Handbook of

Plant Hormones

Actions and Applications

Handbook of Plant Hormones
Actions and Applications

Contributors :
William M. Gray,
F. Odabaşoğlu, *et al.*

Volume 1

KOROS PRESS
KOROS PRESS LIMITED
London, UK

Handbook of Plant Hormones : *Actions and Applications*
Contributors : William M. Gray *and* F. Odabaşoğlu, *et al.*

Published by Koros Press Limited

www.korospress.com

United Kingdom

Printed in 2017 for Sale in the Indian Subcontinent

Handbook of Plant Hormones : *Actions and Applications*

ISBN: 978-1-78163-475-2

British Library Cataloguing in Publication Data
A CIP record for this book is available from the British Library

Exclusively distributed by CBS Publishers & Distributors Pvt. Ltd.

Sales & Distribution Rights only for India, Pakistan, Bangladesh, Sri Lanka, Nepal and Bhutan.This book is not to be sold outside these territories.

Preface

Plant hormones are chemicals that regulate plant growth, which, in the UK, are termed 'plant growth substances'. Plant hormones are signal molecules produced within the plant, and occur in extremely low concentrations. Hormones regulate cellular processes in targeted cells locally and, when moved to other locations, in other locations of the plant. Hormones also determine the formation of flowers, stems, leaves, the shedding of leaves, and the development and ripening of fruit. Plants, unlike animals, lack glands that produce and secrete hormones. Instead, each cell is capable of producing hormones. Plant hormones shape the plant, affecting seed growth, time of flowering, the sex of flowers, senescence of leaves, and fruits. They affect which tissues grow upward and which grow downward, leaf formation and stem growth, fruit development and ripening, plant longevity, and even plant death. Hormones are vital to plant growth, and, lacking them, plants would be mostly a mass of undifferentiated cells. So they are also known as growth factors or growth hormones. The term 'Phytohormone' was coined by Thimann in 1948. Phytohormones are found not only in higher plants, but in algae too, showing similar functions, and in microorganisms, like fungi and bacteria, but, in this case, they play no hormonal or other immediate physiological role in the producing organism and can, thus, be regarded as secondary metabolites.

The word hormone is derived from Greek, meaning set in motion. Plant hormones affect gene expression and transcription levels, cellular division, and growth. They are naturally produced within plants, though very similar chemicals are produced by fungi and bacteria that can also affect plant growth. A large number of related chemical compounds are synthesized by humans. They are used to regulate the growth of cultivated plants, weeds, and in vitro-grown plants and plant cells; these manmade compounds are called Plant Growth Regulators or PGRs for short. Early in the study of plant hormones, "phytohormone" was the commonly used term, but its use is less widely applied now. Plant hormones are not nutrients, but chemicals that in small amounts promote and influence the growth, development, and differentiation of cells and tissues. The biosynthesis of plant hormones within plant tissues is often diffuse and not always localized. Plants lack glands to produce and store hormones, because, unlike animals-which have two circulatory systems powered by a heart that moves fluids around the body-plants use more passive means to move chemicals around the plant. Plants utilize simple chemicals as hormones, which move more easily through the plant's tissues. They are often produced and used on a local basis within the plant body. Plant cells produce hormones that affect even different regions of the cell producing the hormone. Hormones are transported within the plant by utilizing four types of movements. For localized movement, cytoplasmic streaming within cells and slow diffusion of ions and molecules between cells are utilized. Vascular tissues are used to move hormones from one part of the plant to another; these include sieve tubes or phloem that move sugars from the leaves to the roots and flowers, and xylem that moves water and mineral solutes from the roots to the foliage. Not all plant cells respond to hormones, but those cells that do are programmed to respond at specific points in their growth cycle. The greatest effects occur at specific stages during the cell's life, with diminished effects occurring before or after this period. Plants need hormones at very specific times during plant growth and at specific locations. They also need to disengage the effects that hormones have when they are no longer needed. The production of hormones occurs very often at sites of active growth within the meristems, before cells have fully differentiated. After production, they are sometimes moved to other parts of the plant, where they cause an immediate effect; or they can be stored in cells to be released later. Plants use different pathways to regulate internal hormone quantities and moderate their effects; they can regulate the amount of chemicals used to biosynthesize hormones. They can store them in cells, inactivate them, or cannibalise already-formed hormones by conjugating them with carbohydrates, amino acids, or peptides. Plants can also break down hormones chemically, effectively destroying them. Plant hormones frequently regulate the concentrations of other plant hormones. Plants also move hormones around the plant diluting their concentrations. The concentration of hormones required for plant responses are very low. Because of these low concentrations, it has been very difficult to study plant hormones, and only since the late 1970s have scientists been able to start piecing together their effects and relationships to plant physiology. Much of the early work on plant hormones involved studying plants that were genetically deficient in one or involved the use of tissue-cultured plants grown in vitro that were subjected to differing ratios of hormones, and the resultant growth compared. The earliest scientific observation and study dates to the 1880s; the determination and observation of plant hormones and their identification was spread-out over the next 70 years.

In general, it is accepted that there are five major classes of plant hormones, some of which are made up of many different chemicals that can vary in structure from one plant to the next. The chemicals are each grouped together into one of these classes based on their structural similarities and on their effects on plant physiology. Other plant hormones and growth regulators are not easily grouped into these classes; they exist naturally or are synthesized by humans or other organisms, including chemicals that inhibit plant growth or interrupt the physiological processes within plants. Each class has positive as well as inhibitory functions, and most often work in tandem with each other, with varying ratios of one or more interplaying to affect growth regulation. Plant stress hormones activate cellular responses, including cell death, to diverse stress situations in plants. Researchers have found that some plant stress hormones share the ability to adversely affect human cancer cells. For example, sodium salicylate has been found to suppress proliferation of lymphoblastic leukemia, prostate, breast, and melanoma human cancer cells. Jasmonic acid, a plant stress hormone that belongs to the jasmonate family, induced death in lymphoblastic leukemia cells. Methyl jasmonate has been found to induce cell death in a number of cancer cell lines.

This page left intentionally blank.

Contents

List of Contributors

F. Odabaşoğlu

Atatürk University, Faculty of Pharmacy, Department of Biochemistry, 25240 Erzurum -TURKEY

Ö.İ. Küfrevioğlu

Atatürk University, Faculty of Pharmacy, Department of Biochemistry, 25240 Erzurum -TURKEY

William M. Gray

William M. Gray is in Plant Biology at the University of Minnesota, St. Paul, Minnesota, United States of America. E-mail: grayx051@tc.umn.edu

This page left intentionally blank.

Chapter 1

Plant : Growth and Development

New cells form continually in the apical meristems. Cells enlarge slowly in the apical meristem and more rapidly in the subapical regions. The resulting increase in cell volume can range from several folds to one hundredfold, depending on the species and environmental conditions. Classically, plant growth has been analysed in terms of cell number or overall size (*i.e.*, mass). However, these measures tell only part of the story. Tissue growth is neither uniform nor random. The derivatives of the apical meristems expand in predictable and site-specific ways, and the expansion patterns in these subapical regions largely determine the size and shape of the primary plant body. The total growth of the plant can be thought of as the sum of the local patterns of cell expansion. The analysis of the motions of cells or "tissue elements" (and the related problem of cell expansion) is called *kinematics*. In this section we will discuss both the classical definitions of growth and the more modern, kinematic approach. The advantage of the kinematic approach is that it allows one to describe the growth patterns of organs mathematically in terms of the expansion patterns of their component cells.

PLANT GROWTH IN DIFFERENT WAYS

Growth in plants is defined as an irreversible increase in volume. The largest component of plant growth is cell expansion driven by turgor pressure. During this process, cells increase in volume manyfold and become highly vacuolate. However, size is only one criterion that may be used to measure growth. Growth also can be measured in terms of change in fresh weight—that is, the weight of the living tissue—over a particular period of time. However, the fresh weight of plants growing in soil fluctuates in response to changes in water status, so this criterion may be a poor indicator of actual growth. In these situations, measurements of dry weight are often more appropriate. Cell number is a common and convenient parameter by which to measure the growth of unicellular organisms, such as the green alga *Chlamydomonas*. In multi-cellular plants, however, cell number can be a misleading growth measurement because cells can divide without increasing in volume.

Growth is assessed by a count of the number of cells per milliliter at increasing times after the cells are placed in fresh growth medium. Temperature, light, and nutrients provided are optimal for growth. An initial lag period during which cells may synthesise enzymes required for rapid growth is followed by a period in which cell number increases exponentially. This period of rapid growth is followed by a period of slowing growth in which the cell number increases linearly. Then comes the stationary phase, in which the cell number remains constant or even declines as nutrients are exhausted from the medium. For example, during the early stages of embryogenesis, the zygote sub-divides into progressively smaller cells with no net increase in the size of the embryo. Only after it reaches the eight-cell stage does the increase in volume begin to mirror the increase in cell number. Because the zygote is an especially large cell, this lack of correspondence between an increase in cell number and growth may be unusual, but it points out the potential problem in equating an increase in cell number with growth.

Although cell number may not always be a reliable measure of plant growth, under most circumstances dividing cells, particularly in meristems, double in volume during their cell cycle. Therefore, an increase in cell number, such as the increase brought about by the activity of the apical meristems, does contribute to plant growth. However, the largest component of plant growth is the rapid cell expansion that occurs in the subapical region after cell division ceases. Because all the cells of the plant axis elongate under normal conditions, the greater the number of cells produced by the apical meristem, the longer the axis will be. For example, when Arabidopsis plants are transformed with a gene that encodes cyclin, a key component of the cell cycle regulatory machinery, the cells of the apical meristem progress through their cell cycles more rapidly, so more cells form per unit time.

As a result, the roots of these transgenic plants have more cells and are substantially longer than the roots of wild-type plants grown under similar conditions. New cells form continually in the apical meristems. With each new round of cell division and associated cell expansion, the older derivatives are displaced a small distance from the apex. As the cells recede farther from the apex, the rate of displacement is greatly accelerated. By viewing plant growth as a process of cell displacement from the apex, we can apply the principles of kinematics.

PRIMARY GROWTH OF PLANT

The primary growth of a plant body is brought about by the activity of primary meristems, particularly by apical meristem. Details about primary growth occurring in stems, leaves and roots, may be studied by observing a transverse structure is differentiated into outer epidermis, the inner ground tissue and vascular tissue.

- *Dicotyledonous stem* : Example : Sunflower (Helianthus annus). The transverse section is almost circular in outline.

- *Epidermis* : This is the outermost layer, uniseriate, covered with thin cuticle, and bears a few stomata and many multi-cellular trichomes.
- *Ground Tissue* : This fundamental tissue is differentiated into outer cortex and central pith.
- *Cortex* : This is many layered and differentiated into hypodermis and inner cortex. Hypodermis is found below the epidermis, and is constituted with angular collenchyma which is 3-5 layered.

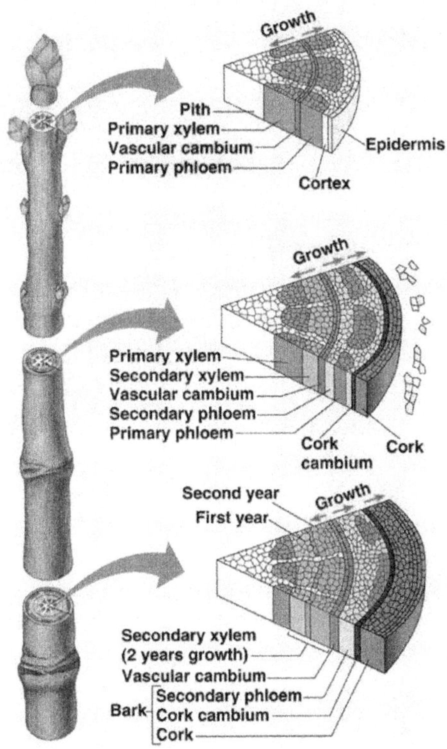

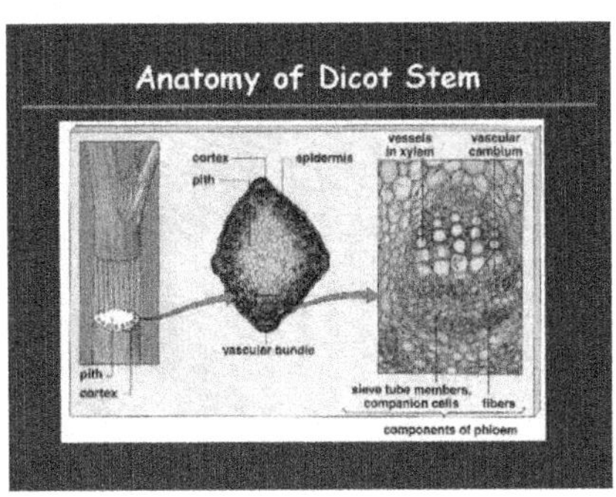

The inner part of cortex is constituted with chlorenchyma and parenchyma; large numbers of intercellular spaces are present between these cells. The inner most cortical layer is wavy in outline, rich in starch grains, with barrel shaped cells and is called endodermoid layer. (Endodermis is generally absent in stem; endodermoid layer differs from endodermis in lacking casparian thickenings).

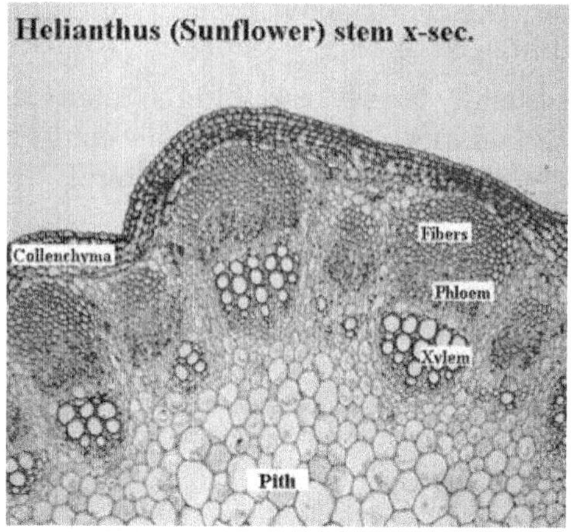

- *Pericycle* : This is the outermost region of vascular tissue. This consists of alternately arranged patches of sclerenchyma and parenchyma. The sclerenchyma is also known as hard bast fibres. Since it is found external to vascular bundle, the patch of sclerenchyma is also called bundle cap.
- *Vascular tissue* : This consists of many vascular bundles arranged regularly in the form of broken ring (Eustele). In each vascular bundle the xylem and phloem elements are in the same line; the presence of one to few stripped intrafascicular cambium is found in between xylem and phloem; the cambial cells are thin walled and are almost rectangular phloem is only external to xylem : the protoxylem is nearest to centre. The vascular bundles are conjoint, collateral, endarch, open and wedge shaped.
- *Pith or Medulla* : This is inner to vascular tissue and occupies the central region. It is composed of cells of parenchyma, with intercellular spaces. Extensions of these cells into the areas in between vascular bundles are called medullary rays.
- *Monocotyledonous stem* :

  Example : Zea mays (Maise). The outline is circular.

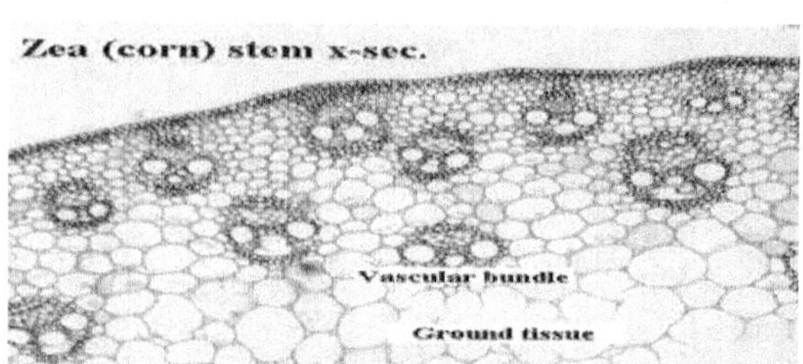

- *Epidermis* : This is the uniseriate, outermost layer and is covered with thick cuticle. Trichomes are absent.

- *Ground tissue* : The ground tissue is not differentiated; into cortex and pith, since there is scattered arrangement of vascular bundles. Even endodermoid layer and pericycle are also not differentiated. Hypodermis constitutes the exterior ground tissue, and is composed of 2 - 3 layers of sclerenchyma. The major part of the ground tissue is parenchymatous.
- *Vascular tissues* : This consists of numerous, oval shaped vascular bundles scattered throughout the ground tissue, hence the arrangement of vascular tissue is called atactostele. Peripheral vascular bundles are small-sised but many in numbers. The central vascular bundles are large sised but a few in numbers in all the vascular bundles the xylem and phloem are in the same line; cambium is absent; phloem is only external to xylem; protoxylem faces the centre.

Hence the vascular bundles are described as conjoint, collateral, endarch and closed. Each vascular bundle is surrounded by a bundle sheath of sclerenchyma, which is prominent at the base and tip of vascular bundle. Xylem comprises usually 2 oppositely arranged Metaxylem elements (pitted, facing periphery) and a few protoxylem elements (facing the centre of stem) in a linear row. If an imaginary line is drawn connecting the two Metaxylem elements with that of protoxylem, xylem appears 'Y' shaped. Usually the lower protoxylem elements break down and form a lysogenous cavity. Above the metaphloem, a small band of obliterated protophloem is seen. Metaphloem consists of sieve tubes and companion cells but not phloem parenchyma.

Comparison between me stems of Dicot and Monocot :

	Dicot stem **Example : Sunflower**	**Monocot stem** **Example : Maize**
1. Epidermal trichomes	Present multi-cellular	Absent
2. Hypodermis	Collenchymatous	Sdererchymatous
3. Ground tissue	Differentiated into cortex and Pith	Undifferentiated Absent
4. Endodermoid layer and Peri-cycle	Present	Absent
5. Stele	Eustle	Atactostele
6. Vascular bundles	Many open, wedge-shaped	Numerous, closed, oval-shaped
7. Lysigenous cavity	Absent	Present
8. Bundle cap	Present	Absent
9. Bundle sheath	Absent	Present
10. Secondary Growth	Occurs later	Doesnot occur

Dicotyledonous root : Example : Lablab purpureus (Beans) :

- *Epiblema* : The external protective layer is called Epiblema or Piliferous layer (the term epidermis is generally not applied to roots). It is uniseriate. Cells are thin walled. Some of these cells extend into unicellular root hairs.

- *Cortex* : The region inner to epiblema is called cortex and it is homogenous. The cortex is composed of thin walled, spaciously arranged parenchyma cells that constitute many layers.

With increase in age, the exterior cortical region becomes suberised and dead. This exterior cortex is known as exodermis.

The innermost cortical layer is called endodermis. Its cells are barrel-shaped and are closely packed. Many cells get suberised on their radial and inner walls. These thickenings are called casparian stripes. Such endodermal cells may become totally suberised on all wall surfaces and are called stone cells.

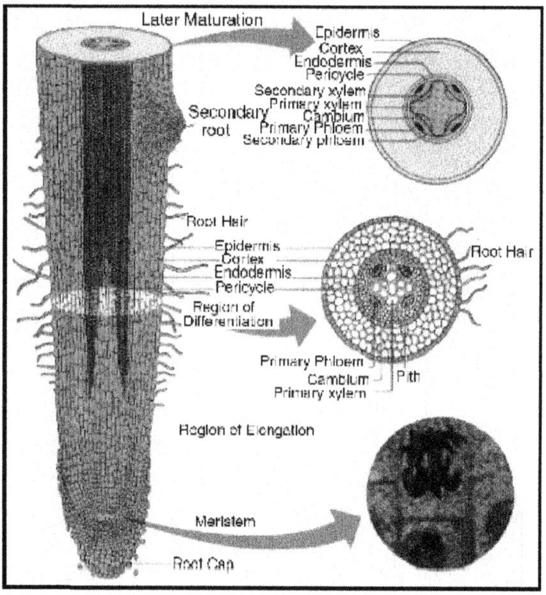

These cells do not allow conduction. The non-suberised endodermal cells are called passage cells. Inner to endodermis, one or two layers of parenchyma are present, constituting pericycle.

- *Stele* : Xylem and phloem patches are in equal number and alternate with each other. Xylem or phloem patches are usually four in number. In Xylem, the protoxylem elements face pericycle; between xylem and phloem elements, parenchyma cells are present constituting conjunctive tissue, which later becomes a secondary meristem.

  Hence vascular bundles are described as radial, exarch and open. Pith is very small, with closely packed parenchyma cells.

  Monocotyledonous root :

  Example : Canna

  Outline almost circular.

The nature of epiblema, exodermis, cortex, endodermis and pericycle is similar to that of Dicot root. However in cortex, large air spaces of schizogenous origin may be seen. The nature and arrangement of vascular bundles are also similar to that of Dicot root. But vascular bundles are usually more than six in number. The Parenchyma present in between xylem and phlegm is called conjunctive tissue, with thin cell walls. This tissue does not become meristematic and thus differs from Dicot root. Pith is relatively large and parenchymatous. Pith and conjunctive tissue, become lignified and Sclerenchymatous. The vascular bundles are described as polyarch, radial, exarch and closed. An endogenous lateral branch may be seen developing from pericycle and passing through cortex. The leaf is horizontal and dorsiventrally differentiated (upper = ventral, lower = dorsal).

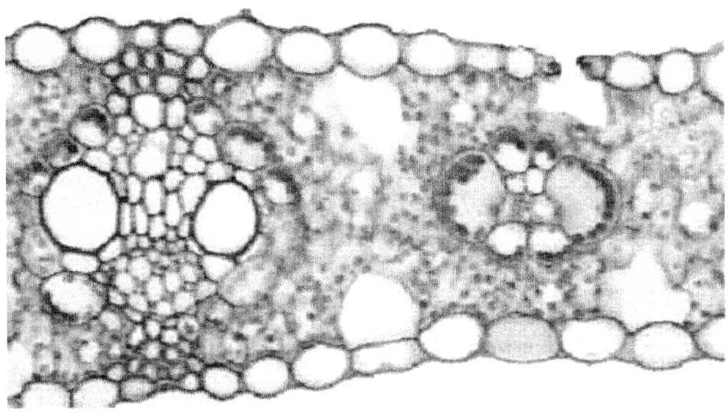

- *Epidermis* : The two surfaces are bounded by their respective epidermal layers. The epidermal cells are barrel-shaped, compactly arranged; upper epidermis is covered with thick cuticle and lacks stomata; lower epidermis is light green, covered with thin cuticle and is interrupted by stomata.
- *Mesophyll* : This is the ground tissue. It is differentiated into upper palisade and lower spongy parenchyma. Palisade parenchyma is found immediately below the upper epidermis, 2 to 3 layered, with compactly arranged tubular cells, rich in parietal chloroplasts.
- Spongy parenchyma is found above the lower epidermis; these cells are varied in shape and sizes, very loosely arranged enclosing air spaces some of which open into stomata. Chloroplasts are parietal in the parenchyma cells.
- *Vascular tissue* : Vascular bundles vary in size; each bundle is conjoint, collateral and closed. The vascular bundle is covered by a bundle sheath of parenchyma cells. An extension of bundle sheath is seen above and below the vascular bundle, reaching the epidermal layers. Phloem is towards lower epidermis; xylem is towards upper epidermis, with metaxylem facing phloem. Fibres are absent in both xylem and phloem. The xylem and phloem elements are conspicuous only in large vascular bundles.

Monocotyledonous leaf Example : Maise.

The leaf is Isobilateral in nature and is vertically oriented.

- *Epidermis* : This is uniseriate, with barrel-shaped, compactly arranged cells and is covered with thick cuticle.

  Stomata are found on both upper and lower epidermal layers hence it is called amphistomatic (more on the lower epidermis). Though the leaf is referred to as isobilateral, it is only in upper epidermis, a few large, empty and colourless bulliform or motor cells are present. During dry weather, these motor cells help the leaf to role over, due to the changes in turgidity. This rolling of leaf reduces the rate of stomatal transpiration.
- *Mesophyll* : There is no differentiation of mesophyll into spongy and palisade parenchyma. All the cells of chlorenchyma are alike, isodiametric, almost compactly arranged with numerous parietal chloroplasts.
- *Vascular tissue* : The vascular bundles are numerous, arranged in parallel series (venation is palmate-parallel), conjoint, collateral and closed. Phloem is towards lower epidermis. Each vascular bundle is surrounded by chlorenchymatous bundle; this sheath also serves for temporary storage of starch. A few vascular bundles are larger in size, with more amount of xylem and phloem and with large bundle sheath cells. A patch of sclerenchyma is present above and below the large sised vascular bundles.

Table. Comparison Between the Leaf Structures of Dicot and Monocot.

	Dicot Root Ex Mango	**Monocot Root Ex : Maize**
1. Orientation of leaf	Horizontal, dorsiventrally differentiated	Vertical, isobilateral
2. Stomata	Usually on lower epidermis	On both epidermal layers
3. Cutide	Thick on upper epidermis	Thick on both epidermal layers
4. Motor cells	Absent	Present in upper epidermis
5. Mesophyll	Differentiated into upper palisade parenchyma and lower spongy parenchyma	No differentiation of mesophyll
6. Vascular bundles	All of them are not seen in parallel series and are supported by bundle sheath extension	Vascular bundles are in parallel series and are supported by sclerenchyma patches

PLANT GROWTH INTERACTIONS

Soil is a medium for plant growth, and its structure should not hinder the movement of water, oxygen and nutrients to plant roots, impede the growth of roots, or allow the buildup of toxic substances around roots. Soil structure can be described as the organisation of the particles in the soil—the internal configuration of the soil matrix. Soil structure is formed by many interacting processes in the soil, especially shrinkage with drying, channel formation by plant roots, faunal activity, and cultivation. Well-developed soil structure is represented by distinct stable aggregations (peds) of soil particles separated by interped pores. The interped pores of an aggregated soil are larger, on average, than pores between particles of unaggregated soil, resulting in a less dense condition.

The larger pores are also critical in terms of allowing movement of air and water through the soil. Unfavourable soil structure, on the other hand, can impart the characteristics of greater soil density, decreased aeration, poor percolation, and mechanical impedance to roots, which can restrict plant growth. There are a number of factors that may result in poor soil structure. Soil texture is important; in fact, very sandy soils are usually classified as structureless due to lack of soil matrix (sand is viewed more as the "skeleton" of the soil). In sandy soils, size distribution and packing of the grains are considered rather than structure *per se.* In loamy and finer-textured soil, clay acts as a cementing agent to provide some level of stability to the particle orientation in structured soil; however, clay can also cement particles in a structurally degraded condition ("massive"). Chemical conditions can be responsible for structurally degraded soil; a relatively high sodium content causes dispersion of particles and elimination of aggregation. These "sodic soils" occur naturally but can also be initiated or exacerbated by human activity. The human-imposed factor that degrades soil structure to the greatest extent is compaction. This chapter focuses on the effects of compacted soil on plant growth, including the direct effect of mechanical impedance as well as the indirect effects that usually accompany compaction. Finally, amelioration of compaction is discussed, with emphasis on the ability of plant roots to effect changes in soil structure and hence modify physical properties.

COMPACTION

Hillel described soil compaction as the compression and densification of an unsaturated soil body, reducing the fractional air volume. The degree to which a soil will compress is a function of the soil moisture content, antecedent bulk density, magnitude of the compactive effort, soil texture, and content of organic matter. Forces that compact soil can originate from natural sources, including rainfall, radial root pressure, cycles of wetting/drying and freesing/thawing, or traffic of people and machinery. The potential to degrade soil structure has increased as modern machinery has become larger and heavier, causing compaction to a greater depth. Compounding this problem is the tendency to tread repeatedly over soil while implementing cultural practices such as seeding, fertilization, pest control, harvesting/mowing, or construction during site development. Soil compaction on recreational sites is thought to be one of the most challenging issues faced by turfgrass managers. Cultivation implements are generally designed to overcome the bonds stabilising structure, breaking aggregates into smaller ones for seedbed preparation to maximise seed-soil contact, enhance water infiltration and retention characteristics, and improve soil aeration. Routine tillage of soil, however, can also degrade soil structure by shattering aggregates, eliminating pore networks, and imposing compacting forces, particularly in a sub-surface layer of soil.

Soil Responses

Porosity, bulk density, and water retention characteristics are physical properties characterising soil structure and compaction phenomena. Compaction of a soil decreases soil porosity and subsequently increases soil bulk density. Veihmeyer and Hendrickson showed that the critical soil bulk density needed to inhibit sunflower (Helianthus annus L.) root growth varied with texture; no roots penetrated soil with a bulk density of 1.9 g cm^{-3}. The lowest density where roots failed to penetrate was 1.46 g cm"3 for an Aiken clay loam. Air tends to occupy the larger pores of soil, whereas the affinity of water for mineral surfaces creates water films and wedges (menisci) within the small fissures and capillaries of the soil. A reduction in the total porosity of soil, as a result of compaction, generally occurs at the expense of the air porosity, those relatively large pores not filled with water at a given soil water potential (usually "33 kPa to represent "field capacity").

Swartz and Kardos observed a decrease in air porosity from 19.3 to 12.7 per cent, with increasing levels of compaction on several sand-soil-peat mixes differing in moisture content at the time of compaction, whereas capillary porosity increased from 32.2 per cent at the lowest compaction level to 35.7 per cent at the highest level. In aggregated soils, the loss of porosity due to compaction can be attributed to the collapse of interaggregate pores. Water retention and flow characteristics in soil are influenced by alterations of soil porosity. Relative to non-compacted soil, compacted soil will generally retain a greater amount of water at a given soil matric potential as a result of increased capillary porosity. Soil water permeability, particularly at saturated or nearly saturated conditions, is a function of air porosity; as air porosity decreases, so does water permeability. Compacted soils with lower infiltration rates are prone to problems with stagnant water and anaerobic conditions.

Constraints on Root Growth

Soil Aeration

Soil oxygen movement to plant roots is critical to maintain adequate respiration for growth. Anaerobiosis, or oxygen stress, occurs in soil when the rate of supply falls below the biological demand. Impeded soil aeration can arise from poor drainage and water-logging or from compacted soil. One approach to evaluate soil aeration is to measure the composition of soil air. This technique indicates that gas exchange between the atmosphere and soil air is restricted when oxygen content of the soil air falls significantly below that of the atmosphere. An important challenge with this approach is how to extract a representative sample of soil air that avoids mixing of gases from outside the point of sampling or contamination from the atmosphere. Another approach for assessing soil aeration is to determine the air porosity, or fractional air space, at a standard soil water potential. Work to identify threshold air porosity values below which plant growth is limited has generated values ranging from 5 to 20 per cent. Madison has indicated that 10 per cent air porosity is the threshold level for intensively utilised turf. The

United States Golf Association Green Section (USGA) has suggested 15 per cent air porosity at a 3 kPa water tension as the lower test limit for materials used in the construction of root zones for golf course putting greens. The extent to which air porosity in the root zone changes with time under a perennial turf system is a subject of current studies.

The rate of air exchange, rather than the volume of soil air, is the more important factor in soil aeration and therefore limits the usefulness of air porosity as an index of soil aeration. Gaseous movement in the soil occurs by two main processes : convection and diffusion. Convection, where the moving force is a gradient of total gas pressure, is thought to be a minor mechanism of soil aeration except at shallow depths and in soils with large pores. Although convection is not considered an important factor for more deeply rooted plant species, it could be more influential for shallow-rooted turfs. A large fraction of the root system of moderately to highly maintained turfs is found at surface depths of 5 cm or less. Furthermore, the crown, the major meristematic organ, of turfgrass plants is located near the soil surface. Thus, turfs with a high plant population could have sufficiently high oxygen demands at the soil surface that convection would play a useful role in maintaining adequate aeration for plant crowns and surface rooting. Convection, therefore, could be partially responsible for the observations of improved turfgrass quality after changes in air masses (pressure) associated with passing weather systems. Below the surface, however, diffusion is considered the more important exchange mechanism in soil, and the moving force is a gradient of partial pressure. Gas exchange between the atmosphere and soil is maintained predominantly by diffusion through air-filled pores. In contrast, live tissue is typically hydrated, and the supply of oxygen to roots occurs by diffusion through water films. This is an important distinction, since the diffusivity of oxygen in water is less than its value in air by a factor of 10"4.

The solubility of oxygen in water may be a compounding issue, especially at higher temperatures, where solubility decreases. Methods that measure gaseous diffusion through air-filled pores do not provide information on the impedance to oxygen presented by the water film surrounding a root. It is possible that an inadequate supply of oxygen to roots can occur in a well-aerated soil (*i.e.*, high air-filled porosity and O_2 concentration) if the roots are surrounded by thick water films. Thus, under such conditions, measurement of soil oxygen movement to plant roots would provide a more complete assessment of soil aeration. Lemon and Erickson introduced the method commonly called ODR (oxygen diffusion rate) to measure oxygen diffusion to a root-like electrode inserted into the soil and enveloped in water films of the soil. The ODR technique has been used by a number of researchers to demonstrate the reduction in oxygen supply in compacted soil. Limiting soil aeration conditions are often transient and strongly affected by wetting and drying cycles in compacted soil. Therefore, the timing of gas composition and ODR measurements is critical for accurate detection of anaerobiosis. Generally, 200 ng cm^{-2} min^{-1} has been considered the minimum ODR value, below which plant roots will not grow. Some researchers have reported values ranging from 50 to 200 ng cm^{-2} min^{-1} as limiting ODR values for turfgrass root growth. The interpretation of these results has been questioned because, in addition to the diffusion of oxygen, other soil factors and components could have affected the measured ODR.

Work by Blackwell has helped to reduce errors associated with reactions other than the reduction of oxygen in the ODR technique. Moreover, early studies of limiting ODR were performed in growth chambers at about 20æ per cent C, whereas soil temperatures in the field can vary considerably during the growing season. Blackwell and Wells concluded that the effect of oxygen flux on root growth is related to temperature differences and differences in root respiration. Thus, it is plausible that the oxygen flux limiting root growth under high soil temperatures may be greater than values that have been observed at 20æ per cent C. When oxygen becomes limiting, some species of bacteria shift metabolic pathways so as to utilise other compounds as terminal electron acceptors. Thus, plants may have not only to survive periods without oxygen but also to withstand toxic substances such as hydrogen sulfide, which are produced during anaerobic microbial respiration. Development of aerenchyma tissue (intercellular spaces) that allows diffusion of oxygen from aerial plant tissue down to roots may be an adaptive response that helps certain species of plants to survive periods of soil anaerobiosis. Agnew and Carrow reported increased root porosity in Kentucky bluegrass (Poa pratensis L.) grown under compacted soil conditions. They observed the greatest root porosity under conditions of both compaction and water stress.

Soil Strength

Soil strength can be described as the capacity of soil to resist a force without rupture, fragmentation, or flow. Various methods exist to quantitate soil strength in the laboratory, but the method most used in the field is the penetrometer. It measures the effort necessary to push a thin, cone-tipped probe into the soil. Taylor and colleagues conducted pioneering research demonstrating that the strength of a soil, as measured with a static penetrometer, will increase as soil bulk density increases and water content decreases; consequently, mechanical impedance to root growth is greater. The limiting strength of soil to cotton (Gossypium hirsutum L.) root growth is in the range 2.5 to 3 MPa. For many crops, a 2-MPa penetrometer resistance is commonly encountered under field conditions and can reduce root length and elongation by at least 50 per cent.

Root Morphology

In experiments designed to study the constricted growth of roots, the ability of ryegrass (Lolium perenne L.) roots to enter rigid pores (glass capillaries and sheets of steel) depended on the size of the root cap and the thickness of the stele, which needed to be about one-third of the nominal root thickness. Roots could elongate down long capillaries while constricted, albeit at a reduced rate. Roots can enter pores of diameter smaller than the root tip itself only if the rigidity of the pore structure is weak enough to allow soil displacement. Thus, major root axes typically thicken when soil pores are too small for root penetration. Ethylene production has been linked to this thickening response of roots subjected to mechanical impedance. Abscisic acid, which increases temporarily under conditions of mechanical impedance, has also been shown to induce root thickening as well as increased root-hair number and curling of roots. The thickening of the root compresses soil laterally and minimises friction with soil as the root extends axially. Additionally, root-hair development aids in anchoring the root as it penetrates compacted soil. Thus these responses can be viewed as adaptive, as they mitigate the effects of axial resistance on growth. Additional contributing factors to thicker roots' relative endurance of compact soil might be greater resistance of thicker roots to bending or higher axial pressures exerted by thick roots. Variation in plant genetics exists for tolerance to high mechanical impedance. Interspecific comparisons have shown differences in root dimensions and the ability to penetrate soils. Greater root thickness and the tendency for roots to expand radially in response to mechanical impedance are correlated with the capacity to elongate in hard soil. Intraspecific studies have also demonstrated genetic variation in characteristics associated with tolerance to mechanical impedance. Two tall fescue (Festuca arundinacea Schreb.) lines with large-diameter roots were able to penetrate a hardpan at the 0.4 to 0.6 m soil depth, extracted more soil water from the 0.6 to 1.2 m depth, and yielded 40 per cent more in dry matter compared to fescues with small-diameter roots. Development of efficient techniques to screen plant germplasm for greater tolerance to high mechanical impedance could prove to be highly useful. Iijima *et. al.* described the morphological development of rice (Oryza sativa L.) and maise (Zea mays L.) root systems as affected by soil compaction. Elongation of the main root axes of rice and maise was restricted in compacted soil. Seminal roots of maise were much less restricted than those of rice. The root system of rice was characterised by a larger fraction of long, thick laterals with potential to produce higher-order laterals on their axes. Growth of higher (second- and third-) order lateral roots in compacted soil compensated for the restricted growth of the main root axes in both species. The work of Carrow and colleagues indicates that the effects of compacted soil on root growth varies with plant species and management. Total root mass of Kentucky bluegrass decreased in response to compaction treatment, whereas tall fescue and perennial ryegrass root systems were more tolerant to compaction stress.

Subsequent work indicated that the most detrimental effect of compaction on root growth of tall fescue and perennial ryegrass occurred under higher nitrogen fertilization. Although total root mass of a turf may not change under compacted conditions, a redistribution of the root system can occur where root mass increases in the surface 5 cm of the soil at the expense of rooting below the 10-cm depth. Root mass may not provide the most detailed index of a root system's response to compacted soil, since root mass is affected less by bulk density than root length is, primarily because of larger-diameter roots in high-density soil. Thus, the measurement of the number and length of roots growing in soil may be a more sensitive indicator of compaction stress on a root system than root mass. Root length and number measurements, however, are expensive, particularly for experiments evaluating a large number of treatments and for plant systems such as turf, which have extremely high root lengths. Physical alteration of water and nutrient uptake by a root system can result as the orientation and morphology of roots change. Higher soil bulk density brings more soil, and associated water and nutrients, into contact with the surface area of the root system and thus has the potential to increase availability. This effect is particularly pronounced for nutrients such as phosphorus, which have relatively low mobility in the soil. Greater soil contact with the root system, and thus availability of nutrients and water, could at least partly explain the greater clipping yields of turf growing on heavily compacted plots compared to lightly compacted plots during the spring. The fact that this response was reversed during the summer suggests that the beneficial response of greater soil and root surface contact may be beneficial until environmental conditions become limiting for other essential growth factors under compacted soil conditions. As environmental conditions become more limiting, the reduced total root elongation rate and length in compacted soil ultimately decreases availability of water and nutrients by restricted access.

Shoot Responses

The direct and indirect effects of compaction on plant roots can elicit responses in the plant shoot as well. A number of studies on grasses demonstrate the potential of compacted soil to limit shoot density, verdure, clipping yield, soil cover, and lateral stem growth. Lower dry-matter production and yield under compacted soil conditions is attributed to reduced light interception caused by restricted leaf area development and is not a result of an impaired ability of crops to utilise intercepted radiant energy. Hormonal mechanisms have been proposed through which roots "sense" the mechanical impedance of the soil and elicit anatomical changes in leaf growth. An increase in abscisic acid in the xylem sap appears to be a root-to-shoot signal that lowers stomatal conductance and leaf expansion rates.

THE PRODUCTION OF CELLS

Moving fluids such as waterfalls, fountains, and the wakes of boats can generate specific forms. The study of the motion of fluid particles and the shape changes that the fluids undergo is called kinematics. The ideas and numerical methods used to study these fluid forms are useful for characterising meristematic growth. In both cases, an unchanging form is produced even though it is composed of moving and changing elements. An example of an unchanging form composed of changing and displaced elements in plants is the hypocotyl hook of a dicot, such as the common bean. As the bean seedling emerges from the seed coat, the apical end of the hypocotyl bends back on itself to form a hook. The hook is thought to protect the seedling apex from damage during growth through the soil. During seedling growth (in soil or dim light) the hook migrates up the stem, from the hypocotyl into the epicotyl and then to the first and second internodes, but the form of the hook remains constant.

The hooked form is maintained over time, while different tissues first curve and then straighten as they are displaced from the seedling apex during growth. If a mark is placed at a fixed point on the surface, it will be displaced (indicated by the arrow), appearing to flow through the hook over time. If we mark a specific epidermal cell on the seedling stem located close to the seedling apex, we can watch it as it flows into the hook summit, then down into the straight region below the hook. The mark is not crawling over the plant surface, of course; plant cells are cemented together and do not experience much relative motion during development. The change in position of the mark relative to the hook implies that the hook is composed of a procession of tissue elements, each of which first curves and then straightens as it is displaced from the plant apex during growth. The steady form is produced by a parade of changing cells.

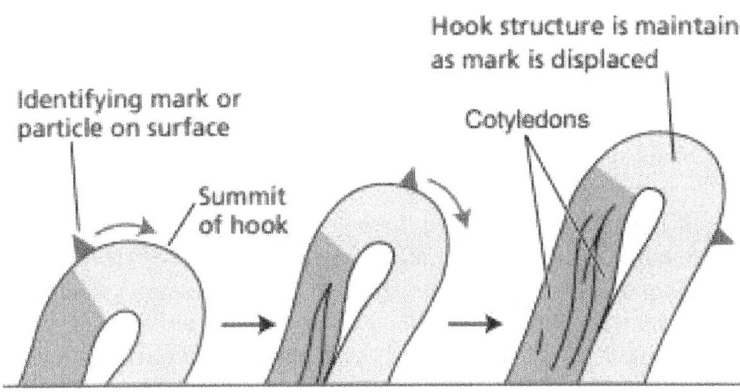

Fig. The dicot Hypocotyl hook is an example of a Constant form Composed of changing Elements.

A root tip is another example of a steady form composed of changing tissue elements. Here, too, the form is observed to be steady only when distance is measured from the root tip. A region of cell division occupies perhaps 2 mm of the root tip. The elongation zone extends for about 10 mm behind the root tip. Phloem differentiation is first observed beginning at 3 mm from the tip, and functional xylem elements may be seen at about 12 mm from the tip. A marked cell near the tip will seem to flow first through the region of cell division, then through the elongation zone and into the region of xylem differentiation, and so on. This shifting implies that developing tissue elements first divide and elongate, and then differentiate. In an analogous fashion, the shoot bears a succession of leaves of different developmental stages. During a period of 24 hours, a leaf may grow to the same size, shape, and biochemical composition that its neighbour had a day earlier. Thus, shoot form is also produced by a parade of changing elements that can be analysed with kinematics. Such an analysis is not merely descriptive; it permits calculations of the growth and biosynthetic rates of individual tissue elements (cells) within a dynamic structure.

Tissue Elements are Displaced During Expansion

As we have seen, growth in shoots and roots is localised in regions at the tips of these organs. Regions with expanding tissue are called growth zones. With time, meristems move away from the plant base by the growth of the cells in the growth zone. If successive marks are placed on the stem or root, the distance between the marks will change, depending on where they are within the growth zone. In addition, all of these marks will move away from the tip of the root or shoot, but their rate of movement will differ depending on their distance from the tip.

From another perspective, if you were to stand at the tip of a root that had marks placed at intervals along the axis, you would see that all marks would move farther away from you with time. The reason is that discrete regions on the plant axis experience displacement as well as expansion during growth and development. As a given region of the plant axis moves away from the apex, its growth velocity increases (the rate of elongation accelerates) until a constant limiting velocity is reached equal to the overall organ extension rate.

The reason for this increase in growth velocity is that with time, progressively more tissue is located between the moving particle and the apex, and progressively more cells are expanding, so the particle is displaced more and more rapidly. In a rapidly growing maise root, a tissue element takes about 8 hours to move from 2 mm (the end of the meristematic zone) to 12 mm (the end of the elongation zone). Beyond the growth zone, elements do not separate; neighbouring elements have the same velocity (expressed as the change in distance from the tip per unit of time), and the rate at which particles are displaced from the tip is the same as the rate at which the tip moves through the soil. The root tip of maise is pushed through the soil at 3 mm h^{-1}. This is also the rate at which the non-growing region recedes from the apex, and it is equal to the final slope of the growth trajectory.

The Growth of Velocity

The velocities of different tissue elements are plotted against their distance from the apex to give the spatial pattern of growth velocity, or growth velocity profile. Growth velocity increases with position in the growth zone. A constant value is obtained at the base of the growth zone. The final growth velocity is the final, constant slope of the growth trajectory equal to the elongation rate of the organ. In the rapidly growing maise root, the growth velocity is 1 mm h^{-1} at 4 mm, and it reaches its final value of nearly 3 mm h^{-1} at 12 mm.

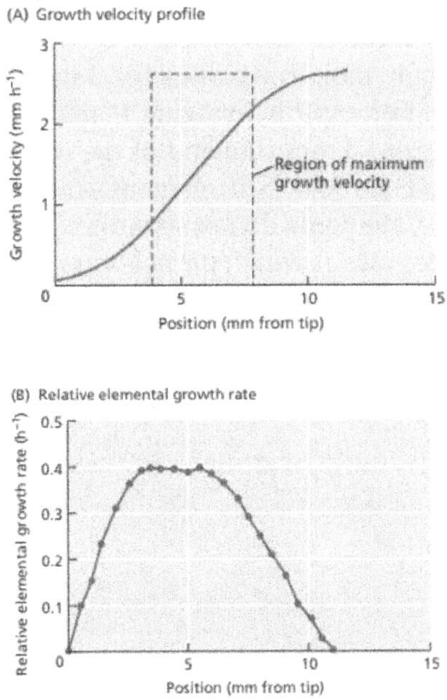

Fig. The Growth of the Primary root of Zea mays (maise) can be represented Kinematically by two related growth curves. (A) The growth velocity profile plots the velocity of movement away from the tip of points at different distances from the tip. This tells us that growth velocity increases with distance from the tip until it reaches a uniform velocity equal to the rate of elongation of the root. (B) The relative elemental growth rate tells us the rate of expansion of any particular point on the root. It is the most useful measure for the physiologist because it tells us where the most rapidly expanding regions are located.

If the growth velocity is known, the relative elemental growth rate, which represents the fractional change in length per unit time, can be calculated. The relative elemental growth rate shows the location and magnitude of the extension rate and can be used to quantify the effects of environmental variation on the growth pattern.

The Relative Elemental Growth Rate

One way to characterise the growth pattern along an axis mathematically is to plot the distance of a tissue element from the apex versus time. The resulting curve is called the growth trajectory. The slope of the curve at any given point is equivalent to the growth velocity at that region of the axis. With the apex as the reference point, the growth trajectory is concave. Marked tissue elements at different locations may be followed simultaneously to generate a family of growth trajectories.

Note that distance is measured from the root tip. This is termed a material specification of growth because real or material particles are followed. As the regions of the plant axis move away from the apex, their growth velocity increases (*i.e.*, the cells accelerate) until a constant limiting velocity is reached equal to the overall organ extension rate. The reason for this increase in growth velocity is that with time, progressively more tissue is located between the moving particle and the apex, and progressively more cells are expanding, so the particle is displaced more and more rapidly. In a rapidly growing maise root, a tissue element takes about 8 hours to move from 2 mm (the end of the meristem) to 12 mm (the end of the elongation zone). The root tissue element accelerates through this region. Beyond the growth zone, elements do not separate; neighbouring elements have the same velocity, and the rate at which particles are displaced from the tip is the same as the rate at which the tip is pushed away from the soil surface.

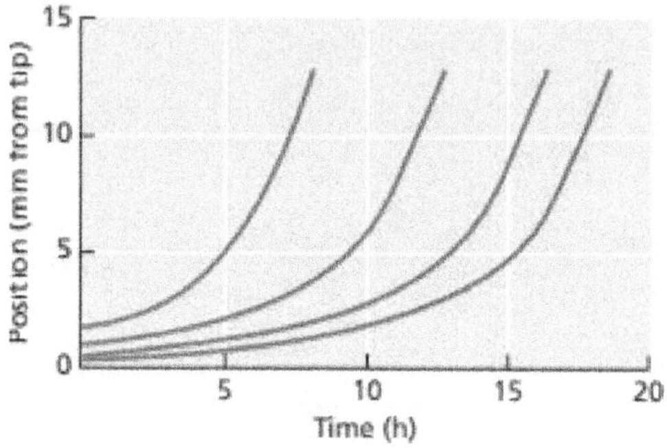

Fig. Growth Trajectories of *Zea mays* (maise) : Growth Trajectories plot the positions of points spaced along the root axis as a Function of time.

The root tip of maise is pushed through the soil at 3 mm h^{-1}. This is also the rate at which the non-growing region recedes from the apex, and it is equal to the final slope of the growth trajectory. The growth trajectory is a material description of growth because an individual element is followed through time. The growth trajectory provides a way to convert spatial information into a developmental time course. For example, if the amount of suberin in the root increases with position, the growth trajectory will give information on the time required for the observed accumulation of suberin in cells moving between any two spatial points. To characterise the local expansion rate, we must consider the growth velocities at the apical and basal ends of a small segment of tissue. If the apical end of the segment is moving faster than the basal end, the segment is elongating. The velocity difference, divided by the segment length, gives the relative growth rate of the segment. Now, imagine that the segment shrinks to a point x. To find the relative growth rate, r, at point x, we can use calculus : The value of r is given by the velocity gradient, dv/dx, where v is the growth velocity. This measure of the local growth rate is called the relative elemental growth rate. It represents the fractional change in length per unit time and has units of h^{-1}.

If the growth velocity is known, the relative elemental growth rate can be evaluated by differentiation of the growth velocity with respect to position. Or we can approximate r by measuring the relative growth rates of closely spaced segments of initial length L. For each segment, $r = (1/L)(dL/dt)$.

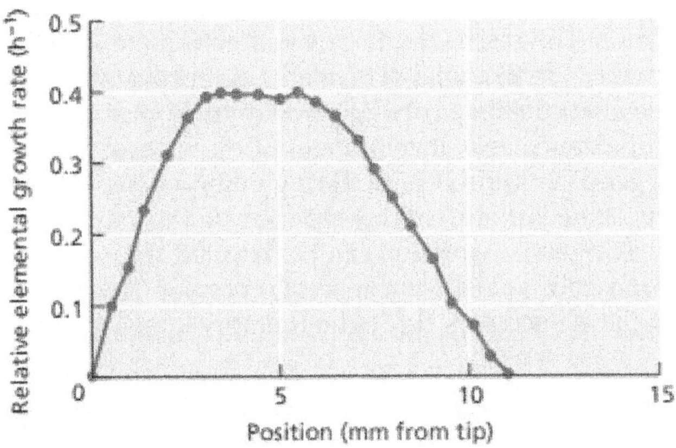

Fig. Relative Elemental Growth rate of *Zea mays* (maise).

The relative elemental growth rate tells us the rate of expansion of any particular point on the root. It is the most useful measure of root growth rate for the physiologist because it tells where the most rapidly expanding regions are located. The relative elemental growth rate profile, $r(x)$, is a spatial description of the expansion pattern in the growth zone. In the maise root, the relative elemental growth rate has a maximum of 0.40 h^{-1} at 4 mm and drops to zero by 12 mm. That is, the tissue element at 4 mm is expanding 40 per cent h^{-1}, while tissue beyond 12 mm has ceased expansion. Again, to understand the profile in physiological terms, we must consider that individual cells expand more and more rapidly as they are displaced to 4 mm and then less and less rapidly as they are displaced through the basal part of the growth zone. The relative elemental growth rate profile shows the location and magnitude of the extension rate and can be used to quantify the effects of environmental variation on the growth pattern. It is also the basis for studying the physiology of growth. The graph of $r(x)$ can be compared to the distribution pattern of any substance thought to be regulating the growth rate. A close correlation between the two curves would provide circumstantial evidence for the role of the substance as a limiting factor for growth.

Embryo Maturation Requires Specific Gene Expression

The Arabidopsis embryo enters dormancy after it has generated about 20,000 cells. Dormancy is brought about by the loss of water and a general shutting down of gene transcription and protein synthesis—not only in the embryo, but also throughout the seed. To adapt the cell to the special conditions of dormancy, specific gene expression is required. For example, the Abscisic Acid Insensitive3 (*ABI3*) and *FUSCA3* genes are necessary for the initiation of dormancy and are sensitive to the hormone abscisic acid, which is the signalling molecule that initiates seed and embryo dormancy. *ABI3* also controls the expression of genes encoding the storage proteins that are deposited in the cotyledons during the maturation phase of embryogenesis.The *LEAFY* COTYLEDON1 (*LEC1*) gene also is active in late embryogenesis. Because *lec1* mutants cannot survive desiccation and do not enter dormancy, the embryos die unless they are rescued through isolation before desiccation occurs. The rescued embryos will germinate in culture and produce fertile plants, which are like wild-type plants except that they lack the 7S storage protein and they have leaflike cotyledons with trichomes on their upper surface. The normal appearance and development of the mature *lec1* mutants indicates that the *LEC1* gene is required only during embryogenesis. Although the most obvious defects of the *lec1* mutants are seen only in the maturation phase embryo, mRNA from *LEC1* gene expression can be detected throughout embryogenesis. It has been proposed that *LEC1* is a general repressor of vegetative development and its expression is necessary throughout embryogenesis.

CELL EXPANSION OF PLANTS

Within 5-10 minutes of application of auxins to isolated epicotyls, or intact tip of the plant or excised segments, growth in their length starts. This early effect is rapid and spontaneous. This rapid growth continues up to 30-45 minutes and then slows down but remains steady. This is also accompanied by an increase in cellular respiration, protein synthesis and the rapid extrusion of protons or in particular Hydrogen ions through the cell walls. Another significant observation of this early rapid growth is insensitive to the inhibition of transcription (RNA synthesis) and translation (protein synthesis) - ex. Actinomycin-D at high concentration inhibits RNA synthesis completely and cycloheximide (CHI) inhibits cytosol protein synthesis. This clearly suggests that the early rapid growth requires neither the synthesis of RNA nor the synthesis of proteins. But this early effect of auxin is totally inhibited by colchicines at 0.5 to 1 per cent concentration. Colchicine, an alkaloid from the tubers of Colchicum autumnale brings about the inhibition of tubulin (TB) polymerisation into long microtubules, which act as the cytoskeleton for mechanical support in the cytoplasm. The colchicines effect on early growth further indicates the important role played by the assembly of microtubules in providing a mechanical structure and the force for all expansion.

Increased metabolic activity, particularly the degradation of osmotically inactive starch etc., leads to the increase of osmotic pressure (they are related to each other) and a steep DPD gradient is created. This facilitates the inward diffusion of water; consequently turgour pressure builds up within the cells. This force acts on all sides of the cell wall and forces the cell to expand in length and breadth. However recent investigations have shown that the cell elongation takes place under negative pressure. The above features suggest that the process or the mechanism of cell elongation is complex and multi-phased, but highly regulated. Various theories have been proposed from time to time and none of them could prove their claim because of one or the other drawbacks. Here we will restrict to a comprehensive account of cell elongation.

Early phase : The auxin's effect in the early phase on cell growth is rapid. This is insensitive to the inhibitors of transcription and translation but insensitive to colchicines and respiratory inhibitors like cyanide, DNP etc. From this it can be deduced that cell elongation does not require new RNA and protein synthesis, but it requires ATP and ATP dependent tubulin polymerisation into microtubular cytoskeleton. Auxin at the early stages activates and increases the rate of respiratory metabolism. With the result ATP pool builds up rapidly. Auxin also activities membrane bound nucleating centre of tubulin polymerisation. These nucleating centres, utilising ATP and tubulin monomers found in the cytoplasm, start polymerising and soon build up long microtubules. These, in turn, start orienting parallel to the long axis of the cell, which probably requires cytochalasin-B sensitive microfilaments. With the rapid increase in microtubules, oriented in longitudinal plane, considerable force is exerted on the cell boundaries.

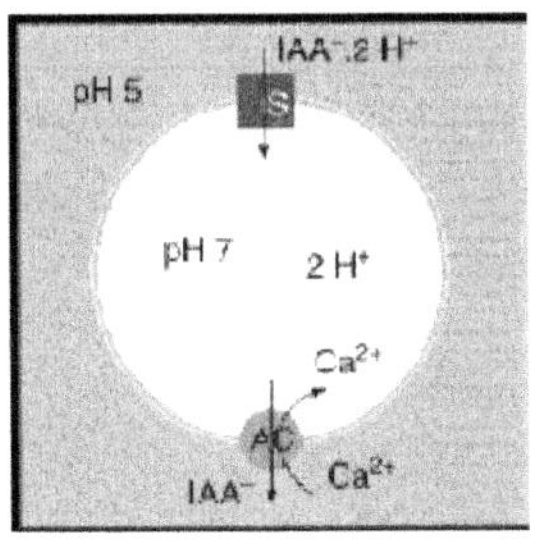

As the microtubular polymerisation is building up, ATP ase/H* gets activated, and H* ions are pumped out into the interspaces of cellulose fibrils acidic pH in turn activates cellulases enzymes, which starts loosening up the cell-wall cellulose fibres. This loosening up of cell wall concomitant with acting as a mechanical force-the cell expands in length. This process starts 4-5 minutes after application of auxins and the elongation continues rapidly for about 20-30 minutes. The late phase starts around 30-40 minutes after auxin application. This is slower than the first phase, but growth is steady. This phase is sensitive to Actinomycin-D, CHI, Colchicine and respiratory inhibitors, indicting that this process requires metabolic energy, tubulin-polymerisation, RNA synthesis as well as protein synthesis. As tubulin pool gets exhausted in the first phase, it has to be replaced. That means it requires the synthesis of RNA and proteins.

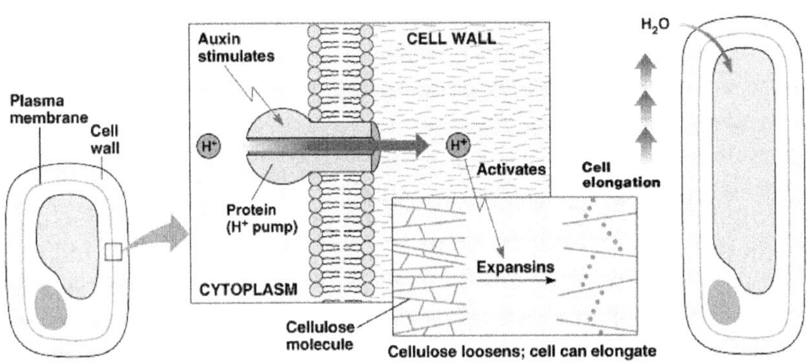

Apart from these proteins, the cytoplasm requires the synthesis of other proteins, for the increase in the mass of protoplasm. It is quite possible that the early activation of early translation may have significant effect on gene activation.

Thus it takes some time for the second phase of growth to set in. The second phase provides all materials for sustained growth.

- *Growth Movements* : The phenomenon of bending movement of plants in response to light stimulus is called Phototropism, the growth of roots towards soil is called Geotropism.

The curvature growth due to the touch is named as Thigmotropism etc. Such differential growth movements are due to differential concentration of auxin. The stem tips and root tips have different optima of their auxin requirement to show maximum growth. The mechanism of these movements will be explained in the chapter on growth movements. Auxin brings about phototropic movement because differential accumulation of auxin at exposed and darker side.

- *Root initiation* : Vegetative propagation of stem cuttings is a normal practice in horticulture, floriculture and agriculture. Out of some stem cuttings which are put into soil, a few produce roots easily but others may or not produce roots. The roots that develop from any part of the plant body other than the radicle are called adventitious roots.

The development of such roots from the stem or any other part is a phenomenon of dedifferentiation. Stem is quite different fro the root with respect to its structure and function.

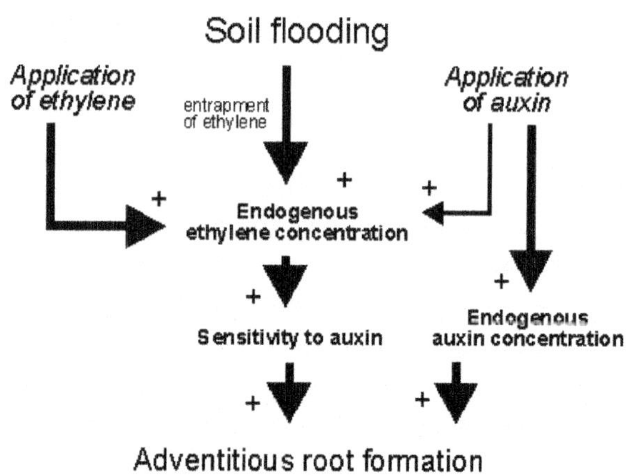

Still some of the tissues found in stem undergo dedifferentiation and develop into roots. This dedifferentiation and growth of roots from the stem is initiated by auxins Indole acetic acid, IBA, NAA, and other hormones. When applied to stem cuttings they induce the formation of roots. IBA has been found to be very good in this regard.This indicates that the pre-existing m RNA protein complex is drawn towards protein synthesis : it shows increased rate of protein synthesis. Some of the proteins thus synthesised in turn activate differential nuclear gene activation. As a result of it, a cascade of events occurs. Generally there is an increase in the synthesis of all cellular proteins. At the same time there is an enhanced rate of synthesis of some specific proteins. Also some new proteins are synthesised.

Among these, tubulin, a membrane bound fraction, is synthesised at a high rate. It is believed that this protein along with microfilaments gets assembled into microtubules which orient themselves in a specific way so as to facilitate the polarity fixation for cell division. In response to auxins, some of the cells in the pericycle region get stimulated and the required macromolecules are synthesised by differential gene activation. Thus these cells endowed with these new potentialities divide and redivide, and soon they organise into root primordial. They in turn develop into new roots which penetrate through cortical cells. The time taken for this new root formation, in response to auxin supply is just 24-36 hours. This effect of auxin induced new root formation has been well exploited by horticulturists and others. The following diagram shows how auxins interact with cytokinins in eliciting growth responses.

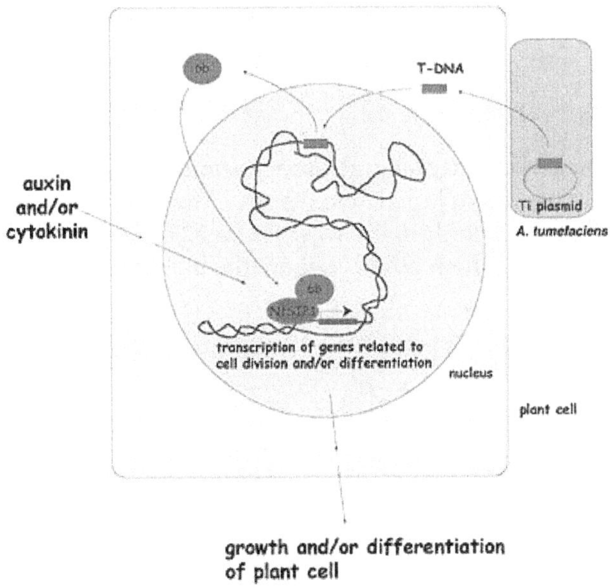

- *Apical dominance* : The active growth of terminal bud inhibits the development of lower axillary bunds. This effect is called apical dominance. Many plants like conifers and other exhibit the apical dominance so predominantly that the plant grows tall with few branches and appear conical in shape. But plants like banyan, mango, tamarind do not show that much of apical dominance, hence these plants spread outwardly. This phenomenon of apical dominance has been attributed to the differential concentration of auxin and nutrient supply to the axillary buds.
- *Abscission* : With the onset of winter season the leaves of deciduous plants turn yellow and fall off. The falling of the leaves is due to the formation of abscission layers at the base of the petioles. This layer consists of thin walled but suberised cells. The middle wall gets disintegrated and the leaves fall down. The development of this layer is stimulated by a plant hormone called ABA. But the application of auxin to such leaves prevent

the formation of abscission layer and thus prevent the falling off the leaves. This has a greater application particularly in preventing the premature falling of fruits like apple, orange etc.

- *Parthenocarpy* : Development of fruits without fertilization is referred to as parthenocarpy and such fruits are seedless. Normally plants do not produce parthenocarpic fruits. Normally plants do not produce parthenocarpic fruits.

But the application of auxin to flowers induces parthenocarpic fruit setting and prevents premature falling of fruits as well. Plants with triploid and many polyploid genomes do not set fruits. If such plants are sprayed with auxins at the time of flowering, fruit setting is enhanced. The fruits thus developed are large, seedless, and sweeter. Seedless oranges, lime, watermelon, sapota, ramphal and seethaphal, if they are produced this way, they have good market at the commercial level.

Gibberllic Acid

Gibberllic Acid is another natural hormone found distributed in higher plants. This was first discovered in Japan, later its structural and functional details have been elucidated by the people in the west. So far 52 or more different kinds have been identified. Among them GA2 2GA1 are the most common ones found in all plants.

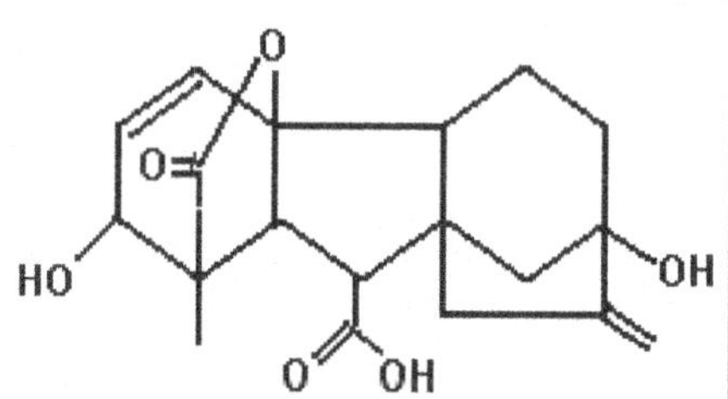

Gibberellic acid (GA_3)

GA is a steroid like molecule in its structure. Young leaves are the sites in which this hormone is mediated by phytochromes and then it is transported to the regions of growth. The most significant effect of GA on the growth of plants is it effects internodal cell elongation profoundly than any other hormones. Even genetic dwarfs (*e.g.* Dwarf Pisum sativum) grow as tall as the natural tall plants. It also brings about a dramatic effect called Bolting and Flowering. Long day plants (which require 12-14 hrs of day light treatment for flowering) under short day conditions produce a cluster of leaves on highly shortened stem and never produce flowers. When such plants are sprayed with GA, the stem starts elongation with enhanced growth of internodes. Simultaneously the terminal buds starts producing floral branches. This phenomenon is called bolting and flowering. However both bolting and flowering are separated by time and space.

Besides these above said effects, GA is essential for the mobilisation of food reserves for the germinating cereal grains like wheat and maise. At the time of germination, the young embryonic axis synthesises this hormone, which is soon released into endosperm and reaches the aleurone layer. It is this layer, GA activates differential gene expression. As a result, specific m.RNAs is transcribed and later specific proteins like amylase are synthesised on translation. The amylase thus synthesised, acts upon the starch found in the endosperm, and the glucose thus released is drawn into the embryonic tissue which is in dire need of energy source. Notwithstanding the above effects, GA also effects parthenocarpy, rooting, germination etc. This hormone also interacts with other natural hormones like auxin and cytokinin and brings about differential growth.

Cytokinins

Another class of growth regulating hormones has been discovered by Miller *et. al.* and they are called Cytokinins. They are the derivatives or nucleotides. 6-Furfuryl amino purines, Benzyl amino purines and such compounds act as cytokinins. Cytokinins are generally synthesised in the root tips, and then they are translocated to other regions. The endosperm of coconut milk, Borassus and of other plants is a rich source of cytokinins. Cytokinin level in plants is further regulated by other hormones like Auxins and GAs. On the contrary cytokinins control the synthesis of other hormones like Abscisic acid, a growth inhibitor. Cytokinin *per se* has an important role in the cytokinesis. But along with auxins, it effects differentiation of plant organs from the callus. Cytokinin by itself inhibits root formation, but in the presence of auxins, it induces organ formation like roots and shoots. If the ratio between auxin and cytokinin is high, roots are differentiated, but the lower ratio induces the formation of shoots.

The intermediate ratio stimulates the development of these hormones in differentiation at the molecular level is very interesting by unfortunately it is least understood. Cytokinins are also found to break dormancy of winter buds as well seeds. The exact mechanism by means of which they over come the dormancy is not known, still it has been thought, that they effect through the suppression of IBA which is responsible for dormancy. Another interesting effect that the cytokinins produce is the prolongation of senescence (Ageing). This effect is often referred to as Richmond and Lang's effect.

Chapter 2

PLANT CELLS

THE PLANT KINGDOM

Since the time of Aristotle, biologists have sought to classify organisms. At first the purpose was ease of identification ("artificial" classification schemes). Carolus Linnaeus (1707–1778), arguably the greatest of the pre-modern Naturalists, sought to classify plants and other organisms according to affinity groups that reflected the mind of the Creator. Later, after Darwin, the goal of classification was to show evolutionary relationships ("natural" classification schemes). For the past 150 years, biologists have emphasised natural systems of classification and have attempted to define morphological criteria that reveal evolutionary relationships. We now know that morphology, the form and structure of organisms, is the end product of the actions of genes. Virtually all of the information needed to form a complete organism is encoded in its DNA sequences, both nuclear and cytoplasmic (mitochondria and chloroplasts). DNA sequence analysis has thus provided evolutionary biologists with a powerful new tool for arriving at a truly natural classification system.

On the basis of phylogenetic analyses of highly conserved DNA sequences, living organisms have been divided into three major domains : Bacteria, Archaea, and Eucarya.

- The common ancestor of all the organisms first gave rise to the Bacteria and the common ancestor of the Archaea and the Eucarya.
- The Archaea branch off from the Eucarya lineage.
- The Eucarya common ancestor acquires mitochondrial endosymbiont (an alpha-proteobacterium-like cell).
- A heterogeneous group of eukaryotes called protists branch off the lineage leading to plants, fungi, and animals.
- The common ancestor of fungi and animals form a branch, followed by a divergence into the fungal and animal lineages.

- The common ancestor of plants, green algae, red algae, and glaucophytes acquires chloroplast endosymbiont (a cyanobacterium).
- The three lineages of glaucophytes, red algae, and green algae diverge.
- Various lineages of protists acquire chloroplasts via green or red algal endosymbionts.
- The earliest branch of green algae (Green Algae I) diverge.
- The later branch of green algae (Green Algae II, Characeae, Coleochaetales) diverge.
- The remaining lineage leads to plants.

Bryophytes are small (rarely more than 4 cm in height), very simple land plants, and the least abundant in terms of number of species and overall population. Bryophytes do not appear to be in the direct line of evolution leading to the vascular plants; rather, they seem to constitute a separate minor branch. Bryophytes include mosses, liverworts, and hornworts. These small plants have life cycles that depend on water during the sexual phase. Water facilitates fertilization, the fusion of gametes to produce a diploid zygote, a feature also seen in the algal precursors of these plants. Bryophytes are like algae in other respects as well : They have neither true roots nor true leaves, they lack a vascular system, and they produce no hard tissues for structural support. The absence of these structures that are important for growth on land greatly restricts the potential size of bryophytes, which, unlike algae, are terrestrial rather than aquatic. The ferns represent the largest group of spore-bearing vascular plants. In contrast to the bryophytes, ferns have true roots, leaves, and vascular tissues, and they produce hard tissues for support. These architectural features enable ferns to grow to the size of small trees. Although ferns are better adapted to the drying conditions of terrestrial life than bryophytes are, they still depend on water as a medium for the movement of sperm to the egg. This dependence on water during a critical stage of their life cycle restricts the ecological range of ferns to relatively moist habitats.

The most successful terrestrial plants are the seed plants. Seed plants have been able to adapt to an extraordinary range of habitats. The embryo, protected and nourished inside the seed, is able to survive in a dormant state during unfavourable growing conditions such as drought. Seed dispersal also facilitates the dissemination of the embryos away from the parent plant. Another important feature of seed plants is their mode of fertilization. Fertilization in seed plants is brought about by wind- or insect-mediated transfer of pollen, the gamete-producing structure of the male, to the sexual structure of the female, the pistil. Pollination is independent of external water, a distinct advantage in terrestrial environments. Many seed plants produce massive amounts of woody tissues, which enable them to grow to extraordinary heights. These features of seed plants have contributed to their success and account for their wide range. There are two categories of seed plants: gymnosperms (from the Greek for "naked seed") and angiosperms (based on the Greek for "vessel seed," or seeds contained in a vessel). Gymnosperms are the less advanced type; about 700 species of gymnosperms are known. The largest group of gymnosperms is the conifers ("cone-bearers"), which include such commercially important forest trees as pine, fir, spruce, and redwood.

Two types of cones are present : male cones, which produce pollen, and female cones, which bear ovules. The ovules are located on the surfaces of specialised structures called cone scales. After wind-mediated pollination, the sperm reaches the egg via a pollen tube, and the fertilized egg develops into an embryo. Upon maturation, the cone scales, which are appressed during early development, separate from each other, allowing the naked seeds to fall to the ground. Angiosperms, the more advanced type of seed plant, first became abundant during the Cretaceous period, about 100 million years ago. Today, they dominate the landscape, easily outcompeting their cousins, the gymnosperms. About 250,000 species are known, but many more remain to be characterised. A typical angiosperm life cycle, that of *Zea mays* (corn). The major innovation of the angiosperms is the flower; hence they are referred to as *flowering plants*. There are other anatomical differences between angiosperms and gymnosperms, but none so crucial and far-reaching as the mode of reproduction.

FLOWER STRUCTURE AND THE ANGIOSPERM LIFE CYCLE

The flower consists of several leaflike structures attached to a specialised region of the stem called the receptacle. Sepals and petals are the most leaflike. Petals have the primary function of attracting insects to serve as pollinators, accounting for their often showy and brightly coloured appearance. The stamen is the male sexual structure, and the pistil is the female sexual structure. The pistil is composed of one or more united carpels; the pistil, or in some flowers a whorl of pistils, is sometimes referred to as the gynoecium. The stamen consists of a narrow stalk called the filament and a chambered structure called the anther. The anther contains tissue that gives rise to pollen grains. The pistil consists of the stigma (the tip where pollen lands during pollination), the style (an elongated structure), and the ovary. The ovary, the hollow basal portion of the pistil, completely encloses one or more ovules. Each ovule, in turn, contains an embryo sac, the structure that gives rise to the female gamete, the egg.

After landing on the stigma, the pollen grain germinates to form a long pollen tube, which penetrates the tissues of the style and ultimately enters the cavity of the ovary, which houses the ovule. Within the ovary, the pollen tube enters the ovule and deposits two haploid sperm cells in the embryo sac. One sperm cell fuses with the egg to produce the zygote; the other typically fuses with the two polar nuclei to produce a specialised storage tissue termed the endosperm, which provides nutrients to the growing embryo. Endosperm tissue also provides the bulk of the worldrquote s food supply in the form of cereal grains. As in conifers, in angiosperms the outer tissues of the ovule harden into a protective seed coat. Angiosperm seeds have a second layer of protective tissues, the fruit. The fruit consists of the ovary wall and, in some cases, receptacle tissue.

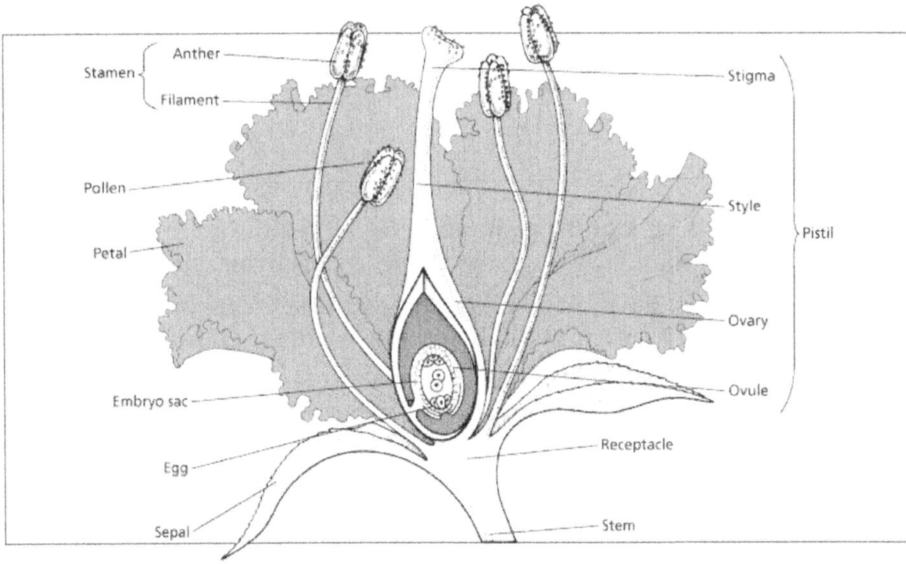

Fig. Schematic Representation of an Idealised flower of the Angiosperms.

Angiosperms are divided into two major groups, dicotyledons (dicots) and monocotyledons (monocots). This distinction is based primarily on the number of cotyledons, or seed leaves. In addition, the two groups differ with respect to other anatomical features, such as the arrangement of their vascular tissues, and their floral structure. As the dominant plant group on Earth, and because of their great economic and agricultural importance, angiosperms have been studied much more intensively than other types of plants, and they are discussed extensively in this book. Plant physiologists have focused on a relatively small number of species that represent convenient experimental systems for the study of specific phenomena. Therefore, while we focus on these famous few, it is important to keep in mind the tremendous diversity of form and function that exists within the angiosperms, and the even greater diversity of form and function that is found within the plant kingdom as a whole.

Plant Tissue Systems : Dermal, Ground, and Vascular

Dermal Tissue

The epidermis is the dermal tissue of young plants undergoing primary growth. It is generally composed of specialised, flattened polygonal cells that occur on all plant surfaces. Shoot surfaces are usually coated with a waxy cuticle to prevent water loss and are often covered with hairs, or trichomes, which are epidermal cell extensions. Pairs of specialised epidermal cells, the guard cells, are found surrounding microscopic pores in all leaves. The guard cells and pores are called stomata (singular stoma), and they permit gas exchange (water loss, CO_2 uptake, and O_2 release or uptake) between the atmosphere and the interior of the leaf. The root epidermis is adapted for absorption of water and minerals, and its outer wall surface typically does not have a waxy cuticle. Extensions from the root epidermal cells, the root hairs, increase the surface area over which absorption can take place.

Ground Tissue

Making up the bulk of the plant are cells termed the ground tissue. There are three types of ground tissue : parenchyma, collenchyma, and sclerenchyma.

- Parenchyma, the most abundant ground tissue, consists of thin-walled, metabolically active cells that carry out a variety of functions in the plant including photosynthesis and storage
- Collenchyma tissue is composed of narrow, elongated cells with thick primary walls. Collenchyma cells provide structural support to the growing plant body, particularly shoots, and their thickened walls are non-lignified so they can stretch as the organ elongates. Collenchyma cells are typically arranged in bundles or layers near the periphery of stems or leaf petioles.
- Sclerenchyma consists of two types of cells, sclereids and fibres. Both have thick secondary walls and are frequently dead at maturity. Sclereids occur in a variety of shapes, ranging from roughly spherical to branched, and are widely distributed throughout the plant. In contrast, fibres are narrow, elongated cells that are commonly associated with vascular tissues. The main function of sclerenchyma is to provide mechanical support particularly to parts of the plant that are no longer elongating.

In the stem, the pith and the cortex make up the ground tissue. The pith is located within the cylinder of vascular tissue, where it often exhibits a spongy texture because of the presence of large intercellular air spaces. If the growth of the pith fails to keep up with that of the surrounding tissues, the pith may degenerate, producing a hollow stem. In general, roots lack piths, although there are exceptions to this rule. In contrast, the cortex, which is located between the epidermis and the vascular cylinder, is present in both stems and roots. At the boundary between the ground tissue and the vascular tissue in roots, and occasionally in stems, is a specialised layer of co rtex known as the endodermis. This single layer of cells originates from cortical tissue at the innermost layer of the root cortex and forms a cylinder that surrounds the central vascular tissue, or stele.

Early in root development, a narrow band composed of the waxy substance suberin is formed in the cell walls circumscribing each endodermal cell. These suberin deposits, called Casparian strips, form a barrier in the endodermal walls to the intercellular movement of water, ions, and other water-soluble solutes to the vascular cells. Leaves have two interior layers of ground tissue that are collectively known as the mesophyll. The palisade parenchyma consists of closely spaced, columnar cells located beneath the upper epidermis. There is usually one layer of palisade parenchyma in the leaf. Palisade parenchyma cells are rich in chloroplasts and are a primary site of photosynthesis in the leaf. Below the palisade parenchyma are i rregularly shaped, widely spaced spongy mesophyll cells.

The spongy mesophyll cells are also photosynthetic, and the large spaces between these cells allow diffusion of carbon dioxide. The spongy mesophyll also contributes to leaf flexibility in the wind, and this flexibility facilitates the movement of gases within the leaf. Vascular tissues : xylem and phloem. The vascular tissue is composed of two major conducting systems : the xylem and the phloem. The xylem transports water and mineral ions from the root to the rest of the plant. The phloem distributes the products of photosynthesis and a variety of other solutes throughout the plant. The tracheids and vessel elements are the conducting cells of the xylem. Both of these cell types have elaborate secondary-wall thickenings and lose their cytoplasm at maturity; that is, they are dead when functional. Tracheids overlap each other, whereas vessel elements have open end walls and are arranged end to end to form a larger unit called a vessel. Other cell types present in the xylem include parenchyma cells, which are important for the storage of energy-rich molecules and phenolic compounds, and sclerenchyma fibres.

The sieve elements and sieve cells are responsible for sugar translocation in the phloem. The former are found in angiosperms; the latter perform the same function in gymnosperms. Like vessel elements, sieve elements are often stacked in vertical rows, forming larger units called sieve tubes, whereas sieve cells form overlapping arrays. Both types of conducting cells are living when functional, but they lack nuclei and central vacuoles and have relatively few cytoplasmic organelles. Substances are translocated from sieve cell to sieve cell laterally through circular or oval zones containing enlarged pores, called sieve areas. In contrast, sieve tubes translocate substances through large pores in the end walls of the sieve elements, called sieve plates. Sugar movement through sieve tubes is more efficient and rapid than through sieve cells and represents a more evolutionarily advanced mechanism. Sieve elements are associated with, and depend on, densely cytoplasmic parenchyma cells called companion cells. The analogous cells adjacent to the sieve cells of gymnosperms are called albuminous cells. Companion cells provide proteins and metabolites necessary for the functions of the sieve tube elements. In addition, the phloem frequently contains storage parenchyma and fibres that provide mechanical support.

CELL-STRUCTURE

CELL

Cell is the unit of structure and function. They are the building blocks of an organism. Irrespective of the nature of organisms (plant or animal) they are either made up of single cell or many cells, the former is called unicellular and later is called multicellular organisms; in the later cells are differentiated into various kinds and they are grouped into tissues, which perform special and special function.

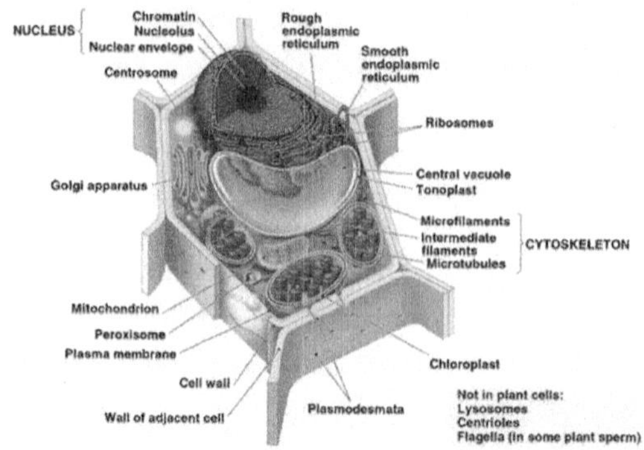

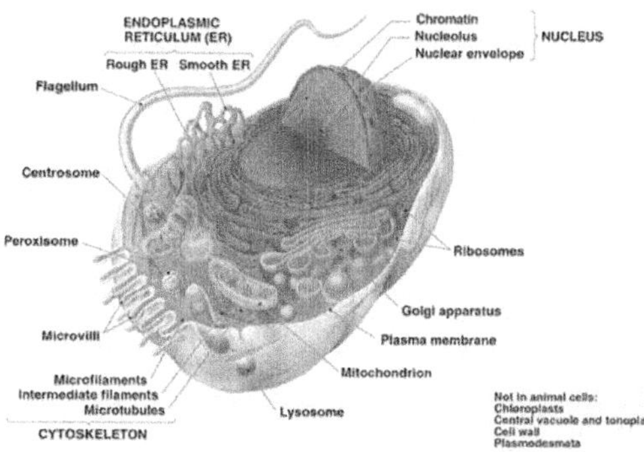

Example meristematic cells perform repeated cell divisions, phloem cells- conduct food material, sclerenchyma- mechanical, xylem-conduction of water and mineral salts and so on. Nevertheless, all these different types of cells are derived from the same embryonic cells. The development of various cell types from a single cell is determined and regulated by a process called differentiation, which in turn is controlled by differential concentrations of plant hormones. Added to this, organ differentiation is another fascinating aspect of development. All these processes are regulated by differential gene regulation in response to environmental stimulus and phytohormones.

Cellular Composition

All cells are made up of a semi-viscous fluid called protoplasm, which is considered as the physical basis of life, for it controls all biochemical reactions of the cell. In fact it is the microcosm of life with many secrets not known to us. The protoplasm is colloidal in nature, because many cell colloidal sised structures and macromolecules are suspended in it. It also exhibits sol and gel properties. The granular nature of the protoplasm is due to the presence of many tiny organelles. Vacuoles of various types are also found, but in plant cells when it is matured, a large central vacuole is present and it is separated from the rest of the protoplasm by a single unit membrane called tonoplast. The fluid present is called cell sap. There are many common features between plant and animal cells; the former is distinguished by the presence of distinct cell wall and plastids, which are totally absent in animal cells. However centrosomes are invariably present in animal cells and rarely in plant cells with exception of some lower plant unicellular algae like Chlamydomonas.

Chemical Composition

When cells are subjected to chemical analysis the following compounds are found (approximate value).

Compound	*Animal*	*Plant Cell*
Carbohydrates	20 per cent	30 per cent
Proteins	45 per cent	40 per cent
Lipids	30 per cent	25 per cent
RNA	1 per cent	0.4 per cent
DNA	0.2 per cent	0.4 per cent
Inorganic &	others 3.8 per cent	3.6 per cent

- *Carbohydrates* : These are organic compounds consisting of C, H and O. The basic structural components of carbohydrates are monosaccharide sugars consisting of 3 to 7 carbons. Example : Glucose (6c), Fructose (6c), Erythrose (4c), Xylose (5c), etc., such monomers by undergoing polymerisation develop into long chained polysaccharides, such as cellulose, cellobioses, starch and glycogen (animal starch). Cellulose is made up of glucose units linked by b1>4 linkages and it is an important component of cell wall. Similarly starch is also a polymer of á1-4 linked glucose units and it is the main source of energy for all living cells.
- *Proteins* : Proteins are the most important organic components of cells, for they act as structural as well as functional molecules. With out proteins life cannot exist, though DNA is the genetic material it is proteins that make it or break it; DNA is master library with information, that cannot be changed. They are made up of basic building blocks called amino acids. The polymers of amino acid residues are called polypeptides (proteins) which exhibit different structural conformation (shape). The 3-D shapes are specific and characteristic for a particular protein, thus they exhibit specific structure and function. Exemple : contractile proteins (muscles), transport proteins (Hemoglobin), enzyme proteins, hormonal proteins (insulin), antibody proteins IgG etc. Almost all biochemical activities, including growth and development are controlled by proteins, without which cell ceases to live.
- *Lipids* : Fatty acids and their derivatives are very important for two reasons, firstly lipids like lecithin, phosphotidyl ethanolamine, sphingolipids, glycolipids, steroids and others are part of the cellular membranes; thus

they contribute to the structure of the membranes. Secondly lipids also act as food reserve and provide energy by oxidation.

- *Nucleic acids* : These are the polymers of nucleotides, consisting of nitrogenous bases, phosphates and pentose sugars. There are two types of nucleic acid - 1) Deoxy-ribose nucleic acid (DNA), 2) Ribose nucleic acid (RNA). DNA is mostly found in chromosomes, it is the repository of genetic information and it provides information for the synthesis of proteins.
- *Inorganic/Organic factors* : Many inorganic metals Fe, Mg^{2+}, Mn^{2+}, Mo, ca^{2+} etc. are very important, for they are either the integral part of some organic molecules or act as co-factor in enzymatic activities. Certain organic factors like important for cellular metabolism, because many of the vitamins act as co-enzymes; these are required for enzymatic activities, without which cellular processes come to stand still.

Size and Shape

The size of cells varies from 10 micron to many centimetres in length. For example cotton fibres are several mm in length The shape ranges from spherical, isodiametric, and hexagonal to tubular. This is genetically predetermined to perform different functions. When the cell is observed through light microscopes, which may have the maximum resolution of about 2000 times, very few details can be made out. On the other hand, if sections of the cells are observed through electron microscope, which has a resolution power ranging from 50,000 to 150,000 times enlargement, even smaller structures stand out clearly. Inspite of high resolution, it is not possible to make out all the structural details.

The following structures are found in the cell :

- Cell wall,
- Plasma membrane,
- Nucleus,
- Plastids,
- Mitochondria,
- Golgi complex,
- Ribosomes,
- Cytoskeleton,
- Micro bodies,
- Centrosomes,
- endoplasmic reticulum,
- central vacuole and
- Non living cell inclusions.

Cell Wall

Only plant cells and bacterial cells posses a protective structure as the cell wall outside the plasma membrane. Bacterial cell wall is firmly adpressed to the underlying plasma membrane.

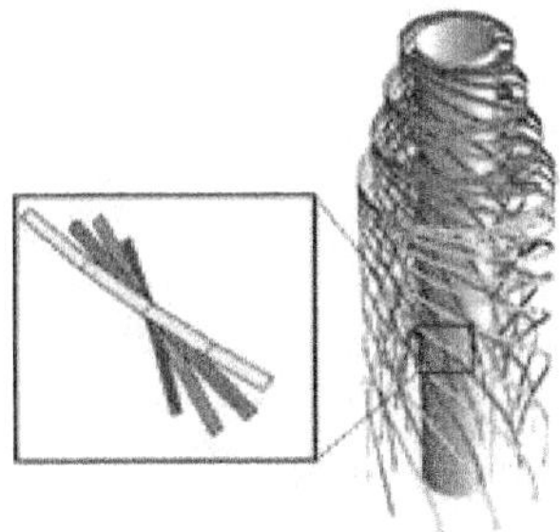

Its chemical composition and structure is more complex. However, it is basically made up of long polymers of glucosamine (NAG) and muramic acids (NAM), which in turn are cross-linked by short-pentamer oligopeptides; thus they form a mat like structural layers around the plasma membrane. Hundreds of such layers are deposited one above the other to form a very tough wall. Many bacterial cells produce a mucilaginous pectose layers outside the cell wall and this layer is called the capsule. But many bacterial cells do contain another lipid layer studded with proteins, oligosaccharides and glycol and proteoglycans.

But higher plants have cell walls mainly made up of cellulose fibres. Addition to these, pectins, hemicelluloses and lignin are also deposited on primary cellulose layer; the thickening is only at the later stages of development.

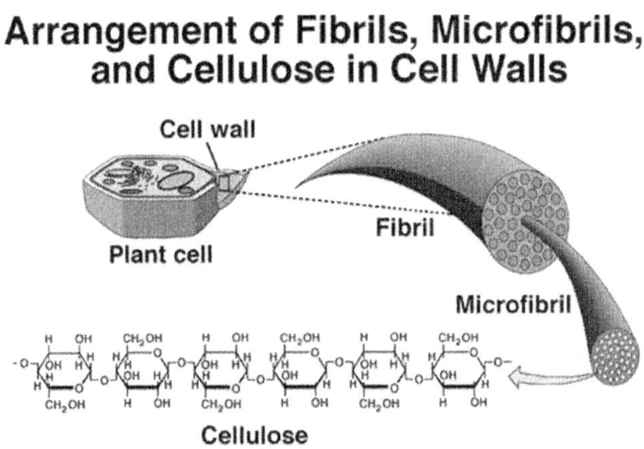

With proper staining, if a group of cells are observed, the cells appear to be held together by a kind of cementing material called middle lamella. It is made up if calcium pectate. Next to calcium pectate layer, the cell wall that is laid later is differentiated into 2 or 3 distinct layers. They are primary wall, secondary wall and tertiary wall.

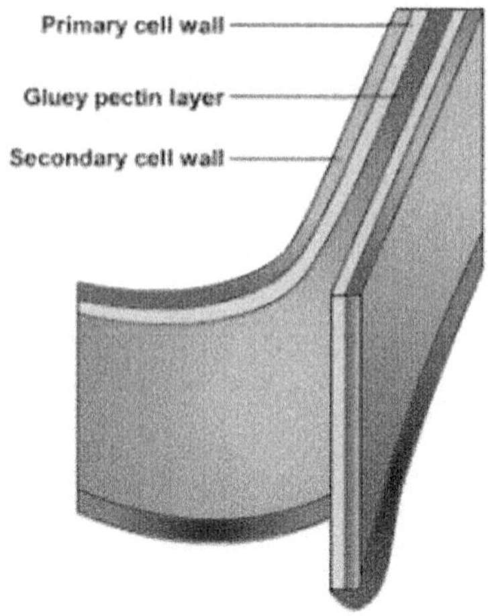

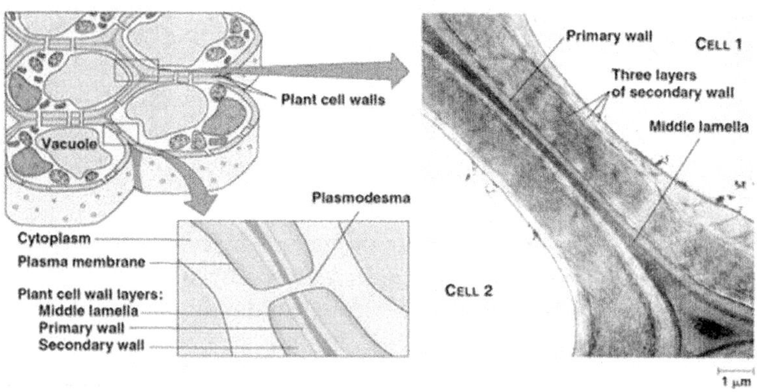

The primary wall is the first true cell wall to be deposited and it is purely made up of cellulose. With the development of the cell, the additional layers are deposited, called secondary and tertiary wall layers, like in the case xylem and sclerenchyma. Certain areas in the cell walls remain without thickening and these are called pit areas, through which fine protoplasmic strands transverse across between two cells and provide a continuum to protoplasm; they are called plasmodesmata. The primary cell wall, which is the first wall layer to be deposited, is made up of cellulose fibres. These fibres are considerably longer and deposited layers after layers; oriented longitudinally, transversely or obliquely.

The deposition and the orientation of these layers are aided by microtubules that are found on the inner face of the plasma membrane. Each cellulose fibre is made up of 8000 to 120,800 D- Glucose units, which are linked to each other by glycosidic bonds to form a long chain of glucose units, which show helical conformation. Hundreds of such cellulose threads are grouped into a bundle called micelles; these in turn are aggregated into micro fibrils which by further aggregation develop in to macro fibrils. Such micro/macro fibrils are deposited regularly either longitudinally or transversely to form uniform layers. These fibres are embedded in a matrix made up of pectate substances and hemi cellulose materials like polyxylose and others.

Primary cell wall	*Secondary cell wall*
Primarily cellulose	Few layers of cellulose
Hemicelluloses	Xylose, mannose
Pectic	Complex lignins
Elastic	Non-elastic
Laid on middle lamellae	Laid over primary wall
1-3 micron thick	>5-10 micron thick

CELL MEMBRANES

These are the most important structures of the cell, for they are responsible for protecting and separating the protoplasm from the external environment. It helps in selective uptake and transport of ions, provides surface area for many biochemical signal transduction reaction and also helps in specialised functions. In fact most of the cell organelles are bounded by membranes. Notwithstanding this, a large number of functional molecules are integrated into these membranes.

Chemical Composition

Almost all membranes are made up of proteins and lipids. The ratio between proteins and lipids may vary in different membranes, but generally it is equal. It is not uncommon to see some carbohydrates as glycoproteins associated with membrane on the outer surface of plasma membranes. Proteins found in membrane are not of the same kind, but differ in their structure, chemical composition and function. Some of them are large and many of them may be smaller. However, most of the proteins are globular in nature having either hydrophilic or hydrophobic or both the characters. Lipids, on the other hand are of various kinds. The common lipids found in the membranes are phosphotidyl choline (Lecithin), phosphotidyl ethanolamine, Phospholipids glycerol, cholesterol, etc., some of the phospholipids exhibit hydrophilic groups at one end and hydrophobic at the other end. Nevertheless, the composition of lipids varies from membrane to membrane, for they have different functions to perform.

Membrane Structure

The structural organisation of various compounds with in the membrane was an enigma for a long time. With the advancement of biological techniques, it has been found that the membranes exhibit fluid-mosaic structure (Singer and Nicholson 1972). Basically lipids form bilayers by organising hydrophobic layers facing each other and the charged region out side. Proteins, depending upon the charged nature, some are found on the surface and some are embedded. But proteins and lipids of various kinds are oriented towards each other in such a way, they exhibit semi-solid (crystalline) and semi-fluid properties. The arrangement of lipids and proteins is of mosaic pattern, where many proteins are half buried in the lipid bilayers and some traverse the entire cross-section of the lipid layer in such a way a part it is buried in the core and some are located at the peripheral surface. The structural and chemical heterogeneity is the hall mark of these membrane structures. This model explains various biological phenomena observed in most of the biological systems. Various membranes found in the cell, like plasma membrane, endoplasmic reticulum; organelle membranes exhibit basically the same structural pattern, but vary in their lipid and protein composition. Some of them are single unit membranes (plasma membrane, tonoplast, Lysosomes and peroxisomes) and some have double membrane systems (endoplasmic reticulum, nuclear membrane, golgi membranes chloroplast membrane and mitochondrial membrane). However they show differences in their chemical composition structural features.

Plasma Membrane

This is the outer most membrane of the cell, within which all protoplasmic structures are included. In plants, cell wall acts as an additional protective layer, but in animal cell, plasma membrane itself is the bounding membrane.

Small Portion of a Plasma Membrane

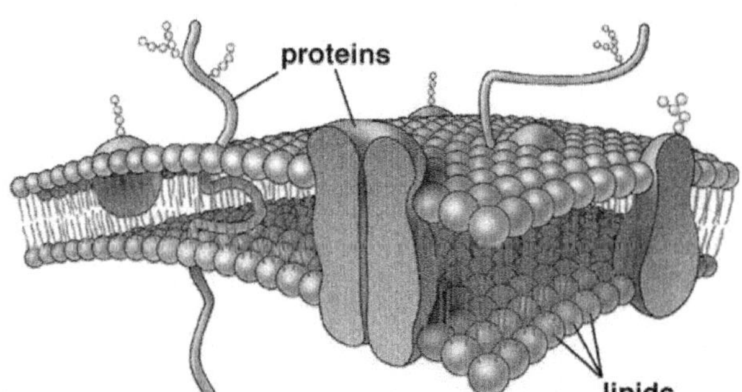

This membrane is in immediate contact with the external environment, and performs various functions like osmosis, selective absorption of various mineral nutrients, accommodates innumerable signal transducing receptors for various stimuli (electrical, light mechanical and chemical) and invariably exhibits dynamic properties. Proteins found in the plasma membrane are vectorial (directional) arranged. At some places, the plasma membrane shows inward projections and they are in continuity with the endoplasmic reticulum. Plasma membrane is also the site for pinocytosis and phagocytosis. Thus the plasma membrane exhibits unique but varied properties of its own. It exhibits dynamic fluidity never to be constant and stagnant.

Endoplasmic Reticulum

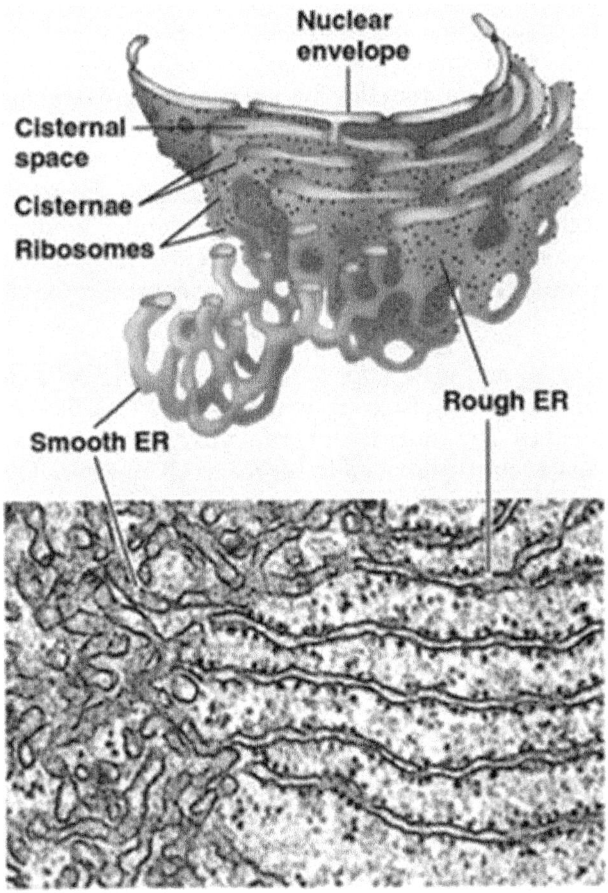

It is a labyrinthine net work of double membrane sheets. They are found in all living cells with the exceptions of mature erythrocytes and cells of bacteria. Endoplasmic reticulum (ER) occupies more than 50 to 90 per cent of total cell volume. It is in contact with the outer plasma lemma and the outer nuclear membrane. ER is made up of two single unit membranes folded to form adpressed sheets, which enclose a channel or cisternae. ER is extensively branched, thus the surface area for reactions is enormously increased.

On the basis of presence or absence of ribosomes on the surface, two classes of ER can be recognised :

- Smooth ER (SER) is without ribosomes,
- Rough ER (RER) has innumerable ribosomes on its outer surface.

These membranes are highly mobile and they undergo rapid flux, changing from SER to RER and RER to SER. Added to this, the entire ER exhibit continuous sweeping movements which help in the distribution of cellular components in the matrix. The ER membranes are supported by microtubule skeletal network. These membranes are highly dynamic and show rapid turnover. Functionally ER exhibits various activities like synthesis of proteins, mechanical support for the fluid protoplasm, synthesis and storage of lipids, synthesis and storage and transport of proteins to different destination through golgi membranes. Some of the functions are detoxification, transport of various cellular components, formation of micro bodies, formation of secretory vesicles, cell plate and others. The ER is at the heart of various cellular activities. Furthermore, during the development of nucleus and other cell organelles like, chloroplast, mitochondria, golgi complex, micro bodies, Lysosomes etc. ER provides membrane fractions to them. Exchange of membrane components among them is pervasive and common.

Golgi Bodies

A group of membrane cisternae, discovered by Camillo Golgi in 1890, are called as Golgi bodies. They are present in all cells except bacterial and blue-green algal cells. These structures vary in number from cell type to cell type. In secretory tissues like thyroid and liver they are present in large numbers than in other type of cells. They are abundant in secretory surfaces like stigma of the pistils.

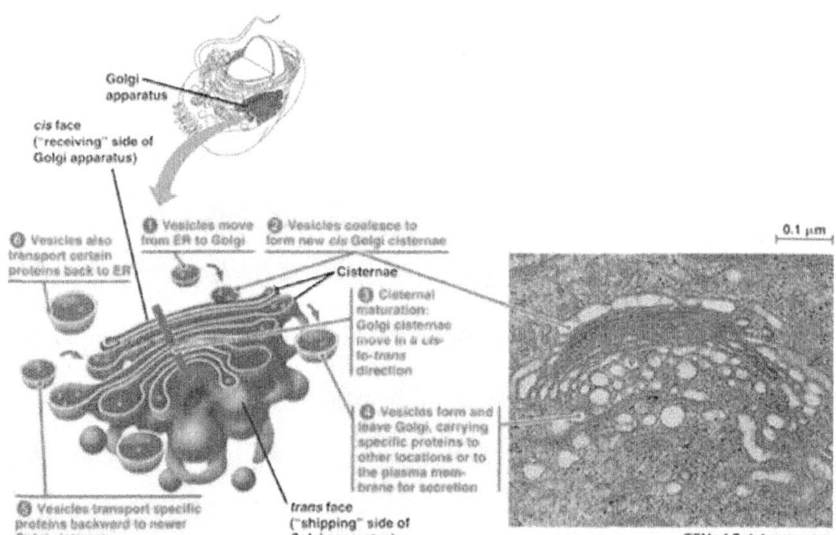

Golgi bodies are made up of a group of stacked membrane cisternae. Some times these membranes structures show extensive reticular network. Generally the distal ends of double membranes are dilated into vesicles and some of the vesicles are in the process of pinching off. An interesting feature is that the Golgi complex is surrounded by ER and at some places they look like in continuity with ER, but these proximal membranes of ER are free from ribosomes called SER (smooth ER). The stacked Golgi membranes have two faces, *i.e.* formation face is called cis face and maturation face as trans face. The formation face has convex surface and has a number of small vesicles pinched off from SER. In fact certain proteins synthesised on the RER, enter into the lumen of ER and hence they are transported in the lumen towards the transitional membranes which are in close association with Golgi complex and then the protein containing ER membranes pinch off as vesicles.

These in turn fuse with one another or with golgi cis membranes and develop into membranous sacs. Within these membranous sacs, proteins and such products get further modified. Later such products are sorted out and get enclosed and pinched off in the form of vesicles from maturation face called Trans surface of the Golgi complex. Similarly many secretary products that are synthesised on RER and transported into Golgi complex, later the matured products get enclosed in vesicles and budded off. The golgi derived vesicles are loaded with proteins that are specifically targeted to various destinations. Thus, Golgi bodies perform various complex processes, like glocolysation of proteins, synthesis of cell wall polysaccharides, maturation of zymogen granules, formation of primary lysosomes, secretion of lipid bodies, acrosome formation, neural secretion etc. Nevertheless, the participation of ER is every essential in the function of Golgi membranes. Golgi membranes are associated with transport proteins and they are responsible for transportation from one site to the other. In all the above mentioned processes, packing, maturation and secretion of specific substances are the most important events of Golgi functions. The golgi complex is at the heart of membrane flow.

WATER AND PLANT CELLS

The water properties of cohesion, adhesion and surface tension, give rise to the phenomenon of capillarity, the movement of water for small distances up a capillary tube. The smaller the tube radius, the higher the capillary rise. How far the water will move may be calculated using the following formula :

$$\text{Capillary rise} = \frac{14.9 \times 10^{-6} \text{m}^2}{\text{radius}}$$

where both capillary rise and radius are expressed in meters.

For a xylem vessel with 25 μm radius, the capillary rise is about 0.6 m. This distance is much too small to be significant for water transport up tall trees. Fibrous materials such as cell walls can act like wicks to draw water by capillarity from nearby xylem. This capillary action ensures that cell wall surfaces that are directly exposed to the air, such as those in leaf mesophyll, remain wetted and do not dry out.

Because the cell wall capillaries have a tiny radius, about 10 m^{-8}, very large physical forces can be generated in the water just below the evaporative surfaces of cell walls. From Fick's law, one can derive an expression for the time it takes for a substance to diffuse a particular distance. If one defines conditions such that all the solute molecules are concentrated at the starting position, then the concentration front moves away from the starting position. As the substance diffuses away from the starting point, the concentration gradient becomes less steep (Δc_s decreases) and thus net movement becomes slower. The time it takes for the substance at any given distance from the starting point to reach one-half of the concentration at the starting point ($t_c = ½$) is given by the following equation :

$$t_{c=1/2} = \frac{(\text{distance})^2}{D_s} K$$

where K is a constant and D_s is the diffusion coefficient. The above equation shows that the time required for a substance to diffuse a given distance increases in proportion to the *square* of that distance. Let us consider two numerical examples. First, how long it would take a small molecule to diffuse across a typical cell? The diffusion coefficient for a small molecule like glucose is about $10^{-9}\ \text{m}^2\ \text{s}^{-1}$, and the cell length may be 50 μm. Thus, for this example :

$$t_{c=1/2} = \frac{\left(50 \times 10^{-6}\text{m}\right)^2}{10^{-9}\text{m}^2\text{s}^{-1}} = 2.5\ \text{s}$$

This calculation shows that small molecules diffuse over cellular dimensions rapidly. What about diffusion over longer distances? Calculating the time needed for the same substance to move a distance of 1 m (*e.g.*, the length of a corn leaf), we find :

$$t_{c=1/2} = \frac{(1\text{m})^2}{10^{-9}\text{m}^2\text{s}^{-1}} = 10^9\ \text{s} \approx 32\ \text{years}$$

a value that exceeds by orders of magnitude the life span of a corn plant, which lives only a few months. This shows that diffusion in solutions can be effective within cellular dimensions but is far too slow for mass transport over long distances. Diffusion is of great importance as a driving force for the water vapour lost from leaves, because the diffusion coefficient for a molecule in air is much greater than in aqueous solutions.

Components of Water Potential

Students planning further study of plant water relations should note that the components of water potential defined in the text are sometimes given different names and symbols. In particular, the equation

$$\Psi_w = \Psi_s = \Psi_p$$

is often replaced by the following equivalent equation :

$$\Psi_w = -\pi + P$$

In this alternative convention, P is the same as Ψ_p. It is the hydrostatic pressure of the solution, and may be positive, as in turgid cells, or negative, as in xylem water. The symbol π is called osmotic pressure and is the negative of Ψ_s. That is, π has positive values, and Ψ_S has negative values. "Osmotic pressure" is the term that physical chemists, zoologists, and many others use to denote the effect of dissolved solutes on the free energy of water. Most handbooks of physics and chemistry use the term "osmotic pressure" and the symbol π. The negative sign in front of π in the equation above accounts for the reduction in water potential (Ψ_w) by dissolved solutes. Thus $\Psi_S = - \pi$. A very interesting, if somewhat unconventional, account of the history and physical meaning of osmotic pressure is given by Hammel and Scholander. Unfortunately, some authors have mixed the conventions for Ψ_S and π, leading to unnecessary confusion about what is meant by the symbol π. Thus, π is sometimes incorrectly called osmotic potential instead of osmotic pressure, and it may be used either as a positive quantity or as a negative quantity. Regarding matric potential, it is usually designated by the symbol t.

Turgor Pressures Exist

One of the challenging aspects of understanding plant water relations is the remarkable range of pressures, both positive and negative, that occur within the bodies of plants. Negative pressures (or tensions), which depend upon the cohesive strength of water coupled with the strength of lignified cell walls to resist deformation, play an important role in water transport through the xylem. Positive pressures, which depend upon the semi-permeable nature of the plasma membrane and the elastic nature of primary cell walls, occur in all hydrated living plants cells but can be especially large in sieve tubes and guard cells. Living plant cells are typically assumed to have only positive pressures. However, there appears to be no reason that negative pressures could not also occur within the cytoplasm of living plant cells. This topic explores the existence and potential role of negative turgor pressures in plants. When a cell loses water in air, turgor declines and solute concentrations increase. At the turgor loss point ($\Psi_p = 0$), the hydrostatic pressure in the cell sap is equal to atmospheric pressure, meaning that no net force is exerted on the cell wall. If water continues to be lost from the cell, the pressure within the cytoplasm drops below atmospheric pressure, resulting in a force imbalance that collapses the cell wall. The deformation of living cells upon desiccation is called cytorrhysis. Note that the plasma membrane is pressed against the cell wall throughout desiccation (*i.e.*, plasmolysis does not occur) because the hydrostatic pressure in cytoplasm remains greater than the hydrostatic pressure in apoplast.

In the living cells described above, a decrease in cell water potential below the turgor loss point is balanced by an increase in solute concentration as the volume of the cell decreases. Thus, true "negative" pressures do not develop. In contrast, xylem conduits have rigid cell walls that resist deformation, allowing them to sustain negative pressures without imploding. This raises the question of what happens when water is lost from living cells that have thick walls or are embedded in a rigid matrix of cells (*e.g.*, xylem parenchyma). Might these cells resist deformation and thus develop negative turgor pressures? It is important to emphasise that this is a controversial area, due in large part to the absence of direct measurements of negative turgor pressures within cells. However, before reviewing the evidence for negative turgor pressures, it is worthwhile to consider what physiological effects might result from the development of such negative pressures in living cells. The major outcome of negative turgor is that it allows stiff-walled cells to decrease in water potential without undergoing major changes in cell volume or osmotic concentration. Because cytorrhysis might cause physical damage to the wall and/or cell membranes, while the high concentrations of solutes resulting from the reduction in cell volume might adversely affect the conformation of membranes and proteins, it is possible to imagine a physiological role or benefit for the development of negative turgor pressures in plant cells. We now turn this around and ask if there are any downsides to the generation of negative turgor pressures in living cells? In the xylem, the primary issue associated with negative pressures is cavitation; could this also be a concern for living cells? The primary mechanism for cavitation in plants is air seeding, reflecting the fact that the probability of the *de novo* formation of gas voids in water (either by homogeneous or heterogeneous nucleation) is extremely low. In air seeding, air is sucked in through the cell wall, where it then "nucleates" the transition to the vapour phase. For air seeding to occur in a living cell experiencing negative pressures, air would have to be pulled through the very small pores of the cell wall. While one can imagine this happening, the movement of air across the cell wall would result in plasmolysis and thus the immediate release of any tension in the cytoplasm because the plasma membrane is not capable of withstanding any significant pressure.

Other costs associated with the ability of living cells to generate negative pressures include the metabolic costs of producing rigid cell walls, as well as any limitations lignified walls might place on physiological function. In addition, a strategy for avoiding cell damage due to desiccation via the generation of negative pressures might impose limitations on cell size. The strength of a cell to withstand collapse (and thus generate negative pressures) is inversely proportional to cell size. Evidence for negative turgor pressure in plants is limited, in part reflecting the fact that few researchers have devoted much attention to this issue. Living cells with flexible (unlignified) cell walls deform relatively easily upon desiccation and thus do not appear to support negative pressures. However, measurements of the forces needed to collapse cell walls suggest that living cells with thick walls can withstand forces in the range of 1.0 MPa. Visual examination of tissues adapted to withstand very cold temperatures also provides indirect evidence for negative turgor. For example, frozen ray parenchyma cells often do not exhibit significant deformation, despite the very strong desiccatory effects (low water potential) of extracellular ice. One can hypothesise that the existence of negative pressures within these cells acts to balance the water potential gradient across the cell membrane (*i.e.*, between the cytoplasm and ice formed within the apoplast). In summary, negative turgor pressure remains an intriguing but little-studied area of plant water relations. While it does not appear to form in cells such as the leaf mesophyll, that can deform easily, its existence in living cells with either lignified cell walls or where the cell is embedded in a matrix of lignified tissues (*e.g.*, living cells within wood) cannot be ruled out. The potential benefits of negative turgor pressures in terms of preventing mechanical and osmotic damage associated with severe desiccation makes this topic worthy of further study.

The Matric Potential

In discussions of soils, seeds, and cell walls, one often finds reference to yet another component of Ψ_w : the matric potential (Ψ_m). The matric potential is used to account for the reduction in free energy of water when it exists as a thin surface layer, one or two molecules thick, adsorbed onto the surface of relatively dry soil particles, cell walls, and other materials. The matric potential does not represent a new force acting on water, because the effect of surface interactions can theoretically be accounted for by an effect on Ψ_s and Ψ_p. In dry materials, however, this surface interaction effect often cannot easily be separated into Ψ_p and Ψ_s components in dry materials, so they are frequently bulked together and designated as the matric potential. It is generally not valid to add Ψ_m to independent measurements of Ψ_s and Ψ_p to arrive at a total water potential. This is particularly true for water inside hydrated cells and cell walls, where matric effects are either negligible or they are accounted for by a reduction in Ψ_p. For instance, the negative pressure in water held by cell wall microcapillaries at the evaporative surfaces of leaves, discussed on page 45 of the textbook, is sometimes described as a wall matric potential. Care is needed to avoid inconsistencies when accounting for this physical effect in definitions of Ψ_p, Ψ_s, and Ψ_m.

Measuring Water Potential

Plant scientists have expended considerable effort in devising accurate and reliable methods for evaluating the water status of a plant. Four instruments that have been used extensively to measure Ψ_w, Ψ_s, and Ψ_p are described here: psychrometer, pressure chamber, cryoscopic osmometer, and pressure probe.

Psychrometer (Y_w Measurement)

Psychrometry (the prefix "psychro-" comes from the Greek word *psychein*, "to cool") is based on the fact that the vapour pressure of water is lowered as its water potential is reduced. Psychrometers measure the water vapour pressure of a solution or plant sample, on the basis of the principle that evaporation of water from a surface cools the surface.

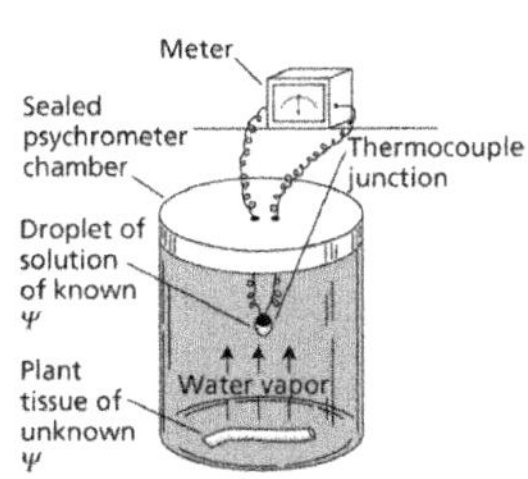

Fig. Diagram illustrating the use of Isopiestic Psychrometry to Measure the Water Potential of a Plant Tissue.

One psychrometric technique, known as *isopiestic psychrometry*, has been used extensively by John Boyer and coworkers. Investigators make a measurement by placing a piece of tissue sealed inside a small chamber that contains a temperature sensor (in this case, a thermocouple) in contact with a small droplet of a standard solution of known solute concentration (known Ψ_s and thus known Ψ_w). If the tissue has a lower water potential than that of the droplet, water evaporates from the droplet, diffuses through the air, and is absorbed by the tissue. This slight evaporation of water cools the drop. The larger the difference in water potential between the tissue and the droplet, the higher the rate of water transfer and hence the cooler the droplet. If the standard solution has a lower water potential than that of the sample to be measured, water will diffuse from the tissue to the droplet, causing warming of the droplet. Measuring the change in temperature of the droplet for several solutions of known Ψ_w makes it possible to calculate the water potential of a solution for which the net movement of water between the droplet and the tissue would be zero signifying that the droplet and the tissue have the same water potential.

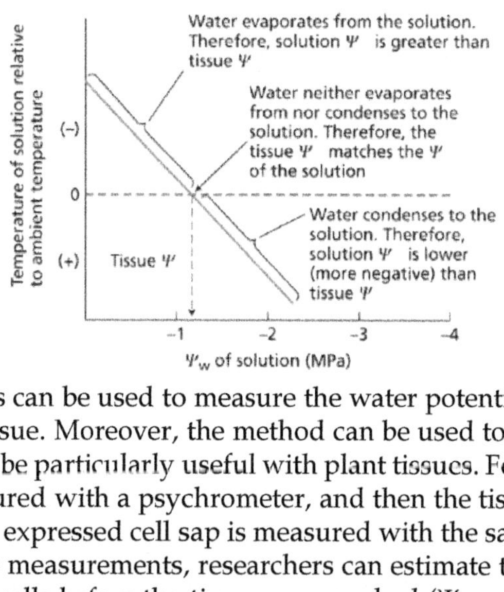

Psychrometers can be used to measure the water potentials of both excised and intact plant tissue. Moreover, the method can be used to measure the Ψ_s of solutions. This can be particularly useful with plant tissues. For example, the Ψ_w of a tissue is measured with a psychrometer, and then the tissue is crushed and the Ψ_s value of the expressed cell sap is measured with the same instrument. By combining the two measurements, researchers can estimate the turgor pressure that existed in the cells before the tissue was crushed ($\Psi_p = \Psi_w - \Psi_s$). A major difficulty with this approach is the extreme sensitivity of the measurement to temperature fluctuations. For example, a change in temperature of 0.01°C corresponds to a change in water potential of about 0.1 MPa. Thus, psychrometers must be operated under constant temperature conditions. For this reason, the method is used primarily in laboratory settings.

Pressure Chamber (Y_w Measurement)

A relatively quick method for estimating the water potential of large pieces of tissues, such as leaves and small shoots, is by use of the pressure chamber. This method was pioneered by Henry Dixon at Trinity College, Dublin, at the beginning of the twentieth century, but it did not come into widespread use until P. Scholander and coworkers at the Scripps Institution of Oceanography improved the instrument design and showed its practical use. In this technique, the organ to be measured is excised from the plant and is partly sealed in a pressure chamber. Before excision, the water column in the xylem is under tension. When the water column is broken by excision of the organ (*i.e.*, its tension is relieved allowing its Ψ_p to rise to zero), water is pulled rapidly from the xylem into the surrounding living cells by osmosis. The cut surface consequently appears dull and dry. To make a measurement, the investigator pressurises the chamber with compressed gas until the distribution of water between the living cells and the xylem conduits is returned to its initial, pre-excision, state. This can be detected visually by observing when the water returns to the open ends of the xylem conduits that can be seen in the cut surface. The pressure needed to bring the water back to its initial distribution is called the *balance pressure* and is readily detected by the change in the appearance of the cut surface, which becomes wet and shiny when this pressure is attained.

The diagram at left shows a shoot sealed into a chamber, which may be pressurised with compressed gas. The diagrams at right show the state of the water columns within the xylem at three points in time : (A) The xylem is uncut and under a negative pressure, or tension. (B) The shoot is cut, causing the water to pull back into the tissue, away from the cut surface, in response to the tension in the xylem. (C) The chamber is pressurised, bringing the xylem sap back to the cut surface. The pressure chamber is often described as a tool to measure the tension in the xylem. However, this is only strictly true for measurements made on a non-transpiring leaf or shoot (for example, one that has been previously enclosed in a plastic bag). When there is no transpiration, the water potential of the leaf cells and the water potential in the xylem will come into equilibrium.

The balancing pressure measured on such a non-transpiring shoot is equal in magnitude but opposite in sign to the pressure in the xylem (Ψ_p). Because the water potential of our non-transpiring leaf is equal to the water potential of the xylem, one can calculate the water potential of the leaf by adding together Ψ_p and Ψ_s of the xylem, provided one collects a sample of xylem sap for determination of Ψ_s. Luckily Ψ_s of the xylem is usually small (> -0.1 MPa) compared to typical midday tensions in the xylem (Ψ_p of –1 to –2 MPa). Thus, correction for the Ψ_s of the xylem sap is frequently omitted. Balancing pressure measurements of transpiring leaves are more difficult to interpret. The fact that water is flowing from the xylem to the leaf means that differences in water potential must exist. When the transpiring leaf or shoot is cut off, the tension in the xylem is instantly relieved and water is drawn into the leaf cells until the water potentials of the xylem and the leaf cells come into equilibrium.

Because the total volume of the leaf cells is much larger than the volume of sap in the xylem, this equilibrium water potential will be heavily weighted towards that of the leaf. Thus, any measurement of the balancing pressure on such a leaf or shoot will result in a value that is approximately the water potential of the leaf, rather than the tension of the xylem. (To be exact, one would have to add the Ψ_S of the xylem sap to the negative of the balancing pressure to get the leaf water potential.) One can explore the differences between the water potential of the xylem and the water potential of a transpiring leaf by comparing balancing pressures measured on covered (*i.e.*, non-transpiring) versus uncovered (transpiring) leaves. Pressure chamber measurements provide a quick and accurate way of measuring leaf water potential. Because the pressure chamber method does not require delicate instrumentation or temperature control, it has been used extensively under field conditions. For a more complete description of the theory and operation of the pressure chamber.

ORIGIN AND BIOGENESIS OF MITOCHONDRIA

Earlier views about the origin of mitochondria from the plasma membranes about nuclear membranes are now discarded, for there is neither substance nor facts to substantiate such claims.

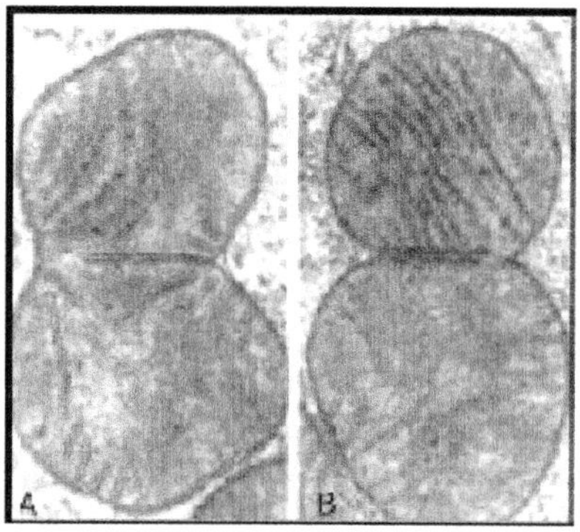

Fig. Mitochondrial Division

Recent studies have unequivocally showed that mitochondria take their origin in the cells from either pre-existing mitochondria or promitochondria. During mitochondrial biogenesis, genes of both nuclear genome and mitochondrial genome products interact. Though mitochondrial DNA codes for some 13 mitochondrial proteins, tRNA & rRNA, other components are coded for by nuclear genes, and they are synthesised in cytoplasm. Then they are transported into mitochondria through mitochondrial membrane transport complex, to form mitochondrial functional structures.

Vacuoles

Vacuoles are the spaces in the cells surrounded by single unit membranes. Don't consider them as empty spaces. In plants, mature cells have a single large centrally located vacuole occupies nearly 50-70 per cent of the cellular volume. It is filled with cell sap. It is separated from the rest of the cytoplasm by a single unit membrane called tonoplast.

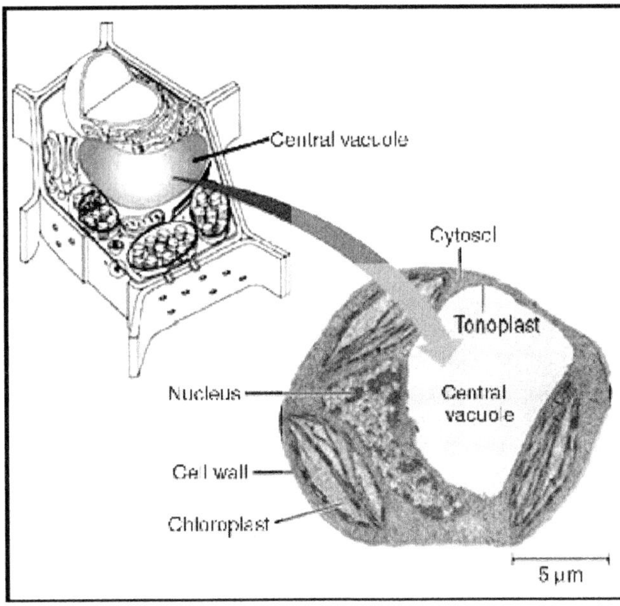

In meristematic cells, vacuoles are absent to being with, but as the cell expands, a number of small vacuoles arise from the endoplasmic reticulum; later as the cells further enlarge, smaller vacuoles fuse with one another to form a single large vacuole. The content of the cell vacuole is important in maintaining the turgidity of the cells. Furthermore, these vacuoles have been found to show lysosomal activities too. Nevertheless, vacuoles found in animal cells have drawn the attention, because they perform some important functions. Contractile vacuoles and food vacuoles are notable among them. Contractile vacuoles are helpful in the excretion and others help in digesting the ingested food materials. Thus vacuoles play different roles in different organisms.

Cytoskeleton

Microfilaments and Microtubules

Organisms, whether they are plants or animals, are made up of various types of cells, which have various shapes and functions. But all these cells are derived from the same mother cell called zygote. The shapes and functions of these cells is defined and determined at the time of differentiation (the topic is more complex). Within these cells, cytoplasm with its components like Intermediate filaments (IF), microtubules (MT and microfilaments (MF) play important roles in determining cellular shape, structure and function. These structures act as the internal cytoskeleton and give a kind of mechanical strength to the cytoplasm and provided directions for the movement of cytoplasm, which makes the cytoplasm as a dynamic fluid.

Microtubules

Microtubules are fine tubular protein structures having dimensions of 200 A0 in diametre and many microns in length. The wall of the individual tubular structure is made up of 13 protein sub-units called tubulins which exist in two forms called α and β sub-units; they are arranged alternatively. Hence the basic building blocks of microtubules are Tubulins of molecular weight 55000-58000 daltons. These sub-units alternatively get polymerised at the membrane bound nucleating centres into tubular structures. This polymerisation can be inhibited by colchicine, an alkaloid extracted from the tubers of Colchicum autumnale. Such microtubules are dispersed and distributed in different patterns in the cytoplasm. Some form a kind of network in the cytoplasm, some are oriented towards nuclear membrane, and some run parallel to the plasma membrane in longitudinal direction of the cell.

These structures being quite rigid, they are able to provide a structural support to various cells. Microtubule are associated with Endoplasmic reticulum for its sustainability and structural stability, so also golgi membranes. Not withstanding being the cytoskeleton structures, they are also involved in the structural organisation of cilia, flagella, centrioles, basal granules, mitotic apparatus, neurotubules etc.

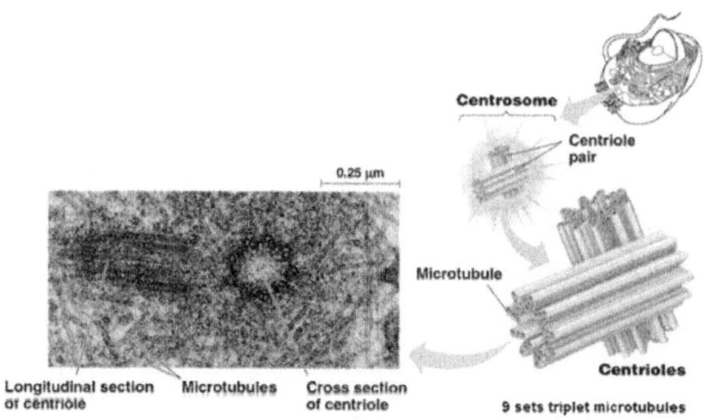

Microtubule Structure

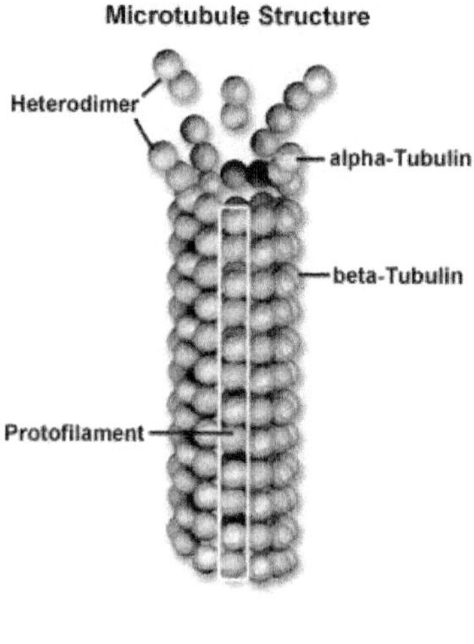

Such being the case, the function of not just being cytoskeleton support to the aqueous cytoplasm, but they are also involved in various functions like maintenance of the cell shape, cytoplasmic fluidity, membrane movement, chromosomal movement, cytokinesis, cell wall deposition, flagellar movement, sensory transduction, transport of cellular materials, cell polarity, morphogenesis and others to say the cytosolic goods to respective destinations and they aided by motor proteins such as kinesins, dyeinins and few more. They are involved in transport of goods along the MTs to their respective destinations. Tubulins are the most abundant proteins in the organic world next to RuBP carboxylase.

Microfilaments

Along with microtubules, cytoplasm also contains category of thin protein fibres called microfilaments. They represent the contractile system of the cytoplasm. Microfilaments are very fine protein filaments of 6-10 nm thickness and 10-100nm in length. These are made up of actin sub-units and they can be associated with myosin and troponin proteins. Actin accounts for more than 20 per cent of the total cellular proteins. Actin consists of basic building blocks called globular proteins called G-actin, of molecular weight 45 Kd. These globular units, by undergoing polymerisation, form into fibrous actin called F-actin. Normally two such F-actin proteins are helically coiled to each other and such filaments are associated with troponin.

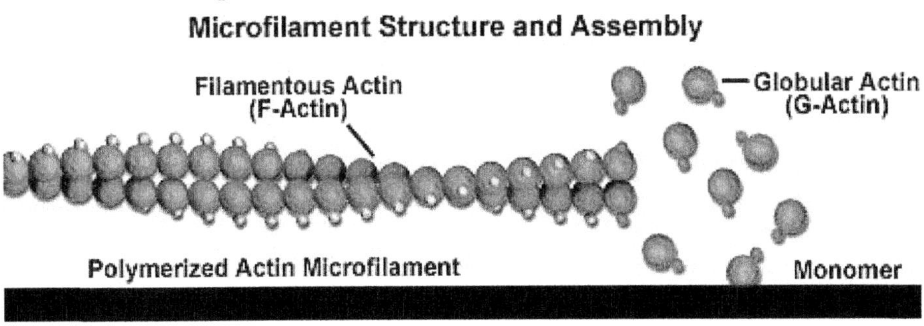

These filaments are further decorated with : Myosin proteins with calcium binding sites and ATPase proteins. These proteins are not just restricted to muscles, but they are also found in cytoplasm of almost all kinds of cells. The interaction between myosin and actin proteins generates the force for movement.

The transformation of G-Actin to F-Actin and *vice versa* represents the interconversion of sol-gel system of the cytoplasm is the dynamic force for cytoplasmic fluidity. The above said microfilaments are often located just beneath the plasma membranes in the form of bundles. These in turn are interconnected by a network of similar filaments which pervade the entire cytoplasm. These are also in contact with small membranous vesicles, microtubules, nucleus, polysomes, root lets of flagella etc. However mitochondria appear to be free from such filaments. These microfilaments, particularly actin filaments undergo depolymerisation in the presence of an alkaloid called cytochalasin B. If this drug is supplied to living cells, it inhibits many cellular activities like cytoplasmic streaming, migration of cells, endocytosis, exocytosis and even cell-polarity fixation. The above said features clearly suggest the involvement of microfilament is vital in various cellular activities. For example, protoplasmic streaming in many plasmodia, plant cells and amoeboid movement of many protozoans is controlled by the activity of these microfilaments. Recent investigations have shown that in cancer cells the microfilaments are in a disorganised state, which certainly indicates the involvement of microfilaments the induction of cancer.

Centrioles

Centrioles are the characteristic organelles of animal cells. Invariably they are present in all animal cells, but they are almost absent in plant cells with the exception of some lower unicellular algae. They are found in the cytoplasm at one pole of the nucleus. It is made up of a pair of dark granular structures called centrioles, which in turn are surrounded by an amorphous region called centrosphere. Each centriole is made up of two open barrel shaped structure measuring 0.2 um μ 0.5 um, sometimes they may be as long as 2um. These two structures are oriented at right angle to each other.

Structure

The wall of the open barrel shaped structure is made up of nine groups of microtubules arranged in a circle. Each group has three microtubules and skewed arrangement, towards the centre. These three microtubules are named A, B & C starting from the centre towards periphery. Compared to the proximal structure of the flagella, the central pair of tubules are absent. However, in the case of centrioles, which develop into cilia, the wall is made up of 9 groups of tubules and each has two microtubules and there are two centrally located microtubules. Chemically centrioles are made up of proteins like tubulin, dynein (ATPase) and other contractile proteins. Added to this, the report about the presence of DNA is very interesting. However the chemical nature of the peripheral amorphous region is not known.

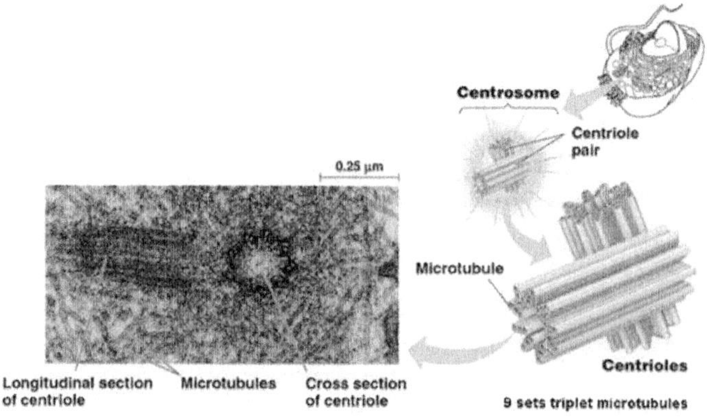

Centrioles have dual functions *viz* :

- Cilia and flagella in lower organisms and other cells take their origin from centrioles.
- During cell division, centrioles help in the organisation of mitotic apparatus in a particular plane.

During cell division, particularly in animal cells, centrioles undergo duplication, so that centrioles produce procentrioles which are oriented at right angles to the mother centrioles. The organisation of microtubules during the formation of daughter centrioles is not clear. Whether the message for the synthesis of these proteins resides in the DNA found in centrioles or in the nucleus is not clear. How the peripheral cytoplasmic ribosomes are involved in the synthesis and secretion of these proteins is again a mystery. Nevertheless, during cell division the two centrioles separate, at the same time a number of microtubular structures emerge from each of the centrioles and radiate in all directions and inter-link with the two centrioles. Thus, these centrioles move away from one another towards predetermined polar directions. However, the mitotic fibres that develop during cell division do not take their origin from the centrioles.

PLANT CELL DIAGRAM AND FUNCTIONS

It is a commonly known fact that plants are highly eukaryotic organism. Their cells are membrane bound and are with many cell organelles. Even animals belong to eukaryoutic groups, but there are certain distinct differences between plant cell parts and animal cell parts. Plant cells possess very well developed cell walls, while an animal cell does not have that developed cell wall. Plant cell have many such distinct differences, which can be observed on studying the plant cell model. Cytoplasm and other organelles are very typical plant cell parts. Many scientific researches have been made to gain knowledge about various aspects of plants cell parts. Like every mechanical machine, which has dedicated parts for specific functions, plant cell parts also have highly specific and dedicated function.

A plant cell is actually a large microscopic 'fully functioning' city, literally and figuratively, housing miniature structures known as organelles. It is fun to teach about a plant cell diagram and definitions, when your kid is already intrigued with the subject. Trouble is when he really isn't and would rather be doing anything else in the world than study biology. Yet, learning and teaching plant cell diagrams to children can actually be fun. You can use plant cell diagram for kids to make miniature models and the project will be fun as well as help your child learn. Kids love making and building things. So a plant cell diagram model set up on a sheet of thermoses, using interesting things to make up the internal organelles should really excite your child. Before you do that though, brush up your own knowledge of the labelled plant cell diagram and functions from this article. Know more on the plant cell structure and parts.

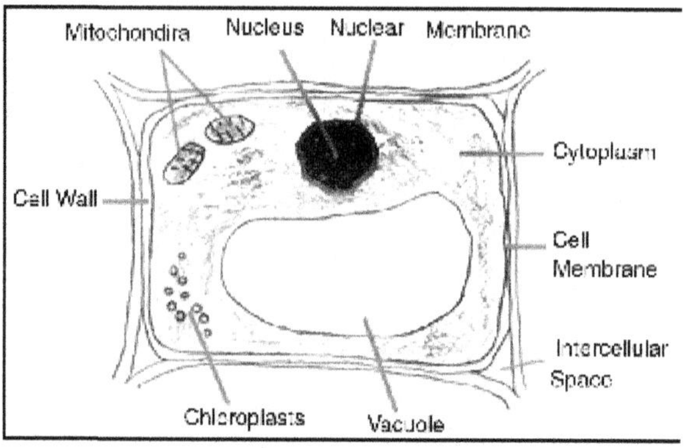

Fig. Labelled Plant Cell Diagram Sketch by Abhishake Sharma.

Functions of Various Organelles

Let us begin with taking each internal organelle found in the plant cell diagram and highlighting its function and purpose within the plant cell. The most important thing to note are the differences that make a plant cell so different from an animal cell. A typical plant cell is distinguished from a typical animal cell as it contains things like the cell wall, vacuoles, chloroplasts and plasmodesmata, that the animal cells do not contain. Here's a look at a plant cell diagram with definitions of all the internal structures. Also know more on the similarities between plants and animal cells. This is a plant cell organelle, that store starch and is not found in all plant cells. It is usually only found in plants that are starchy in nature, like tubers and some fruits. These plastids are non-pigmented organelles that synthesise starch granules to convert into sugar, when the plant requires energy.

Alternately referred to as the ATP, this is a high energy molecule that stores up energy. ATP is produced by the plant cell in the cristae of the mitochondria and chloroplasts and supports the important function of energy transfer within the plant cells. This multifunctional nucleotide provides the energy for cellular processes like biosynthetic reactions and cell division. Cell Membrane This is a thin wall that is found on the inside of the cell wall. It is a layer made up of protein, fats and cellular fibre, which provides the plant cell organelles with support and structure. Also known as the plasma membrane or plasmalemma, the cell membrane is the semi-permeable biological separation, between the cell insides and the cell outsides. This is the think and rigid outer cover that lies above the cell membrane and surrounds the entire plant cell. This cell wall bonds with other plant cell walls and forms the structure of the plant that we know of. Made up of cellulose fibre, the cell wall is tough and acts as a filtering mechanism for the plant cell. Its most important function is to maintain internal plant cell pressure and prevent over-expansion when water enters the cell. Also known as the microtubule organising centre, the centrosome is 'centriole free' structure that has radiating tubules originating from a dense centre. Found near the nucleus of the plant cell, the centrosomes produce the microtubules that regulate the cell-cycle progression. When the plant cell divides (mitosis), the centrosome also divides into two parts that move in the opposite directions. If you draw a coloured plant cell diagram, you will have to make a green coloured molecular structure within, to show chlorophyll. Chlorophyll is a molecule that carries on the process of photosynthesis. This is the process of producing sugar and oxygen using light energy, water and carbon dioxide. This usually green organelle is magnesium based and is found in many different molecular structures.

- Chloroplast : This is usually the elongated or disc shaped photosynthesis site that contains the chlorophyll. Also a part of the 'plastids' group, chloroplasts are similar to mitochondria, but are only found in plants and protista. Chloroplasts have their own DNAs and are protected by the surrounding two lipid-bilayer membranes. Chromatin is a combination of DNA and protein that is highly complex in nature and is central to the makeup of the chromosomes. Found inside the nuclei of eukaryotic cells, the chromatin plays important roles to pack DNA into smaller volumes (to fit in the cell), control DNA replications and allow mitosis and meiosis. Cristae are the finger-like projections that form the folded inner membranes of the mitocondria. ATP is generated in the walls of the cristae and they also help in cellular respiration. Made up of proteins, ATP synthase and various cytochromes, the cristae increase the surface area of the plant cell, on which the various reactions can actually take place. The cytoplasm is essentially the jelly-like substance outside the plant cell nucleus. This is where all the plant cell organelles are situated and it is the substance that is entirely encircled by the cell wall. The contents of the nucleus are called the nucleoplasm and are not part of the cytoplasm, but all the other plant cell organelles are are indeed a part of it. Know more about the structure and functions of cytoplasm.
- Golgi Body : Also known as the golgi apparatus or the golgi complex, the golgi bodies are flat, layered sac-like organelles, that are located near the nucleus. These bodies do the important work of packing proteins and carbohydrates into membrane-bound vesicles. These vesicles are then exported from the cell.
- Mitochondrian : A mitochondrian is a round or rod shaped organelle with a double layered membrane. The inner membrane is double folded and forms projections known as the cristae. The mitochondrian performs the essential task of converting energy stored in glucose form into ATP. Know more on mitochondria : structure and functions. This is spherical body that contains various organelles like the nucleolus (where ribosomal RNA is produced) and surrounded by a nuclear membrane. The nucleus is the control room that controls various cell functions by controlling the protein synthesis of the plant cell. The nucleus contains DNA within the chromosomes. Know more on cell nucleus : structure and functions. These are the sites that see protein synthesis and are basically nothing but small cell organelles made up of RNA-rich cytoplasmic granules. These mol-

ecules make proteins out of amino acids. The word 'ribosome' is derived from the name 'ribonucleic acid' that is biologically very important for the plant cell.

- Rough Endoplasmic Reticulum : Also known as rough ER, this is a vast interconnected system of membranes, infolded tubular structures and convulted sacs that are found in the cell's cytoplasm. This is a network that is covered with ribosomes and these are what give it a rough appearance. These networks transport materials through the cell and produce proteins in sacs known as cisternae. These are in turn sent to the golgi bodies and deposited into the cell membrane.
- Smooth Endoplasmic Reticulum : Also known as smooth ER, this is the same thing as the rough ERs, but with the main difference that they are smooth. They bud off from the rough Ers and also transport materials within the cell. The ER channels, known as the ER lumen, are used to transport proteins and lipids to the golgi bodies and the membranes. These are parts of the plant cell chloroplasts and are located between the grana within their inner cell membranes. A stroma is essentially the fluid matrix that surrounds the thylakoids. These are chloropyll containing disk shaped membrane structures found inside the chroloplasts. Chloroplasts are actually made up of a stack of thylakoid discs that aid in the process of photosynthesis. The stack of thylakoid disks in the chloroplasts are known as grana (singular granum) Vacuole. These are large fluid-filled, membrane-bound spaces within the plant cell. They help in maintaining the cell shape and most plant cells have just one, single vacuole that represents upto 90 per cent of the total plant cell. They contain ions, sugar, secondary metabolites and enzymes. A vacuole is surrounded by a membrane called the tonoplast. There are many other small structures found within the plant cell like the intercellular air spaces (gap between two plant cells) and the peroxisome (has a chrystalline core and helps in removing hydrogen from substrates), but I will conclude my 'plant cell diagram' article here. Hope my overview of the plant cell diagram helps you with your task of educating your child with a plant cell model.

PLANT CELL ORGANELLES

Plants are highly evolved eukaryotic organisms that comprise of membrane bound cell organelles. Even though plants and animals belong to eukaryotic groups, they differ in certain characteristic features. For example, a plant cell possesses a well-developed cell wall and large vacuoles, while an animal cell lacks such structural parts. In addition to these, a plant cell lacks centrioles and intermediate filaments, which are present in an animal cell. The cells are the basic units of life that work together to perform life sustaining functions in both animal and plant worlds. Plant cells are eukaryotic cells having thick and rigid cell walls. Eukaryotic cells are cells that contain complex structures enclosed within membranes called nucleus or nucleus envelop, within which the genetic material is present. These cells are present in almost all living organisms including animals, plants and fungi.

Plant Cell Model and Parts

Plant cells differ from animal cells in that they have three different structures known as cell wall, vacuoles and plastids. However, they lack centrioles and intermediate filaments which are present in animal cells. Read more on plant cell structure and parts :

- *Cell Membrane* : It permits the waste material to exit the cell and regulates the movement of materials in and out of the cell. It also acts as a barrier between the inside of the cell and the outside, so that the chemical conditions on both sides can be different.
- *Cell Wall* : It provides rigid structural support to the cell from which structures, like, leaves and stems are produced. It allows the circulation and distribution of water, minerals and nutrients in and outside the cell. The wall also controls the presence of pathogen microbes and development of tissues in the cell.
- *Golgi Body* : It stores, packages, and circulates the lipids and proteins made in the endoplasmic reticulum. It stores the proteins into packages called vesicles.
- Rough Endoplasmic Reticulum : It synthesises and exports proteins and glycoproteins throughout the cell.
- *Lysosomes* : These organelles digest food by breaking down larger molecules into smaller ones, using special proteins.
- *Cytoplasm* : It helps in maintaining the cell shape. Along with this, it plays an important role in the internal movement of cell organelles, cell mobility and muscle fibre contraction. It also distributes oxygen and nutrients to different parts of the cell.
- *Nucleolus* : It is the most prominent structure in the nucleus wherein ribosomes are made.
- *Vacuole* : The plant cell contains a large, single vacuole (an enclosed compartment) which is used to store water, compounds and minerals that help in plant growth.
- *Ribosomes* : These are packets of RNA (Ribonucleic Acid) and are called as protein builders or synthesisers of the cell.
- *Chloroplasts* : They are the food producers of the cell, containing chlorophyll, the green pigment essential for photosynthesis. Their main function is to generate sugars and starches.
- *Nucleus* : It's the most important part of the cell, comprising chromosomes, *i.e.* structures made up of genetic information that helps in cell growth and reproduction. Read more on cell nucleus : structure and functions.
- *Nuclear Envelope* : It's an enclosure that surrounds nucleus and its contents. Unlike cell membrane which has pores and spaces for RNA and proteins to pass through, it keeps the chromatin and nucleolus inside the nucleus.

- *Smooth Endoplasmic Reticulum* : It's main function is to package proteins for transport, synthesise membrane phosolipids, and secrete calcium. It also performs transformation of bile pigments, glycogenolysis (the breakdown of glycogen), and detoxification of different drugs and chemical agents.
- *Mitochondria* : It provides energy to the cell by combining sugar molecules with oxygen to generate carbon dioxide and water.
- *Amylosplast* : It's an organelle present in some plant cells that stores starch.
- *Druse Crystal* : It's a granular type of crystal found in plant vacuoles. It is composed of calcium oxalate and is considered to deter herbivory.
- *Centrosome* : Also known as Microtubule Organising Centre, it's a region in the cell where microtubules are produced that perform a variety of functions ranging from transportation to structural support. Though both plant and animal cell centrosomes play similar roles during cell division, plant cell centrosomes are simpler and do not contain centrioles.
- *Peroxisomes* : These are membrane bound packets of oxidative enzymes that convert fatty acids into sugar and assist chloroplasts in photo-respiration.

A typical plant cell is made up of cytoplasm and organelles. Scientific studies have been done regarding plant cell organelles and their functions. Each of the organelles of a plant cell has specific functions, without which the cell cannot operate properly. Cytoplasm is the thick, gel-like semi-transparent fluid that is found in both plant and animal cell. It is bounded by the plasma membrane, and contains many organelles in an eukaryotic cell (cell containing membrane bounded nucleus). In the eukaryotic cell, the nucleus is separated from the cytoplasm by a double membrane, known as nuclear membrane. Cytoplasm was discovered in 1835. Though, discovery of the different organelles found in cytoplasm can be attributed to different scientists, no single scientist can be credited for discovering cytoplasm.

Structure of Cytoplasm

Basically, cytoplasm is the fluid, where the organelles remain suspended. So, it fills up the cell, especially the spaces not occupied by any organelle. The constituent parts of cytoplasm are cytosol, organelles and cytoplasmic inclusions.

- *Cytosol* : Cytosol is the part of the cytoplasm that is not occupied by any organelle. It is a gelatinous substance. It mainly consists of cytoskeleton filaments, organic molecules, salt and water. Cytoskeleton filaments are made up of protein filaments, and they are responsible for giving the shape to the cell. It also contains enzymes, fatty acid, sugar and amino acid. Besides, ribosomes, proteasomes and soluble proteins can also be found in cytoplasm. Cytosol accounts for almost 70 per cent of the total cell volume.
- *Organelles* : Cytoplasm also contains some important organelles like mitochondria, endoplasmic reticulum, lysomes and Golgi apparatus. Besides,

it also contains chloroplast in plant cells. Each organelle is bounded by a fatty membrane, and has some specific functions.

- *Cytoplasmic Inclusions* : Some insoluble suspended substances found in cytosol are known as cytoplasmic substances. They are basically granules of starch and glycogen, and they can store energy. Besides, crystals of some minerals and lipid droplets can also be found in cytoplasm. Lipid droplets are composed of lipids and proteins, and they act as a storage of fatty acid and steroids.
- *Functions of Cytoplasm* : Cytoplasm is the site of many vital biochemical reactions crucial for maintaining life. It is the place where cell expansion and growth take place. It provides a medium in which the organelles can remain suspended. Besides, cytoskeleton found in cytoplasm gives the shape to the cell, and facilitates its movement. It also assists the movement of different elements found within the cell. The enzymes found in the cytoplasm breaks down the macromolecules into small parts so that it can be easily used by the other organelles like mitochondria. For example, mitochondria cannot use glucose present in the cell, unless it is broken down by the enzymes into pyruvate. They act as catalysts in glycolysis, as well as in the synthesis of fatty acid, sugar and amino acid. Cell reproduction, protein synthesis, anaerobic glycosis, cytokinesis are some other vital functions that are carried out in cytoplasm. Organelles present in cytoplasm have some specific vital functions; for example, the function of mitochondria is to produce and store energy, while endoplasmic reticulum facilitates synthesis and transport of protein, production of steroids, production as well as storage of glycogen, etc. The important functions of Golgi apparatus include modification, packaging, transportation and processing of macromolecules, like proteins, and lipids. On the other hand, lysosomes contain digestive enzymes and hence, they digest food particles, damaged or worn-out organelles and also virus and bacteria. However, the smooth operation of all these functions depend on the existence of cytoplasm, as it provides the medium for carrying out these vital processes.

List of Plant Cell Organelles

When it comes to plant cell organelles, they are more or less similar to animal cells, except that the latter lacks chloroplast organelles, that are responsible for photosynthesis. Following is a list of organelles found in plant cell :

- *Nucleus* : Nucleus (plural nuclei) is a highly specialised cell organelle, which stores the genetic component (chromosomes) of the particular cell. It serves as the main administrative centre of the cell, by co-ordinating the metabolic processes like cell growth, cell division and protein synthesis. Read more on cell nucleus.

- *Plastids* : Plastids are collective terms for organelles that carry pigments. In a plant cell, chloroplasts are the most prominent forms of plastids, that contain the green chlorophyll pigment. Because of these chloroplast plastids, a plant cell has the ability to undergo photosynthesis in the presence of sunlight and synthesise its own food. Read more on the importance of photosynthesis.
- *Ribosomes* : Ribosomes are plant cell organelles that comprise of proteins (40 per cent) and ribonucleic acid or RNA (60 per cent). They are important organelles responsible for the synthesis of proteins. Each ribosome consists of two parts, a larger subunit and a smaller sub-unit.
- *Mitochondria* : Mitochondria (singular mitochondrion) are spherical to rod-shaped organelles present in the cytoplasm of the plant cell. They break down the complex carbohydrates and sugars into usable forms, for the plant. As mitochondria indirectly supply energy for the plant cell, they are also called as the powerhouse of the cell.
- *Golgi Body* : Golgi body is also referred to as golgi complex or golgi apparatus. It plays a major role in transporting chemical substances in and out of the cell. After the endoplasmic reticulum synthesises lipids and proteins, golgi body alters and prepares them for exporting outside the cell.
- *Endoplasmic Reticulum* : Endoplasmic reticulum (ER) is the connecting link between the nucleus and cytoplasm of the plant cell. Basically, it is a network of interconnected and convoluted sacs that are located in the cytoplasm. Based on the presence or absence of ribosomes, ER can be of smooth or rough types. The former type lacks ribosomes, while the latter is covered with ribosomes. Overall, endoplasmic reticulum serves as a manufacturing, storing and transporting structure for glycogen, proteins, steroids and other compounds.
- *Vacuoles* : Vacuoles are the storage organelles that help in regulating turgor pressure of the plant cell. In a plant cell, there can be more than one vacuole. However, the centrally located vacuole is larger than others, which stores all sorts of chemical compounds. Vacuoles also assist in intracellular digestion of complex molecules and excretion of waste products.

Chapter 3

PLANT MITOCHONDRIA

Mitochondria in plants, as in other eukaryotes, play an essential role in the cell as the major producers of ATP via oxidative phosphorylation. However, mitochondria also play crucial roles in many other aspects of plant development and performance, and possess an array of unique properties which allow them to interact with the specialised features of plant cell metabolism. The two main themes running through the book are the interconnection between gene regulation and protein function, and the integration of mitochondria with other components of plant cells.

MITOCHONDRIAL DNA

Plant cells have three sets of DNA to code for proteins : one set in the chromosomes of the nucleus, another in the chloroplasts, and a third genome in mitochondria. The mitochondrial genomes of higher plants are larger than those of animals and form a complexseries of linear and circular molecules of different sizes. Mitochondria play an essential role in the generation of energy in eukaryoticcells. Mitochondria are the organelles that are the main "chemical factories" of the cell where cellular aerobic respiration—using the Krebs (citric acid) cycle and respiratory electron transport to produce NADH (nicotinamide adenine dinucleotide) and ATP (adenosine triphosphate)—occurs.

In the light microscope, mitochondria look like short rods or thin filaments about 0.5 to 2 microns long. A mitochondrion is made up of a smooth outer membrane and an inner membrane that is folded into tubular shapes called cristae. Many aerobic respiration reactions are catalysed by enzymes that are bound to mitochondrial membranes. Other reactions occur in the space between the inner and outer mitochondrial membranes. Cells may contain several hundred mitochondria. Cells that are dividing and cells that are metabolically active need larger amounts of ATP and usually have large numbers of mitochondria.

Size and Structure

All eukaryotic cells except some primitive protozoans contain mitochondria. All mitochondria contain their own DNA(genomes). There are typically between twenty and one hundred copies of the mitochondrial genome per mitochondrion. The mitochondria of multi-cellular animals contain genomes of 14 to 20 kilobases (kb), present as single circles. The mitochondrial DNA of some organisms, such as some protozoa, algae, and fungi, is organised in linear molecules with ends of chromosomes (telomeres) much like nuclear chromosomes. In contrast, the mitochondrial DNA of higher plants is larger and more complex—from 200 to 2,500 kb—and is present in many different molecules. The size and organisation of the mitochondrial genome vary widely from one plant species to another. Electron micrographs of mitochondrial DNA show linear and circular DNAs of a variety of sizes and complex, branched molecules that are larger than the size of the genome.

Cloning the mitochondrial DNA and comparing the sequences of the clones show that the entire complexity of a plant mitochondrial genome can be represented as a "master circle." Also, it has been learned that sequences are repeated on the master chromosome. The repeated sequences differ for different plant species. A series of recombination events between these identical repeated sequences results in a series of rearrangements of mitochondrial DNA and forms the complex, multiple molecules of varying sizes that are the physical structure of the plantmitochondrial genome. Adding to the complexity of mitochondrial DNAs in higher plants is the fact that some plants, such as corn, contain extra chromosomal mitochondrial nucleic acids. Plasmid-like DNAs (circular double-stranded molecules) and double-stranded and single-stranded RNAs have been found in some corn strains.

Genes Encoded by Mitochondrial DNA

In addition to containing their own genomes, mitochondria contain enzymes for DNA replication and transcription, and ribosomes and transfer RNAs for protein synthesis. (Transfer RNA, or tRNA, carries the building blocks of proteins, called amino acids, to the ribosome, where they are assembled according to the instructions found in messenger RNA.) The ribosomes of mitochondria are different from those of chloroplasts and the cytoplasm, using a slightly different genetic code (a sequence of three bases that codes for a particular amino acid). Mitochondrial genomes code for all of the ribosomal RNAs found in mitochondria and for most of the tRNAs. Mitochondria make only a small number of proteins that are needed for electron transport and ATP production.

The other proteins needed in mitochondria are coded by nuclear DNA, translated in the cytoplasm of the cell, then transported into the mitochondria. Plant mitochondria do not encode a full set of tRNAs, and some are imported from the cytoplasm. Even though the mitochondrial genome of higher plants is much larger than that of animals, the plant mitochondrial genome codes for only a few more genes. Themitochondrial genome of Arabidopsis has been sequenced and contains thirty-two protein-coding genes, twenty-two tRNA genes, and three ribosomal RNA genes.

- Exchange of DNA : Mitochondrial DNA from plants also differs from that of animals in that mitochondrial DNAs contain segments of DNA that originally were in nuclear and chloroplast DNAs. There appear to have been exchanges of DNAs between all three of the higher plant genomes. There is evidence that mitochondrial genes have been transferred to the nucleus and some mitochondrial tRNAs appear to be of chloroplastorigin. Changes in nuclear genes have been shown to lead to changes in the copy number of the different mitochondrial DNA configurations.
- RNA Editing : Mitochondria and chloroplasts contain the biochemical machinery to alter the sequence of the final messenger RNA (mRNA) product in a process called RNA editing. The most common editing is changing a cytosine to a uracil (two of the bases found on the "rungs" of DNA molecules and which are responsible for determining the nucleotide sequences that form the genetic code).
- Inheritance of Mitochondrial DNA : Given the complex branched network of plant mitochondrial DNA, it is difficult to see how the inheritance of a complete genome is ensured. It is still not clearly understood how this complex network of DNAs is passed to daughter cells in away that assures that all of the genetic information is maintained.

TRANSPORT INTO AND OUT OF PLANT MITOCHONDRIA

Plant mitochondria must exchange metabolites and information (signal transduction) with the cytosol. To give some central examples, the end products of glycolysis, pyruvate and malate, need to be taken up, and citric acid cycle intermediates exported to be used as building blocks in biosynthetic processes. Likewise, ADP and P_i must be taken up and ATP exported for the mitochondrion to fulfill its role as the powerhouse of the cell. To carry out the transport of this large number of molecules, plant mitochondria possess a family of mitochondrial carrier proteins.

These are transmembrane proteins with a common general structure and they typically transport relatively small charged molecules, like metabolites. In Arabidopsis there are almost 60 members, most of which are not functionally characterised, giving a hint to the complexity of the mitochondrial transport processes. However, mitochondria exchange more than metabolites, and therefore also other types of transporters are involved. The outer membrane is permeable to molecules of less than 10 kDa, (*i.e.*, all non-polymeric compounds). The permeability results from the presence of pores, which appear to be open in most *in vitro* experiments. Recent studies in animals have characterised pore-forming proteins in the mitochondrial outer membrane, called porins, which appear to regulate transport across the membrane.

Plant mitochondria also have porins but their characterisation awaits further research. In our discussion here we will focus on the regulation of mitochondrial transport across the inner mitochondrial membrane. We will first consider the relatively few molecules that can move across the inner membrane without the help of a specific carrier protein. We will then discuss the various metabolite transporters and, finally, briefly look at the transport of proteins and RNA.

Hydrophobic Molecules Diffuse Across the Inner Membrane

Irrespective of size, molecules that either are charged or are very hydrophilic can only cross a lipid bilayer with the help of a specific protein carrier. Small, *neutral* molecules can diffuse across the lipid bilayer and do not require protein transporters. Among those are O_2 and CO_2, two important molecules for respiration. O_2 is needed on the inner surface of the inner membrane where the active sites of both cytochrome *c* oxidase and the alternative oxidase are located.

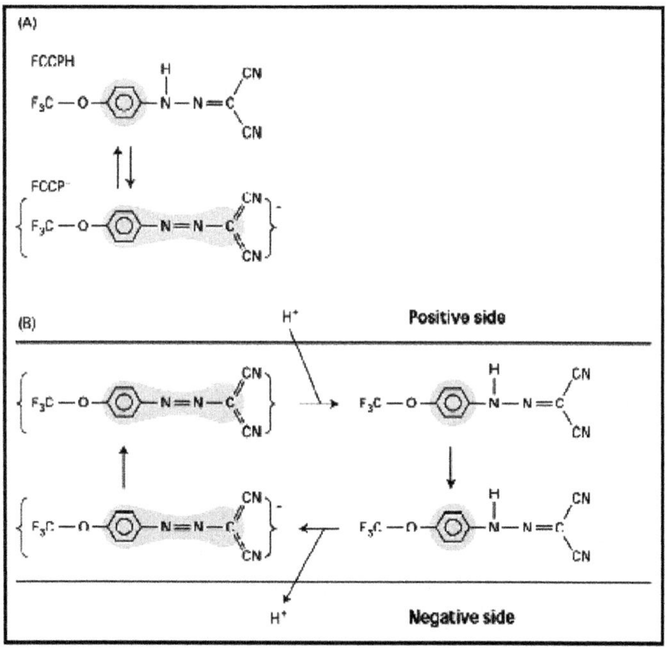

Fig. The Mechanism of action of Uncouplers (also known as protonophores). (A) The protonation/deprotonation of FCCP (trifluoromethoxy carbonyl cyanide phenylhydrazone), which is a weak acid. (B) In the presence of an electrochemical proton gradient across the membrane, FCCP will become protonated and thus pick up a proton on the positive side of the membrane, move across the membrane in the neutral form, and lose the proton on the negative side. This is driven by the proton gradient. In the negative form, FCCP will be driven back out across the membrane by the membrane potential. The net result is the movement of one proton from the positive to the negative side of the membrane, which results in the dissipation of the proton gradient, and the movement of one electrical charge from the negative to the positive side of the membrane, which results in the dissipation of the electrical gradient. The shaded area represents the system of conjugated double bonds (p-orbital system).

The CO_2, produced as an end product in the citric acid cycle, needs to diffuse out of the mitochondrion, ultimately to be lost to the surroundings or re-fixed in the chloroplasts. In addition, however, it is possible that CO_2 can also be transported out of the mitochondrion as bicarbonate (HCO_3^-). Mitochondria are osmotically active and swell in a hypotonic medium. This means that the inner membrane is also permeable to H_2O, but it is not clear whether this occurs by diffusion or if aquaporins may be involved.

For technical reasons, it is of importance that several well-known inhibitors of mitochondrial function can diffuse into mitochondria because they are uncharged and hydrophobic. These are rotenone, the specific inhibitor of complex I, antimycin, the specific inhibitor of complex III, HCN, the protonated form of the cyanide (CN^-) inhibitor of complex IV (and other heme-containing enzymes), and oligomycin, the specific inhibitor of the F_oF_1-ATP synthase (complex V). Chemical uncouplers are very valuable to study membrane bioenergetics because their structure with conjugated double bonds allow them to diffuse across the membrane both in the protonated form and the unprotonated form. A case in point is FCCP (trifluoromethoxy carbonyl cyanide phenylhydrazone), a very efficient mitochondrial uncoupler.

Transport of Inorganic Ions

Inorganic phosphate ($H_2PO_4^-$) is taken up into the matrix in exchange for a hydroxyl ion (HO^-). The energy for the uptake is supplied by the proton gradient ("pH). The phosphate gradient, in turn, drives transport of other metabolites. Several other inorganic ions – Mg^{2+}, Ca^{2+}, K^+, and NH_4^+ – are involved in mitochondrial metabolism, but the mechanism by which they pass the inner membrane of plant mitochondria is not well understood. In mammalian mitochondria, uptake of K^+ through a K^+-specific channel, and extrusion via a K^+/H^+ antiporter, is involved in matrix volume regulation and cellular signalling. A similar system of K^+ transport is present in plant mitochondria.

Both mammalian and plant mitochondria contain an uncoupling protein (UCP), as discussed briefly in Topic. This protein facilitates the movement of protons across the inner membrane and therefore uncouples electron transport and decreases ATP yield.

Transport of Adenylates

One of the important functions of mitochondria is to supply the cell with ATP. To do this, mitochondria have an adenine nucleotide transporter that facilitates the exchange of ADP^{3-} (inward) for ATP^{4-} (outward). This exchange is driven by the membrane potential (inside negative), since there is a net outward movement of one negative charge. In combination with the phosphate carrier reaction, this means that the uptake of one ADP and one phosphate and the export of one ATP (*i.e.*, the transport event connected to mitochondrial ATP synthesis) costs one vectorial proton – the electrical component for adenylate exchange and the "pH component for phosphate uptake.

All of these transporters are directly or indirectly driven by the electrochemical proton gradient. P_i uptake is driven by the proton gradient (hydroxyl anion in the opposite direction) and the resulting P_i gradient is used to drive the uptake of dicarboxylate anions on the dicarboxylate transporter. The dicarboxylates can, in turn, exchange for 2-oxoglutarate or citrate. In the latter case, citrate is protonated on one of its acid groups to maintain electroneutrality. The exchange of glutamate (in) for aspartate (out) is electrogenic in mammals and driven by the electrochemical proton gradient since glutamate is transported as a monovalent anion against aspartate as a divalent anion. The latter transporter is important for the operation of the malate/aspartate shuttle.

Citric Acid Cycle Intermediates

Pyruvate and malate, the end products of glycolysis, both have transporters in the inner membrane. Pyruvate is taken up in exchange for a hydroxyl ion, and malate^{2-} in exchange for $H_2PO_4^{2-}$. The transporter for malate also takes succinate^{2-} and is therefore called the dicarboxylate transporter. Malate^{2-} or succinate^{2-} can, in turn, exchange for citrate^{2-} on the tricarboxylate transporter. Malate^{2-} can also exchange for 2-oxoglutarate^{2-} via a special transporter. An interesting transporter, the oxaloacetate transporter, exchanges oxaloacetate^{2-} for malate^{2-} (or citrate^{2-}), which permits the shuttling of reducing equivalents.

The joint activity of these transporters allows plant mitochondria to import or export most citric acid intermediates. Finally, the glutamate/aspartate transporter facilitates the uptake of glutamate. In mammalian mitochondria, glutamate is taken up together with a proton, and it is thus driven by both the electrical and the chemical proton gradient. A similar transporter is present in plant mitochondria, but a proton co-transport has not been shown, so the transport may be passive in plants. Recently, several additional transporters for metabolites have been identified, and it is likely that certain setups of transporters are only expressed under special developmental conditions, like the mobilisation of storage nutrients in developing seedlings. In several cases the involvement of the mitochondrion in cellular metabolism implies that certain metabolites are taken up across the inner membrane, but the transporter has not yet been described. This is the case for the massive flux of glycine into—and CO_2, serine and NH_4^+ out of—mitochondria in C_3 leaves during photorespiration. This is also the case for proline, which accumulates to high concentrations in the cytosol during episodes of plant water stress, and it is then broken down during the rehydration phase by a pathway involving two mitochondrial enzymes.

Transport of Coenzymes

Coenzymes are either synthesised inside the mitochondrion or imported. There is some evidence that, in addition to ATP, at least NAD^+, thiamine pyrophosphate, and coenzyme A are imported across the inner membrane. The final step in ascorbic acid biosynthesis has recently been shown to be on the outer surface of the inner mitochondrial membrane. In view of the accumulating evidence that ascorbate has an important function in removing ROS inside the mitochondrion, we should expect to find an ascorbate transporter in the inner mitochondrial membrane. The situation with folate (involved in 1C transfers) is the opposite to that with ascorbate : The only site of synthesis of many of the folate metabolites in the plant cell is the mitochondrial matrix. We can therefore expect that these molecules are exported via specific transporters.

Transport of Reducing Equivalents—Metabolic Shuttles

The two shuttles shown have both been demonstrated to work in isolated mitochondria. The malate/aspartate shuttle uses isoforms of two enzymes, malate dehydrogenase (1), and aspartate aminotransferase (2), to interconvert malate and aspartate, which are then exchanged using two transporters, the glutamate/aspartate (A) and the 2-oxoglutarate (B). The electrogenic glutamate/aspartate transporter provides directionality so that reducing equivalents are transferred into the mitochondrial matrix. The malate/OAA shuttle uses malate dehydrogenase isoforms (1) in the matrix and the cytosol to interconvert malate and OAA, which are then exchanged by the OAA transporter (C). This shuttle can transfer reducing equivalents in either direction depending on the requirements of the cell.

Mammalian mitochondria cannot oxidise cytosolic NADH or NAD(P)H directly, so they make use of "shuttle mechanisms." An exchange of metabolites by inner membrane transporters takes place through these shuttles so that a reduced compound moves into the matrix in exchange for a more oxidised compound moving out into the cytosol. In this way, reducing equivalents are transported into the matrix ultimately to be oxidised by the respiratory chain. The shuttles can also be used to transport reducing equivalents in the opposite direction. Although plant mitochondria can oxidise cytosolic NADH and NAD(P)H directly by the two external NAD(P)H dehydrogenases, they can also use these two metabolic shuttles. The malate/oxaloacetate shuttle uses malate dehydrogenase in the cytosol and in the matrix to catalyse the interconversion of malate (reduced) and oxaloacetate (oxidised). These two compounds are exchanged on the oxaloacetate transporter. This exchange, however, is not driven by the electrochemical proton gradient across the inner membrane, so it can only move reducing equivalents from a relatively reduced compartment to a relatively oxidised compartment. Since the mitochondrial matrix is more reduced than the cytosol, the oxaloacetate transporter is likely to work primarily in the export of reducing equivalents.

Under photorespiratory conditions, large amounts of glycine are oxidised in the mitochondria to give NADH while, at the same time, NADH is required in the peroxisomes to reduce hydroxypyruvate to glycerate. It is thought that as much as 50 per cent of the matrix NADH is exported, probably via the malate/oxaloacetate shuttle. The more complex malate/aspartate shuttle involves two matrix enzymes and two cytosolic enzymes as well as two inner membrane transporters. In mammals, it is powered by the electrochemical proton gradient in the direction of the import of reducing equivalents. However, since the glutamate/aspartate exchanger may be passive in plants, this shuttle in plants may similarly be independent of the electrochemical proton gradient. If so, it would only be able to function as an exporter of mitochondrial reductant.

Transport of Proteins—Protein Import

Plant mitochondria contain their own DNA (mtDNA) and ribosomes and are capable of synthesising the proteins encoded in the mtDNA. However, plant mtDNA encodes only a few of the proteins found in the mitochondrion. The great majority of the mitochondrial proteins have to be imported from the cytosol where they are synthesised on free ribosomes. These are synthesised with a presequence or targeting peptide in the N-terminal end, which is recognised by receptors on the outer mitochondrial membrane. The import of proteins requires a complex machinery with two multi-sub-unit protein complexes—one in the outer membrane (TOM, transporter of the outer membrane) and one in the inner membrane (TIM, transporter of the inner membrane). Concomitant with, or immediately afterwards the import of the protein into the correct mitochondrial compartment, the targeting peptide is often cleaved off to yield the mature and active protein.

Transport of tRNA

Mammalian mtDNA encodes the full complement of tRNA required for mitochondrial protein synthesis. In contrast, plant mtDNA only encodes some of the tRNAs required for protein synthesis within the mitochondrion. The missing tRNA must be imported by a yet-to-be-elucidated import mechanism.

ELECTRON TRANSPORT CHAIN OF MITOCHONDRIAL INNER MEMBRANE

The electron transport chain (Etc.) in the mitochondrial inner membrane of most eukaryotes consists of four large protein complexes, complex I to complex IV. The electron transport activity of complexes I, III and IV (the latter two, often jointly called the cytochrome pathway) is coupled to extrusion of protons across the inner mitochondria. The electrochemical proton gradient that is formed and maintained in the process is used by the ATP synthase (also called complex V) to make ATP.The plant Etc., however, is highly branched. As a consequence, the plant oxidative phosphorylation system has several alternative pathways of electron transport via type II NAD(P)H dehydrogenases and the alternative oxidase (AOX), and additionally, an alternative proton transporter called the uncoupling protein (UCP). These proteins mediate bypasses around the proton-translocating multi-protein complexes (I–V) : type II NAD(P)H dehydrogenases bypass complex I, AOX bypasses complexes III and IV, and UCP bypasses the ATP synthase. The type II NAD(P)H dehydrogenases and the AOX transport electrons without pumping protons, whereas the UCP allows proton flow from the intermembrane space to the matrix without ATP synthesis. The consequence of this is that flux through the energy conservation bypasses will neither contribute to respiratory ATP production, nor will they be controlled by cellular adenylate status. This Topic will describe the special features of these energy bypass proteins and discuss the consequences of their existence to the plant.

In addition to the four standard protein complexes found in nearly all other mitochondria, the electron transport chain of plant mitochondria contains five additional enzymes. None of these additional enzymes pumps protons. Additionally, the uncoupling protein directly bypasses the ATP synthase by allowing passive proton influx. This multiplicity of bypasses in plants, where animals have only the uncoupling protein, gives a greater flexibility to plant energy coupling. Specific inhibitors—rotenone for complex I, antimycin for complex III, cyanide for complex IV, and salicylhyroxamic acid (SHAM) for the alternative oxidase—are important tools used to investigate mitochonondrial electron transport. Some also have commercial uses, such as rotenone—which is used as an insecticide and to remove unwanted fish from lakes. Because plants have the alternative pathways, they can survive exposure to inhibitors of the respiratory complexes.

NAD(P)H Dehydrogenases in Plant Mitochondria

Except for yeast, the Etc. of all eukaryotes—including plants—contains a proton-pumping NADH dehydrogenase, complex I, which is inhibited specifically by rotenone. Additionally, plant, fungal and protist mitochondria contain different setups of rotenone-insensitive enzymes capable of oxidising NADH and/or NAD(P)H and passing the electrons on to ubiquinone. These NAD(P)H dehydrogenases are classified as type II, whereas complex I belongs to type I. The inner mitochondrial membrane in plants contains such NAD(P)H dehydrogenases, facing either the matrix or the intermembrane space, thus oxidising NAD(P)H either from the matrix or the cytoplasm. The outer membrane of all mitochondria contains NADH-ferricyanide and NADH-cytochrome *b* reductase activities that are insensitive to rotenone and to the complex III inhibitor, antimycin A. However, the outer membrane activities are not linked to the inner membrane ETC and will not be treated further here.

Inner Matrix Surface of the Inner Membrane

Much of the information about the number of matrix-facing NAD(P)H dehydrogenases has come from experiments using inside-out inner membrane vesicles. Such vesicles have the matrix surface facing the medium and the external NAD(P)H dehydrogenases hidden inside the vesicles and unable to interfere with the assays. Recently, this method has been complemented with a technique where one uses a peptide called alamethicin to make a channel for the substrate through the inner membrane. Experiments with these methods show the presence of three NAD(P)H dehydrogenase activities on the matrix side of the inner membrane :

- Complex I pumps protons, is Ca^{2+}-independent and is inhibited by rotenone and diphenylene iodonium (DPI). Complex I has a high affinity for NADH [low Km(NADH)].

- ND_{in}(NADH) does not pump protons, is Ca^{2+}-independent and DPI insensitive. The enzyme has a tenfold lower affinity for NADH than complex I [*i.e.*, a tenfold higher Km(NADH)].
- ND_{in}(NADPH) does not pump protons, is Ca^{2+}-dependent and very sensitive to DPI.

Because of the contrasting affinities of complex I and ND_{in}(NADH) for NADH, only complex I is engaged at low matrix concentrations of NADH. However, when the NADH concentration in the matrix increases (*e.g.*, if the Etc. is restricted by lack of ADP in the cell), ND_{in}(NADH) can become engaged. When electrons are transported through the type II NAD(P)H dehydrogenases, only the two proton pumping sites of the cytochrome pathway are involved in the flow of electrons from matrix NADH to oxygen. As a consequence, the theoretical ADP/O yield is 1.5 unless the AOX or the UCP are active, in which case it will be even lower.

The External NAD(P)H Dehydrogenases are Ca^{2+}-Dependent

Plant mitochondria can oxidise cytosolic NADH and NAD(P)H directly via external dehydrogenases. The NADH and NAD(P)H dehydrogenase activities are catalysed by separate enzymes. None of the external dehydrogenases pumps protons, which means that, at most, two sites of proton-pumping are involved when the electrons move from the dehydrogenases to oxygen. So, as for ND_{in}(NADH), the theoretical maximum ADP/O yield is 1.5. The oxidation of exogenous NADH or NAD(P)H by most plant mitochondria requires Ca^{2+}, and as a result, is inhibited by EDTA and EGTA, chelators of di- and trivalent cations.

In vitro, the Ca^{2+}concentration needed for half-maximal activity of NADH oxidation is 0.1–0.5 μM, depending on ionic environment and respiratory state This is in the same range as the 0.1–0.2 μM free Ca^{2+} thought to be present in the cytosol of a resting plant cell. Therefore, the external NAD(P)H oxidation may become activated by the increased Ca^{2+}-concentration associated with cellular excitation due to external conditions. Mammalian mitochondria lack external NAD(P) H dehydrogenases. Instead, they have a Ca^{2+}-dependent glycerol-3-phosphate dehydrogenase on the outer surface of the inner membrane, which donates electrons directly to the ubiquinone pool. The dihydroxyacetone phosphate formed is converted back into glycerol-3-phosphate by a cytosolic NAD-dependent glycerol-3-phosphate dehydrogenase. In this way, the two low-molecular-weight compounds, glycerol-3-phosphate and dihydroxyacetone phosphate, shuttle redox equivalents from the cytosol, across the outer membrane and to the inner membrane to yield the same amount of ATP as is produced by the plant mechanism.

Yeast have both external NADH dehydrogenases and a glycerol-3-phosphate shuttle, and since a mitochondrial glycerol-3-phosphate dehydrogenase has been found in plant mitochondria, plants may also have both systems. Another way to oxidise cytosolic NAD(P)H that is found in mammalian mitochondria is to use the malate/aspartate shuttle to tranfer reductant (a collective term for high-energy electrons carried by, for example, NAD(P)H, ferredoxin or an organic acid redox couple like malate/oxaloacetate) into the matrix as NADH that can be oxidised by the internal NAD(P)H dehydrogenases. However, it is not clear whether this shuttle can work for import of cytosolic reductant into plant mitochondria. In the opposite orientation, NADH is shuttled from the matrix to the cytosol via the malate/oxaloacetate shuttle.

In this way, the external NADH dehydrogenase can also work as a bypass around complex I. Support for this bypass has been seen in the form of an up-regulation of external NADH dehydrogenase activity in mutants where complex I or an internal type II NAD(P)H dehydrogenase has been inactivated.

Rotenone-Insensitive NAD(P)H Dehydrogenases

The type II NADH dehydrogenases from *E. coli* and yeast are small, single polypeptide proteins of around 50 kDa. They oxidise only NADH and contain FAD as an electron-accepting prosthetic group. Attempts to purify the type II NAD(P)H dehydrogenases have been unsuccessful in identifying the plant proteins. However, the characterisation of potato homologues to the type II NADH dehydrogenases in *E. coli* and yeasts have provided candidate plant proteins to try and assign to the different reported activities. One protein, NDA1, is internal and the other, NDB1, external of the inner mitochondrial membrane. The latter was shown to be an external calcium-dependent NAD(P)H dehydrogenase. Like a related fungal NAD(P)H dehydrogenase, NDB1 contains a motif for Ca-binding, indicating that in the activation of plant and fungal NAD(P)H dehydrogenase by Ca^{2+}, the ion binds directly to the enzyme.

In Arabidopsis, a weed for which the nuclear genome sequence is known, small families of both NDA-type and NDB-type proteins are encoded by nuclear genes. One of the NDA-type proteins has been shown to be an internal NAD(P)H dehydrogenase. Additionally, a third type of type II NAD(P)H dehydrogenase was found in Arabidopsis, the NDC1. This protein is most closely related to cyanobacterial enzymes, suggesting that the gene entered the eukaryotic domain with the plastid, later to be transferred to the nucleus.

The Alternative Oxidase

The presence of the AOX is one of the features that sets plant (and fungal) mitochondria apart from mammalian mitochondria. This enzyme has therefore received a lot of attention from plant scientists and we now have a fairly good understanding of the regulation of its structure and activity. The AOX is a quinol-oxygen oxidoreductase and it does not pump protons. The oxygenase transfers electrons from ubiquinol (the reduced form of ubiquinone) to oxygen and generates water as the end-product of the reaction, so four electrons must be transferred to oxygen. The precise mechanism is not known, but the protein has been shown to contain a catalytical di-iron centre. The functional form of the enzyme is a dimer with the two polypeptides either covalently or non-covalently bound to each other. The predicted structure places the protein embedded into the inner leaflet of the inner mitochondrial membrane, with the active site and the regulatory cysteines exposed to the matrix.

When electrons from NADH oxidation flow through the AOX, at least two of the three sites of proton translocation (complexes III and IV) are bypassed, so energy conservation in the form of ATP is much smaller when the AOX is active. The activity of the AOX must therefore be carefully regulated. There are often several AOX isoforms in each plant species—for example, five genes encoding AOX proteins are present in Arabidopsis. The relative amounts of three isoforms of AOX vary with development, and cellular redox status, as well as external factors.

The AOX is present in the inner membrane as a dimer in which the two sub-units can be covalently linked through a disulfide bridge between cysteine residues. The reduced non-covalently linked dimer is the active form, and this form appears to be the predominant one *in vivo*. It was recently shown that a mitochondrial thioredoxin system regulates AOX. Thioredoxins regulate enzymes by reducing disulfide bridges, and it is likely that the regulation involves the redox-active cysteines. The presence of thioredoxin regulation connects the AOX activation level to the mitochondrial NAD(P)H pool and the oxidation of citric acid cycle metabolites by NADP-utilising enzymes.

The activity of AOX is also stimulated by 2-oxo organic acids—in particular, pyruvate, which activates by interacting with the same cysteine that is involved in dimerisation. This activation may be a feed-forward mechanism that upregulates the capacity of the Etc. when the glycolytic end-product is available. Both the cytochrome pathway and the AOX pathway use ubiquinol as the substrate. In the absence of pyruvate, the cytochrome pathway is saturated at 40 per cent Q_{red}/Q_{tot} (40 per cent of the UQ is reduced), at which point the AOX pathway is not engaged at all. The AOX pathway starts accepting electrons at 50 per cent reduction and is not fully saturated at 80 per cent.

In the presence of pyruvate, the AOX becomes engaged at a much lower Q_{red}/Q_{tot} (20 per cent), whereas the cytochrome pathway is unaffected. It is puzzling, however, that the normal *in vivo* concentration of pyruvate appears to be higher than the concentration of pyruvate required to give maximal stimulation of AOX activity *in vitro*. It thus appears that there is little regulation of AOX activity via pyruvate under physiological conditions. Q_r is ubiquinol (reduced ubiquinone) and Q_t is total ubiquinone (reduced + oxidised). Myxothiazol is an inhibitor of complex III and, in its presence, only the AOX is active.

AOX Activity Can Be Measured *In Vivo*

For many years, benzhydroxamic acid derivatives, [*e.g.*, salicylhydroxamic acid (SHAM)], were used to quantify the capacity and activity of the AOX pathway in both isolated mitochondria and intact tissues. AOX capacity was determined from the rate of respiration in the presence of KCN (which blocks the cytochrome pathway), and corrected for residual respiration in the presence of both KCN and SHAM. AOX *in vivo* activity was determined from the decrease in the rate of respiration when SHAM was added in the absence of KCN. The assumption was that the cytochrome pathway was always used first and that the AOX would be engaged only when the cytochrome pathway was saturated. However, as we have seen above, this assumption is not correct. The consequence of this is that *in vivo* activity of AOX cannot be measured using inhibitors alone.

A better method for measuring AOX *in vivo* activity is now available. It makes use of the fact that cytochrome *c* oxidase and AOX discriminates differently between ^{18}O relative to ^{16}O. The so-called oxygen discrimination of the sample (tissue or mitochondria) can be measured by mass spectrometry : (1) with KCN to resolve the oxygen discrimination by AOX; (2) with SHAM to resolve the oxygen discrimination by cytochrome *c* oxidase; and (3) without inhibitors to resolve the discrimination value resulting from the activity of both enzymes. From these values, the percentage contribution by AOX and cytochrome *c* oxidase to the total uninhibited rate of respiration can be calculated. An example of the use of the oxygen discrimination technique with an intact tissue is given below. A remaining problem in the estimation of AOX activity *in vivo* is that measurements can only be made in darkness (to avoid interference from photosynthetic oxygen exchange).

PLANT MITOCHONDRIA ALSO CONTAIN AN UNCOUPLING PROTEIN

Mammalian mitochondria contain neither type II NAD(P)H dehydrogenases nor AOX. However, they have a UCP that increases the proton permeability of the inner mitochondrial membrane and, in that way, dissipates the proton gradient. This is another mechanism for reducing the ATP production and increasing heat production, for example, for thermoregulation. Surprisingly, plant mitochondria also contain a UCP. We do not know for certain why plant mitochondria should require two different mechanisms for achieving the same end result. The UCP is related to anion carriers. Its *in vitro* transport activity is dependent on free fatty acids and is inhibited by nucleotides like ATP and GTP but stimulated by superoxide. The latter result indicates that the UCP may protect against oxidative stress.

Energy Bypass Proteins

The respiratory rates of some thermogenic flowers are the highest among plants and, in fact, exceed even those of warm-blooded animals. The external NADH dehydrogenase, AOX and UCP appear particularly active in thermogenic flowers where they are primarily responsible for heat production. At the same time it is clear that the respiratory rate of most plant tissues is insufficient to generate an appreciable increase in tissue temperature. AOX is up-regulated by cold stress, although this is clearly not to heat the tissue. However, during cold stress, reactive oxygen species (ROS) are formed, and also other stress treatments that lead to elevated ROS will induce AOX. ROS (*e.g.*, superoxide and hydrogen peroxide) are formed as by-products of electron transport under aerobic conditions. ROS can cause damage to proteins, lipids, and DNA and the cell must therefore limit their formation. AOX helps prevent overreduction of the ETC and thus lowers ROS production. Supporting this, Maxwell *et. al.* (1999) demonstrated the importance of the AOX for limiting ROS formation in living plant cells. Overexpression of AOX lowered the steady state level of ROS, whereas suppression increased it. This result also points to the mitochondrion as an important site of ROS generation in plant cells. With the recent discovery of ROS stimulation of plant UCP and the increased ROS resistance in plants overexpressing UCP, the picture has become more complex.

By decreasing the resistance for electron transport through the proton-pumping complexes, UCP activity may, like AOX, lead to decreased overreduction and ROS formation. Another factor speaking in favour of UCP having a role in ROS protection is that the AOX is inhibited by an ROS product. For further information on mitochondrial involvement in ROS metabolism. Both type II NAD(P)H dehydrogenases and AOX may act as reductant overflow enzymes by reoxidising surplus NADH to oxygen. This can prevent inhibition of the citric acid cycle under conditions in which the ADP concentration is low, and thus the proton-pumping complexes are restricted by a high membrane potential.

In this way, the citric acid cycle can supply carbon skeletons for biosynthesis (*e.g.*, 2-oxoglutarate for nitrogen assimilation), even if the ATP level in the cell is high. The concept of overflow is also relevant for photorespiratory conditions, under which there is a massive flow of glycine that has to be oxidised by the mitochondria. As much as half of the NADH produced in this process is probably exported, but the other half has to be oxidised promptly. There are several reasons to believe that a reductant overflow via energy bypass proteins is involved. The ND_{in}(NADH) and AOX capacity in leaf mitochondria is higher than in mitochondria from heterotrophic tissue, and both proteins are down-regulated by dark treatment. Consistent with these observations, AOX activity *in vivo* is up-regulated by light, AOX capacity is induced by greening of etiolated tissue, and expression of the *nda*1 gene [encoding an internal NAD(P)H dehydrogenase] in potato and Arabidopsis is light-dependent. Mutant tobacco plants, lacking the proton-pumping complex I can still survive, apparently aided by the action of ND_{in}(NADH), which oxidises matrix NADH with a lower ATP yield. These plants, however, show decreased photosynthetic rates, indicating that the ND_{in}(NADH) cannot fully compensate for the missing complex I.

Also, external NAD(P)H dehydrogenases have been suggested to support photosynthesis. These enzymes may reoxidise NAD(P)H that is shuttled out of the chloroplast in order to balance the NAD(P)H to ATP ratio in the stroma. In summary, there are many indications that mitochondria—especially the energy bypass proteins—have roles in supporting photosynthesis, for example by oxidising excess NAD(P)H. The AOX pathway is involved in the acidification/ deacidification cycle of CAM plants. CAM plants fix CO_2 in the form of malate during the night. This is done by cytosolic PEP carboxylase, and the malate produced is stored in the vacuole. The CO_2 fixed is released during the daytime by malic enzymes, located in cytosol and mitochondria. Robinson *et. al.* (1992) used the oxygen discrimination technique to measure the contribution of the AOX pathway to whole-leaf respiration in the CAM-plant, *Kalanchoë daigremontiana.* They found that the respiratory rate of the leaves increased by 35 per cent in the early hours of the day. This increase was due to increased AOX activity. A more detailed time course showed that the AOX activity increased two- to threefold during the early day at the same time as the malate was metabolised. As soon as the malate had disappeared by mid-day, the AOX activity returned to the low level observed during the night. A reasonable conclusion is that a substantial amount of the malate from the vacuole is oxidised via malic enzyme in the mitochondria. This generates NADH in the matrix, which is oxidised to a large extent via the AOX. The released CO_2 can then be refixed by the Calvin cycle in the chloroplast.

Table. Respiration, oxygen isotope discrimination and partitioning of respiratory pathways in kalanchoe leaf discs during acidification adn deacidification (in the first 6th on the light period).

Parameter	Acidification	Deacicification
Measured		
Unhibited respiration		
Rate (μmol O^2 uptake $m^{-2}s^{-1}$)	1.81	2.56
Discrimination (o/oo)	22.38	25.00
Calculated*		
Cytochrome pathway flux (μmol O_2 uptake $m^{-2}s^{-1}$)	1.33	1.40
Alternative pathway flux (μmol O_2 uptake $m^{-2}s^{-1}$)	0.52	1.16

Respiration, oxygen isotope discrimination and partitioning of respiratory pathways in *Kalanchoë* leaf discs during acidification and deacidification (in the first 6 h of the light period)

FoF1-ATP Synthases : The World2 s Smallest Rotary Motors

F_oF_1-ATP synthases (also called F-ATPases) are present in the inner membranes of mitochondria, chloroplasts, and bacteria. These large, multisubunit enzymes consist of a water-soluble catalytic complex (F_1) attached to an integral membrane protein complex (F_o) that transports protons across the membrane. The F_1 complex is composed of at least five different types of sub-units (three α, three β, one γ, one δ, and one å). When the F_1 complex is dissociated from the membrane, it is active as an ATPase. In fact, under the appropriate conditions the intact F_oF_1-ATP synthase can run in reverse and act as a proton pump, using the energy of ATP hydrolysis to move H^+ across the membrane. Our understanding of how ATP is synthesised was advanced by Paul Boyer at the University of California, Los Angeles, who proposed the binding-change mechanism for catalysis by F-ATPases. From the detailed understanding of how the ATPases function, he proposed that the binding-change mechanism for ATP synthesis contains three important components :

There are three catalytic sites in three different conformations : loose, open, and tight. (For clarity, only the three b sub-units are shown.) Substrate (ADP $+P_i$) initially binds to the open site and is converted to ATP at the tight site. In step 1, rotation of the g sub-unit causes a conformational change resulting in a change in the formation of the sites. As a result, ATP is released from the enzyme. In step 2, substrate again binds to the open site, and another ATP is synthesised at the tight site.

- The major energy-requiring step is not the synthesis of ATP from ADP and P_i, but the release of ATP from the enzyme.
- Substrate is bound and products are released at three separate but interacting catalytic sites, corresponding to the three catalytic sub-units (β sub-units). Each catalytic site can exist in one of three conformations : tight, loose, or open.
- The binding changes are coupled to proton transport by rotation of the γ sub-unit. That is, the flow of protons down their electrochemical gradient through the F_o complex causes the γ sub-unit to rotate.

Rotation of the γ sub-unit then brings about the conformational changes in the catalytic complex that allow the release of ATP from the enzyme, and the reaction is driven forward. The reverse occurs when the enzyme functions as an ATP-driven proton pump. The first two predictions of the binding-change model are supported by many lines of evidence, mainly kinetic studies, and are now generally accepted. However, the prediction of a rotary mechanism for coupling proton flow to ATP synthesis has been more difficult to demonstrate. Two major breakthroughs have led to confirmation of the third prediction, as well. The first breakthrough was the determination of the crystal structure of the F_1-part of bovine mitochondrial ATPase by the laboratory of John Walker in Cambridge, England. The crystal structure showed that the three catalytic β sub-units differ in their conformations and in the nucleotide bound to them, consistent with the binding-change mechanism.

Even more exciting was the discovery that the γ sub-unit is inserted like a shaft through the centre of the catalytic complex, which consists of three α sub-units and three β sub-units arranged alternately in a doughnutlike structure. Moreover, the interface between the γ sub-unit and the α and β sub-units is highly hydrophobic. The hydrophobicity of the interface minimises the interactions between the sub-units, consistent with the rotation of the γ sub-unit within the hole formed by the catalytic complex. In other words, the γ sub-unit looks like a molecular bearing lubricated by a hydrophobic interface. Although many questions were answered by elucidation of the crystal structure of the F_1-ATP synthase, the rotational model cannot be tested by a static "snapshot" of the enzyme. Definitive demonstration of rotation requires a video recording of the spinning of the enzyme in real time.

But although they are large for proteins, F_1-ATP synthases are still far too small to be visualised in a light microscope.

To visualise the rotation of the enzyme, Masasuke Yoshida and his colleagues at the Tokyo Institute of Technology came up with an ingenious method to make the enzyme much larger than it is. They attached an actin filament labelled with a fluorescent dye to the base of the γ sub-unit using another protein as a "glue." They then attached the F_1 complex upside down to a glass surface. If the γ sub-unit rotates with respect to the catalytic complex, the actin filament should swing around with it. Since the filament is very long compared to the ATP synthase (about 1 μm), its rotation should be visible in a fluorescence microscope. In other words, the fluorescently tagged actin filament, which is large enough to visualise in a light microscope, reports the rotation of the γ sub-unit.

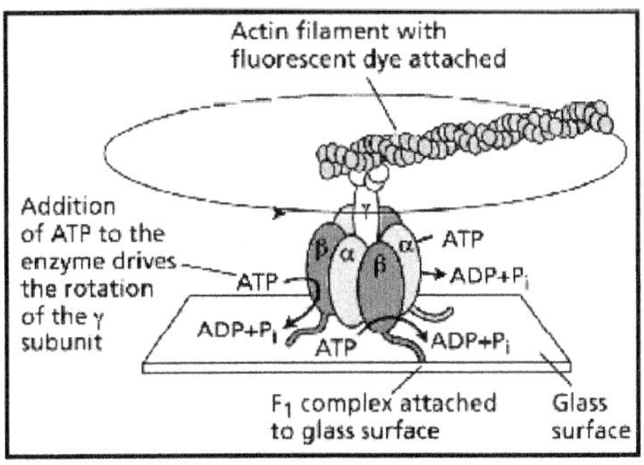

Fig. A Method for Visualising Rotation of the γ sub-unit. A Fluorescently labelled Actin Filament was attached to one Protruding end of the γ sub-unit.

The F_1 complex was then attached upside down to a coverslip. When ATP was added to the coverslip, the actin filament rotated. The results were spectacular! When ATP was added to the modified enzyme, the actin filaments were seen to swing around in a circle at as much as 4 revolutions per second in a fluorescence microscope. To give some idea of scale, if you were a γ sub-unit, this rotation rate would be equivalent to swinging a several hundred meter-long rod around your head at 4 revolutions per second in water! However, the measured velocity is undoubtedly an underestimation of the actual velocity *in vivo*, because of the enormous torque required to swing such a large mass. Demonstration of the rotary motion of the γ sub-unit made it possible to put together a model of how the ATP synthase works. For their contributions to elucidation of the mechanism of ATP synthesis, Paul Boyer and John Walker shared half the Nobel prize in physiology or medicine in 1997.

The other half went to Jens Skou for his pioneering work on the K^+,Na^+-ATP synthase, the mammalian counterpart of the plant plasma membrane H^+-ATP synthase. When the reaction runs in reverse (ATP synthesis), protons diffuse through the F_o complex down their electrochemical gradient. The movement of protons through the channel drives the rotation of the entire F_o complex within the membrane. The γ sub-unit, which is attached to the F_o complex, then turns within the catalytic complex, causing the conformational changes that are required for ATP synthesis. It is assumed that the catalytic complex itself does not rotate, but is anchored to the membrane. The δ sub-unit is located on the outside of the β sub-unit and serves as the site of attachment of the β sub-unit, which anchors the catalytic complex to the membrane and prevents it from spinning. In mechanical terms, the F_1 complex and its membrane anchor act as a stationary housing, or "stator," while γ sub-unit (and possibly the F_o complex) serves as the "rotor."

GENETIC SYSTEM IN PLANT MITOCHONDRIA

As in plastids, the genetic system in mitochondria differs both structurally and functionally from the nuclear genome and its gene expression. On the other hand, mitochondria share several genetic features with bacteria. For example, several genes are transcribed together and the ribosomes are of a prokaryotic type. This is in line with the evolutionary origin of the mitochondrion as an endosymbiont of the α-subclass of Proteobacteria, a group including the intracellular pathogen *Ricketsia prowazekii*. Plants are also distinctly different from other eukaryotes in their mitochondrial DNA (mtDNA) and its expression. For example, the plant mtDNA is very large and highly variable in size between species. The presence of several repeats enables the genome to recombine, changing the gene order or even causing a split into smaller sub-genomic molecules, each carrying a sub-set of the genes. Though the size of mtDNA varies between plants, the number of genes appears to be similar in all plants. In the fully sequenced 367 kilobasepair (kb) mtDNA of *Arabidopsis thaliana*, the known genes (constituting 65 kb) are scattered in a background of non-coding sequence. A similar situation can be found in the sequenced mtDNA of sugarbeet. Plant mtDNA is quite different from the much smaller (16 kb) single mtDNA molecule of mammals in which every base pair has a function in coding for protein or mediating transcription and replication, and neither introns nor repeats exist.

The expression of the mitochondrial genes in plants is carried out by RNA polymerases that are unlike the nuclear polymerases, but related to the enzymes in bacteriophages. For several mitochondrial genes, these phage-type polymerases start up transcription from multiple promoters. Many genes in plant mtDNA are broken up into exons by introns. The introns are removed in the processing of the primary RNA into a mature mRNA that can be translated into the correct polypeptide. The introns in mitochondria belong to the so-called self-splicing type II group, in which splicing is defined by the intron structure. The splicing can also work in the *trans* mode, in which it can join segments from separate RNA molecules, and thus assemble protein-coding reading frames even if the exons are distributed on different primary transcripts. This allows individual exons of a gene to reside independently in the genome, even in different sub-genomic DNA molecules. Because mitochondrial genes are often co-transcribed with unrelated genes into a polycistronic primary transcript, RNA processing also includes endonucleolytic reactions, so that each mature mRNA encodes only one protein. Like mRNAs for nuclear encoded genes, mitochondrial mRNAs are polyadenylated. However, instead of stabilising (the case for nuclear encoded mRNAs), the mitochondrial mRNA polyadenylation mediates degradation. Instead, stem-loops stabilise the mRNA in plant mitochondria.

The mitochondrial *nad2* gene in *Arabidopsis thaliana* encodes the NAD2 sub-unit of respiratory complex I. The RNA processing steps needed for this gene to be expressed into an mRNA that will produce the correct polypeptide. Exons a and b are cotranscribed with a gene for a ribosomal protein (*rps4*), so an endonuclease must cut the RNA. The fate of the *rps4* RNA has been omitted. The *nad2* exons must be joined from two separate primary transcripts by both *cis*- and *trans*-splicing. In this gene, editing of the sequence at the RNA level changes 31 bases (symbolised by green dots) exclusively in exons. For clarity, the different RNA maturation processes are depicted here as consecutive events, whereas *in vivo* they appear to take place in an unordered sequence or simultaneously. In plant mitochondria, an additional RNA processing event takes places—RNA editing. In this process, certain cytidines (C) are deaminated into uridines (U). Rarely, U is aminated to C instead. This means that the coding sequence is changed at the RNA level, in order to restore the correct information for translation where the DNA has diverged. Editing sites mainly occur in DNA encoding protein sequences and functional RNA, and are rare in non-coding mtDNA. In principle, the higher the importance of an individual base, the higher the probability that it constitutes an editing site. Despite the considerable amount of bases that can be edited (up to 15 per cent of the bases of a gene), the functional relevance of this process and the consequent divergence in mtDNA sequence is not understood. Mitochondrial RNA editing has been found in land plants, but not in algae. It also occurs, though to a lesser extent, in plastids.

The mitochondrial genome encodes only a small proportion of the mitochondrial proteins; less than 40 as compared to up to thousands of nuclear-encoded mitochondrial proteins. Through evolution, genes have been transferred from the mitochondrion to the nucleus. This is a process that still takes place, at least in plants. For example, sub-unit 2 of cytochrome oxidase is encoded by the mitochondrial genome in all investigated eukaryotes except for some legumes and protists. Within the legume family, several intermediates of the gene transfer have been found. For example, a functional gene can be found in the mitochondrial genome and an unexpressed one in the nuclear genome, or *vise versa*. In order for a mitochondrial gene to be successfully transferred to the nucleus, the protein-coding frame must be integrated into the chromosome, and several other modifications must be made. For example : (1) a promoter for nuclear transcription must be inserted at the beginning of the gene, (2) the mRNA should acquire sequences specifying the addition of a poly-A tail, and (3) a targeting sequence is necessary for the cytosolically synthesised protein to be imported into the mitochondrion. For proper function, elements regulating the expression in different cell types and under different conditions must also be included.

A mitochondrial gene can be functionally transferred if several rare events take place. The mitochondrial mRNA must be reverse-transcribed and the resulting

DNA integrated into a nuclear chromosome. DNA elements for expression of the gene and for directing the protein to the mitochondrion must then be assembled with it. If this takes place, the mitochondrial gene copy can be deleted without loss of function. Since the coding sequence for most mitochondrial genes in plants is edited at the RNA level, the mitochondrial genomic sequence will not code for a functional protein if inserted into the nucleus. Therefore, the DNA that is integrated into the nuclear genome must in most cases be made by reverse transcription from mitochondrial mRNA. Recently, upon sequencing the nuclear genome of *Arabidopsis thaliana*, a copy of most of the mtDNA was found integrated in a nuclear chromosome, indicating that large segments of DNA can move between cellular compartments. However, most of these mitochondrial genes transferred to the nuclear genome cannot produce functional proteins unless bases corresponding to editing sites are corrected by mutation. However, this is unlikely to take place before other mutations have further disrupted the sequence. As opposed to other mitochondrial genomes, plant mitochondria utilise the universal genetic code. Thus, this factor will not prevent or delay the transfer of genes between plant organelles as it may do in other eukaryotes.

The special features of the mitochondrial genetic system in plants also have consequences for the study of mitochondrial genes. Due to RNA editing and *trans*-splicing, a reading frame is more difficult to identify based only on genomic sequence. Also, protein sequences in databases are automatically translated from genomic DNA sequences without compensating for RNA editing in many cases. This protocol would not work for mitochondrial genes.

Plant Respiration System

Plant respiration can consume an appreciable amount of the carbon fixed each day during photosynthesis over and above the losses due to photorespiration. To what extent can changes in a plant2 s respiratory metabolism affect crop yields? Attempts to establish a quantitative relationship between respiratory energy metabolism and the various processes going on in the cell have led to a break-up of respiration into two components. In growth respiration, reduced carbon is processed to bring about the addition of new biomass. The other component, maintenance respiration, is needed to keep existing, mature cells in a viable state. Utilisation of energy by maintenance respiration is not well understood, but estimates indicate that it can represent more than 50 per cent of the total respiratory flux.

Although numerous questions remain regarding these issues, there are several empirical examples of relations between plant respiration rates and crop yield. In the forage crop, perennial ryegrass (*Lolium perenne*), yield increases of 10 per cent to 20 per cent were correlated with a 20 per cent decrease in the leaf respiration rate. Similar correlations have been found for other plants, including corn (*Zea mays*) and tall fescue (*Festaca arundinacea*). However, later investigations have shown that "selection for low rates of mature leaf respiration is not an appropriate method to select for high-yielding cultivars in perennial grasses". Although there is a potential for increasing crop yield through reduction of respiration rates, a better understanding of the sites and mechanisms that control plant respiration is needed before such changes can be exploited commercially in a systematic fashion by plant physiologists, geneticists, and molecular biologists. Furthermore, much remains to be learned about the general applicability of such observations and the conditions under which slower-respiring lines could be at a disadvantage, causing a reduction in crop yields rather than an increase.

Two recent studies illustrate the present difficulty of predicting the effect of directed changes at the molecular level on plant productivity : Nunes-Nesi *et. al.* found that the productivity of transgenic tomato plants with a reduced activity of mitochondrial malate dehydrogenase was increased as compared to the wild type. Although the respiratory activity of mitochondria isolated from the transgenic plants was unchanged or higher than in the wild type, the rate of leaf respiration was reduced in the transgenic plants. Photosynthesis was markedly increased in the transgenic lines, possibly linked to increased levels of ascorbate. Sieger *et. al.* studied the effect of the alternative oxidase on the growth of tobacco cell cultures. Considering the energy-wasteful nature of the alternative oxidase one might expect that a cell line lacking the enzyme would grow faster than the wild type. Secondly, alternative oxidase has a role in the response of plants to oxidative stress, so a cell line lacking the alternative oxidase would also be expected to handle abiotic stress less well. The results were surprising—the transgenic cell line with a very low expression of alternative oxidase grew as fast as the wild-type cells under normal, nutrient-sufficient conditions, and faster than the wild type under conditions of macronutrient limitation (low phosphate or low nitrogen). It appears that the alternative oxidase is an important factor in modulating the growth rate in response to nutrient availability.

Plant respiration involves an intricate metabolic network of interacting pathways and a complex regulation of gene expression and enzymatic activities. We clearly need to know much more about these interactions before we can predict the effects of changing the expression of single genes on plant productivity.

Lipid Composition of Membranes

The characterisation of lipid mutants and the experimental modification of lipid composition by molecular-genetic means have now revealed some broad generalisations about the roles of membrane lipids in the cell biology and physiology of plants. However, the biochemical and physical bases for the phenotypes that are observed remain uncertain. Changes in fatty acid composition have been shown to alter chloroplast size and architecture. Mutations that eliminate polyunsaturated fatty acids block photosynthesis and prevent the plant from growing autotrophically. Surprisingly, such mutant plants can grow on sucrose, indicating that photosynthesis is the only plant function that absolutely requires a polyunsaturated membrane. However, polyunsaturated fatty acids are also required as precursors of signalling compounds which are necessary for such diverse functions as pollen development and plant defence, such as jasmonic acid.

One of the most extensively studied issues in membrane biology is the relationship between lipid composition and the ability of organisms to adjust to temperature changes. For example, chill-sensitive plants experience sharp reductions in growth rate and development at temperatures between 0 and 12°C. The physical and physiological changes in chill-sensitive plants that are induced by exposure to low temperature, together with the subsequent expression of stress symptoms, are termed chilling injury. Many economically important crops, such as cotton, soybean, maise, rice, and many tropical and subtropical fruits, are classified as chill sensitive. In contrast, most plants of temperate origin are able to grow and develop at chilling temperatures and are classified as chill-resistant plants.

In attempts to link the biochemical and physiological changes associated with chilling injury with a single "trigger" or site of damage, it has been suggested that the primary event of chilling injury is a transition from a liquid-crystalline phase to a gel phase in the cellular membranes. According to this proposal, the transition from liquid-crystalline phase to gel phase would result in alterations in the metabolism of chilled cells and lead to injury and death of the chill-sensitive plants. The degree of unsaturation of the fatty acids would determine the temperature at which such damage occurs. A similar hypothesis has been proposed for chloroplast membranes in which the levels of disaturated phosphatidylglycerol (molecules containing no *cis* double bonds in the fatty acid chains) would determine the chilling sensitivity of plant species. However, the most recent research indicates that the relationship of membrane unsaturation to plant temperature responses is more subtle and complex than is suggested in these earlier hypotheses. On the one hand, studies of five different *Arabidopsis* mutants have demonstrated that reduced unsaturation can result in plants that grow well at 22°C but are less robust than wild-type plants when grown at 2 to 5°C. These results were observed even though the lipid changes in most of the mutants are insufficient to cause a lipid phase transition. In addition, the responses of these mutants to low temperature appear quite distinct from classic chilling sensitivity, suggesting that normal chilling injury may not be related to the level of unsaturation of membrane lipids.

In one particular mutant, *fab1*, saturated forms of phosphatidylglycerol account for 43 per cent of the total leaf phosphati-dylglycerol—a higher percentage than is found in many chill-sensitive plants. Nevertheless, the mutant was completely unaffected (when compared with wild-type controls) by a range of low-temperature treatments that quickly led to the death of cucumber and other chill-sensitive plants. A complementary series of experiments has been carried out in tobacco, which is a chill-sensitive plant. The transgenic expression of exogenous genes in tobacco has been used specifically to decrease the level of saturated phosphatidylglycerol or to bring about a general increase in membrane unsaturation. In each case, damage caused by chilling was alleviated to some extent. These new findings make it clear that the extent of membrane unsaturation or the presence of particular lipids, such as disaturated phosphatidylglycerol, can affect the responses of plants to low temperature. However, membrane lipid composition is not the major determinant of chilling sensitivity in plants.

Chapter 4

Plant Hormone

Plant hormones are chemicals that regulate plant growth, which, in the UK, are termed 'plant growth substances'. Plant hormones are signal molecules produced within the plant, and occur in extremely low concentrations. Hormones regulate cellular processes in targeted cells locally and, when moved to other locations, in other locations of the plant. Hormones also determine the formation of flowers, stems, leaves, the shedding of leaves, and the development and ripening of fruit. Plants, unlike animals, lack glands that produce and secrete hormones. Instead, each cell is capable of producing hormones. Plant hormones shape the plant, affecting seed growth, time of flowering, the sex of flowers, senescence of leaves, and fruits. They affect which tissues grow upward and which grow downward, leaf formation and stem growth, fruit development and ripening, plant longevity, and even plant death. Hormones are vital to plant growth, and, lacking them, plants would be mostly a mass of undifferentiated cells. So they are also known as growth factors or growth hormones. The term 'Phytohormone' was coined by Thimann in 1948.

Phytohormones are found not only in higher plants, but in algae too, showing similar functions, and in microorganisms, like fungi and bacteria, but, in this case, they play no hormonal or other immediate physiological role in the producing organism and can, thus, be regarded as secondary metabolites.

CHARACTERISTICS

The word hormone is derived from Greek, meaning *set in motion*. Plant hormones affect gene expression and transcription levels, cellular division, and growth. They are naturally produced within plants, though very similar chemicals are produced by fungi and bacteria that can also affect plant growth. A large number of related chemical compounds are synthesised by humans. They are used to regulate the growth of cultivated plants, weeds, and in vitro-grown plants and plant cells; these manmade compounds are called Plant Growth Regulators or

PGRs for short. Early in the study of plant hormones, "phytohormone" was the commonly used term, but its use is less widely applied now.

Plant hormones are not nutrients, but chemicals that in small amounts promote and influence the growth, development, and differentiation of cells and tissues. The biosynthesis of plant hormones within plant tissues is often diffuse and not always localised. Plants lack glands to produce and store hormones, because, unlike animals – which have two circulatory systems powered by a heart that moves fluids around the body – plants use more passive means to move chemicals around the plant. Plants utilise simple chemicals as hormones, which move more easily through the plant's tissues. They are often produced and used on a local basis within the plant body. Plant cells produce hormones that affect even different regions of the cell producing the hormone.

Hormones are transported within the plant by utilising four types of movements. For localised movement, cytoplasmic streaming within cells and slow diffusion of ions and molecules between cells are utilised. Vascular tissues are used to move hormones from one part of the plant to another; these include sieve tubes or phloem that move sugars from the leaves to the roots and flowers, and xylem that moves water and mineral solutes from the roots to the foliage.

Not all plant cells respond to hormones, but those cells that do are programmed to respond at specific points in their growth cycle. The greatest effects occur at specific stages during the cell's life, with diminished effects occurring before or after this period. Plants need hormones at very specific times during plant growth and at specific locations. They also need to disengage the effects that hormones have when they are no longer needed. The production of hormones occurs very often at sites of active growth within the meristems, before cells have fully differentiated. After production, they are sometimes moved to other parts of the plant, where they cause an immediate effect; or they can be stored in cells to be released later. Plants use different pathways to regulate internal hormone quantities and moderate their effects; they can regulate the amount of chemicals used to biosynthesise hormones. They can store them in cells, inactivate them, or cannibalise already-formed hormones by conjugating them with carbohydrates, amino acids, or peptides. Plants can also break down hormones chemically, effectively destroying them. Plant hormones frequently regulate the concentrations of other plant hormones. Plants also move hormones around the plant diluting their concentrations.

The concentration of hormones required for plant responses are very low. Because of these low concentrations, it has been very difficult to study plant hormones, and only since the late 1970s have scientists been able to start piecing together their effects and relationships to plant physiology. Much of the early work on plant hormones involved studying plants that were genetically deficient in one or involved the use of tissue-cultured plants grown *in vitro* that were subjected to differing ratios of hormones, and the resultant growth compared. The earliest scientific observation and study dates to the 1880s; the determination and observation of plant hormones and their identification was spread-out over the next 70 years.

CLASSES OF PLANT HORMONES

In general, it is accepted that there are five major classes of plant hormones, some of which are made up of many different chemicals that can vary in structure from one plant to the next. The chemicals are each grouped together into one of these classes based on their structural similarities and on their effects on plant physiology. Other plant hormones and growth regulators are not easily grouped into these classes; they exist naturally or are synthesised by humans or other organisms, including chemicals that inhibit plant growth or interrupt the physiological processes within plants. Each class has positive as well as inhibitory functions, and most often work in tandem with each other, with varying ratios of one or more interplaying to affect growth regulation.

Abscisic Acid

Abscisic acid is one of the most important plant growth regulators. It was discovered and researched under two different names before its chemical properties were fully known, it was called *dormin* and *abscicin II*. Once it was determined that the two compounds are the same, it was named abscisic acid. The name "abscisic acid" was given because it was found in high concentrations in newly abscissed or freshly fallen leaves.

This class of PGR is composed of one chemical compound normally produced in the leaves of plants, originating from chloroplasts, especially when plants are under stress. In general, it acts as an inhibitory chemical compound that affects bud growth, and seed and bud dormancy. It mediates changes within the apical meristem, causing bud dormancy and the alteration of the last set of leaves into protective bud covers. Since it was found in freshly abscissed leaves, it was thought to play a role in the processes of natural leaf drop, but further research has disproven this. In plant species from temperate parts of the world, it plays a role in leaf and seed dormancy by inhibiting growth, but, as it is dissipated from seeds or buds, growth begins. In other plants, as ABA levels decrease, growth then commences as gibberellin levels increase. Without ABA, buds and seeds would start to grow during warm periods in winter and be killed when it froze again. Since ABA dissipates slowly from the tissues and its effects take time to be offset by other plant hormones, there is a delay in physiological pathways that provide some protection from premature growth. It accumulates within seeds during fruit maturation, preventing seed germination within the fruit, or seed germination before winter. Abscisic acid's effects are degraded within plant tissues during cold temperatures or by its removal by water washing in out of the tissues, releasing the seeds and buds from dormancy.

In plants under water stress, ABA plays a role in closing the stomata. Soon after plants are water-stressed and the roots are deficient in water, a signal moves up to the leaves, causing the formation of ABA precursors there, which then move to the roots. The roots then release ABA, which is translocated to the foliage through the vascular system and modulates the potassium and sodium uptake within the guard cells, which then lose turgidity, closing the stomata. ABA exists in all parts of the plant and its concentration within any tissue seems to mediate its effects and function as a hormone; its degradation, or more properly catabolism, within the plant affects metabolic reactions and cellular growth and production of other hormones. Plants start life as a seed with high ABA levels. Just before the seed germinates, ABA levels decrease; during germination and early growth of the seedling, ABA levels decrease even more. As plants begin to produce shoots with fully functional leaves, ABA levels begin to increase, slowing down cellular growth in more "mature" areas of the plant. Stress from water or predation affects ABA production and catabolism rates, mediating another cascade of effects that trigger specific responses from targeted cells. Scientists are still piecing together the complex interactions and effects of this and other phytohormones.

Auxins

Auxins are compounds that positively influence cell enlargement, bud formation and root initiation. They also promote the production of other hormones and in conjunction with cytokinins, they control the growth of stems, roots, and fruits, and convert stems into flowers. Auxins were the first class of growth regulators discovered. They affect cell elongation by altering cell wall plasticity. They stimulate cambium, a subtype of meristem cells, to divide and in stems cause secondary xylem to differentiate. Auxins act to inhibit the growth of buds lower down the stems, and also to promote lateral and adventitious root development and growth. Leaf abscission is initiated by the growing point of a plant ceasing to produce auxins. Auxins in seeds regulate specific protein synthesis, as they develop within the flower after pollination, causing the flower to develop a fruit to contain the developing seeds. Auxins are toxic to plants in large concentrations; they are most toxic to dicots and less so to monocots. Because of this property, synthetic auxin herbicides including 2,4-D and 2,4,5-T have been developed and used for weed control. Auxins, especially 1-Naphthaleneacetic acid (NAA) and Indole-3-butyric acid (IBA), are also commonly applied to stimulate root growth when taking cuttings of plants. The most common auxin found in plants is indole-3-acetic acid or IAA. The correlation of auxins and cytokinins in the plants is a constant (A/C = const.).

Cytokinins

Cytokinins or CKs are a group of chemicals that influence cell division and shoot formation. They were called kinins in the past when the first cytokinins were isolated from yeast cells. They also help delay senescence or the aging of tissues, are responsible for mediating auxin transport throughout the plant, and affect internodal length and leaf growth. They have a highly synergistic effect in concert with auxins, and the ratios of these two groups of plant hormones affect most major growth periods during a plant's lifetime. Cytokinins counter the apical dominance induced by auxins; they in conjunction with ethylene promote abscission of leaves, flower parts, and fruits. The correlation of auxins and cytokinins in the plants is a constant (A/C = const.).

Ethylene

Ethylene is a gas that forms through the breakdown of methionine, which is in all cells. Ethylene has very limited solubility in water and does not accumulate within the cell but diffuses out of the cell and escapes out of the plant. Its effectiveness as a plant hormone is dependent on its rate of production *versus* its rate of escaping into the atmosphere. Ethylene is produced at a faster rate in rapidly growing and dividing cells, especially in darkness. New growth and newly germinated seedlings produce more ethylene than can escape the plant, which leads to elevated amounts of ethylene, inhibiting leaf expansion. As the new shoot is exposed to light, reactions by phytochrome in the plant's cells produce a signal for ethylene production to decrease, allowing leaf expansion. Ethylene affects cell growth and cell shape; when a growing shoot hits an obstacle while underground, ethylene production greatly increases, preventing cell elongation and causing the stem to swell. The resulting thicker stem can exert more pressure against the object impeding its path to the surface. If the shoot does not reach the surface and the ethylene stimulus becomes prolonged, it affects the stem's natural geotropic response, which is to grow upright, allowing it to grow around an object. Studies seem to indicate that ethylene affects stem diameter and height : When stems of trees are subjected to wind, causing lateral stress, greater ethylene production occurs, resulting in thicker, more sturdy tree trunks and branches. Ethylene affects fruit-ripening : Normally, when the seeds are mature, ethylene production increases and builds-up within the fruit, resulting in a climacteric event just before seed dispersal. The nuclear protein Ethylene Insensitive2 (EIN2) is regulated by ethylene production, and, in turn, regulates other hormones including ABA and stress hormones.

Gibberellins

Gibberellins, or GAs, include a large range of chemicals that are produced naturally within plants and by fungi. They were first discovered when Japanese researchers, including Eiichi Kurosawa, noticed a chemical produced by a fungus called *Gibberella fujikuroi* that produced abnormal growth in rice plants. Gibberellins are important in seed germination, affecting enzyme production that mobilises food production used for growth of new cells. This is done by modulating chromosomal transcription. In grain seeds, a layer of cells called the aleurone layer wraps around the endosperm tissue. Absoption of water by the seed causes production of GA. The GA is transported to the aleurone layer, which responds by producing enzymes that break down stored food reserves within the endosperm, which are utilised by the growing seedling. GAs produce bolting of rosette-forming plants, increasing internodal length. They promote flowering, cellular division, and in seeds growth after germination. Gibberellins also reverse the inhibition of shoot growth and dormancy induced by ABA.

Other Known Hormones

Other identified plant growth regulators include :

- Brassinosteroids, are a class of polyhydroxysteroids, a group of plant growth regulators. Brassinosteroids have been recognised as a sixth class of plant hormones, which stimulate cell elongation and division, gravitropism, resistance to stress, and xylem differentiation. They inhibit root growth and leaf abscission. Brassinolide was the first identified brassinosteroid and was isolated from extracts of rapeseed pollen in 1979.
- Salicylic acid – activates genes in some plants that produce chemicals that aid in the defence against pathogenic invaders.
- Jasmonates – are produced from fatty acids and seem to promote the production of defence proteins that are used to fend off invading organisms. They are believed to also have a role in seed germination, and affect the storage of protein in seeds, and seem to affect root growth.
- Plant peptide hormones – encompasses all small secreted peptides that are involved in cell-to-cell signaling. These small peptide hormones play crucial roles in plant growth and development, including defence mechanisms, the control of cell division and expansion, and pollen self-incompatibility.
- Polyamines – are strongly basic molecules with low molecular weight that have been found in all organisms studied thus far. They are essential for plant growth and development and affect the process of mitosis and meiosis.
- Nitric oxide (NO) – serves as signal in hormonal and defence responses. NO can be produced by a yet undefined NO synthase, a special type of nitrite reductase, nitrate reductase, mitochondrial cytochrome c oxidase or non- enzymatic processes and regulate plant cell organelle functions.
- Strigolactones, implicated in the inhibition of shoot branching.
- Karrikins, a group of plant growth regulators found in the smoke of burning plant material that have the ability to stimulate the germination of seeds.

Brassinosteroid

Brassinosteroids (BRs) are a class of polyhydroxysteroids that have been recognised as a sixth class of plant hormones. These were first explored nearly 40 years ago, when Mitchell *et. al.* reported promotion in stem elongation and cell division by the treatment of organic extracts of rapeseed pollen. Brassinolide was the first isolated brassinosteroid in 1979, when pollen from *Brassica napus* was shown to promote stem elongation and cell divisions, and the biologically active molecule was isolated. The yield of brassinosteroids from 230 kg of *Brassica napus* pollen was only 10 mg. Since their discovery, over 70 BR compounds have been isolated from plants.

The BR is biosynthesised from campesterol. The biosynthetic pathway was elucidated by Japanese researchers and later shown to be correct through the analysis of BR biosynthesis mutants in *Arabidopsis thaliana*, tomatoes, and peas. The sites for BR synthesis in plants have not been experimentally demonstrated. One well-supported hypothesis is that all tissues produce BRs, since BR biosynthetic and signal transduction genes are expressed in a wide range of plant organs, and short distance activity of the hormones also supports this. Experiments have shown that long distance transport is possible and that flow is in an acropetal direction, but it is not known if this movement is biologically relevant. Brassinosteroids are recognised at the cell membrane, although they are membrane-soluble.

BRs have been shown to be involved in numerous plant processes :

- Promotion of cell expansion and cell elongation; works with auxin to do so.
- It has an unclear role in cell division and cell wall regeneration.
- Promotion of vascular differentiation; BR signal transduction has been studied during vascular differentiation.
- Is necessary for pollen elongation for pollen tube formation.
- Acceleration of senescence in dying tissue cultured cells; delayed senescence in BR mutants supports that this action may be biologically relevant.
- Can provide some protection to plants during chilling and drought stress.

Extract from the plant *Lychnis viscaria* contains a relatively high amount of Brassinosteroids. Lychnis viscaria increases the disease resistance of surrounding plants. In Germany, extract from the plant is allowed for use as a "plant strengthening substance."

24-Epibrassinolide (EBL), a brassinosteroid isolated from *Aegle marmelos* Correa (Rutaceae), was further evaluated for the antigenotoxicity against maleic hydrazide (MH)-induced genotoxicity in *Allium cepa* chromosomal aberration assay. It was shown that the percentage of chromosomal aberrations induced by maleic hydrazide (0.01 per cent) declined significantly with 24-epibrassinolide treatment.

BRs have been reported to counteract both abiotic and biotic stress in plants. Application of brassinosteroids to cucumbers was demonstrated to increase the metabolism and removal of pesticides, which could be beneficial for reducing the human ingestion of residual pesticides from non-organically grown vegetables. In all Type of brassinosteroids 28-homoBL is the most effective type of brassinosteroids. Brassinosteroids increased tolerance to high temperature in Brassica juncea L. The ability of 28-homobrassinolide to confer resistance to stress in Brassica juncea L. has also established. Application of 24-epiBL have any protective role on shoot, root length, soluble protein, proline content and peroxidases along with proline content PPO and IAA in seedlings of B. juncea L. under seasonal stress

BRs have also been reported to have a variety of effects when applied to rice seeds. Seeds treated with BRs were shown to reduce the growth inhibitory effect of salt stress. When the developed plants fresh weight was analysed the treated seeds outperformed plants grown on saline and non-saline medium however when the dry weight was analysed BR treated seeds only outperformed untreated plants that were grown on saline medium. When dealing with tomatoes under salt stress the concentration of cholophyll a and cholophyll b were decreased and thus pigmentation was decreased as well. BR treated rice seeds considerably restored the pigment level in plants that were grown on saline medium when compared to non-treated plants under the same conditions.

BR might reveal to have a prominent interest in the role of horticultural crops. Based on extensive research BR has the ability to improve the quantity and quality of horticultural crops and protect plants against many stresses that can be present in the local environment. With the many advances in technology dealing with the synthesis of more stable synthetic analogues and the genetic manipulation of cellular BR activity, using BR in the production of horticultural crops has become a more practical and hopeful strategy for improving crop yields and success.

BR could also help bridge the gap of the consumers' health concerns and the producers need for growth. A major benefit of using BR is that it does not interfere with the environment because they act in natural doses in a natural way. Since it is a "plant strengthening substance" and it is natural, BR application would be more favourable than pesticides and does not contribute to the co-evolution of pests.

BRs can be detected by gas chromatography mass spectrometry and bioassays.

Salicylic Acid

Salicylic acid is a monohydroxybenzoic acid, a type of phenolic acid and a beta hydroxy acid. This colourless crystalline organic acid is widely used in organic synthesis and functions as a plant hormone. It is derived from the metabolism of salicin. In addition to being an important active metabolite of aspirin, which acts in part as a prodrug to salicylic acid, it is probably best known for its use in anti-acne treatments. The salts and esters of salicylic acid are known as salicylates.

Chemistry

Salicylic acid has the formula $C_6H_4(OH)COOH$, where the OH group is *ortho* to the carboxyl group. It is also known as 2-hydroxybenzoic acid. It is poorly soluble in water (2 g/L at 20 °C). Aspirin (acetylsalicylic acid or ASA) can be prepared by the esterification of the phenolic hydroxyl group of salicylic acid with the acetyl group from acetic anhydride or acetyl chloride.

Plant Hormone

Salicylic acid (SA) is a phenolic phytohormone and is found in plants with roles in plant growth and development, photosynthesis, transpiration, ion uptake and transport. SA also induces specific changes in leaf anatomy and chloroplast structure. SA is involved in endogenous signaling, mediating in plant defence against pathogens. It plays a role in the resistance to pathogens by inducing the production of pathogenesis-related proteins. It is involved in the systemic acquired resistance (SAR) in which a pathogenic attack on one part of the plant induces resistance in other parts. The signal can also move to nearby plants by salicylic acid being converted to the volatile ester, methyl salicylate.

Production

Salicylic acid is biosynthesised from the amino acid phenylalanine. In *Arabidopsis thaliana* it can also be synthesised via a phenylalanine-independent pathway.

Sodium salicylate is commercially prepared by treating sodium phenolate with carbon dioxide at high pressure (100 atm) and high temperature (390K) -a method known as the Kolbe-Schmitt reaction. Acidification of the product with sulfuric acid gives salicylic acid :

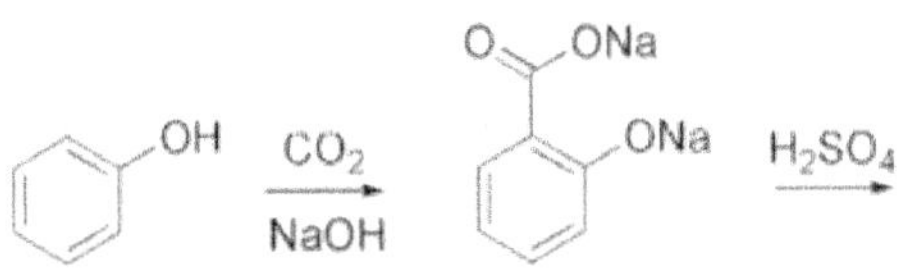

It can also be prepared by the hydrolysis of aspirin or methyl salicylate with a strong acid or base.

History

The Greek physician Hippocrates wrote in the 5th century B.C. about a bitter powder extracted from willow bark that could ease aches and pains and reduce fevers. This remedy was also mentioned in texts from ancient Sumer, Lebanon, and Assyria. The Cherokee and other Native Americans used an infusion of the bark for fever and other medicinal purposes for centuries. The medicinal part of the plant is the inner bark and was used as a pain reliever for a variety of ailments. The Reverend Edward Stone, a vicar from Chipping Norton, Oxfordshire, England, noted in 1763 that the bark of the willow was effective in reducing a fever.

The active extract of the bark, called *salicin*, after the Latin name for the white willow (*Salix alba*), was isolated and named by the German chemist Johann Andreas Buchner in 1828. A larger amount of the substance was isolated in 1829 by Henri Leroux, a French pharmacist. Raffaele Piria, an Italian chemist, was able to convert the substance into a sugar and a second component, which on oxidation becomes salicylic acid.

Salicylic acid was also isolated from the herb meadowsweet by German researchers in 1839. While their extract was somewhat effective, it also caused digestive problems such as gastric irritation, bleeding, diarrhea, and even death when consumed in high doses.

Dietary Sources

Unripe fruits and vegetables are natural sources of salicylic acid, particularly blackberries, blueberries, cantaloupes, dates, raisins, kiwi fruits, guavas, apricots, green pepper, olives, tomatoes, radish and chicory; also mushrooms. Some herbs and spices contain quite high amounts, while meat, poultry, fish, eggs and dairy products all have little to no salicylates. Of the legumes, seeds, nuts, and cereals, only almonds, water chestnuts and peanuts have significant amounts.

Medicinal and Cosmetic Uses

Salicylic acid is known for its ability to ease aches and pains and reduce fevers. These medicinal properties, particularly fever relief, have been known since ancient times, and it is used as an anti-inflammatory drug.

In modern medicine, salicylic acid and its derivatives are used as constituents of some rubefacient products. For example, methyl salicylate is used as a liniment to soothe joint and muscle pain, and choline salicylate is used topically to relieve the pain of mouth ulcers.

As with other beta hydroxy acids, salicylic acid is a key ingredient in many skin-care products for the treatment of seborrhoeic dermatitis, acne, psoriasis, calluses, corns, keratosis pilaris, and warts. The standard treatment for calluses is a 6 per cent aspirin suspension in petroleum jelly, applied on the callus for one hour and then removed with washing. It works as a keratolytic, bacteriocide and comedolytic agent by causing the cells of the epidermis to shed more readily, opening clogged pores and neutralising bacteria within, preventing pores from clogging up again by constricting pore diameter, and allowing room for new cell growth. Because of its effect on skin cells, salicylic acid is used in several shampoos to treat dandruff. Use of concentrated solutions of salicylic acid may cause hyperpigmentation on unpretreated skin for those with darker skin types, as well as with the lack of use of a broad spectrum sunblock.

Bismuth sub-salicylate, a salt of bismuth and salicylic acid, is the active ingredient in stomach relief aids such as Pepto-Bismol, is the main ingredient of Kaopectate, and "displays anti-inflammatory action and also acts as an antacid and mild antibiotic."

A 2004 article in *New Scientist* discussing the controversial idea of treating salicylates as micro-nutrients, "akin to vitamins and antioxidants," suggested that "perhaps in future we might even call salicylate 'vitamin S'."

Mechanism of Action

Salicylic acid has been shown to work through several different pathways. It produces its anti-inflammatory effects via suppressing the activity of cyclooxygenase (COX), an enzyme that is responsible for the production of pro-inflammatory mediators such as the prostaglandins. It does this not by direct inhibition of COX like most other non-steroidal anti-inflammatory drugs (NSAIDs) but instead by suppression of the expression of the enzyme. Salicylic acid has also been shown to activate adenosine monophosphate-activated protein kinase (AMPK), and it is thought that this action may play a role in the anti-cancer effects of the compound and its prodrugs aspirin and salsalate. In addition, the antidiabetic effects of salicylic acid are likely mediated by AMPK activation primarily through allosteric conformational change that increases levels of phosphorylation. Salicylic acid also uncouples oxidative phosphorylation, which leads to increased ADP : ATP and AMP : ATP ratios in the cell. As a consequence, salicylic acid may alter AMPK activity and subsequently exert its anti-diabetic properties through altered energy status of the cell. Even in AMPK knock-out mice, however, there is an anti-diabetic effect, demonstrating that there is at least one additional, yet-unidentified action of the compound.

Other Uses

Although toxic in large quantities, salicylic acid is used as a food preservative and as bactericidal and an antiseptic. For some people with salicylate sensitivity, even these small doses can be harmful. Sodium salicylate is a useful phosphor in the vacuum ultraviolet, with nearly flat quantum efficiency for wavelengths between 10 to 100 nm. It fluoresces in the blue at 420 nm. It is easily prepared on a clean surface by spraying a saturated solution of the salt in methanol followed by evaporation.

Safety

As a topical agent, and as a beta-hydroxy acid, salicylic acid is capable of penetrating and breaking down fats and lipids, making it capable of causing moderate chemical burns of the skin if at very high concentrations. It is capable of damaging the lining of pores in such cases if the solvent is alcohol, acetone, or an oil. Over-the-counter limits are set at 2 per cent for topical left on the face and 3 per cent for those expected to be washed off, such as acne cleansers or shampoo. Caution should be exercised when handling large volumes of salicylic acid, and protective gloves are recommended for any repeat, prolonged exposure. 17 per cent and 27 per cent salicylic acid, which is often sold for wart removal, should not be applied to the face and should not be used for acne treatment. Even for wart removal, such a solution should be applied only twice a day - more frequent use may lead to an increase in side-effects without an increase in efficacy.

When ingested, salicylic acid has a possible ototoxic effect by inhibiting prestin. It can induce transient hearing loss in zinc-deficient individuals. This finding is based on clinical studies with rats. An injection of salicylic acid induced hearing loss in zinc-deficient rats, while a simultaneous injection of zinc reversed the hearing loss. An injection of magnesium in the zinc-deficient rats did not reverse the salicylic acid-induced hearing loss.

There are no studies specifically looking at topical salicylic acid in pregnancy. Oral salicylic acid has not been associated with an increase in malformations if used during the first trimester, but use in late pregnancy has been associated with bleeding, especially intracranial bleeding. The risks of aspirin late in pregnancy are probably not relevant for a topical exposure to salicylic acid, even late in the pregnancy, because of its low systemic levels. Topical salicylic acid is common in many over-the-counter dermatological agents, and the lack of adverse reports suggests a low teratogenic potential.

Salicylic acid overdose can lead to salicylate intoxication, which often presents clinically in a state of metabolic acidosis with compensatory respiratory alkalosis. In patients presenting with an acute overdose, a 16 per cent morbidity rate and a 1 per cent mortality rate are observed.

Some people are hypersensitive to salicylic acid and related compounds.

The United States Food and Drug Administration recommends the use of sun protection when using skincare products containing salicylic acid on sun-exposed skin areas.

There is data that supports an association between exposure to salicylic acid and Reye's Syndrome. The National Reye's Syndrome Foundation cautions against the usage of these substances, and other substances similar to aspirin, on children and adolescents.

Epidemiological research has shown an association between the development of Reye's Syndrome and the use of aspirin for treating the symptoms of influenza-like illnesses, chicken pox, colds, etc.

The U.S. Surgeon General, the Food and Drug Administration, the Centers for Disease Control and Prevention, and the American Academy of Pediatrics recommend that aspirin and combination products containing aspirin not be given to children under 19 years of age during episodes of fever-causing illnesses, because of a concern about Reye's Syndrome.

Jasmonate

Jasmonate (JA) and its derivatives are lipid-based hormone signals that regulate a wide range of processes in plants, ranging from growth and photosynthesis to reproductive development. In particular, JAs are critical for plant defence against herbivory and plant responses to poor environmental conditions and other kinds of abiotic and biotic challenges. Some JAs can also be released as volatile organic compounds (VOCs) to permit communication between plants in anticipation of mutual dangers.

The isolation of methyl jasmonate from jasmine oil derived from *Jasminum grandiflorum* led to the discovery of the molecular structure of jasmonates and their name.

Structure of Jasmonates and Its Derivatives

Jasmonates (JA) are an oxylipin, or oxygenated fatty acid, synthesised from linolenic acid residing in chloroplast membranes. Synthesis is initiated with the conversion of linolenic acid to 12-oxo-phytodienoic acid (OPDA), which then undergoes a reduction and three rounds of oxidation to form (+)-7-iso-JA, jasmonic acid proper. Only the conversion of linolenic acid to OPDA occurs in the chloroplast; all subsequent reactions occur in the peroxisome.

JA itself can be further metabolised into active or inactive derivatives. Methyl JA (MeJA) is a volatile compound that is potentially responsible for interplant communication. JA conjugated with amino acid isoleucine (Ile) results in JA-Ile, which is currently the only known JA derivative needed for JA signaling.

Mechanism of Jasmonate Signaling

In general, the steps in jasmonate (JA) signaling mirror that of auxin signaling : the first step comprises E3 ubiquitin ligase complexes, which tag substrates with ubiquitin to mark them for degradation by proteasomes. The second step utilises transcription factors to effect physiological changes. One of the key molecules in this pathway is JAZ, which serves as the on-off switch for JA signaling. In the absence of JA, JAZ proteins bind to downstream transcription factors and limit their activity. However, in the presence of JA or its bioactive derivatives, JAZ proteins are degraded, freeing transcription factors for expression of genes needed in stress responses.

Because JAZ did not disappear in null *coi1* mutant plant backgrounds, protein COI1 was shown to mediate JAZ degradation. COI1 belongs to the family of highly conserved F-box proteins, and it recruits substrates for the E3 ubiquitin ligase SCF^{COI1}. The complexes that ultimately form are known as SCF complexes. These complexes bind JAZ and target it for proteasomal degradation. However, given the large spectrum of JA molecules, not all JA derivatives activate this pathway for signaling, and the range of those participating in this pathway is unknown. Thus far, only JA-Ile has been shown to be necessary for COI1-mediated degradation of JAZ11. JA-Ile and structurally related derivatives can bind to COI1-JAZ complexes and promote ubiquitination and thus degradation of the latter.

This mechanistic model raises the possibility that COI1 serves as an intracellular receptor for JA signals. Recent research has confirmed this hypothesis by demonstrating that the COI1-JAZ complex acts as a co-receptor for JA perception. Specifically, JA-Ile binds both to a ligand-binding pocket in COI1 and to a 20 amino-acid stretch of the conserved Jas motif in JAZ. This JAZ residue acts as a plug for the pocket in COI1, keeping JA-Ile bound in the pocket. Additionally, co-purification and subsequent removal of inositol pentakisphosphate ($InsP_5$) from COI1 suggest $InsP_5$ is a necessary component of the co-receptor and plays a role in potentiating the co-receptor complex.

Once freed from JAZ, transcription factors can activate genes needed for a specific JA response. The best-studied transcription factors acting in this pathway belong to the MYC family of transcription factors, which are characterised by a basic helix-loop-helix (bHLH) DNA binding motif. These factors tend to act additively. For example, a plant that has only lost one myc becomes more susceptible to insect herbivory than a normal plant. A plant that has lost all three will be as susceptible to damage as coil mutants, which are completely unresponsive to JA and cannot mount a defence against herbivory. However, while all these MYC molecules share functions, they vary greatly in expression patterns and transcription functions. For instance, MYC2 has a greater effect on root growth compared to MYC3 or MYC4.

Additionally, MYC2 will loop back and regulate JAZ expression levels, leading to a negative feedback loop. These transcription factors all have different impacts on JAZ levels after JA signaling. JAZ levels in turn affect transcription factor and gene expression levels. In other words, on top of activating different response genes, the transcription factors can vary JAZ levels to achieve specificity in response to JA signals.

Roles of Jasmonate

Although jasmonate (JA) regulates many different processes in the plant, its role in wound response is best understood. Following mechanical wounding or herbivory, JA biosynthesis is rapidly activated, leading to expression of the appropriate response genes. For example, in the tomato, wounding produces defence molecules that inhibit leaf digestion in the insect's gut. Another indirect result of JA signaling is the volatile emission of JA-derived compounds. MeJA on leaves can travel airborne to nearby plants and elevate levels of transcripts related to wound response. In general, this emission can further upregulate JA synthesis and signaling and induce nearby plants to prime their defences in case of herbivory.

Following its role in defence, JAs have also been implicated in cell death and leaf senescence. JA can interact with many kinases and transcription factors associated with senescence. JA can also induce mitochondrial death by inducing the accumulation of reactive oxygen species (ROSs). These compounds disrupt mitochondria membranes and compromise the cell by causing apoptosis, or programmed cell death. JAs' roles in these processes are suggestive of methods by which the plant defends itself against biotic challenges and limits the spread of infections.

JA and its derivatives have also been implicated in plant development, symbiosis, and a host of other processes included in the list below :

- By studying mutants overexpressing JA, one of the earliest discoveries made was that JA inhibits root growth. The mechanism behind this event is still not understood, but mutants in the COI1-dependent signaling pathway tend to show reduced inhibition, demonstrating that the COI1 pathway is somehow necessary for inhibiting root growth.
- JA plays many roles in flower development. Mutants in JA synthesis or in JA signaling in Arabidopsis present with male sterility, typically due to delayed development. Interestingly, the same genes promoting male fertility in Arabidopsis promote female fertility in tomatoes. Overexpression of 12-OH-JA can also delay flowering.
- JA and MeJA inhibit the germination of non-dormant seeds and stimulate the germination of dormant seeds.

- High levels of JA encourage the accumulation of storage proteins; genes encoding vegetative storage proteins are JA responsive. Specifically, tuberonic acid, a JA derivative, induces the formation of tubers.
- JAs also play a role in symbiosis between plants and micro-organisms; however, its precise role is still unclear. JA currently appears to regulate signal exchange and nodulation regulation between legumes and rhizobium. On the other hand, elevated JA levels appear to regulate carbohydrate partitioning and stress tolerance in mycorrhizal plants.

Jasmonate and Pathogenesis

Pseudomonas syringae causes bacterial speck disease in tomatoes by hijacking the plant's jasmonate (JA) signaling pathway. This bacteria utilises a type III secretion system to inject a cocktail of viral effector proteins into host cells.

One of the molecules included in this mixture is the phytotoxin coronatine (COR). JA-insensitive plants are highly resistant to *P. syringae* and unresponsive to COR; additionally, applying MeJA was sufficient to rescue virulence in COR mutant bacteria. Infected plants also expressed downstream JA and wound response genes but repressed levels of pathogenesis-related (PR) genes. All these data suggest COR acts through the JA pathway to invade host plants. Activation of a wound response is hypothesised to come at the expense of pathogen defence. By activating the JA wound response pathway, *P. syringae* could divert resources from its host's immune system and infect more effectively.

Cross Talk with Other Defence Pathways

While the jasmonate (JA) pathway is critical for wound response, it is not the only signaling pathway mediating defence in plants. To build an optimal yet efficient defence, the different defence pathways must be capable of cross talk to fine-tune and specify responses to abiotic and biotic challenges.

One of the best studied examples of JA cross talk occurs with salicylic acid (SA). SA, a hormone, mediates defence against pathogens by inducing both the expression of pathogenesis-related genes and systemic acquired resistance (SAR), in which the whole plant gains resistance to a pathogen after localised exposure to it.

Wound and pathogen response appear to be interact negatively. For example, silencing phenylalanine ammonia lyase (PAL), an enzyme synthesising precursors to SA, reduces SAR but enhances herbivory resistance against insects. Similarly, overexpression of PAL enhances SAR but reduces wound response after insect herbivory. Generally, it has been found that pathogens living in live plant cells are more sensitive to SA-induced defences, while herbivorous insects and pathogens that derive benefit from cell death are more susceptible to JA defences. Thus, this trade-off in pathways optimises defence and saves plant resources.

Cross talk also occurs between JA and other plant hormone pathways, such as those of abscisic acid (ABA) and ethylene (ET). These interactions similarly optimise defence against pathogens and herbivores of different lifestyles. For example, MYC2 activity can be stimulated by both JA and ABA pathways, allowing it to integrate signals from both pathways. Other transcription factors such as ERF1 arise as a result of JA and ET signaling. All these molecules can act in combination to activate specific wound response genes.

Finally, cross talk is not restricted for defence : JA and ET interactions are critical in development as well, and a balance between the two compounds is necessary for proper apical hook development in *Arabidopsis* seedlings. Still, further research is needed to elucidate the molecules regulating such cross talk.

Plant Peptide Hormone

Peptide signaling plays a significant role in various aspects of plant growth and development and specific receptors for various peptides have been identified as being membrane-localised receptor kinases, the largest family of receptor-like molecules in plants. Signaling peptides include members of the following protein families.

- Systemin – is a small polypeptide functioning as a long-distance signal to activate chemical defences against herbivores. It was the first plant hormone proven to be a peptide. Systemin induces the production of protein defence compound called protease inhibitors. Systemin was first identified in tomato leaves. It was found to be an 18-amino acid peptide processed from the C-terminus of a 200-amino acid precursor, which is called prosystemin.
- CLV3/ESR-related ('CLE') peptide family – *CLV3* encodes a small secreted peptide that functions as a short range ligand to the membrane-bound CLV1 receptor like kinase that together with CLV2 function to maintain stem cell homeostasis in *Arabidopsis* shoot apical meristems. Although the maise embryo-surrounding region protein (ESR). and CLV3 are very different, they are both members of the CLE peptide family given that they share a short conserved 14-amino acid sequence at the carboxy terminal region. To date, more than 150 CLE signaling peptides are identified. This proteolytically processed bioactive region is important for both promoting and inhibiting cellular differentiation in both apical and cambial meristems.
- ENOD40 – is an early nodulin gene, hence ENOD, that putatively encodes two small peptides, one of 12 and the other of 18 amino acid residues. Controversy exists on whether the mRNA or peptides themselves are responsible for bioactivity. Both peptides have been shown "in vivo" to bind to the 93 kDa subunit of sucrose synthase, an essential component in sucrose metabolism. Sucrose degradation is a key step in nitrogen fixation, and is a pre-requisite for normal nodule development.
- Phytosulfokine (PSK) – was first identified as a "conditioning factor" in asparagus and carrot cell cultures. The bioactive five amino acid peptide (PSK) is proteolytically processed from an ~80 amino acid precursor

secreted peptide. PSK has been demonstrated to promote cellular proliferation and transdifferentiation. It has been demonstrated that PSK binds to a membrane bound LRR receptor like kinase (PSKR).

- POLARIS (PLS) – The PLS peptide has a predicted length of 36 amino acids however possesses no secretion signal, suggesting that it functions within the cytoplasm. The PLS peptide itself has not yet been biochemically isolated, however loss-of-function mutants are hypersensitive to cytokinin with reduced responsiveness to auxin. Developmentally it is involved in vascularisation, longitudinal cell expansion and increased radial expansion.
- Rapid Alkalinisation Factor (RALF) – is 49 amino acid peptide that was identified whilst purifying systemin from tobacco leaves, it causes rapid medium alkanalisation and does not activate defence responses like systemin. Tomato RALF precursor cDNA encodes a 115 amino acid polypeptide containing an amino-terminal signal sequence with the bioactive RALF peptide encoded at the carboxy terminus. It is not known how mature RALF peptide is produced from its precursor, but a dibasic amino acid motif is located two residues upstream from the amino terminus of mature RALF. RALF has been identified to bind to potential membrane bound receptors complex containing proteins 25 kDa and 120 kDa in size.
- SCR/SP11 – are small polymorphic peptides produced by the tapetal cells of anthers and is involved in self-incompatibility of *Brassica* species. This secreted polypeptide is between 78 and 80 amino acid residues in length. Unlike other peptide hormones, no further post-translational processing occurs, except for the removal of the N-terminal signal peptide. SCR/SP11 like other small peptide hormones binds to a membrane bound LRR receptor like kinase (SRK).
- *ROTUNDIFOLIA4/DEVIL1* (*ROT4/DVL1*) – The ROT4 and DVL1 are peptides of 53 and 51 amino acids respectively, which have a high degree of sequence homology. They are two members of 23 member peptide family. ROT4 and DVL1 are involved in regulating polar cell proliferation on the longitudinal axis of organs.
- *Inflorescence deficient in abscission* (IDA) – a family of secreted peptides identified to be involved in petal abscission. The peptides are 77 amino acids in length and possess an amino-terminal secretions signal. Like the CLE peptide family these proteins have a conserved carboxy-terminal domain that is bordered by potentially cleavable basic residues. These proteins are secreted from cells in the floral abscission zone. Studies suggest that the HAESA membrane-associated LRR-RLK is likely to be this peptide's receptor as it too is expressed in the zone of floral organ abscission.

Nitric Oxide

Nitric oxide, or nitrogen oxide, also known as nitrogen monoxide, is a molecule with chemical formula NO. It is a free radical and is an important intermediate in the chemical industry. Nitric oxide is a by-product of combustion of substances in the air, as in automobile engines, fossil fuel power plants, and is produced naturally during the electrical discharges of lightning in thunderstorms.

In mammals including humans, NO is an important cellular signaling molecule involved in many physiological and pathological processes. It is a powerful vasodilator with a short half-life of a few seconds in the blood. Long-known pharmaceuticals such as nitroglycerine and amyl nitrite were discovered, more than a century after their first use in medicine, to be active through the mechanism of being precursors to nitric oxide.

Low levels of nitric oxide production are important in protecting organs such as the liver from ischemic damage.

Nitric oxide should not be confused with nitrous oxide (N_2O), an anaesthetic and greenhouse gas, or with nitrogen dioxide (NO_2), a brown toxic gas and a major air pollutant. However, nitric oxide is rapidly oxidised in air to nitrogen dioxide. Humphry Davy discovered this to his discomfort, when he inhaled the gas early in his career.

Despite being a simple molecule, NO is an important biological regulator and is a fundamental component in the fields of neuroscience, physiology, and immunology, with discovery of its key roles leading to Nobel Prize-winning research in these areas. It was proclaimed "Molecule of the Year" in 1992.

Strigolactone

Strigolactones are plant hormones that stimulate the branching and growth of symbiotic arbuscular mycorrhizal fungi, increasing the probability of contact and establishment of a symbiotic association between the plant and fungus. Strigolactones also inhibit plant shoot branching, and trigger germination of parasitic plant seeds. Strigolactones are carotenoid-derived and contain a labile ether bond that is easily hydrolysed in the rhizosphere, meaning that there is a large concentration gradient between areas near the root and those further away. In 2009, strigolactone biosynthesis was found to be DWARF27-dependent.

Karrikin

Karrikins are a group of plant growth regulators of the butenolide class found in the smoke of burning plant material. For many years scientists have known that smoke from forest fires had the ability to stimulate the germination of seeds. In 2004, after studying the thousands of chemical compounds found in smoke, it was discovered that a series of butenolides were responsible for this effect. Karrikins bind to receptor protein KAI2 on seeds to stimulate germination. The relation of seeds' response to karrikins is theorised to be an evolutionary adaptation in response to wildfires. Currently, there are four known karrikins which are designated KAR_1, KAR_2, KAR_3, and KAR_4. Though it is clear that karrikins are found in the smoke of burning plant material, there are competing theories about how they are formed. One hypothesis is that they are formed by Maillard reactions between carbohydrates and amino acids. Another suggests that they are derived from cellulose.

POTENTIAL MEDICAL APPLICATIONS

Plant stress hormones activate cellular responses, including cell death, to diverse stress situations in plants. Researchers have found that some plant stress hormones share the ability to adversely affect human cancer cells. For example, sodium salicylate has been found to suppress proliferation of lymphoblastic leukemia, prostate, breast, and melanoma human cancer cells. Jasmonic acid, a plant stress hormone that belongs to the jasmonate family, induced death in lymphoblastic leukemia cells. Methyl jasmonate has been found to induce cell death in a number of cancer cell lines.

HORMONES AND PLANT PROPAGATION

Synthetic plant hormones or PGRs are commonly used in a number of different techniques involving plant propagation from cuttings, grafting, micropropagation, and tissue culture.

The propagation of plants by cuttings of fully developed leaves, stems, or roots is performed by gardeners utilising auxin as a rooting compound applied to the cut surface; the auxins are taken into the plant and promote root initiation. In grafting, auxin promotes callus tissue formation, which joins the surfaces of the graft together. In micropropagation, different PGRs are used to promote multiplication and then rooting of new plantlets. In the tissue-culturing of plant cells, PGRs are used to produce callus growth, multiplication, and rooting.

Seed Dormancy

Plant hormones affect seed germination and dormancy by acting on different parts of the seed.

Embryo dormancy is characterised by a high ABA : GA ratio, whereas the seed has a high ABA sensitivity and low GA sensitivity. In order to release the seed from this type of dormancy and initiate seed germination, an alteration in hormone biosynthesis and degradation towards a low ABA/GA ratio, along with a decrease in ABA sensitivity and an increase in GA sensitivity, must occur.

ABA controls embryo dormancy, and GA embryo germination. Seed coat dormancy involves the mechanical restriction of the seed coat. This, along with a low embryo growth potential, effectively produces seed dormancy. GA releases this dormancy by increasing the embryo growth potential, and/or weakening the seed coat so the radical of the seedling can break through the seed coat. Different types of seed coats can be made up of living or dead cells, and both types can be influenced by hormones; those composed of living cells are acted upon after seed formation, whereas the seed coats composed of dead cells can be influenced by hormones during the formation of the seed coat. ABA affects testa or seed coat growth characteristics, including thickness, and effects the GA-mediated embryo growth potential. These conditions and effects occur during the formation of the seed, often in response to environmental conditions. Hormones also mediate endosperm dormancy : Endosperm in most seeds is composed of living tissue that can actively respond to hormones generated by the embryo. The endosperm often acts as a barrier to seed germination, playing a part in seed coat dormancy or in the germination process. Living cells respond to and also affect the ABA/GA ratio, and mediate cellular sensitivity; GA thus increases the embryo growth potential and can promote endosperm weakening. GA also affects both ABA-independent and ABA-inhibiting processes within the endosperm.

Chapter 5

AUXIN : THE GROWTH HORMONE

The first plant hormone we will consider is auxin. Auxin deserves pride of place in any discussion of plant hormones because it was the first growth hormone to be discovered in plants, and much of the early physiological work on the mechanism of plant cell expansion was carried out in relation to auxin action. Moreover, both auxin and cytokinin differ from the other plant hormones and signaling agents in one important respect : They are required for viability. Thus far, no mutants lacking either auxin or cytokinin have been found, suggesting that mutations that eliminate them are lethal. Whereas the other plant hormones seem to act as on/off switches that regulate specific developmental processes, auxin and cytokinin appear to be required at some level more or less continuously.

We begin our discussion of auxins with a brief history of their discovery, followed by a description of their chemical structures and the methods used to detect auxins in plant tissues. A look at the pathways of auxin biosynthesis and the polar nature of auxin transport follows. We will then review the various developmental processes controlled by auxin, such as stem elongation, apical dominance, root initiation, fruit development, and oriented, or *tropic*, growth. Finally, we will examine what is currently known about the mechanism of auxin-induced growth at the cellular and molecular levels.

THE EMERGENCE OF THE AUXIN CONCEPT

During the latter part of the nineteenth century, Charles Darwin and his son Francis studied plant growth phenomena involving tropisms. One of their interests was the bending of plants towards light. This phenomenon, which is caused by differential growth, is called **phototropism**. In some experiments the Darwins used seedlings of canary grass (*Phalaris canariensis*), in which, as in many other grasses, the youngest leaves are sheathed in a protective organ called the **coleoptile**.

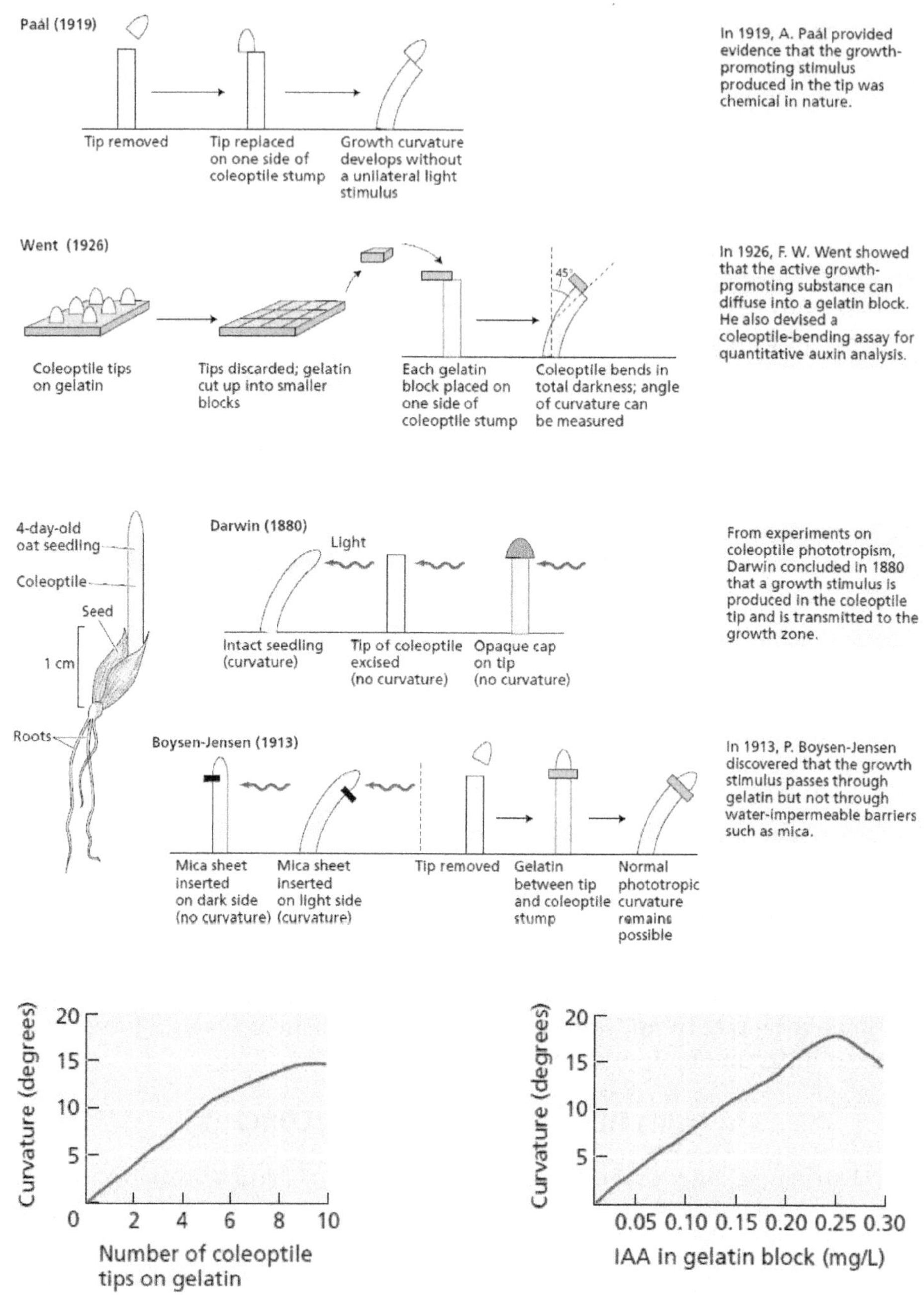

Fig. Summary of early experiments in auxin research.

Coleoptiles are very sensitive to light, especially to blue light. If illuminated on one side with a short pulse of dim blue light, they will bend (grow) towards the source of the light pulse within an hour. The Darwins found that the tip of the coleoptile perceived the light, for if they covered the tip with foil, the coleoptile would not bend. But the region of the coleoptile that is responsible for the bending towards the light, called the **growth zone**, is several millimeters below the tip.

Thus they concluded that some sort of signal is produced in the tip, travels to the growth zone, and causes the shaded side to grow faster than the illuminated side. The results of their experiments were published in 1881 in a remarkable book entitled *The Power of Movement in Plants*. There followed a long period of experimentation by many investigators on the nature of the growth stimulus in coleoptiles. This research culminated in the demonstration in 1926 by Frits Went of the presence of a growth-promoting chemical in the tip of oat (*Avena sativa*) coleoptiles. It was known that if the tip of a coleoptile was removed, coleoptile growth ceased. Previous workers had attempted to isolate and identify the growth-promoting chemical by grinding up coleoptile tips and testing the activity of the extracts. This approach failed because grinding up the tissue released into the extract inhibitory substances that normally were compartmentalised in the cell.

Went's major breakthrough was to avoid grinding by allowing the material to diffuse out of excised coleoptile tips directly into gelatin blocks. If placed asymmetrically on top of a decapitated coleoptile, these blocks could be tested for their ability to cause bending in the absence of a unilateral light source. Because the substance promoted the elongation of the coleoptile sections, it was eventually named **auxin** from the Greek *auxein*, meaning "to increase" or "to grow."

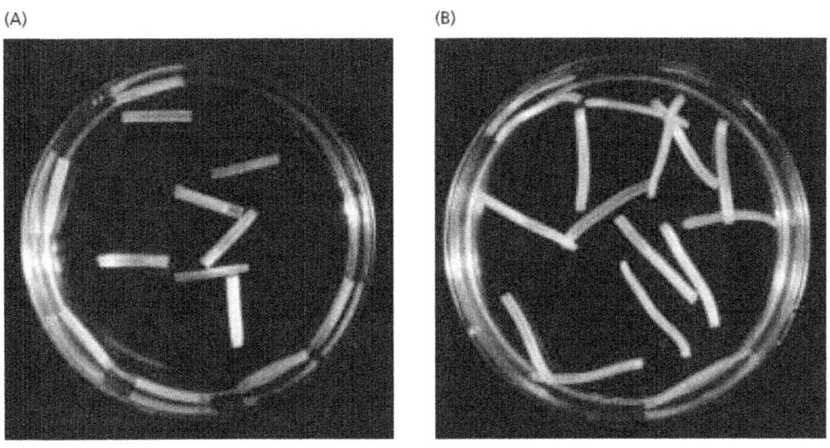

Fig. Auxin stimulates the elongation of oat coleoptiles sections. These coleoptiles sections were incubated for 18 hours in either water (A) or auxin (B). The yellow tissue inside the translucent coleoptiles is the primary leaves.

BIOSYNTHESIS AND METABOLISM OF AUXIN

Went's studies with agar blocks demonstrated unequivocally that the growth-promoting "influence" diffusing from the coleoptile tip was a chemical substance. The fact that it was produced at one location and transported in minute amounts to its site of action qualified it as an authentic plant hormone. In the years that followed, the chemical identity of the "growth substance" was determined, and because of its potential agricultural uses, many related chemical analogs were tested. This testing led to generalisations about the chemical requirements for auxin activity. In parallel with these studies, the agar block diffusion technique was being applied to the problem of auxin transport. Technological advances, especially the use of isotopes as tracers, enabled plant biochemists to unravel the pathways of auxin biosynthesis and breakdown.

Our discussion begins with the chemical nature of auxin and continues with a description of its biosynthesis, transport, and metabolism. Increasingly powerful analytical methods and the application of molecular biological approaches have recently allowed scientists to identify auxin precursors and to study auxin turnover and distribution within the plant.

The Principal Auxin in Higher Plants Is Indole-3-Acetic Acid

In the mid-1930s it was determined that auxin is **indole-3-acetic acid (IAA)**. Several other auxins in higher plants were discovered later, but IAA is by far the most abundant and physiologically relevant.

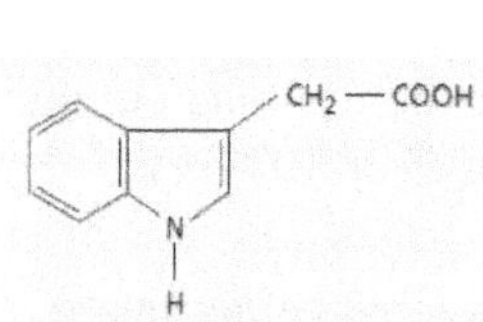

Indole-3-acetic acid (IAA)

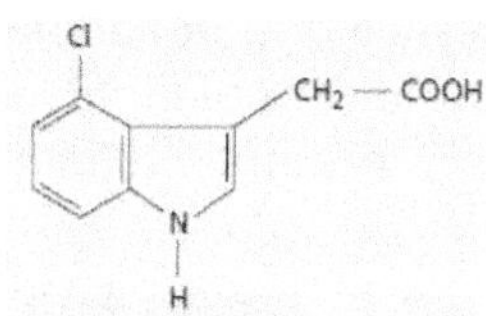

4-Chloroindole-3-acetic acid (4-Cl-IAA)

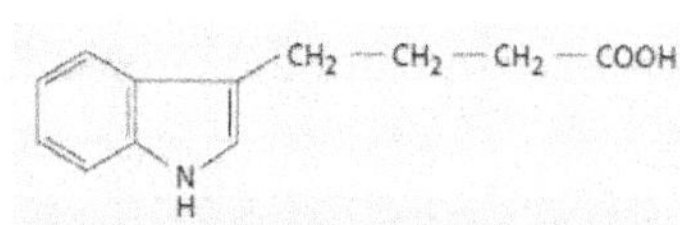

Indole-3-butyric acid (IBA)

Fig. Structure of three natural auxins. Indole-3-acetic acid (IAA) occurs in all plants, but other related compounds in plants have auxin activity. Peas, for example, contain 4-chloroindole-3-acetic acid. Mustards and corn contain indole-3-butyric acid (IBA).

Because the structure of IAA is relatively simple, academic and industrial laboratories were quickly able to synthesise a wide array of molecules with auxin activity. Some of these are used as herbicides in horticulture and agriculture.

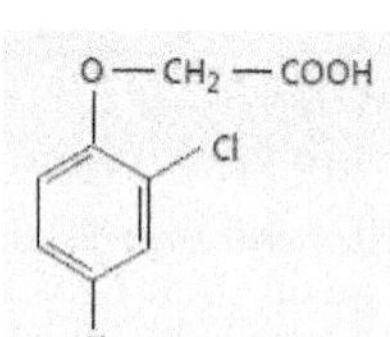

2,4-Dichlorophenoxyacetic acid (2,4-D)

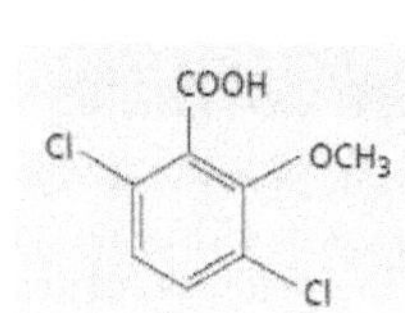

2-Methoxy-3, 6-dichlorobenzoic acid (dicamba)

Fig. Structures of two synthetic auxins. Most synthetic auxins are used as herbicides in horticulture and agriculture.

An early definition of auxins included all natural and synthetic chemical substances that stimulate elongation in coleoptiles and stem sections. However, auxins affect many developmental processes besides cell elongation. Thus auxins can be defined as compounds with biological activities similar to those of IAA, including the ability to promote cell elongation in coleoptile and stem sections, cell division in callus cultures in the presence of cytokinins, formation of adventitious roots on detached leaves and stems, and other developmental phenomena associated with IAA action. Although they are chemically diverse, a common feature of all active auxins is a molecular distance of about 0.5 nm between a fractional positive charge on the aromatic ring and a negatively charged carboxyl group.

Auxins in Biological Samples Can Be Quantified

Depending on the information that a researcher needs, the amounts and/ or identity of auxins in biological samples can be determined by bioassay, mass spectrometry, or enzyme-linked immunosorbent assay, which is abbreviated as ELISA. A **bioassay** is a measurement of the effect of a known or suspected biologically active substance on living material. In his pioneering work more than 60 years ago, Went used *Avena sativa* (oat) coleoptiles in a technique called the ***Avena* coleoptile curvature test**. The coleoptile curved because the increase in auxin on one side stimulated cell elongation, and the decrease in auxin on the other side (due to the absence of the coleoptile tip) caused a decrease in the growth rate—a phenomenon called **differential growth**. Went found that he could estimate the amount of auxin in a sample by measuring the resulting coleoptile curvature. Auxin bioassays are still used today to detect the presence of auxin activity in a sample. The *Avena* coleoptile curvature assay is a sensitive measure of auxin activity (it is effective for IAA concentrations of about 0.02 to 0.2 mg L^{-1}). Another bioassay measures auxin-induced changes in the straight growth of *Avena* coleoptiles floating in solution. Both of these bioassays can establish the presence of an auxin in a sample, but they cannot be used for precise quantification or identification of the specific compound.

Mass spectrometry is the method of choice when information about both the chemical structure and the amount of IAA is needed. This method is used in conjunction with separation protocols involving gas chromatography. It allows the precise quantification and identification of auxins, and can detect as little as 10^{-12} g (1 picogram, or pg) of IAA, which is well within the range of auxin found in a single pea stem section or a corn kernel. These sophisticated techniques have enabled researchers to accurately analyse auxin precursors, auxin turnover, and auxin distribution within the plant.

IAA Is Synthesised in Meristems, Young Leaves, and Developing Fruits and Seeds

IAA biosynthesis is associated with rapidly dividing and rapidly growing tissues, especially in shoots. Although virtually all plant tissues appear to be capable of producing low levels of IAA, shoot apical meristems, young leaves, and developing fruits and seeds are the primary sites of IAA synthesis. In very young leaf primordia of *Arabidopsis*, auxin is synthesised at the tip. During leaf development there is a gradual shift in the site of auxin production basipetally along the margins, and later, in the central region of the lamina. The basipetal shift in auxin production correlates closely with, and is probably causally related to the basipetal maturation sequence of leaf development and vascular differentiation.

By fusing the *GUS* (β-glucuronidase) reporter gene to a promoter containing an auxin response element, and transforming *Arabidopsis* leaves with this construct in a Ti plasmid using *Agrobacterium*, it is possible to visualise the distribution of free auxin in young, developing leaves. Wherever free auxin is produced, *GUS* expression occurs— and can be detected histochemically. By use of this technique, it has recently been demonstrated that auxin is produced by a cluster of cells located at sites where hydathodes will develop.

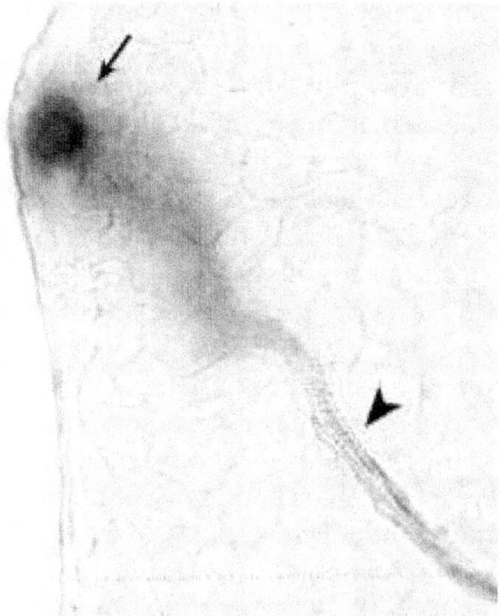

Fig. Detection of sites of auxin synthesis and transport in a young leaf primordium of DR5 Arabidopsis by means of a GUS reporter gene with an auxin-sensitive promoter. During the early stages of hydathode differentiation, a center of auxin synthesis is evident as a concentrated dark blue *GUS* stain (arrow) in the lobes of the serrated leaf margin. A gradient of diluted *GUS* activity extends from the margin toward a differentiating vascular strand (arrowhead 1, which functions as a sink for the auxin flow originating in the lobe.

Hydathodes are glandlike modifications of the ground and vascular tissues, typically at the margins of leaves, that allow the release of liquid water (guttation fluid) through pores in the epidermis in the presence of root pressure. During early stages of hydathode differentiation a center of high auxin synthesis is evident as a concentrated dark blue *GUS* stain (arrow) in the lobes of serrated leaves of *Arabidopsis*. A diffuse trail of *GUS* activity leads down to differentiating vessel elements in a developing vascular strand. This remarkable micrograph captures the process of auxin-regulated vascular differentiation in the very act !

Multiple Pathways Exist for the Biosynthesis of IAA

IAA is structurally related to the amino acid tryptophan, and early studies on auxin biosynthesis focused on tryptophan as the probable precursor. However, the incorporation of exogenous labelled tryptophan into IAA by plant tissues has proved difficult to demonstrate. Nevertheless, an enormous body of evidence has now accumulated showing that plants convert tryptophan to IAA by several pathways, which are described in the paragraphs that follow.

The IPA Pathway

The **indole-3-pyruvic acid (IPA)** pathway is probably the most common of the tryptophan-dependent pathways. It involves a deamination reaction to form IPA, followed by a decarboxylation reaction to form indole-3-acetaldehyde (IAld). Indole-3- acetaldehyde is then oxidised to IAA by a specific dehydrogenase.

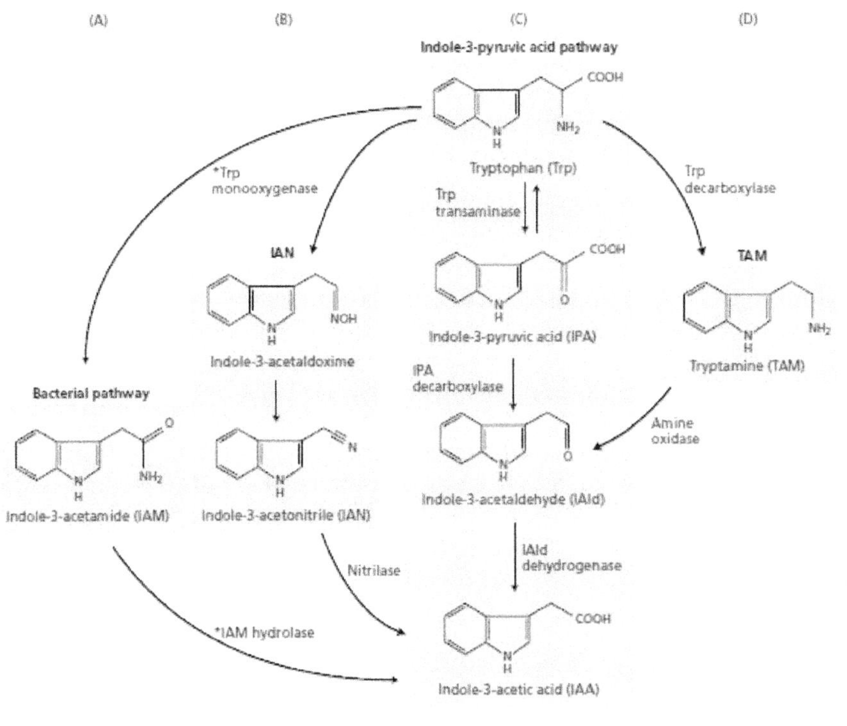

Fig. Tryptophan-dependent pathways of IAA biosynthesis in plants and bacteria. The enzymes that are present only in bacteria are marked with an asterisk.

The TAM Pathway

The **tryptamine (TAM) pathway** is similar to the IPA pathway, except that the order of the deamination and decarboxylation reactions is reversed, and different enzymes are involved. Species that do not utilise the IPA pathway possess the TAM pathway. In at least one case, there is evidence for both the IPA and the TAM pathways.

The IAN Pathway

In the **indole-3-acetonitrile (IAN)** pathway, tryptophan is first converted to indole-3-acetaldoxime and then to indole-3-acetonitrile. The enzyme that converts IAN to IAA is called *nitrilase*. The IAN pathway may be important in only three plant families : the Brassicaceae (mustard family), Poaceae (grass family), and Musaceae (banana family). Nevertheless, nitrilase-like genes or activities have recently been identified in the Cucurbitaceae (squash family), Solanaceae (tobacco family), Fabaceae (legumes), and Rosaceae (rose family). Four genes (*NIT1* through *NIT4*) that encode nitrilase enzymes have now been cloned from *Arabidopsis*. When *NIT2* was expressed in transgenic tobacco, the resultant plants acquired the ability to respond to IAN as an auxin by hydrolysing it to IAA. Another tryptophan-dependent biosynthetic pathway— one that uses **indole-3-acetamide (IAM)** as an intermediate—is used by various pathogenic bacteria, such as *Pseudomonas savastanoi* and *Agrobacterium tumefaciens*. This pathway involves the two enzymes tryptophan mono-oxygenase and IAM hydrolase. The auxins produced by these bacteria often elicit morphological changes in their plant hosts.

In addition to the tryptophan-dependent pathways, recent genetic studies have provided evidence that plants can synthesise IAA via one or more tryptophan-independent pathways. The existence of multiple pathways for IAA biosynthesis makes it nearly impossible for plants to run out of auxin and is probably a reflection of the essential role of this hormone in plant development.

IAA Is Also Synthesised from Indole or from Indole-3-Glycerol Phosphate

Although a tryptophan-independent pathway of IAA biosynthesis had long been suspected because of the low levels of conversion of radio labelled tryptophan to IAA, not until genetic approaches were available could the existence of such pathways be confirmed and defined.

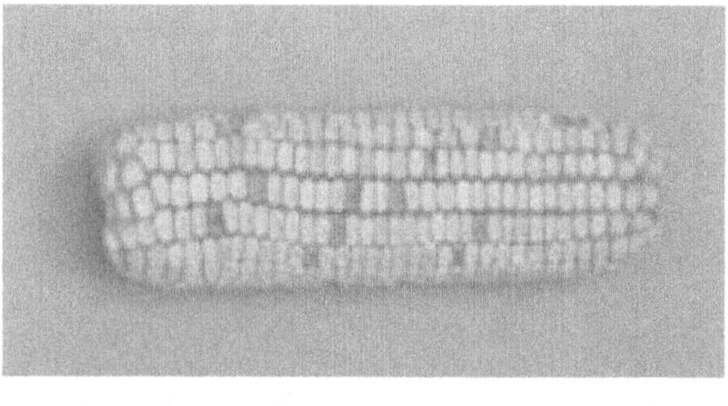

Fig. The orange pericarp (*orp*) mutant of maize is missing both sub-units of tryptophan synthase. As a result, the pericarps surrounding each kernel accumulate glycosides of anthranitic acid and indole. The orange colour is due to excess indole.

Perhaps the most striking of these studies in maise involves the *orange pericarp* (*orp*) mutant, in which both sub-units of the enzyme tryptophan synthase are inactive.

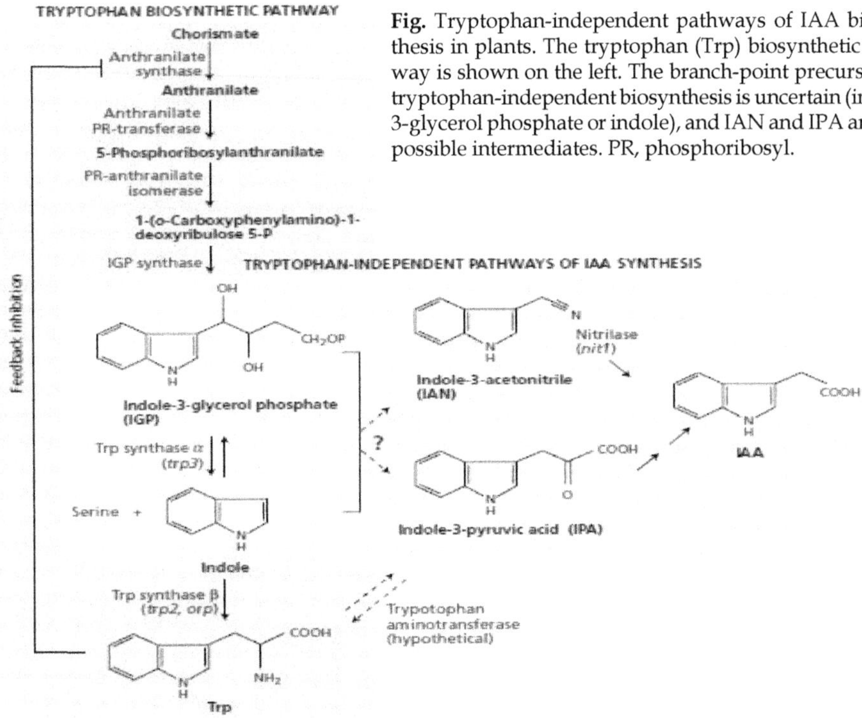

Fig. Tryptophan-independent pathways of IAA biosynthesis in plants. The tryptophan (Trp) biosynthetic pathway is shown on the left. The branch-point precursor for tryptophan-independent biosynthesis is uncertain (indole-3-glycerol phosphate or indole), and IAN and IPA are two possible intermediates. PR, phosphoribosyl.

The *orp* mutant is a true tryptophan auxotroph, requiring exogenous tryptophan to survive. However, neither the *orp* seedlings nor the wild-type seedlings can convert tryptophan to IAA, even when the mutant seedlings are given enough tryptophan to reverse the lethal effects of the mutation. Despite the block in tryptophan biosynthesis, the *orp* mutant contains amounts of IAA 50-fold higher than those of a wild-type plant. Signficantly, when *orp* seedlings were fed [15N] anthranilate, the label subsequently appeared in IAA, but not in tryptophan. These results provided the best experimental evidence for a tryptophan-independent pathway of IAA biosynthesis. Further studies established that the branch point for IAA biosynthesis is either indole or its precursor, indole-3- glycerol phosphate. IAN and IPA are possible intermediates, but the immediate precursor of IAA in the tryptophan-independent pathway has not yet been identified.

The discovery of the tryptophan-independent pathway has drastically altered our view of IAA biosynthesis, but the relative importance of the two pathways (tryptophan-dependent *versus* tryptophan-independent) is poorly understood. In several plants it has been found that the type of IAA biosynthesis pathway varies between different tissues, and between different times of development. For example, during embryogenesis in carrot, the tryptophan-dependent pathway is important very early in development, whereas the tryptophan-independent pathway takes over soon after the root-shoot axis is established.

Most IAA in the Plant is in a Covalently Bound Form

Although free IAA is the biologically active form of the hormone, the vast majority of auxin in plants is found in a covalently bound state. These conjugated, or "bound," auxins have been identified in all higher plants and are considered hormonally inactive. IAA has been found to be conjugated to both high- and low-molecular-weight compounds.

- Low-molecular-weight conjugated auxins include esters of IAA with glucose or *myo*-inositol and amide conjugates such as IAA-*N*-aspartate.
- High-molecular-weight IAA conjugates include IAA glucan (7–50 glucose units per IAA) and IAA-glycoproteins found in cereal seeds.

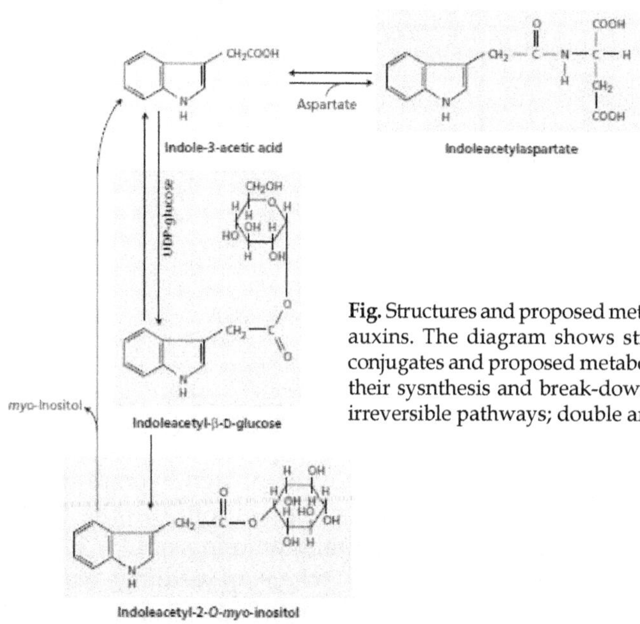

Fig. Structures and proposed metabolic pathways of bound auxins. The diagram shows structures of various IAA conjugates and proposed metabolic pathways involved in their sysnthesis and break-down. Single arrows indicate irreversible pathways; double arrows reversible.

The compound to which IAA is conjugated and the extent of the conjugation depend on the specific conjugating enzymes. The best-studied reaction is the conjugation of IAA to glucose in *Zea mays*. The highest concentrations of free auxin in the living plant are in the apical meristems of shoots and in young leaves because these are the primary sites of auxin synthesis. However, auxins are widely distributed in the plant. Metabolism of conjugated auxin may be a major contributing factor in the regulation of the levels of free auxin. For example, during the germination of seeds of *Zea mays*, IAA-*myo*-inositol is translocated from the endosperm to the coleoptile via the phloem. At least a portion of the free IAA produced in coleoptile tips of *Zea mays* is believed to be derived from the hydrolysis of IAA-*myo*-inositol.

In addition, environmental stimuli such as light and gravity have been shown to influence both the rate of auxin conjugation (removal of free auxin) and the rate of release of free auxin (hydrolysis of conjugated auxin). The formation of conjugated auxins may serve other functions as well, including storage and protection against oxidative degradation.

IAA is Degraded by Multiple Pathways

Like IAA biosynthesis, the enzymatic breakdown (oxidation) of IAA may involve more than one pathway. For some time it has been thought that peroxidative enzymes are chiefly responsible for IAA oxidation, primarily because these enzymes are ubiquitous in higher plants and their ability to degrade IAA can be demonstrated in vitro.

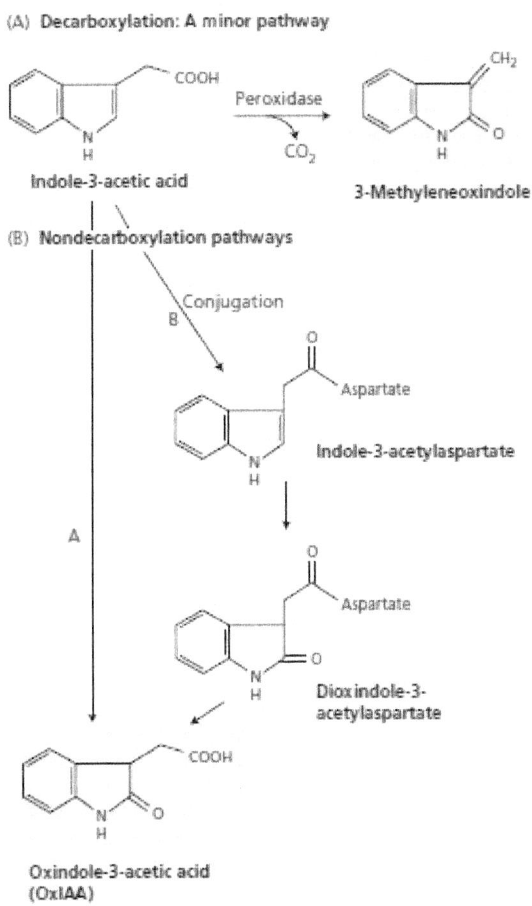

However, the physiological significance of the peroxidase pathway is unclear. For example, no change in the IAA levels of transgenic plants was observed with either a tenfold increase in peroxidase expression or a tenfold repression of peroxidase activity. On the basis of isotopic labelling and metabolite identification, two other oxidative pathways are more likely to be involved in the controlled degradation of IAA. The end product of this pathway is oxindole- 3-acetic acid (OxIAA), a naturally occurring compound in the endosperm and shoot tissues of *Zea mays*. In one pathway, IAA is oxidised without decarboxylation to OxIAA. In another pathway, the IAA-aspartate conjugate is oxidised first to the intermediate dioxindole-3-acetylaspartate, and then to OxIAA.

In vitro, IAA can be oxidised non-enzymatically when exposed to high-intensity light, and its photodestruction in vitro can be promoted by plant pigments such as riboflavin. Although the products of auxin photooxidation have been isolated from plants, the role, if any, of the photo-oxidation pathway in vivo is presumed to be minor.

Two Subcellular Pools of IAA Exist : The Cytosol and the Chloroplasts

The distribution of IAA in the cell appears to be regulated largely by pH. Because IAA^- does not cross membranes unaided, whereas IAAH readily diffuses across membranes, auxin tends to accumulate in the more alkaline compartments of the cell. The distribution of IAA and its metabolites has been studied in tobacco cells. About one-third of the IAA is found in the chloroplast, and the remainder is located in the cytosol. IAA conjugates are located exclusively in the cytosol. IAA in the cytosol is metabolised either by conjugation or by non-decarboxylative catabolism. The IAA in the chloroplast is protected from these processes, but it is regulated by the amount of IAA in the cytosol, with which it is in equilibrium. The factors that regulate the steady-state concentration of free auxin in plant cells are diagrammatically summarised in **Web Topic 19.5**.

AUXIN TRANSPORT

The main axes of shoots and roots, along with their branches, exhibit apex-base structural polarity, and this structural polarity has its origin in the polarity of auxin transport. Soon after Went developed the coleoptile curvature test for auxin, it was discovered that IAA moves mainly from the apical to the basal end (*basipetally*) in excised oat coleoptile sections. This type of unidirectional transport is termed **polar transport**. Auxin is the only plant growth hormone known to be transported polarly. Because the shoot apex serves as the primary source of auxin for the entire plant, polar transport has long been believed to be the principal cause of an auxin gradient extending from the shoot tip to the root tip. The longitudinal gradient of auxin from the shoot to the root affects various developmental processes, including stem elongation, apical dominance, wound healing, and leaf senescence. Recently it has been recognised that a significant amount of auxin transport also occurs in the phloem, and that the phloem is probably the principal route by which auxin is transported *acropetally* (*i.e.*, towards the tip) in the root. Thus, more than one pathway is responsible for the distribution of auxin in the plant.

Polar Transport Requires Energy and Is Gravity Independent

To study polar transport, researchers have employed the *donor–receiver agar block method* : An agar block containing radioisotope-labelled auxin (donor block) is placed on one end of a tissue segment, and a receiver block is placed on the other end. The movement of auxin through the tissue into the receiver block can be determined over time by measurement of the radioactivity in the receiver block.

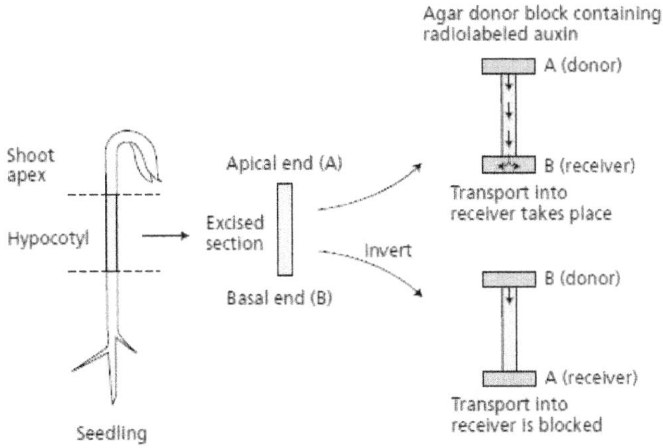

Fig. The standard method for measuring polar auxin transport. The polarity of transport is independent of orientation with respect to gravity.

From a multitude of such studies, the general properties of polar IAA transport have emerged. Tissues differ in the degree of polarity of IAA transport. In coleoptiles, vegetative stems, and leaf petioles, basipetal transport predominates. Polar transport is not affected by the orientation of the tissue (at least over short periods of time), so it is independent of gravity. A demonstration of the lack of effect of gravity on basipetal auxin transport is shown in Figure.

When stem cuttings, in this case grape hardwood, are placed in a moist chamber, adventitious roots form at the basal ends of the cuttings, and shoots form at the apical ends, even when the cuttings are inverted. Roots form at the base because root differentiation is stimulated by auxin accumulation due to basipetal transport. Shoots tend to form at the apical end where the auxin concentration is lowest. Polar transport proceeds in a cell-to-cell fashion, rather than via the symplast. That is, auxin exits the cell through the plasma membrane, diffuses across the compound middle lamella, and enters the cell below through its plasma membrane. The loss of auxin from cells is termed *auxin efflux*; the entry of auxin into cells is called *auxin uptake* or *influx*. The overall process requires metabolic energy, as evidenced by the sensitivity of polar transport to O_2 deprivation and metabolic inhibitors. The velocity of polar auxin transport is 5 to 20 cm h^{-1} – faster than the rate of diffusion, but slower than phloem translocation rates. Polar transport is also specific for active auxins, both natural and synthetic. Neither inactive auxin analogs nor auxin metabolites are transported polarly, suggesting that polar transport involves specific protein carriers on the plasma membrane that can recognise the hormone and its active analogs.

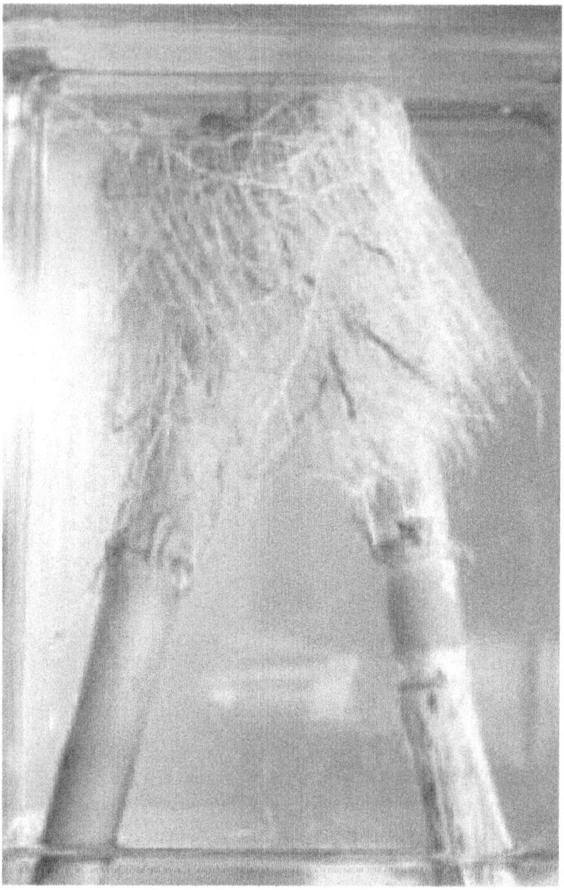

Fig. Adventitious roots grow from the basal ends, and shoots grow from the apical ends, of grape hardwood cuttings, whether they are maintained in the inverted (the two cuttings on the left) or upright orientation (the cuttings on the right). The roots always form at the basal ends because polar auxin transport is independent of gravity.

The major site of basipetal polar auxin transport in stems and leaves is the vascular parenchyma tissue. Coleoptiles appear to be the exception in that basipetal polar transport occurs mainly in the non-vascular tissues. Acropetal polar transport in the root is specifically associated with the xylem parenchyma of the stele. However, most of the auxin that reaches the root tip is translocated via the phloem. A small amount of basipetal auxin transport from the root tip has also been demonstrated. In maise roots, for example, radiolabelled IAA applied to the root tip is transported basipetally about 2 to 8 mm. Basipetal auxin transport in the root occurs in the epidermal and cortical tissues, and as we shall see, it plays a central role in gravitropism.

A Chemiosmotic Model has been Proposed to Explain Polar Transport

The discovery of the chemiosmotic mechanism of solute transport in the late 1960s led to the application of this model to polar auxin transport. According to the now generally accepted **chemiosmotic model** for polar auxin transport, auxin uptake is driven by the proton motive force ($\Delta E + \Delta pH$) across the plasma membrane, while auxin efflux is driven by the membrane potential, ΔE. Acrucial feature of the polar transport model is that the auxin efflux carriers are localised at the basal ends of the conducting cells.

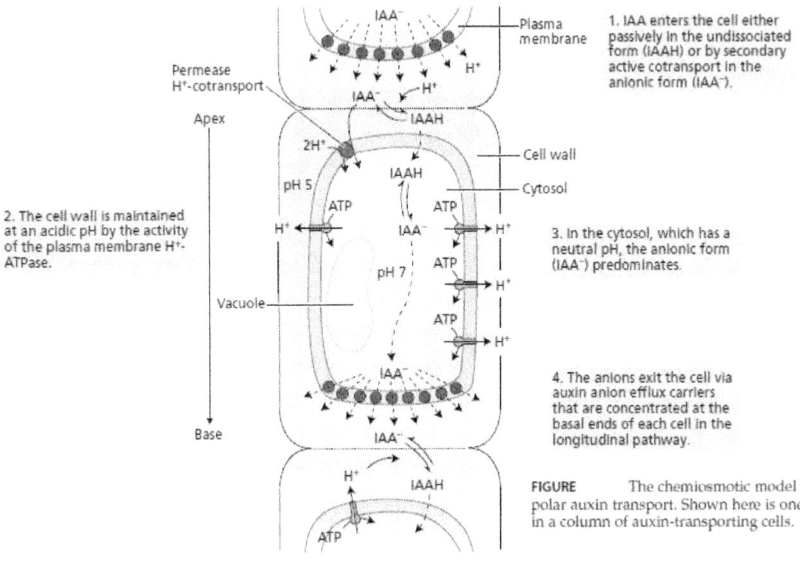

FIGURE The chemiosmotic model for polar auxin transport. Shown here is one cell in a column of auxin-transporting cells.

The evidence for each step in this model is considered separately in the discussion that follows :

Auxin Influx

The first step in polar transport is auxin influx. According to the model, auxin can enter plant cells from any direction by either of two mechanisms :

- Passive diffusion of the protonated (IAAH) form across the phospholipid bilayer
- Secondary active transport of the dissociated (IAA^-) form via a $2H^+$–IAA^- symporter

The dual pathway of auxin uptake arises because the passive permeability of the membrane to auxin depends strongly on the apoplastic pH. The undissociated form of indole-3-acetic acid, in which the carboxyl group is protonated, is lipophilic and readily diffuses across lipid bilayer membranes. In contrast, the dissociated form of auxin is negatively charged and therefore does not cross membranes unaided. Because the plasma membrane H^+-ATPase normally maintains the cell wall solution at about pH 5, about half of the auxin (pKa = 4.75) in the apoplast will be in the undissociated form and will diffuse passively across the plasma membrane down a concentration gradient. Experimental support for pH-dependent, passive auxin uptake was first provided by the demonstration that IAA uptake by plant cells increases as the extracellular pH is lowered from a neutral to a more acidic value. Acarrier-mediated, secondary active uptake mechanism was shown to be saturable and specific for active auxins. In experiments in which the ΔpH and ΔE values of isolated membrane vesicles from zucchini (*Cucurbita pepo*) hypocotyls were manipulated artificially, the uptake of radiolabelled auxin was shown to be stimulated in the presence of a pH gradient, as in passive uptake, but also when the inside of the vesicle was negatively charged relative to the outside. These and other experiments suggested that an H^+–IAA^- symporter cotransports two protons along with the auxin anion. This secondary active transport of auxin allows for greater auxin accumulation than simple diffusion does because it is driven across the membrane by the proton motive force. Apermease-type auxin uptake carrier, AUX1, related to bacterial amino acid carriers, has been identified in *Arabidopsis* roots. The roots of *aux1* mutants are agravitropic, suggesting that auxin influx is a limiting factor for gravitropism in roots. As predicted by the chemiosmotic model, AUX1 appears to be uniformly distributed around cells in the polar transport pathway. Thus in general, the polarity of auxin transport is governed by the efflux step rather than the influx step.

Auxin Efflux

Once IAA enters the cytosol, which has a pH of approximately 7.2, nearly all of it will dissociate to the anionic form. Because the membrane is less permeable to IAA^- than to IAAH, IAA^- will tend to accumulate in the cytosol. However, much of the auxin that enters the cell escapes via an *auxin anion efflux carrier*. According to the chemiosmotic model, transport of IAA^- out of the cell is driven by the inside negative membrane potential.

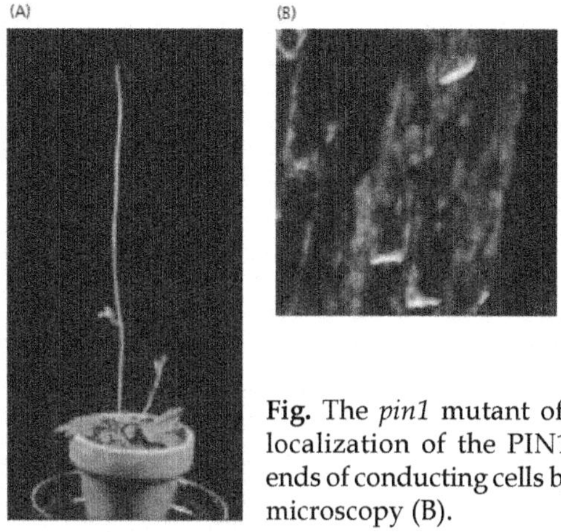

Fig. The *pin1* mutant of Arabidopsis (A) and localization of the PIN1 protien at the basal ends of conducting cells by immunofluorescence microscopy (B).

As noted earlier, the central feature of the chemiosmotic model for polar transport is that IAA^- efflux takes place preferentially at the basal end of each cell. The repetition of auxin uptake at the apical end of the cell and preferential release from the base of each cell in the pathway gives rise to the total polar transport effect. A family of putative auxin efflux carriers known as **PIN proteins** are localised precisely as the model would predict—that is, at the basal ends of the conducting cells.

PIN proteins have 10 to 12 transmembrane regions characteristic of a major superfamily of bacterial and eukaryotic transporters, which include drug resistance proteins and sugar transporters.

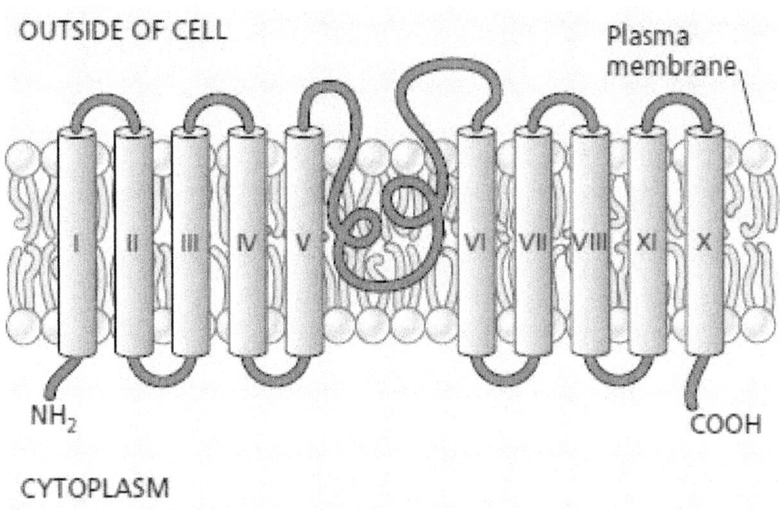

Fig. The topology of the TIN1 protein with ten transmembrane segments and a large hydrophilic loop in the middle.

Despite topological similarities to other transporters, recent studies suggest that PIN may require other proteins for activity, and may be part of a larger protein complex.

Inhibitors of Auxin Transport Block Auxin Efflux

Several compounds have been synthesised that can act as **auxin transport inhibitors (ATIs)**, including NPA (1-*N*naphthylphthalamic acid) and TIBA (2,3,5-triiodobenzoic acid).

These inhibitors block polar transport by preventing auxin efflux. We can demonstrate this phenomenon by incorporating NPA or TIBA into either the donor or the receiver block in an auxin transport experiment. Both compounds inhibit auxin efflux into the receiver block, but they do not affect auxin uptake from the donor block. Some ATIs, such as TIBA, that have weak auxin activity and are transported polarly, may inhibit polar transport in part by competing with auxin for its binding site on the efflux carrier. Others, such as NPA, are not transported polarly and are believed to interfere with auxin transport by binding to proteins associated in a complex with the efflux carrier. Such NPA-binding proteins are also found at the basal ends of the conducting cells, consistent with the localisation of PIN proteins. Recently another class of ATIs has been identified that inhibits the AUX1 uptake carrier. For example, 1-naphthoxyacetic acid (1-NOA) blocks auxin uptake into cells, and when applied to *Arabidopsis* plants it causes root agravitropism similar to that of the *aux1* mutant. Like the *aux1* mutation, neither 1- NOAnor any of the other AUX1-specific inhibitors block polar auxin transport.

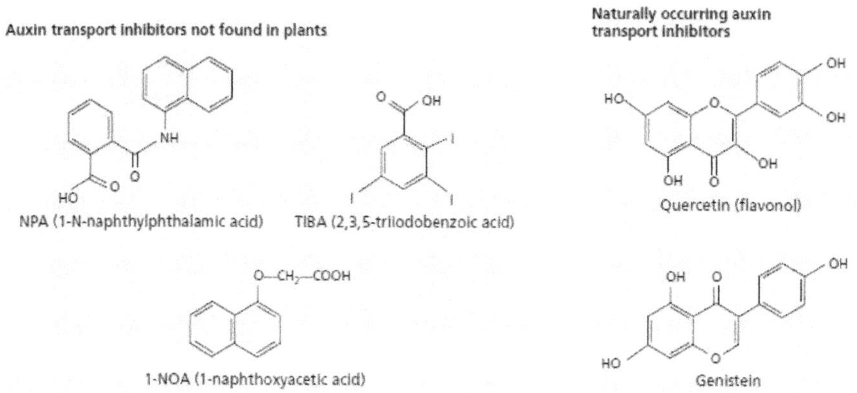

Fig. Structures of auxin transport inhibitors.

PIN Proteins Are Rapidly Cycled to and from the Plasma Membrane

The basal localisation of the auxin efflux carriers involves targeted vesicle secretion to the basal ends of the conducting cells. Recently it has been demonstrated that PIN proteins, although stable, do not remain on the plasma membrane permanently, but are rapidly cycled to an unidentified endosomal compartment via endocytotic vesicles, and then recycled back to the plasma membrane. Prior to treatment, the PIN1 protein is localised at the basal ends (top) of root cortical parenchyma cells.

Treatment of *Arabidopsis* seedlings with brefeldin A(BFA), which causes Golgi vesicles and other endosomal compartments to aggregate near the nucleus, causes PIN to accumulate in these abnormal intracellular compartments. When the BFA is washed out with buffer, the normal localisation on the plasma membrane at the base of the cell is restored. But when cytochalasin D, an inhibitor of actin polymerisation, is included in the buffer washout solution, normal relocalisation of PIN to the plasma membrane is prevented. These results indicate that PIN is rapidly cycled between the plasma membrane at the base of the cell and an unidentified endosomal compartment by an actin-dependent mechanism.

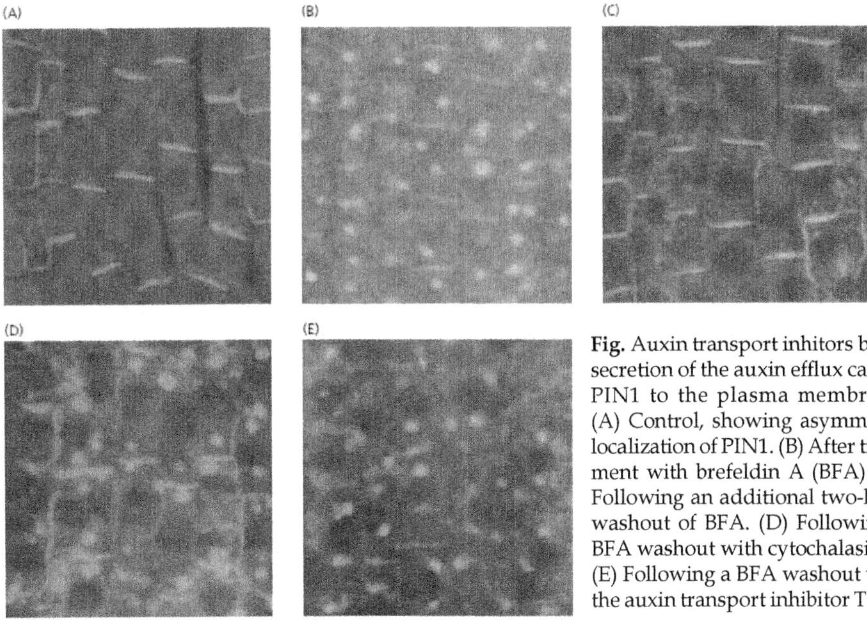

Fig. Auxin transport inhitors block secretion of the auxin efflux carrier PIN1 to the plasma membrane. (A) Control, showing asymmetric localization of PIN1. (B) After treatment with brefeldin A (BFA). (C) Following an additional two-hour washout of BFA. (D) Following a BFA washout with cytochalasin D. (E) Following a BFA washout with the auxin transport inhibitor TIBA.

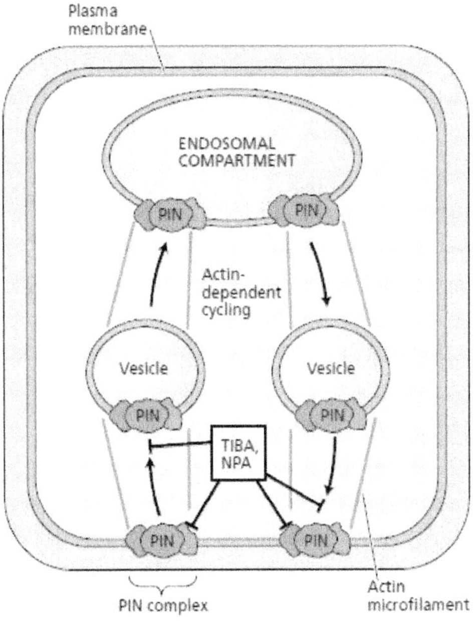

Although they bind different targets, both TIBA and NPA interfere with vesicle traffic to and from the plasma membrane. The best way to demonstrate this phenomenon is to include TIBA in the washout solution after BFAtreatment. Under these conditions, TIBA prevents the normal relocalisation of PIN on the plasma membrane following the washout treatment. The effects of TIBA and NPAon cycling are not specific for PIN proteins, and it has been proposed that ATIs may actually represent general inhibitors of membrane cycling. On the other hand, neither TIBA nor NPA alone causes PIN delocalisation, even though they block auxin efflux. Therefore, TIBA and NPAmust also be able to directly inhibit the transport activity of PIN complexes on the plasma membrane—by binding either to PIN or to one or more regulatory proteins (as NPAdoes). A simplified model of the effects of TIBA and NPA on PIN cycling and auxin efflux is shown in Figure.

Flavonoids Serve as Endogenous ATIs

There is mounting evidence that flavonoids can function as endogenous regulators of polar auxin transport. Indeed, naturally occurring aglycone flavonoid compounds (flavonoids without attached sugars) are able to compete with NPA for its binding site on membranes and are typically localised on the plasma membrane at the basal ends of cells where the efflux carrier is concentrated. In addition, recent studies have shown that the cells of flavonoid-deficient *Arabidopsis* mutants are less able to accumulate auxin than wild-type cells, and the mutant seedlings that lack flavonoid have altered auxin distribution profiles. Many of the flavonoids that displace NPAfrom its binding site on membranes are also inhibitors of protein kinases and protein phosphatases. An *Arabidopsis* mutant designated *rcn1* (*r*oots *c*url in *NPA1*) was identified on the basis of an enhanced sensitivity to *NPA*. The *RCN1* gene is closely related to the regulatory subunit of protein phosphatase 2A, a serine/threonine phosphatase (Garbers *et. al.* 1996).

Protein phosphatases are known to play important roles in enzyme regulation, gene expression, and signal transduction by removing regulatory phosphate groups from proteins. This finding suggests that a signal transduction pathway involving protein kinases and protein phosphatases may be involved in signaling between NPA-binding proteins and the auxin efflux carrier.

Auxin is Also Transported Non-polarly in the Phloem

Most of the IAA that is synthesised in mature leaves appears to be transported to the rest of the plant non-polarly via the phloem. Auxin, along with other components of phloem sap, can move from these leaves up or down the plant at velocities much higher than those of polar transport. Auxin translocation in the phloem is largely passive, not requiring energy directly. Although the overall importance of the phloem pathway *versus* the polar transport system for the long-distance movement of IAA in plants is still unresolved, the evidence suggests that long-distance auxin transport in the phloem is important for controlling such processes as cambial cell divisions, callose accumulation or removal from sieve tube elements, and branch root formation. Indeed, the phloem appears to represent the principal pathway for long-distance auxin translocation to the root (Aloni 1995; Swarup *et. al.* 2001). Polar transport and phloem transport are not independent of each other. Recent studies with radiolabeled IAA suggest that in pea, auxin can be transferred from the non-polar phloem pathway to the polar transport pathway. This transfer takes place mainly in the immature tissues of the shoot apex.

Asecond example of transfer of auxin from the non-polar phloem pathway to a polar transport system has recently been documented in *Arabidopsis*. It was shown that the AUX1 permease is asymmetrically localised on the plasma membrane at the upper end of root protophloem cells (*i.e.*, the end distal from the tip).

It has been proposed that the asymmetrically oriented AUX1 permease promotes the acropetal movement of auxin from the phloem to the root apex (Swarup *et. al.* 2001). This type of polar auxin transport based on the asymmetric localisation of AUX1 differs from the polar transport that occurs in the shoot and basal region of the root, which is based on the asymmetric distribution of the PIN complex. That AUX1 is also strongly expressed in a cluster of cells in the columella of the root cap, as well as in lateral root cap cells that overlay the cells of the distal elongation zone of the root. These cells form a minor, but physiologically important, basipetal pathway whereby auxin reaching the columella is redirected backward towards the outer tissues of the elongation zone. The importance of this pathway will become apparent when we examine the mechanism of root gravitropism.

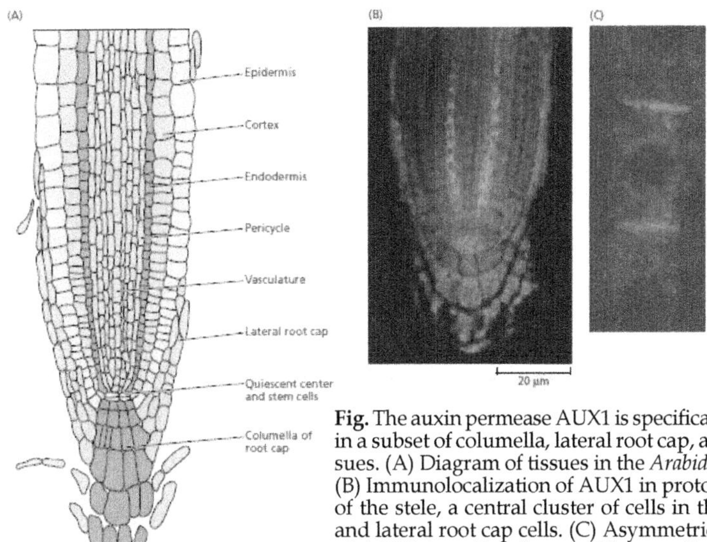

Fig. The auxin permease AUX1 is specifically expressed in a subset of columella, lateral root cap, and stellar tissues. (A) Diagram of tissues in the *Arabidopsis* root tip. (B) Immunolocalization of AUX1 in protophloem cells of the stele, a central cluster of cells in the columella, and lateral root cap cells. (C) Asymmetric localization of AUX1 in a file of protophloem cells. Scale bar is 2 µm in C.

PHYSIOLOGICAL EFFECTS OF AUXIN : CELL ELONGATION

Auxin was discovered as the hormone involved in the bending of coleoptiles towards light. The coleoptile bends because of the unequal rates of cell elongation on its shaded versus its illuminated side. The ability of auxin to regulate the rate of cell elongation has long fascinated plant scientists.

Auxins Promote Growth in Stems and Coleoptiles, While Inhibiting Growth in Roots

As we have seen, auxin is synthesised in the shoot apex and transported basipetally to the tissues below. The steady supply of auxin arriving at the subapical region of the stem or coleoptile is required for the continued elongation of these cells. Because the level of endogenous auxin in the elongation region of a normal healthy plant is nearly optimal for growth, spraying the plant with exogenous auxin causes only a modest and short-lived stimulation in growth, and may even be inhibitory in the case of darkgrown seedlings, which are more sensitive to supraoptimal auxin concentrations than light-grown plants are. However, when the endogenous source of auxin is removed by excision of sections containing the elongation zones, the growth rate rapidly decreases to a low basal rate. Such excised sections will often respond dramatically to exogenous auxin by rapidly increasing their growth rate back to the level in the intact plant. In long-term experiments, treatment of excised sections of coleoptiles or dicot stems with auxin stimulates the rate of elongation of the section for up to 20 hours.

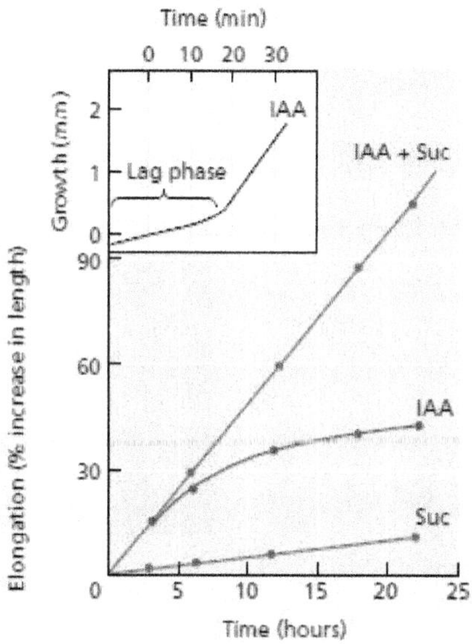

Fig. Time course for auxin-induced growth of *Aueaa* (oat) coleoptile sections. Growth is plotted as the percent increase in length. Auxin was added at time zero. When sucrose (Suc) is included in the medium, the response can continue for as long as 20 hours. Sucrose prolongs the growth response to auxin mainly by providing osmotically active solute that can be taken up for the maintenance of turgor pressure during cell elongation. KC1 can substitute for sucrose. The inset shows a short-term time course plotted with an electronic position-sensing transducer, in this graph, growth is plotted as the absolute length in millimeters versus time, The curve shows a lag time of about 15 minutes for auxin-stimulated growth to begin.

The optimal auxin concentration for elongation growth is typically 10^{-6} to $10^{-5}M$.

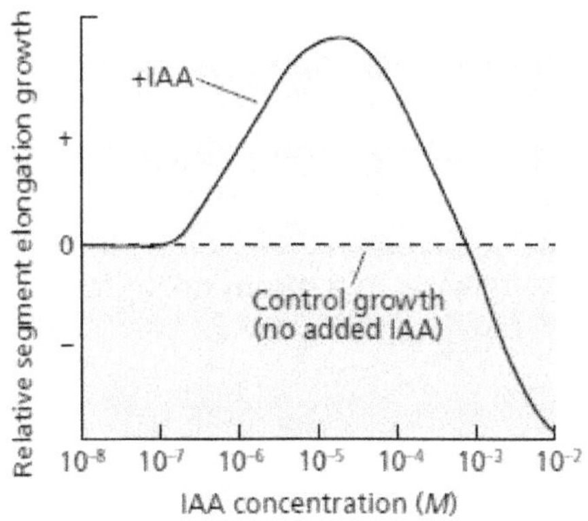

The inhibition beyond the optimal concentration is generally attributed to auxin-induced ethylene biosynthesis. The gaseous hormone ethylene inhibits stem elongation in many species. Auxin control of root elongation growth has been more difficult to demonstrate, perhaps because auxin induces the production of ethylene, a root growth inhibitor. However, even if ethylene biosynthesis is specifically blocked, low concentrations (10^{-10} to 10^{-9} *M*) of auxin promote the growth of intact roots, whereas higher concentrations (10^{-6} *M*) inhibit growth. Thus, roots may require a minimum concentration of auxin to grow, but root growth is strongly inhibited by auxin concentrations that promote elongation in stems and coleoptiles.

The Outer Tissues of Dicot Stems are the Targets of Auxin Action

Dicot stems are composed of many types of tissues and cells, only some of which may limit the growth rate. This point is illustrated by a simple experiment. When stem sections from growing regions of an etiolated dicot stem, such as pea, are split lengthwise and incubated in buffer, the two halves bend outward. This result indicates that, in the absence of auxin the central tissues, including the pith, vascular tissues, and inner cortex, elongate at a faster rate than the outer tissues, consisting of the outer cortex and epidermis. Thus the outer tissues must be limiting the extension rate of the stem in the absence of auxin. However, when the split sections are incubated in buffer plus auxin, the two halves now curve inward, demonstrating that the outer tissues of dicot stems are the primary targets of auxin action during cell elongation.

The observation that the outer cell layers are the targets of auxin seems to conflict with the localisation of polar transport in the parenchyma cells of the vascular bundles. However, auxin can move laterally from the vascular tissues of dicot stems to the outer tissues of the elongation zone. In coleoptiles, on the other hand, all of the non-vascular tissues (epidermis plus mesophyll) are capable of transporting auxin, as well as responding to it.

The Minimum Lag Time for Auxin-Induced Growth is Ten Minutes

When a stem or coleoptile section is excised and inserted into a sensitive growth-measuring device, the growth response to auxin can be monitored at very high resolution. Without auxin in the medium, the growth rate declines rapidly. Addition of auxin markedly stimulates the growth rate after a lag period of only 10 to 12 minutes. Both *Avena* (oat) coleoptiles and *Glycine max* (soybean) hypocotyls (dicot stem) reach a maximum growth rate after 30 to 60 minutes of auxin treatment.

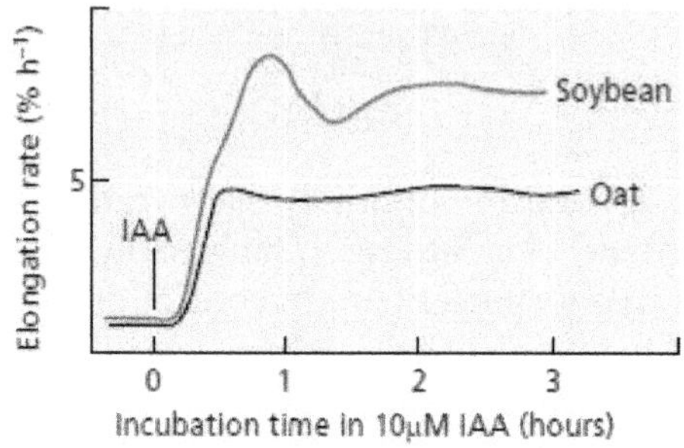

Fig. Comparison of the growth kinetics of oat coleoptile and soybean hypocotyls sections, incubated with 10 μM IAA and 2% sucrose. Growth is plotted as the rate at each time point, rather than the rate of the absolute length. The growth rate of the soybean hypocotyls oscillates after 1 hour, whereas that of the oat coleoptile is constant.

This maximum represents a five- to tenfold increase over the basal rate. Oat coleoptile sections can maintain this maximum rate for up to 18 hours in the presence of osmotically active solutes such as sucrose or KCl. As might be expected, the stimulation of growth by auxin requires energy, and metabolic inhibitors inhibit the response within minutes. Auxin-induced growth is also sensitive to inhibitors of protein synthesis such as cycloheximide, suggesting that proteins with high turnover rates are involved. Inhibitors of RNA synthesis also inhibit auxininduced growth, after a slightly longer delay (Cleland 1995). Although the length of the lag time for auxin-stimulated growth can be increased by lowering of the temperature or by the use of sub-optimal auxin concentrations, the lag time cannot be shortened by raising of the temperature, by the use of supraoptimal auxin concentrations, or by abrasion of the waxy cuticle to allow auxin to penetrate the tissue more rapidly. Thus the minimum lag time of 10 minutes is not determined by the time required for auxin to reach its site of action. Rather, the lag time reflects the time needed for the biochemical machinery of the cell to bring about the increase in the growth rate.

Auxin Rapidly Increases the Extensibility of the Cell Wall

How does auxin cause a five- to tenfold increase in the growth rate in only 10 minutes? To understand the mechanism, we must first review the process of cell enlargement in plants. Plant cells expand in three steps :

- Osmotic uptake of water across the plasma membrane is driven by the gradient in water potential ($\Delta\psi_w$).
- Turgor pressure builds up because of the rigidity of the cell wall.
- Biochemical wall loosening occurs, allowing the cell to expand in response to turgor pressure.

The effects of these parameters on the growth rate are encapsulated in the growth rate equation :

$$GR = m\ (\psi_p - Y)$$

where GR is the growth rate, ψp is the turgor pressure, Y is the yield threshold, and m is the coefficient (*wall extensibility*) that relates the growth rate to the difference between ψ_p and Y.

In principle, auxin could increase the growth rate by increasing m, increasing Ψp, or decreasing Y. Although extensive experiments have shown that auxin does not increase turgor pressure when it stimulates growth, conflicting results have been obtained regarding auxininduced decreases in Y. However, there is general agreement that auxin causes an increase in the wall extensibility parameter, m.

Auxin-Induced Proton Extrusion Acidifies the Cell Wall and Increases Cell Extension

According to the widely accepted **acid growth hypothesis**, hydrogen ions act as the intermediate between auxin and cell wall loosening. The source of the hydrogen ions is the plasma membrane H^+-ATPase, whose activity is thought to increase in response to auxin. The acid growth hypothesis allows five main predictions :

- Acid buffers alone should promote short-term growth, provided the cuticle has been abraded to allow the protons access to the cell wall.
- Auxin should increase the rate of proton extrusion (wall acidification), and the kinetics of proton extrusion should closely match those of auxin-induced growth.
- Neutral buffers should inhibit auxin-induced growth.
- Compounds (other than auxin) that promote proton extrusion should stimulate growth.
- Cell walls should contain a "wall loosening factor" with an acidic pH optimum.

All five of these predictions have been confirmed. Acidic buffers cause a rapid and immediate increase in the growth rate, provided the cuticle has been abraded. Auxin stimulates proton extrusion into the cell wall after 10 to 15 minutes of lag time, consistent with the growth kinetics.

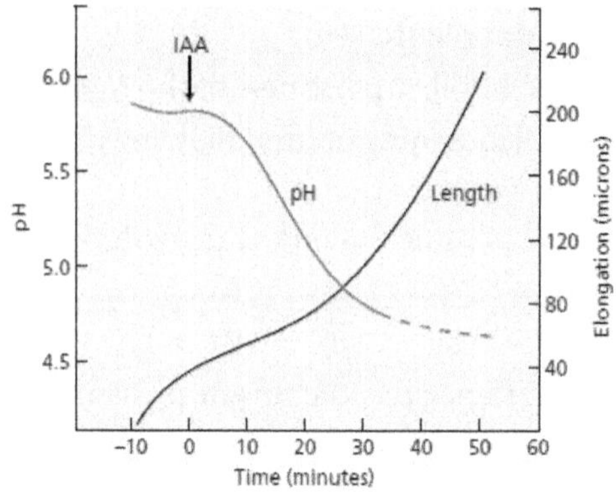

Fig. Kinetics of auxin-induced elongation and cell wall acidification in maize coleoptiles, The pH of the cell wall was measured with a pH micro-etectrode, Note the similar lag times (10 to 15 minutes) for both cell wall acidification and the increase in the rate of elongation.

Auxin-induced growth has also been shown to be inhibited by neutral buffers, as long as the cuticle has been abraided. **Fusicoccin**, a fungal phytotoxin, stimulates both rapid proton extrusion and transient growth in stem and coleoptile sections (see **Web Topic 19.6**). And finally, wallloosening proteins called **expansins** have been identified in the cell walls of a wide range of plant species. At acidic pH values, expansins loosen cell walls by weakening the hydrogen bonds between the polysaccharide components of the wall

Auxin-Induced Proton Extrusion May Involve Both Activation and Synthesis

In theory, auxin could increase the rate of proton extrusion by two possible mechanisms :

- Activation of pre-existing plasma membrane H^+- ATPases
- Synthesis of new H^+-ATPases on thc plasma membrane

H^+-ATPase activation

When auxin was added directly to isolated plasma membrane vesicles from tobacco cells, a small stimulation (about 20 per cent) of the ATP-driven proton-pumping activity was observed, suggesting that auxin directly activates the H^+-ATPase. A greater stimulation (about 40 per cent) was observed if the living cells were treated with IAA just before the membranes were isolated, suggesting that a cellular factor is also required (Peltier and Rossignol 1996). Although an auxin receptor has not yet been unequivocally identified, various auxin-binding proteins (ABPs) have been isolated and appear to be able to activate the plasma membrane H^+-ATPase in the presence of auxin (Steffens *et. al.* 2001).

Recently an ABP from rice, ABP_{57}, was shown to bind directly to plasma membrane H^+-ATPases and stimulate proton extrusion—but only in the presence of IAA (Kim *et. al.* 2001). When IAA is absent, the activity of the H^+- ATPase is repressed by the C-terminal domain of the enzyme, which can block the catalytic site. ABP_{57} (with bound IAA) interacts with the H^+-ATPase, activating the enzyme. A second auxin-binding site interferes with the action of the first, possibly explaining the bell-shaped curve of auxin action. This hypothetical model for the action of ABP_{57} is shown in Figure.

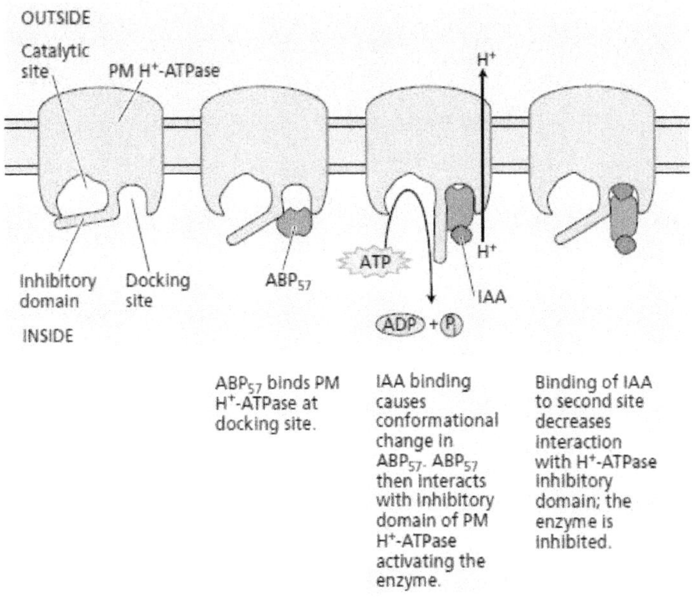

Fig. Model for the activation of the plasma membrane (PM) H^+-ATPase by ABP_{57} and auxin.

H^+-ATPase Synthesis

The ability of protein synthesis inhibitors, such as cycloheximide, to rapidly inhibit auxininduced proton extrusion and growth suggests that auxin might also stimulate proton pumping by increasing the synthesis of the H^+-ATPase. An increase in the amount of plasma membrane ATPase in corn coleoptiles was detected immunologically after only 5 minutes of auxin treatment, and a doubling of the H^+-ATPase was observed after 40 minutes of treatment. A threefold stimulation by auxin of an mRNA for the H^+-ATPase was demonstrated specifically in the non-vascular tissues of the coleoptiles. In summary, the question of activation versus synthesis is still unresolved, and it is possible that auxin stimulates proton extrusion by both activation and stimulation of synthesis of the H^+-ATPase. Figure summarises the proposed mechanisms of auxin-induced cell wall loosening via proton extrusion.

Activation hypothesis: Auxin binds to an auxin-binding protein (ABP1) located either on the cell surface or in the cytosol. ABP1-IAA then interacts directly with plasma membrane H^+-ATPase to stimulate proton pumping (step 1). Second messengers, such as calcium or intracellular pH, could also be involved.

Synthesis hypothesis: IAA-induced second messengers activate the expression of genes (step 2) that encode the plasma membrane H^+-ATPase (step 3). The protein is synthesized on the rough endoplasmic reticulum (step 4) and targeted via the secretory pathway to the plasma membrane (steps 5 and 6). The increase in proton extrusion results from an increase in the number of proton pumps on the membrane.

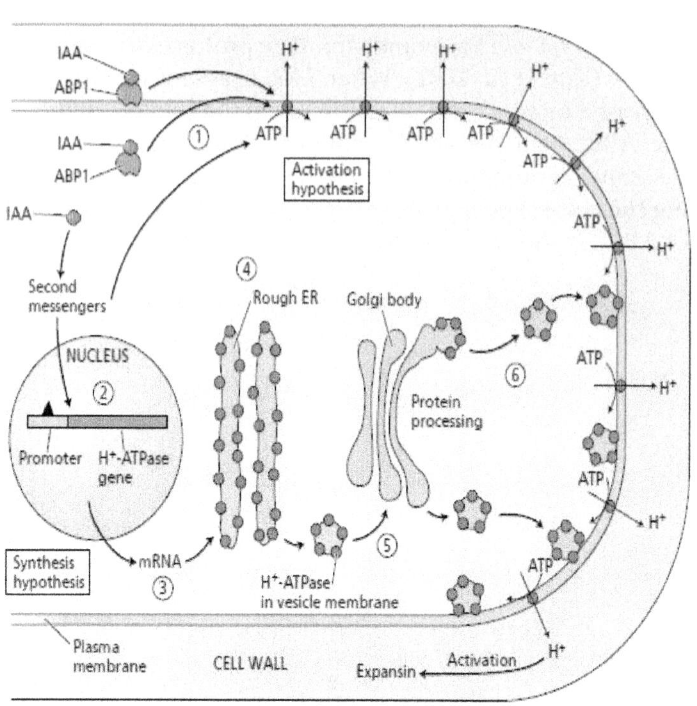

Fig. Current models for IAA-induced H^+ extraction. In many plants, both of these mechanisms may operate. Regardless of how H^+ pumping is increased, acid-induced wall loosening is thought to be mediated by expansins.

PHYSIOLOGICAL EFFECTS OF AUXIN : PHOTOTROPISM AND GRAVITROPISM

Three main guidance systems control the orientation of plant growth :

- **Phototropism**, or growth with respect to light, is expressed in all shoots and some roots; it ensures that leaves will receive optimal sunlight for photosynthesis.
- **Gravitropism**, growth in response to gravity, enables roots to grow downward into the soil and shoots to grow upward away from the soil, which is especially critical during the early stages of germination.
- **Thigmotropism**, or growth with respect to touch, enables roots to grow around rocks and is responsible for the ability of the shoots of climbing plants to wrap around other structures for support.

In this section we will examine the evidence that bending in response to light or gravity results from the lateral redistribution of auxin. We will also consider the cellular mechanisms involved in generating lateral auxin gradients during bending growth. Less is known about the mechanism of thigmotropism, although it, too, probably involves auxin gradients.

Phototropism Is Mediated by the Lateral Redistribution of Auxin

As we saw earlier, Charles and Francis Darwin provided the first clue concerning the mechanism of phototropism by demonstrating that the sites of perception and differential growth (bending) are separate : Light is perceived at the tip, but bending occurs below the tip. The Darwins proposed that some "influence" that was transported from the tip to the growing region brought about the observed asymmetric growth response. This influence was later shown to be indole-3-acetic acid—auxin.

When a shoot is growing vertically, auxin is transported polarly from the growing tip to the elongation zone. The polarity of auxin transport from tip to base is developmentally determined and is independent of orientation with respect to gravity. However, auxin can also be transported laterally, and this lateral movement of auxin lies at the heart of a model for tropisms originally proposed separately by the Russian plant physiologist, Nicolai Cholodny and Frits Went from the Netherlands in the 1920s. According to the Cholodny–Went model of phototropism, the tips of grass coleoptiles have three specialised functions :

- The production of auxin
- The perception of a unilateral light stimulus
- The lateral transport of IAA in response to the phototropic stimulus

Thus, in response to a directional light stimulus, the auxin produced at the tip, instead of being transported basipetally, is transported laterally towards the shaded side. The precise sites of auxin production, light perception, and lateral transport have been difficult to define. In maise coleoptiles, auxin is produced in the upper 1 to 2 mm of the tip. The zones of photosensing and lateral transport extend farther, within the upper 5 mm of the tip. The response is also strongly dependent on the light fluence.

Two flavoproteins, *phototropins* 1 and 2, are the photoreceptors for the blue-light signaling pathway that induces phototropic bending in *Arabidopsis* hypocotyls and oat coleoptiles under both high- and low-fluence conditions (Briggs *et. al.* 2001). Phototropins are autophosphorylating protein kinases whose activity is stimulated by blue light. The action spectrum for **blue-light** activation of the kinase activity closely matches the action spectrum for phototropism, including the multiple peaks in the blue region. Phototropin 1 displays a lateral gradient in phosphorylation during exposure to low-fluence unilateral blue light. According to the current hypothesis, the gradient in phototropin phosphorylation induces the movement of auxin to the shaded side of the coleoptile (see **Web Topic 19.7**). Once the auxin reaches the shaded side of the tip, it is transported basipetally to the elongation zone, where it stimulates cell elongation. The acceleration of growth on the shaded side and the slowing of growth on the illuminated side (differential growth) give rise to the curvature towards light.

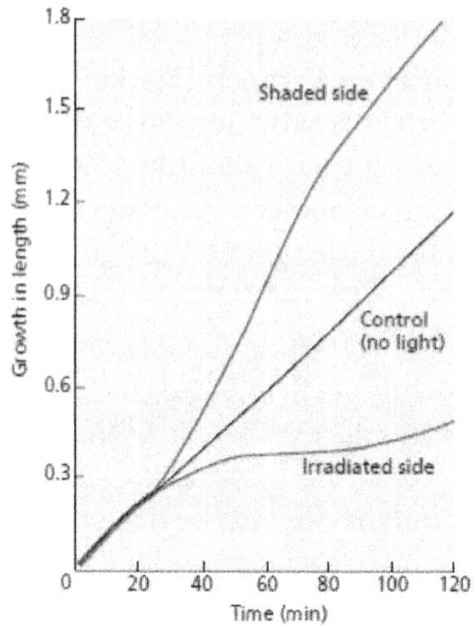

Fig. Time course of growth on the illuminated and shaded sides- of a coleoptile responding to a. 30-second pulse of unidirectional blue light. Control coleoptiles were not given a light treatment.

Direct tests of the Cholodny–Went model using the agar block/coleoptile curvature bioassay have supported the model's prediction that auxin in coleoptile tips is transported laterally in response to unilateral light.

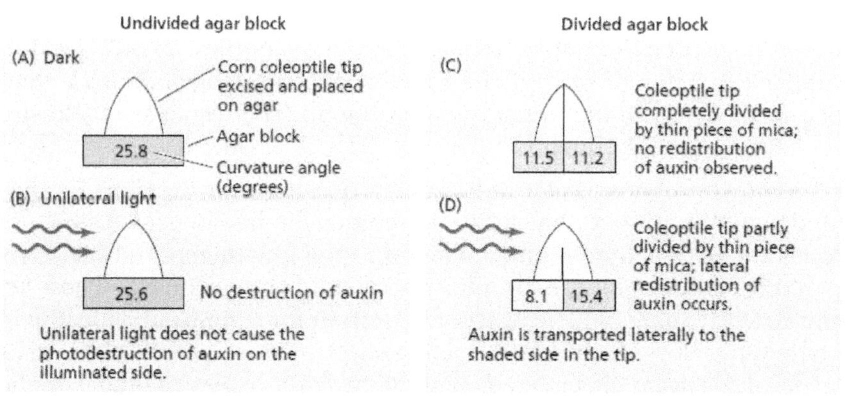

Fig. Evidence that the lateral redistribution of auxin is stimulated by unidirectional light in corn coleoptiles.

The total amount of auxin diffusing out of the tip (here expressed as the angle of curvature) is the same in the presence of unilateral light as in darkness. This result indicates that light does not cause the photo-destruction of auxin on the illuminated side, as had been proposed by some investigators. Consistent with both the Cholodny–Went hypothesis and the acid growth hypothesis, the apoplastic pH on the shaded side of a phototropically bending stem or coleoptile is more acidic than the side facing the light (Mulkey *et. al.* 1981).

Gravitropism Also Involves Lateral Redistribution of Auxin

When dark-grown *Avena* seedlings are oriented horizontally, the coleoptiles bend upward in response to gravity. According to the Cholodny–Went model, auxin in a horizontally oriented coleoptile tip is transported laterally to the lower side, causing the lower side of the coleoptile to grow faster than the upper side. Early experimental evidence indicated that the tip of the coleoptile can perceive gravity and redistribute auxin to the lower side. For example, if coleoptile tips are oriented horizontally, a greater amount of auxin diffuses from the lower half than the upper half.

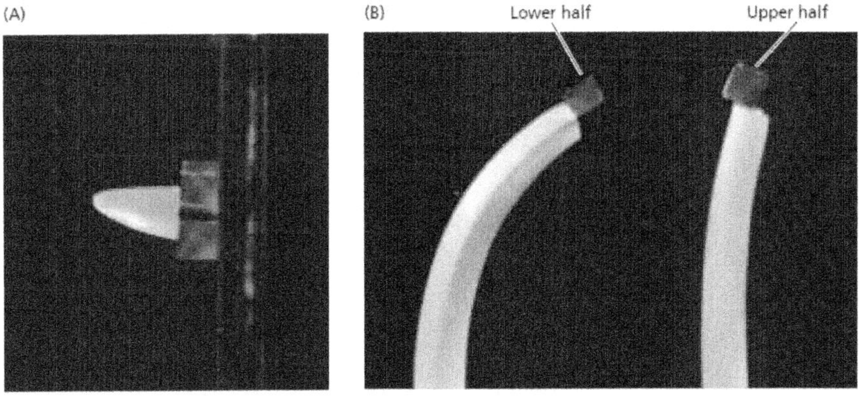

Fig. Auxin is transported to the lower side of a horizontally oriented oat coleoptile tip. (A) Auxin from the upper and lower halves of a horizontal tip is allowed to diffuse into two agar blocks. (B) Tine agar block from the lower half (left) induces greater curvature in a decapitated coleoptile than the agar block from the upper half (right).

Tissues below the tip are able to respond to gravity as well. For example, when vertically oriented maise coleoptiles are decapitated by removal of the upper 2 mm of the tip and oriented horizontally, gravitropic bending occurs at a slow rate for several hours even without the tip. Application of IAA to the cut surface restores the rate of bending to normal levels. This finding indicates that both the perception of the gravitational stimulus and the lateral redistribution of auxin can occur in the tissues below the tip, although the tip is still required for auxin production. Lateral redistribution of auxin in shoot apical meristems is more difficult to demonstrate than in coleoptiles because of the presence of leaves. In recent years, molecular markers have been widely used as reporter genes to detect lateral auxin gradients in horizontally placed stems and roots. In soybean hypocotyls, gravitropism leads to a rapid asymmetry in the accumulation of a group of auxin-stimulated mRNAs called SAURs (small *a*uxin *u*p-regulated *R*NAs) (McClure and Guilfoyle 1989). In vertical seedlings, SAUR gene expression is symmetrically distributed. Within 20 minutes after the seedling is oriented horizontally, SAURs begin to accumulate on the lower half of the hypocotyl. Under these conditions, gravitropic bending first becomes evident after 45 minutes, well after the induction of the SAURs. The existence of a lateral gradient in SAUR gene expression is indirect evidence for the existence of a lateral gradient in auxin detectable within 20 minutes of the gravitropic stimulus.

The *GH3* gene family is also up-regulated within 5 minutes of auxin treatment and has been used as a molecular marker for the presence of auxin. By fusing an artificial promoter sequence based on the *GH3* promoter to the *GUS* reporter gene, it is possible to visualise the lateral gradient in auxin concentration that occurs during both photo- and gravitropism.

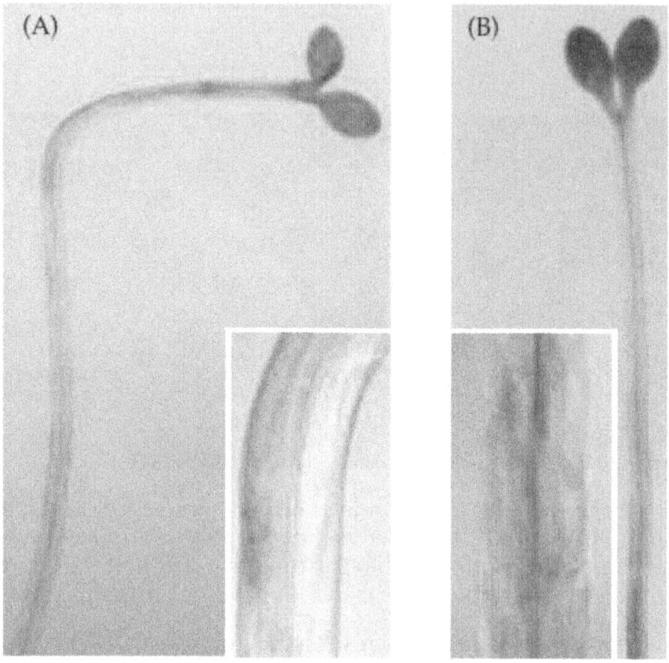

Fig. Lateral auxin gradients are formed in Arabidopsis hypocotyls during the differential growth responses to light (A). Treatment with the auxin efflux inhibitor NPA blocks auxin redistribution and bending (B). The plants were transformed with the *DR5 : : CUS* reporter gene. Auxin accumulation on the shaded side of the hypocotyls is indicated by the blue staining shown in the insets (A). A similar redistribution of auxin occurs in gravitropism.

Statoliths Serve as Gravity Sensors in Shoots and Roots

Unlike unilateral light, gravity does not form a gradient between the upper and lower sides of an organ. All parts of the plant experience the gravitational stimulus equally. How do plant cells detect gravity? The only way that gravity can be sensed is through the motion of a falling or sedimenting body. Obvious candidates for intracellular gravity sensors in plants are the large, dense amyloplasts that are present in many plant cells. These specialised amyloplasts are of sufficiently high density relative to the cytosol that they readily sediment to the bottom of the cell.

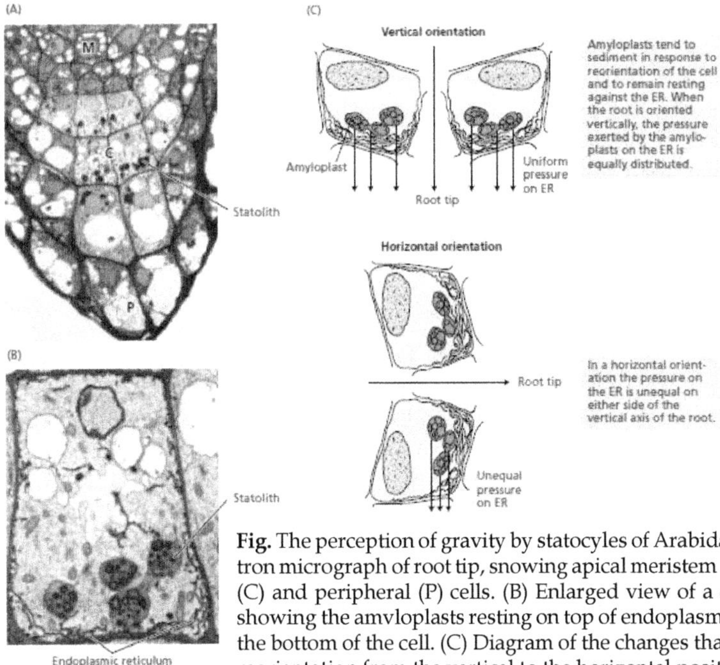

Fig. The perception of gravity by statocyles of Arabidapsis. (A) Electron micrograph of root tip, snowing apical meristem (M). columella (C) and peripheral (P) cells. (B) Enlarged view of a columella cell, showing the amvloplasts resting on top of endoplasmic reticulum at the bottom of the cell. (C) Diagram of the changes that occur during reorientation from the vertical to the horizontal position.

Amyloplasts that function as gravity sensors are called **statoliths**, and the specialised gravity-sensing cells in which they occur are called **statocytes**. Whether the statocyte is able to detect the downward motion of the statolith as it passes through the cytoskeleton or whether the stimulus is perceived only when the statolith comes to rest at the bottom of the cell has not yet been resolved.

Shoots and Coleoptiles

In shoots and coleoptiles, gravity is perceived in the **starch sheath**, a layer of cells that surrounds the vascular tissues of the shoot. The starch sheath is continuous with the endodermis of the root, but unlike the endodermis it contains amyloplasts. *Arabidopsis* mutants lacking amyloplasts in the starch sheath display agravitropic shoot growth but normal gravitropic root growth. The *scarecrow* (*scr*) mutant of *Arabidopsis* is missing both the endodermis and the starch sheath. As a result, the hypocotyl and inflorescence of the *scr* mutant are agravitropic, while the root exhibits a normal gravitropic response. On the basis of the phenotypes of these two mutants, we can conclude the following :

- The starch sheath is required for gravitropism in shoots.
- The root endodermis, which does not contain statoliths, is not required for gravitropism in roots.

Roots

The site of gravity perception in primary roots is the root cap. Large, graviresponsive amyloplasts are located in the statocytes in the central cylinder, or **columella**, of the root cap. Removal of the root cap from otherwise intact roots abolishes root gravitropism without inhibiting growth. Precisely how the statocytes sense their falling statoliths is still poorly understood. According to one hypothesis, contact or pressure resulting from the amyloplast resting on the endoplasmic reticulum on the lower side of the cell triggers the response. The endoplasmic reticulum of columella cells is structurally unique, consisting of five to seven rough-ER sheets attached to a central nodal rod in a whorl, like petals on a flower. This specialised "nodal ER" differs from the more tubular cortical ER cisternae and may be involved in the gravity response (Zheng and Staehelin 2001).

The **starch–statolith hypothesis** of gravity perception in roots is supported by several lines of evidence. Amyloplasts are the only organelles that consistently sediment in the columella cells of different plant species, and the rate of sedimentation correlates closely with the time required to perceive the gravitational stimulus. The gravitropic responses of starch-deficient mutants are generally much slower than those of wild-type plants. Nevertheless, starchless mutants exhibit some gravitropism, suggesting that although starch is required for a normal gravitropic response, starch-independent gravity perception mechanisms may also exist. Other organelles, such as nuclei, may be dense enough to act as statoliths. It may not even be necessary for a statolith to come to rest at the bottom of the cell. The cytoskeletal network may be able to detect a partial vertical displacement of an organelle.

Recently Andrew Staehelin and colleagues proposed a new model for gravitropism, called the **tensegrity model** (Yoder *et. al.* 2001). *Tensegrity* is an architectural term—a contraction of *tensional integrity*—coined by the innovative architect R. Buckminster Fuller. In essence, *tensegrity* refers to structural integrity created by interactive tension between the structural components. In this case the structural components consist of the meshwork of actin microfilaments that form part of the cytoskeleton of the central columella cells of the root cap. The actin network is assumed to be anchored to stretch-activated receptors on the plasma membrane. Stretch receptors in animal cells are typically mechanosensitive ion channels, and stretch-activated calcium channels have been demonstrated in plants. According to the tensegrity model, sedimentation of the statoliths through the cytosol locally disrupts the actin meshwork, changing the distribution of tension transmitted to calcium channels on the plasma membrane, thus altering their activities. Yoder and colleagues (2001) have further proposed that the nodal ER, which is also connected to channels via actin microfilaments, may protect the cytoskeleton from being disrupted by the statoliths in specific regions, thus providing a signal for the directionality of the stimulus.

Gravity Perception without Statoliths?

An alternative mechanism of gravity perception that does not involve statoliths has been proposed for the giant-celled freshwater alga *Chara*.

Auxin is Redistribution Laterally in the Root Cap

In addition to functioning to protect the sensitive cells of the apical meristem as the tip penetrates the soil, the root cap is the site of gravity perception. Because the cap is some distance away from the elongation zone where bending occurs, a chemical messenger is presumed to be involved in communication between the cap and the elongation zone. Microsurgery experiments in which half of the cap was removed showed that the cap produces a root growth inhibitor.

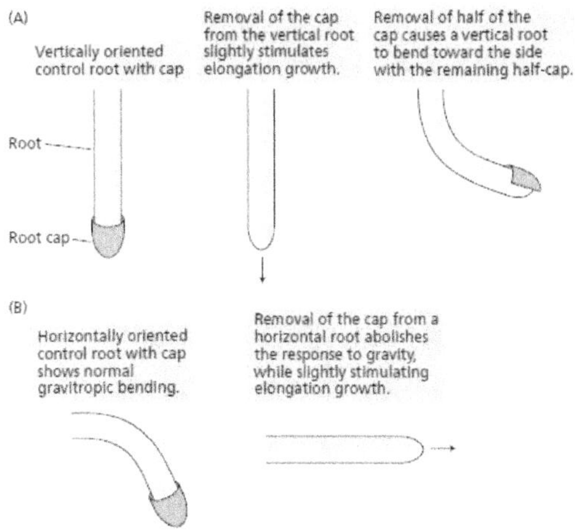

Fig. Microsurgery experiments demonstrating that the root cap produces an inhibitor that regulates root gravitropism.

This finding suggests that the cap supplies an inhibitor to the lower side of the root during gravitropic bending. Although root caps contain small amounts of IAA and abscisic acid (ABA), IAA is more inhibitory to root growth than ABAwhen applied directly to the elongation zone, suggesting that IAA is the root cap inhibitor. Consistent with this conclusion, ABA-deficient *Arabidopsis* mutants have normal root gravitropism, whereas the roots of mutants defective in auxin transport, such as *aux1* and *agr1*, are agravitropic (Palme and Gälweiler 1999). The *agr* mutant lacks an auxin efflux carrier related to the PIN proteins (Chen *et. al.* 1998; Müller *et. al.* 1998; Utsuno *et. al.* 1998). The AGR1 protein is localised at the basal (distal) end of cortical cells near the root tip in *Arabidopsis.*

How do we reconcile the fact that the shoot apical meristem is the primary source of auxin to the root with the role of the root cap as the source of the inhibitory auxin during gravitropism? Auxin from the shoot is translocated from the stele to the root tip via protophloem cells. Asymmetrically localised AUX1 permeases on the protophloem parenchyma cells direct the acropetal transport of auxin from the phloem to a cluster of cells in the columella of the cap. Auxin is then transported radially to the lateral root cap cells, where AUX1 is also strongly expressed.

The lateral root cap cells overlay the distal elongation zone (DEZ) of the root— the first region that responds to gravity. The auxin from the cap is taken up by the cortical parenchyma of the DEZ and transported basipetally through the elongation zone of the root. This basipetal transport, which is limited to the elongation zone, is facilitated by auxin anion carriers related to the PIN family (called AGR1), which are localised at the basal ends of the cortical parenchyma cells. The basipetally transported auxin accumulates in the elongation zone and does not pass beyond this region. Flavonoids capable of inhibiting auxin efflux are synthesised in this region of the root and probably promote auxin retention by these cells.

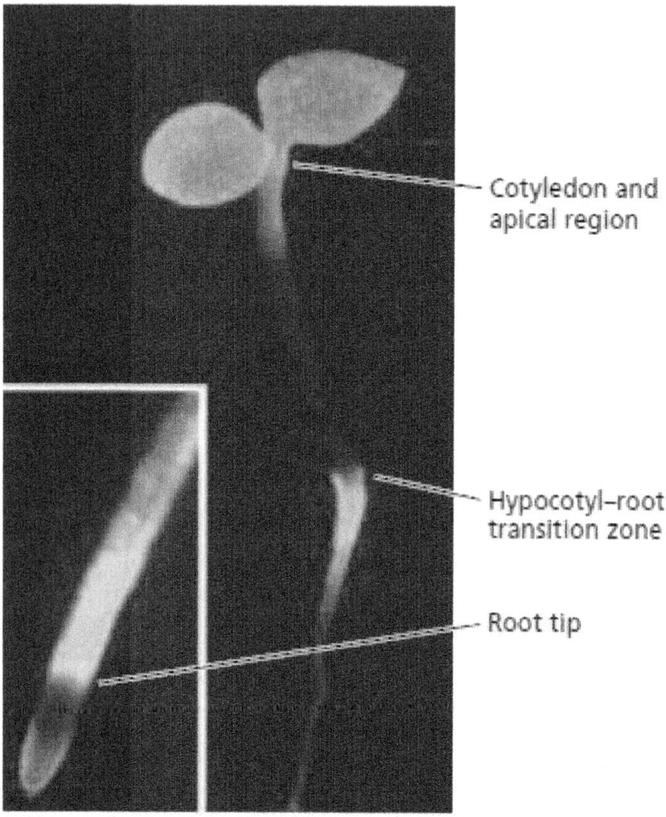

Fig. Flavonoid localization in a 6-day-old *Arabidopsis* seedling. The staining procedure used causes the flavonoids to fluoresce. Flavonoids are concentrated in three regions : the cotyledon and apical region, the hypocotyl-root transition zone, and the root tip area (inset). In the root tip, flavonoids are localized specifically in the elongation zone and the cap, the tissues involved in basipetal auxin transport.

According to the model, basipetal auxin transport in a vertically oriented root is equal on all sides.

When the root is oriented horizontally, however, the cap redirects the bulk of the auxin to the lower side, thus inhibiting the growth of that lower side. Consistent with this idea, the transport of [^{3}H]IAA across a horizontally oriented root cap is polar, with a preferential downward movement.

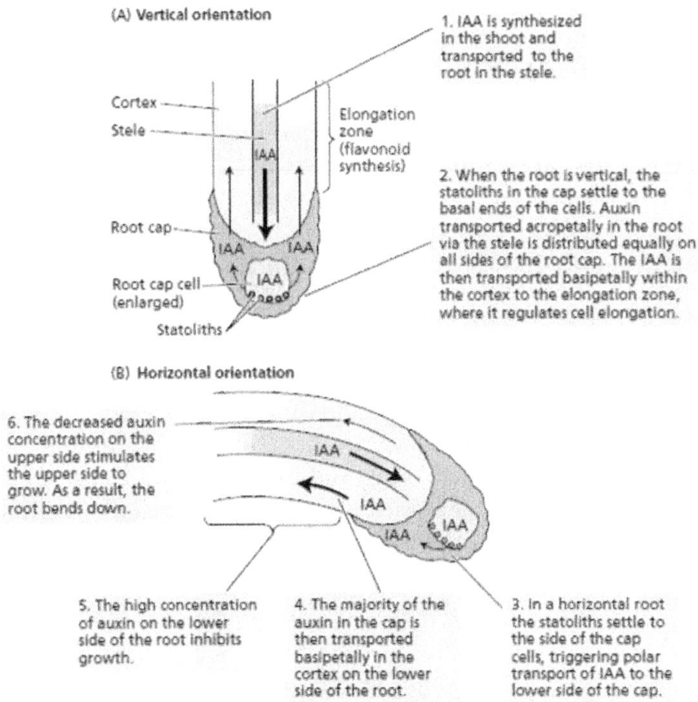

Fig. Proposed model for the redistribution of auxin during gravtitropism in maize roots.

PIN3 is Relocated Laterally to the Lower Side of Root Columella Cells

Recently the mechanism of lateral auxin redistribution in the root cap has new been elucidated (Friml *et. al.* 2002). One of the members of the PIN protein family of auxin efflux carriers, PIN3, is not only required for both photo-and gravitropism in *Arabidopsis*, but it has been shown to be relocalised to the lower side of the columella cells during root gravitropism.

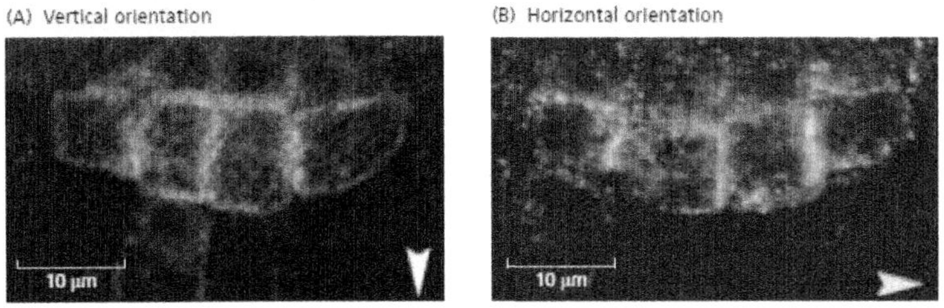

Fig. Relocalization of the auxin efflux carrier PIN3 during root gravitropsm in Arabirfopsis. (A) In a vertically oriented root, PIN3 is uniformly distributed around the columella cells. (E) After being oriented horizontally for 10 minutes, PIN3 has been reloealized to the lower side of the columella cells. The photo in (B) has been reorientated so that the lower side is on the right. (The direction of gravity is indicated by the arrows.)

As noted previously, PIN proteins are constantly being cycled between the plasma membrane and intracellular secretory compartments. This cycling allows some PIN proteins to be targeted to specific sides of the cell in response to a directional stimulus. In a vertically oriented root, PIN3 is uniformly distributed around the columella cell. But when the root is placed on its side, PIN3 is preferentially targeted to the lower side of the cell. As a result, auxin is transported polarly to the lower half of the cap.

Gravity Sensing may Involve Calcium and pH as Second Messengers

A variety of experiments have suggested that calcium– calmodulin is required for root gravitropism in maise. Some of these experiments involve EGTA(ethylene glycol-bis(®- aminoethyl ether)-N,N,N2,N2-tetraacetic acid), a compound that can chelate (form a complex with) calcium ions, thus preventing calcium uptake by cells. EGTAinhibits both root gravitropism and the asymmetric distribution of auxin in response to gravity (Young and Evans 1994). Placing a block of agar that contains calcium ions on the side of the cap of a vertically oriented corn root induces the root to grow towards the side with the agar block.

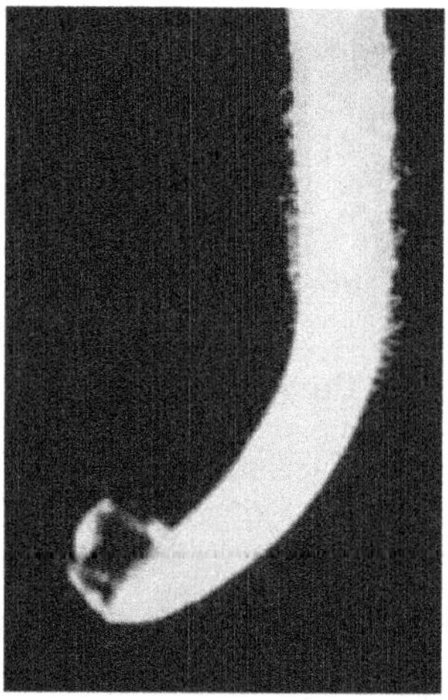

Fig. A corn root bending toward an agar block containing calcium placed on the cap.

As in the case of [$^{+}$H]IAA, $^{45}Ca^{2+}$ is polarly transported to the lower half of the cap of a root stimulated by gravity. However, thus far no changes in the distribution of intracellular calcium have been detected in columella cells in response to a gravitational stimulus. Recent evidence suggests that a change in intracellular pH is the earliest detectable change in columella cells responding to gravity. Fasano *et. al.* (2001) used pH-sensitive dyes to monitor both intracellular and extracellular pH in *Arabidopsis* roots after they were placed in a horizontal position. Within 2 minutes of gravistimulation, the cytoplasmic pH of the columella cells of the root cap increased from 7.2 to 7.6, and the apoplastic pH declined from 5.5 to 4.5. These changes preceded any detectable tropic curvature by about 10 minutes. The alkalinisation of the cytosol combined with the acidification of the apoplast suggests that an activation of the plasma membrane H^+-ATPase is one of the initial events that mediate root gravity perception or signal transduction.

DEVELOPMENTAL EFFECTS OF AUXIN

Although originally discovered in relation to growth, auxin influences nearly every stage of a plant's life cycle from germination to senescence. Because the effect that auxin produces depends on the identity of the target tissue, the response of a tissue to auxin is governed by its developmentally determined genetic programme and is further influenced by the presence or absence of other signaling molecules. Interaction between two or more hormones is a recurring theme in plant development. In this section we will examine some additional developmental processes regulated by auxin, including apical dominance, leaf abscission, lateral-root formation, and vascular differentiation. Throughout this discussion we assume that the primary mechanism of auxin action is comparable in all cases, involving similar receptors and signal transduction pathways.

Auxin Regulates Apical Dominance

In most higher plants, the growing apical bud inhibits the growth of lateral (axillary) buds—a phenomenon called **apical dominance**. Removal of the shoot apex (decapitation) usually results in the growth of one or more of the lateral buds. Not long after the discovery of auxin, it was found that IAA could substitute for the apical bud in maintaining the inhibition of lateral buds of bean (*Phaseolus vulgaris*) plants. This classic experiment is illustrated in Figure.

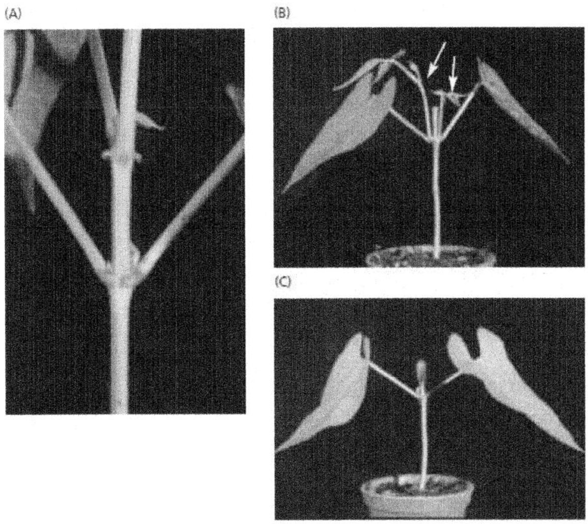

Fig. Auxin suppresses the growth of axillary buds in bean (Phaseolus vulgaris) plants. (A) The axillary buds are suppressed in the intact plant because of apical dominance. (B) Removal of the terminal bud releases the axillary buds from apical dominance (arrows). (C) Applying 1AA in lanolin paste (contained in the gelatin capsule) to the cut surface prevents the outgrowth of the axillary buds.

This result was soon confirmed for numerous other plant species, leading to the hypothesis that the outgrowth of the axillary bud is inhibited by auxin that is transported basipetally from the apical bud. In support of this idea, a ring of the auxin transport inhibitor TIBA in lanolin paste (as a carrier) placed below the shoot apex released the axillary buds from inhibition. How does auxin from the shoot apex inhibit the growth of lateral buds? Kenneth V. Thimann and Folke Skoog originally proposed that auxin from the shoot apex inhibits the growth of the axillary bud directly—the so-called *directinhibition model*. According to the model, the optimal auxin concentration for bud growth is low, much lower than the auxin concentration normally found in the stem. The level of auxin normally present in the stem was thought to inhibit the growth of lateral buds.

If the direct-inhibition model of apical dominance is correct, the concentration of auxin in the axillary bud should decrease following decapitation of the shoot apex. However, the reverse appears to be true. This was demonstrated with transgenic plants that contained the reporter genes for bacterial luciferase (*LUXA* and *LUXB*) under the control of an auxin-responsive promoter (Langridge *et. al.* 1989). These reporter genes allowed researchers to study the level of auxin in different tissues by monitoring the amount of light emitted by the luciferase-catalysed reaction. When these transgenic plants were decapitated, the expression of the *LUX* genes increased in and around the axillary buds within 12 hours. This experiment indicated that after decapitation, the auxin content of the axillary buds *increased* rather than decreased. Direct physical measurements of auxin levels in buds have also shown an increase in the auxin of the axillary buds after decapitation. The IAA concentration in the axillary bud of *Phaseolus vulgaris* (kidney bean) increased fivefold within 4 hours after decapitation (Gocal *et. al.* 1991). These and other similar results make it unlikely that auxin from the shoot apex inhibits the axillary bud directly. Other hormones, such as cytokinins and ABA, may be involved. Direct application of cytokinins to axillary buds stimulates bud growth in many species, overriding the inhibitory effect of the shoot apex. Auxin makes the shoot apex a sink for cytokinin synthesised in the root, and this may be one of the factors involved in apical dominance.

Finally, ABAhas been found in dormant lateral buds in intact plants. When the shoot apex is removed, the ABA levels in the lateral buds decrease. High levels of IAA in the shoot may help keep ABA levels high in the lateral buds. Removing the apex removes a major source of IAA, which may allow the levels of bud growth inhibitor to fall.

Auxin Promotes the Formation of Lateral and Adventitious Roots

Although elongation of the primary root is inhibited by auxin concentrations greater than 10^{-8} *M*, initiation of lateral (branch) roots and adventitious roots is stimulated by high auxin levels. Lateral roots are commonly found above the elongation and root hair zone and originate from small groups of cells in the pericycle. Auxin stimulates these pericycle cells to divide. The dividing cells gradually form into a root apex, and the lateral root grows through the root cortex and epidermis.

Adventitious roots (roots originating from non-root tissue) can arise in a variety of tissue locations from clusters of mature cells that renew their cell division activity. These dividing cells develop into a root apical meristem in a manner somewhat analogous to the formation of lateral roots. In horticulture, the stimulatory effect of auxin on the formation of adventitious roots has been very useful for the vegetative propagation of plants by cuttings. Aseries of *Arabidopsis* mutants, named *alf* (*a*berrant *l*ateral root *f*ormation), have provided some insights into the role of auxin in the initiation of lateral roots. The *alf1* mutant exhibits extreme proliferation of adventitious and lateral roots, coupled with a 17-fold increase in endogenous auxin.

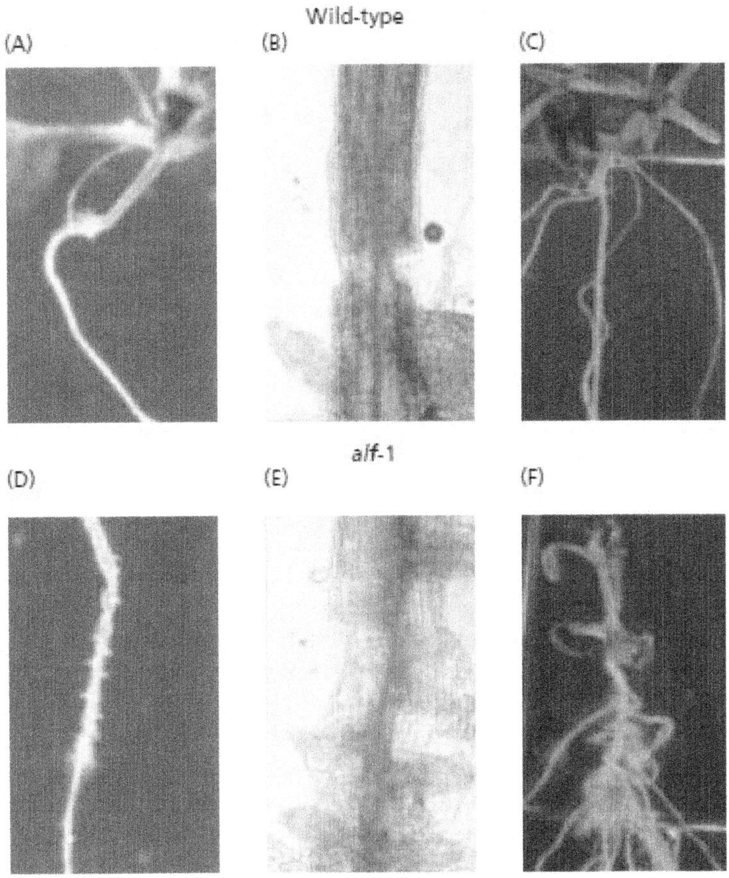

Fig. Root morphology of *Arabidopsis* (A-C) wild-type and *alfa* seedlings (D-F) on hormone-free medium. Note the proliferation of root primordia growing from the pericycle in the *alfa* seedlings (D and E).

Another mutant, *alf4*, has the opposite phenotype : It is completely devoid of lateral roots. Microscopic analysis of *alf4* roots indicates that lateral-root primordia are absent. The *alf4* phenotype cannot be reversed by application of exogenous IAA. Yet another mutant, *alf3*, is defective in the development of lateral-root primordia into mature lateral roots. The primary root is covered with arrested lateral-root primordia that grow until they protrude through the epidermal cell layer and then stop growing. The arrested growth can be alleviated by application of exogenous IAA. On the basis of the phenotypes of the *alf* mutants, a model in which IAA is required for at least two steps in the formation of lateral roots has been proposed :

- IAA transported acropetally (towards the tip) in the stele is required to initiate cell division in the pericycle.
- IAA is required to promote cell division and maintain cell viability in the developing lateral root.

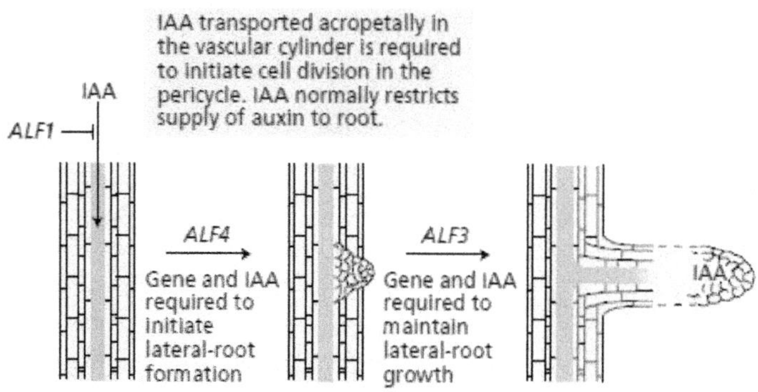

Fig. A model for the formation of lateral roots, based on the *alf* mutants of *Arabidopsis.*

Auxin Delays the Onset of Leaf Abscission

The shedding of leaves, flowers, and fruits from the living plant is known as **abscission**. These parts abscise in a region called the **abscission zone**, which is located near the base of the petiole of leaves. In most plants, leaf abscission is preceded by the differentiation of a distinct layer of cells, the **abscission layer**, within the abscission zone. During leaf senescence, the walls of the cells in the abscission layer are digested, which causes them to become soft and weak. The leaf eventually breaks off at the abscission layer as a result of stress on the weakened cell walls. Auxin levels are high in young leaves, progressively decrease in maturing leaves, and are relatively low in senescing leaves when the abscission process begins. The role of auxin in leaf abscission can be readily demonstrated by excision of the blade from a mature leaf, leaving the petiole intact on the stem. Whereas removal of the leaf blade accelerates the formation of the abscission layer in the petiole, application of IAA in lanolin paste to the cut surface of the petiole prevents the formation of the abscission layer. (Lanolin paste alone does not prevent abscission.) These results suggest the following :

- Auxin transported from the blade normally prevents abscission.
- Abscission is triggered during leaf senescence, when auxin is no longer being produced.

Auxin Transport Regulates Floral Bud Development

Treating *Arabidopsis* plants with the auxin transport inhibitor NPA causes abnormal floral development, suggesting that polar auxin transport in the inflorescence meristem is required for normal floral development. In *Arabidopsis*, the "pin-formed" mutant *pin1*, which lacks an auxin efflux carrier in shoot tissues, has abnormal flowers similar to those of NPA-treated plants. Apparently the developing floral meristem depends on auxin being transported to it from subapical tissues. In the absence of the efflux carriers, the meristem is starved for auxin, and normal phyllotaxis and floral development are disrupted (Kuhlemeier and Reinhardt 2001).

Auxin Promotes Fruit Development

Much evidence suggests that auxin is involved in the regulation of fruit development. Auxin is produced in pollen and in the endosperm and the embryo of developing seeds, and the initial stimulus for fruit growth may result from pollination. Successful pollination initiates ovule growth, which is known as **fruit set**. After fertilization, fruit growth may depend on auxin produced in developing seeds. The endosperm may contribute auxin during the initial stage of fruit growth, and the developing embryo may take over as the main auxin source during the later stages. Figure shows the influence of auxin produced by the achenes of strawberry on the growth of the receptacle of strawberry.

Fig. (A) The strawberry "fruit" is actually a swollen receptacle whose growth is regulated by auxin produced by the "seeds," which are actually achenes— the true fruits, (B) When the achenes are removed, the receptacle fails to develop normally. (C) Spraying the achene-less receptacle with IAA restores normal growth and development.

Auxin Induces Vascular Differentiation

New vascular tissues differentiate directly below developing buds and young growing leaves, and removal of the young leaves prevents vascular differentiation (Aloni 1995). The ability of an apical bud to stimulate vascular differentiation can be demonstrated in tissue culture. When the apical bud is grafted onto a clump of undifferentiated cells, or *callus*, xylem and phloem differentiate beneath the graft. The relative amounts of xylem and phloem formed are regulated by the auxin concentration : High auxin concentrations induce the differentiation of xylem and phloem, but only phloem differentiates at low auxin concentrations. Similarly, experiments on stem tissues have shown that low auxin concentrations induce phloem differentiation, whereas higher IAA levels induce xylem (Aloni 1995). The regeneration of vascular tissue following wounding is also controlled by auxin produced by the young leaf directly above the wound site.

Removal of the leaf prevents the regeneration of vascular tissue, and applied auxin can substitute for the leaf in stimulating regeneration. Vascular differentiation is polar and occurs from leaves to roots. In woody perennials, auxin produced by growing buds in the spring stimulates activation of the cambium in a basipetal direction. The new round of secondary growth begins at the smallest twigs and progresses downward towards the root tip.

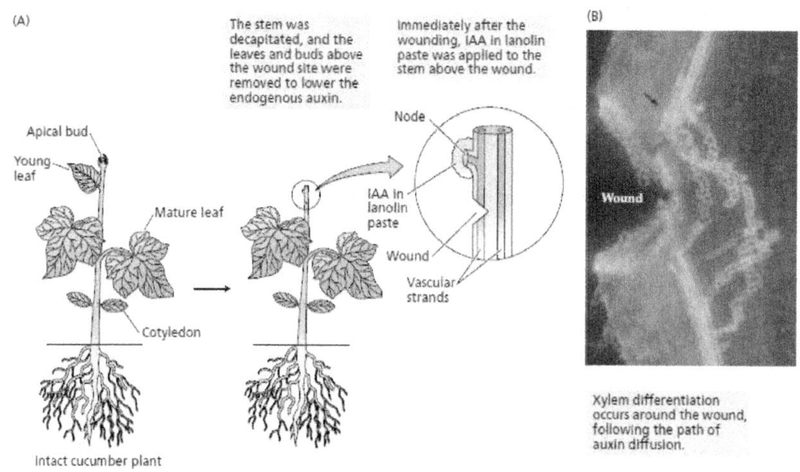

Fig. IAA-induced xylem regeneration around the wound in cucumber (***Cucumis sativus***) stem tissue. (A) Method for carrying out the wound regeneration experiment. (B) Fluorescence micrograph showing regenerating vascular tissue around the wound.

Further evidence for the role of auxin in vascular differentiation comes from studies in which the auxin concentration is manipulated by the transformation of plants with auxin biosynthesis genes through use of the Ti plasmid of *Agrobacterium*. When an auxin biosynthesis gene was overexpressed in petunia plants, the number of xylem tracheary elements increased. In contrast, when the level of free IAA in tobacco plants was decreased by transformation with a gene coding for an enzyme that conjugated IAA to the amino acid lysine, the number of vessel elements decreased and their sizes increased (Romano *et. al.* 1991). Thus the level of free auxin appears to regulate the number of tracheary elements, as well as their size. In *Zinnia elegans* mesophyll cell cultures, auxin is required for tracheary cell differentiation, but cytokinins also participate, perhaps by increasing the sensitivity of the cells to auxin. Whereas auxin is produced in the shoot and transported downward to the root, cytokinins are produced by the root tips and transported upward into the shoot. Both hormones are probably involved in the regulation of cambium activation and vascular differentiation.

Synthetic Auxins have a Variety of Commercial Uses

Auxins have been used commercially in agriculture and horticulture for more than 50 years. The early commercial uses included prevention of fruit and leaf drop, promotion of flowering in pineapple, induction of parthenocarpic fruit, thinning of fruit, and rooting of cuttings for plant propagation. Rooting is enhanced if the excised leaf or stem cutting is dipped in an auxin solution, which increases the initiation of adventitious roots at the cut end. This is the basis of commercial rooting compounds, which consist mainly of a synthetic auxin mixed with talcum powder. In some plant species, seedless fruits may be produced naturally, or they may be induced by treatment of the unpollinated flowers with auxin. The production of such seedless fruits is called **parthenocarpy**. In stimulating the formation of parthenocarpic fruits, auxin may act primarily to induce fruit set, which in turn may trigger the endogenous production of auxin by certain fruit tissues to complete the developmental process.

Ethylene is also involved in fruit development, and some of the effects of auxin on fruiting may result from the promotion of ethylene synthesis. In addition to these applications, today auxins are widely used as herbicides. The chemicals 2,4-D and dicamba are probably the most widely used synthetic auxins. Synthetic auxins are very effective because they are not metabolised by the plant as quickly as IAA is. Because maise and other monocotyledons can rapidly inactivate synthetic auxins by conjugation, these auxins are used by farmers for the control of dicot weeds, also called *broad-leaved weeds*, in commercial cereal fields, and by home gardeners for the control of weeds such as dandelions and daisies in lawns.

AUXIN SIGNAL TRANSDUCTION PATHWAYS

The ultimate goal of research on the molecular mechanism of hormone action is to reconstruct each step in the signal transduction pathway, from receptor binding to the physiological response. We will examine candidates for the auxin receptor and then discuss the various signaling pathways that have been implicated in auxin action. Finally we will turn our attention to auxinregulated gene expression.

ABP1 Functions as an Auxin Receptor

In addition to its possible direct role in plasma membrane H^+-ATPase activation (discussed earlier), the auxin-binding protein ABP1 appears to function as an auxin receptor in other signal transduction pathways. ABP1 homologs have been identified in a variety of monocot and dicot species (Venis and Napier 1997). Knockouts of the *ABP1* gene in *Arabidopsis* are lethal, and less severe mutations result in altered development (Chen *et. al.* 2001). Recent studies indicate that, despite being localised primarily on the endoplasmic reticulum (ER), a small amount of ABP1 is secreted to the plasma membrane outer surface where it interacts with auxin to cause protoplast swelling and H^+- pumping (Venis *et. al.* 1996; Steffens *et. al.* 2001). However, it is unlikely that ABP1 mediates all auxin response pathways because expression of a number of auxin-responsive genes is not affected when protoplasts are incubated with anti-ABP1 antibodies. It is also unclear what role the ABP1 in the ER plays in auxin-responsive signal transduction. Finally, it remains to be determined whether ABP_{57}, the soluble and unrelated ABP from rice that activates the H^+-ATPase is involved in a signal transduction pathway.

Calcium and Intracellular pH are Possible Signaling Intermediates

Calcium plays an important role in signal transduction in animals and is thought to be involved in the action of certain plant hormones as well. The role of calcium in auxin action seems very complex and, at this point in time, very uncertain. Nevertheless, some experimental evidence shows that auxin increases the level of free calcium in the cell. Changes in cytoplasmic pH can also serve as a second messenger in animals and plants. In plants, auxin induces a decrease in cytosolic pH of about 0.2 units within 4 minutes of application. The cause of this pH drop is not known. Since the cytosolic pH is normally around 7.4, and the pH optimum of the plasma membrane H^+-ATPase is 6.5, a decrease in the cytosolic pH of 0.2 units could cause a marked increase in the activity of the plasma membrane H^+-ATPase. The decrease in cytosolic pH might also account for the auxin-induced increase in free intracellular calcium, by promoting the dissociation of bound forms. MAP kinases that play a role in signal transduction by phosphorylating proteins in a cascade that ultimately activates transcription factors have also been implicated in auxin responses. When tobacco cells are deprived of auxin, they arrest at the end of either the G_1 or the G_2 phase and cease dividing; if auxin is added back into the culture medium, the cell cycle resumes. Auxin appears to exert its effect on the cell cycle primarily by stimulating the synthesis of the major cyclin-dependent protein kinase (CDK) : Cdc2 (*c*ell *d*ivision *c*ycle 2).

Auxin-Induced Genes Fall into Two Classes : Early and Late

One of the important functions of the signal transduction pathway(s) initiated when auxin binds to its receptor is the activation of a select group of transcription factors. The activated transcription factors enter the nucleus and promote the expression of specific genes. Genes whose expression is stimulated by the activation of pre-existing transcription factors are called **primary response genes** or **early genes.** This definition implies that all of the proteins required for auxin-induced expression of the early genes are present in the cell at the time of exposure to the hormone; thus, early-gene expression cannot be blocked by inhibitors of protein synthesis such as cycloheximide. As a consequence, the time required for the expression of the early genes can be quite short, ranging from a few minutes to several hours (Abel and Theologis 1996).

In general, primary response genes have three main functions : (1) Some of the early genes encode proteins that regulate the transcription of **secondary response genes**, or **late genes**, that are required for the long-term responses to the hormone. Because late genes require de novo protein synthesis, their expression can be blocked by protein synthesis inhibitors. (2) Other early genes are involved in intercellular communication, or cell-to-cell signaling. (3) Another group of early genes is involved in adaptation to stress. Five major classes of early auxin-responsive genes have been identified :

- Genes involved in auxin-regulated growth and development :
  - The *AUX/IAA* gene family
  - The *SAUR* gene family
  - The *GH3* gene family.
- Stress response genes :
  - Genes encoding glutathione S-transferases
  - Genes encoding 1-aminocyclopropane-1-carboxylic acid (ACC) synthase, the key enzyme in the ethylene biosynthetic pathway.

Early genes for growth and development

Members of the *AUX/IAA* gene family encode short-lived transcription factors that function as repressors or activators of the expression of late auxin-inducible genes. The expression of most of the *AUX/IAA* family of genes is stimulated by auxin within 5 to 60 minutes of hormone addition All the genes encode small hydrophilic polypeptides that have putative DNA-binding motifs similar to those of bacterial repressors. They also have short half-lives (about 7 minutes), indicating that they are turning over rapidly. Auxin stimulates the expression of *SAUR* genes within 2 to 5 minutes of treatment, and the response is insensitive to cycloheximide. The five *SAUR* genes of soybean are clustered together, contain no introns, and encode highly similar polypeptides of unknown function. Because of the rapidity of the response, expression of *SAUR* genes has proven to be a convenient probe for the lateral transport of auxin during photo- and gravitropism.

GH3 early-gene family members, identified in both soybean and *Arabidopsis*, are stimulated by auxin within 5 minutes. Mutations in *Arabidopsis GH3*-like genes result in dwarfism (Nakazawa *et. al.* 2001) and appear to function in light-regulated auxin responses (Hsieh *et. al.* 2000). Because *GH3* expression is a good reflection of the presence of endogenous auxin, a synthetic *GH3*-based **reporter** gene known as *DR5* is widely used in auxin bioassays.

Early Genes for Stress Adaptations

Auxin is involved in stress responses, such as wounding. Several genes encoding glutathione-S-transferases (GSTs), a class of proteins stimulated by various stress conditions, are induced by elevated auxin concentrations. Likewise, ACC synthase, which is also induced by stress and is the rate-limiting step in ethylene biosynthesis is induced by high levels of auxin. To be induced, the promoters of the early auxin genes must contain response elements that bind to the transcription factors that become activated in the presence of auxin. Alimited number of these response elements appear to be arranged combinatorily within the promoters of a variety of auxin-induced genes.

Auxin-Responsive Domains Are Composite Structures

Aconserved **auxin response element (AuxRE)** within the promoters of the early auxin genes, like *GH3*, is usually combined with other response elements to form **auxin response domains (AuxRDs)**. For example, the *GH3* gene promoter of soybean is composed of three independently acting AuxRDs (each containing multiple AuxREs) that contribute incrementally to the strong auxin inducibility of the promoter.

Early Auxin Genes are Regulated by Auxin Response Factors

As noted previously, early auxin genes are by definition insensitive to protein synthesis inhibitors such as cycloheximide. Instead of being inhibited, the expression of many of the early auxin genes has been found to be stimulated by cycloheximide. Cycloheximide stimulation of gene expression is accomplished both by transcriptional activation and by mRNA stabilisation. Transcriptional activation of a gene by inhibitors of protein synthesis usually indicates that the gene is being repressed by a short-lived repressor protein or by a regulatory pathway that involves a protein with a high turnover rate.

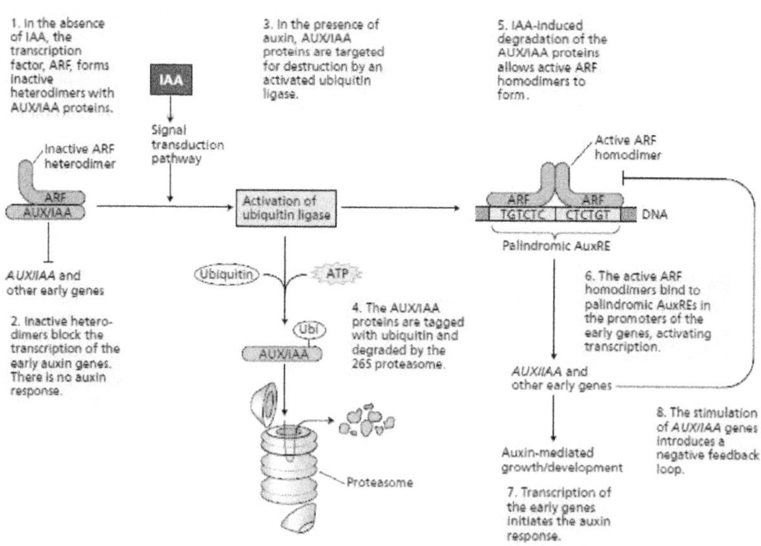

Fig. A model for auxin regulation of transcriptional activation of early response genes by auxin.

A family of **auxin response factors** (**ARFs**) function as transcriptional activators by binding to the auxin response element TGTCTC, which is present in the promoters of *GH3* and other early auxin response genes. Mutations in ARF genes result in severe developmental defects. To bind the AuxRE stably, ARFs must form dimers. It has been proposed that ARF dimers promote transcription by binding to two AuxREs arranged in a palindrome (Ulmasov *et. al.* 1997). Recent studies also indicate that proteins encoded by the *AUX/IAA* gene family (itself one of the early auxin response gene families) can inhibit the transcription of early auxin response genes by forming inactive heterodimers with ARFs. These inactive heterodimers may act to inhibit ARF–AuxRE binding, thereby blocking either gene activation or repression. AUX/IAA proteins may thus function as ARF inhibitors. It is now believed that auxin induces the transcription of the early response genes by promoting the proteolytic degradation of the inhibitory AUX/IAA proteins so that active ARF dimers can form. The precise mechanism by which auxin causes AUX/IAA turnover is unknown, although it is known to involve ubiquitination by a ubiquitin ligase and proteolysis by the massive 26S proteasome complex. Note that a negative feedback loop is introduced into the pathway by virtue of the fact that one of the gene families turned on by auxin is *AUX/IAA*, which inhibits the response.

A model for auxin regulation of the early response genes based on the findings described here is shown in Figure page 150.

SUMMARY

Auxin was the first hormone to be discovered in plants and is one of an expanding list of chemical signaling agents that regulate plant development. The most common naturally occurring form of auxin is indole-3-acetic acid (IAA). One of the most important roles of auxin in higher plants is the regulation of elongation growth in young stems and coleoptiles. Low levels of auxin are also required for root elongation, although at higher concentrations auxin acts as a root growth inhibitor. Accurate measurement of the amount of auxin in plant tissues is critical for understanding the role of this hormone in plant physiology. Early coleoptile-based bioassays have been replaced by more accurate techniques, including physicochemical methods and immunoassay.

Regulation of growth in plants may depend in part on the amount of free auxin present in plant cells, tissues, and organs. There are two main pools of auxin in the cell : the cytosol and the chloroplasts. Levels of free auxin can be modulated by several factors, including the synthesis and breakdown of conjugated IAA, IAA metabolism, compartmentation, and polar auxin transport. Several pathways have been implicated in IAA biosynthesis, including tryptophan-dependent and tryptophan-independent pathways. Several degradative pathways for IAA have also been identified. IAA is synthesised primarily in the apical bud and is transported polarly to the root. Polar transport is thought to occur mainly in the parenchyma cells associated with the vascular tissue. Polar auxin transport can be divided into two main processes : IAA influx and IAA efflux. In accord with the chemiosmotic model for polar transport, there are two modes of IAA influx : by a pH-dependent passive transport of the undissociated form, or by an active H^+ cotransport mechanism driven by the plasma membrane H^+-ATPase.

Auxin efflux is thought to occur preferentially at the basal ends of the transporting cells via anion efflux carriers and to be driven by the membrane potential generated by the plasma membrane H^+-ATPase. Auxin transport inhibitors (ATIs) can interrupt auxin transport directly by competing with auxin for the efflux channel pore or by binding to regulatory or structural proteins associated with the efflux channel. Auxin can be transported non-polarly in the phloem.

Auxin-induced cell elongation begins after a lag time of about 10 minutes. Auxin promotes elongation growth primarily by increasing cell wall extensibility. Auxin-induced wall loosening requires continuous metabolic input and is mimicked in part by treatment with acidic buffers. According to the acid growth hypothesis, one of the important actions of auxin is to induce cells to transport protons into the cell wall by stimulating the plasma membrane H^+-ATPase. Two mechanisms have been proposed for auxin-induced proton extrusion : direct activation of the proton pump and enhanced synthesis of the plasma membrane H^+-ATPase. The ability of protons to cause cell wall loosening is mediated by a class of proteins called expansins. Expansins loosen the cell wall by breaking hydrogen bonds between the polysaccharide components of the wall. In addition to proton extrusion, long-term auxin-induced growth involves the uptake of solutes and the synthesis and deposition of polysaccharides and proteins needed to maintain the acid-induced wall-loosening capacity.

Promotion of growth in stems and coleoptiles and inhibition of growth in roots are the best-studied physiological effects of auxins. Auxin-promoted differential growth in these organs is responsible for the responses to directional stimuli (*i.e.*, light, gravity) called tropisms. According to the Cholodny–Went model, auxin is transported laterally to the shaded side during phototropism and to the lower side during gravitropism. Statoliths (starch-filled amyloplasts) in the statocytes are involved in the normal percepton of gravity, but they are not absolutely required. In addition to its roles in growth and tropisms, auxin plays central regulatory roles in apical dominance, lateralroot initiation, leaf abscission, vascular differentiation, floral bud formation, and fruit development. Commercial applications of auxins include rooting compounds and herbicides. The auxin-binding soluble protein ABP1 is a strong candidate for the auxin receptor. ABP1 is located primarily in the ER lumen. Studies of the signal transduction pathways involved in auxin action have implicated other signaling intermediates such as Ca^{2+}, intracellular pH, and kinases in auxin-induced cell division.

Auxin-induced genes fall into two categories : early and late. Induction of early genes by auxin does not require protein synthesis and is insensitive to protein synthesis inhibitors. The early genes fall into three functional classes : expression of the late genes (secondary response genes), stress adaptation, and intercellular signaling. The auxin response domains of the promoters of the auxin early genes have a composite structure in which an auxin-inducible response element is combined with a constitutive response element. Auxin-induced genes may be negatively regulated by repressor proteins that are degraded via a ubiquitin activation pathway.

Chapter 6

GIBBERELLIN

Gibberellins (GAs) are plant hormones that regulate growth and influence various developmental processes, including stem elongation, germination, dormancy, flowering, sex expression, enzyme induction, and leaf and fruit senescence.

Gibberellin was first recognised in 1926 by a Japanese scientist, Eiichi Kurosawa, studying *bakanae*, the "foolish seedling" disease in rice. It was first isolated in 1935 by Teijiro Yabuta and Sumuki, from fungal strains (*Gibberella fujikuroi*) provided by Kurosawa. Yabuta named the isolate as gibberellin.

Interest in gibberellins outside of Japan began after World War II. In the United States, the first research was undertaken by a unit at Camp Detrick in Maryland, via studying seedlings of the bean *Vicia faba*. In the United Kingdom, work on isolating new types of gibberellin was undertaken at Imperial Chemical Industries. Interest in gibberellins spread around the world as the potential for its use on various commercially important plants became more obvious. For example, research that started at the University of California, Davis in the mid-1960s led to its commercial use on Thompson seedless table grapes throughout California by 1962. A known antagonist to gibberellin is paclobutrazol (PBZ), which in turn inhibits growth and induces early fruitset as well as seedset.

CHEMISTRY

All known gibberellins are diterpenoid acids that are synthesised by the terpenoid pathway in plastids and then modified in the endoplasmic reticulum and cytosol until they reach their biologically-active form. All gibberellins are derived via the *ent*-gibberellane skeleton, but are synthesised via *ent*-kaurene. The gibberellins are named GA1 through GAn in order of discovery. Gibberellic acid, which was the first gibberellin to be structurally characterised, is GA3.

As of 2003, there were 126 GAs identified from plants, fungi, and bacteria.

Gibberellins are tetracyclic diterpene acids. There are two classes based on the presence of either 19 or 20 carbons. The 19-carbon gibberellins, such as gibberellic acid, have lost carbon 20 and, in place, possess a five-member lactone bridge that links carbons 4 and 10. The 19 carbon forms are, in general, the biologically active forms of gibberellins. Hydroxylation also has a great effect on the biological activity of the gibberellin. In general, the most biologically active compounds are dihydroxylated gibberellins, which possess hydroxyl groups on both carbon 3 and carbon 13. Gibberellic acid is a dihydroxylated gibberellin.

BIOACTIVE GAS

The bioactive GAs are GA1, GA3, GA4, and GA7. There are three common structural traits between these GAs : 1) a hydroxyl group on C-3β, 2) a carboxyl group on C-6, and 3) a lactone between C-4 and C-10. The 3β-hydroxyl group can be exchanged for other functional groups at C-2 and/or C-3 positions. GA5 and GA6 are examples of bioactive GAs that do not have a hydroxyl group on C-3β. The presence of GA1 in various plant species suggests that it is a common bioactive GA.

BIOLOGICAL FUNCTION

Gibberellins are involved in the natural process of breaking dormancy and various other aspects of germination. Before the photosynthetic apparatus develops sufficiently in the early stages of germination, the stored energy reserves of starch nourish the seedling. Usually in germination, the breakdown of starch to glucose in the endosperm begins shortly after the seed is exposed to water. Gibberellins in the seed embryo are believed to signal starch hydrolysis through inducing the synthesis of the enzyme α-amylase in the aleurone cells. In the model for gibberellin-induced production of α-amylase, it is demonstrated that gibberellins (denoted by GA) produced in the scutellum diffuse to the aleurone cells, where they stimulate the secretion α-amylase. α-Amylase then hydrolyses starch, which is abundant in many seeds, into glucose that can be used in cellular respiration to produce energy for the seed embryo. Studies of this process have indicated gibberellins cause higher levels of transcription of the gene coding for the α-amylase enzyme, to stimulate the synthesis of α-amylase.

Gibberellins are produced in greater mass when the plant is exposed to cold temperatures. They stimulate cell elongation, breaking and budding, seedless fruits, and seed germination. They do the last by breaking the seed's dormancy and acting as a chemical messenger. Its hormone binds to a receptor, and Ca^{2+} activates the protein calmodulin, and the complex binds to DNA, producing an enzyme to stimulate growth in the embryo.

A major effect of gibberellins is the degradation of DELLA proteins, the absence of which then allows phytochrome interacting factors to bind to gene promoters and regulate gene expression. Gibberellins are thought to cause DELLAs to become polyubiquitinated and, thus, destroyed by the 26S proteasome pathway.

BIOSYNTHESIS

GAs are usually produced from the methylerythritol phosphate (MEP) pathway in higher plants. In this pathway, bioactive GA is produced from trans-geranylgeranyl diphosphate (GGDP). In the MEP pathway, three classes of enzymes are used to yield GA from GGDP :

- terpene synthases (TPSs),
- cytochrome P450 monooxygenases (P450s), and
- 2-oxoglutarate–dependent dioxygenases (2ODDs).

There are 8 steps in the methylerythritol phosphate pathway :

- GGDP is converted to ent-copalyl diphosphate (ent-CPD) by ent-copalyl diphosphate synthase –
- etn-CDP is converted to ent-kaurene by ent-kaurene synthase –
- ent-kaurene is converted to ent-kaurenol by ent-kaurene oxidase (KO) –
- ent-kaurenol is converted to ent-kaurenal by KO –
- ent-kaurenal is converted to ent-kaurenoic acid by KO –
- ent-kaurenoic acid is converted to ent-7a-hydroxykaurenoic acid by ent-kaurene acid oxidase (KAO) –
- ent-7a-hydroxykaurenoic acid is converted to GA12-aldehyde by KAO –
- GA12-aldehyde is converted to GA12 by KAO. GA12 is processed to the bioactive GA4 by oxidations on C-20 and C-3, which is accomplished by 2 soluble ODDs : GA 20-oxidase and GA 3-oxidase.

Sites of Biosynthesis

Most bioactive GAs are located in actively growing organs on plants. Both GA20ox and GA3ox genes (genes coding for GA 20-oxidase and GA 3-oxidase) and the SLENDER1 gene (a GA signal transduction gene) are found in growing organs on rice, which suggests bioactive GA synthesis occurs at their site of action in growing organs in plants. During flower development, the tapetum of anthers is believed to a primary site of GA biosynthesis.

Differences between Biosynthesis in Fungi and Lower Plants

Arabidopsis, a plant, and Gibberella fujikuroi, a fungus, possess different GA pathways and enzymes. P450s in fungi perform functions analogous to the functions of KAOs in plants. The function of CPS and KS in plants is performed by a single enzyme, CPS/KS, in fungi. In fungi, the GA biosynthesis genes are found on one chromosome, but in plants, they are found randomly on multiple chromosomes.

GIBBERELLIN METABOLISM GENES

One or two genes encode the enzymes responsible for the first steps of GA biosynthesis in Arabidopsis and rice. The null alleles of the genes encoding CPS,

KS, and KO result in GA-deficient Arabidopsis dwarves. Multigene families encode the 2ODDs that catalyse the formation of GA12 to bioactive GA4.

AtGA3ox1 and AtGA3ox2, two of the four genes that encode GA3ox in Arabidopsis, affect vegetative development. Environmental stimuli regulate AtGA3ox1 and AtGA3ox2 activity during seed germination. In Arabidopsis, GA20ox overexpression leads to an increase in GA concentration.

DEACTIVATION

Several mechanisms for inactivating GAs have been identified. 2β-hydroxylation deactivates GA, and is catalysed by GA2-oxidases (GA2oxs). Some GA2oxs use C19-GAs as substrates, and other GA2oxs use C20-GAs.

Cytochrome P450 mono-oxygenase, encoded by elongated uppermost internode (eui), converts GAs into 16α,17-epoxides. Rice eui mutants amass bioactive GAs at high levels, which suggests cytochrome P450 mono-oxygenase is a main enzyme responsible for deactivation GA in rice.

The Gamt1 and gamt2 genes encode enzymes that methylate the C-6 carboxyl group of GAs. In a gamt1 and gamt2 mutant, concentrations of GA is developing seeds is increased.

REGULATION

Regulation by Other Hormones

The auxin indole-3-acetic acid (IAA) regulates concentration of GA1 in elongating internodes in peas. Removal of IAA by removal of the apical bud, the auxin source, reduces the concentration of GA1, and reintroduction of IAA reverses these effects to increase the concentration of GA1. This phenomenon has also been observed in tobacco plants. Auxin increases GA 3-oxidation and decreases GA 2-oxidation in barley. Auxin also regulates GA biosynthesis during fruit development in peas. These discoveries in different plant species suggest the auxin regulation of GA metabolism may be a universal mechanism.

Ethylene decreasing the concentration of bioactive GAs.

Regulation by Environmental Factors

Recent evidence suggests fluctuations in GA concentration influence light-regulated seed germination, photomorphogenesis during de-etiolation, and photoperiod regulation of stem elongation and flowering. Microarray analysis showed about one fourth cold-responsive genes are related to GA-regulated genes, which suggests GA influences response to cold temperatures. Plants reduce growth rate when exposed to stress. A relationship between GA levels and amount of stress experienced has been suggested in barley.

Role in Seed Development

Bioactive GAs and abcisic acid levels have an inverse relationship and regulate seed development and germination. Levels of FUS3, an Arabidopsis transcription factor, are upregulated by ABA and downregulated by GA, which suggests that there is a regulation loop that establishes the balance of GA and ABA.

GIBBERELLIN HOMEOSTASIS

Feedback and feedforward regulation maintains the levels of bioactive GAs in plants. Levels of AtGA20ox1 and AtGA3ox1 expression are increased in a GA deficient environment, and decreased after the addition of bioactive GAs, Conversely, expression of AtGA2ox1 and AtGA2ox2, GA deactivation genes, is increased with addition of GA.

IMPACT ON THE "GREEN REVOLUTION"

A chronic food shortage was feared during the rapid climb in world population in the 1960s. This was averted with the development of a high-yielding variety of rice. This variety of semi-dwarf rice is called IR8, and it has a short height because of a mutation in the sd1 gene. Sd1 encodes GA20ox, so a mutant sd1 is expected to exhibit a short height that is consistent with GA deficiency.

RECEPTORS

GA receptors must have four traits : (1) reversibly bind GA, (2) GA saturability (3) high affinity for bioactive GA, and (4) reasonable ligand specificity for bioactive GAs.

Cereal Aleurone Cells

The first reported GA-binding protein (GBP) activity was discovered in wheat seed aleurone homogenates by using the GA-dependent induction of aleuron hydrolytic enzymes. Alpha-amylase induction in aleurone protoplasts is dependent on GA levels, but induction can still occur if GA does not pass through the plasma membrane. The existence of GA perception outside the cell and a plasma membrane GA receptor in aleurone cells can be concluded from these observations. Trimeric G proteins are involved in GA signaling, which supports the previous theory for plasma membrane GA receptors. Four different GBPs have been discovered in the plasma membrane of aleuron cells. Cereal aleurone cells could be unique in GA perception because they are some of the few plant cells that cannot synthesise bioactive GA.

GID1 Receptor

The rice GID1 gene encodes an unknown protein that only has a high affinity for bioactive GAs. In yeast, GID1 binds SLR1, GA signaling repressor, depending on GA levels. GID1 is a soluble GA receptor mediating GA signaling. The GA-GID1 complex interacts with SLR1 to transduce the GA signal to SLR1. SLR1 increased the GA-binging activity of GID1 threefold. Arabidopsis has three GID1 homologs that bind GA and interact with DELLAs : AtGID1a, b, and c.

GIBBERELLIN AND DELLA PROTEINS

DELLA Proteins : Repressors of GA-dependent Processes

DELLA proteins act as intracellular repressors of GA responses, DELLAs inhibit seed germination, seed growth, and other GA-dependent pathways, but GA can reverse these effects.

GA-GID1-DELLA Complex

For GA to bind to the GID1 receptor, the C3-hydroxyl on GA must form a hydrogen bond to the tyrosine 31 residue of GID1, which induces a conformational change in GID1 to enclose GA. Once GA is trapped in the GID1 pocket, the lid of the pocket binds to DELLA to form the GA-GID1-DELLA complex.

GA Promotes Proteasome-dependent Degradation of DELLAs

In the absence of GA, many DELLAs are present to repress GA responses, but the formation of the GA-GID1-DELLA complex increases the degradation of DELLAs. The SCF complex is composed of SKP1, CULLIN, and F-BOX proteins. F-box proteins catalyse the formation of polyubiquitin on target proteins to be degraded by the 26S proteasome. The formation of the GA-GID1-DELLA complex is believed to cause a conformational change in DELLA, which enhances recognition of DELLAs by F-box proteins. Next, the SCF complex promotes ubiquitylation of DELLAs, which are then degraded by the 26S proteasome. Degradation of DELLAs allows GA regulated growth to resume. Thus, GA stimulates growth by activating the degradation of DELLAs.

Chapter 7

CYTOKININ

Cytokinins (CK) are a class of plant growth substances (phytohormones) that promote cell division, or cytokinesis, in plant roots and shoots. They are involved primarily in cell growth and differentiation, but also affect apical dominance, axillary bud growth, and leaf senescence. Folke Skoog discovered their effects using coconut milk in the 1940s at the University of Wisconsin–Madison.

There are two types of cytokinins : adenine-type cytokinins represented by kinetin, zeatin, and 6-benzylaminopurine, and phenylurea-type cytokinins like diphenylurea and thidiazuron (TDZ). Most adenine-type cytokinins are synthesised in roots. Cambium and other actively dividing tissues also synthesise cytokinins. No phenylurea cytokinins have been found in plants. Cytokinins participate in local and long-distance signalling, with the same transport mechanism as purines and nucleosides. Typically, cytokinins are transported in the xylem.

Cytokinins act in concert with auxin, another plant growth hormone.

MODE OF ACTION

The ratio of auxin to cytokinin plays an important role in the effect of cytokinin on plant growth. Cytokinin alone has no effect on parenchyma cells. When cultured with auxin but no cytokinin, they grow large but do not divide. When cytokinin is added, the cells expand and differentiate. When cytokinin and auxin are present in equal levels, the parenchyma cells form an undifferentiated callus. More cytokinin induces growth of shoot buds, while more auxin induces root formation.

Cytokinins are involved in many plant processes, including cell division and shoot and root morphogenesis. They are known to regulate axillary bud growth and apical dominance. The "direct inhibition hypothesis" posits that these effects result from the cytokinin to auxin ratio. This theory states that auxin from apical buds travels down shoots to inhibit axillary bud growth. This promotes shoot growth, and restricts lateral branching. Cytokinin moves from the roots into the shoots, eventually signaling lateral bud growth. Simple experiments support this theory. When the apical bud is removed, the axillary buds are uninhibited, lateral growth increases, and plants become bushier. Applying auxin to the cut stem again inhibits lateral dominance.

While cytokinin action in vascular plants is described as pleiotropic, this class of plant hormones specifically induces the transition from apical growth to growth via a three-faced apical cell in moss protonema. This bud induction can be pinpointed to differentiation of a specific single cell, and thus is a very specific effect of cytokinin.

Cytokinins have been shown to slow aging of plant organs by preventing protein breakdown, activating protein synthesis, and assembling nutrients from nearby tissues. A study that regulated leaf senescence in tobacco leaves found that wild-type leaves yellowed while transgenic leaves remained mostly green. It was hypothesised that cytokinin may affect enzymes that regulate protein synthesis and degradation.

BIOSYNTHESIS

Adenosine phosphate-isopentenyltransferase (IPT) catalyses the first reaction in the biosynthesis of isoprene cytokinins. It may use ATP, ADP, or AMP as substrates and may use dimethylallyl diphosphate (DMAPP) or hydroxymethylbutenyl diphosphate (HMBDP) as prenyl donors. This reaction is the rate-limiting step in cytokinin biosynthesis. DMAPP and HMBDP used in cytokinin biosynthesis are produced by the methylerythritol phosphate pathway (MEP).

Cytokinins can also be produced by recycled tRNAs in plants and bacteria. tRNAs with anticodons that start with a uridine and carrying an already-prenylated adenosine adjacent to the anticodon release on degradation the adenosine as a cytokinin. The prenylation of these adenines is carried out by tRNA-isopentenyltransferase.

Auxin is known to regulate the biosynthesis of cytokinin.

USES

Because cytokinin promotes plant cell division and growth, produce farmers use it to increase crops. One study found that applying cytokinin to cotton seedlings led to a 5–10 per cent yield increase under drought conditions.

Cytokinins have recently been found to play a role in plant pathogenesis. For example, cytokinins have been described to induce resistance against *Pseudomonas syringae* in *Arabidopsis thaliana* and *Nicotiana tabacum*.

6-BENZYLAMINOPURINE

6-Benzylaminopurine, benzyl adenine or BAP is a first-generation synthetic cytokinin that elicits plant growth and development responses, setting blossoms and stimulating fruit richness by stimulating cell division. It is an inhibitor of respiratory kinase in plants, and increases post-harvest life of green vegetables. Influence of cytokinin as 6-benzylaminopurine (BAP) in combination with other methods on postharvest green colour retention on broccoli heads and asparagus spears, showed positive results for quality retention. Treatment with 10 and 15 ppm BAP can be used to extend shelf life of fresh-cut broccoli florets and shredded cabbage during storage at 6±1°C at commercial level.

6-Benzylaminopurine was first synthesised and tested in the laboratories of plant physiologist Folke K. Skoog.

FORCHLORFENURON

Forchlorfenuron is a plant growth regulator. It has been approved for use on kiwi fruit and grapes in the USA, and it has been associated with exploding watermelons in China.

KINETIN

Kinetin is a type of cytokinin, a class of plant hormone that promotes cell division. Kinetin was originally isolated by Miller and Skoog *et. al.* as a compound from autoclaved herring sperm DNA that had cell division-promoting activity. It was given the name kinetin because of its ability to induce cell division, provided that auxin was present in the medium. Kinetin is often used in plant tissue culture for inducing formation of callus (in conjunction with auxin) and to regenerate shoot tissues from callus (with lower auxin concentration).

For a long time, it was believed that kinetin was an artifact produced from the deoxyadenosine residues in DNA, which degrade on standing for long periods or when heated during the isolation procedure. Therefore, it was thought that kinetin does not occur naturally, but, since 1996, it has been shown by several researchers that kinetin exists naturally in the DNA of cells of almost all organisms tested so far, including human and various plants. The mechanism of production of kinetin in DNA is thought to be via the production of furfural — an oxidative damage product of deoxyribose sugar in DNA — and its quenching by the adenine base's converting it into N6-furfuryladenine, kinetin.

Since 1994, kinetin has been thoroughly tested for its powerful anti-aging effects in human skin cells and other systems. At present, kinetin is one of the widely used components in numerous skin care cosmetics and cosmeceuticals, such as Valeant products kinerase. There are some reports published on other biological effects of kinetin in human beings, for example its effects as anti-platelet aggregation factor reducing thrombosis formation. In addition, it has been shown to be capable of correcting RNA mis-splicing in the disease of familial dysautonomia, in which exon 20 of IKBKAP is skipped instead of included in the disease, leading to insufficient levels of IKBKAP protein product as a result of frame-shift-induced non-sense-mediated decay.

Kinetin is currently sold commercially under the trade name "Bonide Tomato and Blossom Set Spray", and can be used to increase yields of various fruits and vegetables, produce seedless fruits, and increase 'budding' of various herbs.

Kinetin is also widely used in producing new plants from tissue cultures.

History

In 1939 P. A. C. Nobécourt (Paris) began the first permanent callus culture from root explants of carrot (*Daucus carota*). Such a culture can be kept forever by successive transplantations onto fresh nutrient agar. The transplantations occur every three to eight weeks. Callus cultures are not cell cultures, since whole tissue associations are cultivated. Though many cells keep their ability to divide, this is not true for all. One reason for this is the aneuploidy of the nuclei and the resultant unfavourable chromosome constellations.

In 1941 J. van Overbeek (Rijksuniversiteit Utrecht) introduced coconut milk as a new component of nutrient media for callus cultures. Coconut milk is liquid endosperm. It stimulates the embryo to grow when it is supplied with food at the same time. Results yielded from callus cultures showed that its active components stimulate the growth of foreign cells too.

In 1954 F. Skoog (University of Wisconsin, Madison) developed a technique for the generation and culture of wound tumour tissue from isolated shoot parts of tobacco (*Nicotiana tabacum*). The developing callus grows when supplied with yeast extract, coconut milk, or old DNA preparations. Freshly prepared DNA has no effect but becomes effective after autoclaving. This led to the conclusion that one of its breakdown products is required for cell growth and division. The substance was characterised, was given the name *kinetin*, and classified as a phytohormone.

ZEATIN

Zeatin is a plant hormone derived from the purinebase called adenine. Zeatin belongs to the family of plant-growth hormones called cytokinins and was discovered in immature corn kernels from the genus *Zea*. It promotes growth of lateral buds and when sprayed on meristems stimulates cell division to produce bushier plants. Zeatin and its derivatives are the active ingredient in coconut milk, which causes plant growth. Zeatin has several anti-aging effects on human skin fibroblasts.

Toxicology

The acute transperitoneal toxicity in mice (LD_{50}) is 2200 mg/kg.

Application

Zeatin has a variety of effects including :

1. Promotes callus initiation when combined with auxin, concentration 1 ppm.

2. Promotes fruit set. Zeatin 100 ppm + GA3 500 ppm + NAA 20 ppm, sprayed at 10th, 25th, 40th day after blossom.
3. Retards yellowing for vegetables, 20 ppm, sprayed.
4. Causes auxiliary stems to grow and flower.

Zeatin can also be applied to stimulate seed germination and seedling growth.

Chapter 8

ETHYLENE

Ethylene is a hydrocarbon with the formula C_2H_4 or $H_2C=CH_2$. It is a colourless flammable gas with a faint "sweet and musky" odour when pure. It is the simplest alkene (a hydrocarbon with carbon-carbon double bonds), and the simplest unsaturated hydrocarbon after acetylene (C_2H_2).

Ethylene is widely used in chemical industry, and its worldwide production (over 109 million tonnes in 2006) exceeds that of any other organic compound. Ethylene is also an important natural plant hormone, used in agriculture to force the ripening of fruits.

STRUCTURE AND PROPERTIES

This hydrocarbon has four hydrogen atoms bound to a pair of carbon atoms that are connected by a double bond. All six atoms that comprise ethylene are coplanar. The H-C-H angle is 117.4°, close to the 120° for ideal sp^2 hybridised carbon. The molecule is also relatively rigid : rotation about the C-C bond is a high energy process that requires breaking the π-bond.

The π-bond in the ethylene molecule is responsible for its useful reactivity. The double bond is a region of high electron density, thus it is susceptible to attack by electrophiles. Many reactions of ethylene are catalysed by transition metals, which bind transiently to the ethylene using both the π and π^* orbitals.

Being a simple molecule, ethylene is spectroscopically simple. Its UV-vis spectrum is still used as a test of theoretical methods.

USES

Major industrial reactions of ethylene include in order of scale : (1) polymerisation, (2) oxidation, (3) halogenation and hydrohalogenation, (4) alkylation, (5) hydration, (6) oligomerisation, and (7) hydroformylation. In the United States and Europe, approximately 90 per cent of ethylene is used to produce ethylene oxide, ethylene dichloride, ethylbenzene and polyethylene.

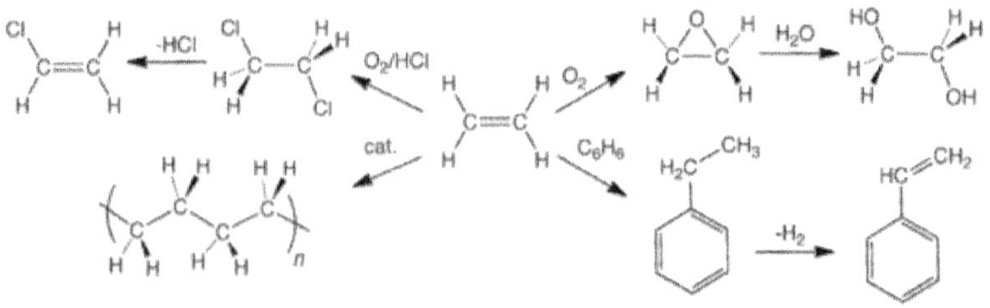

Fig. Main industrial uses of ethylene. Clockwise from the upper right : its conversions to ethylene oxide, precursor to ethylene glycol; to ethylbenzene, precursor to styrene; to various kinds of polyethylene; to ethylene dichloride, precursor to vinyl chloride.

Polymerisation

Polyethylene consumes more than half of world ethylene supply. Polyethylene, also called *polythene*, is the world's most widely used plastic, being primarily used to make films used in packaging, carrier bags and trash liners. Linear alpha-olefins, produced by oligomerisation (formation of short polymers) are used as precursors, detergents, plasticisers, synthetic lubricants, additives, and also as co-monomers in the production of polyethylenes.

Oxidation

Ethylene is oxidised to produce ethylene oxide, a key raw material in the production of surfactants and detergents by ethoxylation. Ethylene oxide is also hydrolysed to produce ethylene glycol, widely used as an automotive antifreeze as well as higher molecular weight glycols, glycol ethers and polyethylene terephthalate.

Ethylene undergoes oxidation by palladium to give acetaldehyde. This conversion remains a major industrial process (10M kg/y). The process proceeds via the initial complexation of ethylene to a Pd(II) center.

Halogenation and Hydrohalogenation

Major intermediates from the halogenation and hydrohalogenation of ethylene include ethylene dichloride, ethyl chloride and ethylene dibromide. The addition of chlorine entails "oxychlorination," *i.e.* chlorine itself is not used. Some products derived from this group are polyvinyl chloride, trichloroethylene, perchloroethylene, methyl chloroform, polyvinylidene chloride and copolymers, and ethyl bromide.

Alkylation

Major chemical intermediates from the alkylation with ethylene is ethylbenzene, precursor to styrene. Styrene is used principally in polystyrene for packaging and insulation, as well as in styrene-butadiene rubber for tires and footwear. On a smaller scale, ethyltoluene, ethylanilines, 1,4-hexadiene, and aluminium alkyls. Products of these intermediates include polystyrene, unsaturated polyesters and ethylene-propylene terpolymers.

Oxo Reaction

The hydroformylation (oxo reaction) of ethylene results in propionaldehyde, a precursor to propionic acid and n-propyl alcohol.

Hydration

Ethylene has long represented the major non-fermentative precursor to ethanol. The original method entailed its conversion to diethyl sulfate, followed by hydrolysis. The main method practiced since the mid-1990s is the direct hydration of ethylene catalysed by solid acid catalysts :

$$C_2H_4 + H_2O \rightarrow CH_3CH_2OH$$

Dimerisation to n-Butenes

Ethylene can be dimerised to n-butenes using processes licensed by Lummus or IFP. The Lummus process produces mixed n-butenes (primarily 2-butenes) while the IFP process produces 1-butene.

Niche Uses

An example of a niche use is as an anesthetic agent (in an 85 per cent ethylene/15 per cent oxygen ratio). It can also be used to hasten fruit ripening, as well as a welding gas.

PRODUCTION

Global ethylene production was 107 million tonnes in 2005, 109 million tonnes in 2006. NNFCC Renewable Chemicals Factsheet : Ethanol, 138 million tonnes in 2010 and 141 million tonnes in 2011. By 2010 ethylene was produced by at least 117 companies in 55 countries. To meet the ever increasing demand for ethylene, sharp increases in production facilities are added globally, particularly in the Mideast and in China.

Ethylene is produced in the petrochemical industry by steam cracking. In this process, gaseous or light liquid hydrocarbons are heated to 750–950 °C, inducing numerous free radical reactions followed by immediate quench to stop these reactions. This process converts large hydrocarbons into smaller ones and introduces unsaturation. Ethylene is separated from the resulting complex mixture by repeated compression and distillation. In a related process used in oil refineries, high molecular weight hydrocarbons are cracked over zeolite catalysts. Heavier feedstocks, such as naphtha and gas oils require at least two "quench towers" downstream of the cracking furnaces to recirculate pyrolysis-derived gasoline and process water. When cracking a mixture of ethane and propane, only one water quench tower is required.

The areas of an ethylene plant are :

1. Steam cracking furnaces :
2. Primary and secondary heat recovery with quench;
3. A dilution steam recycle system between the furnaces and the quench system;
4. Primary compression of the cracked gas (3 stages of compression);
5. Hydrogen sulfide and carbon dioxide removal (acid gas removal);
6. Secondary compression (1 or 2 stages);
7. Drying of the cracked gas;
8. Cryogenic treatment;
9. All of the cold cracked gas stream goes to the demethaniser tower. The overhead stream from the demethaniser tower consists of all the hydrogen and methane that was in the cracked gas stream. Cryogenically (−250 °F (−157 °C)) treating this overhead stream separates hydrogen from methane. Methane recovery is critical to the economical operation of an ethylene plant.
10. The bottom stream from the demethaniser tower goes to the deethaniser tower. The overhead stream from the deethaniser tower consists of all the C 2,'s that were in the cracked gas stream. The C 2 stream contains acetylene, which is explosive above 200 kPa (29 psi). If the partial pressure of acetylene is expected to exceed these values, the C 2 stream is partially hydrogenated. The C 2's then proceed to a C 2 splitter. The product ethylene is taken from the overhead of the tower and the ethane coming from the bottom of the splitter is recycled to the furnaces to be cracked again;
11. The bottom stream from the de-ethaniser tower goes to the depropaniser tower. The overhead stream from the depropaniser towerconsists of all the C 3's that were in the cracked gas stream. Before feeding the C 3's to the C 3 splitter, the stream is hydrogenated to convert the methylacetylene and propadiene (allene) mix. This stream is then sent to the C 3 splitter. The overhead stream from the C 3 splitter is product propylene and the bottom stream is propane which is sent back to the furnaces for cracking or used as fuel.
12. The bottom stream from the depropaniser tower is fed to the debutaniser tower. The overhead stream from the debutaniser is all of the C 4's that were in the cracked gas stream. The bottom stream from the debutaniser (light pyrolysis gasoline) consists of everything in the cracked gas stream that is C 5 or heavier.

Since ethylene production is energy intensive, much effort has been dedicated to recovering heat from the gas leaving the furnaces. Most of the energy recovered from the cracked gas is used to make high pressure (1200 psig) steam. This steam is in turn used to drive the turbines for compressing cracked gas, the propylene refrigeration compressor, and the ethylene refrigeration compressor. An ethylene plant, once running, does not need to import steam to drive its steam turbines. A typical world scale ethylene plant (about 1.5 billion pounds of ethylene per year) uses a 45,000 horsepower (34,000 kW) cracked gas compressor, a 30,000 hp (22,000 kW) propylene compressor, and a 15,000 hp (11,000 kW) ethylene compressor.

Laboratory Synthesis

Although of great value industrially, ethylene is rarely used in the laboratory and is ordinarily purchased. It can be produced via dehydration of ethanol with sulfuric acid or in the gas phase with aluminium oxide.

ETHYLENE AS A PLANT HORMONE

Ethylene serves as a hormone in plants. It acts at trace levels throughout the life of the plant by stimulating or regulating the ripening of fruit, the opening of flowers, and the abscission (or shedding) of leaves. Commercial ripening rooms use "catalytic generators" to make ethylene gas from a liquid supply of ethanol. Typically, a gassing level of 500 to 2,000 ppm is used, for 24 to 48 hours. Care must be taken to control carbon dioxide levels in ripening rooms when gassing, as high temperature ripening (20 °C; 68 °F) has been seen to produce CO_2 levels of 10 per cent in 24 hours.

History of Ethylene in Plant Biology

Ethylene has been used since the ancient Egyptians, who would gash figs in order to stimulate ripening (wounding stimulates ethylene production by plant tissues). The ancient Chinese would burn incense in closed rooms to enhance the ripening of pears. In 1864, it was discovered that gas leaks from street lights led to stunting of growth, twisting of plants, and abnormal thickening of stems. In 1901, a Russian scientist named Dimitry Neljubow showed that the active component was ethylene. Sarah Doubt discovered that ethylene stimulated abscission in 1917. It wasn't until 1934 that Gane reported that plants synthesise ethylene. In 1935, Crocker proposed that ethylene was the plant hormone responsible for fruit ripening as well as senescence of vegetative tissues.

Ethylene Biosynthesis in Plants

Ethylene is produced from essentially all parts of higher plants, including leaves, stems, roots, flowers, fruits, tubers, and seeds. Ethylene production is regulated by a variety of developmental and environmental factors. During the life of the plant, ethylene production is induced during certain stages of growth such as germination, ripening of fruits, abscission of leaves, and senescence of flowers. Ethylene production can also be induced by a variety of external aspects such as mechanical wounding, environmental stresses, and certain chemicals including auxin and other regulators.

Ethylene is biosynthesised from the amino acid methionine to *S*-adenosyl-L-methionine (SAM, also called Adomet) by the enzyme Met Adenosyltransferase. SAM is then converted to 1-aminocyclopropane-1-carboxylic acid (ACC) by the enzyme ACC synthase (ACS). The activity of ACS determines the rate of ethylene production, therefore regulation of this enzyme is key for the ethylene biosynthesis. The final step requires oxygen and involves the action of the enzyme ACC-oxidase (ACO), formerly known as the ethylene forming enzyme (EFE). Ethylene biosynthesis can be induced by endogenous or exogenous ethylene. ACC synthesis increases with high levels of auxins, especially indole acetic acid (IAA) and cytokinins. ACC synthase is inhibited by abscisic acid.

Ethylene Perception in Plants

Ethylene is perceived by a family of five transmembrane protein dimers such as the ETR_1 protein in *Arabidopsis*. The gene encoding an ethylene receptor has been cloned in *Arabidopsis thaliana* and then in tomato. Ethylene receptors are encoded by multiple genes in the *Arabidopsis* and tomato genomes. Mutations in any of the gene family, which comprises five receptors in *Arabidopsis* and at least six in tomato, can lead to insensitivity to ethylene. DNA sequences for ethylene receptors have also been identified in many other plant species and an ethylene binding protein has even been identified in Cyanobacteria.

Environmental and Biological Triggers of Ethylene

Environmental cues such as flooding, drought, chilling, wounding, and pathogen attack can induce ethylene formation in plants. In flooding, roots suffer from lack of oxygen, or anoxia, which leads to the synthesis of 1-aminocyclopropane-1-carboxylic acid (ACC). ACC is transported upwards in the plant and then oxidised in leaves. The ethylene produced causes nastic movements (epinasty) of the leaves, perhaps helping the plant to lose water.

List of Plant Responses to Ethylene

- Seedling triple response, thickening and shortening of hypocotyl with pronounced apical hook.
- In pollination, when the pollen reaches the stigma, the precursor of the ethene, ACC, is secreted to the petal, the ACC releases ethylene with ACC oxidase.
- Stimulates leaf and flower senescence.
- Stimulates senescence of mature xylem cells in preparation for plant use.

- Induces leaf abscission.
- Induces seed germination.
- Induces root hair growth – increasing the efficiency of water and mineral absorption.
- Induces the growth of adventitious roots during flooding.
- Stimulates epinasty – leaf petiole grows out, leaf hangs down and curls into itself.
- Stimulates fruit ripening.
- Induces a climacteric rise in respiration in some fruit which causes a release of additional ethylene.
- Affects gravitropism.
- Stimulates nutational bending.
- Inhibits stem growth and stimulates stem and cell broadening and lateral branch growth outside of seedling stage.
- Interference with auxin transport (with high auxin concentrations).
- Inhibits shoot growth and stomatal closing except in some water plants or habitually flooded ones such as some rice varieties, where the opposite occurs (conserving CO_2 and O_2).
- Induces flowering in pineapples.
- Inhibits short day induced flower initiation in *Pharbitus nil* and *Chrysanthemum morifolium.*

Commercial Issues

Ethene shortens the shelf life of many fruits by hastening fruit ripening and floral senescence. Ethene will shorten the shelf life of cut flowers and potted plants by accelerating floral senescence and floral abscission. Flowers and plants which are subjected to stress during shipping, handling, or storage produce ethene causing a significant reduction in floral display. Flowers affected by ethene include carnation, geranium, petunia, rose, and many others.

Ethene can cause significant economic losses for florists, markets, suppliers, and growers. Researchers have developed several ways to inhibit ethene, including inhibiting ethene synthesis and inhibiting ethene perception. Aminoethoxyvinylglycine (AVG), Aminooxyacetic acid (AOA), and silver salts are ethylene inhibitors. Inhibiting ethene synthesis is less effective for reducing post-harvest losses since ethene from other sources can still have an effect. By inhibiting ethene perception, fruits, plants and flowers don't respond to ethene produced endogenously or from exogenous sources. Inhibitors of ethene perception include compounds that have a similar shape to ethene, but do not elicit the ethene response. One example of an ethene perception inhibitor is 1-methylcyclopropene (1-MCP).

Commercial growers of bromeliads, including pineapple plants, use ethene to induce flowering. Plants can be induced to flower either by treatment with the gas in a chamber, or by placing a banana peel next to the plant in an enclosed area.

Chrysanthemum flowering is delayed by ethene gas and growers have found that carbon dioxide 'burners' and the exhaust fumes from inefficient glasshouse heaters can raise the ethene concentration to 0.05 vpm causing delay in flowering of commercial crops.

LIGAND

Ethylene is a ligand in organometallic chemistry. One of the first organometallic compounds, Zeise's salt is a complex of ethylene. Useful reagents containing ethylene include $Pt(PPh_3)_2(C_2H_4)$ and $Rh_2Cl_2(C_2H_4)_4$. The Rh-catalysed hydroformylation of ethylene is conducted on industrial scale to provide propionaldehyde.

HISTORY

Many geologists and scholars believe that the famous Greek Oracle at Delphi (the Pythia) went into her trance-like state as an effect of ethylene rising from ground faults.

Ethylene appears to have been discovered by Johann Joachim Becher, who obtained it by heating ethanol with sulfuric acid; he mentioned the gas in his *Physica Subterranea* (1669). Joseph Priestley also mentions the gas in his *Experiments and observations relating to the various branches of natural philosophy : with a continuation of the observations on air* (1779), where he reports that Jan Ingenhousz saw ethylene synthesised in the same way by a Mr. Enée in Amsterdam in 1777 and that Ingenhousz subsequently produced the gas himself. The properties of ethylene were studied in 1795 by four Dutch chemists, Johann Rudolph Deimann, Adrien Paets van Troostwyck, Anthoni Lauwerenburgh and Nicolas Bondt, who found that it differed from hydrogen gas and that it contained both carbon and hydrogen. This group also discovered that ethylene could be combined with chlorine to produce the *oil of the Dutch chemists*, 1,2-dichloroethane; this discovery gave ethylene the name used for it at that time, *olefiant gas* (oil-making gas.)

In the mid-19th century, the suffix *-ene* (an Ancient Greek root added to the end of female names meaning "daughter of") was widely used to refer to a molecule or part thereof that contained one fewer hydrogen atoms than the molecule being modified. Thus, *ethylene* (C2H4) was the "daughter of ethyl" (C2H5). The name ethylene was used in this sense as early as 1852.

In 1866, the German chemist August Wilhelm von Hofmann proposed a system of hydrocarbon nomenclature in which the suffixes -ane, -ene, -ine, -one, and -une were used to denote the hydrocarbons with 0, 2, 4, 6, and 8 fewer hydrogens than their parent alkane. In this system, ethylene became *ethene*. Hofmann's system eventually became the basis for the Geneva nomenclature approved by the International Congress of Chemists in 1892, which remains at the core of the

IUPAC nomenclature. However, by that time, the name ethylene was deeply entrenched, and it remains in wide use today, especially in the chemical industry.

Nomenclature

The 1979 IUPAC nomenclature rules made an exception for retaining the non-systematic name ethylene, however, this decision was reversed in the 1993 rules so the IUPAC name is now *ethene*.

Safety

Like all hydrocarbons, ethylene is an asphyxiant and combustible. It has been used as an anesthetic. It is listed as an IARC class 3 carcinogen.

Chapter 9

ABSCISIC ACID

Abscisic acid (ABA), also known as abscisin II and dormin, is a plant hormone. ABA functions in many plant developmental processes, including bud dormancy. It is degraded by the enzyme (+)-abscisic acid 8′-hydroxylase into phaseic acid.

FUNCTION

ABA was originally believed to be involved in abscission. This is now known to be the case only in a small number of plants. ABA-mediated signalling also plays an important part in plant responses to environmental stress and plant pathogens. The plant genes for ABA biosynthesis and sequence of the pathway have been elucidated. ABA is also produced by some plant pathogenic fungi via a biosynthetic route different from ABA biosynthesis in plants.

Abscisic acid owes its names to its role in the abscission of plant leaves. In preparation for winter, ABA is produced in terminal buds. This slows plant growth and directs leaf primordia to develop scales to protect the dormant buds during the cold season. ABA also inhibits the division of cells in the vascular cambium, adjusting to cold conditions in the winter by suspending primary and secondary growth.

Abscisic acid is also produced in the roots in response to decreased soil water potential and other situations in which the plant may be under stress. ABA then translocates to the leaves, where it rapidly alters the osmotic potential of stomatal guard cells, causing them to shrink and stomata to close. The ABA-induced stomatal closure reduces transpiration, thus preventing further water loss from the leaves in times of low water availability. A close linear correlation was found between the ABA content of the leaves and their conductance (stomatal resistance) on a leaf area basis.

Seed germination is inhibited by ABA in antagonism with gibberellin. ABA also prevents loss of seed dormancy.

Several ABA-mutant *Arabidopsis thaliana* plants have been identified and are available from the Nottingham Arabidopsis Stock Centre – both those deficient in ABA production and those with altered sensitivity to its action. Plants that are hypersensitive or insensitive to ABA show phenotypes in seed dormancy, germination, stomatal regulation, and some mutants show stunted growth and brown/yellow leaves. These mutants reflect the importance of ABA in seed germination and early embryo development.

Pyrabactin (a pyridyl containing ABA activator) is a naphthalene sulfonamide hypocotyl cell expansion inhibitor, which is an agonist of the seed ABA signaling pathway. It is the first agonist of the ABA pathway that is not structurally related to ABA.

BIOSYNTHESIS

Abscisic acid (ABA) is an isoprenoid plant hormone, which is synthesised in the plastidal 2-C-methyl-D-erythritol-4-phosphate (MEP) pathway; unlike the structurally related sesquiterpenes, which are formed from the mevalonic acid-derived precursor farnesyl diphosphate (FDP), the C_{15} backbone of ABA is formed after cleavage of C_{40} carotenoids in MEP. Zeaxanthin is the first committed ABA precursor; a series of enzyme-catalysed epoxidations and isomerisations via violaxanthin, and final cleavage of the C_{40} carotenoid by a dioxygenation reaction yields the proximal ABA precursor, xanthoxin, which is then further oxidised to ABA.

Abamine has been designed, synthesised, developed and then patented as the first specific ABA biosynthesis inhibitor, which makes it possible to regulate endogenous level of ABA.

CATABOLISM

ABA can be catabolised to phaseic acid via AtCYP707A{1,2,3,4} (a group of P450 enzymes) or inactivated by glucose conjugation (ABA-glucose ester) via the enzyme AOG. Catabolism via the CYP707As is very important for ABA homeostasis, and mutants in those genes generally accumulate higher levels of ABA than lines overexpressing ABA biosynthetic genes. In soil bacteria, an alternative catabolic pathway leading to dehydrovomifoliol via the enzyme vomifoliol dehydrogenase has been reported.

LOCATION AND TIMING OF ABA BIOSYNTHESIS

- Released during desiccation of the vegetative tissues and when roots encounter soil compaction.
- Synthesised in green fruits at the beginning of the winter period.
- Synthesised in maturing seeds, establishing dormancy.
- Mobile within the leaf and can be rapidly translocated from the roots to the leaves by the transpiration stream in the xylem.

- Produced in response to environmental stress, such as heat stress, water stress, salt stress.
- Synthesised in all plant parts, *e.g.*, roots, flowers, leaves and stems.

EFFECTS

- Antitranspirant — Induces stomatal closure, decreasing transpiration to prevent water loss.
- Inhibits fruit ripening.
- Responsible for seed dormancy by inhibiting cell growth - inhibits seed germination.
- Inhibits the synthesis of Kinetin nucleotide.
- Downregulates enzymes needed for photosynthesis.
- Acts on endodermis to prevent growth of roots when exposed to salty conditions.

IN FUNGI

Like plants, some fungal species (for example, Botrytis cinerea) have an endogenous biosynthesis pathway for ABA. In fungi, it seems to be the MVA biosynthetic pathway that is predominant (rather than the MEP pathway that is responsible for ABA biosynthesis in plants).

IN ANIMALS

ABA has also been found to be present in metazoans, from sponges up to mammals including humans. Currently, its biosynthesis and biological role in animals is poorly known. ABA has recently been shown to elicit potent anti-inflammatory and anti-diabetic effects in mouse models of diabetes/obesity, inflammatory bowel disease, atherosclerosis and influenza infection. In mammalian cells ABA targets a protein known as lanthionine synthetase C-like 2 (LANCL2), triggering an alternative mechanism of activation of peroxisome proliferator-activated receptor gamma (PPAR gamma).

Chapter 10

Non-traditional Plant Growth Regulators

MATERIALS AND METHOD

Plant Materials and Growth Regulators

Medicinally important plant species, *Mentha piperita,* were selected for the present investigation. Them stem cutting were obtained from the Horticulture Department, Faculty of Agriculture, Annamalai University Tamilnadu. The plant growth regulators Gibberellic acid were purchased from sigma chemicals Bangalore, Difenoconazole (DIZ) a triazolic group of fungicide having PGR properties is obtained from syngenta, India Ltd., Mumbai used for the study. The experimental part of this work was carried out in Botanical garden and Stress physiology lab, Department of Botany, Annamalai University, Tamilnadu. During this study the average temperature was ± 32/26°C and relative humidity (RH) varied between 6075 per cent.

Treatments and Samplings

23 nodes of stems were selected and planted separately in 40 pots. 10mg L^{-1} DIZ, 5 μm L^{-1}GA3 and 5μM L^{-1} ABA concentrations were used for the treatments and control plants, irrigated with tap water. The treatments were given on 35, 50, 65 and 80 days after planting (DAP) by soil drenching. The plants were taken randomly on 45, 60, 75 and 90 DAP and separated into root, stem and leaves and used for determining growth and biochemical contents.

GROWTH PARAMETERS

Height of the Plant and Root Length

The plant height was measured from the soil level to the tip of the shoot and expressed in cm. The plant root length was measured from nodal initiation of the shoot to the tip of longest root and expressed in cm.

Number of Leaves and Total Leaf Area

The total leaf area of the plants was measured using L1COR Photo Electric Area Meter and expressed in cm 2 per plant.

Determinations of Fresh and Dry Weight

After washing the plants in the tap water, fresh weight was determined by using an electronic balance and the values were expressed in grams. After taking fresh weight, the plants were dried at 60 °C in hot air oven for 24 hours. After drying, the weight was measured and the values were expressed in grams.

Photosynthetic Pigment Content

Chlorophyll a and b, Carotenoids were extracted from the leaves and estimated by the method of Arnon (1949). Absorbance was measured at 645, 663 and 480 nm with a spectrophotometer against 80 percent acetone as blank. Chlorophyll content was calculated using the formula of Arnon.

RESULTS AND DISCUSSION

Effect of Plant Growth Regulators on Plant Height

The total height of the plant increased with the age in the control, ABA and GA3 treated *Mentha* plants, but it decreased under DIZ treatments. The increase was higher in gibberellic acid treated when compared to ABA. Highest plant height was noted in 90 DAP under GA3 treatments and it was nearly 136.66 per cent over control. The lower plant height was observed in DIZ treated plants on 90 DAP and it was 94.35 per cent over control. The plant height reduced under treatments with DIZ and ABA in *Mentha* plants. The gibberellic acid treated plants increased the plant height. Triazole treatments reduced stem elongation and plant height in *Plectranthus forskholii,* in *Cassava,* and *Catharanthus roseus.* The growth retarding effect of triazole is caused by the inhibition of GA3 as observed in *Cucurbita maxima.* ABA induced growth inhibition was resulted from signal transduction at the single cell level and thereby induces closure of stomata. The growth retarding effect of triazole is caused by the inhibition of GA3. GA3 exerts profound effects on fundamental process of plant growth and development. GA3 is widely regarded as a growth promoting compound that positively regulates processes such as seed germination, stem elongation and leaf expansion..

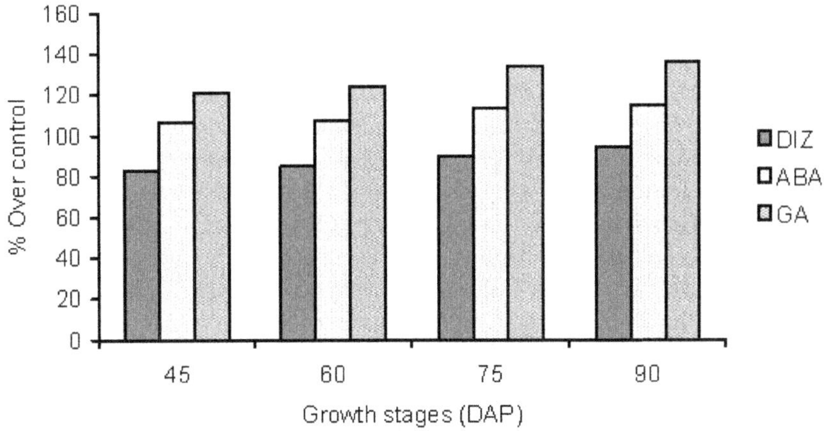

Fig. Effect of difenoconazole, abscisic acid and gibberellic acid, on Plant height of Mentha piperita on different growth stages

Effect of Plant Growth Regulators on Root Length

The total root length of the *Mentha* plant increased with the age in control and DIZ treated plants, but a decrease was observed in ABA and GA3 treatments. The increase was higher in DIZ treated on 90 DAP and it was 129.41 per cent over control. The root length was increased with DIZ treated plants to a higher extent on the other hand, GA3 inhibited root growth in *Mentha piperita*. Triadimefon treatment increased the root growth in mungbean. An increase in root length was reported in paclobutrazol and triadimefon treated in *C. roseus*. Paclobutrazol increased the root length and enhanced the lateral roots in tomato plants.

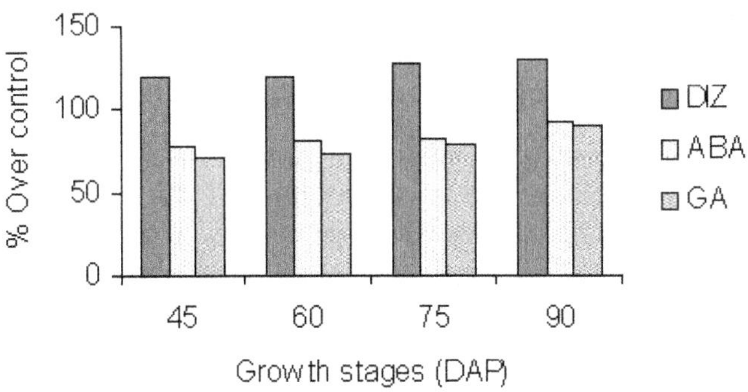

Fig. Effect of difernoconazole, abscisic acid and gibberellic acid, on Root length of Mentha piperita on different growth stages.

Effect of Plant Growth Regulators on Total Leaf Area

The total leaf area of the plant decreased with the age in all the treatments. The more decrease was prominent in ABA treated plants and it was 69.00 per cent over control on 90 DAP when compared to DIZ and GA3 treatments in *Mentha piperita* plants. DIZ, ABA and GA3 treatments were reduced in total leaf area when compared to control in *Mentha* plants. Paclobutrazol reduced the leaf area in tomato and barley. The leaf production is reduced in *Catharanthus* plants under PBZ treatment. Uniconazole reduced the leaf number in *Pyrecantha* species. Application of ABA at room temperature results in reduced leaf production in many plants. The leaf production is reduced in strawberry and soybean by triazole treatment. Uniconazole reduce the leaf number in *Pyrecantha* species GA3 caused a significant decrease in leaf area in all stages of growth *Mentha* plant. The application of gibberellic acid has the potential to control growth and flowering by reduced leaf area and induce earliness in strawberry. The response of strawberry to exogenous GA3 is similar to that caused by certain natural environmental factors.

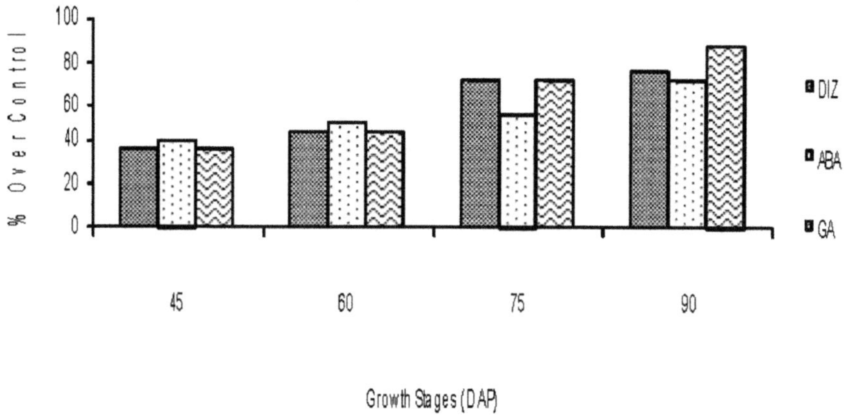

Fig. Effect of difenoconazole, abscisic acid and gibberellic acid, on Total Leaf Area of Mentha piperita on different growth stages.

Effect of Plant Growth Regulators on Fresh and Dry weight

The whole plant fresh weight of the plant increased in age on all of the treatments. But there is increases to a larger extent on 90 DAP in DIZ treatments and it was 130.12 per cent over control. The least increase was observed on 45 DAP in GA3 treatments and it was 102.79 per cent over control. The whole plant dry weight increased with DIZ, ABA and GA3 treatments when compared with control. The whole plant dry weights increased in a larger extent on 90 DAP in DIZ treatment and it was 120.00. When compared to ABA and GA3 treatments. Among the treatment ABA and GA3 slightly increased and it was 118.27 and 112.11 respectively on 90 DAP. DIZ, ABA and GA3 treatments in *Mentha piperita* increased the whole plant fresh weight to a large extent. Similar results were reported in PBZ treated

Catharanthus plants under salt stress. Triazole compounds inhibited gibberellin biosynthesis, cytokinin and abscisic acid stimulated tuberisation and reduced stolon length in potato by counteracting gibberellin action content induced by these triazoles might be the cause for increase root growth.

GA3 application was reported to increase weight of aerial parts in *Viola*. Exogenously applied, gibberellin promoted stolon elongation and inhibited tuber formation and increased fresh weight in potato DIZ, ABA and GA3 treatments increased the dry weight considerably in *Mentha* plants when compared to control plants. ABA plays a critical role in regulating plant water status through guard cells and growth as well as by induction of genes that encodes enzymes and other protein involved in cellular dehydration tolerance, which might be the reason for increased dry weight. Triadimefon and paclobutrazol altered the shoot dry weight in peanut and tomato. It was apparent that treating plants with GA3 increased height of the plant over the control, which may be attributed to the growth promotion effect of GA3 in stimulating and accelerating cell division, increasing cell elongation and enlargement which in turn increased the dry weight of the plants. GA3 application was reported to increase fresh and dry weights in *Viola*.

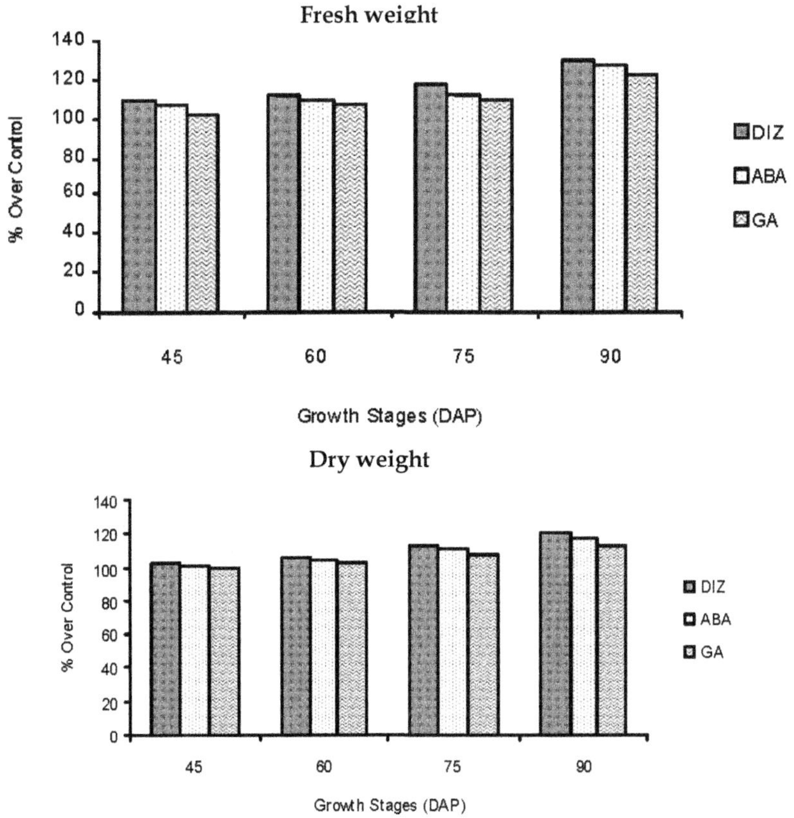

Fig. Effect of difenoconazole, abscisic acid and gibberellic acid, on fresh and dry weight of Mentha piperita on different growth stages.

Effect of Plant Growth Regulators on Chlorophyll Content

The total chlorophyll contents of the leaves increase with the age in control and treated *Mentha* leaves. The maximum increase was found on 90 DAP in DIZ treatments and it was 132.91 per cent over control. Among the treatments ABA and GA3 slightly increased and it was 123.00 and 122.11 respectively on 90 DAP. Treatments with DIZ, ABA and GA3 significantly increased the total chlorophyll contents in *Mentha* plants. Similar results were observed in Paclobutrazol treated barley carrot and tomato. Paclobutrazol treated leaves were dark green due to high chlorophyll content in potato. Sebastian *et. al.*, 2002 reported enhanced chlorophyll synthesis in *Dianthus caryophyllus* treated with Palobutrazol. Gibberellic acid increased the vegetative growth and pigment concentration in maise. In *Lotus tenuis* low photosynthetic photon flux density induced an ortotropic growth of stems with greater supply of GA1 and GA3. Foliar application of GA3 improved the chlorophyll levels in salinity stressed maise plants.

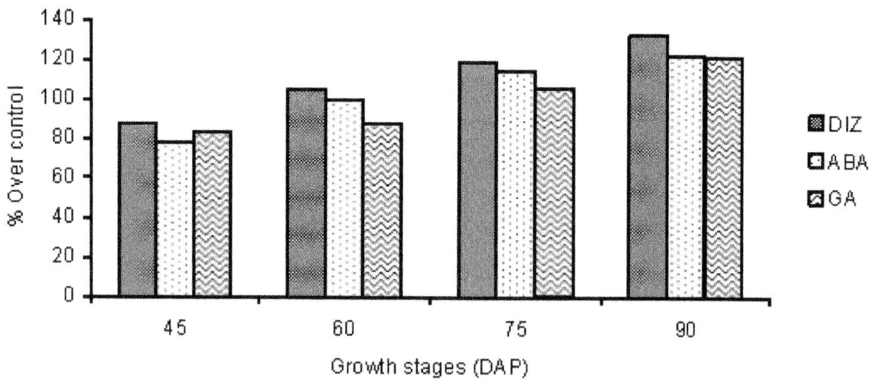

Fig. Effect of difenoconazole, abscisic acid and gibberellic acid on chlorophyll content of Mentha piperita.

Effect of Plant Growth Regulators on Carotenoid Content

The carotenoid content increased with the age in control and treated plants on all sampling days. The higher carotenoid contents were observed under DIZ treatments on 90 DAP and it was 133.48 per cent over control. The other treatments increased the carotenoid contents on 90 DAP and it was 125.00 and 125.89 respectively on ABA and GA3 The carotenoid content of the *Mentha piperita* leaves increased with age in control and treated plants. Treatment increased the carotenoid content in *catharanthus* plants. Triadimefon treatment increased the carotenoid content to a higher level in cucumber. An increase in carotenoid content was reported in maise plants. Plant growth of wheat decreased with increasing salinity levels, but was increased by seed treatment with GA3, which accompanied increased photosynthetic pigment contents.

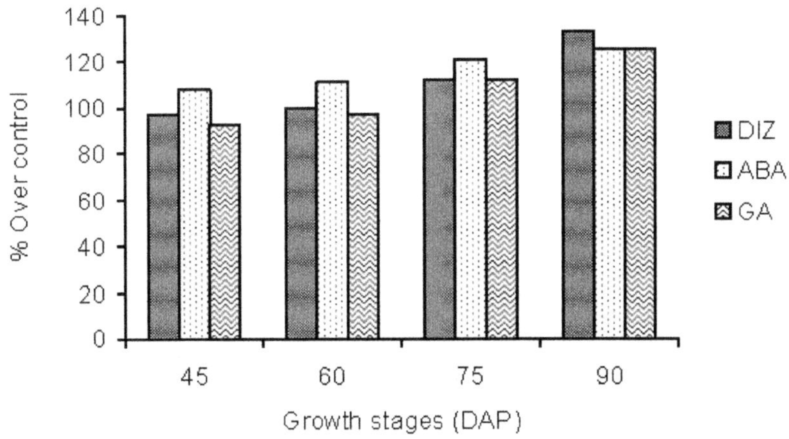

Fig. Effect of difenoconazole, abscisic acid and gibberellic acid on carotenoids content of Mentha piperita.

CONCLUSION

The experiments showed that the traditional and non-traditional (DIZ) plant growth regulators significantly altered the growth and photosynthetic pigment content in *Mentha* plants. The increasing in quantity of growth and photosynthetic pigments were in the order DIZ > ABA > GA3. Our results have good significance, as these increase the growth and photosynthetic pigments of this kind of traditional medicinal plant.

Chapter 11

Cellular Metabolism and Growth

It is generally recognised that the energy necessary for cellular activities, whether it be used to maintain the steady state of adult life, the "dynamic equilibrium" of Schoenheimer, or the state of growth and synthesis which predominates in dividing and growing cells, is provided by enzymatic reactions largely of the oxidation-reduction type. This has been amply demonstrated by the increased O_2 utilisation during growth and by the necessity of aerobic processes for the synthesis of carbohydrates, fats, and proteins which together make the bulk of the cells. It will be therefore necessary before discussing the role of enzyme reactions in cellular growth to review some fundamental aspects of the properties of the oxidation-reduction enzymes.

GENERAL PROPERTIES OF ENZYMES

Enzymatic oxidation-reduction reactions are distinguished from chemical reactions in general by the facts that they are reversible, that electron exchange occurs in steps, and that they have a great tendency to form coupled reactions.

The reversibility of enzyme reactions can be accepted as a general property, since more and more of those reactions hitherto considered irreversible are being shown to be reversible. Witness the enzymatic synthesis of phosphopyruvic acid from pyruvic acid, a reversal of the dephosphorylation of phosphopyruvic acid, which had been reported as an irreversible process. In many instances reversibility of oxidation-reduction reactions is made easy by the entrance of P into the reactions and generation of energy-rich phosphate bonds, as was shown by Warburg and Christian in the compulsory coupling between the oxidation of phosphoglyceraldehyde and the phosphorylation of adenylic acid, a finding extensively generalised to other reactions by Lipmann.

The first step in an enzymatic reaction is the formation of the enzymesubstrate complex. Evidence in favour of such a concept, introduced by Michaelis and Menten, is abundant. Knowledge of the nature of this complex formation awaits clarification of the structure of proteins, since it is probably the protein moiety which combines with the substrate. The substrate to be oxidised must fit quite closely to the side chains of the protein molecule where combination occurs. No lock, however, remains immune to the prying of false keys; and substances with chemical structure close to that of the substrate will compete for union with the protein, thus affecting the activity of the enzyme. This is strikingly demonstrated by the inhibition of acetate oxidation by fluoroacetate discovered in animal tissues and in yeast. In yeast, acetate oxidation, completely inhibited by monofluoroacetate, was not affected at all by the other halogen acids, bromo-, iodo-, chloro-acetic acids, as well as by trifluoroacetic acid.

This inhibition is due to competition of fluoroacetate with acetate in the protein of the enzyme, for inhibition occurs in the first step of acetate metabolism, namely citric acid formation. Of all these acids it is monofluoroacetic acid which approaches most closely in size acetic acid, as the interatomic distance between C and F (1.41 A) is only 0.32 A more than the interatomic distance between C and H.

An increase in the interatomic distance of 0.35 A as in C and Cl (1.76 A) was enough to avoid penetration of the halogen acid into the space reserved for acetic acid in the protein moiety. The protein-substrate complex formation resembles thus strikingly the antibody-antigen complex formation in the specificity.

Some proteins require the presence of –SH groups for enzymatic activity. Furthermore, in some cases only a portion of the –SH groups present in the protein are necessary for enzyme activity, as demonstrated with urease by Hellerman and coworkers.

Yeast cells, 8.5 mgs.; Mg acetate, 0.077 *M*; fluoroacetate, 0.005 *M*. Duration of experiments, 4 hours. Blank values for oxygen uptake and citric acid have been subtracted.

A survey made by Barton and Singer showed that a large number of enzymes which catalyse the metabolism of carbohydrates, fats and protein derivatives need the electronegative–SH groups for activity. Doubtless there are still other enzymes, not yet studied, which belong to this class. An indication in favour of this assumption is given by the inhibition of mitosis produced by small doses of x-rays and by nitrogen mustards. Up to the time of Dale's experiments little attention was paid to enzyme inhibition by x-rays. In 1940 Dale demonstrated that enzyme inhibition could be produced by small doses of x-rays provided protein concentration was kept low.

Barron, Dickman, and Singer have shown that x-rays, α, and β particles and γ radiations inhibit sulfhydryl enzymes at small doses by oxidation of the–SH groups of the protein. Inhibitions produced by x-rays and α particles could be released on addition of glutathione. This inhibition is undoubtedly produced by the OH and OH_2 radicals and H_2O_2 formed by the ionisation of water during irradiation.

In fact, protection of the –SH groups by mercaptide formation with Cl-Hg benzoic acid abolished inhibition of urease by γ radiation. Destruction of H_2O_2

by catalase, diminished the degree of inhibition of phospho-glvceraldehyde dehydrogenase by a particles, and suppressed the inhibition by β particles. The combination of nitrogen mustards with – SH groups is known.

However, as this combination is irreversible, it is difficult to establish definitely whether enzyme inhibition is due to this reaction or to combination with some other group of the protein. Very little is known about the role of these – SH groups.

Protection of the – SH groups of the protein moiety of succinoxidase by succinate and by malonate would suggest that the substrate is anchored in close vicinity to the – SH groups.

The proximity of the carboxyl electronegative groups would decrease the activity of the – SH groups and thus render them less susceptible to the action of sulfhydryl reagents. Some other proteins require the presence of a metal (Mg, Mn, Zn, Co) for enzyme activity; the metal does not participate in the reaction.

Their role is still more obscure than that of the – SH groups. It is known, however, that – SH groups and metals contribute to increase the stability of the native protein.

The combination of the substrate with the protein brings forth the "activation" of the substrate. It is not yet clear what is meant by activation of the substrate. It is plausible, however, to postulate with Michaelis that activation consists in the formation of a radical, the first step in the process of oxidation of organic substances with a loss of two electrons.

This is implicit in Michaelis' theory of compulsory univalent oxidation. In condensation reactions where two substrates take part in the reaction, as in the formation of acetylcholine, it seems that only one of the substrates combines with the protein and becomes activated.

Evidence for this assumption is provided by inhibitions with competitive inhibitors. For example, methyl bis (dichloroethyl) amine HCI inhibits choline oxidase because of the similarity of its first transformation product, ethylene imonium, to choline; it also inhibits the synthesis of acetylcholine and its hydrolysis. It has no effect on the hydrolysis of tributyrin.

Fluoroacetate inhibits the oxidation of acetate while it has no effect on the synthesis of acetylcholine and acetylations in general. It must be concluded that in acetylations the protein combines with the acetylated compound and not with the acetyl group.

A number of proteins, components of enzyme systems, have been isolated and prepared in crystalline form. Furthermore our knowledge of the physical and chemical properties of proteins has considerably increased during the last years. The time has come to use these techniques for a serious study of the protein moiety of the enzyme.

It is necessary to know the architecture of the protein moiety of the enzyme, the spacing of the side groups to which the substrates attach themselves, and the nature of the activation of the substrate.

THE OXIDATION-REDUCTION SYSTEMS

In 1930 Barron and Hoffmanfound that the catalytic power of reversible oxidation-reduction dyes on cellular respiration was conditioned by two factors: the oxidation-reduction potential of the dye, and the permeability of the cell membrane.

The speed of this catalysis was correlated to :

- The speed at which the dye was reduced by the cell,
- The speed at which the leucodye was oxidised by atmospheric oxygen.

Such a correlation between free energy (as determined by the oxidation-reduction potential of the dye) and rate of oxidation (as determined by the increase in the O_2, uptake of the cell) was naturally received with caution. At the suggestion of W.

Mansfield Clark and with his generous co-operation these experiments were repeated in homogenous systems by studying the rate of oxidation of reduced dyes by atmospheric oxygen. The same relation between free energies and rates of reaction was found.

The theory of absolute reaction rates, developed mainly in Princeton University by Eyring and his coworkers, which focuses its attention on the thermodynamic probability of molecules entering the transition state, has shown that the speed of reaction depends not on the heat of activation alone but also on the entropy of activation :

Where 3per centF is the free energy of formation of the activated state, and k and h the Boltzman and Plank constants, respectively. This equation means that the Fe^{++} porphyrin forms the more stable compound; if the Fe^{+++} form associates more strongly than the Fe^{++} form the trend of the potential will be in the negative direction.

This is of great significance for an understanding of the varied catalytic effects of the different oxidation-reduction systems which form part of the enzyme systems.

In the same manner as the Fe-porphyrins, and its nitrogenous complexes give a series of oxidation-reduction systems, the potentials of the pyridine nucleotides and flavins in combination with different proteins may change, at constant *p*H, according to the value of the dissociation constants of the protein complex.

The breakdown of foodstuffs in stepwise manner is facilitated by the interposition of reversible oxidation-reduction enzymatic systems, and the stepwise transfer of electrons is facilitated by the property of the three main electron-transfer systems (pyridine nucleotides, flavins, Feporphyrins) of forming reversible complex compounds with proteins with a wide variety of oxidation-reduction potentials.

It must be recalled, however, that electron transfer between two reversible enzymatic systems does not occur but through the mediation of electroactive systems as shown by the experiments of Borsook.

THE PYRUVATE OXIDATIVE PATHWAY

Of all the foodstuffs, carbohydrate provides most free useful energy in its breakdown. It is mainly carbohydrate-provided energy which is used during the process of embryonic growth and development. In general, carbohydrate breakdown occurs in two phases : the anaerobic phase which ends in pyruvic acid formation, and the oxidative phase which involves the utilisation of pyruvate via the oxidative pathway. In 1932 Barron and Miller found that pyruvate was oxidised to acetate and CO_2 by gonococci.

Although formation of acetate from pyruvate by animal tissues has also been reported, there is a tendency to ignore this oxidative pathway of pyruvate metabolism and to give as the main oxidative pathway that of pyruvate carboxylation to oxaloacetate and condensation between pyruvate and oxaloacetate, as indicated by Krebs in his "tricarboxylic acid cycle". The powerful inhibiting effect of fluoroacetate on the oxidation of acetate was used by Bartlett, Ahrens, and Barron to test the oxidative pathway of pyruvate metabolism. To discover this oxidative pathway fluoroacetate is indeed the ideal agent, since if acetate is formed during the metabolism of pyruvate, and it normally disappears in condensation reactions (acetylations, citrate formation, fatty acid synthesis), it will not be detected unless some of these reactions are inhibited.

Experiments where the metabolism of pyruvate was studied in the presence of fluoroacetate in kidney, liver, and muscle showed accumulation of acetate and inhibition of pyruvate utilisation due to accumulation of its oxidation product, acetate.

The other pathways of pyruvate metabolism (amination to alanine, reduction to lactate, and dismutation to lactate and acetate) were not affected. Furthermore none of the oxidative steps of the tricarboxylic acid cycle—oxidation of isocitric, α-ketoglutaric, succinic, and malic acid—were affected by fluoroacetate.

The experiments with fatty acids showed an inhibition in the oxidation and utilisation of these substances with an inhibition of acetoacetate formation in liver slices, but with acetate as substrate there was accumulation of acetoacetate. This inhibition of acetoacetate formation in the presence of fluoroacetate supports the view, now generally accepted, that the oxidation of fatty acids occurs by steps until the formation of acetate. In the metabolism of acetate there are then two pathways :

Oxidation inhibited by fluoroacetate, and condensation to acetoacetate not affected by it. Since acetate accumulates in the presence of fluoroacetate, it is plausible to postulate that acetate is the end product in the oxidation of fats and carbohydrates. Furthermore the high toxicity of fluoroacetate is indication of the essential role of acetate metabolism in higher animals and in some plants. The oxidative pathway, however, is not the only avenue of utilisation of pyruvate. Pyruvate, the most highly reactive molecule among the intermediates found during the breakdown of foodstuffs, is in fact utilised in a variety of ways in the different cells, and even in the same cell the utilisation of pyruvic acid will depend on the oxygen tension, thiamin concentration, glutathione concentration, etc. The multiple pathway of pyruvate metabolism postulated in this labouratory for several years may thus be summarised as follows, all these reactions being reversible. How much of the oxidative removal is performed via acetate and how much *via* carboxylation to oxaloacetate has not been determined in all tissues. In the experiments of Bartlett and Barron with kidney slices, at least 45 per cent of the pyruvate utilised could be accounted for as due to direct oxidation.

The interposition of enzymatic reversible oxidation-reduction systems in the various steps during the burning of foodstuffs, already shown in the scheme of pyruvate metabolism, allows the production of a series of coupled reactions whose speed and direction will be determined not only by the value of the equilibrium constants but also by the environment (temperature, O_2 tension, CO_2 tension, electrolytes, hormones, etc.).

Ochoa has thus shown a number of coupled reactions which in *in vitro* experiments end in carboxylations :

- α-ketoglutarate + CO_2 + glucose 6-phosphate $\rightleftharpoons$ 1-isocitrate + phosphogluconate;
- Malate + TPN + pyruvate + CO_2 + $TPNH_2$

These reactions do not occur in the presence of pure isolated enzyme systems. Carboxylation of α-ketoglutarate and pyruvate, difficult of attainment because of the values of the equilibrium constants, was rendered possible by coupling carboxylation with an oxidation-reduction reaction. Whether such reactions do occur in living cells is problematic.

OXIDATIONS AND PHOSPHORYLATIONS

Whenever an important discovery is made, there spring forth applications, similarities, extensions. Years have to pass by before harvesting from the melee the useful crop. The coupling between phosphorylation and oxidation observed by Meyerhof in his studies on the mechanism of glycolysis was explained in 1939 by the striking demonstrations in Warburg's labouratory which culminated in 1942 in the isolation and crystallisation of the specific transphosphorylase. Warburg and his coworkers showed that the aldehyde group in phosphoglyceraldehyde is not oxidised as aldehyde hydrate but as aldehyde phosphate. By a non-enzymatic reaction glyceraldehyde 3-phosphate takes up another phosphate group to give 1,3-diphosphoglyceraldehyde, which is oxidised by DPN to 1,3-diphosphoglyceric acid :

The phosphate transfer enzyme, a Mg-protein, transfers the unstable phosphate group into adenosinediphosphate to give ATP. This reaction has been taken as a model of phosphate transfer from aceylphosphates :

Lipmann has applied and extended these concepts to the oxidation of pyruvate and α-keto acids in general, and indeed has found that certain bacteria produce acetylphosphate on oxidation of pyruvic acid, and that animal tissues hydrolyse acetylphosphate rather rapidly. These important contributions of Lipmann were promptly accepted, but little has been accomplished towards further clarification of the mechanism of aerobic phosphorylations.

The fact that they exist was demonstrated by Belitzer and Tsibakowa, Ochoa and Colowick *et. al.* From Ochoa's work it is clear that on the oxidation of pyruvate there are about 15 phosphorylations per mole of pyruvate completely oxidised, *i.e.* 3 per atom of oxygen. Of these 15 phosphorylations, Ochoa has reported that during the oxidation of α-ketoglutaric acid to succinic acid *in the presence of glucose and adenylic acid,* 3 phosphate molecules are simultaneously esterified to the sugar forming hexose diphosphate. The transfer of 2 electrons to oxygen was thus accompanied by the release of energy in three successive steps in the form of high-energy phosphate bonds. How they are formed remains unknown. The postulated succinylphosphate formed as the first product of oxidation has not been demonstrated.

Ball's suggestion that the step by step transfer of electrons from oxidisable substrate to pyridine nucleotide, from this to flavin, from flavoprotein to cytochrome, and from cytochrome to oxygen might release the energy (0.25 V.) necessary to form a high-energy phosphate bond seems reasonable, but again is still in the realm of speculation. To complicate the problem, Stumpf *et. al.* maintain that α-ketoglutarate is oxidised to succinate by preparations from pigeon breast muscle without the necessity of inorganic phosphate, adenosinetriphosphate, and magnesium, all components considered by Ochoa essential for the oxidation process.

The phosphorylations in the first step of the cycle—pyruvate + $\frac{1}{2}O_2 \rightarrow$ acetate + CO_2 —have not been established in animal tissues in spite of the repeated efforts of Lipmann. Here again, Stumpf *et. al.* maintain that this oxidation occurs in the absence of phosphate and with no formation of acetyl phosphate. The second oxidation step, isocitrate + TPN $\rightleftharpoons$ oxalosuccinate + TPN H_2, is postulated to give no high-energy phosphate bonds; however in the experiments of Ochoa there was but one step in the transfer of electrons. If Ball's suggestion is verified, here as well as in the oxidation of pyruvate, the step by step release of energy might be utilised for the formation of highenergy phosphate bonds.

In the fourth step, succinate + $\frac{1}{2}O_2 \rightleftharpoons$ fumarate + H_2O, there has been found evidence of one phospho rylation. This is probably due to the high potential of the system (0 V. at *p*H 7) and the fewer steps in electron transport to molecular oxygen (0.8 V.). The final step, malic + $\frac{1}{2}O_2 \rightleftharpoons$ oxaloacetic + H_2O, was found by Kalckar to produce at least two phosphorylations. It seems as if the insistence on finding analogs to Warburg's reaction, which after all is a fermentation reaction, has stopped progress and hindered the search for other channels of energy transfer.

Perhaps Ball's suggestion ought to be studied experimentally. Aerobic phosphorylations may be produced by the release of energy from the different oxidation-reduction catalysts to adenylic acid, for it is accepted that adenylic acid enhances completeness of oxidation. Lack of phosphorylation in isolated enzyme systems where substrate is oxidised by only the first oxidation-reduction system and failure of phosphorylation in some tissue extracts when cytochrome C is not present favour this opinion. The ATP thus produced would represent the free useful energy ready to be used.

Energy transfer from electrons and its utilisation by the cell through adenosinetriphosphate occurs not only on the oxidation of carbohydrates but also on that of fats, as the studies of Lehninger and Muñoz and Leloir have demonstrated.

THE OXIDATION AND THE SYNTHESIS OF FATS

Our knowledge of the mechanism of oxidation and synthesis of fatty acids has advanced considerably in the last years through the efforts of a number of investigators, although no single enzyme has yet been isolated. The classical concept of Knoop and Dakin with the modifications suggested by MacKay —oxidation at their βC atoms with subsequent removal of two C units which may condense to acetoacetate or to citrate—seems now generally accepted.

As in the complete oxidation of carbohydrate, the series of oxidations which take place during the breakdown of fatty acids require the presence of ATP or of C_4 dicarboxylic acids, which in their oxidation provide energy-rich phosphate bonds. Whether there is a transphosphorylation process with the formation of unstable acyl phosphates is not known, in spite of some efforts to prove that such substances are formed. The acyl phosphates (monopalmityl phosphoric acid and monooctanoylphosphoric acid), although hydrolysed much more slowly than acetyl phosphate, were physiologically inactive. When octanoic acid was oxidised by the liver in the absence of C_4 dicarboxylic acids, acetoacetate was the end product. In the presence of C_4 dicarboxylic acid, the C_2 fragments gave citric acid and its oxidation products, α-ketoglutaric acid, and succinic acid.

The interconversion of acetic acid to acetoacetic acid has been demonstrated by direct interconversion as well as by the synthesis of citric acid from acetoacetate presumably by its previous breakdown to "active" acetate. This acetic acid, the end product of carbohydrate oxidation and of fat oxidation, is the connecting link of carbohydrate and fat metabolism. The further oxidation of acetic acid, we have already shown, occurs through the tricarboxylic acid cycle. The synthesis of fat starts with the condensation of acetate to acetoacetate. Indeed, Rittenberg and Bloch demonstrated with isotope experiments that acetic acid is utilised for the synthesis of fats.

THE OXIDATION AND SYNTHESIS OF PROTEINS

Proteins in the oxidation of their fundamental units, the amino acids, contribute greatly to the formation of the carboxylic acids which by their reversible coupled oxidation-reductions form the common pathway for the final steps of complete oxidation of foodstuffs, whether it be carbohydrate, fat or protein.

The rapid and continuous breakdown and synthesis of proteins so well demonstrated by Whipple and his coworkers and by Schoenheimer and his coworkers, as well as the great specificity of proteinases and peptidases, brought forth the suggestion that protein synthesis could be performed by coupled reactions between the hydrolytic enzymes.

Bergmann and Fruton, in fact, presented models of peptide bond synthesis by selecting suitable amino acid derivatives and by producing the conditions necessary for reversal of hydrolysis. Whether this mechanism actually takes place in living cells is not known. The synthesis of proteins, like that of glycogen, must be due to the co-operation of a large number of enzymes.

In cytochrome containing living cells it does not occur in the absence of oxygen, and it is plausible to postulate that proteins are, like carbohydrates and fats, synthesised after their breakdown to amino acids and that peptide bond formation requires the utilisation of energy produced during the oxidation of carbohydrates and fats. Support for this assumption is given by the experiments of Cohen and McGilvery on the formation of p-amino-hyppuric acid from p-aminobenzoic acid and glycine by rat tissue :

$$NH_2 COOH + NH_2 CH COOH \rightarrow NH_2 CO\text{-}NHCH_2COOH + H_2$$

This may be taken as a model of peptide bond synthesis. This reaction could not proceed anaerobically nor in the presence of oxidative inhibitors (HCN, NaN_3). Inhibition of this reaction by arsenite, iodoacetate and malonate favours the view that the energy for the synthesis is provided by oxidations in the Krebs cycle where there are a number of —SH enzymes. Whether this energy is furnished by the high-energy phosphate bonds produced in these oxidations is not yet established.

It must be emphasised however that proteins possess a great specificity and complexity. Besides enzymes forming peptide bonds there must be others which add these bonds together to form the complicated and varied structures of the different proteins. Casperson demonstrated that ribonucleic acid is present in abundance in growing cells and gave some evidence in favour of the view that it is essential for protein synthesis.

The mechanism of action of nucleic acids is not yet known. They must act in the last steps when the architecture of the molecule acquires its specific characteristics. Spiegelman and Kamen believe that they are "the specific energy donators which make possible reactions leading to protein and enzyme synthesis." They found that new protein formation in yeast cells was accompanied by a marked transfer of phosphate from the nucleoprotein. However the phosphate groups in nucleoproteins are phosphate esters which on hydrolysis provide small amount of energy. Northrop has dealt extensively on this subject of the last stages of protein synthesis.

From this brief review of some enzyme reactions it can be seen that what distinguishes biochemical reactions from chemical reactions in general is the means provided for the rapid attainment of equilibrium, so strikingly demonstrated by the enzymatic reversible oxidation-reduction systems. Furthermore, although on a thermodynamic basis a very large number of reactions are theoretically possible in a given biologic system, the presence of enzymes will result in the selective catalysis of only a limited number of these reactions.

But the isolation, purification, and study of the catalytic action of individual enzymes is only the beginning of the study of the biochemical reactions which occur in the cell, whether it be in the state of "dynamic equilibrium" or in the state of growth.

Living matter is the result of the co-ordinated interplay of substance and structure, as Pauling has remarked, and the advance of biology requires equal devotion to the work of isolation and identification of active chemical substances, to the study of the macromolecular stromatic structures present in the living cell, and to the investigation of the multiple regulating mechanisms which contribute to the cellular dynamic equilibrium.

The rate of reaction of isolated enzyme systems is extremely high when compared with the rate of reaction in the living cell. Enzyme reactions are controlled in the living cell by a variety of regulating mechanisms. To the usual physico-chemical factors which govern the rate of reactions in solutions there must be added the effect of macromolecular structures which constitute the framework of the cell, such as the genes, and which can orient or fix the position of the intracellular enzymes by means of loose unions.

As the cell increases in efficiency and becomes more differentiated, the activity of the enzymes becomes more and more controlled by specific regulating mechanisms, which in multi-cellular organisms freely circulate through the intercellular fluid. In vertebrates these regulatory mechanisms increase in number and specificity, and they control and orient the rate of enzymatic reactions by interaction among each other.

These regulatory mechanisms (hormones, nucleoproteins, genes, electrolytes, certain oxidation-reduction systems) do not act by forming part of enzyme components; they act by controlling the rate of enzyme reactions or by changing the orientation of these reactions. They act as the traffic policemen in the crowded streets of a city. Thus the steady state of the adult cell is maintained, and the synthetic processes required in growing cells are brought forth.

An example of the presence of regulating mechanisms is seen in the effect of dilution on the respiration of sea urchin sperm. The respiration of sea urchin sperm is maintained at a low level under the regulating influence of some substances possessing —SH groups; on dilution or on addition of iodoacetate or malonate (usual respiratory inhibitors), the regulatory system becomes inactivated and the respiration is increased.

A striking demonstration of the role of hormones as regulators of enzyme reactions, a theory postulated in 1943, is found in the experiments of Price Colowick, and Cori of inhibition of muscle hexokinase by a substance (hormone?) contained in the anterior pituitary gland and the release of this inhibition by insulin. In the living cell enzymes may be present which are so located as to be inactive in the intact cell.

Physiological alterations, rupture of the cell, destruction of the normal architecture, may bring these enzymes into action. An example of physiological alteration is given by the experiments of Korr who found that the respiration of the salivary glands, which is insensitive to azide when the gland is in the resting state, becomes azide sensitive when the gland is stimulated.

An example of enzyme liberation on damage to the cells is given by the marked and rapid inhibition of the respiration of lung tissues after the tissue has been ground. The QO_2 value goes from 7.5 down to 1.0.

This inhibition is due to destruction of the pyridine nucleotides by the nucleosidase present in lung tissue in high concentration but kept inactive in the intact cell. Evans has dealt with the properties of the anterior hypophyseal growth hormone isolated in his labouratory which plays an important role in the synthesis of proteins, a role which in our opinion is that of a regulating agent.

CELL DIVISION

From the extensive investigations on the mechanism of cell division it must be concluded that there is as yet no full understanding of the physical and chemical mechanisms which bring forth this process. There is agreement however that we are confronted with a series of events which together culminate in the phenomena of membrane formation and cell division.

The initial process, formation of fertilization membrane and entrance of the spermatozoon into the ovum with no nuclear changes, may be called the *anaerobic phase,* because it may take place under strictly anaerobic conditions. If energy is required for this process it must be obtained from glycolysis.

The nuclear changes which follow membrane formation and end in cell division constitute the aerobic phase and must require the energy provided by oxidative reactions. The increased utilisation of glycogen accompanied by the increased utilisation of pyruvate may provide the energy necessary for the process of cell division.

Furthermore, the introduction during the cell division process of three of four oxidative steps (the series of oxidations carried on by the cytochrome system) in the series of reactions from substrate to molecular oxygen will increase the efficiency of the reactions so that for the same amount of O_2 utilised there will be a more efficient utilisation of energy.

Whether carbohydrate is metabolised entirely via glycolysis, or is oxidised through the pathway of hexosemonophosphate oxidation as suggested by Örström and Lindberg is not known. Neither is the nature of the acid formed on cell division known. Some of it may be acetic acid formed on oxidation of pyruvic acid.

Whether acetate is utilised or not is not known. If it is, the pathway of its oxidation must be different from that of most animal tissues because neither succinate nor α-keto-glutarate is oxidised by sea urchin eggs. Fluoroacetate had no effect on the O_2 uptake.

CELL GROWTH

Our knowledge of the enzymatic reactions which take part in cellular growth is meagre. In general cellular growth, new protein formation, takes place at the expense of aerobic oxidations except in those cells. deprived of cytochromes.

Voegtlin, Hammett, and Rapkine have for a long time championed the importance of sulfhydryl groups in cell growth, and evidence in support of this opinion is good. They have been found increased in concentration in growing cells and growth was inhibited on addition of substances which combine with –SH groups. It has already been pointed out that there are a large number of enzymes which require the presence of –SH groups in their protein moiety for activation. Destruction of these groups by oxidation (x-rays, O_2, Cu), alkylations (CH_2ICOOH, $CH_2ICOONH_2$, mustard, nitrogen mustard), or mercaptide formation (Hg, Pb, Cd, Zn, Bi, As, V, Sb) will produce enzyme inhibitions, an inhibition which affects such high-energy yielding reactions as pyruvate oxidase, α-ketoglutarate oxidase, succinoxidase, triosephosphate oxidase, adenosinetriphosphatase, malate oxidase. Sluggish oxidation-reduction systems, mostly in their reduced state (glutathione, ascorbic acid), have also been found abundant in cells during growth and development. These systems must have for function that of keeping the–SH groups of the enzymes in the reduced, physiologically active state.

It must be recalled that ionising radiations, halogenated alkylamines, and mustard gas inhibit mitosis. Furthermore Hevesy has found that the synthesis of nucleoproteins (which take part in mitosis) is inhibited by x-rays. Under certain conditions enzymes like adenosinetriphosphatase from aged myosin are so susceptible to ionising radiations that amounts as small as 1r produce inhibition. The inhibiting effect of x-rays may well indicate that some enzyme necessary for the synthesis of nucleoproteins requires –SH groups for its activity, groups which are oxidised by the OH and O_2 H radicals, and H_2O_2 produced during the irradiation of water. The inhibiting effect of mustard gas and of nitrogen mustard might also be due to combination of these substances with the–SH groups. The fact that lewisite produces no effect is no serious objection, because Barron *et. al.* have shown that d-amino acid oxidase, a sulfhydryl enzyme, was not inhibited by lewisite.

The close interrelation between the metabolism of carbohydrates, fats, and proteins has shown that for the synthesis of protoplasm energy is obtained by the aerobic oxidation of carbohydrate. The preponderant part of synthesis reactions over breakdown reactions in growing cells must be due not only to the increased formation of enzymes but also to the distribution of the regulating mechanisms which not only modify the rate of reactions but are also able to change the direction of these reactions.

The study of these regulating mechanisms which channel the multiple enzymatic reactions towards different pathways will clarify the role of enzymes in growth processes and will help us in understanding the change from normal controlled growth to the uncontrolled growth of cancer.

DEFFERENTIAL GROWTH BY BAUL WEISS

The appearance of different properties in initially equivalent cells is called "differentiation." Growth being only one cellular attribute among many, "differential growth" is to be subsumed under the larger topic of "differentiation," and our discussion therefore will have to centre on the latter. Since the accent of this book lies more on prospects than on retrospects, It is intend to concentrate on the gaps in our concepts and knowledge of differentiation and point to possible ways of filling them rather than dwell on past achievements, which, though impressive in themselves, are dwarfed by the task that remains to be accomplished.

Biology has discovered the strength it can derive from a more rigourous analysis of biological phenomena by the tools and standards of the physical sciences, and if the study of development is to share the new light with its more alert sister branches, such as physiology, genetics and immunology, to name only a few, it will have to submit to the same reorientation that has proved so beneficial with the others. This reorientation is in the direction of greater precision, objectivity, and consistency in the description and interpretation of phenomena, preferably in terms of physical and chemical order.

It implies the adoption not only of the tools of the physical sciences, but above all of their disciplined methodology. Analytical embryology is still full of problems which it can only settle within its own right and by its own techniques. But in doing this it must apply standards, criteria, and methods of procedure as rigourous as those of the physical sciences. It must strive to replace the abstract formalism and verbalism of its groping infancy by a mature realism, in which a term stands as symbol for a known or knowable entity or relation, rather than as a soothing device of convenient elasticity to pretend knowledge where there is none.

We must not allow its analytical principles to be contaminated by relapses into animistic mythology, as we are doing when we charge the unresolved residue of meticulous determinations of physico-chemical properties to "organisers," "inductors," and similar ill-descript agents. Labels must no longer be allowed to pass for content, nor generalities for explanations. What we need most in entering the era ahead is the will to drive the analysis of developmental phenomena far beyond the mark where we used to stop to coin a term, and the determination to become as exact, specific, objective, and realistic in the formulation of our problems and in the description of our observations and results as has become the rule in the physical sciences. This is what we mean by reorientation. It is just as important as is the gathering of more data. With this in mind, we turn to the question which forms the core of my assignment : What is differentiation?

WHAT IS DEFFERENTIATION?

By the standards just advocated, this question can immediately be recognised as far too indefinite to be answered by anything but another generality. For "differentiation" refers not to a single circumscribed natural phenomenon, but is a summary name for a heterogeneous collection of notions about a variety of events associated with development. Consequently, the various statements and generalisations which have been made in the past concerning "differentiation" and "dedifferentiation" are equally heterogeneous and full of contradictions. Only a few authors have deemed it necessary to be explicit about terms of such general currency. A definition of "differentiation" is partly a matter of objective description, partly of plain logic. We shall deal with the logical portion first, because it has been more persistently ignored.

CHARACTERS *VS.* PROCESSES

One source of confusion is the tradition of confounding the process of differentiation with its criteria. We usually tell differentiation by its more conspicuous products; *e.g.* melanin granules in a pigment cell, myofibrils in a muscle fibre, secretion granules in a gland cell. Our criteria are predominantly morphological and we judge mostly by appearances. In the examples just mentioned, our judgement happens to be well founded.

It becomes dubious when based merely on superficial differences of shape, arrangement, or size, and it misleads us completely when we try to reverse the argument and take the absence of obvious differences for evidence of lack of differentiation. To appraise a cell correctly we must take into account all of its properties, not just the ones that impress us optically, and to appreciate the development of these properties we must bear in mind that we are dealing with material systems in space and time, in which whatever is present at any one stage is the result of antecedent *processes*.

Thus the conventional "characters" by which we distinguish differentiation are in reality only incidental manufacturing samples of the cellular production plant. And just as one kind of industrial plant differs from another not only by its different end products but by the whole machinery and processing method turning out those products, so a differentiated cell of one kind differs from that of another kind not only in its final equipment (pigment, fibrils, secretions, etc.) but in the basic mechanisms through which that equipment has been produced.

Tracing pigment cells, muscle cells, or blood cells back into earlier phases of their development, we find, or postulate, that they contain then, instead of the terminal products of melanin, myosin, or hemoglobin, half-finished "precursors." But if we go back still further, beyond the "precursors of precursors," the future products lose their identity and we are faced with nothing but the plant capable of turning them out, and perhaps some of the ingredients.

Now, evidently a plant designed to turn out chiefly melanin must be different from one designed to make contractile myofibrils. We express this fact in the names "melanoblast" and "myoblast" given to those early stages, and we generally concede that they are intrinsically different, although the differences are no longer demonstrable by ordinary microscopic or chemical techniques. By further refinements of technique we can extend the range of our discriminative powers; structurally, by the ultramicroscope, polarisation microscope, electron microscope and x-ray diffraction; chemically, by microtests (including cytological stains), spectroscopy, microincineration, etc.

These techniques will undoubtedly reveal criteria of differentiation much earlier in the embryonic history than was possible with the cruder classical tools of observation. This is a line of investigation worthy of vigourous prosecution. But we must expect that it too will come to an end short of retracing the whole ontogenetic course of a given cell strain, for we are still essentially concerned with tracing "products" rather than the production processes that lie behind. Once we can detect the "product," most of the story of its production is over and we have certainly missed its essential beginnings. It looks as if the analytical methods at hand are of no avail.

Preformistic theory would deny the validity of our dilemma. It would assert that all distinctions of later stages have existed, though less manifestly, from the beginnings of ontogeny. But experimental embryology has thoroughly discredited this concept by proving that mature cells which differ profoundly in their physico-chemical constitution may be derived from cells that are demonstrably equivalent and, to all practical purposes, identical. When and by what means have their developmental courses then become divergent? This is the crucial question of cytodifferentiation. Fortunately, though analytical techniques fail us, we can answer it by a test that might be called "behavioural." This is where we must draw on logics rather than gadgets.

Suppose we are to compare two closed material systems, whether they be molecules or cells or organisms, about whose character and composition we know nothing. How are we to tell whether they are the same or different? The only way of telling is by watching their behaviour. If both behave identically under identical conditions, we decide that they are alike. However, if both, again *under identical conditions, behave differently,* then we must conclude that *they have been intrinsically different* from the very beginning of the test.

Against the cogency of this postulate even the most persistent failures to detect more tangible signs of difference would prevail nothing. Two seeds giving rise to different plants in the same soil, or two eggs giving rise to different animals in the same pond, would have to be acknowledged as intrinsically different even if they were otherwise indistinguishable.

Similarly, if two cells under identical conditions behave and respond differently in any respect (*e.g.* movement, growth rate, chemical activity, sensitivity to radiation or drugs, etc.), then they are different, which implies that if they have descended from a common source of identical cells, they must have become different, *i.e.* have gone through a process of differentiation, no matter whether they show it by morphological signs or not. The behavioural test is as compelling as any morphological test, and far more pertinent; for morphological criteria are but residues and convenient indicators of prior activities.

Early differences of this type, which are not immediately recognisable and can only be deduced from later formative activities, are commonly described as differences of "potency." If this were to imply that they are virtual rather than real (in the sense of Driesch), then it would mean reinstating the old "vital spirits" under another name. Actually there can be no doubt that we are dealing with differences of equipment and conditions which are real, yet too subtle to be detectable by present techniques. In conclusion, we propose to define "differentiation" as the unidirectional changes in the character of cells and cell generations during their life history, transformations by which they become increasingly different from their own former selves, their parent cells, and the cells of other strains which have taken divergent courses. And, as we have already intimated, "character" is to include the totality of reactions of which the cell is capable at a given stage. In any given situation, only a fraction of these possible reactions will be realised, and only a small part of this fraction will become externally recognisable.

MODULATION

We have said that cells that look alike may be intrinsically very different. The opposite is equally true, namely, cells may look very different and yet be essentially alike in character. Let me cite a striking example from my own experience, the vagaries of the adult sheath cell of Schwann of adult mammalian nerve. In the embryonic stage these cells are long spindles. They then envelop the nerve axons as thin protoplasmic cylinders. When axons degenerate, as after nerve section, the sheath cells regain mass, become spindly again, and glide out from the nerve tubes. After nerve regeneration they resume their original sheathing position and corresponding cylindrical shape.

But they can vary even much more than just shuttling back and forth between these two states. Embryonic Schwann cells, reared in tissue culture along the interface of cover glass and fluid medium, tend to transform from the spindly type into a flat round form of the appearance and behaviour of macrophages. A repetition of the experiments with adult nerve proved that the fully mature Schwann cell is capable of the same transformations and more. Plate illustrates a series of cell forms obtained *in vitro* from Schwann cells of peripheral nerve.

They include tandem strands of filamentous sheath cells, ramified astrocyte-like forms, giant macrophages with ruffled membrane in active phagocytosis, monocyte-like round cells with clear ectoplasm and eccentric nucleus, and foamy round cells whose vitality is proven by their mitotic activity. Each one of these phases occurs under certain definable conditions, among which the fine-structural constitution of the medium seems to play a predominating role. Some of the transformations are easily reversible, others have as yet been observed in one direction only.

Since they can be followed directly as they take place in the cultures, valuable insight into the processes underlying the morphological results has been obtained. There was evidence of significant changes in the distribution of the cell content, size of the ectoplasmic border, constitution, elasticity, rigidity, mobility, etc., of the cell surface, average number and shape of pseudopodia, mode and rate of locomotion, viscosity, vacuolisation, adhesiveness, orientation of fibrillar plasma constituents (as indicated by silver impregnation), fat production, distribution of nuclear materials, shape and hydration of nucleus and nucleoli, phagocytic power, and other associated properties.

In short, these cells, in response to different environments, have undergone rather profound re-organisations of the kind unhesitatingly classified as "differentiation" or "metaplasia" by a nomenclature based on sheer appearances. But is this type of change directly comparable to the progressive cytodifferentiation during ontogeny of which we have spoken before? It evidently is not, and it is in this place that facts supplement logic. As we mentioned before, many of the observed changes are reversible if the cell is returned to its prior environment. There are many more cell types of the mature organism which behave similarly in that their behaviour and shape change reversibly with their environment.

Changes in the nutrient and hormone composition of the internal milieu, sub-lethal toxic agents, inflammatory processes, seasonal fluctuations, denervation, and in fact, on a more moderate scale, practically all ordinary physiological stimulations, cause a more or less thorough re-shuffling of the cell content with consequent change in cell behaviour, and often in cell form. The important point to keep in mind is that in all these instances the cell can revert to its original state when the original environmental conditions are restored.

Evidently the cell passes through these transmutations with no alteration, either gain or loss, of its basic equipment. At the end of a cyclic change it turns out to be the same cell that it was before, and it is quite beside the point that sometimes the reversal may not occur in the same cell, but in one of its descendants which then prove to be "chips off the old block." There is no reason why the reversible changes observed in our Schwann cells should be regarded in any different light than, let us say, the contractions and expansions of a melanophore in response to cyclic nervous or humoral stimuli, or the change in the secretory state (and associated morphology) of a hormone-sensitive cell in response to varying concentrations of the respective hormone.

They all are merely expressions of how a given cell can react to a variety of conditions—its "reaction repertory," based on its material constitution such as it is at the time. The living insight into the processes underlying the morphological results has been obtained. There was evidence of significant changes in the distribution of the cell content, size of the ectoplasmic border, constitution, elasticity, rigidity, mobility, etc., of the cell surface, average number and shape of pseudopodia, mode and rate of locomotion, viscosity, vacuolisation, adhesiveness, orientation of fibrillar plasma constituents (as indicated by silver impregnation), fat production, distribution of nuclear materials, shape and hydration of nucleus and nucleoli, phagocytic power, and other associated properties. In short, these cells, in response to different environments, have undergone rather profound reorganisations of the kind unhesitatingly classified as "differentiation" or "metaplasia" by a nomenclature based on sheer appearances.

But is this type of change directly comparable to the progressive cytodifferentiation during ontogeny of which we have spoken before? It evidently is not, and it is in this place that facts supplement logic. As we mentioned before, many of the observed changes are reversible if the cell is returned to its prior environment. There are many more cell types of the mature organism which behave similarly in that their behaviour and shape change reversibly with their environment. Changes in the nutrient and hormone composition of the internal milieu, sub-lethal toxic agents, inflammatory processes, seasonal fluctuations, denervation, and in fact, on a more moderate scale, practically all ordinary physiological stimulations, cause a more or less thorough re-shuffling of the cell content with consequent change in cell behaviour, and often in cell form. The important point to keep in mind is that in all these instances the cell can revert to its original state when the original environmental conditions are restored.Evidently the cell passes through these transmutations with no alteration, either gain or loss, of its basic equipment. At the end of a cyclic change it turns out to be the same cell that it was before, and it is quite beside the point that sometimes the reversal may not occur in the same cell, but in one of its descendants which then prove to be "chips off the old block."

There is no reason why the reversible changes observed in our Schwann cells should be regarded in any different light than, let us say, the contractions and expansions of a melanophore in response to cyclic nervous or humoral stimuli, or the change in the secretory state (and associated morphology) of a hormone-sensitive cell in response to varying concentrations of the respective hormone. They all are merely expressions of how a given cell can react to a variety of conditions—its "reaction repertory," based on its material constitution such as it is at the time. The living ments, in each other's places. This can be done by the techniques of explantation (tissue culture) and of transplantation, respectively. The answer has been unequivocal.

Cells taken from different tissues of late embryonic or mature organisms and transferred into standard extraneous media offering nutrition, protection, and support lose many of their specialised aspects, *i.e.* criteria by which we used to tell their "differentiation." Spectacular as their differences may have been during residence in the body, they soon come to look and behave much more alike, in adaptation to the common environment. In the early days of tissue culture this simplification of appearance was actually often interpreted as a sign of "dedifferentiation," in the sense of a return not only to a more primitive but to a veritably primordial state.

This was again judgement by appearances with all its pitfalls. Yet later work showed clearly that the explanted cells, in spite of their similar disguise, retain distinguishing features of fundamental nature which they pass on unaltered to their descendant cells for indefinite numbers of generations. As a glance at Plate will illustrate, Schwann cells and endoneurial connective tissue cells, while assuming both the aspect of common "mesenchym," *i.e.* spindle shape and amoeboid activity, can be well distinguished by their sizes, nuclear characteristics, staining properties, affiliations, and other criteria revealed only by subtler analysis.

In other words, the parallel modifications which the cell strains undergo on transfer to tissue culture are merely modulations carried out by variously "differentiated" cells that otherwise preserve their basic distinctions. Their distinctiveness persists even during such further modulations in culture as the transformation to the macrophage stage. This transformation has thus far been observed in cultures of various connective tissues, Schwann cells, and muscle. They all can give rise under similar conditions to a highly phagocytic amoeboid cell with hydrated nucleus, dense entoplasm, and a more hyaline ruffled border, *i.e.* macrophage-like characters.

But in doing so they also retain the signs of their descendancy, so that a "macrophage" resulting from the conversion of an endoneurial cell can usually be clearly distinguished from one that has come from a Schwann cell. Tissue culture thus does not really abolish character differences established during ontogeny. This conclusion is fully borne out by the critical litreature on the subject. It is strikingly confirmed by experiments in which cells that had lost their more specialised aspects during intensive proliferation in tissue culture were again given opportunity to produce their specialties, either by suppressing further proliferation *in vitro* or by transplanting them back into an organism.

Under such conditions they were essentially true to the kind from which they stemmed : those derived from pigment epithelium would resume the production of pigment granules; those from thyroid, of colloid; those from gut, of digestive enzymes. Cultured cells returned to an organism, but into atypical locations, moreover, showed no tendency to differentiate in conformity with their new environment, proving that the culture period had not restored to them the ability of younger cells to switch into a new local course of differentiation. All this adds up to the realisation that the measure of true differentiation is the absence of true dedifferentiation, if dedifferentiation connotes not the mere assumption of a semblance of primitivity but the actual return to a state of wider potency, *i.e.* the recuperation of something that had ostensibly been lost. By the testimony of the reported and similar experiments, vertebrate cells are incapable of such true dedifferentiation; hence, for them, true differentiation is the rule.

TISSUE DEFFERENTIATION

All our statements thus far refer to cytodifferentiation. Histodifferentiation and organ differentiation require some additional comment. Tissues and organs consist of cells and matrix in definite mutual relations and groupings. What happens during the differentiation of a tissue is in large measure a direct reflection of the differentiation of its constituent cells. Their order of proliferation, their movements, transformations, affinities, secretions, etc., and the spatial arrangements and restraining conditions resulting from group occupancy of a common space determine, for the main part, the character and organisation of the tissue or organ.

By their collective activity the cells set up conditions for one another that would not develop if they remained single and independent. For this reason, what is commonly called "histodifferentiation" implies an added element of order and novelty over the mere "cytodifferentiation" of the unit cells. This is particularly striking in the formation of composite organs where the combination of cells from diverse sources leads to novel results which neither of the contributing components could have produced by itself. In addition to material contributions, there are dynamic influences of mechanical, chemical, and electrical nature acting from one tissue or organ upon another. But on close inspection there remains no property of a tissue that could not be ultimately traced to the activity of either its own or other cells. It is by virtue of properties acquired in cytodifferentiation that cells can combine, interact, and arrange themselves in certain specific ways, conditions permitting. "Histodifferentiation" implies merely the existence of conditions under which cytodifferentiated cells can realise their individual and group faculties.

This helps to clarify the meaning of the term "tissue dedifferentiation." Evidently cells may lose their typical associations and orderly arrangements just as we have seen before that they may lose internal differentiation products, without giving up their essential character. Destruction of intercellular matrix, for instance, can set cells free, and such cells may resume motility, scatter, and modulate in various ways. Their community has been dissolved, but they themselves have retained their identity.

To this extent, what is called "tissue dedifferentiation" is merely tissue disintegration. Since mobilisation of cells is frequently attended by the dissolution of internal differentiation products (*e.g.* myofibrils), the superficial picture becomes one of general simplification and regression. There is no harm in continuing to speak of tissue dedifferentiation, so long as one bears in mind that it need involve no true dedifferentiation of cells.

WHICH CELL PARTS DEFFERENTIATE ?

A cell is a highly complex and heterogeneous system. Then where specifically do the changes occur that constitute differentiation? In the cytoplasm, in the nucleus, or in both? Or rather, which ones of their sub-divisions undergo change, and which do not? It is customary in speaking of cytodifferentiation to emphasise mainly the concomitant profound changes in the content of the cytoplasm. But the nucleus likewise undergoes patent modifications, which become evident in the divergence of nuclear characters and behaviour among different cell types. Some of these nuclear distinctions can be deduced from differential reactions to histological stains, which in essence constitute microchemical tests. In special instances even more direct signs of nuclear differentiation have been observed, as in the extrusion of formed secretion bodies from the nuclei of certain nerve cells, or in the transfer of the major mineral content from nucleus to cytoplasm during the maturation of neuroblasts.

On the other hand, it is a basic tenet of genetics that the chromosomes, or at least those parts of them that constitute the genes, remain unaltered in the process of somatic differentiation. Accordingly each nucleus contains both "differentiated" and "undifferentiated" portions.

The question arises whether the cytoplasm might not be similarly sub-divided into a differentiated fraction and an undifferentiated residue. Such is generally conceded to be the case for the cytoplasm in the "germ track," *i.e.* of those cells that lie in direct ascendancy of the primordial germ cells. The same must be assumed for those "totipotent" somatic cells of the lower invertebrates which can give rise to a new individual by budding, regeneration, or other forms of asexual reproduction.

The specialised somatic cells of higher forms, on the other hand, have never given any evidence of such totipotence. This could be interpreted to indicate that *all* of their cytoplasm has been diverted into a specialised course, has become liver-specific for the liver cell, kidney-specific for the kidney cell, and so forth. Yet again, there is undeniably some stock of cytoplasmic equipment that practically all cells have in common (*e.g.* respiratory enzymes, mitochondria, centrosomes, spindle fibres, etc.) and which, thus, seems to have escaped the progressive changes of cytodifferentiation.

In short, any attempts at making general statements for the nucleus as a whole or the cytoplasm as a whole end up in contradictions. The more concretely and specifically we try to state our problems, the more clearly we realise the inadequacy and ambiguity of our traditional vocabulary. So long as we address our questions not to a real living cell, but to a vague and abstract symbol, we can expect no more than oracular replies. Those general concepts of cell and cytoplasm with which we conventionally operate, with their sundry textbook attributes of contractility, excitability, adaptability, and so forth, have just about lived out the term of their usefulness.

We no longer view contractility and excitability vaguely as "fundamental attributes of protoplasm," but have recognised them as defined properties of circumscribed physicochemical systems, protein chains and surface membranes, respectively. Differentiation deserves a similarly specific treatment. This will be feasible only if we replace the abstract notion of the cell by a much more specific picture. Such a picture is gradually emerging from the combined efforts of cytology, general physiology, and biochemistry. According to this picture, a cell is made up of known or knowable molecular elements, and cellular activities are resolvable into elementary molecular processes.

Study of these elementary processes is well under way and has yielded spectacular results. However, preoccupation with the elements has also at times tended to obscure the fact that in order to constitute a living cell the constituent molecular processes must not just happen, but must take place according to a highly ordered plan of interrelations based on orderly spatial distributions, orderly chronological sequences, and orderly ratios of rates. No model of a cell can be pertinent unless it takes into account both the elementary processes and their organisational frame.

It is imperative, therefore, that we match research on the former by an equally vigourous search for the physical basis of the latter. To facilitate this task we have tried to sketch a sort of crude molecular blueprint of the cell, in which *the molecular phenomena are placed in relation to the organisation of the cell as a unit*. This picture is of necessity diagrammatic, oversimplified, tentative, and full of blank spaces, but it does serve to crystallise the problems. Most of it has already been presented on an earlier occasion.

MOLECULAR ECOLOGY

Science advances not only by discovering new truths but also by eliminating old errors. Most older cell concepts held to the view that cellular organisation is based on rigid structural frameworks of microscopic order, but this view has been ruled out by modern cell research. Cytoplasm behaves as a viscous liquid. It can be thoroughly mixed up by stirring or centrifugation, and yet its countless chemical workshops will continue to operate in good order, which implies proper spatial segregation. No orderly microscopic array could have survived in the commingled content; what has survived is a fixed set of conditions—the rule, as it were—according to which the original order can be restored.

And where do those conditions reside? Evidently they must be looked for in those parts of the system which can preserve organisation in the shuffle. There are two categories of cell constituents that qualify under this title, and they are of quite different order of magnitude, one molecular and the other supramolecular. The former consists of the large organic molecules or molecular aggregates with their specific structure and faculty to perpetuate this structure in the presence of the necessary ingredients and appropriate sources of metabolic energy. The latter consists of the existing surfaces and interfaces of microscopic or sub-microscopic order, whose molecular linings are anchored by adsorption or chemical fixation and thus better protected against random disruption.

Organisation within any sample space of a cell is then to be regarded as a result of the counterplay between these two poles : the specific faculties for selective combination and segregation of the molecular constituents on the one hand; and the marginal reference frames of surfaces on the other, which, by their power of adsorbing and orienting the mobile elements, force them into a supraelemental collective order. It seems impossible to account for cellular organisation otherwise than by such a dual concept. Its essence can be summarised as follows : Manifest cell organisation results from the response of organised elements to fields of organised (*i.e.* non-random) physical and chemical conditions, here tentatively identified with conditions prevailing along interfaces.

The nearest simile to this type of organisation is found in the behaviour of human or animal populations, which likewise consists of the reaction of organised individuals to an organised environment. In both cases, the order is one of *ecological* conditions. Thus, cell biology, in molecular terms, becomes not merely molecular physics and chemistry of the cellular constituents, but *molecular ecology*, a science not of molecular individuals or pairs, but of *molecular populations and their existential conditions*. Some day we may hope to have a fairly complete list of the chemical compounds present in any given cell, from simple elements to the highly complex organic macromolecules, and thus develop a taxonomic catalogue of molecular species. But just as a museum collection of animals is never lifelike unless it shows them in their natural environment and mutual ecological relations, so this chemical catalogue would not be representative of the cell unless it were supplemented by information on the groupings, behaviour, interdependence, and general operating conditions that tie the various molecular species into a durable, and indeed "viable," community.

Biochemistry is producing an ever-growing wealth of evidence for the complexity of molecular interdependence in living systems. Even very elementary physiological processes, as for instance the energy transfer in respiration or photosynthesis, involve a great number of co-ordinated steps, each of which requires the presence of a definite set of properly dosed reactants, mediators (enzymes), and physical conditions (*p*H, temperature, pressure).

Undisturbed operation of these energy-yielding systems, in turn, is a prerequisite for the synthesis, among others, of the complex protein molecules, some of which, as enzymes, are again fed back as indispensable links into the energy-producing systems. Thus neither system can maintain itself without the other. Protein synthesis itself proceeds in series of steps, each of which can occur only under specified conditions which will vary for different compounds. The various chemical systems of the cell thus form "symbiotic" interdependences, and the integrity of the cellular system as a whole is contingent on their coexistence and co-operation.

For each particular chemical system we can thus postulate a definite set of "existential and operational prerequisites," that is, conditions indispensable for the synthesis, preservation, and operation of specific organised products. The character of a given animal population is determined by the physical environment (climate, physiographic factors, etc.) and by the organic environment (food, competition among species, biocoenosis, etc.) in that particular area.

Similarly, the character of the molecular population occupying a given site in a cell will depend on the physical conditions (electric potentials, surface tensions, etc.) and the chemical interactions among the various molecular species in that locality.

The ecological simile is primarily a convenient device to describe the behaviour of the molecular populations of cells. It permits us, for instance, to answer in principle the question of how the morphological segregation and orderly distribution of different chemical systems in the cell is brought about, preserved, and restored after disturbance, despite the absence of static internal frameworks. If each complex biochemical system requires highly specific conditions for its formation and operation, then obviously the probability of finding that kind of system will be high at sites where those conditions are satisfied, and low at others.

And after commingling, the content will again sort itself out according to the frame of marginal conditions, and if the latter have remained undisturbed, the re-sorting will likewise follow the old order. We will thus find preferential sites for the synthesis of certain compounds, namely, sites where the necessary ingredients, templates, energy sources, and physical factors for that particular synthesis coexist. There must also be preferential sites for the settlement of the finished products; for as synthesis continues, the accumulating products are driven further and further from their production centers and thus brought under conditions not necessarily compatible with their continued existence. It is suggested on a previous occasion that such preferential settlement might be a matter of the configuration of the molecule, as a stable layer of molecules with specifically shaped ends firmly adsorbed to an interface could readily trap roaming molecules of complementarily fitting configuration.

The concept of molecular ecology gives a more truly representative picture of cell life than the older morphological concepts. It gives expression to the fact that what is "determined" in a given cell is only the general statistical norm of the proportions, arrangement, and distribution of its constituent elements, but not an absolutely fixed and stereotyped pattern exactly repeated in each individual cell specimen. No two cells are ever precisely congruous. It is in the very nature of the cell that we can predict no more than the degree of *probability* of the occurrence of a particular process in a given place, and this only with reference to what is happening in the rest of the cell.

In animal ecology the sizes of the mobile units vary greatly. Some animals move independently, others in flocks, and still others as heterogeneous groups. Similarly, in the molecular populations of the cell some molecules move singly, others combine into larger bodies, sub-microscopic or microscopic particulates, and still others form mixed aggregates, with members of different species compounded in definite proportions (*e.g.* lipoprotein complexes, coacervates, etc.).

The formation of any such larger unit introduces a new element of complexity (new specific surface, new conditions for adsorption, restraints of mobility, etc.), and if one thinks the matter through to its logical conclusion, one realises that cytodifferentiation effects its manifest order through a series of steps of progressive complication, in which the basic determinants are the properties and mutual affinities of the different molecular species, and a frame of "conditions" prevailing in their common space, favouring the various species differentially. This frame or ground plan, in relation to which the mobile elements will become ordered and sorted, must evidently possess a greater measure of physical stability than the rest of the system. In a relatively fluid system, such as a cell, the entities to answer this demand are the surfaces, in which molecular mobility is restrained. Surfaces thus assume foremost rank as ecological niches for the assembly and segregation of different segments of the molecular population of the cell. Their mode of operation can be briefly sketched as follows.

SURFACE ORGANISATION

Let us consider the interfaces which set off a cell from its surroundings, or a cytoplasmic inclusion from the matrix, or a chromosome from the nuclear medium. Heterogeneous systems meeting along a common border immediately affect each other across the border.

This is due to the fact that the molecular population of one system begins to react differentially to the conditions introduced by the new contact system (electric charges, surface tensions, combinations and reactions between border molecules). The resulting interactions lead to :

- The segregation of a selected fraction of the molecular population into a surface layer,
- The immobilisation, hence, solidification, of this portion.

The mechanisms involved will vary, but for our present purpose they can simply be lumped under a common term as "*selective adsorption*." Their common denominator is the fact that a given interface of given constitution will intercept and retain certain member species of a mixed molecular population more easily and more strongly than others. Consequently, the favoured species will become concentrated at the surface and the less favoured ones correspondingly diluted. A change in the outer conditions may force the replacement of the former border population by new and more appropriate species from the interior.

A complex system of this kind will have a labile phase, during which the molecular surface grouping responds to changes in the physical state and chemical composition of the adjacent outer system, and a secondary stable phase, which follows the consolidation of the surface due to chemical bonding, gelation, precipitation, etc. If the outer conditions are not uniform over the entire surface, but vary locally, the adsorbed molecular populations will be correspondingly variegated.

The same cell can thus develop different types of surface zones on different sides exposed to different environments. Moreover, each of the manifold types of interfaces inside the cell will also acquire its own appropriate type of covering. The segregative power of specific surfaces can thus be readily understood.

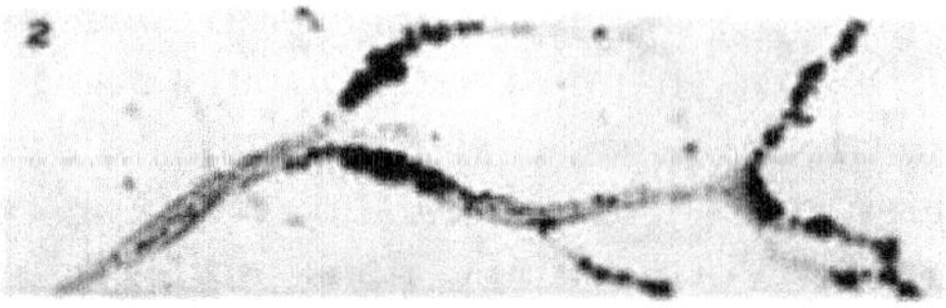

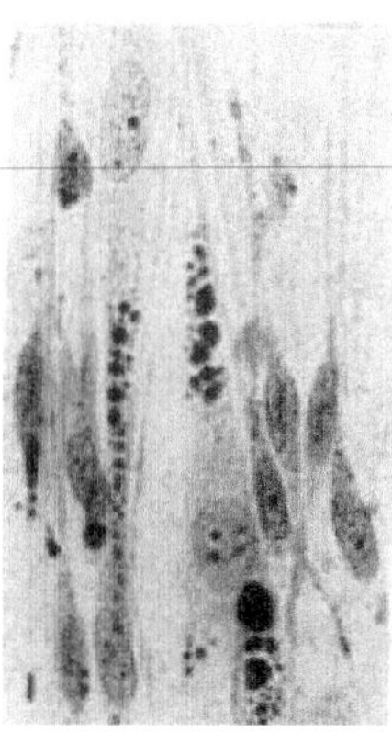

In becoming adsorbed to an interface, polar molecules assume a common orientation. Their exposed ends thus constitute a new free surface to which further molecular species of fitting configuration may be built on in a second layer. This, in turn, may serve for the deposition of a third group, and so layer upon layer may be stacked up in a specific sequence, the molecules of each interlocking with those of the next. Although factual information is scarce, it might be surmised that the apposition of new layers in living systems is governed by highly specific relations among the combining molecules, perhaps due to the steric fitting, key-lock fashion, of the specifically shaped ends of the conjugating molecules.

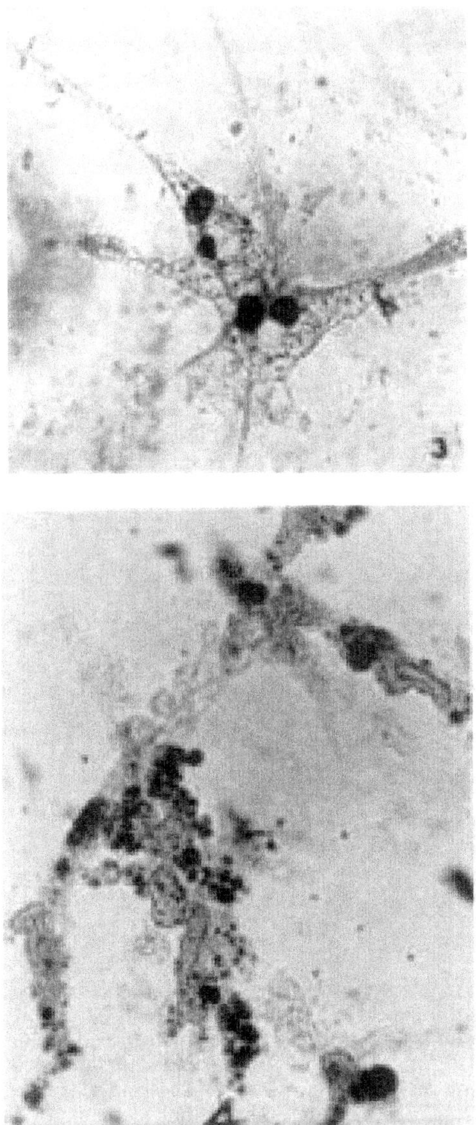

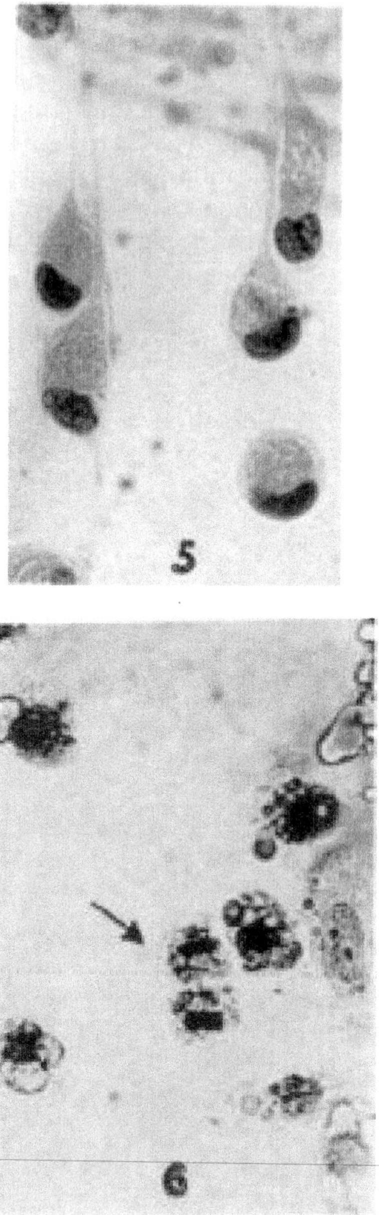

PLATE I

Transformations of Schwann cells from peripheral nerves of adult rats in tissue culture. All figures at identical magnification (X 710). *Panel I* shows the original slender spindle shape of Schwann cells freshly emigrating from nerve. The picture also contains two large endoneurial fibroblasts (large, oval, lightlystained nuclei; dark, sudan-stained fat droplets in cytoplasm). *Panel 2* shows beginning ramification of Schwann cells in plasma clot (note rod-shaped nuclei and fat droplets). *Panel 3* shows giant multi-nucleate "macrophage" developed through transformation and coalescence of a group of Schwann cells at interphase between cover glass and capillary exudate below plasma clot.

Panel 4 shows extensively arborised Schwann cells with polymorphic nuclei at fringe of culture (note active extrusion of fat bodies). *Panel 5* shows transformation stages from spindle type through monopolar "flask" shape to spherical "monocytoid" form with sickle-shaped eccentric nucleus and distinct zone of hyaline exoplasm. *Panel 6* shows small macrophages formed from Schwann cells in liquid areas of medium. (Note mitosis in telophase indicated by arrow.)

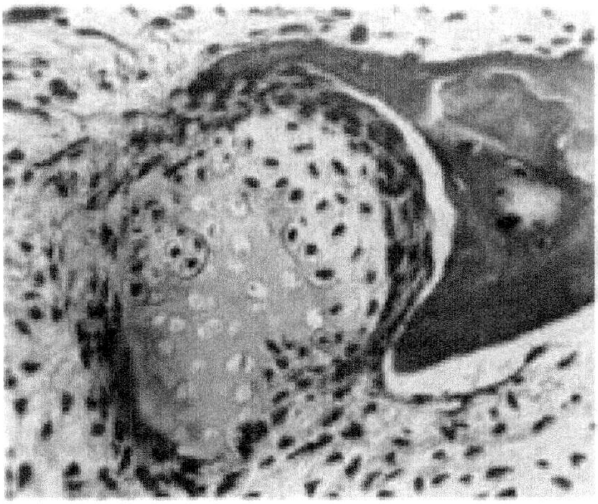

PLATE II

Induction of cartilage formation by dead cartilage graft in Amblystoma tigrinum (X 185). A metatarsal bone, quick-frozen in isopentane at-159°C, dehydrated in vacuo at-40°C, rehydrated in Ringer's solution, and transplanted into the tarsus of another premetamorphic specimen of the same species, has healed in; new cartilage and a new perichondrium have formed along most of the surface of the epiphysis of the graft, which has become detached from the bony shaft, seen in the right half of the picture. The new cartilage can be clearly distinguished by its darkly stained nuclei from the dead graft with unstained ghosts of dead nuclei in its capsules. As complex macromolecules will mostly have different configurations at opposite ends, successive layers will consist of different molecular species. Here we have, therefore, a process that, once it has started from a given surface, could lead to a progressive segregation with stratification of molecular species which were originally intermingled. In this process molecular orientation is of paramount importance, since it exposes the maximum number of specific receptive ends to the interior.

This consideration gains added significance in the case of those polar molecules that have enzymatic properties, as parallel orientation and packing might increase their effectiveness by turning all active groups towards the substrate. Then selective surface adsorption would become a prime factor not only in the segregation of existing molecular species, but also in the creation of new ones. Moreover, the molecular fringe which settles along the surface acquires master control over substance transfer across the cell boundary in that it can selectively facilitate or prevent the exchange of materials between the areas it divides.

The extent to which a surface layer grows inward by apposition, will vary with the circumstances. For example, in some cells the outermost layers form a thick and distinctive coat (myelin sheath of nerve fibre, envelope of red blood cell, vitelline membrane of egg), while in others they remain an inconspicuous "plasma membrane."

The subjacent layers (crust of egg, ectoplasms) likewise vary in thickness as well as in sharpness of demarcation. It must be remembered, however, that stratified molecular segregation need not be so massive as to become microscopically discernible.

The picture of surface organisation here presented is incomplete and greatly oversimplified. For instance the specific key molecules, which were given such prominence, obviously lie interspersed with many more, smaller, simpler and trivial molecules of less specific behaviour, a mosaic feature which we have not duly taken into account. It has been suggested on the basis of studies on drug action, immunological reactions, and radiation effects, that target points of specific response are present in isolated patches occupying only a small fraction of the cell surface.

A similar discontinuity characterises the structure of the chromosomes. Accordingly, many of the specific effects assigned above to surfaces in general will actually be confined to certain portions of the total surface only. An extension of this thought leads to the realisation that complex molecular aggregates (particulates) may be subject to specific adsorption and fixation, with one of their constituent key patches effecting the specific attachment, and the rest remaining uninvolved. But to discuss these and other qualifications in detail, would exceed the scope of this paper.

DIFFERENTIATION IN MOLECULAR TERMS

We return from this excursion into the molecular realm better equipped to phrase questions about differentiation and growth in tangible form. We can now ask, for instance, whether differentiation really signifies a change in the composition of the molecular population of the cell or merely a change in the distribution of pre-existing molecular species. The latter view has been implicit in those embryological and pathological theories which hold that the various cells of a given organism are all composed of essentially the same basic and immutable "protoplasm," and that their diverse appearances and performances are merely the overt responses to a variety of external situations or "stimuli."

A given response display (called "differentiation") is thought to last only as long as the given stimulus situation lasts. Change the latter and the display will change accordingly ("dedifferentiate" and "redifferentiate") in a new direction.[5] Yet, when we view the facts from the molecular level, there seems to be no doubt that cytodifferentiation does not leave the chemical composition of "protoplasm" basically unaltered. The ubiquitous species of small and simple molecules going in and out of cells remain about the same at all stages except for differences in concentration. Some species of large and complex molecules, including many of the proteins, are also present throughout development. All these comprise what we have called above the stock in common to all cells. But in addition to these, differentiation brings with it, or virtually consists of, the continuous elaboration of molecular novelties which have not been present from the beginning.

Thus, a liver cell, a nerve cell, a pigment cell, a mucus cell, a thyroid cell, etc., each produces its own chemical specialties. Can this really be taken as evidence that the respective "protoplasms" are chemically different? Might it not be that all mature cells have the synthetic faculty for the whole array of cell products known to the body, and that a given cell is merely prevented by local external conditions from materialising any but the appropriate one? This contention is contradicted by the experimental evidence quoted above, showing that cells transferred into a different environment (explantation or transplantation), though they may cease to produce their customary products, never reacquire the faculty to give rise to any but the products of their original line, when given another chance.

Thus the conclusion is inescapable that the various cell types mentioned do not just temporarily exhibit different production processes, but have themselves become chemically converted into different production plants. That is, each cell type develops, in addition to the molecular species which it contains in common with all or several other cell types, its own peculiar and distinctive array of molecular species. These are the ones that account for the differential behaviour of the various protoplasms; they are the true objects of differentiation.

One is tempted to identify them with specific proteins, but factual knowledge is still far too inadequate for any generalisation. We need much more precise information on the time and mode of appearance of various cytochemical constituents in the ontogeny of various cell strains. Such an inventory is bound to contain the answers to numerous questions about which we can as yet do no more than speculate.

In contrast to differentiation, cell modulation implies no essential change in the composition of the molecular population, but merely a re-grouping of the existing species. According to our previous discussion, any change in the environment of the cell that alters the conditions in the surface could cause a re-shuffling of the content and replacement of the surface population with marked changes in the behaviour and morphology of the cell, but without necessarily affecting its basic composition. However, whether such a modulation is fully reversible, *i.e.* whether or not the re-shuffled population can return once more to its initial distribution when the original environmental conditions are restored, depends on the degree of consolidation the surface has undergone in the interim.

These considerations suggest the possibility that all differentiations start out as modulation, as mere molecular re-grouping, in response to changes in the cellular environment. Let us consider an embryonic cell shifting from a site *A* to another site *B*, or, what amounts to the same, a change in the environment of a stationary cell from a condition *A* to a condition *B*. Exposure to the new contact conditions brings molecules with affinities to *B* to the surface. Now, suppose this molecular coat initiates a chain of chemical reactions, through which the composition of the interior is materially changed.

This implies the appearance of novel molecular types as well as the disappearance of some of the existing species as they become converted into new ones. Such change cannot take place all at once throughout the cell. We must assume that it starts from a localised area (in our concept, the surface) and spreads at a measurable rate. Therefore if the cell were returned to its original condition *A* during the early phases of this process, it would still contain sufficient amounts of all the species it had before to resume its original state.

To the observer this would appear as reversible modulation. But continued residence under condition *B* would permit continued progress of the changes indicated, until eventually the cell will have lost a major segment of its earlier constituents. It would thus have undergone an irreversible change of its constitution, *i.e.* true differentiation. The lability of differentiation in its incipient stages is a matter of record. If a cell that has assumed character *B* then moves on to a site *C* (or if its neighbour cells or the intercellular milieu changes to a condition *C*), its surface population will again become reconditioned and then initiate a new chain of processes, marking a further step of differentiation, the earliest phase of which would again be labile. A cell transferred from *A* to *C* directly may or may not respond in the same fashion as when coming by way of *B*, depending on how decisively its molecular population has been altered during the intermediary stay at *B*.

We are portraying "differentiation" as a chain of events consisting of alternate physical re-grouping and chemical alteration of the molecular populations, the latter phenomenon involving the emergence of novel species of compounds. It can be seen that according to this concept the potentialities for differentiation are strictly limited by the initial chemical endowment of the cell and that they become further restricted as the cell passes through the various phases *A*, *B*, *C*. What we conventionally call "loss of potency" is therefore merely the counterpart of the positive increase in definite chemical specialisation incurred by the "differentiating" segments of the molecular population. We need assume no separate inhibitory agents.

This concept contains also, in principle, the answer to the problem of divergent cytodifferentiation, that is, of how two initially identical cells may become the source of two qualitatively different strains. Let us assume that of two cells with identical molecular populations, including key species *a*, ^, *y*, *s*, *e*, etc., one is exposed to an environment *E*, and the other to a different environment *F*. Given some agitation and sufficient mobility of the cell contents, we may further assume that in each cell a selected fraction of key species will be concentrated at the surface, namely, those species which best conform to the adjacent medium; let us say, compounds *e* in the cell exposed to the *E* surface, and compounds <*t*> in that at the *F* surface, where *e* and </>may be entirely unrelated. According to the preceding section, such adsorbed and oriented border species can exert key effects in turning the chemical reactions, segregations, and syntheses among the rest of the population into a definite course.

Consequently the chemical fates of the two cells, coated by qualitatively different key species, will likewise become qualitatively different. Thus a difference which started out as merely one of distribution gradually develops into one of composition and character. This would be the fundamental step in the dichotomy of cell fates. This concept remains, for the present, hypothetical. But being more concrete and specific than the conventional verbal concepts, it has the advantage over the latter of being amenable to experimental verification. Tentatively, it fits the known facts of development very satisfactorily, as will be shown in the following.

The principle of discreteness refers to the fact that "differentiation produces a definite number of discrete, distinct, discontinuous and more or less sharply delimited cell types which are not connected by intergradations." The principle of exclusivity states that once one type of differentiation has taken hold of a cell, no fraction of such a cell can ever become engaged concurrently in any other type of differentiation; the cell in differentiation acts as an entity in all-or-none fashion. For instance, there is a sharp dividing line at the rim of the optic cup separating optic retina from tapetum; one cell is a typical retinal cell and the next cell is just as typical a pigmented tapetum cell.

Although we know from experimental evidence that originally both have the endowment to transform into either type, they never form a half-way blend. Examples of this kind could be multiplied at will. It is evident that these principles, derived from empirical facts, can be logically deduced from the postulated discreteness of the molecular key species <<, *p*, *y*, *8*, *e*, and their mutual exclusiveness in the competition for surface positions. Thus, a given cell must follow either the *y*-determined or the *a*-determined or the /3-determined, etc., course, and follow it wholly and with no compromises.

The third principle, genetic limitation, stating that each cytodifferentiation occurs according to the special methods characteristic of the parent species, reflects the fact that the numbers and kinds of molecular key species available for differentiation are limited by the inherited, hence, genetically controlled, initial endowment. It would seem premature, however, to speculate about the relation between our postulated key species and genes.

Phenomena of dominance and recessivity of cellular activities can likewise be explained in terms of our hypothesis. Any molecular species assuming a monopolistic surface position thereby gains "dominance" over competitively inferior species which, being barred from the surface, are in no position to express themselves effectively.

INDUCTION

After this consideration of cellular responses in differentiation, some comment should be given to the factors calling them forth in the organised play of development.

According to the testimony of experimental embryology, development proceeds as follows. Cleavage sub-divides the egg into blastomeres, each of which receives in addition to a nucleus a certain fraction of the original cytoplasmic system. While the nuclei are essentially equivalent, the cytoplasmic allotments vary, depending on the degree of specific molecular segregation attained in the undivided egg.

The role of the most nearly stabilised part of the egg, namely its surface, must be especially taken into account. Different blastomeres inherit different parts of the original egg surface. This initial differential among egg parts is then further elaborated and amplified by the subsequent reactive transformations which the various cleaved-off cells undergo in response to the organised conditions of the system which they constitute.

Descriptively, these conditions have been referred to as "fields"; but we shall not concern ourselves here with their character and operation. We need only keep in mind that in the early phases of development any area of the cleaved egg is endowed with dual capacities. On the one hand, it is composed of a large number of individual cells which in many cases, according to experimental evidence, are practically equivalent ("equipotential") and hence interchangeable as far as their response faculties are concerned.On the other hand, the totality of these cells, as a collective unit, exercises specific functions not exhibited by the individual elements as such. To mention a familiar though crude example, the differential exposure to environmental factors of cells in the core of a given group, as contrasted with those near the surface, is a property of the collective not owned by its members.

Cell groups under the influence of such collective factors, but prior to the appearance of a manifest response on their part, are usually referred to as "determined." There follow material transformations of the kind discussed in the preceding chapter, introducing novel relations among parts. Stepwise, in assembly-line fashion, the diversity of activities increases. As initial time differences become accentuated, some parts of the germ soon outstrip others in their tempo of differentiation. From then on, influences of the more advanced ("determined") upon the less advanced ("undetermined") parts become increasingly prominent. These influences find essentially two forms of expression :

- Influences spreading within a primary cell continuum,
- Influences exerted from one cell continuum upon another contiguous one, usually after secondary junction.

Influences of this type are customarily referred to as "inductions." The term is purely descriptive and covers a considerable variety of heterogeneous phenomena which have often no more in common than the label. There is not the least justification for the supposition that a single operative mechanism underlies them all, much less so a single chemical entity. Much more pertinent than the illusory search for a master solution to the "induction" problem is an objective analysis of each and every one of the various concrete instances of induction that have thus far been recognised. Therefore, instead of discussing induction in general, let us concentrate on two specific examples, one for each of the categories mentioned above.

The first example concerns what might be called "homoeo-induction," that is, the progressive recruiting of cells for a given type of differentiation, spreading from a focal area like an infectious wave (*e.g.* the antero-posterior progression of myotome formation). One gets the impression that cells which have attained a certain differentiated character can communicate their state to their neighbours, which then pass it on further, and so on down the line. The experimental test lies in transplanting a differentiated fragment amidst less far advanced cells. In many embryonic operations of this type, completion of the fragmentary graft structures by recruitment from surrounding cell sources has been observed, but in no case has the precise mechanism been revealed.

Technically it might be easier to start with post-embryonic stages, particularly regeneration. For instance, the fact that transplants of mature cartilage may induce new cartilage formation lends itself to a much more detailed analysis than has as yet been undertaken. Nageotte reported that the implantation of alcohol-fixed cartilage into the rabbit's ear stimulates the formation of new cartilage in the adjacent connective tissue.

In preliminary experiments in which transplanted pieces of salamander skeleton which had been devitalised by quick-freesing and drying. It is found that cells of the larval host built a new shell of cartilage onto the dead model. Since new cartilage has formed only in direct continuity with the graft, we must conclude that the influence in question is transmitted only by contact and not by diffusion. It seems that objects of this kind would be very favourable for a detailed physical and chemical analysis of communicable differentiation.

In terms of a molecular population concept, this type of inductive effect could be described as follows : The exposed surface of the cartilage contains certain cartilage-specific key compounds. Any potentially chondrogenic cell would contain similar key species in its interior. On making contact with cartilage, these key species would become concentrated and oriented along the surface by virtue of their affinity to the cartilage. After saturating the surface, they would then direct the subsequent chemical transformations in the cell interior into a cartilage-specific course.

Thereupon the cell can, in its turn, act on the next one, and so forth. In the case of cartilage, the selective effect would be exerted by the secreted ground substance rather than the cell body itself, but in other instances it would pass directly from cell to cell. This concept has been developed in greater detail in an earlier publication. Its main thesis is that adjacent cells influence each other's development by means of selective attraction among specifically configurated molecules, key-lock fashion. An alternative assumption would be an actual transfer of specific master units from cell to cell, in the manner of a spreading infection. Present evidence is insufficient for a final decision.

The second category to be illustrated here is "heteroinduction," in which one tissue induces another tissue, with which it has secondarily come in contact, to differentiate in a direction different from that of the inducing part. Inductions of neural plate by chorda-mesoderm and of lens by retina are familiar examples. According to our hypothesis the mechanism might be as follows : The cells of the inductive layer have a certain specific surface organisation, that is, they possess already a surface layer of specific key molecules in relatively stable configuration and orientation.

Corresponding molecules of complementary configuration would be present in the ectoderm cells; but during the earlier labile phase of these cells, these would still lie scattered in the interior, mixed with an assortment of other species. As the organised substratum then doubles under the ectoderm, its surface molecules would attract their molecular counterparts in the ectoderm cells into the contact surface. From their oriented surface positions, these molecular layers in the basal face of the ectoderm cells would then control the further chemical developments in their cell bodies, as outlined above. It can readily be seen that the type of response to be obtained from such an inductive action depends both on the character of the substratum and the types of molecular species present in the reacting cells at the time of exposure. The progressive transformation of the molecular equipment of the cell explains why there is often a limited period of "competence" during which cells can appropriately conform to a given inductive stimulus.

In contrast to previous hypotheses, which for the most part have ascribed the inductive effect to diffusible "evocator" substances, the theory here advanced presumes a definite spatial organisation of the molecular population as essential for inductive success. Induction of this type could be transmitted only by direct contact. This is in agreement with experience. Lens induction requires intimate adhesion between retina and epidermis. Neural plate arises in direct contact with the chorda plate. Many more examples could be cited that would seem to discount the diffusible nature of the inductive agent.

One might object that neural inductions have been obtained in explants floating in body fluids, but it is conceivable that in these cases the active role is played by a protein film adsorbed to existing surfaces to which the explant has temporarily adhered, or a similar film settling along the outer surface of the cells. Since there is a direct relation between the presence of disintegrating cells and the occurrence of neural inductions, it is plausible to assume that some neural key species, set free from the destroyed cells, collect and become oriented along the outside of the surviving cells and thereby become an inductive film in the sense discussed under homoeoinduction.

These comments are conjectural. There are not enough facts known on which to decide the issue of contact versus distance action conclusively. It stands to reason that a precise and objective description of the cytological changes attending induction, together with experimental tests of their relevance for the process, could provide some of the wanting clues. A suggestive instance is the following: In observing tissues in inductive interaction we have often been struck by the remarkable correspondence of their cell and nuclear orientation just during the crucial phase. This phenomenon is very marked, for instance, in the formation of the lens. When the eye cup makes contact with the overlying epidermis, the irregular cells of the latter, within an area exactly coextensive with the area of contact, turn into a prevalent radial orientation, each elongate nucleus lining up in the direction of the axis of the opposite retinal cell. Precisely these oriented cells will later constitute the lens placode.

The orientation effect is transitory, being obliterated by the subsequent transformations of the placode. While experimental tests remain to be applied, the cell-sharp coincidence between area of contact, orientation, and lens determination makes it almost certain that the observed cell changes are directly related to the primary inductive effect. Evidently, what happens is that the content of the epidermal cells undergoes a thorough re-shuffling, with the underlying retinal cells acting as attractive foci. In short, the elongation and orientation of cell structure would be visible expressions of the oriented shifts of the molecular population, which would have to be expected from our theory. A systematic investigation of this orientation phenomenon is under way.

The foregoing comments on induction were presented mainly as illustrations of how problems of development gain in sharpness when resolved into terms of molecular behaviour. The special solutions suggested should be considered merely as a first crude and wholly tentative effort to replace the vague, non-committal, and unrealistic notion of "inductor substances" by a specific, concrete, and testable concept. The limitations of this concept are self-evident. Therefore our insistence lies more on the merits of the particular manner of viewing the phenomena than on the tentative views presented.

As for the limitations, one must not lose sight of the fact that the concept pertains only to the instrumentalities of biological processes and not to the overall order which these processes follow and which we usually refer to as organisation. It can explain the manifest differentiation of a given molecular population, provided only that there is a regular frame of organised surface conditions to start with. The organisation of the cell content is thus referred back to a prior topographical organisation of the various surfaces of cell, nucleus, chromosomes, etc., and the principle, *omnis organisatio ex organisatione*, still holds. The "field" character of the organising conditions in development, which cannot be reduced to molecular terms, has not been touched at all in our discussion.

Another point is that our concept is perhaps too exclusive in assigning the major ordering role to the surfaces. Some faculty of self-sorting of molecular mixtures will perhaps have to be brought into the picture to make it complete. Also our assumption that selective transmission effects and affinities are essentially explicable in terms of steric conformance of specifically shaped molecules remains to be verified. It is still possible that ultimately we may have to take recourse to some other sort of "resonance" mechanism, "resonance-like" interrelations among tissues being clearly recognisable on the biological level.

GROWTH

The problem of growth likewise resolves itself now into a number of tangible issues. The definition of growth gains precision. Growth is often defined as any increase in the mass of an organic system. Since such an increase may occur by the mere intussusception of water—and even a dead seed can swell—this popular definition is scientifically of no value. The other part of the population, consisting of inorganic substances like water and salts, passing in and out of the system, is left out of consideration, and so are the smaller organic compounds that enter as ready-made building blocks, *e.g.* sugars and amino acids, until they have become assimilated and thereby lost their identity.

Since, as far as we know, the synthesis of the complex high molecular systems in a cell occurs only in the presence and with the intervention of other similarly complex and cell-generated systems, after their own image, growth in our definition is essentially synonymous with reproduction. In population terminology, we might say that immigration of molecules into the cell space as such does not constitute growth, but their assimilation into complex systems after the model of pre-existing indigenous systems does.

We can thus divide the cell content roughly into two categories : systems capable of turning out more of their own kind, and others devoid of that faculty – reproductive and non-reproductive ones. The non-reproductive group includes all the elementary ingredients for the higher compounds, as well as the sterile terminal cell products – membranes, fibres, secretions, pigments, etc. In thinking about these matters it is rather distressing to note that the proteins about which we have the most direct information, such as collagen, keratin, melanin, myosin, pepsin, etc., belong in the latter class and hence can divulge nothing about the reproductive systems, which are still very much of a mystery.

At any rate the distinction here proposed enucleates the core of the growth problem, as it were. We shall say of a pigment cell that it grows not when it synthesises more melanin granules, but only when it synthesises more of that peculiar part of its protoplasm which, other conditions favourable, can synthesise pigments. Similarly, growth of a myoblast is not the synthesis of more myosin and actin, but of more of the basic muscle protoplasm possessing, among others, the faculty to synthesise myosin and actin; and so forth.

For the sake of simplicity let us ignore for the moment the fact that "more" refers not to absolute production, but to the net excess of production over destruction in opposite metabolic phases. It would also be premature to regard the "basic" systems in question as nothing but enzyme mixtures, although enzymatic activity is one of their prime characteristics.

The fundamental feature of growth, therefore, is the capacity of a reproductive system to procreate more systems with similar reproductive faculty. In comparison with this fundamental phenomenon of multiplication of the molecular key population, the often superimposed phenomenon of cell division is as incidental as the setting up of new administrative districts in a growing country. The relation between cell division and growth is by no means as close as the widespread habit of treating them interchangeably would make it appear. During cleavage, an egg undergoes repeated cell divisions with practically no growth at all, while conversely, certain cells, such as mature neurons or the single-celled kidneys of some nematodes, grow to comparatively enormous sizes without division of either cytoplasm or nucleus.

The growth process is being explored from many directions and with increased momentum. Pertinent information comes now from genetics, now from virus research, now from embryology. Views begin to converge, and concepts, to become unified. Yet the more diversified the contributions, the more we must guard against the obvious danger of confusing a common label with a common object. If growth means cell division to one, increase in weight to another, elongation to a third, metabolism to a fourth, protein synthesis to a fifth, their mutual communication on "growth" is apt to become deceptive unless they are fully explicit.

With that definition in mind we can now ask a number of specific questions pertaining to the mode, the rate, and the site of growth. First, do cells ever stop growing? Superficially, cells fall into two categories, those that keep on proliferating, and those that do not. The ratio of the former declines steadily during ontogeny. Yet even in the mature body proliferation is still very active in many tissues, such as skin, blood-forming organs, many glands, sex cords, and others. Other cell groups become stationary. Since incremental growth can be observed only if there is a net surplus of production over breakdown, it is quite conceivable that the stationary cells are likewise engaged in growth, but at a rate just sufficient to balance the metabolic losses. In this case, protoplasm would be continually regrown in all cells.

What evidence is there for this view? Our habit of lumping all synthetic processes in one class and all degradation processes in another tends to make incremental growth appear simply as a shift in the balance towards the former; in other words, simply as more of the same kind of activity that is going on in any metabolising cell. But if we distinguish, as we must, between different levels and types of synthesis, the problem loses its aspect of simplicity. To crystallise the issue let us focus on the most complex molecular systems, the proteins. We know from isotope tracer work that proteins are in a state of continuous renovation.

Now, this may mean either that the individual molecule merely interchanges constituent parts with its environment while preserving its identity and individuality, or that the molecule breaks down completely while another whole one, freshly delivered from the reproductive system, appears in its stead. The difference is as between keeping an automobile in repair by replacing the worn-out parts one by one, or scrapping it and replacing it by a new one. If the latter interpretation is correct, protein replacement and reproduction, *i.e.* true growth, would definitely be two different processes. To the best of my knowledge, the matter has not been finally settled.

Let us then confine ourselves to the cells which are known to grow perpetually (basal skin, lymph, intestine, etc.). Each of these cell types is known to produce each more of its own kind. What does this imply at the molecular level? Again we are faced with alternatives that we can only present, without deciding between them. The question is whether the basic systems of reproduction are the same for all cell strains or whether they change in character with progressive differentiation.

Perhaps all cell strains retain a certain germinal core population which remains exempted from differentiation and constitutes the sole source of new protoplasmic compounds; in that case the synthesis of each specific compound in skin, blood, gut, etc. would have to recapitulate the whole course of the ontogenetic differentiation of that cell strain. The new units would be turned out in a primordial mold common to all cells and then gradually pass through a series of transformations after the pattern of the already present skin-, blood-, gut-specific units, as the case may be.

The other possibility is that differentiation gradually changes the reproductive apparatus of the various cell strains itself so that the skin-, blood-, gut-specific systems would each be enabled to reproduce themselves directly. If we were to postulate that the genes represent the only self-reproducing systems, and assume, on the basis of genetic evidence, that genes do not change in character during ontogeny, the former alternative would become axiomatic. Some new experimental contributions bearing on this problem will be reported later.

GROWTH RATE

The fact that we do not know just what fraction of the molecular population of a cell constitutes the actively reproductive growth apparatus makes it impossible to determine "true" growth rates, *i.e.* the rate at which the reproductive fraction procreates more of its kind. This places serious limitations on our interpretation of relative growth, for in all our measurements of growth the reproductive and non-reproductive fractions of the cell are lumped. As long as their ratio remains constant the error in estimating true growth from measurements of total growth is constant and can be discounted for most purposes.

For example, if identical cells, kept in different conditions (temperature, nutrients, etc.), grow at different rates, the difference can be properly ascribed to effects on the true growth process. However, if cells of different character grow unequally, even under identical external conditions, this does not necessarily reflect differences of their true growth rates; it may merely be an expression of the fact that different fractions of their total mass are in reproductive activity.

Since the measured growth rate will be smaller than the true growth rate by an amount varying inversely with the unknown constant D. Consequently, of two cells with different proportions of R and D but identical fundamental growth rates of the portions R, the one with the larger D will appear to grow more slowly. The fraction D is itself heterogeneous, including water, inorganic compounds, and all organic compounds that have either not yet been incorporated in the reproductive systems or have become detached from them as metabolites and differentiation products. Accordingly the ratio between D and R will be significantly different in different cell types, and hence different actual growth rates must be expected simply as a corollary of differentiation. Whether there is also a concurrent change in basic growth rate remains undecided.

Moreover, the ratio R/D does not remain constant for a given cell strain over its whole ontogenetic history. As differentiation proceeds, the proportion of "differentiated" terminal products, making up a large fraction of D, increases progressively at the expense of *R*. This increasing drain on the reproductive fraction of the cell is reflected in the wellknown decline of measured overall growth rate with age. After our earlier discussion of differentiation it seems hardly necessary to reiterate that the *D*-fraction in question includes not merely the formed cell products (pigments, secretions, fibrils, etc.) but also that part of their formative apparatus that is not self-reproductive. Even after the loss of the formed products, as in tissue culture, cells of different strains of differentiation exhibit different growth rates.

Comparative evaluation of differential growth rates among different cell types is further complicated by the fact that in some cells the formed differentiation products stay within the cell and are therefore included in all measurements, while in other cell types they are extruded and lost. For instance myofibrils, neurofibrils, most pigment granules, membranes, and the like stay with the protoplasms from which they arose, while formed secretions, intercellular ground substances, and the like do not. Comparisons based on visible increments are therefore quite deceptive.

Not until most of these matters have received much more precise formulation and investigation will it be possible to develop a rational theory of differential growth. Attempts have been made to account for differential growth rates by different partition coefficients in the appropriation of nutrients, by competition for "growth substrate," by differences in the efficiency of utilisation of foodstuffs, and by differential growth stimulators and growth inhibitors. Plainly, many of these theories miss the point in that they try to explain differences which only appear to be differences of basic growth but in reality are natural results of differentiation. Further compelling evidence on this point will be presented below, where we shall deal with tissue growth.

THE SITE OF GROWTH

The distinction between a growing and non-growing fraction in protoplasm raises the question of the sites of growth. Evidently, the and fractions might be either segregated or interspersed. They might be mingled as mosaics of molecular, sub-microscopic, or microscopic (particulate) dimensions; they might be uniformly dispersed or each concentrated in certain localities, which might or might not correspond to the nucleus, the chromosomes, nucleolus, cell surface, or other topographical features.

Some unexpected light has been shed on these problems from recent observations on nerve growth. A nerve fibre is the long cytoplasmic extension of a nucleated cell body, surrounded by sheath cells. The fibre grows enormously in length and width during ontogeny, and even more dramatically during regeneration. When a fibre is cut, the distal stump, severed from its nucleated base, perishes, while the proximal stump, still connected with its central cell, grows out anew.

The rapid elongation of the regenerating fibre is a process of protoplasmic movement rather than of growth in the proper sense. But true growth also supervenes and becomes most conspicuous in the increase in width of the fibre (from less than 1 micron to often more than 10 micra, *i.e.* more than a hundred times in mass). Since the fibre is surrounded by accessory sheath cells and well supplied with blood, one could well imagine that its growth is sustained from local sources. The experimental evidence, however, is that growth takes place exclusively in the nucleated central part of the cell and that all the specific cytoplasm for the whole axon is synthesised there. The evidence comes from constriction experiments as follows.

A local constriction (by a ring of artery) reduces the diameters of the neurilemmal tubes which contain the nerve fibre. Fibres which have regenerated through such "bottlenecks" remain permanently undersised from that point on distally. In contrast, these same fibres swell to excessive dimensions just proximal to the constriction. Plate III shows cross sections through an experimental and adjacent control nerve at the indicated levels, located one centrally and the other distally to the bottleneck.

A detailed analysis of this phenomenon in a large series of varied experiments has led consistently to the same conclusion, namely, that new axoplasm is constantly produced in the nucleated part of the cell body and that the whole column of axoplasm is in steady slow motion in proximo-distal direction. If this translatory movement is partially throttled by constriction, the supply to more peripheral levels is correspondingly reduced, while the surplus piles up at the entrance to the narrows. If the constriction is later removed, the effect is much as that of opening flood gates; that is, the dammed up material moves downward, its front advancing, according to preliminary determinations, at the order of one millimeter per day.

There is evidence that this proximo-distal growth of the axoplasm occurs not only during regeneration, when it adds directly to the width of the fibre, but actually goes on continuously in all axons, even the mature ones, which have reached stationary size. The central cell body seems to turn out a steady supply of new axoplasm, most of which is then conveyed distad. This raises the puzzling question as to where the material goes, if it no longer adds to the size of the fibre. It simply serves to replace the protein systems of the cell as they wear off in a steady natural degradation process, and that the nucleated cell body is, in fact, the sole source of such replacement.

We know that any isolated fragment of nerve can exhibit the synthetic phases of respiratory metabolism with the aid of the enzyme systems it contains. But in the light of our experiments, it would seem that the enzyme systems themselves cannot be synthesised anywhere except in the central cell body, whence they are then distributed over the whole neuron.

As supporting evidence for this view we can quote the intensive production of nucleoproteins at the nucleus of the nerve cell, the identification of nucleoproteins in the axis cylinder, and finally what might be taken to be the external sign of their peripheral breakdown, namely the liberation of nitrogen in the form of ammonia from peripheral nerve. As a matter of fact, a preliminary estimate of the rate of protein breakdown from the known values of ammonia production leads to a figure which is satisfactorily close to the rate of proximodistal replacement determined from the damming-and-release experiments reported above. While details of this concept remain to be worked out, it leaves no doubt that the site of growth and replacement of the neuron is strictly confined to the nucleated portion. It is probable, though not proven, that the same would then hold for other cell types.

These facts have an immediate bearing on efforts, currently in the foreground of interest, to bring concepts of genetics and growth in line. Our experiments place the site of growth near the nucleus, not necessarily in the nucleus. We mentioned before that if the reproductive sources should prove to be identical in all types of cells (as is assumed for the chromosomal genes), we would have to assume that their products are instantly recast into specific patterns corresponding to the surrounding differentiated cytoplasm.

In our present case the nucleus may continuously reproduce basic protoplasmic systems of the general character dictated by the genic constitution, which, however, under the molding influence of the existing specific neural population to which they are added, would assume specific nerve-character. The alternative would be that the intranuclear extrachromosomal systems themselves have become specifically transformed in the course of differentiation, and in that case the proliferated new compounds (nucleoproteins, etc.) could acquire their full nerve-specific character at their intranuclear site of origin *in statu nascendi*. Our data are non-committal on this point. They do prove, however, that growth is not a ubiquitous process in cytoplasm and should not be treated at par with the type of synthesis commonly met with in other types of metabolism.

TISSUE GROWTH

Analogous considerations, as applied in the foregoing to cell growth, apply to the growth of tissues and organs. Just as the molecular population of the cell consists of a reproductive and a non-reproductive fraction, so the cellular population of a tissue is composed of proliferative and non-proliferating components. In most organs proliferation becomes gradually confined to certain restricted areas, so-called germinal zones, arranged in layers (*e.g.* central nervous system, skin), cords (*e.g.* sex cords) or foci (*e.g.* buds). Of the newly formed cells, some remain in the germinal zone, while the rest move away from it. The former continue to proliferate, while the latter cease to multiply and usually undergo some terminal specialisation.

This dichotomy has sometimes been interpreted as signifying an antagonistic relation between proliferation and differentiation.9 Statistically speaking, it is correct that as differentiation progresses, proliferation declines. But this does not imply that differentiation in the strict sense and proliteration are mutually exclusive. From our earlier discussion, it should be plain that the proliferating sources of epidermis, bone, blood, muscle, glands, etc. are differentiated to the extent that each can reproduce only its own kind and none of the others. Yet they are still actively multiplying.

Only the full transformation into terminal stages seems to be held up in those cells that are kept in mitotic agitation. Proliferation does not interfere with differentiation, but it does impede the elaboration of certain manifest products of differentiation. Conversely, the general impairment of proliferative capacity in cells undergoing terminal specialisation can be ascribed to the fact that most of their substance, including mitotic prerequisites, is diverted into the building of differentiation products (*e.g.* intracellular fibre systems).

In early developmental stages practically all cells proliferate. With progressive differentiation, however, the outlined segregation into reproductive and sterile groups takes place. Since the latter do not further contribute to the increase in mass, the total growth of a tissue or organ will vary directly with the ratio of reproductive to non-reproductive cells. This ratio in turn depends on the geometric configuration of the proliferating zone and the rate at which newly proliferated cells desert it.

Further variability is introduced by the fact that most organs receive secondary additions from foreign sources (*e.g.* blood vessels, nerves, pigment cells), and, conversely, lose some of their own cells by dispersion or destruction. Since each organ has its own rule, it is evident that bulk measurements, which fail to separate the actual sources of the measured materials from the "dead weight," can furnish no valid basis for comparing "differential" growth of parts differing in constitution. If percentage increase is faster in one organ than in another, this could be due to either :

- A larger initial proportion of reproductive cells,
- A larger quota of retention of new cells in the germinal zone,
- Greater infiltration of foreign elements,
- Less dissipation of cells and cell products,
- An intrinsically higher rate of reproduction in the respective germinal cells.

Only differences not accounted for by are to be ascribed to that is, to a higher "growth rate."

For an illustration let us choose the growing vertebrate eye. Due to its simple form and accessibility it has been a favourite object for comparative studies on growth, with the increase in dimensions serving as index. How heterogeneous a collection of events this index covers can be readily seen from the following account. The original eye vesicle consists of a certain initial allotment of cells from the embryonic brain wall. At first, all of these cells divide.

The growth function at this stage is therefore a volume function. In the cup stage the retina becomes multi-layered, with a sharp division into a germinal and a sterile zone. Only the cell layer in contact with the outer surface, corresponding to the ventricular (ependymal) layer of the brain, continues to proliferate, while the cells released into deeper layers differentiate the various retinal strata without further multiplication.

The source of growth thus has become reduced to a two-dimensional one, causing a marked decline in the relative growth rate taken over the whole organ (*e.g.* from measurements of diameter). Later, the cells of the germinal layer themselves cease to proliferate and transform into sensory cells, a process which starts from the centre (macula) and spreads rapidly towards the periphery (ciliary zone) of the retina. Eventually, only the cells at the rim retain residual capacity to multiply.

Further growth is then essentially by apposition from this rim; that is, the growth source has shrunk from planar to linear extension. Meanwhile some of the neuroblasts, though no longer multiplying, grow in size as they sprout nerve processes, which, grouped into plexiform layers, add to the thickness of the retina. During the later stages a gelatinous secretion, supposed to come from cells of both retina and lens, fills the interior with vitreous humour, thereby progressively distending the eyeball. In addition, blood vessels and other mesenchym penetrate into the eye from the surroundings.

This diversity and complexity of the component processes contributing to eye size makes the search for a single "growth-controlling" principle appear utterly unrealistic, and comparisons between different eyes on the sole basis of size must be fallacious.

First, what causes the confinement of proliferation to the single cell layer at the convexity of the retina? The answer is unknown. But a clue lies in the fact that in this and most similar instances the site of proliferation corresponds to some prominent geometrical feature of the system, such as a surface or fold or tip or crest. For the cells concerned this means unique physical conditions (*e.g.* tensions, pressures, unique exposure to surrounding agencies) not shared by the rest of the tissue.

In neural tube and retina, for instance, the direct exposure to the fluid of the central canal and ventricular spaces might be the crucial factor. The fact that injury can activate cells of the interior to take part in proliferation could be explained by the penetration of the central fluid to the deeper layers as a result of the break in the internal limiting membrane. This is purely conjectural, but capable of an experimental test.

Next, what determines the further fate of the cells proliferated in the germinal layer? Of the two daughter cells of a germinal cell, either :

- Both stay in the germinal layer
- Both move away
- One stays and the other moves off. The same alternatives recur at subsequent divisions. Continuation of procedure
- Would lead to exponential increase of the proliferative source with no cells left for terminal differentiation, while procedure
- Would rapidly deplete the proliferative source. The actual growth of the eye at this phase is therefore determined by the relative incidence of events.

But again, factual information on this point is practically wholly lacking. In anuran amphibians the number of dividing cells in the retina has been reported to remain remarkably constant, which would mean that all divisions follow scheme.

But this is definitely not true of the chick, for instance, in which the number of proliferating cells increases with the expansion of the retina. Here the rate of growth is therefore largely dependent on the statistical apportionment of the newly formed cells according to :

- This in turn seems to resolve itself into a matter of the relative orientation assumed by the daughter cells, either during mitosis or immediately after, tangential orientation leading
- Radial orientation
- A plausible hypothesis is the following.

It has been demonstrated that dividing cells separate in the direction of maximum cell elongation, which corresponds to the direction of maximum tension. Hertwig's rules of cleavage are an illustration of this principle. Assuming that retinal cells follow the same principle, we may conclude that distribution of type :

- Will result whenever the locally prevailing tension is tangential, while types
- Will result from radial tension or lateral pressure. According to this concept the proliferating layer can expand only as long as the retina is under tangential tension. The source of such force is found in the distension of the eyeball by the growing lens and the vitreous body. Retina, lens, and vitreous body would thereby form a self-regulating system operating as follows.

Accumulation of vitreous fluid would stretch the retina, producing a prevalence of tangential tensions, which, in turn, would cause a large proportion of newly proliferated cells to stay in the germinal surface according to :

- As these additional cells are intercalated, the lateral stretch is reduced and may eventually even be reversed into pressure if the cells begin to crowd laterally; in that case, radial movement.
- Meanwhile, however, further secretion of vitreous humour has produced further distension so that more new cells can be accommodated in the surface expanse, and the cycle repeats itself. Because of widely varying local conditions, these events will not cause a regular alternation of phases of tangential and radial divisions as in cleavage, but will merely statistically increase the probability of the occurrence of one or the other. There are some experimental results suggesting such a mechanism of growth regulation. When the vitreous humour of an embryonic chick eye is drained, the collapsed retina remains greatly reduced in size, but increases in thickness. The loss of a progressively distending core would properly account for the result.

It has also been demonstrated that when incongruous eye-lens combinations are produced by heteroplastic transplantation, a mutual adjustment of size occurs. This could be ascribed to the fact that an undersised lens, both because of its smallness and because it will permit vitreous body to flow out, will not be able to exercise the distensive function which promotes retinal growth until it has caught up with the size of the eye and sealed the borders. Reciprocally, an oversised lens will produce greater distension, hence, expansion of the germinal layer, resulting in intensified retinal growth.

In a similar capacity, distension of vesicles, glands, ducts, etc. by their own or foreign discharges may be a major factor in regulating the extent of their proliferating layers, and consequently of the amount of growth. Perhaps the excessive size of the gut in amphibians fed on a bulky vegetable diet as compared with meat-fed ones is due to the same principle. One could even envisage that the air in the larval lungs of gill-breathing aquatic amphibians is there for a growth-supporting rather than respiratory purpose.

These examples may suffice to illustrate the dependence of overall "growth rate" on the special mode of growth of each organ, particularly the geometric configuration of its germinal zone.

ORIENTATION OF GROWTH

Growth as such, *i.e.* the mere increase of the specific molecular population of the cell, is a scalar process. It has no intrinsic direction. Its directiveness is given to it by vector components of the physical frame-work in which it occurs. Yet growth itself plays a decisive role in creating those physical conditions which, in turn, will orient its further course. A few examples may illustrate this principle of reverberation.

In the preceding section reference was made to the effect of tension on the form of growth in epithelial tissues. Even more spectacular is its effect on the architecture of tissues of mesenchymal origin. Connective tissue, tendon, fascia, cartilage, bone, blood vessels, muscles, all respond to mechanical tensions by orientation of the component cells and intercellular fibre systems along the lines of stress. Experimental analysis of this response in tissue culture showed that the observed cell orientation is a secondary product following a primary orientation of the colloidal matrix in which the cells lie embedded. The sequence is as follows :

- Tension produces alignment of aggregates of the anisodiametric molecules of the matrix into chains.
- Such chains join to build up submicroscopic fibrils, which eventually may merge into microscopic fibres.
- Cytoplasmic filaments (filopodia) from cells inhabiting the matrix extend along the interfaces between these solid micellar threads and the interstitial liquid.
- The rest of the cell body follows the advancing filopodia.

In this manner the cell becomes greatly elongated and is made to trace the ultrastructural pattern of its surroundings. Nerve processes advance in the same fashion, except that their cell body remains anchored.

In the original tissue culture experiments the matrix consisted of blood plasma, with fibrin constituting the orientable ultrastructural reticulum. Later experiments revealed that even in so-called cultures in liquid medium, cell orientation is brought about by a similar reticulum, namely, coagulable fibrous material exuding from the cells themselves, spreading along the surfaces of the liquid and constituting a trellis ("ground mat") on which the cells then advance. Tensional fibrin organisation in blood clots has likewise been disclosed as the mechanism orienting cell growth in wound healing and nerve regeneration in the body. In embryonic tissues, other fibre proteins can be expected to operate in like manner. All pertinent observations and experiments made thus far confirm the conclusion that tension, by way of its organising effect on the colloidal cellular milieu, can determine :

- Orientation of cell movement,
- Cell elongation,
- Cell shape,
- Orientation of mitotic spindles, hence, direction of division,
- Rate of movement,
- Rate of multiplication,
- Deposition of fibrous cell products.

Figure illustrates diagrammatically how different degrees of stretch applied to a fibrous colloid translate themselves into the morphology of inhabitant, as well as immigrating, cells. Potent as these effects are, they could not be a major factor in development if only tensions of extraneous origin were involved.

But tensional stresses also arise within the developing organism as a result of shifts, growth, and chemical activity of its parts.

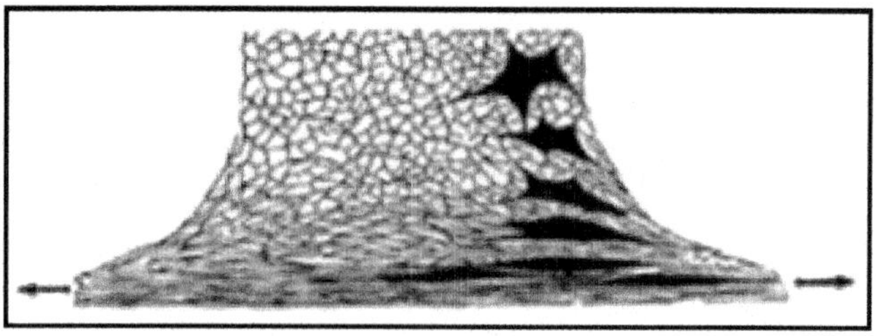

Fig. Diagram illustrating the effect of graded stretching (in the direction of the arrows) of a reticular matrix on the shape of mesenchymal cells.

Let us consider the two main intrinsic sources of morphogenetic tensions – expansion and contraction. On the left side (A_O B_O) are sectors of a tissue possessing a continuous reticulated matrix and containing a circumscribed area (dark circle) in which some localised developmental process goes on, as follows. The lower half represents a case in which the marked area expands relative to its surroundings, either by rapid proliferation or by deposition of large amounts of ground substance or, in the case of hollow organs, by distension.

This local expansion creates in the surrounding continuum tensions that are oriented circumferentially, declining in strength with increasing distance from the border. The fibrous matrix assumes a corresponding pattern as depicted, and this in turn imparts itself to the cells. Cell movements and cell divisions will follow tangential courses. It is evident that the familiar concentric architecture of the mesenchymal sheaths, coats, tunics, and membranes around both solid organs and hollow ducts and vesicles finds a natural explanation on this basis. Essentially the same effect is observed if the surrounding tissue contracts against a rigid core, as in the formation of foreign-body capsules.

A wholly different pattern results if a tissue contains a focal area that contracts. In this case, illustrated in the top row of diagrams, the matrix adhering to the shrinking area is subject to radial tensions, and the surrounding tissue will thus be gathered into a starshaped pattern. Cell movements and divisions will be directed at the area, instead of around it as in the preceding example. Many embryonic invaginations (*e.g.* blastopore, neural tube, nasal pits) seem to take their origin from such localised contractile centers in the surface coat of the germ.

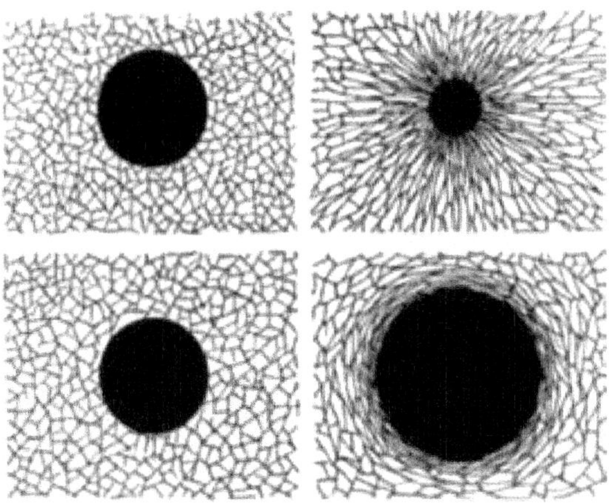

Fig. Diagrams illustrating the effect of local contraction (top) and local expansion (bottom) on the structure of a reticular matrix.

It was very illuminating to find that certain biochemical activities attending growth and differentiation may lead to the same type of expansion and contraction patterns. The expansion type, for instance, is observed around areas with proteolytic activity. When a local process dissolves part of an erstwhile coherent protein network, the remaining portion, owing to its elastic tension, retracts from the hole and assumes a pattern like that shown in Figure, much like a pierced cobweb. A contraction pattern such as in Figure, on the other hand, arises around actively proliferating centers through the following chain of events. Cells in proliferation, and apparently only then, liberate agents with strong dehydrating potency.

The result is a progressive condensation of the surrounding fibrous phase of the medium. Such local condensation in a rigid tissue frame generates tensions, which in turn orient the condensing system towards the centre of shrinkage. The familiar ray-shaped growth of ordinary tissue cultures in blood plasma is due to this effect. Analogous configurations can be recognised in many embryonic formations.

By its radial orientation such a contraction pattern can evidently guide peripheral cells towards the centre. This is one of the major mechanisms by which cells are drawn towards distant destinations, as if "attracted." Its morphogenetic importance becomes particularly impressive when several such centers act on a common matrix, where they create interconnections of great regularity among the otherwise independent centers. As is readily understood from Figure, two

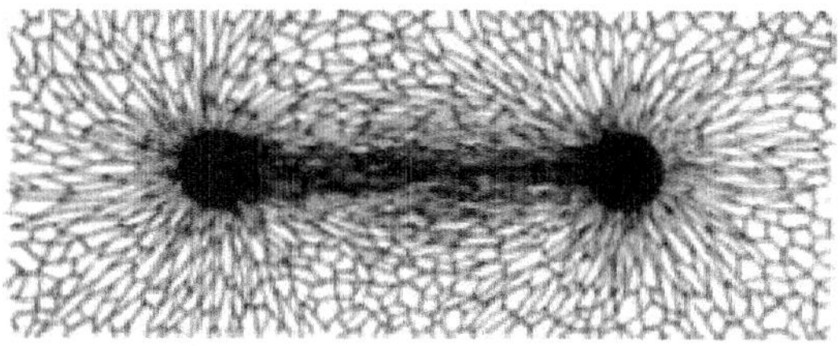

Fig. Diagram illustrating the orientation of a reticular matrix along the connecting line between two centers of contraction.

contractile centers produce the greatest amount of shrinkage, hence tension, and thereby orientation of the matrix along their connecting line; this automatically establishes a bridge for cell traffic between the centers. Three centers determine a triangle, and so forth. In principle, these experiments have solved the problem of orientation in cell movement and growth. In acknowledging mechanical stress as a major morphogenetic factor, two reservations must be borne in mind. Firstly, tension is not the only agent capable of orienting tissue structure. Any other physical force that is capable of affecting the orientation and aggregation of polar molecules (electrostatic fields, electrophoresis, streaming, etc.) may have comparable effects.

Secondly, stress is decidedly not a determining factor incytodifferentiation. Though it seems to be prerequisite for some cell types in order that they may be able to carry out terminal transformations for which they are already predisposed by previous differentiation, it is not the cause of such differentiation. Without stretch, myoblasts cannot transform into muscle fibres, but stretch itself neither makes a cell into a myoblast nor ever transforms a non-myoblast into a muscle fibre, as has once been claimed.

An interesting combination effect of both contractile tension and proteolysis has recently been observed in the healing of nerve stumps. Tension transmitted from stump to stump through an intervening blood clot, intensified by the syneretic contraction of the clot itself, orients the fibrin network in the latter in a prevailingly longitudinal direction. The oriented strands tend to fuse and thus become heavier than their irregularly arranged cross-links. Then proteolytic enzymes appear on the scene and rout out the cross-links, while leaving the straight linear bundles standing. Thereby preferentially linear order becomes strictly linear.

The main value of these varied results is that they demonstrate in principle how chemical activity (proteolysis, dehydration) can engender spatial order of a high degree, although it has itself no space order. Through its physical and chemical by-products, plain growth thus leads to ordered growth. As the order, arrangement, and consequent exposure of the growing tissue reciprocally affect the condition, manner, and hence, rate of its further growth, we realise how complex and diverse are the factors that contribute to the growth of a given tissue. Some of the factual sequences are summarised in Figure. Many more are still unexplored.

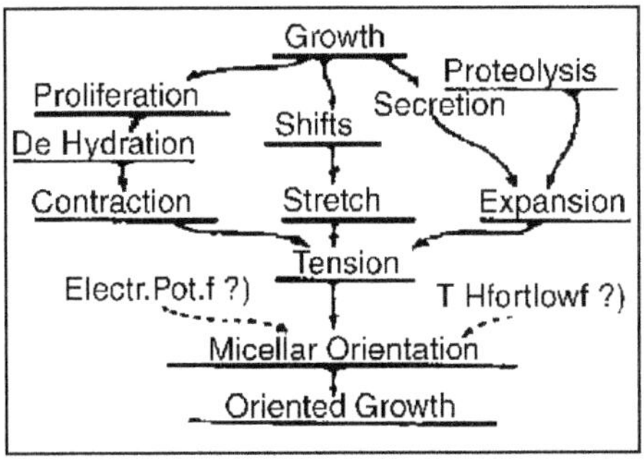

Carrying this exploration of the "microtechnology" of development, *i.e.* of "Entwicklungs-mechanik" in its original sense, for-ward on an intensified scale will go far in rectifying the current rather abstract and often patently unrealistic concepts of "differential growth."

But even the existing knowledge, of which we have presented some samples, is sufficiently compelling to discredit the growing fashion of reducing, on paper at least, problems of differential growth to terms of simple chemical reaction rates, with "growth stimulators" and "growth inhibitors" called upon to do the "regulating."

ELABORATION OF GROWTH PATTERNS

According to the preceding sections, differential growth is simply a corollary of differentiation, and differences in growth rates result from a great variety of causes. As such, we have singled out the unequal depletion of reproductive protoplasm in the building of specialised differentiation products in different cell types, the various inequalities arising from shifts, aggregations and dispersals of cell groups in accordance with properties acquired during differentiations, the resultant differences of geometric configuration, and hence, exposure of different parts of a tissue, with the ensuing confinement of proliferative activity to circumscribed zones of varying sizes, the different fates of differentiation products, which are retained in some tissues but extruded in others, and other related disparities contingent upon the divergent courses of differentiation.

Such inequalities in the growth pattern can, in turn, secondarily influence the setting for subsequent steps of differentiation, and the resulting changes will further modify the growth pattern. The actual elaboration of the mature body thus appears as a complex sequence of transforming patterns in which the contributions of differentiation and differential growth are intimately interlocked. Basically, however, differential growth is not an agent but merely a product and index of differentiation.

This realisation is of the utmost importance when it comes to interpreting the manifold general non-localised influences which modify the growth process during later developmental stages, and which are impressively illustrated by the diversity of phenotypic variants that can be obtained from genetically identical germs : giants, dwarfs, duplications, defects, disproportions, excesses, and suppressions. Since these can often be directly related to aberrations of nutrition, hormones, oxygen supply, physical factors (radiations, pressure), and even climatic changes, all of these agents have at one time or other been claimed as the "dominating" factors in the growth process. We realise now that any factor that has an effect on the physico-chemical setting in which growth occurs may thereby alter the manifest form which the growing process will take. But it is evident that such actions will be in the nature of merely modifying already established patterns rather than instituting wholly new ones. Consequently an agent applied systemically (non-topically) may accentuate or reduce inherent differentials but cannot create differentials not yet in existence. A few illustrations may be in order.

As the developmental realisation of the primordial growth patterns of the germ progresses, small initial differences among parts become greatly amplified. Minor deviations from normal become exaggerated. Therefore, if each part were to keep on growing independently of the other parts, the harmony of development would soon be endangered. The setting up of new systemic growth controls co-ordinating the separate parts thus becomes vital for harmonious development. This is achieved through the humoral, more specially the circulatory, systems for effective and rapid diffusion of substances.

But we understand now that a humoral pool could exercise no discriminative control over the various parts supplied by it unless these parts had already previously acquired different response qualities. For instance, the fraction of a common supply that a given part will be able to secure is a function of the accessibility of the part to the circulatory channels; this relation, however, is entirely a matter of previous differentiation and growth of both the part and the vessels. Similarly, the capacity of a part electively to remove, concentrate, utilise, or store certain components from a common pool only reflects affinities of its cells acquired during their prior biochemical specialisation; affinity for iodine in the thyroid cell, for vitamin A in the sensory cell of the retina, etc.

Or if one cell group responds to a circulating hormone in a manner or degree different from that of another cell group, equally exposed, the two must have been already different constitutionally at the time the hormone appeared on the scene.

The case is no different from that of selective drug response of mature cells, which is likewise based on the specific sensitivity of the reacting targets.

Since no nutrient or hormone or other chemical of ubiquitous distribution can make a cell render any contribution for which that cell has not been predisposed by its previous differentiation, none of those agents can be rated as primary causes of differential cell behaviour. But they do affect the subsequent elaboration and final proportions of the growth pattern in a variety of ways. A general deficiency of nutrients, oxygen, or specific growth accessories, for instance, hits different growing parts differentially, depending on their respective metabolic rates, margins of tolerance, reserves, substitutive capacities, and the like.

Faced with deficiencies, they compete for the available requisites in what Roux has described as the "struggle among parts." One part's loss may either be another part's gain, or may doom other parts, depending on the mode and degree of their mutual dependence. The production of differential growth defects by different vitamin deficiencies may be pointed out as a most instructive example. But the lessons from such deficiencies are not reversible. While increasing the supply of growth essentials from deficient to adequate levels will benefit growth correspondingly, a further increase to levels of overabundance does not keep on enhancing growth indefinitely. A limit will be reached when the production plant works at full capacity, the maximum rate of production being determined by the intrinsic rates of the component processes, which, because of the assembly-line character of growth, are limited by the slowest links in the chains.

Of course, since what we call the "normal" level of nutrients, vitamins, hormones, and other growth requisites still falls short of the attainable optimum, the growth potentialities of the organism are "normally" not fully realised. Raising such growth factors to supranormal dosages can therefore occasionally have the spectacular effect of bringing out latent formations which under "normal" circumstances would have remained unrealised. For example, overdosing a female opossum with sex hormones causes the formation of prostate glands, a capacity inherent in the urogenital tract but failing to come to expression at "normal" hormone levels.

To the various humoral principles thus far recognised as secondary regulators of growth will perhaps have to be added an even more general category, namely, systems of the antigen-antibody type consisting of organ-specific discharges and their molecular counterparts. The case for this assumption has been presented on an earlier occasion, together with supporting experimental evidence. In the most condensed version, it is as follows. The specific key molecules which arise progressively in the course of differentiation act as templates ("positives") in the intracellular reproduction of more of their kind. At the same time they cause the formation of molecular species of complementary shape ("negatives"), which may or may not be an integral step in the reproductive process itself (double reversal). Complementary pairs could inactivate each other by conjugation. The amount of free active growth catalysts in a cell would thus vary with the excess of one type over the other.

Now, if we assume that the "positives" are retained in the cell (*e.g.* because of large dimensions), while the "negatives" can escape, circulate, and enter other cells (*e.g.* because of smaller size), we can see the general outlines of a highly specific and sensitive mechanism of growth regulation, which would co-ordinate all cells of common character into a single operational system, no matter how much they are scattered. Local reduction (or augmentation) of growth in any part of the system would imply a corresponding decline (or increase) of both "positives" and "negatives" in that locality, and because of their easier diffusibility, of the "negatives" only throughout the system. Since systemic reduction of the "negatives" means an increased ratio of active over conjugated "positives," a compensatory increase of growth throughout the rest of the system will ensue; and conversely, enhancement of growth in one part will automatically entail depression of growth in all other members of the system. Thus the total growth of a given system would be held under joint control of all its components.

That some such mechanisms exist is becoming increasingly evident from a study of compensatory growth reactions. That these mechanisms operate, at least in part, after the fashion of immune reactions is indicated by experiments in which organ-specific antibodies were shown to have selective effects on the growth of the homologous organs. But the precise mode of operation here sketched by way of illustration is sheer supposition.

Whatever its nature may eventually turn out to be, one must remember that it constitutes merely another contributing factor to the highly intricate system of dependencies and interactions which control the growth process. It is no more the master key to differential growth than are any of the other factors discussed before.

What our whole discussion adds up to is that there is no single master clue to the problems of differential growth or of growth in general. The measurable growth of different parts is so intimately dependent on their peculiar configurations and cellular differentiations, which themselves differ sharply and qualitatively, that a purely quantitative comparison of different growth processes is a deliberate abstraction. There are instances where such abstraction has led to the recognition of striking regularities capable of mathematical formulation *e.g.* allotropic growth.

Whether these regularities are due to the fact that the compared systems resemble each other closely enough in their mode of growth to be essentially commensurable, or to the existence of some undefined principle of organismic nature, exercising overall control over the multiplicity of heterogeneous components involved in growth, can only be decided after a detailed analysis of the underlying phenomena. At any rate, a purely formal treatment of growth, as is often attempted through the interpretation of growth curves, is only a valuable guide to and supplement of, but never a substitute for, a precise analysis of the different forms in which growth manifests itself.

There can be no research on growth as such. We can only study growing objects. And different growing objects follow different methods. The growth of a differentiating metazoan introduces aspects not presented by the growth of a bacterial culture, and the growth of a higher plant has still other peculiarities.

As we have pointed out repeatedly, each individual tissue and organ has its peculiar mode of growth. To know growth we must therefore first break down each one of its manifestations into its constituent elementary processes and then study these and describe them in objective terms. This is a long way to go, but there is no short cut.

In our present primitive stage of knowledge impatient attempts to formulate a general and universal theory of growth seem to have little chance of success. Those general theories that are sporadically being advanced are usually "general" only in the sense that they are first derived from some rather special segment of the growth problem and then broadly generalised by proclamation or mere implication. Consequently they may be sound and pertinent within the area of their derivation but gratuitous and invalid in their illegitimate extensions. Over-simplification of facts and overgeneralisation in theory are the main causes of the existing discordance among the several contending concepts of growth. Unless we desist from those practices the confusion is bound to grow, in spite of all the excellent experimental work being done in various areas.

Growth is not a simple and unitary phenomenon. Growth is a word, a term, a notion, covering a variety of diverse and complex phenomena. It is not even a scientific term with defined and constant meaning, but a popular label that varies with the accidental traditions, predilections, and purposes of the individual or school using it. It has come to connote all and any of these : reproduction, increase in dimensions, linear increase, gain in weight, gain in organic mass, cell multiplication, mitosis, cell migration, protein synthesis, and perhaps more. It is gravely inconsistent to apply the most exacting standards of precision to our research data and then proceed to mix into their description and interpretation such vague terminology as this.

The mixture can be no more precise than its vaguest ingredient. To deal with growth as an entity, which can be "activated," "stimulated," "retarded," or "suppressed," is only part science, and for the other part, fiction. The less we let our work and thoughts be misled by the delusion that "growth" is basically but a simple elementary process, like a "bimolecular reaction," the faster will be our progress towards true insight into the real mechanisms of development.

Chapter 12

Genetic Resources and Plant Breeding

In order for genetic resources to be efficiently utilised in plant breeding programmes, it is first necessary to determine whether useful genetic variation exists in the material and secondly to develop the most cost-effective method of introducing the potentially useful genes into commercially acceptable material. Until recently, genetic analysis has relied on conventional segregation analysis in controlled crosses for qualitative traits determined by major genes, and on standard biometrical methods for quantitative traits controlled by polygenes. Now, however, the use of DNA techniques has opened up new possibilities for genetic analysis and breeding, and in this chapter the general principles and applications of these molecular marker-based procedures will be discussed.

In its specific sense, marker-assisted or marker-aided selection (MAS) is the use of appropriate easily recognisable genetic markers to facilitate or accelerate the selection of linked genes controlling useful agronomic traits. The nature of the marker loci themselves is not important and they are used simply to indicate the presence of useful alleles of genes of commercial or practical importance. However, the term has also come to include the wider use of markers to improve breeding programmes.

USES OF MARKER-AIDED SELECTION

In an ideal world, selection should be aimed directly at those genes which control the trait to be improved, but this assumes that genotypes at these target genes can be recognised easily, unambiguously and at an appropriate time without impeding the breeding programme. There are several reasons for using marker genes to improve selection efficiency.

Difficulties in Identifying Trait Genotype

The genotypes at a single-trait locus may be difficult to recognise because the desired alleles may have poor penetrance, be recessive, or interact with other genes or the environment. Environmental variation itself may impede accurate genotyping by causing the phenotype of different genotypes to overlap; this is particularly a problem with polygenic traits.

Some phenotypes such as resistance to a particular pest, pathogen or abiotic factor, like drought or salinity, may only appear under particular conditions which are difficult to define or control. These difficulties may be more severe during the early stages of a breeding programme when plant numbers are too few to allow adequate replication or the breeder does not want to place his valuable yet scarce breeding material at hazard.

Earlier Selection

Another major reason why it may be valuable to use marker or indicator genes is to reduce the time from sowing to selection. Many traits of economic importance are only apparent in the mature plant and so may not occur for months or even years after sowing. This is particularly a problem with biennial species and long-lived crops such as trees. Indicator genes, on the other hand, may be detectable in seedlings within days of sowing, if not in the seeds themselves, so avoiding the waste of valuable resources involved in raising and harvesting plants, most of which may prove to be of no value to the breeding programme.

More Intense Selection

Selection at a juvenile stage, particularly among seedlings, may also allow much larger populations to be studied and hence more intense selection to be applied. Indeed, selection may be possible in tissue culture without raising plants or running trials.

Non-destructive Scoring

Many characters are scored before maturity and the act of scoring may prevent seed being collected. Imposition of selection for pests and diseases may result in plants which are less able to produce seed. The use of certain types of marker loci may only require the removal of small quantities of leaf or other material to permit full genotyping of the plant.

Linkage Drag

Conventionally plant breeders have used backcrossing with selection to introduce a useful allele, such as one conferring disease resistance, from one strain into another. The source of the allele is often an alien or exotic species, but may be a related cultivar. The F1 is backcrossed to the recurrent parent, and backcrossing is repeated for five to ten generations, the intention being to introduce the desired donor allele whilst returning to the recurrent parent's genotype at all other loci. At every backcross generation, the breeder selects for the phenotype of the allele to be introgressed whilst simultaneously selecting for the phenotype of the recurrent parent for all other traits.

It was shown by Stam and Zeven (1981) that this procedure results in the leaving of very large amounts of donor genome associated with the selected alien allele. On average, about 20 per cent of an average chromosome will consist of a donor segment even after ten backcross generations, while in any given line the actual size could be much more. Thus, as well as introducing useful alleles, the breeder will also be introducing very many undesirable alleles from the donor and so the gain in one trait may be at the expense of losses elsewhere. The use of marker loci can enable the breeder to reduce the size of this unwanted associated region as well as to accelerate the speed of return to the desired genotype of the recurrent parent.

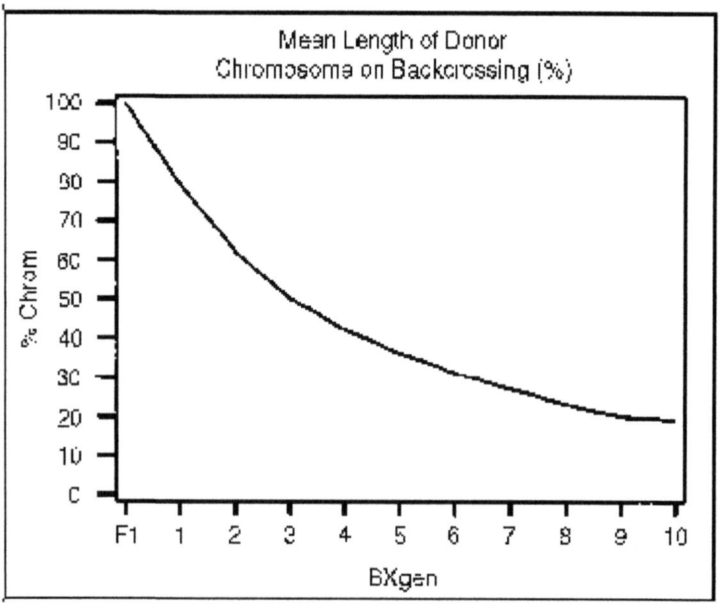

Fig. Linkage Drag. the Average Amount of Chromosome Surrounding a Selected Gene Following Back crossing. Bxgen = Generation of Backcrossing from F.

Heterosis Potential

Heterosis, or hybrid vigour, is due mainly to the presence of dispersed dominant alleles in the parental cultivars. Breeders concerned with developing lines for hybrid production are often concerned to identify lines which differ at as many gene loci as possible. It is often assumed that this implies that the lines also have very different genotypes at marker loci and hence such loci can be used to assess the genetic distance between potential parents, but this can be a dangerous assumption unless a wide range of different types of markers are used.

Quality Control

A major problem faced by most conservationists and breeders is to maintain the integrity of the material they are working with and to keep it free from contamination. Contamination can occur in various ways. It may arise through errors due to carelessness in labelling or sorting seed or plant material, mistakes in maintaining the pedigree, or as a result of outcrossing due to stray pollen from unrelated material. Such errors are difficult to recognise when only the gross phenotype is available as a guide, but genetic markers can provide a clear genetic fingerprint which can be used at any time to confirm the origin of the material. It can also be used to confirm that the intended cross or self has occurred, or to separate the desired progeny from a particular parent which might be a mixture of selfs and crosses.

Other Uses

Marker information can be used in other situations also : distinguishing haploids, diploids and aneuploids, for example, particularly following wide interspecific crosses or tissue culture. Doubled haploid lines produced by microspore or ovule culture can be recognised in this way and possible aneuploids arising during chromosome doubling can be identified and removed. Not only can genetic markers identify contaminant material but they also allow closely related or identical material to be identified in genebank collections, so reducing unnecessary duplication.

TYPES OF GENETIC MARKERS

Up to the mid-1960s the only easily usable genetic markers were those that produced clearly visible effects on the plant's phenotype such as colour, size, shape, disease resistance, etc.. All these were the result of mutations to alleles which had, obviously, a major effect on the development of the plant. Such mutants were rare and generally not desirable in commercial material although they may be selected by breeders of decorative plants for their novelty value. As genetic markers, however, they have many disadvantages. They often affect the fitness of the individual and could well have effects on important agronomic traits, *i.e.* they exhibit pleiotropy. Because they are rare, it would be necessary to bring several such mutants together in the same material in order to use them, and they would not normally be present in commercial cultivars or, indeed, in wild material. This not only takes time, but the combined effects of many such alleles would mitigate against their value as a genetic tool.

In the 1960s through to the 1980s, naturally occurring genetic polymorphisms at the protein level came to be easily recognised. Variation in enzymes and storage proteins was detectable on an electrophoretic gel for a very high proportion of gene loci which could be studied in this way and, unlike major mutants, such naturally occurring allelic variants had much smaller effects on fitness and hence were more common and existed naturally in populations. The number of such polymorphisms was, however, insufficient to provide the coverage of the genome desired by breeders and conservationists. The development of techniques during the late 1980s up to the present revealed the vast amounts of genetic polymorphism which exists at the DNA level and has revolutionised genetic analysis, opening up a whole new range of genetic tools.

As described elsewhere in this book, this variation arises through the existence of occasional base changes in the DNA which can be recognised by restriction enzymes or by primers used in analysis involving the polymerase chain reaction (PCR). Much of the DNA, possibly as much as 90 per cent in many species, is non-coding and hence reliance on morphological or biochemical variants was only using the restricted variation that plants could tolerate within the coding parts of their DNA.

Clearly, there are limits to the tolerance of such variation. Variation in the non-coding regions, whether in the intergenic regions or within introns, is probably under far less constraint by natural selection, and hence the number of polymorphic sites is potentially enormous. Because most of these sites are outside coding genes they are referred to as loci rather than genes; genes are coding regions whilst most marker loci are not.

Restriction fragment length polymorphisms (RFLPs) have been the most widely used until now. They have the advantage of being codominant and hence all genotypes in a cross can be identified. On the other hand they require quite large quantities of DNA, and hence plant material, and generally rely on radioactive probes, although fluorescent techniques are becoming more popular.

As a result, RFLPs are expensive to use. The use of radioactive probes incurs safety considerations, as well as taking longer to visualise the bands. PCR based markers are now being used more widely.

Randomly amplified polymorphic DNAs (RAPDs), for example, require far less material because only small amounts of DNA are needed and the required sequences are then amplified. They do not involve radioactive probes and are both quick and cheap. On the other hand RAPDs are dominant although it is possible to distinguish heterozygotes from homozygotes by the amount of product. Amplified fragment length polymorphisms (AFLPs) offer even cheaper and more easily identified and extensive PCR based polymorphism.

Because of the number of polymorphisms which exist on individual gels and the speed of production, computer software is necessary to scan the banding patterns on the gels and to analyse and assimilate the information that is generated. Other popular markers are microsatellites (short sequence repeats) and cleaved amplified polymorphisms (CAP)-based primers.

All the genetic variants discussed above will be subsumed under the broad title of marker loci. They may be of little interest in their own right, and this is particularly true of the molecular polymorphisms; their value lies in their use as markers of more useful genes.

GENE MAPPING

An important, though by no means an essential, step in genetic analysis is to produce genetic maps of the marker loci. Such maps are often referred to as 'framework maps' because they provide a framework within which important genes can be located, as well as providing a means of comparing chromosome organisation in other closely or distantly related species.

There are two stages to mapping. Firstly, to arrange the markers in a linear sequence separated by an appropriate map distance, *i.e.* to construct a linkage map. The second is to relate the linkage maps to particular recognisable chromosomes. The latter is often the most difficult and generally not essential for breeding or conservation work. We will, therefore, concentrate on the former.

Chromosomes contain a single linear molecule of DNA and hence the markers on that chromosome occur at particular positions along that molecule. Typically, each chromosome contains 107 to 108 base pairs (bp), *i.e.* 104 to 105 kilo-base pairs (kbp) of DNA, while a typical structural gene, coding for a polypeptide chain, would be between 1 and 2 kbp long. Assuming only 10 per cent of the genome is coding DNA, then a chromosome probably contains something of the order of 1000 to 10 000 genes. Fortunately, the fact that the chromosome is a linear molecule means that the genes need to be mapped in only one dimension.

Currently, the only useful method of gene mapping, at least as far as breeders and conservationists are concerned, relies on recombination between homologous chromosomes resulting from genetic exchange, which can be seen as chiasmata at diplotene through first metaphase of meiosis. A single chiasma on a particular chromosome results in half the gametes from such a meiosis being recombinant for that chromosome : half the gametes will contain a copy of that chromosome that is part maternal and part paternal in origin, while the remaining gametes will be entirely parental. Very few plant species have a sufficiently low number of clearly recognisable chromosomes that the number of chiasmata on a particular chromosome can be compared in different nuclei.

However, in many species the total number of chiasmata in each diplotene nucleus can be counted, at least in pollen meiosis, and the average number of chiasmata per nucleus calculated. The number and position of chiasmata on a particular pair of homologous chromosomes will vary from nucleus to nucleus; it is as though there are a large number of potential sites of exchange, but that in any given meiosis only a few sites are actually involved.

The longer the chromosome the more potential sites there might be. There will invariably be one chiasma and there may be two, three, four or more; a typical set of results is shown in Table. A chromosome that has one chiasma on average is said to be 50 centimorgans (cM) long – a map length unit named after the American geneticist Thomas Hunt Morgan.

This relates to the fact mentioned above that one chiasma results in 50 per cent recombinant chromosomes. By extension, a chromosome with an average of 2.5 chiasmata is said to be 125 cM (*i.e.* 2.5 350 cM) long. It appears that, as a rough rule of thumb, each chromosome has on average two chiasmata and so is 100 cM long. It follows, therefore, that providing the haploid chromosome number is known (*n*), the total map length will be approximately 2 3 *n* 3 50 cM. The average number of chiasmata per nucleus has been calculated in many species and this rule generally holds.

Table. Chiasma Fraquency and Mapping.Chiasma Frequency in Chromosomes of Secale Cereale.

Number of chiasmata per chromosome	*Frequency*	*Percentage unrecombined Chromosomes*
1	0.267	50.0
2	0.716	25.0
≥3	0.017	12.5
Mean	–	22.0

This rough rule is important because it gives the geneticist an idea of the total genetic map length that should be expected as more genes or markers are mapped. It provides a guide to the extent to which the currently mapped markers cover the full genetic map. Because the map is based in chiasma units, it is not the same as a physical map. The distribution of chiasmata is not uniform over a given chromosome in a species and will vary with the chromosome and the species.

Although a knowledge of the number of chromosomes and, ideally, the mean chiasma frequency per nucleus for the species, indicates the total map length, the actual map has to be constructed from examining the frequencies of progeny in crosses; *i.e.* maps are constructed from information based on the consequences of chiasmata, namely recombination of genetic markers. Recombination of genetic markers is identified by the presence of gametes that contain recombined markers, and such gametes can be recognised from the phenotypes of progeny of a cross.

The frequency of such gametes is an estimate of the recombination frequency between the two markers concerned. The closer two markers are to each other on the chromosome, the less likely a chiasma will occur between them and the less likely they are to recombine. Providing the two markers are sufficiently close that either none or one chiasma occurs between them but never more, then the recombination frequency as a percentage is equal to the map distance in centimorgans as described above. The problem arises when the markers are sufficiently far apart for two or more chiasmata to occur between them in some meioses. It can be shown that not only does a single chiasma between a pair of markers result in 50 per cent recombination between them, but so also do two, three or more chiasmata, on average.

It is necessary to say 'on average', because there are a variety of possible consequences with two, three or more chiasmata in any given meiosis but, in practice, when one is looking at the progeny of a cross or self, each progeny will be the result of gametes from different meioses.This fact implies that although map distance increases linearly with the number of chiasmata, the same is not true with the frequency of recombination, which can never be more than 50 per cent for a pair of markers even though they might be 100 cM apart at opposite ends of a chromosome.

These relationships are illustrated in Fig.. In order to overcome this problem, mapping functions have been devised to correct for multiple chiasmata. The most common of these are the Haldane (1919) and Kosambi (1943) mapping functions. Haldane's assumes that the probability of none, one, two or more chiasmata in a given interval follows a Poisson distribution, *i.e.* that the chiasmata are independent, random events. Kosambi's method allows for a certain degree of non-independence in chiasma occurrences. It has long been known by cytogeneticists that chiasma interference occurs over quite large regions of a chromosome.

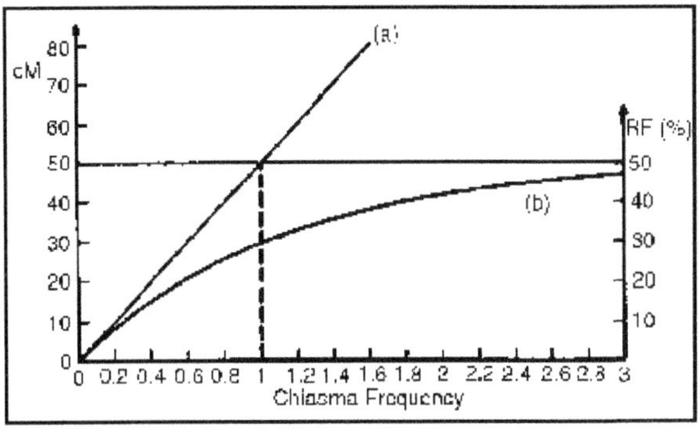

Fig. The Relationship of Chiasma Frequency with (a) Genetic Map Length and (b) Recombination Frequency.

Interference means that if a chiasma occurs at a particular point on the chromosome, the next one will never occur closer than some fixed distance, the interference distance, from it. Beyond this distance, there is a short distance in which the interference disappears and the next chiasma can then occur randomly outside this range.

Observational data in various species of plants are confirming this and suggest that the interference distance is generally between 15 and 20 cM, *i.e.* 15 to 20 per cent of a typical chromosome on either side of the chiasma. This has important and useful consequences for genetic and breeding work as will be shown later.

Recombination frequencies between pairs of markers can be scored in a wide variety of different crosses. The simplest to use are generations derived from an F1 because only two alleles are segregating, and their distribution in the chromosomes of the parents of the F1 can be determined. The generations that can be used are F2, backcrosses (Bc), recombinant inbred lines (RILs) or doubled haploid (DH) lines, with F2s being the most informative.

General formulae for calculating recombination frequencies are given by Allard (1956). In outbreeding species where F1s are not available, a given individual can be considered as an F1 and its selfed progeny as being an F2, even though the inbred parents do not exist. If it cannot be selfed but controlled crossing to another individual is possible, then again those genes segregating in the cross can be mapped although the situation is more complex.

There could be from two to four alleles segregating at each locus and, with respect to any one locus, the cross could represent an F2, Bc1 or Bc2. Moreover, the distribution of alleles in the two parents cannot be known and has to be inferred from the progeny. This complexity is compounded by the large number of marker loci that may be segregating in any given cross.

Fortunately, a range of software packages are available to estimate these recombination frequencies, identify linkage groups, assign the markers to the most likely order and space them in map units (cM) on these linkage groups; examples are MAPMAKER and JOINMAP. Ideally the number of linkage groups should be equal to the chromosome number in the gametes but, unless the markers available provide a good coverage of the genome, it is likely that those markers on a given chromosome may appear as two or more separate linkage groups simply because the subsets are not sufficiently close for them to be recognised as being together on one chromosome.

A typical marker framework map is shown in Fig. for the chromosomes of *Brassica oleracea*. Clearly these maps provide a very detailed coverage of the chromosome. Many of the markers are very close, *i.e.* less than 10 cM, and over this distance it matters little which of the two mapping functions is used. With the more widely spaced markers, *i.e.* recombination frequency >15 per cent, Haldane's function will exaggerate their distance apart because it will allow for more double crossovers than actually occur.

As more markers are mapped, the total length should converge on the value predicted by the chiasma frequency, *i.e.* approximately 2 3 n 3 50 cM. It is important to remember that genetic map distances have standard errors which are dependent on the size and type of population used to construct the map, the mapping population. The map obtained is that which best fits, given the data and the assumptions underlying it, but there may well be other maps which also fit the data well, though are slightly less likely.

Any particular map always develops a greater aura of respectability once it is published, frequently without stating its reliability, and subsequent yet different maps are treated with suspicion. A recombination frequency of true value p has a standard error of ?[p(1 2 p)/N], where N is the size of the gamete population. For example, if p is 0.1 (*i.e.* RF = 10 per cent) and N = 100, the standard error is 0.03 or 3 per cent. In other words the 95 per cent confidence interval of the estimate lies approximately between 4 per cent and 16 per cent.

There are other reasons, apart from statistical sampling, why maps could be wrong. The most obvious is the accuracy with which the data are scored and recorded. Autoradiographs and banding patterns on gels can easily be misread and research workers are loath to discard data even though its interpretation is ambiguous. All data should be checked independently by two or more people and any ambiguities removed. Wrong scores will bias the data and often exaggerate map lengths as they suggest double recombination; wherever double recombination appears to have occurred in two short, adjacent intervals, the data should be checked. Changes in methylation rather than scoring errors could cause a restriction site to appear or disappear so giving the false impression of double crossing- over, or it could simply be an error.

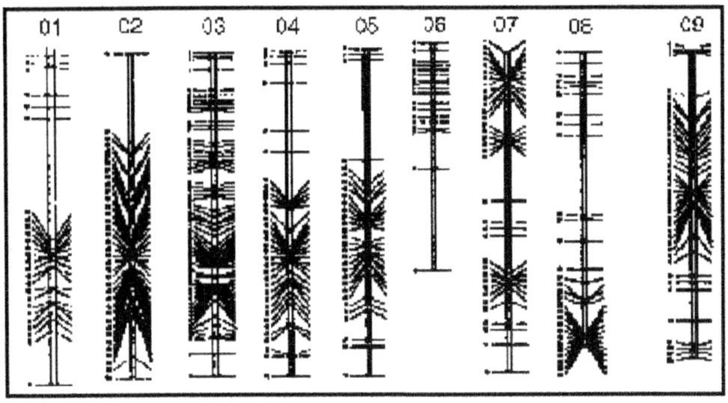

Fig. Typical Molecular Marker Framework Map. The Nine Groups of Brassica Oleracea Based on RFLP, Isozyme and Morphological Markers.

Where maps appear to have major inconsistencies in different crosses, different chromosomal structural arrangements may be responsible, such as translocations or inversions. There is considerable evidence that chiasma frequencies, and hence recombination frequencies, may be quite different in male and female meioses even on the same plant although in other studies this may not be so.

Thus although the order of markers on a linkage group may remain the same, the relative distances between them may be quite different if the map derives from male versus female meioses. For example, in *Brassica* it appears that the genetic map obtained from a backcross where the F1 is the female parent is 60 per cent longer than when the F1 is the male parent. It is also clear that environmental factors during meiosis, particularly temperature, can affect the number and distribution of chiasmata, and this could be very critical if the crosses are set up at different times of the year.

Two or more different populations will almost certainly be segregating for different combinations of polymorphic markers, although some at least should be in common. These common ones can be used to overlay the two or more maps and a consensus map produced from an amalgamation of these. Again software is available to produce these consensus maps. Despite the various error-causing factors discussed above a considerable degree of consensus can be produced from such populations and, considerable similarity of chromosome sequence is conserved between even distantly related species, allowing the potential for cross-species genetic transfer.

It was stated above that the total map length should be dictated by the chiasma frequency, in so far as this can be accurately measured. When the first few genes were initially put onto genetic maps in the first 40 years of this century, they only covered a small part of the genome - except in wellstudied species such as *Drosophila* and maise. As more and more genes were placed on the map, so the total map lengths increased towards the asymptote dictated by the chiasma frequency. However, during the first decade following the use of molecular markers, the map lengths of some species appeared to exceed that expected, and it was even argued by some that the chiasma theory of recombination might be wrong.

No one would wish to argue that chiasma frequency data are free from error but there would have had to have been quite excessive underscoring to result in the disparity which was occasionally found. It would appear, however, that the excessive lengths were due in part to errors in scoring and to the use of small mapping populations, because subsequent map sizes have shown a progressive decrease towards the predicted asymptote. Scoring errors bias estimates of RF generally upwards, while RFs have to be large to be detectable in small populations. Our own experience with mapping in cereals and brassicas has also shown a progressive decrease in estimated map length as more markers can be found to fill some of the large gaps in the map and as errors are removed from the mapping data.

LOCATING GENES OF MAJOR EFFECT

The use of extensive molecular and other markers as described above provides a general framework map. Although this has value in its own right for comparing linkage groups in different species as well as possible structural variation within a species, the main value lies in providing a set of markers to locate genes of economic or special scientific interest.

Locating individual major genes is a straightforward development of the procedures above. Let us assume that a cultivar is identified which contains an allele of interest which changes the phenotype in a clearly recognisable fashion and which appears to segregate as a single gene in crosses to other cultivars. Using a knowledge of the positions of existing marker loci, a small subset of markers can be chosen which provide a reasonably even coverage of the genome, say every 20 to 30 cM, *i.e.* approximately five per chromosome. These then have to be shown to differ between the two cultivars and any which are monomorphic replaced by polymorphic markers.

A rough position of the gene of interest can then be obtained by bulked segregant analysis. This involves taking the mapping population, *e.g.* a Bc, and dividing the plants into two groups depending on whether or not they show the phenotype of interest. Bulk samples of DNA are taken from each group, and the two samples assayed for the marker systems chosen as in the paragraph above.

Any marker which is unlinked to the gene to be located will produce similar banding patterns in both samples because it will be segregating independently of the target gene. Conversely, should one of the markers be very closely linked to the target gene, it will co-segregate and the two samples will show quite different patterns for that marker.

In practice, complete co-segregation with any one marker is not very likely but, given a reasonable spread of markers over the genome, one or two are likely to show a strong association with the two groups. Thus although bands associated with both marker alleles will be found in each sample, their relative intensities will be quite different because of the association with the target gene.

Having located the chromosome and the region on the chromosome, the actual position can be identified in a segregating population using the principal marker associated in the bulked segregant analysis together with other polymorphic markers situated about 10 cM on either side of the principal marker.

The target gene is then mapped with respect to the three markers. This approach can be used for a wide variety of populations and traits. Even if the trait has poor penetrance, the affected group will clearly show the marker banding pattern even if the non-affected group consists of a mixture due to misclassification.

MAPPING GENES CONTROLLING QUANTITATIVE TRAITS

Quantitative traits such as yield, quality, height and flowering time, which are controlled by several genes and are greatly influenced by the environment, create particular difficulties for gene mapping. The problems arise because the genotype for the trait concerned can never be clearly identified from the phenotype; many different genotypes could produce the same or very similar phenotypes whilst the same genotype can result in different phenotypes depending on what may be elusive factors in the environment.

Whereas with a single gene difference controlling a major effect, two or more Mendelian phenotypic classes might be recognised in ratios of 3 : 1, etc., quantitative traits resulting from the joint segregation of many genes show a continuous, often normally distributed range of phenotypes. The total phenotypic variation in an F2, for example, VP, is made up of genetic and environmental components, VG and VE. The genes underlying such traits are variously called polygenes, effective factors or, more recently, quantitative trait loci or QTL.

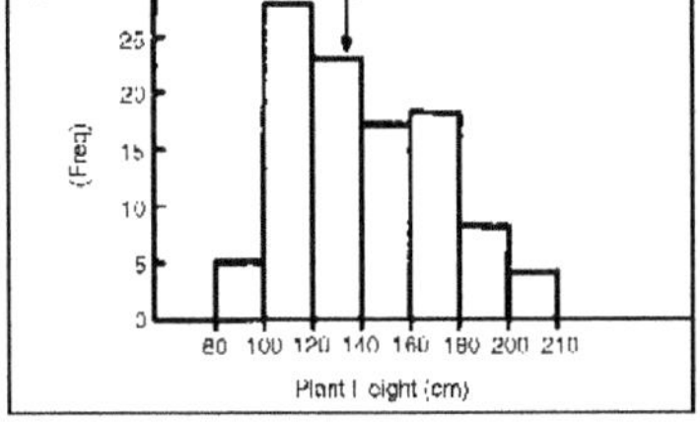

Fig. Contrast Between Distribution of a Qualitive and Quantitative trait in an F2. Data for NIcotiana Rustica, Ovary Colour and Height.

However, it is still possible to map and measure the effects of the QTL by techniques analogous to those used for single major genes, *i.e.* by looking for the effects of co-segregation of particular marker loci with differences in the quantitative trait. Various mapping populations can be used as before, F_2, Bc, RIL, DH or open-pollinated full sib (FS) or half sib (HS) families, and the individuals or families are scored both for their phenotype for the quantitative trait(s) and for their genotype at the marker loci. The type of population that is used will depend on the traits studied and the breeding system of the plant species but, other things being equal, an F_2 population will normally provide the most information for a given size. Unfortunately, most traits of economic importance such as yield, quality and disease resistance are not usefully scored in an F_2 population because one is interested in how the material performs in something similar to the high-density monoculture that it will encounter in agricultural practice. Measuring yield in an F_2 population with spaced plants is straightforward and statistically very powerful but it may well not provide any insight into how yield is controlled in agricultural practice.

It is therefore necessary to use a plot structure for most traits in an attempt to simulate commercial conditions. For this reason most plant breeders will prefer to work with genotypes that can be extensively and easily replicated such as RILs, DHs or F_3s. It is also easier to explain the principles underlying QTL mapping using a population of DH lines, and so for both reasons the methodology will be illustrated with DH lines, although the same principles apply to all populations. Because several trait loci are concerned, probably on different chromosomes, bulk segregant analysis is of limited applicability, at least initially.

THEORY OF QTL MAPPING IN DOUBLED HAPLOID LINES

Doubled haploid (DH) lines can be produced parthenogenetically from an F1 by microspore or ovule culture and, in some species, such as wheat and barley, by wide outcrossing. Each original DH plant is derived from a single gamete, which subsequently becomes homozygous and disomic either naturally or by treatment with the drug colchicine, which inhibits spindle formation, and hence chromosome disjunction, at metaphase of mitosis.

Selfed seeds from such a plant produce a DH line of identical individuals so the line becomes effectively immortal. If we consider an F1 that is heterozygous for a marker locus M1/M2 and for a linked QTL locus Q1/Q2, where the 1/2 indicates the allele with an increasing or decreasing effect on the trait, this will produce the gametes and hence the DH lines shown in Table.

If the DH lines are scored for some trait for which the Q+/Q+ homozygotes increase the trait above the overall mean, *m*, by *a* units while the Q2/Q2 homozygotes decrease the mean by *a*, then the means of the four gametic types are as shown in Table, where *R* is the recombination frequency between the QTL and the marker. From this it follows that the mean trait score of all those DH lines which are M1/M1 (M1M1) is m 1 a(1 2 2R), while the mean of all those DH lines which are M2/M2 (M2M2) is m 2 a(1 2 2R). In other words, half the difference between these two means (M1M1 2 M2M2) is *a*(1 2 2*R*). Clearly, if the marker is so close to the QTL that it never recombines, *R* will be 0 and the marker difference will represent the QTL effect, *a*. Conversely if the two loci are unlinked, *i.e.* *R* = 0.5, then the difference will be zero. This effect is, of course, the basis of bulked segregant analysis. In general, the magnitude of the marker effect will decline linearly with the distance of the marker from the QTL in terms of recombination frequency, and so with several linked markers, their individual trait effects will be as shown in Fig.. Similar Arguments Apply to Other Populations Such as F_2s and RILs.

Table. QTL model for market analysis in doubles haploid (DH) lines from and F_1.(a)The F_1 constitution; M=Market and Q = QLT genotype.(b)DH genotypes, frequencies and genetic values. (c) Expertations of marker means and differences. R=recombination frequency between M and Q; m and a are the mean and additive genetic effects of the QLT;δ is the additive genetic.

(a)

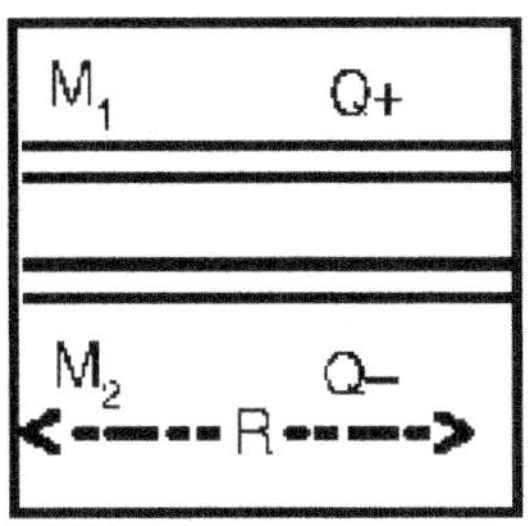

(b)

	F_1 gametes			
	M_1Q^+	M_1Q^-	M_2Q^+	M_2Q^-
Frequency	a(1−R)	a(R)	a(R)	a(1−R)
DH genotype	$M_1M_1Q^+Q^+$	$M_1M_1Q^-Q^-$	$M_2M_2Q^+Q^+$	$M_2M_2Q^-Q^-$
Genetic value	m+a	m−a	m+a	m−a

(c)

$$M_1M_1 = \{a(1-R)(m+a)+a(R)(m-a)\}a$$
$$= m+(1-R)a$$
$$M_2M_2 = \{a(R)(m+a)+a(R)(1-R)(m-a)\}a$$
$$= m-(1-R)a$$
$$\delta = (\overline{M_1M_1}-\overline{M_2M_2})/2$$
$$= (1-2R)a$$

If the trait means of the markers and their map are known very accurately, and only one QTL existed on the chromosome, then as Fig. shows, it would be relatively easy to locate the QTL. Figure shows the distribution of flowering time among a number of doubled haploid lines of *Brassica oleracea* containing alternative alleles at a marker located very close to a QTL.

In practice neither the map positions of the markers nor their means are known with very great accuracy because the number of DH lines and replicated plots are generally few and the heritability of quantitative traits is generally low. Moreover, linked QTLs create difficulties because their individual effects can combine either to create the impression of an intermediate, ghost QTL, or to conceal their effects altogether.

Various analytical procedures have been developed to maximise the efficiency and accuracy of locating individual QTLs and to separate linked QTLs. They all rely on the relationships described in the previous paragraph and use either maximum likelihood or weighted least squares regression procedures to estimate the QTL locations and effects that best fit the observed trait scores for each genotype.

The main difficulty is that there are two unknown parameters with respect to each marker score; the QTL effect, *a*, and map position. This means that it is necessary, in all methods, to use an iterative approach which involves trying all possible QTL locations along each chromosome and identifying the position or positions of the QTL which best fit the observed data as indicated by the size of the likelihood or the residual regression variance. As a result, a very large number of statistical tests are performed with the concomitant risk of false positive results.

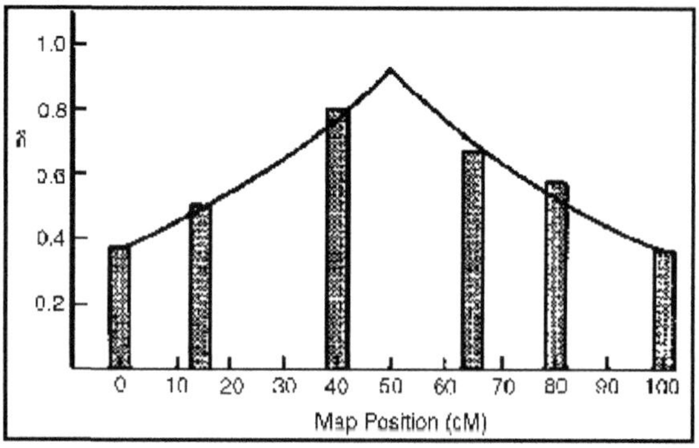

Fig. Quantitative Trait Effects (δ) Associated with Marker Loci.The Bars Indicate the Observed Effects,While the Curve Illustrates the Expected Relationship if a QTL of δ = 1.0 Existed at 50 cM.

All methods provide estimates of QTL locations and effects with similar precision in terms of confidence intervals, and this precision drops rapidly as the heritability of individual QTLs declines, *i.e.* the estimates are accurate only when there are just two or three unlinked QTLs with large effects. For example, Hyne *et. al.* (1995) showed that the 95 per cent confidence interval for the location of a QTL contributing 10 per cent to the phenotypic variance of an F2 could be as large as 35 cM. However, conservationists and breeders who are concerned with introducing QTL into commercial cultivars from other cultivars or more distantly related species, would normally only be interested in those cases where a few QTLs of major effect are to be manipulated. There is no evidence that dense maps are required for initial.

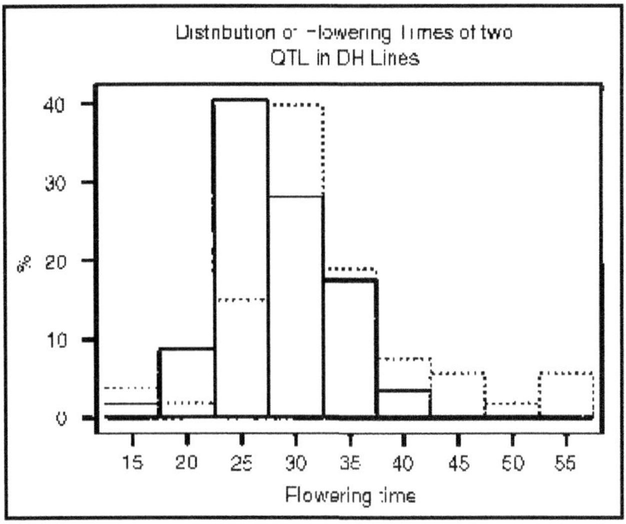

Fig. Distribution of Flowering Times Among Doubled Haploid (DH) Lines in Brassica Oleracea Containing alternative Alleles of a Marker Linked to a QLT. The two sets Significantly Different Means Indicating a QLT with an Effect of Approximately ± 2 Days.

QTL location. It is better to have large populations and a few (5–10) markers per chromosome.

There is evidence that many of the larger QTLs located so far map closely to previously known major genes. This is often used as evidence that there are, in fact, very few QTLs, but this may well be misleading for several reasons. The QTL map locations are sufficiently imprecise that they could well appear spuriously close to candidate loci. Only QTLs of large effect will stand much chance of being located and, by definition, they are likely to be major genes. It is likely that many cases where QTLs have very large effects could be due to the chance association of alleles of like effect at several QTLs along a chromosome, which cannot be separated as individual effects.

Much effort has been devoted to increasing the precision of QTL location, in particular in improving the power of the test to reduce the failure to detect genuine QTL. A long-established problem with breeding for quantitative traits in plants is the genotype 3 environment interaction. Using QTL location techniques, it is now possible to explore these effects at the level of individual QTLs although it is often difficult to know whether genotype 3 environment or poor repeatability is responsible.

ALIEN GENE TRANSFER

Once useful genes have been located on a framework map, it is then possible to use the molecular markers to facilitate the transfer of the useful genes between species and strains in an efficient manner. As stated earlier, recombination is not a frequent event along a chromosome and, furthermore, two recombination events rarely occur within a distance of 15 to 20 cM of each other.

It therefore follows that if a useful target gene (a major gene or a QTL) is known to be located within a region of chromosome flanked by two markers that are no more than 15 cM apart, then these markers can be used to follow and control the progression of the target gene through successive stages of a breeding programme. Clearly, the alleles at the marker loci have to differ in the donor and recipient cultivars, but any chromosome that contains A1 and B1 at the marker loci will also contain T1, the target allele to be transferred.

Should single recombination occur close to T, then the chromosome will contain A1 or B1 but not both. Providing the markers are less than 15 cM apart double recombination resulting in A1T2B1, and hence loss of T1, can be safely ignored. If the location of T is not accurately known, then three or more marker loci may be needed to be sure of safely bracketing the region containing T without fear of double recombination.

There are conflicting requirements in marker-assisted gene transfer : the need to keep the bracketed region large enough to be sure of holding the gene to be transferred whilst not having it so large that too many other linked but undesirable alleles are transferred with it. Let us now consider the basic procedure of gene transfer.

Traditionally, breeders have introduced a useful allele into their cultivars by backcrossing the original F1 to the commercial cultivar whilst selecting at each generation for the desired allele. As was shown earlier, this can result in a very large region of chromosome around the target gene surviving even after ten generations of backcrossing – a phenomenon known as linkage drag. With markers, however, it is possible to reduce linkage drag considerably whilst simultaneously reducing the number of backcrossing generations to two or three instead of the normal six to eight. Moreover, the amount of plant material raised at each generation can be reduced because much of the molecular genotyping can be achieved at the juvenile stage.

The aim is to hold the target allele, T1, heterozygous throughout the backcrossing process, by selecting for the flanking markers, whilst simultaneously selecting for the genotype of the recurrent, commercial cultivar at all other loci. The latter can readily be achieved by having as few as four or five well-spaced markers on all chromosomes and selecting for recurrent parent alleles at each generation.

Depending on the chromosome number, some individuals in the first backcross generation will be homozygous for several whole chromosomes whilst still being heterozygous for the target gene. Those which have most of the recurrent parent chromosomes are selected and backcrossed again, and if necessary the process is repeated for a third generation. For each generation, fewer markers need to be screened as those found to be homozygous in the previous backcross no longer have to be checked.

Unless the position of the target gene is very accurately defined, it is prudent to have several markers covering its likely position, again with the proviso that they need to be separated by no more than 15 cM. At the final stage of the introgression process, a backcross individual which is homozygous for most of the recurrent parent alleles is selfed or, if this is not possible, intercrossed with a similar genotype. From among the progeny, individuals are chosen which have different combinations of markers bracketing the target region.

These can be selfed again and homozygotes for the different sequences identified and multiplied for reassessment for the target trait. Some will fail to show the target trait because of recombination, but of those that do, the lines with the shortest sequence of donor markers are selected for further trials. The approach can obviously be adapted to the simultaneous introgression of several genes : no new principles are involved but the screening process becomes more elaborate. It is always possible at a later stage to reduce the length of the introgressed region around T by looking at other, more closely linked, markers. Similarly, it is wise to check the origins of the other chromosomes to avoid having introgressed unwanted regions through missing double recombinants in the central regions of individual chromosomes or single recombinants towards their ends.

If such errors are found they can easily be corrected by another round of backcrossing to the recurrent parent. When attempting to reduce the length of donor chromosome surrounding the target gene, it is important to realise that this should be done in two successive generations. If the sites of the required recombination events are less than 15 cM apart, then the simultaneous double event will not occur.

Even if they are far enough apart for the double event to occur, the probability of such an event may be too low to be practicable. If two recombination events each have a probability of 0.05 (*i.e.* 5 per cent), their combined probability is at best 0.0025. Therefore, it would be necessary to genotype approximately 1200 individuals to be 95 per cent sure of having at least one double recombinant. If it was tackled in two stages then it would require only the genotyping of 86 (= 28 in the first round plus 58 in the second; less being needed in the first round because the recombination could occur on either side of T), a very considerable saving in effort of 93 per cent, or 1114 plants!

Marker-aided selection has been successfully tried for quantitative traits in several crops while the efficiency of the procedure has been examined by Lande and Thompson (1990).

MAP-BASED GENE CLONING

Providing a target gene can be located to within a pair of marker loci, it should be possible to use standard cloning techniques to walk along the chromosome between the markers and to find the target gene. The feasibility of this depends on the distance between the markers and the ease with which the target gene can be recognised from among the clones.

Depending on the species, a genetic length of 1 cM can, on average, consist of as few as 280 kbp of DNA in rice, or 2220 kbp in maise. It is not always easy to locate a target gene with sufficient accuracy to be able to say that it is between two markers as close as 1 cM, and hence the actual lengths of DNA between them can often be much greater. With QTLs one can seldom achieve anything approaching this accuracy by existing methods of conventional mapping. Furthermore, these are average distances per centimorgan; some chromosomal regions could be much longer or, indeed, less.

Increases in mapping reliability can be obtained by using what are called near isogenic lines (NILs), which are lines obtained by selfing or backcrossing and are known to differ from some standard genotype by just a short defined section of chromosome. Providing this section is delineated by accurately mapped flanking markers, and the NILs which do and do not contain this region can clearly be shown to differ for the target gene, then the target gene has to be in that region. The genetic and physical proximity of the markers can be established by conventional mapping and by gene cloning.

Map-based gene cloning involves starting from one of the markers and walking to the next by means of a series of partially overlapping cosmid clones. Cosmid clones are necessary because they can be up to 50 kb in length and hence require fewer cloning steps in order to cover the DNA between the markers (280 to 2200 kb per cM in the examples cited above). Identifying the target gene is more difficult. Clearly, if the gene is between the two flanking markers it has to be on one of the cloned regions and could possibly be on two overlapping clones.

It could be identified by transforming plants with each clone separately and identifying which transformant expresses the effect. If the gene concerned exhibits classical dominance it would be necessary to transform the homozygous recessive with the dominant allele or the transformant would not be recognised. Alternatively, the gene has to be introduced into a species that does not normally express the gene. Once the effective cloned fragment is identified, the actual location of the gene within the fragment could be identified by successively transforming sub-fractions of the clone or, following sequencing, by looking for potential open reading frames. In those cases where the target gene produces a known product, cDNA clones derived from tissue likely to be rich in the appropriate mRNA of the target gene can be used to identify likely sequences in the previous overlapping clones.

Once the gene has been identified by one of these methods, its structure and activity can be studied in detail. Disease resistance genes are obvious candidates for such studies as their action is very specific and their location on the genome of many species is well documented. However, unless suitable candidate loci.

Chapter 13

Culturing of Plants

CELL CULTURE

Plant tissue culture encompasses culturing of plant parts on an artificial medium. The plant parts can be a single cell, tissue or an organ. It is also referred to as micropropagation. Plant tissue culture was practically implemented for the first time by Haberlandt, a German scientist, in 1902. Later in 1934, Gautheret found successful results on in-vitro culture of plants. The basic key used in plant tissue culture is the totipotency of plant cells, meaning that each plant cell has the potential to regenerate into a complete plant. With this characteristic, plant tissue culture is used to produce genetically identical plants (clones) in the absence of fertilization, pollination or seeds.

Plant tissue culture encompasses culturing of plant parts on an artificial medium. The plant parts can be a single cell, tissue or an organ. It is also referred to as micropropagation. Plant tissue culture was practically implemented for the first time by Haberlandt, a German scientist, in 1902. Later in 1934, Gautheret found successful results on in-vitro culture of plants. The basic key used in plant tissue culture is the totipotency of plant cells, meaning that each plant cell has the potential to regenerate into a complete plant. With this characteristic, plant tissue culture is used to produce genetically identical plants (clones) in the absence of fertilization, pollination or seeds.

Methods and Techniques

Raupp *et. al.* evaluated the relative yield, product quality and soil life after long-term (17 years) organic or mineral fertilization. They found that production yields varied : yields of potatoes and rye were lower and yields of spring wheat were similar in the organic system. However, their evaluation also showed that humus content and biological activity in the soil were greater, that products from the organic system had a better storage quality and that vegetables had lower nitrate contents.

Apart from the quality of the product from different production systems, the production potential of organic in comparison with conventional production systems is important. Research findings seem contradictory as production yields are found to be lower, equivalent, or higher. Simulating the production potential after long-term differing management systems showed the following : Droogers *et. al.* compared the production potential of two farming systems (biodynamic and conventional) by converting 'static' basic soil properties into a 'dynamic' assessment using simulation modelling. Soil conditions on two farms — one having been managed biodynamically for 70 years — were investigated with morphological and physical methods. A simulation model including 30 years climatic data was used to predict water-limited potato yields, showing that the simulated yields were significantly higher on the biodynamic fields.

Rhizosphere — Interaction with Plant Growth

No published research was identified in New Zealand on this topic. The international work published emphasised the influence that soil flora and fauna have on plant growth in an organic system. Scullion *et. al.*, (1998), for example, described how management before conversion can have an impact on soil fungi, that in turn will influence plant growth under the organic system. Yeates *et. al.*, described soil microbe and faunal diversity in Welsh soils, and while the soil types may differ, the message to enhance diversity is the same. This paper also described indicators that could be used to define soil "health" in an organic system. Eason *et. al.* and Cook *et. al.*, both described how management can impact on soil microbe and fungi populations and how this in turn can influence plant growth.

Trace Elements

Condron *et. al.* presented a comparison of soil quality under conventional and organic management in New Zealand. One area of focus in the paper was on trace elements in soils under organic and conventional pasture and their effects on animal health. Trace element impacts on animal health were also mentioned in some of the more generic animal health papers described earlier. Other than this work, there was very little information on trace elements in soils under organic pastures.

Microbial Activities and Biomass

Synthetic water soluble fertilizers are known to be detrimental to soil microorganisms and therefore to influence biological and physical soil characteristics negatively in the long-term. Mader *et. al.*studied the effect different fertilization intensities on different crops have on soil micro-organisms at different soil depths. They compared organic and conventional farming systems in Switzerland and reported that soil microbial biomass was significantly higher under the biodynamic system. Wood reported on studies in the UK, that showed an increase in earthworm populations with additions of manure; and comparisons of conventional and biodynamic farms in New Zealand showed that the biodynamically farmed soils had better structure, lower bulk density, higher organic matter content and respiration rates, and higher earthworm populations.

Pfiffner *et. al.* described a long-term trial studying the effect of different farming systems (biodynamic, organic and conventional) on ground beetles. The number of species was consistently higher (193 per cent) in the biodynamic than in the conventional (100 per cent) and organic (188 per cent) plots. The number of species differed from 18-24 in biodynamic plots, compared to 19-22 in organic and 13-16 in conventional plots. In 1998 Pfiffner *et. al.*, (1998) reported on a long-term trial into the effect of different farming systems (biodynamic, organic and conventional) on earthworm populations. The earthworm biomass and density, and the number of juveniles and earthworm species were significantly higher in the biological than the conventional or organic plots.

Ryan reviewed studies with regard to the question 'Is an enhanced soil biological community a consistent feature of alternative agricultural systems?' Studies that examined four groups of soil organisms, comparing the soil biological community in conventional, organic and biodynamic management system were reviewed, and a case study of biodynamic and conventional dairy farms in Australia included. Ryan's conclusion was that the enhanced soil community found on the biodynamic farms relative to conventional neighbours should not be considered as a definitive feature of alternative agricultural systems, but rather as an effect of the higher input of organic matter.

Lytton-Hitchins *et. al.* came to the same conclusion after comparing the physical and chemical properties of biodynamically and conventionally managed pastures in NE Victoria, Australia. The more favourable properties found on the biodynamically managed soils are attributed to decreased grasing pressure, longer intervals between irrigations, reduced tractor traffic and intermittent applications of compost and hornmanure preparations.

Ryan *et. al.*, undertook a glasshouse experiment with soil samples from 3 biodynamic (without conventional fertilizers for 17 years) and conventional (regular inputs of soluble P and N fertilizers) dairy pastures. Plant nutrient uptake was examined by assessing the response of white clover, perennial rye grass and indigenous VAM fungi to the addition of 4 levels of soluble P and N. Their findings indicated that the soils had not developed substantially different processes to enhance plant nutrient uptake. The experiments did not examine the response of plant nutrient uptake to non-water-soluble fertilizer, which might have produced different results.

PLANT CELL MODEL

Biotechnology is name given to the methods and techniques that involve the use of living organisms like bacteria, yeast, plant cells etc. or their parts or products as tools (for example, genes and enzymes). They are used in a number of fields : food processing, agriculture, pharmaceutics, and medicine, among others. Plant tissue culture can be defined as culture of plant seeds, organs, explants, tissues, cells, or protoplasts on nutrient media under sterile conditions.

The science of plant tissue culture takes its roots from path breaking research in botany like discovery of cell followed by propounding of cell theory. In 1839, Schleiden and Schwann proposed that cell is the basic unit of organisms. They visualised that cell is capable of autonomy and therefore it should be possible for each cell if given an environment to regenerate into whole plant. Based on this premise, in 1902, a German physiologist, Gottlieb Haberlandt developed the concept of *in vitro* cell culture. He isolated single fully differentiated individual plant cells from different plant species like palisade cells from leaves of Laminum purpureum, glandular hair of Pulmonaria and pith cells from petioles of Eicchornia crassiples etc and was first to culture them in Knop's salt solution enriched with glucose. In his cultures, cells increased in size, accumulated starch but failed to divide. Therefore, Haberlandt's prediction failed that the cultured plant cells could grow, divide and develop into embryo and then to whole plant.

This potential of a cell is known as totipotency, a term coined by Steward in 1968. Despite lack of success, Haberlandt made several predictions about the requirements in media in experimental conditions which could possibly induce cell division, proliferation and embryo induction. G Haberlandt is thus regarded as father of tissue culture. Taking cue from Haberlandt's failure, Hannig chose embryogenic tissue to culture. He excised nearly mature embryos from seeds of several species of crucifers and successfully grew them to maturity on mineral salts and sugar solution. In 1908, Simon regenerated callus, buds and roots from Poplar stem segments and established the basis for callus culture. For about next 30 years, there was very little further progress in cell culture research. Within this period, an innovative approach to tissue culture using meristematic cells like root and stem tips was reported by Kolte and Robbins working independently.

All these research attempts involving culture of isolated cells, root tips or stem tips ended in development of calluses. There were two objectives to be achieved before putting Haberlandt's prediction to fruition. First, to make the callus obtained from the explants to proliferate endlessly and second to induce these regenerated calluses to undergo organogenesis and form whole plants. It was in 1930s, when progress in plant tissue culture accelerated rapidly owing to an important discovery that vitamin B and natural auxins were necessary for the growth of isolated tissues containing meristems. This breakthrough came from White who reported that not only could cultured tomato root tips grow but could be repeatedly subcultured to fresh medium of inorganic salts supplemented with yeast extract. He later replaced YE by vitamin B namely pyridoxine, thiamine and proved their growth promoting effect.

Description

Plant cells are the cells found in plants comprising sub cellular organelles. Following is a plant cell model for kids that will give your child a visual of the plant cell and a better understanding of its various parts.

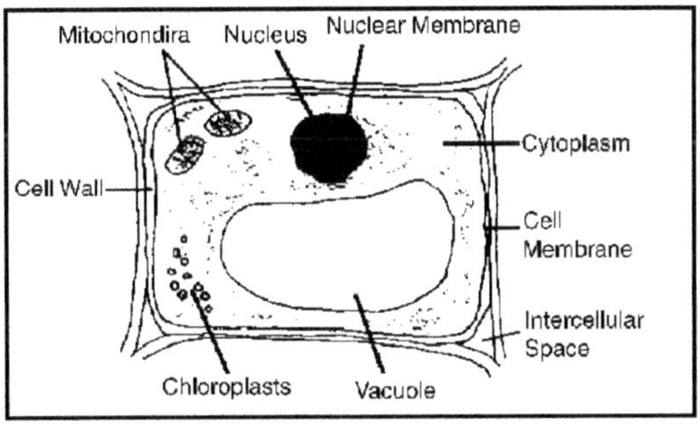

The cells are the basic units of life that work together to perform life sustaining functions in both animal and plant worlds. Plant cells are eukaryotic cells having thick and rigid cell walls.

Eukaryotic cells are cells that contain complex structures enclosed within membranes called nucleus or nucleus envelop, within which the genetic material is present. These cells are present in almost all living organisms including animals, plants and fungi.

Plant Cell Model and Parts

Plant cells differ from animal cells in that they have three different structures known as cell wall, vacuoles and plastids. However, they lack centrioles and intermediate filaments, which are present in animal cells. Read more on plant cell structure and parts.

- **Cell Membrane :** It permits the waste material to exit the cell and regulates the movement of materials in and out of the cell. It also acts as a barrier between the inside of the cell and the outside, so that the chemical conditions on both sides can be different.
- **Cell Wall :** It provides rigid structural support to the cell from which structures, like, leaves and stems are produced. It allows the circulation and distribution of water, minerals and nutrients in and outside the cell. The wall also controls the presence of pathogen microbes and development of tissues in the cell.
- **Golgi Body :** It stores, packages, and circulates the lipids and proteins made in the endoplasmic reticulum. It stores the proteins into packages called vesicles.
- **Rough Endoplasmic Reticulum :** It synthesises and exports proteins and glycoproteins throughout the cell.
- **Lysosomes :** These organelles digest food by breaking down larger molecules into smaller ones, using special proteins.
- **Cytoplasm :** It helps in maintaining the cell shape. Along with this, it plays an important role in the internal movement of cell organelles, cell mobility and muscle fibre contraction. It also distributes oxygen and nutrients to different parts of the cell.

- **Nucleolus :** It is the most prominent structure in the nucleus wherein ribosomes are made.
- **Vacuole :** The plant cell contains a large, single vacuole (an enclosed compartment), which is used to store water, compounds and minerals that help in plant growth.
- **Ribosomes :** These are packets of RNA (Ribonucleic Acid) and are called as protein builders or synthesisers of the cell.
- **Chloroplasts :** They are the food producers of the cell, containing chlorophyll, the green pigment essential for photosynthesis. Their main function is to generate sugars and starches.
- **Nucleus :** It's the most important part of the cell, comprising chromosomes, *i.e.* structures made up of genetic information that helps in cell growth and reproduction. Read more on cell nucleus : structure and functions.
- **Nuclear Envelope :** It's an enclosure that surrounds nucleus and its contents. Unlike cell membrane, which has pores and spaces for RNA and proteins to pass through, it keeps the chromatin and nucleolus inside the nucleus.
- **Smooth Endoplasmic Reticulum :** It's main function is to package proteins for transport, synthesise membrane phosolipids, and secrete calcium. It also performs transformation of bile pigments, glycogenolysis (the breakdown of glycogen), and detoxification of different drugs and chemical agents.
- **Mitochondria :** It provides energy to the cell by combining sugar molecules with oxygen to generate carbon dioxide and water.
- **Amylosplast :** It's an organelle present in some plant cells that stores starch.
- **Druse Crystal :** It's a granular type of crystal found in plant vacuoles. It is composed of calcium oxalate and is considered to deter herbivory.
- **Centrosome :** Also known as Microtubule Organising Center, it's a region in the cell where microtubules are produced that perform a variety of functions ranging from transportation to structural support. Though both plant and animal cell centrosomes play similar roles during cell division, plant cell centrosomes are simpler and do not contain centrioles.
- **Peroxisomes :** These are membrane bound packets of oxidative enzymes that convert fatty acids into sugar and assist chloroplasts in photo-respiration.
- **Golgi Vesicles :** These vesicles transfer proteins and lipids to different parts of the cells.

Plant Cell Model

Following are some simple instructions or plant cell model ideas, to construct a model of a plant cell :

- Enumerate and collect all possible materials required for the model and look out for the different cell parts in a biology textbook.
- Use a heavy sturdy wooden base to construct the base of the model so that it can hold the weight and facilitate presentation of the model.
- Construct the cell nucleus and ensure that it's round in shape. Paint the nucleus in one colour so that it can be identified as a distinct cell part.
- With the help of a dowel rod or thick, straight stick, connect the nucleus to the base of the model. For this, drill a hole into the wooden base, stick the dowel rod into the hole and connect the nucleus to the rod.
- The cell wall and cell membrane of the plant cell model should be rectangular in shape. The outer wall is the cell wall, which should be made of a strong material like hard plastic or wood, and the cell membrane can be made of cellophane or thin plastic. Now attach both the walls to the nucleus and the model base.
- Add various other key components of the plant cell to the model. Like, for chlorophyll, use a piece of green fabric stuffed with cotton. Sew it and attach it to the model.
- Finally, prepare a report on the various plant cell parts and their corresponding functions.

PLANT CELL ORGANELLES

Plants are highly evolved eukaryotic organisms that comprise of membrane bound cell organelles. Even though plants and animals belong to eukaryotic groups, they differ in certain characteristic features. For example, a plant cell possesses a well-developed cell wall and large vacuoles, while an animal cell lacks such structural parts. In addition to these, a plant cell lacks centrioles and intermediate filaments, which are present in an animal cell.

A typical plant cell is made up of cytoplasm and organelles. Scientific studies have been done regarding plant cell organelles and their functions. Each of the organelles of a plant cell has specific functions, without which the cell cannot operate properly.

List of Plant Cell Organelles

When it comes to plant cell organelles, they are more or less similar to animal cells, except that the latter lacks chloroplast organelles, that are responsible for photosynthesis. Following is a list of organelles found in plant cell :

- **Nucleus :** Nucleus (plural nuclei) is a highly specialised cell organelle, which stores the genetic component (chromosomes) of the particular cell. It serves as the main administrative center of the cell, by co-ordinating the metabolic processes like cell growth, cell division and protein synthesis.
- **Plastids :** Plastids are collective terms for organelles that carry pigments. In a plant cell, chloroplasts are the most prominent forms of plastids,

which contain the green chlorophyll pigment. Because of these chloroplast plastids, a plant cell has the ability to undergo photosynthesis in the presence of sunlight and synthesise its own food.

- **Ribosomes :** Ribosomes are plant cell organelles that comprise of proteins (40 per cent) and ribonucleic acid or RNA (60 per cent). They are important organelles responsible for the synthesis of proteins. Each ribosome consists of two parts, a larger sub-unit and a smaller sub-unit.
- **Mitochondria :** Mitochondria (singular mitochondrion) are spherical to rod-shaped organelles present in the cytoplasm of the plant cell. They break down the complex carbohydrates and sugars into usable forms, for the plant. As mitochondria indirectly supply energy for the plant cell, they are also called as the powerhouse of the cell.
- **Golgi Body :** Golgi body is also referred to as golgi complex or golgi apparatus. It plays a major role in transporting chemical substances in and out of the cell. After the endoplasmic reticulum synthesises lipids and proteins, golgi body alters and prepares them for exporting outside the cell.
- **Endoplasmic Reticulum :** Endoplasmic reticulum (ER) is the connecting link between the nucleus and cytoplasm of the plant cell. Basically, it is a network of interconnected and convoluted sacs that are located in the cytoplasm. Based on the presence or absence of ribosomes, ER can be of smooth or rough types. The former type lacks ribosomes, while the latter is covered with ribosomes. Overall, endoplasmic reticulum serves as a manufacturing, storing and transporting structure for glycogen, proteins, steroids and other compounds.
- **Vacuoles :** Vacuoles are the storage organelles that help in regulating turgor pressure of the plant cell. In a plant cell, there can be more than one vacuole. However, the centrally located vacuole is larger than others, which stores all sorts of chemical compounds. Vacuoles also assist in intracellular digestion of complex molecules and excretion of waste products.
- **Peroxisomes :** Peroxisomes are cytoplasmic organelles of the plant cell, which contains certain oxidative enzymes. These enzymes are used for the metabolic breakdown of fatty acids into simple sugar forms. Another important function of peroxisomes is to help chloroplasts in undergoing photorespiration process. Well ! This was brief information regarding plant cell organelles, their structure and specific functions. As we have seen, co-ordination of the functions of plant cell organelles is crucial for carrying out the physiological and biochemical functionalities of the plant.
- **Cell Nucleus : Structure :** The structure of a cell nucleus consists of nuclear membrane (nuclear envelope), nucleoplasm, nucleolus and chromosomes. Nucleoplasm, also known as karyoplasm, is the matrix present inside the nucleus. Let's discuss in brief about the several parts of a cell nucleus.

- *Nuclear Membrane :* The nuclear membrane is a double-layered structure that encloses the contents of the nucleus. The outer layer of the nuclear membrane is connected to the endoplasmic reticulum. A fluid-filled space or perinuclear space is present between the two layers of a nuclear membrane. The nucleus communicates with the remaining of the cell or cytoplasm through several openings called nuclear pores. Nuclear pores are the sites for the exchange of large molecules (proteins and RNA) between the nucleus and cytoplasm.
- *Chromosomes :* Chromosomes are present in the form of strings of DNA and histones (protein molecules) called chromatin. Chromatin is further classified into heterochromatin and euchromatin based on the function. The former type is a highly condensed trancriptionally inactive form, mostly present in adjacent to the nuclear membrane. Euchromatin is a delicate, less condensed organisation of chromatin, which is found abundantly in a transcribing cell.
- *Nucleolus :* The nucleolus is a dense, spherical-shaped structure present inside the nucleus. Some of the eukaryotic organisms have nucleus that contains up to four nucleoli. The nucleolus plays an indirect role in protein synthesis by producing ribosomes. Ribosomes are cell organelles made up of RNA and proteins; they are transported to the cytoplasm, which are then attached to the endoplasmic reticulum. Ribosomes are the protein-producing structures of a cell. Nucleolus disappears when a cell undergoes division and is reformed after the completion of cell-division.

Cell Nucleus : Functions

Speaking about the functions of a cell nucleus, it controls the hereditary characteristics of an organism and is responsible for the protein synthesis, cell division, growth and differentiation.

Here is a list of the functions carried out by a cell nucleus :

- Storage of hereditary material, the genes in the form of long and thin DNA (deoxyribonucleic acid) strands, referred to as chromatins.
- Storage of proteins and RNA (ribonucleic acid) in the nucleolus.
- Nucleus is a site for transcription in which messenger RNA (MRNA) are produced for the protein synthesis.
- Exchange of hereditary molecules (DNA and RNA) between the nucleus and rest of the cell.
- During the cell division, chromatins are arranged into chromosomes.
- Production of ribosomes (protein factories) in the nucleolus.
- Selective transportation of regulatory factors and energy molecules through nuclear pores.

As the nucleus regulates the integrity of genes and gene expression, it is also referred to as the control center of a cell. Overall, the cell nucleus stores all the chromosomal DNA of an organism. The environment which we inhabit and live our daily lives, is an incorporation of several kinds of components. These components, although may not share any resemblance, yet are in a mutual dependency with each other. These components not only refer to the living things, but other elements which are classified as the non-living things on the Earth. Both these elements are important in their own domain in the environment, as they work to maintain the delicate balance in the ecosystem. And this very balance helps to regulate all those things which makes life possible on the Earth. Ergo, with any of these things missing in this system, the biological organisation will suffer nothing less than chaotic events. So let's proceed further to understand these types of components and what role do they play in the levels of organisation of life.

Levels of Organisation

In an ascending manner, the levels of biological organisation are as follows :

- **Subatomic Particles :** Any substance is composed mainly of three subatomic particles. They include the protons, neutrons and electrons. As you must be aware, protons refer to positively charged particles, while neutrons are ones which posses no charge. These two kinds reside inside the nucleus of the atom. Now, about the electrons, they posses negative charge and rotate around the nucleus, in different energy shells. Another subatomic particle is what is known as the photon - possessing zero mass and rest energy and traverses with the speed of light. It is actually defined as a quantum of electromagnetic energy.
- **Atoms :** After the subatomic particles, now we come to the next point in the levels of organisation, know as atoms. All objects are made up of matter in this universe. And the basic building blocks of these matter are known as atoms. These components consist of equal number of protons and neutrons. However, a single element might have two different atoms, wherein, there might be a difference in the number of neutrons. The center of the atom is known as the nucleus, and it comprises of protons and neutrons. This structure of the atomic nucleus is orbited by negatively charged electrons.
- **Small Molecules :** By saying small molecules, I am referring to those levels of organisation, which comprises of amino acids, fatty acids, glucose, etc.
- **Macromolecules :** Now these are the large molecules which are contained in a cell, and are required for carrying out the vital processes of life. Examples include carbohydrates, lipids, proteins, nucleic acids, etc.
- **Molecular Assemblies :** A level higher than macromolecules, these assemblies are defined as sets which comprise of one or more molecular entity.
- **Organelle :** Every cell has a specialised part which serve as organs. These are known as organelle. Examples are nucleus, mitochondria, etc.

- **Cells :** This is one of the important levels of organisation, wherein, cells are known to play vital roles. Every living creature on this Earth consists of cells, which are primarily, their structural and functional units. For instance, micro-organisms like bacteria are known to be single-celled organisms. While, we humans and other organisms, are known to have as many as 100,000,000,000,000 cells!
- **Tissues :** Tissue is nothing but a term that is used to refer to a group of cells. There are several functions in the body of living organisms. And different function is assigned to different tissues. For example, muscle tissue, nervous tissue, etc., are found in animals, while plants, have meristematic tissues and permanent tissues.
- **Organs :** Again, when tissues are grouped together, they form what is known as an organ, which also has an important place in the levels of organisation. As you are aware, heart, lungs, kidneys, stomach are organs. And these organs are enabled by the different tissues to carry out their specific duties.
- **Organ System :** As you must have guessed already, this system is a formed by different kinds of organs. The system is assigned to perform higher levels of work in the body. This is more important in multicellular organisms like us and animals. A very simple example of an organ system is the circulatory system. It is a collaboration of heart, blood and blood vessels.
- **Organisms :** Now when all the above levels of organisation are taken in one picture, what is derived is known as organism like plants, animals, humans, bacterium, etc.
- **Population :** It is here that we speak about a broader concept in the levels of organisation of living things. A group of organisms, belonging to the same species may have the capability to inter-breed with each other. And when they inhabit a particular area, they are known as population. Its example can be the population of frogs in a pond.
- **Species :** Species is a term that is used to mark the distinction between different kinds of organisms. For example, human species are different from animal species and species of plants.
- **Community :** Taking it down to the last three of the levels of organisation, we are speaking of community. It is nothing but a reference to a group of organism of different species, which populate a given area and stay in mutual interaction with each other.
- **Ecosystem :** To define it in a layman language, interaction of a group of organism with the environment that they live in, is an ecosystem. For example, a pond ecosystem may consists of organisms such as fish, frogs, insects, etc., who depend on the water, mud, plants, etc., for survival.
- **Biosphere :** Finally, we have reached the last block of the pyramid of the levels of organisation, with what is known as biosphere. It encompasses

all. Meaning, it consists of the regions which lie on, above and below the surface of the Earth, including the atmosphere where the living organisms thrive. So there you are with a quick explanation on the different levels of organisation, which we, along with several other components are living. As we can see, each level serves as a building block for the other, and this is what lays emphasis on the importance of all the components.

CELLULAR TOTIPOTENCY

Totipotency is the ability of all living cells potentially to regenerate whole new individuals – the trick is how to trigger them to do it and so to demonstrate it. If I remember correctly, Dolly the sheep was cloned from a cell from another sheep's udder – a differentiated cell, which in normal circumstances would have remained as such. Similarly in plants, it is possible to trigger single epidermal cells in oilseed rape, sunflower, and I am sure there are others, to develop into embryos and subsequently to whole plants. Even immature pollen grains have this potential – these immature cells can be switched to develop into whole plants, even though they only have one set of chromosomes, and give rise to haploid plants, although sometimes these may diploidise during culture, to generate homozygous diploid plants.

One of the hardest things in plant tissue culture is to get tissues from mature trees to illustrate totipotency, but by forcing a rejuvenation step, it is even possible to do this with some species – I believe quite a bit of work has been carried out on pines.

Simple cuttings, and the SAPS tissue culture system depends on this ability of cells to switch their development – hence root formation from hypocotyls, and root formation from woody cuttings, which allow new plants to be obtained for propagation. We described the different media used for cell/tissue culture, and also discussed culture of individual differentiated cells. However, in plant tissue culture experiments, unless single cell cultures are absolutely necessary, more often we use explants (pieces of differentiated tissues) to initiate their growth in culture.

This is also necessary to obtain callus from which single cell cultures are derived. During the last few decades, even for development of new crops or for improvement in the characteristics of existing crops (*e.g.* to obtain disease free crop, etc.), the techniques of tissue culture in combination with genetic manipulation have become important tools.

For instance, as an alternative to vegetative propagation, shoot tip propagation was used commercially for rapid and consistent reproduction of elite or difficult to propagate genotypes. When an explant from differentiated tissue is used for culture on a nutrient medium, the non-dividing, quiescent cells first undergo certain changes to achieve a meristematic state. The phenomenon of the reversion of mature cells to the meristematic state leading to the formation of callus is called dedifferentiation. The component cells of callus have the ability to form a whole plant, a phenomenon described as redifferentiation.

These two phenomenons of dedifferentiation and redifferentiation are inherent in the capacity described as cellular totipotency, a property found only in plant cells and not in animal cells. In other words, while a differentiated plant cell retains its capacity to give rise to a whole plant, an animal cell loses this capacity of regeneration after differentiation. Although, generally a callus phase is involved before the cells can undergo redifferentiation leading to regeneration of a whole plant, but rarely the dedifferentiated cells can give rise to whole plants directly without an intermediate callus phase. These aspects of tissue culture and its application in clonal propagation and in the general on of hereditary variation called somaclonal variation.

Some plants have cells that are able to convert from a structured, or differentiated, state to a totipotent state. One example is the ZZ plant (Zamioculcas zamiifolia). This plant is so cellularly precocious that a mere piece of leaf or stem can be used to grow a new plant. Other plants and trees can produce spontaneous buds on the stem or trunk. These emerge from the activity of the meristematic tissue of the cambium layer, and since they are not buds that are normally produced in the axils of leaves, they are called adventitious buds. This can happen even when the tree in question is not the kind that will root from cuttings. The formation of adventitious buds begins with the production of a callus, which is essentially an undifferentiated lump of cells resembling a small tumor. Through genetic and hormonal processes, cells differentiate into growing points which then produce new stems with leaves. A similar process can result in adventitious roots as well.

Advances in the understanding of totipotency and how cells can be stimulated to become totipotent has resulted in an entire industry, the business of micropropagation, or what is commonly referred to as "tissue culture". While micropropagation labs do culture tissue (meristematic tissue), this work is not the same as what is done in a scientific tissue culture labouratory. The work done in scientific labs is much more esoteric and amasing. I studied the science of tissue culture when I attended graduate school, and we did things like isolating single cells from a callus and removing the cell wall using enzymes to yield a naked plant cell called a protoplast. These are important in scientific research because genes are more easily inserted into plant cells when they don't have a cell wall around them. We grew pure callus cultures and cell lines that were, essentially, all totipotent cells. Using the right combinations of plant hormones, we could then cause the cells to turn into plant shoots with roots. These and other experiments done in scientific tissue culture labs are not the kinds of activities engaged in by micropropagation operations, but they do demonstrate vividly the power of totipotency in plant growth.

By contrast, commercial tissue culture is the business of producing many thousands of clones of particular plants using meristematic tissue. Different plants require different formulations of media, as the formula that is ideal for one plant will produce no results with some other plants. Conventional micropropagation requires a sterile or aseptic environment and the operating costs of such a facility are substantial. However, without totipotency, none of this would be possible at any cost.

SOMATIC EMBRYOGENESIS

A plant reproduces naturally through the development of zygotic embryos. Formation of the embryo begins with the division of the fertilized eggs or zygote within the embryo sac of the ovule. Through an orderly progression of divisions, the embryo eventually differentiates, matures, and develops into the new plantlet. Alternatively, the plant can be derived from a single somatic cell or a group of somatic cells. This regeneration process, which differs from the natural pathway, is called somatic embryogenesis. All the plantlets produced have the same genetic makeup. Combined with genetic engineering, micropropagation through somatic embryogenesis provides an efficient means of producing a large number of elite or transgenic plants.

The somatic embryo's development is analogous to its zygotic counterpart. The embryo development process is characterised by a series of morphological changes which correspond to its maturition. Research in our lab focuses on the applications of image analysis and pattern recognition systems for studies of somatic embryo development. Carrot embryos have been used as a model system to develop the image analysis and pattern recognition system. Currently, Douglas-fir cultures are being investigated because of their economic importance to the timber industry.

Chapter 14

SEEDS : DEVELOPMENT AND DISEASES

TREATING SEEDS TO PREVENT DISEASES

Sometimes chemicals are applied to seeds, bulbs, corms, tubers, and roots to prevent their decay after planting and to control seed-borne and soil-borne plant diseases.

To be satisfactory, a seed treatment has to be effective yet reasonably safe from seed injury in case of overdosage; economical, readily available, easily applied, and chemically stable; and not overly poisonous or disagreeable to operators or corrosive to metal.

Fungicides may be classified as seed disinfestants, disinfectants, or protectants, just as to the location of the organisms to be combated.

Disinfestants inactivate organisms, such as bunt spores, that are borne on the surface of the seed.

Disinfectants are effective against those located deeper within the seed. Protestants protect the seeds from attack by organisms that are present in the soil.

Practically all effective seed-treatment materials are disinfestants. Many are also disinfectants and protectants. The formaldehyde and hot-water treatments, however, are disinfestants and disinfectants but are not seed protectants. In fact, seeds that have been treated with formaldehyde or hot water frequently are attacked by soil-borne fungi more severely than are untreated seeds and therefore should be treated also with a protestant before planting.

Based on composition, fungicides may be organic or inorganic, mercurial or non-mercurial, and metallic or non-metallic. There are organic mercurials (Ceresan) and inorganic mercurials (calomel); there are non-mercurial metallic organics (Fermate) and non-metallic organics (Spergon); there are metallic inorganics (copper carbonate) and non-metallic inorganics (sulfur). Fungicidal seed treatments may be dry or wet just as to the form in which the fungicide is applied to the seed. In a dry treatment, the fungicide is applied in dust form, usually in a mechanical mixer at rates ranging from 1/2 to 4 ounces or more to the bushel. Wet treatments once meant soaking the seed in a water solution of the fungicide for a certain period, after which the seed was allowed to drain and dry. Wet treatments now are applied mostly by the slurry method or the"quick-wet" method. In the slurry method, the fungicide is applied to the seed as a soup-like water suspension, which is mixed with the seed in a special slurry treater. The seed requires no drying but may be bagged immediately for sowing or storage.

In the "quick-wet" method, a concentrated solution of a volatile fungicide is applied to the seed and thoroughly mixed with it. The dosage may range from 1/2 to 5 fluid ounces to a bushel. As in the slurry treatment, that adds less than 1 per cent of moisture to the seed. The well-known formaldehyde spray treatment of oats is essentially a "quick-wet" treatment. So also is the method recommended for applying Panogen, Mercuran, Setrete, and several other materials.

Inorganic mercurials used for treating seed are limited practically to mercuric chloride, mercurous chloride (calomel), and mercuric oxide. Mercuric chloride, as a 1 to 1,000 solution, may be used for treating potato seed pieces, sweetpotatoes, and rhubarb roots for planting. It is also used for seed of crucifers (plants of the mustard or cabbage family), celery, cucumber, pepper, tomato, watermelon, and certain other vegetables. Most seeds are more or less susceptible to injury by mercuric chloride. Calomel is used on seeds of crucifers, celery, and onion. Mercuric oxide may be used as a dip treatment for sweet-potatoes (1 pound to 30 gallons of water).

Organic mercurials are more numerous and more widely used than the inorganics. They are used on seed of small grains, legumes, grasses, cotton, beets, flax, sorghum, and some other field crops, and also on certain corms, bulbs, tubers, and roots and the seeds of some vegetables.

Ceresan, 2 per cent ethyl mercury chloride, introduced in 1926, was the first organic mercurial widely used in the United States. It is applied at 2 ounces a bushel. It was followed and largely replaced in 1933 by New Improved Ceresan, 5 per cent ethyl mercury phosphate, which is applied at 1/2 ounce per bushel. Both were used mostly on small grains, flax, cotton, peas, hemp, and sugar beets.

Ceresan M, 7.7 per cent ethyl mercury p-toluene sulfonanilide, appeared in 1948. It largely replaced the two already Ceresans because of several advantages over them, including its application as a slurry.

Leytosan and Agrox, 7.2 per cent and 6.8 per cent phenyl mercury urea, respectively, are applied to small grains, peas, rice, and sorghum at 1/2 ounce to the bushel and to flax at 1 1/2 ounces. They may be applied in dust or slurry form.

Mercuran, 3.5 per cent mercury as methoxy ethyl mercuric acetate, is used at the rate of 1/2 ounce per bushel on small grains. It may be applied as a dust, in concentrated solution by the "quick-wet" method, or in a more dilute solution with a slurry machine.

Panogen, 2.2 per cent methyl mercury dicyan diamide, is a concentrated liquid applied at 3/4 fluid ounce per bushel to small grains, 1 1/2 fluid ounces to flax, and 4 fluid ounces per 100 pounds of segregated beet seed. It is applied in a special Panogen treater, but can be successfully applied in a slurry treater if diluted with water.

Setrete, 7 per cent phenyl mercury ammonium acetate, is a concentrated liquid that may be applied as such at 1/2 ounce per bushel, or it may be diluted 1 to 9 with water and applied in a slurry treater. Mersolite, 5 per cent phenyl mercury acetate, is used as a dip treatment (1 pound to 800 gallons) for narcissus corms to combat basal rot. Merthiolate, sodium ethyl mercury thiosalicylate, is used to prevent corm rot and yellows in gladiolus. Sanoseed, 7.9 per cent ethanol mercury chloride, and Corona P. D. 7, 5 per cent mercury in a mercury bromine-phenol compound, are used as dip treatments for seed potatoes. Semesan, 30 per cent hydroxy mercuric chlorophenol (19 per cent Hg), is an excellent mercurial used as a wet soak treatment for bulbs, tubers, and corms and as a dust treatment for seeds of flowers and vegetables. Semesan Bel, a mixture of 2 per cent hydroxy inercurichlorophenol and 12 per cent hydroxymercurinitrophenol, is used as a dip treatment for seed potatoes.

Puratised N-5 E, 10 per cent phenyl mercury triethanol ammonium lactate, is used for treating seed potatoes and lily bulbs. L-224, an experimental mercuryzinc-chromate material, is an excellent treatment for seed corn.

Aagrano, 3.5 per cent ethoxy propyl mercury bromide, is effective against cereal diseases, especially when it is applied in slurry form. Semenon, 2 per cent isopropyl methyl mercury acetate, gave excellent results in controlling diseases of small grains and sorghum. Both Aagrano and Semenon are European products. They were not available in the United States in 1953.

Non-mercurial organic fungicides have increased greatly in number since 1945. Generally they are less effective than the mercurials, but as a rule they are less injurious to seeds and less dangerous to persons using them. The organic sulfurs and quinones are prominent ingredients in these compounds and often are combined with phenol, chlorine, bromine, quinoline, zinc, iron, copper, sodium, or other materials. Spergon, 98 per cent chloranil (tetrachloro-p-benzoquinone), was among the first non-metallic organics to be widely used for treating seed, especially peas and beans. It is used for vegetable seeds, corn, sorghum, peanuts, alfalfa, clover, soybeans, and some other crops. It may be applied as a dust or as a slurry. Arasan, 50 per cent thiram (tetramethylthiuram disulfide), still another early organic fungicide, is used for the same crops as Spergon. Both will also control bunt in wheat, but are not recommended for treating oats or barley.

Arasan SFX, 75 per cent thiram, is the wettable form of Arasan for treating seeds by the slurry method. Tersan, also a wettable form of thiram, is used for the control of diseases of turf and lawn grass.

Phygon (formerly Phygon XL) consists of 50 per cent 2, 3-dichloro-1, 4- napthoquinone and 50 per cent talc. It is an effective seed treatment for corn, peanuts, rice, sorghum, and most vegetables. It controls bunt in wheat, but is not recommended for other small grains.

Zerlate, 70 per cent ziram (zinc di-methyl dithio-carbamate), is effective as a prebedding dip for controlling black rot in sweetpotatoes. It is similar to Zincate, Methasan, Zimate, and Karbam, as all contain ziram as the active ingredient.

Fermate, 70 per cent ferbam (ferric dimethyl dithio-carbamate), like Zerlate, is used as a prebedding dip for sweetpotatoes. Both materials are used also as foliage dusts or sprays.

Dow 9 B, 50 per cent zinc trichlorophenate, has been used to treat gladiolus bulbs and seed of cotton, corn, and sorghum.

Dithane Z-78, 65 per cent zinc ethylene bisdithio-carbamate, has shown promise as a disinfectant and chemotherapeutic fungicide.

Mycon, 7.7 per cent methyl arsenisulfide, in extensive field tests, has been found effective in controlling those seed-borne diseases of wheat, oats, and barley that are amenable to control by fungicides.

SeedOx, 50 per cent 2, 4, 5-trichlorophenyl acetate, has been used to treat cottonseed. Mycotox is similar to Seed-ox. Neither is effective as a seed treatment for small grains. Anticarie, 40 per cent hexachlorobenzene, is effective as a seed treatment for the control of bunt in wheat. When applied to the soil it also prevents infection due to bunt spores in the soil. It is not recommended for treating seeds of other cereals.

Pentachloronitrobenzene (50 per cent) controlled covered kernel smut in kafir and a 20-per cent product controlled bunt in wheat. In Europe this chemical is reported as having controlled infection from soil-borne spores of both common bunt and dwarf bunt when it was applied to the soil at planting time at the rate of about 50 pounds an acre. Copper carbonate, the first dust seed treatment to be widely used in agriculture, and basic copper sulfate are still used on wheat as bunt preventives. Copper sulfate (bluestone) solution, once a popular seed treatment for wheat, now is used for that purpose to a very limited extent.

Cuprous oxide (yellow or red) serves as a seed protectant for vegetable seeds to prevent seed decay and pre-emergence damping-off. It is injurious to seeds of lettuce, crucifers, and onions. Vasco 4, a mixture of zinc oxide and zinc hydroxide, is used on seed of crucifers, spinach, and other vegetables that are sensitive to cuprous oxide.

Other seed-treatment materials, some effective and some experimental. The hot-water treatment remains the standard method for controlling the flower-infecting loose smuts of wheat and barley. It is effective also for treating seed of crucifers, onion, tomato, and some other vegetables.

Some gases, such as chlorine, have been suggested for treating large quantities of seed, but their effectiveness has not been proved.

Hot vapour was described in 1944 as being applied to tons of seed exposed on a moving belt. Ultraviolet and infrared rays, short waves, Hertzian waves, diathermy, X-rays, and other similar devices have been tried as seed disinfectants, but none has been proved practicable. Like hot water, these materials would not performance as seed protectants, and so a supplementary treatment would be necessary to guard against soil-borne fungi.

Testing the effectiveness of fungicides in the control of seed-borne diseases presents two chief difficulties : Obtaining a supply of suitable seed that is sufficiently infected to furnish an adequate test for the fungicides, and obtaining environmental conditions after planting that favour infection in the plants.

In diseases like bunt of wheat, in which the causal spores are located on the surface of the seed, clean seed can be infested artificially if a supply of spores is available. But many disease organisms are located deeper within the seed in a manner that cannot be duplicated artificially. So one has to get seed from a badly infected crop, or, better still, from the seed lot that produced that crop. At times seed obtained from a heavily infected field may be infected too lightly to serve as an adequate test for seed treatment because conditions for infection may have been very unfavourable at the critical period.

We must observe certain precautions in testing fungicides for seed treatment. The seed should be thoroughly cleaned, before treatment, to remove dust, chaff, weed seeds, and other substances, all of which take up much of the fungicide. Proper dosage is important because the seed sample used usually is relatively small and hence the amount of fungicide applied must be carefully weighed or measured. In treating cereal seeds experimentally, 500 cubic centimeters is a convenient sample. This volume, which is 1/70 of a bushel, simplifies the conversion of ounces-per-bushel to grams-per-sample. If the desired rate of application is one ounce (28.34 grams) per bushel, 1/70 of a bushel will require 1/70 of 28.34 grams, or 0.4 grams. Rates of 1/2, 2, 3, and 4 ounces per bushel are easily converted to 0.2, 0.8, 1.2 and 1.6 grams per sample, respectively. Differences in bushel weight among different seeds or seed lots then can be ignored. It also avoids the error involved in treating samples of light, chaffy seed as compared with plump, heavy seed. The light seed should receive more fungicide for each weighed bushel than the heavy seed.

When small samples of seed are treated, the capacity of the container should be such that it is only half filled by the sample. It should be first"conditioned" by treating in it a sample of seed at a rate sufficient to coat the inside with the fungicide. This seed is then discarded.

After the fungicides have been applied to the different samples, the containers should be shaken in some mechanical contrivance so that all receive the same amount of mixing. Thorough mixing is especially essential when applying the non-volatile materials.

Between the treatment and sowing, the treated and untreated samples should be stored at a moderate temperature and preferably at a low humidity. Metal or glass containers are preferable to paper envelopes because if the envelopes are stored in contact with one another, the fumes from a volatile mercurial fungicide in one envelope will treat the seed in the adjoining envelope. If this envelope contains the untreated check sample, it will be rendered useless for that purpose.

The effects of the treatments on germination of the seed, seedling emergence, disease control, and plant growth and yield are among the details usually desired from seed-treatment experiments.

Germination tests, to study any harmful effects of the fungicides on the seed itself, may be made on wet blotters placed in incubators (in which temperature and humidity are controlled) or in steamed wet sand or soil. Relatively disease-free seed and disinfected soil should be used in the tests, because the harmful effects of a fungicide on the seed may be masked by its protective effect against seed-borne or soil-borne fungi that cause seed rot or pre-emergence damping-off.

The use of infested soil is essential for determining the effect of the seed-protectant qualities of treatments on emergence. That may involve the isolation and culture of soil-borne fungi, such as species of Pythium, Fusarium, Helminthosporium, and Rhizoctonia, and using the pure cultures to inoculate soil in which the treated (and untreated) seed is to be planted. Such tests may be made in the greenhouse in pots, flats, or beds. The soil should be steamed or chemically fumigated before being inoculated in order to determine the effectiveness of the fungicides against each specific soil-borne fungus culture.

The effectiveness of fungicides in the control of seed-borne diseases, such as the smuts of cereals. that are not apparent in the seedling stage usually is studied in field plots. The seed should be sown at the proper date so that the soil temperature as well as moisture before emergence are conducive to infection. For the cereals, that calls for soil that is not too moist for aeration and germination of the seed-borne spores.

Along somewhat sub-medium moisture content of the soil, the temperatures considered conducive to infection by cereal diseases are : Bunt of wheat, 41 to 50 F.; barley covered smut, 50 to 68; false loose smut of barley, 59 to 68 the smuts of oats, about 64 to 72; the barley stripe disease, 46 to 59; and kernel smuts of sorghum, 75 to 86. Periods favourable to infection cannot be predicted with certainty. Frequently, because of the absence of such conditions, significant field data are not obtained.

The effect of the seed treatment on yield should be obtained by treating and sowing relatively disease-free seed in replicated plots along with untreated seed. Increases in yield from such treated seed presumably reflect the seed-protestant qualities of the fungicide used.

Cereal seed treatment was rather widely practiced for quite a few years before the treatment of other crop seeds was generally recommended.

The reasons perhaps were that smuts could be seen in cereals and that it was discovered early that some of them could be prevented by seed treatment.

The beneficial effects of the treatment of cereal seeds may result from the elimination of seed-borne diseases, the prevention of seed rot and seedling blight, and the suppression of weeds by better and more even stands.

One of the greatest benefits lies in the elimination of some seed-borne fungi or bacteria that cause primary infection lesions from which the disease spreads to other plants. Outstanding examples are certain helminthosporium diseases of wheat, oats, and barley. This spread by secondary infection may cause heavy loss, although only a small percentage of the seed sown may have been infected. Annual seed treatment of Cereal seed is now considered a wise farm practice because the use of disease-free or treated seed one year does not insure the production of disease-free seed for the next year's crop. Airborne spores from neighbouring fields may contaminate the heads of grain grown from disease-free seed, so that seed from these heads, if sown untreated, may produce a diseased crop the following year. Growers of certified seed have found it wise to guard against this source of infection, as it may disqualify their fields for certification.

Wheat is treated mostly for the control of bunt, which if only seed-borne is the most easily controlled of all the seed-borne cereal smuts. Ceresan M, Agrox, Secrete, Panogen, Leytosan, Mercuran, Aagrano, and some other organic mercurial compounds are generally most effective in bunt control, especially if infection is severe. Most are applied at less than an ounce to the bushel. Many non-mercurials also are effective copper carbonate, basic copper sulfate, Arasan, Spergon, Phygon, Anticarie, Mycon, and several experimental materials. The mercurials by and large are preferable because they eliminate also some of the pathogens borne more deeply within the seed. Loose smut is prevented only by the hot-water treatment.

Rye may be treated to prevent the spread of seed-borne diseases, like stalk smut and bunt. The treatments for wheat may be used also for rye.

Barley is treated largely for the prevention of covered smut, black or false loose smut, and stripe disease. Seed treatment also reduces the amount of primary infection from such diseases as bacterial blight, scab, net blotch, and spot blotch. The fungicides recommended are restricted largely to the organic mercurials such as Ceresan M, Panogen, Leytosan, and Agrox. The non-mercurial organics may improve stands and reduce infection by these diseases to some extent but, with a few exceptions, they seldom control them satisfactorily. The flower-infecting loose smut is controlled only by the hot-water treatment.

Seed of oats, like that of barley, is treated most frequently for the prevention of the smut diseases, which are visible at heading time. The effective treatments are restricted largely to the organic mercurials, although the formaldehyde spray treatment is widely used. It is cheap and effective but may injure the seed. Also it is not a seed protectant. The effective organic mercurials also prevent primary infection from seed-borne halo blight, fusarium blight, anthracnose, and the helminthosporium diseases; they will not prevent these diseases, however, if the causal organisms are present in the soil.

Hybrid corn seed, which constitutes about 80 per cent of the corn seed planted, is treated at the seed houses before it is sold to the growers; thus the work of seed treatment of corn mostly has been taken out of the hands of individual growers.

Treatment of rice seed was not a generally recommended farm practice until about 1947. Experiments proved that the treatments increased stands, especially in the early seedings when the soil was cold and wet. Often yields also were improved. Best results were obtained with Ceresan M, Phygon, Arasan, and Spergon. Ceresan M prevented seedling blight caused by Helminthosporium oryzae. Dow 9 B injured the seed after long storage. Cuprocide (cuprous oxide) seemed best for seed sown in water, but it may injure presprouted seed. Rice may be fumigated with methyl bromide to combat the seed-borne rice nematode. Exposure to a concentration of 1 1/4 pounds of methyl bromide to 1,000 cubic feet of space for 12 to 15 hours will kill the seed-borne nematodes without injuring the seed seriously.

Sorghum, like corn, is benefited most by seed treatment when cold, wet weather follows planting. It reduces seed rot and seedling blight and prevents infection by the kernel smuts. The non-mercurial organics, such as Phygon and Arasan, have been found beneficial in improving stands and controlling smut. In varieties whose seeds have persistent plumes, however, the kernel smuts are controlled more effectively by the use of volatile organic mercurials, such as Ceresan M and Panogen.

Sugar beet seed is treated mostly to combat seed-borne infection by Phoma betae and Cercospora beticola. Seed treatment is effective in preventing pre-emergence damping-off, caused either by seed- or soil-borne fungi. Materials used for beet seed treatment include the organic fungicides N. I. Ceresan, Ceresan M, Panogen, Phygon, and Arasan. Inorganic mercury compounds, cuprous oxide, and various mixtures of mercurials with copper carbonate have proved effective experimentally but never have come into widespread commercial use.

Pre-emergence and post-emergence damping-off have been successfully combated in greenhouse experiments by applying an Arasan-fertilizer mixture to the soil so that the sugar beet seed germinated in the soil impregnated with the mixture and the seedlings grew through it. The Arasan was used at a rate of about 4 pounds an acre. Field experiments to control black root by Arasan-fertilizer mixtures have not given consistently favourable results.

Seed treatments for the control of damping-off caused by Pythium species, Phoma betae, and the Rhizoctonia species are beneficial unless soil infestation is severe or soil moisture conditions are unfavourable. Black root, caused by Aphanomyces cochlioides, however, is not prevented by seed treatment.

Proper soil management helps reduce the soil populations of sugar beet pathogens. Adequate drainage and heavy application of commercial fertilizers, especially phosphate, are important. Of great importance also is a rotation in which sugar beets do not immediately follow a legume sod crop but follow an early fall-plowed legume. Such handling of the legume crop is necessary because clovers, sweetclover, and alfalfa harbor the various pathogens that cause damping-off. Their sods, if spring-plowed, produce peak populations of the fungus at the period corresponding to planting time for sugar beets. If associated with proper soil management, control of excessive soil moisture by drainage, and a good fertilizer practice, seed treatments show value.

Cotton, flax, and hemp respond to seed treatment in the order named. The diseases of cotton that are reduced somewhat by seed treatment are bacterial blight or angular leaf spot (Xanthomonas malvacearum), anthracnose (Colletotrichum gossypii), sore shin (Rhizoctonia solani), and seedling blight caused by species of Aspergillus, Fusarhim, Diplodia, Sclerotium, and other fungi.

Seed of cotton is generally delinted before being treated because seed-borne infection is more easily eliminated in delinted seed. Delinting may be done mechanically by reginning or chemically by acid treatment. Mechanical delinting may injure the seed and impair its germination. Acid-delinted seed germinates better than fuzzy seed, but it rots more easily, especially in cold, wet soil. Effective seed treatments largely prevent that. Organic mercurials (such as the Ceresans) with some exceptions have generally been more effective than the non-mercurial organics (such as Spergon, Dow 9 B, Phygon, Arasan, and Seedox) in eliminating seed-borne infection. Some growers object to the use of poisonous fungicides, such as the mercurials, however, because excess treated seed may become mixed with untreated seed used for making cottonseed meal or oil. The non-mercurials are especially useful as seed protectants for acid-delinted seed.

Treatment of flaxseed is made necessary largely because seed, especially of large-seeded kinds, may be injured in threshing. Many of the fractured seeds rot after planting, particularly in heavy soils, unless they are first treated with an efficient protectant, which prevents invasion by species of Alternaria, Penicillium, Fusarium, and Pythium. Several seed-borne diseases of flax are alleviated by seed treatment. Pasmo (Mycosphaerella linorum), when seed-borne, causes primary infection lesions, which initiate secondary infection in other plants. Browning and stem break (Polyspora lini) and anthracnose (Colletotrichum linicolum) also may be seed-borne.

One of the difficulties in treating flaxseed is the failure of dry fungicides to adhere to the smooth seed coat. The seed therefore requires a much heavier dosage of dusts than is applied to most other seeds. Wet treatments cause gumming of the seed because of the mucillaginous coat. The organic mercury dusts usually are applied at 1 1/2 ounces per bushel. non-mercurials are applied somewhat more heavily. Seed treatment of flax may increase stands, but increased yields do not always follow unless an abundance of weeds prevents sufficient branching to compensate for the thinner stands from untreated seed.

Treatment of hempseed with New Improved Ceresan, Spergon, and Arasan was found to improve stands when planting was followed by unfavourable conditions for germination and growth.

Treatment of seeds of forage crops controls some diseases, such as certain smuts in slender wheatgrass, millet, Canada wildrye, and Sudangrass. Seed treatment sometimes has increased stands in some species of Lespedeza, Lotus, Medicago, Mehlolus, and Trifolium. Other species are injured by certain treatments when the treated seed is sown in dry soil. Seed of winter peas, mung beans, cowpeas, hop clover, hairy vetch, and alfalfa gave better germination, improved stands, and superior plants when treated with Spergon, Arasan, Phygon, or Dow

9 B in extensive field and greenhouse tests in 1949. Nodulation was not inhibited by treatment when the nitrogen-fixation culture was applied to the treated seed immediately before sowing. Some investigators, however, legume seed should not be treated before being inoculated if it is to be sown in soil not already cropped to legumes.

Experiments with treating soybean seeds have been more numerous and extensive than with those of almost any other legume. In general, improved stands were had after the use of Arasan, Spergon, Phygon, and Dow 9 B. Organic mercurials are sometimes injurious. Increased stands due to treatment were not always followed by increased yields, probably because branching of the plants often compensates for thinner stands and because a higher percentage of the soybean flowers in a thin stand of plants will form pods.

Treatment of peanut seed is a profitable farm practice, especially when mechanically shelled seed is used. Increases in stand have ranged from 30 to 100 per cent. Uninjured hand-shelled seed frequently gets no benefit from seed treatment except when unfavourable growing conditions follow planting. Arasan, Spergon, Phygon, Dow 9 B, and Ceresan M are commonly recommended.

Vegetable seeds are treated primarily to prevent seed rot and damping-off. Sometimes the control of seed-borne diseases is a major aim. The materials so used include Arasan, Phygon, Spergon, Fermate, Semesan, Dow 9 B, N. I. Ceresan, Ceresan M, Cuprocide, zinc oxide, zinc hydroxide, mercuric and mercurous chlorides, copper sulfate, phenothiazine, Zerlate, Dithane, and others.

Arasan and Spergon are two widely used fungicides for vegetable seeds. Arasan seems most suitable for seed of beets, chard, and spinach. Spergon seems best for legumes. Some fungicides display differential benefit or injury towards the seed of certain crops. Cuprocide, for example, is injurious to seed of crucifers and lima beans and causes necrosis, delayed absorption, and delayed seedling growth in peas. It is especially beneficial, however, to lettuce seed, which in turn is injured somewhat by Arasan and Fermate. Zinc oxide is injurious to peas but is highly beneficial to seed of spinach and crucifers.

Potato tubers often are treated before planting. Fifteen or more diseases of potatoes may be transmitted in or on the tubers. Few of them are amenable to control by treating the tubers before planting. Scab, rhizoctonia or black scurf, and fusarium seed-piece decay respond to seed treatment if they are seed-borne. The principal treatments recommended are hot formaldehyde dip, cold formaldehyde soak, hot mercuric chloride dip, cold mercuric chloride soak, yellow oxide of mercury dip, and the hydrochloric acid-mercuric dip. Several organic mercury fungicides also are used. Among them are Semesan Bel, Sanoseed, and Corona P. D. 7, all of which are made specifically for treating potato seed pieces. Beneficial results have been obtained also from the use of Fermate, Semesan, Spergon, and Dithane.

Sweetpotatoes used for planting are treated to prevent injury due to seed-borne black rot, scurf, and stem rot and soil-infesting pathogens as species of

Pythium Rhizoctonia, and Sclerotium. The standard treatment is one 10-minute dip in a 1 to 1,000 mercuric chloride solution, or a dip in Semesan Bel solution (1 pound to 7 1/2 gallons of water). Both are effective but sometimes they delay or reduce the production of sprouts. Fairly good results without injury have been had with Spergon, Phygon, Fermate, Zerlate, Tersan, and Puratised N-5 E.

Treating the seeds of ornamentals is a common practice. Semesan has been widely used for this purpose. The non-mercurial organics, such as Arasan and Spergon, also are satisfactory. Ornamentals grown from bulbs, corms, tubers, and roots also are benefited somewhat by the use of fungicides. Gladiolus corms, for example, are helped by a 15-minute dip in a solution of 1 pound of New Improved Ceresan in 50 gallons of water just before planting. The standard mercuric and mercurous chloride solutions also are used. Dipping the corms in slurries of Spergon, Fermate, or Dow 9 B after digging is beneficial.

Tulip bulbs have not responded very well to treatment. Some fungicides have lowered the yield of bulbs. Dipping bulbs in slurries of Spergon or Fermate has increased some yields. Narcissus bulbs may be dipped in a phenyl mercury acetate solution (1 pound to 800 gallons) for 5 minutes, after digging in spring and again before planting in fall, to control fusarium basal rot. Arasan SFX, Dow 9 B, and New Improved Ceresan also are beneficial.

Hormones in seed treatments have been tried often. Results have varied. Of 30 investigators whose work was reviewed, 10 reported beneficial results from the use of growth-promoting substances on seeds. Twenty failed to obtain any benefits. Apparently the conditions under which hormones may or may not be beneficial in seed treatments are not fully understood.

Growth-promoting substances are used commercially to induce root formation in cuttings, prevent fruit drop in apple orchards, induce fruit formation without pollination in some plants, and to prevent sprouting in stored potato tubers. It seems reasonable that under proper conditions the materials may improve seed germination and early growth of the seedlings. Definite and reliable recommendations cannot be made until more extensive research has been carried out. Synergism and antagonism between different fungicides, when mixed together, has been demonstrated often enough to restrain one from mixing fungicides with one another or with insecticides without knowing how they will affect each other and the seeds on which they are to be used.

The addition of New Improved Ceresan to DDT reduced both the fungicidal action of Ceresan and the insecticidal action of DDT. Magnesium oxide, added to copper carbonate or to Spergon, reduced the beneficial effect of those materials on emergence in wheat and on smut control in sorghum. Magnesium oxide also reduced the fungicidal efficiency of cuprous oxide and of Dow 9 B, but seemed to increase the fungicidal effectiveness of sulfur. Pyrophyllite containing 3 per cent DDT when mixed with Dow 9 B reduced the control of sorghum kernel smut from 0.3 per cent to 40 per cent, with 6o per cent infection in the check. Copper compounds in general are reduced in effectiveness when mixed with materials high in protein.

A good fungicide, prepared especially for seed treatment, usually is a well-balanced combination of active ingredients and suitable diluents, perhaps with the addition of wetting and dispensing agents, dyes, and other materials in proper proportion. The addition of other materials, such as insecticides or other fungicides, may cause chemical reactions and the formation of compounds that are ineffective as fungicides or injurious to the seed.

The labels on containers for fungicides used for dusting or spraying vegetation often mention the insecticides with which they are not to be used. Labels for seed-treatment fungicides, however, do not include such directions because, as a rule, those fungicides are not mixed with insecticides or other fungicides. That may change, however, with the growing need for combating insects that attack seeds after they have been planted. Experiments in New York showed that Arasan SFX mixed with chlordane, lindane, or aldrin and applied to lima beans prevented both seed rot caused by fungi and seed injury due to the seed-corn maggot. Mergamma, a treatment for cereal seed, contains phenyl mercury urea for the control of certain cereal diseases and benzene hexachloride for wireworm control. The number of these insecticidal-fungicidal seed-treatment combinations doubtless will increase, but their use in combination should follow careful chemical and biological experiments.

Seed injury following treatment was common when the treatments were mostly copper sulfate solutions, formaldehyde, or mercuric chloride, especially when planting was delayed after treatment.

When copper carbonate dust began to be used to treat wheat, it was found that delayed planting after treatment caused no injury to the seed but actually protected it against rodents and insects in storage. The more volatile organic mercury treatments, however, occasionally lowered the viability of seed after storage periods of more than a few days, especially when the moisture content of the seed was high. Several factors govern the degree of such injury : The moisture content of the seed; the volatility of the fungicide and the rate at which it is applied; the length of the storage period; the temperature, humidity, and aeration during storage; the kind of seed (seed of some genera, species, or varieties are more susceptible to chemical injury than are those of others); and the condition of the seed coat (cracked, chipped, or broken seed coats are conducive to seed injury).

If seed is to be stored for a while after treatment with a volatile fungicide, its moisture content should be relatively low-13 per cent or less for cereals and a lighter rate of application used. Different portions of oats of 12 per cent moisture content were treated with New Improved Ceresan at 1/2 and 1/8 ounce per bushel and either sown at once or stored for several weeks. The seed treated at the 1/2-ounce rate yielded better when sown the day after treatment. The seed that got the lighter application yielded better when souring was delayed for several weeks after treatment. Several experiments proved that sound seed of wheat, oats, and barley of good quality and proper moisture content, treated with one of the better organic mercury disinfectants at the recommended rate and properly stored for a year, was not injured in viability but yielded better than did untreated seed similarly stored. Occasionally in the more humid areas of the Southeast, treated seed is stored with a too-high moisture content and the poor viability is ascribed to the treatment. Subsequent tests often show that the viability of the untreated seed is equally poor.

Pretreatment of seed sold by seed dealers has been advocated for years. Some large seed houses pretreat seed of some field crops, such as cereals, flax, cotton, sugar beets, peas, corn, broomcorn, and some forage crops, either as a general practice or on a buyer's request.

Pretreatment of seeds by all dealers would mean cheaper but more general and more effective seed treatment; fewer outbreaks of preventable diseases; less waste of chemicals; less need of storing large stocks of chemicals in many places; more economical packaging, distribution, and use of seed-treatment chemicals; the use of the proper disinfectant at the proper rate for each type of seed; and many other advantages.

Some objections to general pretreatment of seed are valid enough. There is no general agreement as to what treatment is best for each kind of seed. Some buyers object to planting "poisoned" seed. Some persons might not realise that treated seed is sometimes poisonous and they might suffer injury though, that all the advantages of pretreating seed outweigh the objections.

Continued advances in seed treatment doubtless will bring new and better fungicides, better apparatus, and improved procedures into use. Fungicidal materials that promise to be more effective but less costly and less poisonous and disagreeable are sought. Slurry treaters that are more accurate and less troublesome are promised. A process that will fix the slurry fungicide to the seed and prevent its dusting off when the dried seed is handled will rid the slurry method of its chief shortcoming.

The possibility of systemic fungicides and chemotherapeutic disease prevention has been suggested and has been demonstrated in a few instances. This might eventually lead to the prevention of such Nation-wide calamities as epidemics of stem rust of cereals. Such fungicides would be applied to the soil and, when taken up by the plant, would render it resistant or fatal to the fungus attacking it. The fact that a tiny amount (3 parts per million) of selenium in soil is fatal to aphids and spider mites feeding on plants grown in the treated soil should encourage the search for fungicides equally effective against fungus infection but not poisonous to humans and animals. Such fungicides would be a tremendous advance in our war against plant diseases.

MAKING SURE OF HEALTHY SEED

The primary purpose of seed certification is to maintain and to make available to our farmers crop seeds, tubers, or bulbs of good seeding value and true to name. The factors considered in determining good seeding value or quality include viability, content of weed seeds, and freedom from seed-borne pathogens or viruses.

The production of certified seed of superior varieties of field and forage crops involves the co-operative effort of many State, Federal, and private agencies. Among them are State agricultural experiment stations and extension services, State departments of agriculture, State crop improvement associations, the International Crop Improvement Association, the seed trade, and the United States Department of Agriculture. The work started at a meeting in 1919 of representatives of the States and Canadian certifying agencies.

The determination of eligibility of varieties for certification is the chief responsibility of the State agricultural experiment station in every State. Factors considered in determining the suitability of a crop variety for inclusion in a certified-seed programme include yield, adaptation, and resistance to diseases and insects.

The requirements for eligibility for certification vary as to type of crop. Certification for such cereal crops as wheat, oats, or barley, because of their limited area of adaptation, is relatively simple and frequently is conducted on a local-area basis. But the distribution of good seed of perennial forage crops alfalfa, red clover, the grasses requires the effort of many persons in widely separated areas because most of such seed is not produced in the locality where it is used for hay and pasture seedings.

The farmer may make application for inspection but is under no obligation to do so.

Varieties of field crops must be approved by a State agricultural experiment station before they are eligible to be considered for certification. In general, only one variety of the same crop, for seed production, is permitted on a farm except by prior approval of the certifying agency.

Field inspection is an important phase of the certification procedure. The International Crop Improvement Association has established certain minimum field standards, which form the basis for the regulations adopted by State certifying agencies. The standards take into consideration the type of crop, the degree of isolation necessary to prevent cross-pollination, and the class of seed produced.

Four classes of seed are recognised in seed certification of field and forage crops : Breeder seed, foundation seed, registered seed, and certified seed. The International Crop Improvement Association defines them thus :

"Breeder seed is seed or vegetative propagating material directly con- trolled by the originating, or in certain cases the sponsoring plant breeder or institution, and which provides the source for the initial and recurring increase of foundation seed.

"Foundation seed shall be seed stocks that are so handled as to most nearly maintain specific genetic identity and purity and that may be designated or distributed by an agricultural experiment station. Production must be carefully supervised or approved by representatives of an agricultural experiment station. Foundation seed shall be the source of all other certified seed classes, either directly or through registered seed.

Registered seed shall be the progeny of foundation or registered seed that is so handled as to maintain satisfactory genetic identity and purity and that has been approved and certified by the certifying agency. This class of seed should be of a quality suitable for the production of certified seed.

Certified seed shall be the progeny of foundation, registered, or certified seed that is so handled as to maintain satisfactory genetic identity and purity and that has been approved and certified by the certifying agency."

State and Federal plant breeders for years have been developing superior varieties of forage crops. For a long time, however, less than 1 per cent of the legume and grass seed produced in the United States was of improved varieties.

The great demand for seed of the newer varieties in forage-producing areas meant the draining away of the early generations of seed stocks, which should have been used to increase seed. The small supply of breeder seed has been the principal limiting factor in production of adequate supplies of foundation seed of some varieties of legumes and grasses. But, equally important, no procedure for maintaining and distributing foundation seed of forage crops on a Nation-wide basis was available.

The Foundation Seed Project was initiated in 1948 to set up the organisation and financial procedure whereby foundation seed stocks of improved varieties of legumes and grasses could be rapidly produced from breeder seed and distributed. Methods necessary to maintain genetic purity of the varieties were devised. It is a co-operative effort utilising the facilities of State and Federal agencies and the commercial seed trade. In 1952, 34 States participated. The co-operating agencies include State experiment stations, State extension services, State seed certifying agencies, State foundation seed organisations, the International Crop Improvement Association, the American Seed Trade Association, and the Department of Agriculture.

The operational phases of the programme are concerned with producing, assembling, distributing, and limited stockpiling of breeder and foundation seed. The co-ordination of those functions is the responsibility of the Bureau of Plant Industry, Soils, and Agricultural Engineering of the Department of Agriculture. Financial assistance for contracting for the maintenance of adequate reserves of breeder and foundation seed stocks is provided by the Grain Branch of the Production and Marketing Administration, representing the Commodity Credit Corporation of the Department of Agriculture. Each State is represented by a State foundation seed representative, who is responsible for initiating the foundation seed work and handling operations in his State.

Direction of the programme is the responsibility of a 16-man advisory group. Its members serve without pay. It is known as the Planning Conference and includes two representatives of each of these organisations : Northeastern Experiment Station Region, Southern Experiment Station Region, North Central Experiment Station Region, Western Experiment Station Region, International Crop Improvement Association, American Seed Trade Association, Production and Marketing Administration, and Bureau of Plant Industry, Soils, and Agricultural Engineering.

A superior variety developed by a plant breeder is tested for regional adaptation under the supervision of the Regional Forage Crops Technical Committee. A new variety found to be good in part or all of a region is then recommended by this technical committee to the Planning Conference for inclusion in the Project.

The Planning Conference determines the areas for foundation seed production for each new variety, estimates the requirements for breeder and foundation seed stocks, and co-operates with State workers in the areas of production in the development of the seed-increase programme.

The breeders' seed of a new variety is assigned by the Planning Conference to State foundation seed representatives in the States where it has been determined that foundation seed is to be produced. All foundation seed is grown under contract with individual growers, who are selected by the State foundation seed representative. Because the number of growers needed is small, only the most dependable growers, located in an area with favourable soil and climate, are chosen.

The foundation seed of all varieties included in the programme is allocated by the Planning Conference to State foundation seed representatives for planting to produce registered seed. The registered seed is handled by commercial seedsmen, who sell it to growers for the production of certified seed. The certified seed, which is subsequently used by farmers for forage plantings, is distributed through regular seed trade channels.

The 12 varieties included in the programme in 1952 were Atlantic, Buffalo, Narragansett, and Ranger alfalfas; Kenland red clover; Tift Sudan-grass; Climax lespedeza; and five blend lines of Midland red clover.

The 27 million pounds of certified seed of Ranger alfalfa available for 1953 planting was the largest amount ever produced of an improved alfalfa variety. It was almost double the amount farmers had the previous year. The market had almost twice as much Buffalo alfalfa seed in 1952 as was available in 1951. The supply of certified Atlantic alfalfa for hay and pasture seedings had increased from 150,818 pounds in 1950 to 1,511,000 pounds in 1952. Narragansett alfalfa was added to the list of varieties in the Project in 1951. Yet it was possible to produce more than 5,000 pounds of certified seed the first year in addition to increasing breeder and foundation seed supplies.

Enough stock seed of Kenland red clover was distributed to make available 215,000 pounds of certified seed in 1950, the first large-volume production. Yet in 1952, only 2"years later, there was a tenfold increase in certified seed production, which meant farmers had 2,000,000 pounds of planting. The increase of Tift Sudangrass has been comparable.

The potato tuber is subject to many diseases caused by fungus and bacterial pathogens and viruses. Disease-free seed is important.

The quality of seed potatoes has greatly improved since 1925 through the elimination of tuber-transmitted virus diseases by roguing, tuber-unit planting, tuber indexing, early harvest and pulling or killing vines, the use of winter field test plots; the use of immune or resistant varieties; the enforcement of high standards of certification; the production and use of better foundation seed; and the production of certified seed.

Many tuber-transmitted diseases of potatoes, like mosaic, curly dwarf, spindle tuber, leaf roll, and blackleg (Erwinia atroseptica), cannot be controlled by spraying or dusting. The only way is to remove the diseased plants from the field, a procedure that is commonly termed roguing. Seed pieces, tops, and tubers are removed sufficiently far from the field and destroyed to prevent winged aphid vectors from migrating from the rogued plants or parts back to the potato field. Roguing usually commences when the plants are 4 to 6 inches above the ground. The first roguing is followed by roguing at weekly or 10-day intervals throughout the season.

Foundation seed plots a mile away from other potato fields can become infected with virus diseases transmitted by winged aphids. As the growing season progresses, insect vectors of virus diseases may increase, and infection, contracted by the tops during the current season, is more likely to have reached the tubers in late – than in early-harvested stock. Healthy plants harvested early show a lower percentage of virus diseases than do healthy plants harvested late in the season. Early harvesting of seed plots or foundation seed is now a common practice in some potato seed producing areas.

Hand pulling was one of the first methods employed to destroy potato vines in an effort to produce better disease-free seed. It soon became apparent that the method was impractical because of the great cost and the high labour requirement.

Flame burning destroyed the leaves but left stems standing, from which new growth was produced in late-maturing varieties. A similar situation occurred when the foliage was mascerated with a rotobeater machine.

The use of chemicals to kill the vines has become a common practice in many of the potato seed producing areas. It helps reduce the spread of virus diseases, prevent infection of tubers by the late blight fungus (Phytophthora infestans), complete harvest before freesing weather, regulate the size of seed tubers, and reduce skinning and bruising of tubers. Late-maturing varieties have been found to be more difficult to kill than early-maturing ones. Vascular discolouration of the tuber frequently occurs from the use of chemicals to kill vines.

More work needs to be done to determine the factors that cause the discolouration. Because some weather conditions mask symptoms of some tuber-transmitted virus diseases, detection is sometimes difficult in the northern areas. The diseases can be detected when the plants are grown under field conditions in winter in some of the Southern States and California. Thus it is now mandatory to test all foundation seed (and much of the certified seed) in winter test plots. Through this means some northern growers of foundation seed have been able to maintain relatively disease-free stocks. The winter tests of samples from fields of potatoes grown for certification are completed in time so that the data may be used by certification officials of the seed-producing States. The information also helps growers of certified seed to keep from planting inferior stock.

It is generally agreed that very little progress would have been made in controlling mosaic and other tuber-transmitted virus diseases if winter tests had not been established for guidance in the programme of production of good foundation and certified seed.

The production of the good, new, resistant varieties of potatoes is largely due to the accomplishments of the National Potato Breeding Programme, which began in 1929. It is a Nationwide programme in which State agricultural experiment stations and the Department of Agriculture co-operate.

The Katandin variety was the first distributed under this programme. It has resistance to mild mosaic and some resistance to leaf roll. It is immune to net necrosis. Nearly 13 million bushels of certified Katandin seed were grown in the United States in 1952. In 1952 it represented more than 30 per cent of all certified seed potatoes and led all other varieties. Since then, 42 additional varieties have been introduced, some of which have replaced older varieties to a significant extent. Not all are resistant to any of the major potato diseases but were released because of their reported superior horticultural characters.

Potato seed certification dates from 1914 and the work of J. G. Milward of the University of Wisconsin. It was the object of the work then, as now, to put upon the market the best seed that could be produced. Certification is recognised as a constructive preventive measure in tuber-borne diseases. The industry has expanded greatly. The average production of certified seed potatoes in the United States, for the years between 1949 and 1952, was 44.7 million bushels.

The enforcement of certification standards is the responsibility of the colleges of agriculture and State departments of agriculture in most States. In Nebraska, Utah, and South Dakota, the work has been conducted by growers' organisations.

Certified potato seed is grown under a system of inspection. The plants are inspected twice in the field during the growing season. The tubers are inspected in the bin after they are dug. The first field inspection is usually made early so as to identify and rogue the diseased plants. The second inspection is made in the period between the time of blossoming and just before the vines mature. For each inspection a maximum percentage of affected plants or tubers for each disease is allowed. This disease tolerance varies for disease and by certifying States but is usually between 1 and 5 per cent.

Each State has jurisdiction over its own certification work and sets the tolerances permitted for various diseases. For example, the tolerances allowed for various diseases and varietal mixtures by Maine are : Leaf roll, 2 and 1; mosaic, 3 and 2; spindle tuber, 2 and 2; yellow dwarf, 0-5 and 0.5; total virus diseases, 5 and 3; blackleg, 2 and 1; wilt, 2 and 1; bacterial ring rot, 0 and 0; total for all diseases, 6 and 4; giant hills, 1 (second inspection); varietal mixtures, 1 and 0.25.

The, production of certified seed potatoes depends upon the quality of foundation seed stocks from which they are grown. All plantings produced for certification must be grown from foundation stock of the best quality and should be thoroughly rogued. Improvement of foundation seed stock through field roguing permits the removal of diseased, weak, off-type, or varietal-mixture plants during the growing season. In order to insure freedom from virus diseases, the foundation seed is produced in areas isolated from other potatoes and free from disease-transmission insects. A comparatively small number of tubers indexed one year and planted in the field the next spring will give a sisable increase in seed stock for the following year. After one season's increase, a sufficient quantity would then be produced so that the grower could replace his old stock entirely with disease-free potatoes for the production of certified seed.

The most recent development is the operation of seed-source farms by the certifying agencies. The farms (in Maine and some other States) are operated to produce seed stocks for growers of foundation seed. Seed is carefully grown in isolated areas and is tested in winter field test plots. Seed thus produced is released to selected growers for the production of foundation seed. The varieties grown on the farms are planted by tuber-unit methods and carefully rogued by representatives of the certification agencies.

Vegetable seed production areas have been shifted to areas where the weather is less favourable to the development of plant diseases. Such a shift is possible with crop-seed production because the total acreage of a particular crop is usually relatively small in proportion to the crop for which the seed is used.

Anthracnose (Colletotrichum lindemuthianum) and the bacterial blights (Xanthomonas phaseoli and Pseudomonas phaseolicola) are three major diseases of beans in the United States. They are wet-weather diseases, and their spread and development depend to a great degree on the presence of high moisture and suitable temperatures. Before 1925, the production of bean seed was largely limited to the New England – New York area and Michigan. In those regions the weather conditions are generally favourable for the development of the three diseases. Between 1916 and 1919, losses in some of these localities were as much as 25 per cent. For many subsequent years the diseases were widespread and destructive, wherever beans were grown.

Since the causal organisms of the diseases are disseminated chiefly with the seed and it is imperative to have clean seed for planting, attention was centered on seed production in the West, where weather is unfavourable to the development of the diseases. Such areas were found in Idaho and California. Those centres of certified seed production have elevations of 2,000 to 3,000 feet, exceptionally low humidity, few showers, and little hail during the growing season. As a consequence of producing certified bean seed there, the three diseases have ceased to be the limiting factors in the Production of the crop. At one time pea seed for gardens and canning was grown chiefly in the Northeastern States. Since 1925 or so, the centre of production has shifted to the irrigated and more and parts of some of the Western States. Pea seed for these purposes is now produced in the Snake River Valley and the Twin Falls areas of Idaho, near Bozeman, Mont., the Palouse section of northern Idaho, and eastern Washington. This shift was made in order to produce seed free of bacterial blight (Pseudomonas pisi) and leaf and pod spot.

The growing of seed stocks of cabbage, turnip, rutabaga, and cauliflower used to be limited to the Midwest and East. More recently the ravages of blackleg (Phoma lingam) and black rot (Xanthomonas campestris) have caused the shifting of the seed-producing areas for these crops to localities along the Pacific coast. Because of the low rainfall in those areas during the time the seedling plants are growing in seedbeds, the two diseases do not become established. Hence a crop of disease-free seed can be produced from the seedlings, which are subsequently transplanted to clean fields.

HOW NURSERIES GET VIRUS-FREE FRUIT STOCK

More than 40 virus diseases affect stone fruits. Others attack pome, citrus, avocado, fig, and other fruits. No major fruit crop is free from virus diseases. Some virus diseases have caused the destruction of orchards throughout communities and others have ruined orchards in large areas. Some, more insidious, do not kill trees but take an annual toll by reducing the yield and quality of the crops. Still others produce only mild effects and are important chiefly because they complicate the question of control.

Part of the geographic occurrence of most of the virus diseases affecting stone fruit is traceable to distribution in infected nursery stock. Most fruit trees now are grown from nursery stock produced by budding or grafting the desired variety onto seedlings or rootstocks propagated vegetatively from cuttings. If the variety or the rootstock is infected with a virus, the resulting nursery trees usually will be infected. If the nursery is located near infected trees, viruses may spread naturally into nursery stock during the growing season. When diseased nursery stock is planted in a district where the disease was not already present, spread may take place to other trees and the disease soon becomes established.

The question of nursery improvement is complex. The many factors that are involved vary among areas because different diseases are present, different varieties of fruit are grown, and different conditions exist in each. It seems impossible to devise a programme with provisions that would be entirely applicable to all areas, although some general procedures can be formulated.

Because some virus diseases of stone fruits are known to have been carried in the nursery stock, a logical starting point for improvement is the use of procedures to eliminate them from scion and budwood sources and from rootstock seed sources. Growers of such sources then could be issued certificates indicating the standards that have been met. Certification has value only when it refers to specified definite standards.

Any plan for the production of certified fruit tree nursery stock could well make use of the same principles and procedures developed for certified seed potatoes : Establishing-disease-free foundation stocks true to variety; increasing the stocks in the field under rigid inspection and roguing; passing the stocks for certification if the number of diseased or off-type plants is maintained at or below a standard, which has been determined by practice to be necessary to insure high yields and good quality; and supervising sales to maintain the identity of the certified stock.

Considerable progress has been made towards nursery improvement in the more important fruit-growing States. The programme has been mostly voluntary. The approach has been from different angles. Unknown and variable factors have prevented the formulation of any uniform procedure usable in all States.

In most of the States, the first step has been to inspect the orchard trees desired as a source of budwood and the trees adjacent to them. If no symptoms of virus or virus-like diseases are found, the nurseryman who gets buds from them may obtain a certificate that his trees were propagated from sources that had been inspected and found visibly free of virus diseases. The procedure has helped reduce the prevalence of such virus diseases as the peach yellows group, peach wart, and y certain cherry diseases, which are generally expressed on all the horticultural varieties of the affected host. It has also helped in the elimination of cherry diseases like mottle leaf, twisted leaf, rusty mottle, the necrotic rusty mottle; apricot ring pox; the psorosis of citrus; and other diseases that damage some fruit varieties but are only meagrely expressed on others. Orchard inspection has also materially helped in the elimination of certain virus-like non-transmissible but bud-perpetuated disorders, such as sweet cherry crinkle leaf, sweet cherry deep suture, almond bud failure, and Italian Prune leaf spot and sparse leaf, and has assisted in the selection of fruitful types true to variety.

The usual procedure has been for the nurseryman to apply for the service by a given date to the State department of agriculture in his State. The trees then are inspected in the proper season and given some sort of identifying designation. Standards have to be set, such as minimum age of the trees to be used and the distance from the nearest diseased trees. Because most of the diseases spread in orchards, any certificate based on orchard inspection is good only for the year in which the inspection is made and new inspections must be made each year.

Orchard inspection alone is not enough to determine the presence of all the viruses that affect fruit trees. Some viruses that are destructive to one variety may exist in another without symptoms. Buds from such infected but symptomless trees produce similar infected nursery trees, which carry the virus to orchard locations where the trees are planted. The mottle leaf virus ruins the Napoleon (Royal Ann) and Bing varieties of cherries, but may cause few symptoms or none on Lambert.

Environment, such as high or low temperatures, influences the expression Of symptoms of some diseases. Symptoms of leaf yellowing of the sour cherry yellows disease are expressed on sour cherry trees and damage is accentuated in areas where temperatures following petal fall are relatively low, but no leaf symptoms occur in areas where the temperatures are higher. Nursery stock propagated in warmer climates from vigourous-appearing trees can very innocently carry the sour cherry yellows virus and result in serious losses if planted in such areas as those near the Great Lakes, where summer growing temperatures are low.

The western X-disease virus, conversely, may not produce symptoms, especially on sweet cherries growing on mazzard rootstock in areas of high elevation where temperatures are low. Some virus diseases have long incubation periods; hence, if buds are cut from orchard trees in the early stages of infection before symptoms appear, they may carry the virus to nursery stock. The ring spot virus produces symptoms only during the acute or initial stages of infection on many fruit tree hosts, yet buds taken from trees in the chronic stages and showing no symptoms carry the virus.

The presence of viruses in symptomless trees is determined by indexing them on varieties that express symptoms. That commonly is done by budding healthy nursery trees of a susceptible symptom-expressing variety with buds from the suspect trees. Certain varieties and species are known to produce consistent and characteristic symptoms when infected with particular viruses.

By use of a combination of such hosts in index procedures, a fruit tree can be tested for the presence of any of the known viruses. In order to keep the number of necessary hosts to a minimum, hosts that will express and differentiate a large number of the viruses may be used.

Here is a list of index hosts and the diseases which each may serve to diagnose :

Peach. Elberta : Peach yellows, little peach, red suture, peach rosette, rosette mosaic, phony, peach mosaic, X-disease, western X-disease, yellow bud mosaic, wart, peach mottle, peach necrotic leaf spot, asteroid spot, golden-net, peach calico, peach blotch.

- *Peach. J. H. Hale* : Ring spot, willow twig.
- *Peach. Muir* : Muir peach dwarf.
- *Peach. Seedlings open-pollinated seedlings of Lovell and Halehaven have been used)* : Necrotic ring spot, sour cherry yellows.
- *Sour cherry. Montmorency* : Sour cherry yellows, green ring mottle, necrotic ring spot, pink fruit, peach mottle.
- *Sour cherry. On mahaleb* : Western X-disease wilt and decline.
- *Sweet cherry. Bing* : Buckskin, albino, mottle leaf, rusty mottle, mild rusty mottle, rasp leaf, twisted leaf, tatter leaf, small bitter cherry, western X little cherry, peach mottle.
- *Sweet cherry. Royal Ann* : Black canker, cherry rugose mosaic, pinto leaf.
- *Sweet cherry. Lambert* : Necrotic rusty mottle, little cherry, small bitter cherry, Lambert mottle, Utah Dixie rusty mottle.
- *Prunus serrulata vars. Shirofugen* : Ring spot.
- *Prunus serrulata vars. Kwanzan* : Other latents, rough bark.
- *Plum. Italian Prune* : Prune dwarf.
- *Plum. Shiro* : Line pattern.
- *Plum. French Prune* : Prune diamond canker.

- *Plum. Standard prune* : Standard prune constricting mosaic.
- *Plum. Santa Rosa* : Plum white spot. Apricot.
- *Tilton* : Ring pox.
- *Almond. non-pareil* : Drake almond bud failure.

It may not be necessary to index budwood sources in all areas on all of those hosts. For example, our evidence indicates that some of the cherry viruses are not present in peaches in some sections where only peaches are grown. Also, most peach virus diseases (or at least most of those that seriously damage peaches) affect all varieties of peaches similarly, and their presence generally is determined easily by orchard inspection. But cherries appear to be more commonly affected by viruses than peaches.

Some viruses are ruinous on one variety of cherry but may infect another with only meagre or no symptoms; index procedure to find virus-free trees therefore is needed for cherries more than for peaches.

The question of how much indexing should be done depends on what diseases are present in the area, the destination of the nursery stock, and the variety of the host.

Stone-fruit clones completely free of all viruses are difficult to achieve for a number of reasons. The complete host range of many of the stone-fruit viruses has not yet been determined. Completely satisfactory index hosts for all of the stone fruits have not been determined, particularly because of the variability of reaction caused by different forms of certain viruses. Some stone fruits are nearly universally infected, particularly with some of the latent viruses. Certain viruses (sour cherry yellows and ring spot) are transmitted through seeds. As insect vectors are known for only a few of the diseases, it is not known what measures are necessary to protect healthy stock from outside infection.

The problem of obtaining virus-free stocks by indexing procedure is complex. Index hosts have to be found that will serve with certainty to indicate the presence of a given virus in all its forms. The search for such hosts has been complicated by the fact that stocks of certain of those used were already infected. Orchard trees being indexed often are infected by more than one virus and therefore give confusing results. The long incubation period of some virus diseases, such as diamond canker of French Prune and willow twig of peach, makes the procedure slow and costly.

The climate also favours development of the sour cherry yellows disease, the cause of serious losses. Infected trees or nursery stocks grown in warmer climates may not be seriously affected and may not show leaf symptoms. A simple procedure for testing the sour cherry orchard trees for the presence of the sour cherry yellows virus or freedom from it would be to grow progenies located in an area where, if the virus were present in the progenies, symptoms would be sure to develop on them. Trees of other species could be correspondingly tested by placing buds from them into healthy sour cherry nursery trees. The chief objection to this procedure is that the nursery trees propagated from diseased trees, or healthy nursery trees infected by inoculation from diseased trees, sometimes take 2 years to develop symptoms; thus it takes a long time to get results. Some difficulty has been experienced in obtaining and maintaining disease-free indicator trees. Also, sour cherry does not express symptoms of many of the other viruses which may be present, thus necessitating further indexing on other hosts.

A high percentage of sour cherry trees is infected with the ring spot virus. Build-up of this virus has taken place over the years by indiscriminate propagation from infected trees, in which the virus had become latent; by propagation on infected seedling rootstocks, infected by passage of the virus through seeds; and by orchard spread.

Ring spot is much more prevalent than sour cherry yellows; in fact, all cultures of sour cherry yellows appear to contain the ring spot virus.

The universal occurrence of ring spot with yellows might indicate that yellows is the expression of the combined effect of two or more viruses, of which ring spot is one. Ring spot is known to exist without yellows and may be due to a single virus that is generally a contaminant of yellows. Any index procedure for sour cherry yellows would necessarily have to take ring spot into account.

In Michigan, index procedure for sour cherries has been developed with peach seedlings as indicator plants. Halehaven peach seedlings grown during the current season are budded with cherry buds in late August. If ring spot alone is present in the sour cherry trees from which the peach seedlings were budded, growth of the seedlings the following spring is retarded, buds die on many of the branches, and sometimes the branches die. Subsequent new peach growth from surviving buds assumes a normal appearance.

If sour cherry yellows is present in the cherry trees, the inoculated peach trees will show the retarding and die-back characteristic of and caused by the ring spot virus but in addition subsequent peach growth from surviving buds produces shoots with short inter-nodes and abnormally green leaves crowded into loose rosettes. If neither ring spot nor sour cherry yellows was present in the cherries, the peach seedlings grow normally and should compare with uninoculated checks.

The peach seedling technique is quick and inexpensive. It can be done on a large scale in many areas in the open field. Its shortcomings are that peach does not react well in the greenhouse and cannot be used out-of-doors in regions where peaches are subject to winter injury. In some instances, possibly because of virus forms or individual seedling differences, peach does not give clear-cut reactions. The possibility exists that the dwarfing reaction attributed to yellows is due to a third virus, which is a contaminant commonly associated with sour cherry yellows.

In Wisconsin, indexing procedure has been developed by making use of the fact that ring spot is usually associated with sour cherry yellows. Indexing is done in one of two ways. Scions from the trees are grafted on potted disease-free Montmorency cherry trees in the greenhouse and held at 70 F. for 3 to 4 weeks; if the orchard tree has ring spot, ring spot symptoms will develop on the leaves of the potted tree. The second way is to cut scions from the orchard tree and hold them in cold storage.

The tree is inoculated with a known culture of ring spot and observed for symptoms. If no symptoms develop, the tree is assumed to have had ring spot before the inoculation and the scions in storage are discarded. If the tree develops ring spot symptoms, it is assumed that it was not already infected and the scion wood in storage is used for propagation. By eliminating ring spot, it is reasoned that sour cherry yellows also is eliminated. Our experience thus far has supported this result.

Use has been made in Oregon of two varieties of Prunus serrulata, Kwanzan and Shirofugen, for indexing for the presence of ring spot and possibly other latent viruses. When buds carrying ring spot are inserted into arms of Shirofugen, the buds die without uniting, and gumming lesions are formed around the bud insertion points. The virus apparently moves very slowly, because if the branch is severed below the gumming lesion the virus is removed. By spacing index buds at intervals of 6 inches or less along a branch, a single Shirofugen tree can be used to index a large number of orchard trees. Trees that test negative on Shirofugen are then tested on Kwanzan, because in a few instances a virus has been found that will not affect Shirofugen but will cause a reaction on Kwanzan. Viruses usually spread rapidly through Kwanzan. A Kwanzan tree therefore can be used only for one test.

Prunus tomentosa, Manchu cherry, has been used in Iowa and is reported to be a more sensitive host for ring spot than Lovell peach seedlings. Manchu cherry was found unsatisfactory in California because of variability among seedlings and its failure to give a reaction with forms of the ring spot virus which reacted on Hale peach. Results of tests in Washington indicate that Shirofugen is a much more sensitive test plant than Manchu cherry.

More information is needed before the different index hosts can be evaluated. Where sour cherries are maintained as clones and do not show any symptoms of sour cherry yellows under growing conditions favourable for yellows, it can be presumed with reasonable certainty that they are free of yellows. Peach, sour cherry, and the Shirofugen and Kwanzan varieties of oriental flowering cherry all appear to be of value in indexing for ring spot. Shirofugen can be grown in climates with insufficient chilling requirements for sour cherries and appears to be as sensitive to the ring spot virus as sour cherry or peach.

Programmes have been started in some States to develop certified foundation stocks. In some instances nurserymen were furnished budwood direct from orchard trees that had been determined by index methods to be free of virus. Tree performance was determined by observations directly on the orchard tree. In other instances progenies have been grown from indexed orchard trees and budwood has been supplied to the nurserymen from the progenies. In a few instances enough budwood for direct propagation of nursery stock has been supplied from progeny trees, but mostly nurserymen have increased their own foundation stocks to supply budwood sufficient for their needs.

Peach budwood heated at 122 F. for 5 minutes was used in Michigan on a sufficiently large scale to show that such treatment was practical for nursery procedure. Experiments earlier had shown that the treatment would eliminate viruses of the peach yellows group and X-disease. Certification on the basis of inspection of orchard trees has been used satisfactorily in Michigan for avoiding the yellows group of diseases in nursery stock; the heat treatment adds assurance against any of these diseases getting through.

The foundation stock-mother block procedure has several advantages over use of approved trees in orchards. Mother-block trees properly isolated can be maintained with less risk of becoming infected by natural spread. Pedigreed stocks, centrally located, can be given proper care and can more easily be checked for off-types or diseases. Centralisation also makes for simplification of record keeping, especially as regards progeny performance and tracing troubles which may arise.

It also has limitations. Standards may be hard to determine. It has not been established that certain viruses, especially certain of the latent ones, are sufficiently harmful to warrant exclusion. New diseases are continually being found, and it may be difficult to prevent some of these from getting into stocks. There will be a continuing demand for the new varieties and new strains of varieties, which will pose a continuous problem of what should be stocked.

If the isolation station procedure is generally adopted, much expense can be saved by putting it on a regional basis. In the same way, procedure for interstate shipment of nursery stock could be simplified by interstate agreement on the requirements for certification.

The most serious virus diseases of citrus and pome fruits in most instances can be detected and avoided by orchard inspection. The psorosis disease of citrus occurs in all of the citrus-growing areas of the world. It was distributed in infected budwood before people knew it was caused by a virus. Infected trees do not commonly develop the spectacular scaling lesions and decline symptoms until they are 12 to 16 years old, yet the buds taken from them before the symptoms appear carry the virus. Exocortis, a disease that affects trifoliate orange rootstock, resulting in dwarfing the trees growing on it, can be carried by the top variety without symptoms if it is grown on other rootstocks. Stubborn disease, a third virus disease that reduces the vigour and fruitfulness of sweet orange, also requires several years for recognisable symptoms to develop in infected nursery stock.

All these diseases can be avoided by selecting budwood from vigourous, healthy-appearing orchard trees old enough to show symptoms. For freedom from the exocortis virus, bud-wood must be taken from trees growing on trifoliate orange rootstock. The discovery that psorosis could be diagnosed on the basis of symptoms in young leaves has greatly simplified certification procedure. Sweet orange seedlings can be used to index any species of citrus on which the presence of psorosis cannot be determined by orchard inspection.

None of the three citrus viruses we display appears to have any means of natural spread in North America except by occasional natural root grafts between trees. Production of virus-free nursery stock is therefore an important and efficient way to control them.

The contagious nature of the quick decline disease of citrus, the lack of distinctive symptoms, and the wide occurrence of the causal virus in sweet orange on rootstocks other than sour orange make production of nursery stock free of the quick decline virus impractical as a control procedure within infected areas.

Several virus diseases affect pome fruits, but only one, stony pit of pear, has caused sufficient damage to merit selection of disease-free scion wood. Fruits of the Bose variety on diseased trees are variously misshapen and pitted. Tissue at the base of the pits and around the core becomes hard and stony, making the fruit worthless. The disease can readily be recognised in the orchard just before harvest on fruits of the Bosc variety. Other varieties can be indexed by grafting healthy Bosc on one arm.

One rather serious virus disease of avocado, sun blotch, is the cause of unfruitfulness and misshapen fruits. The expression of the disease is erratic and no good indicator variety is known. The best method of avoiding the disease seems to be by the use of propagation material from trees shown by progeny performance to be free of sun blotch. The disease appears to have been spread chiefly in diseased nursery stock and is the cause of enough damage to warrant efforts to avoid it.

Elimination of viruses and virus-like disorders from nursery stock must precede measures applied in the orchard for effective control. To do this, several general steps are necessary. Viruses and virus-like disorders, which cause obvious symptoms, can be avoided by use of budwood from orchard trees that show no symptoms, are fruitful, and true to type. Such trees preferably should not be in plantings where contagious virus diseases are present and should not in any case be adjacent to virus-infected trees. Screening by index procedure is necessary to avoid viruses that may be latent in orchard trees. When desirable trees are once determined free of virus, they should be propagated on virus-free rootstocks and grown under isolation where they can be maintained under observation and periodic testing to assure virus freedom and desirability of type. Such trees can serve as foundation material from which propagating materials can be supplied to nurserymen for establishing mother tree blocks, which in turn supply budwood for nursery propagation.

It is equally important that orchards producing seeds for growing rootstocks be virus-free and of desirable type.

Specifications for indexing and isolation would necessarily vary with districts, depending on the diseases present and the fruits grown, but effort should be made to devise provisions with enough uniformity to allow for interstate shipment. Growers should demand virus-free trees. Nurserymen need the co-operation and assistance of research, regulatory, and extension men, and growers. Nursery improvement programmes are under development in several States and there is reason for optimism.

THE INSPECTION OF IMPORTED PLANTS

Federal plant quarantines have been in effect in the United States since 1912. They do not prevent the importations of large numbers of plants each year. The plants offered for introduction must be examined to determine that they are free from plant pests that are not present in our country or are not widely established here. This first examination at the ports of entry searches out all plant pests insects, fungi, virus diseases, and nematodes.

The Bureau of Entomology and Plant Quarantine is responsible for the pest-risk problem in the importation of plants. Two States, California and Florida, collaborate with the Federal Government at the ports within their borders. All States co-operate in the follow-up inspections that are given certain genera of plants when they are grown in the field under postentry quarantine.

The inspector's basic tools are a hand lens and a microscope. He examines the imported plant material to see if any plant diseases are present and identify the ones he finds. On the identification depends the decision as to whether those plants should be rejected, treated, or released. If he cannot identify the disease, the inspector holds the shipment and refers a specimen to the Washington office for determination by specialists.

The very nature of plant diseases makes it hard to enforce plant quarantines. Bacteria and the spores of the higher fungi that spread the diseases are so minute that they are usually invisible without magnification unless massed in large numbers. Even when the spores have germinated and invaded the tissues of the plant, evidence of the developing disease ordinarily does not appear at once. Leaves inoculated with Colletotrichum cypripedii show the first symptoms in 15 or 16 days. That is about the normal period for the incubation of many other diseases, but some may appear in as few as 4 days and others only after a month or longer.

Few fungus diseases can be eradicated when they are present in living plants. Therefore the plants on which a new disease is found usually are rejected. There are some exceptions. Hot-water treatments for nematode diseases and a few fungi (such as the leaf smut of rice, Entyloma oryzae, and the mint rust, Puccinia menthae) under favourable conditions may eliminate the parasites. Another example : The citrus seeds may carry the bacteria Xanthomonas citri, which cause the citrus canker. Citrus seeds immersed in a solution of 1 part peroxide of hydrogen to 2 parts of water for 10 minutes are completely cleansed of viable citrus canker bacteria.

The plant quarantine inspector at Houston, Tex., made his customary examination of the stores of a cargo vessel from Japan. The fruits and vegetables in ship's stores, though for use on the ship, may be a serious risk. The crew members may attempt to smuggle fruit ashore. Peelings and spoiled stores may be thrown into harbors and washed ashore.

In the stores the inspector found five citrus fruits on whose rinds were numerous small, round, corky spots. The fruits were confiscated and destroyed, as citrus from Japan is prohibited entry into the United States. Samples of the rind bearing the spots were sent to a Bureau mycologist at Hoboken, N. J. All five fruits were infected with citrus canker, Xanthomonas citri, a highly destructive disease of citrus and one that is believed to have been completely eradicated from the United States after long and expensive effort.

Citrus canker was found in passengers' baggage and ships' stores on 10 occasions in a single month at the port of San Francisco an illustration of the need for continuous vigilance.

At the old inspection house in Washington, D. C., which stood on the corner of Constitution Avenue at Twelfth Street, an inspector was examining some orchids from the wilds of Brazil. Some small patches of dusty, yellow material on a leaf caught his eye.

It seemed harmless at first glance and resembled amorphous materials sometimes found on orchid leaves. Upon turning the leaf over, however, he saw that there was a yellowing of the tissue extending through the leaf. Other leaves with more numerous spots were found. Some of the leaves were dead.

The dusty material proved to be the uredospores of the rust, Hemileia oncidii. All the infected leaves were removed and destroyed. The plants were then disinfected by a dip in bordeaux mixture. Hemileia oncidii and other rusts of orchids, which have also been intercepted many times, are not known to be established in the United States.

A large shipment of lily-of-the-valley pips (Convallaria majalis) arrived on the docks at New York in 1950. The pips originated near Hamburg, Germany. The inspectors at the port gave them the usual thorough examination on the pier, as these plants are a known host of the stem and bulb nematode (Ditylenchus dipsaci).

No Ditylenchus was found, but the roots carried some sandy soil. Soil is a serious hazard in quarantine enforcement. It may carry insects, particularly the larvae and pupal stages, various soil-borne fungi, and plantnematodes. Imported plant material therefore must come with clean roots. The pips were ordered to be sent to the Hoboken Inspection House for cleaning. At Hoboken the washings from the roots were examined carefully for plant pests, and finally processed to recover any nematode cysts that might be present.

The plant pathologist, looking down through the microscope at the debris floating in the dish beneath the lens, observed a smooth, dark-red, spherical body with a short neck, floating around with the debris. The whole object was less than 1 millimeter in diameter. Here was a truly dangerous immigrant the cyst of the golden nematode (Heterodera rostochiensis) filled with living eggs.

Further search revealed that it was not a lone cyst, but that the whole shipment of pips was heavily contaminated with golden nematode cysts, by hundreds at least, and possibly by thousands. Each cyst contained 10 to 400 eggs. It is quite probable that this one shipment of lily-of-the-valley, after being distributed by sale to florists for growing, might have established the nematode in not one but many new areas. Its introduction into our remaining potato-growing areas would be a disaster as it has proved to be on Long Island and in some parts of Europe.

The interception of the golden nematode on these Convallaria was not the first interception of this pest nor by any means the last. The cysts are found frequently not only in soil with plants, but they have also been intercepted adhering to straw and burlap bagging. The menace they present is heightened by the longevity of the eggs within the cysts. H. rostochiensis eggs have been known to hatch after 8 years.

Part of the Convallaria shipment was treated with hot water. The pips were immersed in hot water at 118 F. for 30 minutes, then removed and cooled with water after draining for 5 minutes.

The treatment must be given with care as the margin between a complete kill of the nematodes and serious injury to the plants is very small. The plants not treated were forced in postentry quarantine in isolated greenhouses. After the flowers were harvested the plants were destroyed, and the soil and the benches then sterilised with steam.

Our three examples emphasise the varied nature of the inspector's problem. The virus diseases are especially difficult to detect in dormant plants, the condition in which most nursery stock is imported. It was largely to overcome this difficulty that postentry quarantine was devised.

Post-entry quarantine is the requirement that some genera of plants be grown under observation of State and Federal inspectors, usually for two growing seasons, before they are released for sale or distribution. The plants subjected to this treatment are the ones known to be the hosts of some serious plant disease that does not occur here or is restricted in distribution in the United States. In most instances the plants are prohibited from the countries in which the disease is known to occur. The post-entry provision- then applies only to other countries in which it might be present but unreported.

Rose wilt virus (Marmor flaccumfaciens) occurs in Italy, Australia, and New Zealand. Rose plants or cuttings are prohibited from those countries. If a nurseryman wishes to import the material from any other foreign country he will be required to make a legal agreement that he will grow the material at a designated place, accessible to the State nursery inspector and the Federal plant quarantine inspector until it is released.

In each of the following two growing seasons the State nursery inspector will inspect the plants several times. Sometimes he will be accompanied by an inspector from the postentry section of the division of plant quarantines. The plants will be examined particularly for any evidence of rose wilt virus, but also for other foreign diseases or insects which may have accompanied them.

Postentry inspectors also must gather from the published records information on the foreign plant diseases which we are attempting to exclude. Summaries of the information are then distributed to the State inspection services. In that way the State inspectors can know the proper time to inspect roses, hops, or other postentry plants in their areas, and something of the appearance of the foreign diseases.

Daphne mosaic virus was found in 1950 in a lot of 250 Daphne mezereum plants from Holland. Then the disease was known only in Australia and New Zealand. It was the basis for the inclusion of daphne in the post-entry list. The plants were promptly destroyed under the supervision of the State nursery inspector. At the end of the second growing season those postentry plants which have remained free of diseases and insects new to the United States are released from quarantine.

The postentry inspection makes it possible to examine the plants when they are in leaf and at times favourable for detection of the particular diseases for which they are quarantined. In 1953 there were 50 genera of plants that must be grown in postentry quarantine when imported from certain parts of the world. Besides them, a blanket provision includes fruit and nut plants.

PROTECTION THROUGH QUARANTINES

Modern transportation in the air age brings us close in time to lands far distant in point of miles. Travel for business, recreation, and cultural purposes stimulates interest in exotic plants, fruits, and other plant products. Modern commerce and world economy are conducive to an international flow of plants and plant products together with the plant diseases which these materials may carry across international boundaries, including our own.

Were there no barriers to plant disease entry, the United States could readily become the habitat for a host of plant diseases which are not now known in this country and which, if introduced and established here, could become the cause of untold additional annual losses in field crop, fruit, horticultural, floricultural, and forest production.

We take comfort, however, in the provisions of the Plant Quarantine Act of 1912, as amended, by which many safeguards are erected against entry and establishment of foreign plant diseases as well as against the domestic spread of introduced diseases. The safeguards consist of embargo and regulatory provisions in both foreign and domestic commerce. The following remarks relate only to the provisions of the Act concerning plant diseases. In order to prevent the introduction of any plant disease that is not known in this country or that is not widely prevalent or distributed in the United States, the Act empowers the Secretary of Agriculture to issue a quarantine prohibiting the importation of the host materials of that disease and prescribes certain legal requirements which must precede the issuance of that quarantine.

They include a determination of the necessity for the action, and a public hearing on the subject. The quarantine notice must specify the host materials to be excluded and name the countries where the disease occurs. Thereafter, and until the quarantine is withdrawn, the importation of the named host materials from the specified countries is prohibited regardless of the use for which the materials are intended.

All nursery stock not excluded by such embargo action may enter this country only when a permit has been issued by the Secretary of Agriculture for the importation. If the plants are to come from a country with an official phytosanitary inspection system, the shipment is to be accompanied by a certificate of the proper official of the country of origin that the nursery stock has been thoroughly inspected and is believed to be free from injurious plant diseases and insect pests. In the case of importations from countries without official systems of inspection, the shipment shall meet such conditions as the Secretary of Agriculture may prescribe.

Any plants, fruits, vegetables, roots, bulbs, seeds, or other plant products not defined as nursery stock in the Act may be brought under regulated entry comparable to that for nursery stock by action of the Secretary. To do so he must determine that the unrestricted importation of any such plant materials may result in the entry of injurious plant diseases. Other requirements are similar to those necessary to promulgate a prohibitory quarantine.

An amendment to the Act that was approved July 31, 1947, improves the protection against foreign plant disease entry by making provision for growing nursery stock under postentry quarantine by, or under the supervision of, the Department of Agriculture, for the purpose of determining whether it may be infested with plant pests not discernible by port-of-entry inspection methods. Should there be such infection, the Secretary is authorised to prescribe remedial measures to prevent the spread. This feature is particularly useful in the instance of virus diseases. The various quarantines and orders issued under authority of the Act and now in force cover practically the entire field of foreign host materials plants and plant parts, including fruits, vegetables, seeds, several fibres, and cut flowers likely to carry injurious plant diseases into this country. Several of the quarantines and orders are issued under authority of both the prohibitory and regulatory parts of the Act.

A number of the foreign quarantines contain provisions complementing domestic plant disease control programmes. For example, a feature of Nursery Stock, Plant, and Seed Quarantine No. 37 regulates the entry of barberry seeds and plants in harmony with the provisions of Federal domestic Stem Rust Quarantine No. 38. Another feature of Quarantine No. 37 prohibits the entry of citrus seeds into Florida, where there is a comparable State quarantine against citrus propagating materials from other States.

The quarantines and orders are enforced by plant quarantine inspectors at the principal ports of entry. The inspectors are assisted through co-operative relations by their associates in the Customs, Immigration, Public Health, and Postal Services, and the Bureau of Animal Industry and by collaborating State plant quarantine services. The total effort of those services and the foreign inspection and certification of plants before shipment provide an effective programme for the enforcement of the quarantines and prevention of the entry of plant diseases. Carriers and cargoes, mail, and baggage from foreign countries are examined for unauthorised plant material. The plant material that is permitted entry into the United States is inspected for the presence of plant pests. The more hazardous of the plant propagating materials that are permitted entry are routed, upon arrival, for intensive inspection at specified locations where specialised facilities are available, both for inspection and for whatever treatment is required. These facilities are maintained at Hoboken, N. J.; Miami, Fla.; Laredo, Tex.; San Francisco and San Pedro, Calif.; Seattle, Wash.; Honolulu, T. H.; and San Juan, P. R. Certain designated kinds of plants are then released only for growing by the importer pending further observation for plant pests particularly virus diseases not detectible at the time of entry.

Not all importations are inspected with equal thoroughness because that would be too costly and, just as to experience, unnecessary. The Department therefore places greater emphasis on the inspection of some classes of plants and plant products than on others (just as to known or potential risk of entry of plant diseases) in order to afford the maximum degree of protection possible with the facilities available. The State plant quarantine agencies support the Department in connection with the postentry quarantine work by assuming the primary responsibility for the growing-season surveillance of imported plants held under postentry quarantine conditions. Certain States contribute materially to the entry control programme at ports of arrival.

The Department itself makes numerous importations in connection with its programme for improving field and vegetable crops, the horticulture, floriculture, and forestry of the country, and for kindred experimental and scientific reasons. These introductions, which may legally include material prohibited entry, are subjected to even more rigid control and safeguards on entry than are introductions by the public.

As to domestic commerce, the Act in some respects is more adaptable to the needs of a programme for prevention of the spread of plant diseases because it gives the Secretary authority to embargo or regulate the movement of "stone or quarry products, or any other substance of any character whatsoever, capable of carrying any dangerous plant disease," as well as authority to embargo or regulate the movement of nursery stock and other plants and plant products.

In a manner similar to the procedure prescribed in the field of foreign commerce, the Secretary may quarantine any State, Territory, or District of the United States, or part thereof, whenever he determines it is necessary to do so in order to prevent the spread of a new or not yet widely prevalent or distributed dangerous plant disease. He is directed, when public interest will permit, to provide by regulation for the inspection, treatment, and certification of the regulated materials in order to govern their movement from the quarantine area.

Unless the Secretary has issued a quarantine to prevent the domestic spread of a dangerous plant disease, the Act specifically provides that a State, Territory, or District may enforce a measure for the same purpose. The Secretary has discretionary authority also to co-operate with States to carry out their quarantine measures.

The quarantines and regulations governing the domestic movement of plant materials may be divided roughly into two classes those governing interstate movement on the mainland of the United States and those governing movement between the off-shore territories and possessions and the mainland, or other off-shore territories and possessions. These domestic quarantines on the mainland are usually associated with disease control projects of the Department.

The States help materially in carrying out these projects, including enforcement of related intrastate quarantines. Further assistance in the domestic quarantine enforcement is derived from the inspection by Federal inspectors of shipments of quarantined materials moving interstate through traffic centres. Although the Act provides for such protection, practical enforcement problems resulted in lifting the only quarantine of the mainland for the protection of an off-shore area. Only three quarantines of offshore areas exist which are designed for plant disease reasons to protect the continental United States; namely the quarantine of Hawaii and Puerto Rico on account of sugarcane diseases, and the quarantines of Hawaii on account of citrus canker.

The ACT also makes especial provision with respect to the District of Columbia in order that it may have a plant quarantine inspection service similar to that of the States and Territories. The Secretary of Agriculture is authorised to make regulations to govern the movement of plants and plant products into and out of the District "in order further to control and eradicate and to prevent the dissemination of dangerous plant diseases." This part of the Act also confers authority to compel an owner to take sanitary action in case of a nuisance situation to prevent the dissemination of plant disease.

Handbook of

Plant Hormones

Actions and Applications

Handbook of
Plant Hormones
Actions and Applications

Contributors :
William M. Gray,
F. Odabaşoğlu, *et al.*

Volume 2

KOROS PRESS LIMITED
London, UK

Handbook of Plant Hormones : *Actions and Applications*
Contributors : William M. Gray *and* F. Odabaşoğlu, *et al.*

Published by Koros Press Limited

www.korospress.com

United Kingdom

Printed in 2017 for Sale in the Indian Subcontinent

Handbook of Plant Hormones : *Actions and Applications*

ISBN: 978-1-78163-475-2

British Library Cataloguing in Publication Data
A CIP record for this book is available from the British Library

Exclusively distributed by CBS Publishers & Distributors Pvt. Ltd.
Sales & Distribution Rights only for India, Pakistan, Bangladesh, Sri Lanka, Nepal and Bhutan.This book is not to be sold outside these territories.

Preface

Plant hormones are chemicals that regulate plant growth, which, in the UK, are termed 'plant growth substances'. Plant hormones are signal molecules produced within the plant, and occur in extremely low concentrations. Hormones regulate cellular processes in targeted cells locally and, when moved to other locations, in other locations of the plant. Hormones also determine the formation of flowers, stems, leaves, the shedding of leaves, and the development and ripening of fruit. Plants, unlike animals, lack glands that produce and secrete hormones. Instead, each cell is capable of producing hormones. Plant hormones shape the plant, affecting seed growth, time of flowering, the sex of flowers, senescence of leaves, and fruits. They affect which tissues grow upward and which grow downward, leaf formation and stem growth, fruit development and ripening, plant longevity, and even plant death. Hormones are vital to plant growth, and, lacking them, plants would be mostly a mass of undifferentiated cells. So they are also known as growth factors or growth hormones. The term 'Phytohormone' was coined by Thimann in 1948. Phytohormones are found not only in higher plants, but in algae too, showing similar functions, and in microorganisms, like fungi and bacteria, but, in this case, they play no hormonal or other immediate physiological role in the producing organism and can, thus, be regarded as secondary metabolites.

The word hormone is derived from Greek, meaning set in motion. Plant hormones affect gene expression and transcription levels, cellular division, and growth. They are naturally produced within plants, though very similar chemicals are produced by fungi and bacteria that can also affect plant growth. A large number of related chemical compounds are synthesized by humans. They are used to regulate the growth of cultivated plants, weeds, and in vitro-grown plants and plant cells; these manmade compounds are called Plant Growth Regulators or PGRs for short. Early in the study of plant hormones, "phytohormone" was the commonly used term, but its use is less widely applied now. Plant hormones are not nutrients, but chemicals that in small amounts promote and influence the growth, development, and differentiation of cells and tissues. The biosynthesis of plant hormones within plant tissues is often diffuse and not always localized. Plants lack glands to produce and store hormones, because, unlike animals-which have two circulatory systems powered by a heart that moves fluids around the body-plants use more passive means to move chemicals around the plant. Plants utilize simple chemicals as hormones, which move more easily through the plant's tissues. They are often produced and used on a local basis within the plant body. Plant cells produce hormones that affect even different regions of the cell producing the hormone. Hormones are transported within the plant by utilizing four types of movements. For localized movement, cytoplasmic streaming within cells and slow diffusion of ions and molecules between cells are utilized. Vascular tissues are used to move hormones from one part of the plant to another; these include sieve tubes or phloem that move sugars from the leaves to the roots and flowers, and xylem that moves water and mineral solutes from the roots to the foliage. Not all plant cells respond to hormones, but those cells that do are programmed to respond at specific points in their growth cycle. The greatest effects occur at specific stages during the cell's life, with diminished effects occurring before or after this period. Plants need hormones at very specific times during plant growth and at specific locations. They also need to disengage the effects that hormones have when they are no longer needed. The production of hormones occurs very often at sites of active growth within the meristems, before cells have fully differentiated. After production, they are sometimes moved to other parts of the plant, where they cause an immediate effect; or they can be stored in cells to be released later. Plants use different pathways to regulate internal hormone quantities and moderate their effects; they can regulate the amount of chemicals used to biosynthesize hormones. They can store them in cells, inactivate them, or cannibalise already-formed hormones by conjugating them with carbohydrates, amino acids, or peptides. Plants can also break down hormones chemically, effectively destroying them. Plant hormones frequently regulate the concentrations of other plant hormones. Plants also move hormones around the plant diluting their concentrations. The concentration of hormones required for plant responses are very low. Because of these low concentrations, it has been very difficult to study plant hormones, and only since the late 1970s have scientists been able to start piecing together their effects and relationships to plant physiology. Much of the early work on plant hormones involved studying plants that were genetically deficient in one or involved the use of tissue-cultured plants grown in vitro that were subjected to differing ratios of hormones, and the resultant growth compared. The earliest scientific observation and study dates to the 1880s; the determination and observation of plant hormones and their identification was spread-out over the next 70 years.

In general, it is accepted that there are five major classes of plant hormones, some of which are made up of many different chemicals that can vary in structure from one plant to the next. The chemicals are each grouped together into one of these classes based on their structural similarities and on their effects on plant physiology. Other plant hormones and growth regulators are not easily grouped into these classes; they exist naturally or are synthesized by humans or other organisms, including chemicals that inhibit plant growth or interrupt the physiological processes within plants. Each class has positive as well as inhibitory functions, and most often work in tandem with each other, with varying ratios of one or more interplaying to affect growth regulation. Plant stress hormones activate cellular responses, including cell death, to diverse stress situations in plants. Researchers have found that some plant stress hormones share the ability to adversely affect human cancer cells. For example, sodium salicylate has been found to suppress proliferation of lymphoblastic leukemia, prostate, breast, and melanoma human cancer cells. Jasmonic acid, a plant stress hormone that belongs to the jasmonate family, induced death in lymphoblastic leukemia cells. Methyl jasmonate has been found to induce cell death in a number of cancer cell lines.

This page left intentionally blank.

CONTENTS

LIST OF CONTRIBUTORS

F. Odabaşoğlu

Atatürk University, Faculty of Pharmacy, Department of Biochemistry, 25240 Erzurum -TURKEY

Ö.İ. Küfrevioğlu

Atatürk University, Faculty of Pharmacy, Department of Biochemistry, 25240 Erzurum -TURKEY

William M. Gray

William M. Gray is in Plant Biology at the University of Minnesota, St. Paul, Minnesota, United States of America. E-mail: grayx051@tc.umn.edu

This page left intentionally blank.

Chapter 15

GROWING HEALTHIER PLANTS

THE NATURE OF RESISTANCE TO DISEASE

We can get a double economic gain if we can introduce into our agriculture crops that are naturally immune to diseases and thereby avoid both the loss from disease and the cost of sprays and other ways of combating disease.

To do that as W. A. Orton pointed out in 1908 one has to know the problems of heredity, the nature of a disease, its governing factors, and the type of resistance involved in order to adopt the most promising lines of approach in breeding.

Disease resistance in plants is not an easy subject to understand. It involves the intricate relations between the plant the host that is being attacked and the fungus or bacterial organism the parasite; that is doing the attacking.

The terms "disease resistance" and "immunity" can be used to denote different degrees of the same thing. Various degrees of disease resistance are possible. Immunity means complete resistance to disease; immune means not subject to attack by a pathogenic organism or virus.

We might consider other definitions here at the start : A pathogen is a parasitic organism or virus whose activity causes disease in the host. The host is a living organism that harbors another organism or virus that depends on it for existence. Pathogenic, the adjective, means having the ability to induce disease. A parasite is an organism or virus that lives on the tissues of another living organism. To inoculate is to introduce a micro-organism or virus or a material containing either into an organism, a culture medium, soil, or something like them. To infect is to invade an organism and to bring about infection. A suscept is an organism that is affected or can be affected by a given disease. Resistance is the ability of a plant to withstand, oppose, lessen, or overcome the attack of a pathogen. Susceptibility is the inability of a plant to defend itself against an organism or to overcome the effects of invasion by a pathogenic organism or virus.

Immunity is absolute. Resistance and susceptibility are relative : A plant is either immune or not immune to a pathogen, but it may be more or less susceptible or resistant. A plant may be "slightly susceptible," "moderately resistant, " or "extremely susceptible" but not "moderately immune" or "highly immune."

The ability of a susceptible plant to avoid infection because it possesses some quality (such as earliness of maturity) that prevents successful inoculation is called escape, or klendusity. Escape must be clearly distinguished from resistance. Tolerance is the ability of a plant to endure the invasion of a pathogen without showing many symptoms or much damage. A degree of resistance great enough so that no serious economic loss results (although there might be considerable invasion by the pathogen) is termed "practical resistance." Certain varieties of wheat, for example, for practical purposes are resistant to leaf rust the rust causes little loss although the plants may become heavily rusted as they approach maturity.

We use the term hypersensitiveness to denote such a violent reaction of a plant to an attack by an obligate parasite (a pathogen that depends on living tissue for its nutrition) that the invaded tissues of the host are quickly killed, so that there is no further spread of infection. In essence, hypersensitiveness is extreme susceptibility, but its practical effect, as far as crop loss is concerned, amounts to extreme resistance. This type of reaction is common in cases of infection of many plants by rust fungi and some of the viruses. The definitions help us avoid some wrong ideas about the nature of resistance and help us understand what can and cannot be done in plant breeding. For instance, a variety that escapes a particular disease is not necessarily a resistant variety.

William A. Orton long ago pointed out the differences between what he termed disease-escaping, disease-enduring, disease-resisting, and immune varieties.

He said : "Disease endurance sometimes results from the ability of the plant to grow in spite of an attack, either through exceptional vigour or through a hardier structure, as in the case of certain melons which better survive the attacks of leaf-blight because the leaves do not dry out as quickly as do those of the ordinary melons. Drought-resistant plants are often disease-enduring. Watermelons from semi-arid Russia were for this reason the last to succumb to the wilt disease when planted in our Southern States.

"Finally, we have disease-escaping varieties. Such, for example, are the extra early cowpeas which mature before the season for wilt and root-knot to develop. These varieties which escape disease through earliness or lateness are often really very susceptible. The Early Ohio and other early potatoes, which commonly mature before the appearance of the late blight disease, are among the first to succumb to this disease if planted so late as to be still immature when the moist weather of the late summer or early fall enables late blight to spread."

E. M. Freeman in the University of Minnesota and one of the pioneers in this subject, pointed out that little difference exists in the real resistance powers of oat varieties to the common oat rusts, and when a grower is told that an oat variety is resistant because it usually escapes the rust through earliness of ripening, he is led into a serious mistake. The essential character of true resistance lies in a protoplasmic activity and is independent of inoculation accidents.

A variety may escape a disease through certain peculiarities of the host. To prove that, Dr. Freeman grew a variety of barley in different soils, which varied from the normal garden soil to those containing about 2 per cent of alkaline salts. Plants in the different soils, inoculated by sprays in the greenhouse, showed different amounts of rust. Those in the stronger alkaline soils generally showed less rust. The latter, however, when attacked outside the greenhouse, displayed large and vigourous growths of the rust. That was undoubtedly due to the greater development of "bloom" on the barley foliage when it was grown in the strong alkali; the bloom caused the drops of water to roll off and the inoculating material to be lost. The bluish colour of the plants in the alkaline soils and the greater tendency of the water drops to run off were quite pronounced. There seems to be no reason for assuming that there was any difference in real resistance to rust.

The economic importance of disease-enduring varieties ought not to be overlooked by pathologists and plant breeders. Effective work in selection and breeding may be accomplished in the production of disease-enduring plants. Some of the best varieties of the hard spring wheats commonly used in the North Central States, for example, come under this class in regard to stem rust, since they do not possess any appreciable amount of resistance. They can endure moderate attacks of rust, but all go down in a heavy epidemic of rust. No sharp distinction may exist between disease-enduring and disease-resistant varieties, but for practical purposes well-marked resistance can be detected readily under proper experimental conditions.

Immune varieties have perfect resistance. Their production will probably always remain rare in comparison with those possessing only a partial resistance or high powers of endurance.

rue resistance to disease is different from those instances in that it depends on some structural or physiological characters of the plant that prevent successful invasion of the parasite.

C. Brick of Germany in 1919 pointed out that susceptibility of plants to disease is not due to degeneration, old age, and such causes, but usually is the result of certain differences in the structures of the host itself. He believed that a higher content of acid, sugar, and tannin also has a bearing on the resistance of a variety to parasites. He said that some plants escape disease because their season of blossoming or maturing does not coincide with the development of the parasite.

DIFFERENT TYPES OF DISEASE RESISTANCE EXIST IN PLANTS

To quote Orton again : "The typical form of disease resistance involves a specific reaction on the part of the host cell against a true parasite, a character developed in nature in the evolution of the species and strengthened in cultivated plants through the work of the breeder.

Less important from the breeder's standpoint are plants resistant through :

- Structural differences,
- Disease endurance, and
- Disease avoidance.

The evidence indicates that the resistance is due to a specific protective reaction of the host cell against the parasite. So in plants the evidence leads us to believe that more is involved than the acidity of the cell sap or the chemotactic effect of sugars or other food substances.

"The first group is the most important, relating as it does to diseases due to the most highly developed parasites, such as the rusts, mildews, and other injurious fungi. The evidence indicates that the resistance is due to a specific protective reaction of the host cell against the parasite." Orton was not sure that the reaction between host cell and fungus parasite in plants was the same as that described for some forms of immunity in man and the higher animals, in which substances in the blood serum neutralise the toxin excreted by the invading bacteria and assist in the destruction of the latter. He stated, however, that the evidence in plants leads us to believe that more is involved than the acidity of the cell sap or the chemotactic effect of sugars or other substances.

He said : "The delicacy of the reaction may be better understood if we withdrawal the fact that it is adjusted to repel specific invaders. A plant resistant to one disease may be quite susceptible to another. General hardiness is also another matter. A plant may be resistant to cold and yet extremely susceptible to the attack of some parasite.

"Structural differences do not seem to play much part in enabling plants to resist the true parasite. Satisfactory demonstrations of cases where resistance to highly adapted parasites is due to thickened epidermis, development of hairs, etc., are lacking. It has, on the other hand, been shown; that germinating spores of fungi often penetrate the epidermis of plants they can not parasitise, and are killed forthwith by the cells they attack. It is hard to understand why a thick cell wall should protect from infection a leaf which has many thousand openings as breathing pores through which a fungus might enter.

"Resistance due to structural causes does occur in troubles due to wound parasites, a fruit or a tuber with a thick rind being thereby less liable to bruising; as there may be an indirect connection, a plant of more open habit of growth being thereby less subject to attack by fungi which require moisture for their development."

Observations by other early scientists showed that mechanical immunity may have some significance in certain instances but cannot be accepted as a universal phenomenon; natural immunity does not depend on the anatomical peculiarities of plants, but on properties of their cytoplasmic cell contents and on active resistance of host plant cells, usually accompanied by a complicated physiological reaction in response to penetration by the parasite.

Other workers just as strongly contend that resistance or immunity in plants is due to morphological factors. Still others have confused the "disease-enduring" and "disease-escaping" varieties with resistant or immune varieties.

The causes for apparent resistance or immunity to a disease and tolerance to a disease are rather easily explained. But when we consider the causes for true immunity and resistance, we find ourselves involved in a maze of anatomical, physiological, biochemical, and ecological evidence and theories, which have been offered, some to explain specific cases of resistance and the others to explain the problem in general.

Before entering into a discussion of them, we should have in mind the fact that what we commonly think of as infection consists of two stages. The first is the entrance of the parasite into the tissues of the plants. The second is the establishment of parasitic relationship with the host. Some fungi, such as the rusts, enter many plants from which they cannot obtain nourishment and therefore perish because they cannot accomplish the second stage of infection.

One realises, then, that immunity and resistance may be due to the plant's morphological and anatomical characters that prevent the first stage of infection or entrance of the parasite into the tissues; or to biochemical properties or anatomical characters of the tissues of the host, which prevent the second stage of infection or the establishment of parasitic relationship with the tissues of the plant.

Cases of immunity or resistance due to the anatomical characters of the host preventing entrance by the parasite are frequent. Coffee leaf disease usually starts on the under surface of the leaf because there are few stomata pores or openings on the upper surface.

Young beet leaves are practically immune to attacks by Cercospora beticola because their stomata are so small as to be incapable of opening widely enough to allow for the entrance of the germ tube of the spores of the fungus which can only enter the host through the mature stomata.

Some varieties of plums resist brown rot caused by Sclerotinia cinerea because the stomata soon become plugged with masses of small parenchymatous cells. The toughness of the skin, the firmness of the flesh, and high fibre content also are characteristics that make varieties of plums resistant to brown rot. As ripening progresses, the texture of the resistant varieties remains firm, while that of the susceptible becomes softer. The stomata of Kanred wheat, which is resistant to certain strains of Puccinia graminis tritici, are said to be of such a nature as to shut out most of the fungi. Wheat has a tendency to keep many of its stomata closed. It may be that a secretion of the rust fungus makes the mechanism of the stomata inoperative and they remain closed, thus excluding the fungus. In the Kanred wheat, which has smaller stomata, those peculiarities might be more effective in excluding fungi than in varieties of wheat with larger stomata.

The resistance of Citrus nobilus to citrus canker is due to a broad ridge overarching the outer chamber of the stomata, which practically excludes water from them, thus preventing the entrance of the canker bacteria. The resistance of some carnations to rust may be due to the type of stomata that make it impossible for the rust hyphae to penetrate the leaves.

The resistance of some varieties of barley to rust is due to the bloom waxy coating on the leaves, which keeps drops of water from adhering to them. Consequently the rust spores cannot germinate on them. The more waxy varieties of raspberries and grapes are less damaged than others by cane blight.

Mr. B. F. Lutman, of the Vermont Agricultural Experiment Station, after investigating the resistance of potato tubers to scab (Actinomyces scabies), reported that at the head of the resistant class stand the potatoes of the russet type and that thickness of skin determines resistance of tubers to scab.

The hairs on the leaves may influence resistance of plants. Varieties of potatoes with small, hairy leaves and open habit of growth dry quickly after wetting and are less liable to infection by late blight than other kinds, but perhaps the open habit of growth (and not the abundant hairs) is the chief factor with such varieties.

Some technicians believe that hairs seem to save the under surface of the apple leaves from infection with the apple scab fungus (Venturia inaequalis), while Venturia pyrina readily attacks the smooth under surface of the pear leaves. That belief, however, does not agree entirely with other observations in regard to the scab on apple. Frequently most, if not all, primary infection by apple scab is on the lower surface of the leaves. It is also a matter of common observation that the young fruits, while still covered with down, are very susceptible to scab.

The very hairy common mullein (Verbascum thapsus) is attacked by six leaf-inhabiting parasitic fungi, but only three are reported as occurring on the moth mullein (V. blattaria), which is free of hairs. Variations of form or structure in certain varieties may influence susceptibility. As an example : Varieties of pears with an open channel from calyx to core are most susceptible to Fusarium petrefaciens, a fruit-rotting organism. The immunity to loose smut in barley has been attributed to the closed flowers found in the resistant varieties. In closed flowers the stigmas are not exposed to infection. The rare occurrence of ergot on wheat is said to be due to the brief and irregular openings of the glumes at maturity.

Let us next consider cases of immunity due to incompatibility of the invading organism with the tissues of the plant the failure of the organism to accomplish the second stage of infection, which is the establishment of parasitic relationship with the host.

The prevention of this second stage of infection sometimes is due to the anatomical characters of the host tissue. More often it is due to biochemical properties of the cells.

In making a physiological study of the parasitism of Pythium debaryanum on the potato tuber, Lena Hawkins and R. B. Harvey, of the Department of Agriculture, found that the fungus secretes a toxin that kills the cells and an enzyme that breaks down the middle lamellae but does not affect the secondary thickenings to any extent.

They observed that White McCormick, a potato very resistant to Pythium debaryanum, has a higher crude fibre content than that of susceptible varieties. The higher fibre content they attributed to more secondary thickenings in the cell walls. Resistance to infection is apparently due to the resistance of the cell wall to mechanical puncture by the invading hyphae.

When we come to consider cases of immunity due to biochemical properties of the cells, we are confronted with many complicated theories and speculations, all of which support the theory that such immunity is due to a chemical interaction between the host and the parasite.

E. C. Stakman, of the University of Minnesota, in 1919 showed that resistance of some plants to rust is due to the dying of the cells surrounding the point of infection and the consequent starving of the obligate parasite. That is called supersensitivity, or hypersensitiveness. The cause has not been determined, but it might be of biochemical origin.

J. G. Leach, then of the University of Minnesota, stated that in a variety of bean highly resistant to anthracnose seldom more than one or two cells are attacked and that resistance seems to be due to the inability of the fungus to obtain nourishment from the living protoplasm of the bean.

Tannin, often found in vegetable cells, generally is toxic to fungi. M. T. Cook and J. J. Taubenhaus, of the Delaware Agricultural Experiment Station, pointed out that beans are more susceptible to anthracnose (Colletotrichum lindemuthianum) during the stages of their growth when the enzyme that acts on gallic acid in the cells to form tannin is least abundant. In apples, pears, and persimmons, and other fruits, the enzyme is less abundant when the fruits are ripe than when they are green. That explains the lower resistance to rots in the ripe fruits.

Cook found also that tannin affects differently the various species of the Endothia fungus. It inhibits E. radicalis and E. gyrose. E. parasitica, the cause of chestnut blight, is retarded at first but later is able to feed on the tannin.

Acidity of the cell sap evidently plays some part in resistance of grapes to black rot (Guignardia bidwellii). Varieties that contain less tartaric acid in their cell sap show greater resistance to the rot than those containing greater amounts. Botrytis cinerea, nerea, a parasitic fungus, is repelled by certain acids in plant cells. The resistance of grapes to powdery mildew (Uncinula necator) also has been correlated with the acidity of the cell sap. The resistant varieties contain more tartaric acid than do the less resistant kinds.

O. Comes, an Italian investigator, has drawn attention to the fact that Rieti, a variety of wheat strongly resistant to rust in Italy, has a more acid sap than that of any of the other less resistant kinds he tested. The loss of resistance in Rieti when it is grown in warmer localities than its native area is correlated with reduced acidity of the sap.

M. Popp, a German scientist who investigated the hydrogen-ion concentration and natural immunity of plants, found that infections of pathogenic bacteria cause the plants to respond with variations of the hydrogen-ion concentration.

Immediately after infection the acidity decreases. At the end of the incubation period the acidity rises. If the plant can withstand the infection, the acidity then falls back to normal. If the plant cannot withstand the infection, hydrogen-ion concentration rises to a high level and then falls, usually below normal.

Other cell contents seem to account for some types of resistance. Biologic species of parasitic fungi are sensitive to minute differences in the amounts of albumin in host plants. Maybe that explains why some varieties are resistant to certain biologic strains and not to others.

J. C. Walker, of the University of Wisconsin, observed that the smudge fungus, Colletotrichum circinans, normally attacks only white onions and not those with red or yellow papery scales. But if he removed the coloured scales, the coloured onions then became susceptible to the smudge disease. Acting on the hunch that the pigments were associated with the resistance of the coloured onions, he extracted the pigments with water and found that in the extracts the spores of the smudge fungus would not germinate normally. Nor would the fungus grow, although it developed normally in similar extracts of white-scaled onions.

The extracts of red and yellow pigments were analysed later by K. P. Link and H. R. Angell, Dr. Walker's associates. They found the toxic substances to be protocatechuic acid and catechol. They discovered that the substances also protect the coloured onions from certain other diseases.

Chemical disease resistance in plants offers the biochemist a fertile field for research, and in that field we should search for the causes of disease resistance in plants.

We must not regard the different degrees of disease resistance displayed by plants as fixed and absolute. Environmental factors modify them profoundly. In fact, growers often believe that unfavourable rainfall and temperature are the direct and only reasons for the trouble. Temperature, moisture, fertility, and reaction of the soil markedly affect disease development. Light and the temperature and humidity of the air also may be important.

Other factors, such as the age and maturity of the plant, may affect disease resistance. Thus the observed degree of susceptibility or resistance in any case is a product of many interacting factors of which the inherent susceptibility of the host is only one. The degree of virulence of the pathogen, the age and condition of the plant, and the environment with its many effects on both host and pathogen, all must be suitable before maximum susceptibility can be expressed.

K. Starr Chester, of the Oklahoma Agricultural Experiment Station, has compared infection with the operation of a complicated lock, every tooth and tumbler of which must be in proper alinement before the lock will open; as the failure of a single correspondence between lock and key will prevent the performance of unlocking, just so the failure of any of the many factors required for disease expression may entirely inhibit the infection.

We have seen that plants have many types and grades of resistance to disease. In certain types the resistance is due to a reaction of the cells or tissues of the host to invasion by the parasite. It is natural, therefore, for us to ask whether there is in plants a type of induced or acquired immunity, such as is so important a factor in animal and human pathology. A patient, for instance, recovers from scarlet fever; his blood contains antibodies that destroy the pathogen; he has acquired an immunity from scarlet fever and he will not again take the disease. Also, his blood serum may be used to prevent or cure scarlet fever in other patients. Once a child is vaccinated with the virus of cowpox, a mild form of smallpox, he suffers a mild attack of cowpox, and as a consequence he acquires immunity from the more virulent smallpox disease.

Do plants have the same ability as animals and human beings to recover from one attack of disease and thereby acquire an immunity against subsequent attacks of the same disease? If not, why not? Much thought has been given to those questions. With the exception of some of the virus diseases, however, we have little or no clear evidence of the development of acquired immunity in plants. There appears to be no good evidence of the existence of antibodies elsewhere than in the immediate vicinity of the site of infection by fungus and bacterial parasites. Because infection by the organisms is usually localised, only local immunisation is apt to occur.

The organisation of plants is much simpler than that of animals in the sense that each component part is less closely linked with the well-being of all the other parts and is less dependent for its functioning on them. There is no central nervous system and no blood or lymph streams. Every living part has or may have within it the capacity to regenerate the whole plant.

Therefore we should not draw analogies between what occurs in animal disease (where any local disturbances or localised injury may have rapid repercussions on other parts) and disease in a plant (where, an organ, such as a leaf or branch, may be lost with little or no effect on the rest of the plant). The only cases in which it has been fully demonstrated that infection can lead in plants, as in animals, to immunity from subsequent reinfection occur in the most completely systemic plant diseases that are known the ones that are caused by viruses. The virus diseases of plants are much like the diseases of animals in that the virus can spread through the entire plant and infect all its cells. In several virus diseases a already inoculation, or"vaccination," will confer immunity against a later attack, even when the second inoculation is with a much more virulent strain of the virus.

Tobacco ring spot is an example. Its virus produces many dead lesions on the leaves of young tobacco plants. If the diseased plants are protected and allowed to continue growth, however, the severe phase of the disease passes, and the new leaves show less and less evidence of disease.

Finally leaves are produced that are entirely normal in appearance. If the leaves are inoculated with the ring spot virus, no disease will develop; the plant has thus acquired immunity from ring spot. Cuttings may be taken from the recovered parts of such plants and from the cuttings new plants may be produced. The plants appear entirely normal and they are immune, as they cannot again be made to show symptoms of ring spot by inoculation with the virus. These "recovered" plants, however, still contain the virus. If juice from one of them is inoculated into a normal plant, the latter will develop typical ring spot symptoms. This type of immunity, as in animal medicine, is specific. The recovered ring spot plant is immune to ring spot but not to other virus diseases, just as the person who has recovered from smallpox or has been vaccinated against it is immune to that disease but not to any other disease.

BREEDING FOR RESISTANCE TO DISEASE

In the botanical garden of the National School of Agriculture at Montpellier, France, stands a statue dedicated to G. Foex, commemorating his success in saving the grape industry of France by resistant varieties from America. It symbolises the spirit and aim behind man's effort to breed healthier plants. The figure of the weak, old man in the statue represents the grape industry of southern France, about to die. Supporting and sustaining the man is a sturdy, young woman, America. The statue commemorates the modern application of disease resistance as a control measure that began about 1870. The story of it has great meaning for us today.

Attempts to control powdery mildew of grape in Europe by introducing rooted American vines for breeding purposes had backfired because the American grapes had carried with them the grape root aphis, or phylloxera. The insect, native of United States, did no great damage to our wild and cultivated grapes because it and its host had reached an equilibrium over the centuries. Introduced into Europe some time before 1860, the phylloxera found the vinifera types of grape highly susceptible. Soon the insect spread throughout the entire grape-growing area.

When the vines were first found dying under attacks of the root aphis, the American varieties were remarkable in escaping injury. Attempts were then made to substitute the best American varieties to replace the dead vines. That was not satisfactory. Next, the old French vinifera varieties were grafted to the American varieties that had maintained vigourous growth despite the phylloxera. That procedure was more successful. It saved the grape industry until hybrids between the resistant or nearly immune American sorts and the French varieties could be bred to furnish better adapted, resistant stocks on which to graft the vinifera scions.

Powdery mildew, phylloxera, and downy mildew, introductions from the United States, revolutionised methods of grape culture in Europe. Some of the resistant stocks that were developed to meet the crisis still are being used wherever vinifera types are grown.

All that the statue at Montpellier tells. And more : It underscores the international aspects of the research programme that seeks to control plant diseases by developing disease-resistant varieties.

THAT PROGRAMME HAS MANY FACETS

Men the world over have searched within all species of crop plants subject to serious diseases and their cousins for breeding materials that carry the hereditary factors for resistance. They have made worldwide collections of germ plasm and comprehensive assemblages of varieties and strains of crop plants and their related species.

The crop specialists have sought the wild forebears of the cultivated sorts at the places where the species presumably originated. They have combed the countries where the particular diseases are endemic in the search for strains and individuals which, through operation of natural selection, might have come to bear factors for disease resistance.

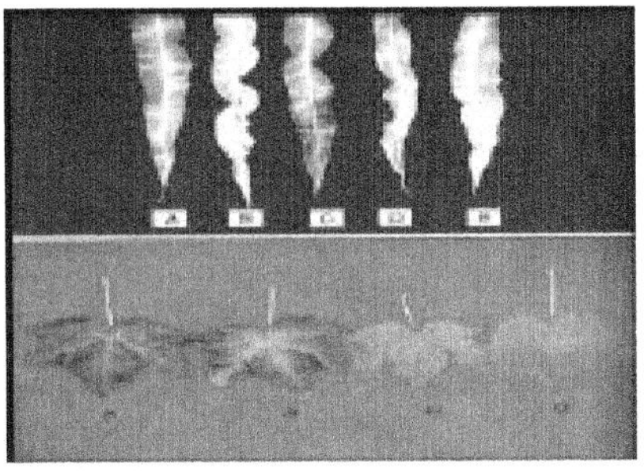

Mineral deficiency disease caused the discolouration of the old corn leaves. A, a normal leaf. B, C, D, and E lack nitrogen, phosphorus, potassium, and magnesium, respectively.

Leaves from a cotton plant showing symptoms of magnesium hunger. The leaves A to D represent progressively the reactions on old to new growth.

For experiments and tests of resistance, they have chosen sites where incidence of disease may be expected to be extremely high, the exposure severe and uniform, and other conditions conducive to efficient research. All in all, then, breeding healthier plants is a long-time programme whose many phases require diversified yet balanced, strong, consistent, and thoroughly coordinated efforts.

The research has to be international and national. It must also be regional in scope : Seasons have to be telescoped; tests of breeders' strains and selections under a range of disease exposures have to be conducted in many areas at the same time. Delays in producing seed and multiplying clones of desirable stocks must be avoided. Increases of seed for commercial purposes have to go forward promptly and efficiently.

Breeding for disease resistance has constituted for many years a major part of the Department of Agriculture's plant pathological programme, in which the Bureau of Plant Industry, Soils, and Agricultural Engineering works in co-operation with State agricultural experiment stations, the seed growers, and farmers.

Outstanding results have come from the research. Among them are fundamental biological discoveries that apply not alone to plants and their reactions but shed light and afford new techniques for solution of problems of human and animal disease. They have added to our knowledge, especially to the concepts concerning man and his relations to his environment. They have had immediate practical significance. The disease-resistant varieties that have been developed and introduced have added greatly to our own wealth and to the wealth of nations.

An attempt was made in 1936 to appraise the contribution of disease-resistant varieties to American agriculture. The crop statistics and the dollar values of the 8-year average, 1928-1935, that were compiled then of Course do not apply to today's farm production and values. But at that time the contribution to farm value for 17 leading farm and vegetable crops was placed at 10 per cent.

In the years since 1935, the degree of disease resistance in the new varieties has been greatly increased, and many other disease problems have been put in line for solution by the introduction of superior types. The acreage of many crops has increased, and on the expanded acreage the disease-resistant varieties have gained wider and wider acceptance among growers. Almost revolutionary changes have occurred as old varieties have been replaced by new productions whose chief superiority derives from disease resistance. The acreage occupied by varieties for which it is proper to apply the term "disease-resistant" has moved from approximately 25 per cent in 1935 to more than 50 per cent. No small part of the increase derives from the almost total swing in many States to varieties of hybrid corn and to the popularity of newer kinds of wheat, oats, and potatoes. Very generally in the areas of greatest production the higher yielding, better adapted, and more disease-resistant sorts have replaced the old varieties.

If we want to put the benefit from disease resistance in terms of dollars, we may use the early, conservative estimate of 10 per cent, which allows for seasons with lessened disease outbreak. But we must take into account the greater utilisation of the superior sorts. We assume that about one-half of the acreage is concerned. Our total farm value of crops ranges from about 12 to 15 billion dollars annually. So we get a figure of 600 to 750 million dollars as the annual benefit that comes from the use of resistant varieties.

One more aspect : The utilisation of disease resistance is not merely the stopping of leaks here and there in an otherwise profitable agriculture. The disease situation with our crops very frequently is serious and, almost without exception, control measures add to farming costs. The farmer, as he pays for fungicides and their application and as he employs other disease-control measures such as sterilisation of seed and soil pays a heavy impost to the plant pathogens. Control of disease by use of a disease-resistant variety has been described as the "painless method" that does not levy on the farmer's pocketbook except as he has to pay for the care and harvest of a larger crop. As increasing crop production costs tend to make more and more crops marginal, the lessened expense for disease control may mean the difference between profit and loss from the farming business.

With some crops, notably the sugar plants, cereals, and potatoes, disease-resistant varieties may spell the difference between success and failure. Without the kinds resistant to curly top, western United States would have abandoned the culture of sugar beets, and this keystone crop would have been lost to irrigation agriculture. Without resistance to mosaic and root rot, sugarcane culture would have disappeared in the Sugar Belt of the South. The citrus industry must depend on trees reworked on resistant stocks if tristeza should develop in the United States at all comparably to its advance in Argentina, Brazil, and other countries. Wheat production has been threatened by a new race of the black stem rust, 15B, and new resistant varieties are imperatively needed.

The contribution of a disease-resistant variety includes other things. Improvement in yield and quality usually has accompanied the resistant variety as a result of better and more nearly normal growth or as intrinsic improvement, irrespective of the factors of resistance. The farmer gives better culture to a crop that shows promise. Certainly the varieties that have gained ready acceptance and have moved into position of standards in our agriculture such as Washington asparagus; Wisconsin yellows-resistant cabbage; the numerous wheat varieties; Bond, Victoria, and other disease-resistant oats : Michigan golden celery; the Robust bean; mosaic and root rot resistant C. P. sugarcane; U. S. 22/3 sugar beet, which is resistant to curly top; U. S. 215 x 216/3 and U. S. 216 x 226, the varieties of sugar beet that are resistant to leaf spot; Katandin and Chippewa potatoes; and mildew-resistant cantaloups 40 (to cite only a few) all have had improved commercial quality and high capacities for yield, along with disease resistance. With the resistant varieties, even if the crop is somewhat reduced under epidemic conditions, there is certainly something to harvest which offsets labour costs of production a striking contrast to crops of some of the old varieties that were not worth harvesting at all. Those intangible contributions stability in rotation systems, permanence in the agricultural programme of an area, and the increased security that comes from lessening of hazards in crop production; permit the farmer to plan with greater confidence.

Almost all of our cultivated plants trace back to primitive man. During the ages, as people wrested the plants from nature and conserved them, they must have improved them whenever disease outbreak retarded or eliminated the less resistant ones.

The early experiences in which resistant host plants were found in regions where a given disease or insect pest is endemic have shaped our thinking and forced recognition of the first and fundamental principle in breeding for disease resistance, namely, that where host and parasite are long associated, then in the evolutionary process resistant forms are developed by natural selection.

Conversely, when an introduced parasite enters a new environment and finds new host plants, conditions may be conducive to its growth and spread; above all, the pathogen may find a host plant in which no resistance has ever been developed. Such plants are attacked with great virulence. Many of our serious outbreaks of plant disease trace to the introduction of parasites to which our non-resistant crop plants immediately succumb. Faced with such emergencies, it is almost axiomatic to apply this basic concept about host and parasite relations that came to light 75 years ago when grape culture in Europe was threatened with extinction.

William a. Orton inaugurated in the Department of Agriculture breeding for disease resistance as an effective and practical means of disease control. His investigations met the specific disease-control problems and gave the guiding principles to this branch of plant pathology. When Erwin F. Smith, famed plant pathologist and bacteriologist of the Department, had completed his studies on the pathology of the fusarium wilts of cotton, watermelon, and cowpea, he assigned to young Orton fresh from the University of Vermont the job of developing control measures. Each disease problem was solved by the application of disease resistance, but each crop required a different approach. To combat the wilt of cowpea, Orton utilised the natural resistance of an existing variety after his comparative tests on infested soil had shown the"Iron" cowpea not only resistant to wilt but nearly immune to the root knot nematode.

Against cotton wilt, Orton employed methods that are operative today in all attempts to improve cotton varieties the selection within desirable strains of individuals that survive under conditions of drastic exposure and the proving of the selections by subsequent tests of the progeny. Co-operating closely with growers and a practical cotton breeder E. L. Rivers, who had started some selections in 1895 at Centerville, S. C. Orton centered his attention on fields of highest infestation with the fusarium wilt fungus. He subjected to further tests the progenies from the individual plants that were selected, because he soon learned that mass selection alone was not effective. In less than to years he produced many highly resistant varieties Rivers, Centerville, Dillon, and Dixie varieties, each, in its day, a successful introduction that grew well where ordinary types failed and each a contributor of germ plasm for the use of cotton breeders to improve their varieties further.

In developing wilt-resistant watermelons, Orton had to go beyond just selection and progeny testing. Failing to find resistance in edible varieties of watermelon, he turned to the highly resistant citron melon, used only for feeding livestock, and incorporated genes for resistance from it into the "Eden" watermelon. Thus he synthesised a disease-resistant variety by hybridisation. Those investigations antedated the rediscovery of Mendel's law, whose disciplines would have been extremely useful to the young scientist as he selected for desirable characteristics from the segregates in the F, generation. By further selection, Orton obtained the wilt-resistant watermelon, Conqueror, capable of giving good crops despite disease. In repeated tests in later years, Conqueror retained its qualities, and the problem apparently was solved, except, as Orton quaintly said,"styles in watermelons changed." Market demand for long melons of the Tom Watson type made the round type of melon unwanted. But the scientific contribution was there, and the genes from Conqueror still are used in research.

H. L. Bolley, another pioneer in plant pathological research, shares with Orton the distinction of bringing to therefore the possibility of meeting serious disease problems by resistance breeding. Bolley discovered in 1900 that flax wilt, most serious of all flax diseases, was caused by a soil-infesting fungus, Fusarium lini. Bolley added a new concept to plant pathology, namely, that of flax-sick soil that is, soil infested with Fusarium lini. Bolley extended this concept of soil infestation to apply to other crops. He pointed out that greatly reduced yields below those of virgin soils and the so-called "running out of soil" might have a biological explanation. He used experimental plots, notably famed Plot 30, that were highly infested with Fusarium lini to develop resistant flax varieties : North Dakota Resistant 52, North Dakota Resistant 114, and Buda, and later, with O. A. Heggeness, produced the variety Bison, which is still in wide commercial use.

The investigations of Lewis R. Jones and his students, J. C. Gilman, J. C. Walker, and W. B. Tisdale, at the Wisconsin Agricultural Experiment Station, in breeding yellows-resistant cabbage firmly established disease-resistance breeding as a control measure. This wilt disease, caused by Fusarium conglutinans, had practically ruined cabbage production in the rich bottom lands near Racine, Wis.

Building on the work of Orton and Bolley, the Wisconsin investigators selected in 1910 some individual plants from fields sustaining an almost complete loss. Only a few of the remaining plants produced heads. The plants to serve as seed bearers were critically selected from them and brought to seed. Then the individual progenies were tested on highly infested soil. By this technique, yellows-resistant strains of Hollander type were obtained. Later, Walker and his associates produced yellows-resistant strains of other standard cabbage varieties. The results with the resistant cabbages were dramatic. The new strains gave almost a full crop where the commercial type failed. Before the work was concluded, resistant types, equivalent in quality and productiveness to any non-resistant types already grown, were made available to growers. This important vegetable crop was saved not alone for Wisconsin but for many other States where the disease had been introduced.

The work in Wisconsin stressed the influence of environment upon the parasitism of Fusarium. Soil temperature particularly was found to be highly significant in determining the ability of the fungus to attack and of the host to withstand the parasite. Here we have the beginning of the concept that disease resistance needs to be defined not alone in terms of the organism-host relation, but as well in terms that include the environmental conditions as they influence both host and parasite.

The work of W. H. Tisdale at the Wisconsin Agricultural Experiment Station with flax wilt developed that idea further. He showed why some varieties of flax resisted Fusarium ling only under certain conditions. Temperature relations determined the infecting powers of the fungus. Wilt-susceptible flax grown with soil temperatures below 60 F. escaped disease; grown at 68 F., it was completely susceptible.

By his experiments, conducted on a field that had grown flax continuously for a decade, H. D. Barker at the Minnesota Agricultural Experiment Station clarified a confused situation with respect to disease resistance by showing that resistance depends on the genotype and is not something acquired by mere association of host and parasite.

The high incidence of infection in an experimental field at St. Paul, Minn., augmented by inoculation of the soil with pure cultures of the flax wilt Fusarium, allowed him to obtain clear-cut reactions with the resistant varieties then available. He showed that lines may be selected that breed true for resistance and that some lines are heterozygous. In his work we find the first research that indicated that the flax fungus itself breaks up into strains.

A basic thing in Orton's work with the wilt diseases was that he centered attention on fields with the highest disease incidence. With the watermelon wilt, Orton insisted on producing practically 100 per cent infection in the test field. That he did by locating the tests in fields whose soil was known to harbor the fungus and by placing manure infested with the causal organism in the melon hills. He thereby subjected the plants to the severest of both natural and artificial exposures. In breeding techniques, disease exposures are of paramount importance.

Although in the breeding of rust-resistant varieties of cereal, the significance of biological races that exist within the rust species was given prominence from almost the beginning of the research, that concept, as we have seen, was rather slow in being recognised in the disease-resistance breeding programmes with vegetables and certain other crops. The occasional failure of resistant types to meet disease situations eventually forced attention to the problem.

From the investigations of the cereal rusts and smuts, we have the most Complete delineation of the problem to be faced in breeding for disease resistance namely, the play of forces within the pathogen and the host, both living entities with inherent capacities for variation.

Writing in 1924, William B. Brierley, a British scientist, took note of the growing evidence of existence of biotypes within common species of parasitic fungi and pointed out that the true relation is between"host-strain" and"parasite-strain." That biotypes exist within our species of pathogens is now commonly accepted. Comparison of isolates on the basis of physiological reactions reveals the species to be composed of many sub-groups that fall within the limits of the species, but differ among themselves; notably, this also is the situation with respect to pathogenicity. By means of the host plant and its varieties and segregates, it has been possible to demonstrate existence of a wide range of biotypes. The complexity of the disease producers, therefore, parallels that of the crop plant. We are thus concerned with the genetic make-up of the host and the parasite.

In seeking to explain the rather broad adaptation of resistant varieties as they were developed, some ot them many decades ago, and their continued value in agriculture, we are led to another principle in disease-resistance breeding.

The first successes were obtained with a group of soil-infesting species. All had the characteristics of producing such severe disease exposures that only those plants survived that were truly resistant. Progeny tests conducted on the same soil eliminated the "escapes" and concentrated attention on truly resistant portions of the population.

But in addition to those characteristics of the early experiments, the research was conducted in the field and against the full range of organisms that would be likely to accumulate in years of intensive agriculture. Thus we may suppose that a fairly large array of biotypes was included in the pathogenic material. A relatively high degree of resistance was sought and, as such, it would be likely to manifest itself similarly against most biotypes of the organism. Too often in subsequent research, similar breadth of exposures has not been sought, nor have criteria for resistance been high enough.

From this background and with such diverse plants as grape, cotton, watermelon, flax, tomato, and the cereals, a system of breeding has evolved. We have learned that wherever a serious disease has occurred in nature, plants carrying factors for resistance may be found, because without such qualities there would have been no survival.

From the research with Fusarium, we are taught to concentrate on fields having nearly total and uniform exposure to a given disease complex. The successful outcome may largely be attributed to the characteristics of the Fusaria to produce almost complete infestation.

Thus the parasite co-operated by pointing out the plants to select and safeguarded against mistake. Should factors for resistance be lacking in the cultivated varieties, then we have learned to seek the genes in related resistant species and to incorporate them into the genotype. Already a body of knowledge has been built up concerning the inheritance of the factors for resistance.

Some form of inoculation procedure has commonly been found necessary as insurance that the work will not receive discouragement and setback by selection of individual plants that are mere"disease-escapes" a bugbear in all such breeding. Here the plant pathologist by learning how to create a localised epidemic in the experimental field, may make a highly significant contribution. Other logical steps along the course set by the pioneer research are the location of test fields at places where environmental factors are conducive to epidemic outbreak, the manipulation of cultural operations to increase disease exposure, the maintenance over the years of plots to insure heavy soil infestation, and other similar measures to insure maximum exposure to diseases.

We have learned the significance of the existence within both host and parasite of species of biotypes that differ with respect to their physiological reactions using this term in its broadest sense.

If the breeding is to succeed, if it is to have wide applicability and meet more than some restricted situation, and if it is to avoid disappointments, then there must be a utilisation of the widest possible array of genotypes of both host and pathogen. The former could provide a broad base for selection and could provide genes for high resistance, and the latter could give as broad a base of exposure as possible.

How a crop plant may be tailored to fit a disease complex created by the multiplicity of biotypes within a pathogen is shown by research on wheat and oat diseases. Even though the parasites are on the move, and a once-prised variety may pass out of the picture because of a disease, a defeatist attitude is not warranted. Nevertheless, the facts are patent that unless the exposure encompasses about the ordinary range of virulence displayed by the gamut of biotypes of the pathogen likely to be encountered, the resistant type as developed may be very restricted as to its area of adaptation and short-lived as an agricultural variety.

The success of these classic researches lead to the application of the principles and techniques with those diseases with which it was possible to produce high incidence of infection. The successful results of the investigations that stemmed from the early research constitute highlights of accomplishment. There has been extensive application of the techniques of disease-resistance breeding. The sugar-plant investigations leads me to draw upon them to show how it has been possible to broaden the scope of this method of disease control. These extensions fall into three groups : Breeding for resistance to virus infection, breeding for resistance to necrotic diseases, and the combining in one variety of resistance against several diseases.

Control of virus diseases by breeding resistant varieties is well exemplified by sugarcane mosaic. When that disease was first definitely identified as the cause of decline of sugarcane production in the sub-tropical cane-growing districts of the Western Hemisphere probably no recent plant disease aroused more public attention. The early work of identification and clarification of the etiology among a welter of speculations and the discovery by E. W. Brandes at the Department of Agriculture that a plant-louse. These discoveries came recently after H. A. Allard's fundamental studies on tobacco mosaic had shown that insects can transmit virus infections. The first control of sugarcane mosaic was by introduction of mosaic-tolerant P. O. J. varieties from Java, where the disease had long been endemic. The P. O. J. varieties were introduced into Louisiana in 1926, when the sugarcane industry was about to fail. They stemmed the tide of bankruptcy and saved the industry.

Then began extensive breeding for the control of sugarcane disease in which lessons learned from experience in Java, where mosaic was endemic, were applied to the American problems. The research centered on hybridising Java and Indian sugarcanes that derived their resistance from wild sugarcane, Saccharum spontaneum, with Saccharum officinarum, the noble cane. It was necessary to find a place in the United States where the sugarcane varieties would flower in order to make the hybrids and backcrosses. Canal Point, Fla., on Lake Okeechobee, was chosen for the experimental work. Then countless seeds had to be grown and Progenies selected in the fields on the basis of agronomic characteristics. After this screening to obtain economic types, the seedlings that remained were subjected to a further Winnowing process under severe exposure to sugarcane mosaic. Of the thousands of seedlings started in the first 10 years of work, only three C. P. 807, C. P. 28/11, and C. P. 28/19 were introduced as economic varieties. These three and two importations Co. 281 and Co. 290 became standard canes and were great advances over old varieties formerly cultivated.

The early breeding work at Canal Point was dominated by the need to find immediate replacements for D 74 and Louisiana Purple, old varieties that had failed because they were susceptible to mosaic, root rot, and red rot. The research was somewhat hit-and-miss. Beginning in 1928, a programme of purposeful crossing, backcrossing, and selling, accompanied by selection, was instituted. The plan sought to combine the favourable factors from two or more varieties of proved value into a single variety. The most important hybridisations have been those that sought to nobilise the interspecies crosses between S. officinarum and S.

spontaneum, as well as S. barberi. These crosses have given rise to disease-resistant varieties that increased the unit yield several fold and are adapted to many soils and climatic conditions.

The success of the breeding programme, which in essence has been the progressive nobilisation of the resistant wild plant by application of genetic techniques while retaining its characteristics for disease resistance, is evidenced by the stream of disease-resistant, highly productive, and high-quality canes that have been introduced. Bearing in mind that the designation "C. P." denotes a selected sugarcane coming from the Canal Point breeding station, and that no cane could succeed under Louisiana conditions if it were not resistant to mosaic and other diseases, the census of varieties in use in Louisiana in 1951 showed that about 90 per cent of the acreage was planted with disease-resistant C. P. varieties.

The early successes in obtaining resistant plants were confined mostly to the diseases in which the reaction between the host and pathogen was, as Orton phrased it,"very delicate." Such a highly specialised relationship would exist with diseases produced by the most highly developed parasites or with pathogens whose activities were intracellular. The rust fungi, together with the powdery mildews, and possibly the viruses, usually are considered as displaying the highest forms of parasitism. With them, therefore, we could expect that breeding investigations whose objective is to upset the relation of pathogen strain to host strain would be likely to be fruitful. The success in breeding varieties resistant to fusariumwilt was with organisms belonging in the sub-section of this genus in which the parasite invades the water-conducting vessels of the stem and induces systemic poisoning a specialised form of parasitism. No cases have been reported in which resistance has been obtained against the Fusaria that cause necrosis of roots, stems, or leaves.

It is therefore of interest to consider diseases of the necrotic type to find if they offer possibilities for control by breeding. It has been more or less accepted that the general run of organisms such as Sclerotinia libertiana, Sclerotium rolfsii, Phymatotrichum omnivorum, Fusarium solani and its allies, Botrytis species, Penicillium species, Rhizopus, and others occur widely in nature as saprophytes; their invasion of plant tissue seems to be almost a function of the environmental conditions, irrespective of the host genotype.

The breeding work with sugarcane may be cited as contributing evidence that it is possible to make selections and obtain sugarcane types resistant to root rot and red rot, both necrotic diseases. As part of the regular screening procedure in selection of sugarcane seedlings, the plants are inoculated with red rot and are planted in fields where root rot is serious. Only resistant plants are considered for retention. In the newer sugarcane introductions, it is difficult to differentiate between resistance to red rot, root rot, and mosaic in their respective contributions to the success of the variety.

Investigations by John O. Gaskill of the Department of Agriculture indicate that resistance to rotting of sugar beet roots by species of Botrytis and by strains of Phoma belie may be found among sugar beet genotypes. In this study, Gaskill inoculated sectors taken from individual sugar beet roots, representative of various lines of breeding, with pure cultures of the fungi. The inoculated pieces were held in storage for more than 2 months under high temperature and humidity, which favour rotting of beet tissue. It was demonstrated that the sugar beet varieties differ in their resistance to decay organisms and that individuals within varieties also differ significantly among themselves. In the inoculation tests only a portion of the root was used for indexing for keeping quality of the mother root. The remaining portion was thus available for planting out if its record warranted its use as a seed bearer. Progenies of roots selected for superior keeping quality have been compared for keeping quality with parent strains, commercial varieties, and with progenies from roots that were very subject to decay. Replicated tests have shown that significant advance has been made by the selections.

Additional evidence that resistance against necrotic diseases may be obtained by breeding is given by the investigations with sugar beet leaf spot (Cercospora beticola). This disease has been controlled by disease-resistance breeding so that the U. S. varieties now introduced into districts subject to the disease do not blight despite epidemics of leaf spot. With the disease, however, resistance manifests itself chiefly with the younger and maturing leaves, but not so strikingly with the older and mature foliage.

In comparison with susceptible varieties, the resistant plants show fewer spots on a leaf often only flecks are produced and the spots tend to be small and not confluent. Leaves of susceptible varieties become densely spotted, however, and recently the leaves turn yellow and die. A plot of resistant plants next to a susceptible variety will be a green strip flanked by one that is dry and brown. The loss caused by leaf spot comes from the death of the leaves and the drain on the stored substance in the root brought about by the constant leaf replacement; resistant varieties retain their foliage to add to root substance, not to waste it.

Mass-selection methods, so effective in breeding varieties resistant to curly top, were not adapted for breeding for resistance to leaf spot. Even under epidemic conditions and with inspections throughout the season, resistant individuals within a variety were not readily recognised. No immune plants were to be found; only degrees of disease involvement and the individual plants were hard to classify. Inbred strains of sugar beets, from earlier breeding work of the Department, were studied and of these a few (14 Out of 250) were outstanding in resistance as compared with the others.

The pedigrees of the outstanding strains revealed that they had been subjected to several selfings as against mass selection operative with less resistant lines. The entire leaf spot breeding programme centered on inbreeding as the technique to segregate genotypes with leaf spot resistance. Inbreds, as they were obtained, were subjected to localised epidemics of leaf spot in the test fields, and selections among inbreds were made under conditions of drastic exposure.

After many years of inbreeding and intensive selection, inbreds highly resistant to leaf spot have been obtained. A few of them have enough vigour to be desirable as breeding stock. Continued inbreeding was found necessary to segregate the plants with adequate resistance, but vigour was lost in the process.

The finding that hybrids made between certain inbred lines tended to regain productivity was exceedingly important. Although it is relatively easy to find hybrids that exceed in yield the mean of the parents, hybrids that exceed an open-pollinated variety in yield are hard to find. Very vigourous inbreds, when mated, give best possibilities for high productivity. Once highly resistant inbreds are found, the job resolves itself into production of a series of hybrids to determine the inbreds that give the best interactions.

As a result of the breeding for resistance to leaf spot, a number of varieties U. S. 217, U. S. 200 × 215, U. S. 215 × 216, and U. S. 216 × 226 have been introduced. They and the varieties produced by research conducted by beet sugar companies now occupy essentially all the sugar beet districts subject to leaf spot and have accomplished such control that spraying or dusting against the disease is not necessary. The U. S. varieties in general are grown in Michigan, Ohio, Indiana, Illinois, Wisconsin, and in eastern Canada.

Varieties of the beet sugar companies occupy the districts of Colorado, Nebraska, Wyoming, Iowa, and Minnesota. An outstanding inbred from this research is U. S. 216, a high-sugar type. It is a component of nearly all recent U. S. hybrids. In the production of sugar beet hybrids, male sterility introduced into one strain is used as a device to enforce the hybridisation with the other strain that supplies the pollen. The beet breeder, to produce male-sterile plants, uses a special breeding technique to incorporate the genetic factor that makes a plant fail to produce pollen.

A highly important development in the breeding research with the sugar plants is the demonstration that it is possible to combine many desirable characteristics in one variety. In a way that had been foreshadowed when yielding capacity, high sucrose quality, and disease resistance were incorporated in one variety.

It has already been noted that sugarcane varieties resistant to mosaic also could be bred for resistance to root rot and red rot. In the production of such strains, the job resolves itself to sorting out from the genotype complex those entities that manifest the desired qualities. Because of the polyploid nature of the sugarcane material, great diversity in biotypes resulted from the hybrids and the backcrosses so that there were many combinations from which to choose.

With the sugar beet, the possibility of finding within one variety factors for resistance against several diseases was first demonstrated with U. S. 15, a variety selected in experiments for curly top resistance in New Mexico. It was not so resistant to curly top as U. S. 22, but it had moderate resistance. It also had a high resistance to beet rust, Urongces betae, and to beetdowny mildew, Peronospora schochiii. Furthermore, when planted in winter plantings in California it was non-bolting that that is, it did not go to seed the first, or vegetative year in strong contrast to other resistant varieties that tended to bolt in high percentage. The non-bolting character and the resistance to rust and downy mildew, coupled with enough resistance to curly top to meet needs with early-planted sorts, made U. S. 15 outyield others with which it was compared.

Development of the sugar beet industry in the Imperial Valley of California is to be credited to U. S. 15. There the sugar beets are planted in the fall, the crop is grown during the winter months, and the harvest is in the spring. Under such conditions ordinary sugar beet varieties bolt. U. S. 33 and U. S. 34 proved to be entirely unsuitable. The non-bolting character of U. S. 15 made it especially adapted to the winter plantings. Its moderate resistance to curly top was adequate for the exposures to curly top that usually are encountered. In the San Joaquin Valley and coastal valleys of California it was possible to plant U. S. 15 in the winter, its non-bolting character again giving it advantage over other varieties. Winter plantings in coastal valleys are not usually subjected to curly top, but rust and downy mildew are serious. For those conditions, the resistance of U. S. 15 to both diseases was highly important.

Studies have been continued in California to combine resistance to many diseases and the non-bolting character in one variety. Already U. S. 56, a variety with more resistance to curly top than U. S. 15, has been released to replace the older variety. More recently, U. S. varieties resistant to leaf spot have been inter-crossed with U. S. strains resistant to curly top and then backcrossed. By repeated selections when leaf spot or curly top occurs, depending on the nature of recurrent parent in backcross, resistance to both diseases essentially equivalent to that of the parents has been combined in one variety.

It has not been possible as yet to achieve immunity or near-immunity to disease in the breeding of sugar beets as with tobacco. Immunity from disease must ultimately be obtained if losses, which are considerable even with highly resistant varieties, are to be avoided. Growers in their enthusiasm when they see the contrast between a disease-resistant variety and the old susceptible type may minimise or overlook the loss that disease still causes. Probably the varieties resistant to curly top are as outstanding in their advantages over susceptible types as any that could be found, short of immune types; nevertheless, with the best resistant varieties, if exposures are severe, the reduction in yield may be as much as 25 per cent of the potential were it possible to remove curly top as a factor. Similarly, varieties resistant to leaf spot make a profitable crop secure from crop failure, but the loss is Still great over what could come from Immunity.

Certain wild species of Beta notably Beta Patellaris, B. procumbens, and B. webbiana are immune to both leaf Spot and curly top. They produce F1 seeds with Beta vulgaris that are viable, but the seedlings only grow a few inches tall and then die. By application of various techniques in breeding, and especially by use of a far wider range of biotypes of both the wild and cultivated species, it may be possible to introduce into the sugar beet the genes for immunity to two of the most serious beet diseases.

The sugar plants as examples of the extension of disease-resistance techniques for control of serious diseases. The investigations have required interspecies crosses when factors for resistance were not found in the cultivated species.

Mass selection was effective with curly top, but to obtain resistance to leaf spot we had to inbreed repeatedly to intensify the factors for resistance, and then utilise hybrid vigour to get productive varieties. Male sterility has been used as a genetic tool to enforce hybridisations. Because of the breadth of its applicability, the demonstration that resistance to many diseases may be combined in one variety is exceedingly important.

Sugarcane, representing polyploids, and sugar beets, representing heterozygous material, have been tailored to meet important disease conditions. Immunity to certain diseases now known for certain wild species of Beta offers possibilities for incorporation of those factors into cultivated beets. Those are goals for future work.

So FAR we have considered breeding for disease resistance from the point of view of its contributions, the techniques employed in obtaining disease-resistant varieties, and the more recent developments that have broadened the scope of this method of disease control. These manipulations of plant material have been made with only meagre understanding of the cellular mechanisms and the reactions involved. It is not, however, a new thing in science for great basic forces to be manipulated with only imperfect knowledge of their nature or even of the basic principles involved. In the play of biological forces, we are often even more in the dark than with purely physical or chemical phenomena.

Our approach to the problems of disease resistance has had to be by trial and error because of the wide gaps in our knowledge of the physiology of both host and pathogen. But it may be of some value to summarise what we know of the causes of disease and to record the hypotheses that have been advanced as to the nature of disease resistance.

Many plant diseases are caused by fungi that are known chiefly in their asexual stage. It has often been possible to find the sexual stage for a particular asexual stage, thus permitting classification of the pathogen as an Ascomycete or Basidiomycete. It is common for the fungus to be a pathogen in its asexual stage and for fruiting to take place on dead tissue in what has been termed the "saprogenic phase." Certain groups of fungi follow a common pattern of behaviour.

Thus the rust fungi, all obligate parasites, evoke a definite train of symptoms and frequently show the phenomenon of heteroecism, that is, two botanically unrelated hosts are involved in the complete life cycle of the rust. Some forms of smuts, after a period of semi-commensalism eating together, as it were within the developing hosts, may then completely occupy the inflorescence. Many smuts, however, show localised invasion and local tissue occupation. Among the plant pathogens the host-parasite relations range from a living together, through obligate parasitism, to rather simple necrotic processes. A classification which places obligate parasitism with conservation of the host as of high order and predatism as of low order seems generally accepted.

Plant pathogens may produce typical effects of underdevelopment. Others, such as the Exoascaceae, cause definite over-development. The most common effects produced are necrotic lesions, the plant as a whole responding in various ways, depending on the physiological disturbance that may ensue. There is also generalised decay, which results from attack by bacteria or fungi on fleshy plant parts.

Sometimes an invading fungus may exert pressure in its penetration of the plant body and in its advance from cell to cell. A common histological picture is that of a cell wall pierced by a strongly constricted hypha, which enlarges when the cell wall is passed. Many organisms form an attachment organ, called an appressorium. As invasion takes place, a peg-like structure passes into the plant. Through this structure the contents of the fungus cell flow; the penetrating thread enlarges to normal size or produces haustorial, or absorbing, apparatus within the cell. The cytological picture as the host cells are invaded is a movement of the nucleus towards the invading thread, followed by discolouration, dissolution of the cell contents, and eventual death of the cell. Certain specialised organisms occupy the vascular tissues and cause wilting and death by production of toxic substances.

The entrance of fungi into plants may be direct; through the unbroken epidermis or at points along the lines of juncture of epidermal cells. The entrance also may be through plant organs such as the stomata and floral parts. In lower types of parasitism the entrance is through wounds. There is also evidence that certain organisms kill in advance of penetration and move into the disintegrating cells.

The great German mycologist, Anton de Bary, showed in 1885 that Sclerotinia liberhana, the cause of a soft rot of vegetables, invades plants by means of a softening enzyme. L. R. Jones, in his studies at the Universities of Michigan and Vermont, greatly extended our knowledge on how parasites attack. He showed that Erwinia carotovora, the cause of bacterial soft rot of carrots, cabbage, and other vegetables, produces an enzyme that dissolves the cementing layer between cells, causing tissues to lose their form and structure. He called the enzyme pectinase and distinguished it from those that dissolve cellulose. Mancel T. Munn, in his investigations at the Michigan Agricultural Experiment Station, resolved the products of the onion neck rot fungus, Botrytis allii, into a pectinase-like enzyme and a toxic substance that worked together in rotting the onion. Such combination effects of an enzyme that dissolves the pectin-like substances in cell walls and a plant poison have been found to occur commonly with necrotic diseases.

Agricultural Experiment Station in 1925 showed that the blackleg organism not only softens potato tissues but turns the cells black ahead of the softening. E. carotovora causes only a limited amount of softening and no blackening. Potatoes were grown in water culture and their uninjured roots were exposed to sterile filtrates from the blackleg bacillus cultures. The potato plants promptly turned canary yellow, matching the disease response frequently seen in diseased plants in the field. E. carotovora did not produce these effects. The two bacterial organisms therefore are alike in producing a softening enzyme, but only the blackleg organism produces the substance toxic to potato.

It is probable that effects characteristic of certain plant diseases on analysis will be found to come from the joint action of enzymes and toxic substances produced by the pathogens.

Death of cells is commonly attributed to toxic substances excreted by the invading organism, rather than to general effects such as plasinolysis or desiccation. Stimulation and other types of host response have been considered comparable to effects produced by minimal doses of plant poisons. The term toxin is avoided in referring to these substances in order that analogy to toxins as recognised in animal pathology may not be implied. Some of the toxic substances are simple metabolic byproducts. L. J. Krakover, at the Michigan Agricultural Experiment Station, working with Stemphylium sarciniforme cause of clover leaf spot produced lesions on red clover leaves similar to those obtained by inoculations with the fungus, by pricking in filtrates from cultures of the organism. He obtained the same effects with ammonia solutions comparable in strength, as shown by Nessler reagent tests, to the filtrates. Erwin F. Smith, of the Department of Agriculture, in seeking an explanation of the mechanism of crown gall formation, attributed the effects produced by Agrobacterium tumefaciens to acetic acid and other chemicals arising from the growth of the bacterium. Smith suggested that they incite excessive cell division or remove natural inhibitors to growth of cells, which, in the normal plant, maintain a balance in cell division. He supported his hypothesis by experimental production of plant tumors by comparable chemical treatments in absence of the organism.

The vascular diseases caused by the Fusaria have received a great deal of study. It is characteristic of these diseases that the shoots of affected plants wilt. At first it was assumed that wilting came from plugging of the water-conducting vessels, comparable to what Erwin F. Smith had demonstrated with the bacterial wilt of cucurbits. R. W. Goss, at the Michigan Agricultural Experiment Station, demonstrated that if cut ends of potato shoots were immersed in filtrates of Fusarium oxysporum, the shoots promptly wilted. E. W. Brandes at Cornell University demonstrated the same thing for Fusarium cubense with buckwheat, bean, and banana leaves. That wilting comes from mechanical blocking was thus brought into question. Filtrates of ordinary saprophytes were shown to produce wilting. But the association of wilting with a specific Fusarium is significant, because it enters the plant and there produces its toxic substances. More recently, Ernst Gaumann in Switzerland has extensively developed this field of research and attempted to relate it to antibiosis.

The defence mechanisms, if they may be so called, that are invoked in plants by infectious organisms appear to be relatively simple; the invader may in some cases stimulate the formation of corky tissues or may produce other growth effects. Automatic walling-off of the parasite occurs with hypersensitive varieties, in which invasion of the fungus is accompanied by prompt collapse of tissue. Thus the paradoxical situation exists in which hypersensitivity limits fungus extension to a mere flecking, and the most susceptible plants are the most resistant.

It might seem simplest to explain disease resistance as an antagonism of the juices of the host cell to the parasitic invader. But antagonistic chemicals of host plants have mostly remained undiscovered except for the catechol and protocatechuic acid in the red- and brown-skinned onions resistant to Colletotrichum circinans and Botrytis allii, as found by K. P. Link and J. C. Walker in their investigations at the University of Wisconsin in 1933. These chemicals are not found in the susceptible white-skinned types. But the chemical antagonism found by Link and Walker did not extend to all fungi, since Aspergillus niger could grow in extracts from the pigmented onions.

Plants, depending on how they are grown, vary in their resistance and in their chemical make-up. Furthermore, it is difficult to determine biochemical compounds, and so far there are no leads to relate any compound or class of compounds to disease resistance. Because of such obstacles, this field of research has remained almost unexplored. In the two decades since the discovery of resistance in pigmented onions, and the determination of the chemicals responsible, no additional information of parallel exactness has been obtained.

There is no accepted evidence of the production of antibodies in plants as a result of invasion by fungi. Localisation in fungus attack, absence in plants of any circulating medium comparable to the blood stream, and the apparently simple chemical reactions involved in many of the lethal effects of fungi upon plant cells may explain why infected plants do not produce protective substances comparable to those known in animals.

Even with the bacterial organism, A. tumefaciens, which (on the basis of the type of cell division, presence of tumor strands, and the profound disturbances in cell morphogenesis) Erwin F. Smith considered as producing a plant cancer, it was not possible to demonstrate antibody formation. In Smith's tests, repeated inoculations of an individual plant did not bring about lessening of reaction.

The virus disease curly top furnishes some evidence of the production of protective substances as a result of infection. This effect differs from saturation phenomena known for tobacco ring spot and interference as known to take place between viruses. J. M. Wallace, working with curly top in the Department of Agriculture in 1944, inoculated tobacco and tomato plants by means of beet leafhoppers, Circulifera tenellus, which had fed on infected plants. The inoculated plants soon showed severe symptoms of curly top.

Tomato plants did not recover but many tobacco plants did. The leaves from such plants showed only mild symptoms in a few weeks. If healthy shoots of tobacco or tomato were grafted onto the tobacco plants that showed recovery, only mild symptoms were produced on the scions. The virus in the recovered plants was demonstrated to be highly virulent by insect transmission tests, but the symptoms produced were mild. When cuttings were taken from recovered plants of tobacco, or from tomato plants protected by the scion grafts from recovered tobacco plants, and these in turn were grafted onto tomato plants, the disease produced on the tomato was mild. Wallace attributed these results to protective substances against curly top developed in the respective tobacco and tomato plants in the course of infection. Once a tomato plant is protected, serial grafts give only a succession of mild curly top reactions, the scions taken from such plants serving as donors of the protective substance. If protective substances were to be sought, it would be logical to look for them with a systemic disease such as curly top.

We have considered so far entrance of fungi into plants, the mechanism of disease production, and the responses provoked. For the higher grades of parasitism, certain other essential requirements in disease production, namely, conditions for establishment of the organism within the host, are important. Suggestions have been made as to osmotic differentials which must prevail in host-parasite relationships. The concentration of the cell sap of the parasite must be higher than that of the crop plant, if the parasite is to take water from the host. No regularity in this respect has been found. To explain resistance of Zea to Puccinia sorghi, the hypothesis was advanced that susceptibility is determined by the presence of a relatively large quantity of a nutritive substance that attracts the fungus after penetration and makes possible abundant development of the rust. In resistant host Plants, this hypothetical substance is Present only in very small quantity; hence the fungus dies of starvation, and that, in turn, leads to necrosis of the host tissue. It has been suggested that the slower development of the pathogen in a resistant host, as compared with abundant growth in a susceptible host, may be ascribed to a more favourable nutritive substance ratio in the latter.

Cercospora apii causes lesions only on celery and closely related umbellifers. It does not attack plants from other families. Diseased celery leaves gave strong nitrite tests, as compared with absence of nitrite in healthy leaves, indicating that breakdown of proteins and other nitrogenous compounds occurred. Now any of the species of Cercospora will grow readily and indiscriminately on cooked plant tissue, but with respect to living tissue they are very choicy. Furthermore, the pathogenic limits apparently tend to disappear as leaf tissue matures and cells are about to die. It does not seem that a parasite such as Cercospora, once it invades the cells, should have difficulty in appropriating water necessary for its growth. It seems probable also that the carbohydrates, such as sucrose, dextrose, or other sugars, and water-soluble substances exist in plants in much the same available state as in synthetic or cooked media. As a result of our investigations, we suggested that the nitrogen phase of fungus nutrition was particularly important and that the protein-dissolving enzymes that a given fungus has may determine its capacity for attacking a given plant species.

Investigations in 1937 by F. G. Larmer, of the Department of Agriculture, on the phoma root rot of sugar beets gave some interesting results. Phoma betae, the causal organism, occurs rather generally in sugar beet tissue. The fungus is apparently seed-borne, and young plants may be killed by severe attack. Under ordinary conditions it appears to live in the beet plant without provoking any visible symptoms. If the sugar beet lags in its growth because of drought or lack of soil nutrients, the fungus may produce extensive and conspicuous decay of tissue. In the fall the fungus causes serious rotting of sugar beets in storage piles. Larmer found that well-nourished sugar beets, grown with adequate water supply, kept better in storage than sugar beets from the check plots that suffered from drought and poor soil conditions. The effects of plant food elements, especially phosphorus, were decisive. Sugar beets, grown with adequate phosphate, showed minimal decay as compared with serious decay in the sugar beets not receiving additional phosphorus. Inoculation tests were made with Phoma betae and showed that the plants grown with adequate phosphate definitely resisted invasion by the fungus, whereas the sugar beet roots grown with limited phosphate decayed. The practical applications of the experiments are obvious and afford an accessory advantage, besides crop increase, to be derived from appropriate applications of fertilizer. But the point to be stressed is that it seems definitely to be shown that plants whose nutrition was adequate, especially with respect to phosphate, were more resistant to Phoma betae. In recent experiments, adequate nitrogen nutrition has given improved keeping quality to sugar beets.

Black root of sugar beet has been found to be a disease complex, consisting of an acute form caused by the ordinary damping-off organisms and a chronic form caused by Aphanomyces cochlioides. Black root is most serious on soils of low available phosphate. Poor stands that result from attack of the soil organisms may be prevented by proper phosphate fertilization. A plausible explanation of the better stands that accompany liberal applications of P2 O 5 fertilizer is that phosphate makes the young sugar beet plants more resistant to the pathogens. The effect must be upon the beet, because abundant trial has failed to indicate any effects upon the fungi from comparable phosphate feeding. Black root control now is based on two things : First, the production of varieties that combine resistance to black root and to leaf spot; second, proper fertilization of the resistant variety with P2O5 and other necessary food elements. A favourable environment for the sugar beet increases the advantages that come from disease resistance.

The associative effects of nutrition and disease are not unique in phytopathology. Pronounced fungus invasion occurs with certain deficiency diseases; for example, the fungus rots or severe outbreaks of mildew that occur on plants short of boron. Existence of fungus disease does not of itself indicate malnutrition, but the association of disease and poor plant feeding is frequent. Sugar beet plants growing with scant phosphate show severe leaf injury following spotting of the leaves by Cercospora beticola, not because the leaf spot fungus is more aggressive but because weak parasites such as Alternaria can now invade and greatly enlarge the cercospora spots. The secondary attacks following leaf spot of this type do not occur with well-nourished plants. With many important plant diseases, the host-parasite relationships are very delicately balanced, and proper feeding of the crop plant may be highly important.

The nature of disease causation and of plant response may at first glance seem to contribute little to the practical problems of disease-resistance breeding. These seem to be moving along, with little reference to the theory. It is not unusual for practical applications to go far beyond the scientific explanation of phenomena just as a driver may run a machine without understanding the mechanism or the source of power. But always in science the theoretical basis is the fruitful source of new concepts and new approaches.

We are concerned with the ways and means of the pathogen's attack and of the host plant's defence. With hundreds of thousands of pathogens, each composed of numerous biotypes, we cannot expect a single pattern of disease production or to find some specific substance that confers resistance. Thus, the theory is important in teaching what not to search for. The theory is important also in its teachings with regard to plant reactions. Disease exposure must be maximum, and knowledge of plant pathology must be put into service to initiate infection, produce disease, and finally to classify affected plants. The theory teaches also the specificity and delicateness of the interactions that exist between host and parasite, and the play of environmental factors as they influence these reactions. Thus light, temperature, length of day, nutrition, and the entire range of physiological forces are concerned. Plant response when subject to the parasite is the sum total of these effects. The goal of breeding for disease resistance is to manipulate host and parasite and the factors playing upon them in such a way that for a given environment disease-safe varieties may be provided.

The fungi, bacteria, and the viruses, each in its own specific manner, establish a food and water relation with the host plant. The invader may be tolerated; it may dwarf the plant or cause overgrowths or decay; it may kill the host. The tools the invader uses are enzymes and toxic substances; in considerable part, mere occupancy of the cells, appropriation of food, and the effects of the metabolic byproducts from the growth of the pathogen may be sufficient to explain the disease signs and symptoms.

Disease resistance as considered here is that which is of protoplasmic origin, as opposed to mechanical walling-off, escape, or other reactions than purely vital ones. We do not know at all what makes one plant less susceptible than another, nor do we know the basis of fungus specificity that makes one species, genus, or family of plants completely immune from a given parasite. Of the two hypotheses commonly advanced, antagonism (presumably chemical) and the so-called starvation hypothesis, the latter seems most tenable, with nitrogen nutrition appearing to be a significant phase.

Despite the common question posed by farmers, we do not know how to feed plants so as to make them more disease-resistant. Until recently, research along this line has been neglected. There are also strong indications that disease-resistant varieties may be made to do better by proper nutrition.

Chapter 16

HELMINTHOSPORIUM

HELMINTHOSPORIUM (GENERAL)

- *Lovegrass (Eragrostis)* : Sometimes seedling blight and common root rot (Southwest and Plains).
- *Orchardgrass (Dactylic)* : Root necrosis (general); seed rot (general); seedling blight (scattered).
- *Panicum grasses* : Seedling blight (Great Plains); root necrosis (Plains and South).
- *Redtop and bents (Agrostis)* : Seed rots (general); rhizoctonia rot (general); damping-off (coastal areas).
- *Ryegrasses (Lolium)* : Common root rot (general); root necrosis (general); rhizoctonia rots (general); seed rots (general).
- *Timothy (Phleum pratense)* : Seedling blight (general); seed rots (general); root necrosis (general).
- *Wheatgrasses (Agropyron)* : Seedling blight (Western States); common root rot and crown rot (Western States).
- *Wild-rye grasses (Elymus)* : Seedling blight (general); common root rot and crown rot (general).
- *Stipa grasses* : Seedling blight (Plains and far West); common root rot and crown rot (Western States).

SEED DISORDERS OF FORAGE PLANTS

Seed diseases of forage plants are relatively few, if we do not count the smuts. The four seed disorders are different from the seed-borne diseases, in which the causative agents primarily infect leaves, stems, or roots, but also may attack seeds.

Seed diseases are particularly important when they reduce the supply of seed needed to plant forage crops, lawns, or turf. In two diseases ergot and grass seed nematode disease the grass seeds are replaced by the sclerotic and galls, which are poisonous to animals.

Blind seed disease of perennial rye-grass, Lolium perenne, apparently became established in the United States about 1940, although it has been a seed production problem in New Zealand since 1932. Poor germination of seed of domestic perennial ryegrass alarmed growers in Oregon in 1943 and led to positive identification of the disease. But three-fourths of the Oregon crop by then had become infested, and more than one-third of the 1943 seed could not be certified.

The causal fungus, Phialea temulenta, was identified in France on cereal rye in 1892, and was identified on perennial ryegrass in New Zealand in 1942. Since then the pathogen has been recognized on perennial ryegrass in England, Ireland, and Scotland; probably it can be found on the grass wherever climate permits infection.

Heavily diseased crops have many seeds that fail to germinate. Such dead seeds were referred to as blind seeds in New Zealand, and the malady was named blind seed disease.

It is hard to tell infected seeds from healthy ones unless the lemma and palea are removed. Then one can see the shriveled, soft, pasty appearance of diseased seeds. Healthy seeds have hard, plump, purple endosperms.

Diseased seeds reach the soil by pre-harvest shattering of the crop, planting diseased seed, feeding infested seed or screenings, harvesting operations, and, in unharvested areas, by the natural dispersal of seed.

The blind seeds remain dormant during the winter. In spring, when perennial ryegrass flowers, the small-stalked, cup-shaped spore-producing organs (apothecia) arise from the overwintered blind seeds and forcibly discharge the primary spores (ascospores). The ascospores are showered on the ryegrass flowers and infect the developing seeds. Asexual spores are produced in a slimy matrix surrounding infected seeds.

These secondary spores can infect other developing seeds when rain and insects spread them from head to head. Badly infected seeds that overwinter at or near the surface of the soil produce spore cups the following spring and repeat the cycle. The entire life cycle is confined to the seed. Infected seeds are not toxic to livestock. The disease is important only when the grass is grown for seed.

To provide a basis for control, each sample of cleaned seed of perennial ryegrass entered for certification in Oregon is examined for disease. Growers then are advised through their county agents of the amount of disease in all fields. Plowing before May 1 is recommended for fields that appear too badly infested to produce a profitable crop of seed. The procedure makes it possible for each seed grower in Oregon to know how much disease is in his field and what to do each year. He can avoid unprofitable crops and use the land for spring grain or leave it fallow to reduce weeds. Thorough burning of the straw and stubble after harvest gives good control for a year.

Disease-free seed is selected after harvest and approved for planting in new seed-production fields. The blind seed fungus dies after 24 months in dry storage; aged seed therefore is also safe to plant. The planting of any seed more than one-half inch deep, with complete soil coverage, prevents emergence of the apothecia.

Much heavily infected seed is spread on the field with the straw during combine harvesting, as such seed is lighter in weight than healthy seed. Because the disease is perpetuated primarily by infected seed that is left on the soil, precautions should be taken during harvest so that light as well as heavy seed is removed from the field.

It is helpful also to destroy infested perennial ryegrass screenings, prevent heading in pastures until after July, plow clean to bury infected seeds deeply, have good soil drainage, and plow all ryegrass on a farm at the same time to prevent spread of disease from old fields to new plantings nearby.

Anther mold was first discovered in Ladino clover in 1947. A seed blight of white clover had been seen in Scotland before 1928, but the identity of the causal fungus was unclear until Mary Noble, of the University of Edinburgh, named it in 1948. By comparing data with Dr. Noble, we found that the seed blight of white clover in Scotland and the anther mold in Ladino clover in Oregon are different stages of the same disease.

Anther mold is caused by Sclerotinia spermophila, a fungus that can live inside the clover stems and leaves. Infected plants look the same as healthy plants. The disease is evident only in the flowers and seeds. Infected flowers bear gray, fuzzy anthers, on which the fungus produces spores that largely replace the pollen; hence the name anther mold. Bees carry the fungus spores to healthy flowers. Infection of young seeds in the flower is the only known way the fungus can enter the plant.

Infected seeds are shriveled and have a dull-brown or gray-pink colour. Healthy seeds are plump and bright yellow or reddish brown. Most infected seeds will not germinate. The fungus therefore appears to be largely self-eliminating, and the disease has not been economically important. If a less lethal strain of the fungus would appear by mutation or other genetic change, however, more diseased seeds could grow and produce infected plants, which could furnish inoculum for infecting healthy seeds. The disease might then rapidly become a serious problem in the production of seed of white and Ladino clover.

The disease seems not to affect the vegetative growth of infected plants. Its dissemination has been retarded by the use of modern seed-cleaning machinery, which removes the lighter infected seed. Grass seed nematode disease, caused by Anguina agrostis, is a main disorder of chewings fescue and Astoria and Seaside bents west of the Cascade Mountains in the Pacific Northwest, where those grasses are grown for seed. A similar nematode infests seeds of buffalograss in the Great Plains. Other related nematode species infest seeds sporadically in species of Calamagrostis, Danthonia, Elymus, Holcus, Sporobolus, and Stipa throughout the United States.

The life antiquity of the grass seed nematodes is like that of Anguina tritici, which causes the eelworm disease of wheat and rye in Southeastern States. Only the seed is infested, so that utilisation of a grass for lawn and turf purposes is not affected by the presence of the grass seed nematode.

Striking symptoms occur in bent-grass. The seeds are transformed into purple to black galls that may be much longer than the healthy seed. In orchardgrass the seed panicles are greatly malformed. In chewings fescue the symptoms are less easily seen.

Each gall contains many nematodes, which leave the galls after fall rains and migrate to grass leaves. Eventually they work their way between the folded leaves near the growing points. When the grass panicles develop, the microscopic roundworms penetrate the ovaries and stimulate the plant to produce the galls, within which many eggs are laid. Thus the cycle is repeated.

Satisfactory methods have been devised for control of the grass seed nematode in chewings fescue in the Pacific Northwest. Clean seed for planting seed-production fields can be had from dry-land districts or by disinfesting seed with a specific gravity separator, using air flotation. Chewings fescue seed weighing 21 pounds per bushel from such seed-cleaning machinery contained no nematode galls. Burning fescue fields after seed harvest is effective because it destroys many of the nematode galls in the grass stubble or in the straw. Vacuum machines, which were developed to pick up seeds on bare soil, could be helpful in removing the lightweight nematode galls from a grass field if they are adapted to work in the thick stubble. Crop rotations that use crops other than susceptible grasses effectively starve out the grass seed nematode.

Several species of ergot fungi attack grasses in the United States. Claviceps cinerea infects curly-mesquite (Hilaria belangeri) and tobosa grass (H. mutica) in the Southwest. Claviceps tripsaci occurs on eastern gamagrass, Tripsacum dactyloides, in the Southeast. Claviceps paspali is restricted to members of the genus Paspalum. The most common and important species, Claviceps purpurea, causes the ergot disease in about 150 different grasses throughout the United States.

Ergot has retarded seed production of grass desirable for regrassing programmes. Seed production of big bluegrass, Poa ampla, was discontinued in Union County, Oreg., because of such heavy infestations that harvesting and cleaning were difficult. Production of seed of some grasses in the Great Plains is often a failure because of sterility due to ergot. Most of the seed often is blighted, although only a few sclerotia may develop. Ergot poisons livestock that graze heavily diseased grass inflorescences and feed on contaminated seed screenings. Substances in the ergot sclerotia cause abortion, nervous disorders, blindness, and paralysis. In gangrenous ergotism there is a sloughing of the hoofs, tips of ears, and tail; shedding of teeth and hair; and death.

Ergotism in man is no less severe. The disease is contracted usually by eating rye bread made from contaminated flour. Epidemics in humans were common in the Middle Ages.

Modern methods of grain cleaning permit the removal of ergot to within the tolerance of 0.3 per cent by weight set by the Federal grades and pure food laws. This has almost eliminated the disease in man, although local epidemics occurred in France and India in 1951.

The ergot fungus attacks only the developing seeds of grasses. It has economic importance when the susceptible grasses are grown for seed or are allowed to flower and seed before grasing or cutting for hay. Ergot is of no importance in seed for planting lawns or other clipped turf.

The first sign of ergot infection appears at flowering time when a sweet exudate,"honeydew," is noticeable. Infested heads feel sticky when drawn through the hand. The exudate, which contains the asexual spores (conidia) of the fungus, attracts flies and other insects. Conidia carried by such insects to healthy heads or spread by rain in the same head start new infections. As the infection progresses, purple or blue-black horny bodies (sclerotia) develop in place of seeds. These sclerotia may fall to the ground or be harvested with the seed. When planted or left on the soil they produce specialised sporebearing organs that eject spores into the air. The spores are then carried by wind to grass flowers and infect the young seeds, thereby repeating the disease cycle the following spring. Many grasses are susceptible to ergot, Claviceps purpurea, and wild grass may serve as sources of infection for grass seed or grain crops.

Control of ergot involves avoidance of the airborne primary spores or the secondary spores spread by rain and insects. This can be accomplished by planting non-susceptible crops, deep plowing to bury the sclerotia too deeply for emergence of the spore organs, clipping grass before seed heads bloom, and eradication or preventing the heading of grasses in fence rows or other adjacent waste areas. Deep planting, permissible with the larger grass seeds, will prevent emergence of the sporebearing organs from the ergot sclerotia. However, selection of ergot-free seed, removal of ergot sclerotia by seed cleaning with specific gravity separators, or aging the seed for 2 years would be preferable.

Some grasses, such as the bents, produce their flowers in midsummer when the soil is too dry in many areas for development of spore organs from sclerotia. Under such conditions, infection depends entirely on spores carried by insects that have visited the spore slime on grasses infected earlier; control of the offending insect carriers could prevent ergot infection. Ergot in perennial ryegrass has been materially reduced in western Oregon by burning the fields after the seed harvest. The fire destroys many of the ergot sclerotia in the straw and stubble and also prevents regrowth head formation in the late summer and fall. Such late-formed spikes are usually abundantly infected even in the years that are dry.

ALTHOUGH infection starts in the spring on the early-flowering grasses, the heaviest ergot infestations occur in the fall in many areas. The phenomenon is due to recurrent heading of many grasses, which become infected by insect-carried spores.

This is important, because of the increase in sclerotia that furnish inoculum for infection of grass-seed crops the following spring. Such heavy infestation in late-grass heads is especially dangerous to livestock, because leaf growth is at a low ebb and grasing is heavy on the infected inflorescences.

SOME OF THE 125 RUSTS OF GRASSES

About 125 different species of rusts attack grasses in the United States. Nearly 400 species of grasses are among the hosts of the rusts. Some of the rusts attack only one or a few grasses. Others can attack a great many. Stem rust, Puccinia graminis, for example, has been recorded on nearly 200 species of grasses in the United States. Among the rusts, which have such large numbers of hosts, we find innumerable strains or races, which look alike, even under the microscope, but differ in their comparative ability to attack species or varieties of grasses. Some of the rusts of grasses are often destructive on the cereals. Most of the so-called cereal rusts have numerous grass hosts; probably the cereal rusts originally were grass rusts that found susceptible hosts among the cereals during the centuries they have been cultivated by man.

However, since the cereals wheat, oats, barley are really only grasses whose seeds are large enough and nutritious enough to warrant intensive cultivation as food, it is only to be expected that there would be a great deal of similarity between the rusts that attack grasses and those that attack cereals. The rusts are microscopic fungi and are strictly parasites. They are among that parasitic classification known as obligate parasites; that is, they can grow only on a living plant. Indeed, many of the rust fungi require not only a certain species of grass or cereal but frequently a definite strain or variety of that species a phenomenon known as host specialisation, host specificity, or physiologic specialisation. It has an important part in breeding crop plants for resistance to rust.

Because the rust fungi are parasites, their development on a host plant is at the expense of that plant and the nutriments they take would otherwise go into seed, forage, or both. A light infestation of rust is not likely to cause noticeable effects on yield of seed or forage, but a heavy infestation definitely will. In grasses grown for seed, a heavy infestation will result in low test weight of the seed because of the direct effect of parasitism in sapping nutrients from the host and because of water loss through the numerous open rust pustules on the leaves and stems. Rust likewise affects the production of grasses for forage, chiefly in lower yields as a result of reduced vigour. That, however, is partly offset by the somewhat higher protein content of rust-infested grass over rust-free grass.

Secondary or indirect adverse effects should not be overlooked. Heavy attacks of rusts on grasses will make them more likely to succumb to other factors that are always more severe on the already weakened plant drought, winter injury, root rot, snow mold, and perhaps other diseases.

Reproduction and dissemination in the rust fungi, as in fungi generally, is by microscopic cells or groups of cells known as spores. Many of the rusts produce several kinds of spores during their life cycle. Very common are the familiar red uredospores, whose mass appearance accounts for the name rust. Those spores are borne throughout the summer and spread the disease rapidly during the growing season. As the plants mature, however, the red pustules gradually change to a black type. The black spores (teliospores) usually cannot germinate until the following spring; thus they insure survival of the rust fungus through fall, winter, and early spring.

In many of the rusts, especially the grass and cereal rusts, the black spores cannot infect any grass or cereal host but only some entirely different kind of plant, as the examples given later will show. Such a plant is known as the alternate host and usually is infected in the spring or early summer. The spores produced on the alternate host cannot reinfect that host; instead, they give rise to the initial red spore stage again on grasses or cereals, and thus the life cycle is completed.

Few of the 125-odd species of rust fungi that attack grasses in the United States are of economic importance : Many of the rusts are so limited geographically that they have not presented any general problem. Many are specialised to a few grasses, frequently of limited distribution and importance in the over-all range and forage picture. Another circumstance : In many wild grasses, especially of the bunch type, individual plants of a species often are separated by many other types of vegetation and by resistant grasses of other species. Susceptible individuals of one species have to be close by before rusts and leaf diseases in general can develop into an epidemic. That has been demonstrated countless times and often with disastrous results in cultivated fields of cereals and grasses.

Nevertheless, several rusts often reach economic proportions in important forage grasses and merit specific mention here. They are stem rust, leaf rust, stripe rust, crown rust, and bluegrass rust.

Stem rust of grasses (Puccinia graminis) is better known as a destructive disease of small-grain cereals than as an important disease of numerous grasses. It is evident primarily as small red pustules, chiefly on the stems, but sometimes also on the leaves, particularly on young grass plants. As the summer draws to a close and the plants mature, the red pustules gradually are replaced by black ones.

The six varieties of the stem rust fungus in the United States are differentiated mostly by the cereals and grasses they can attack. Thus the wheat variety of stem rust (Puccinia graminis tritici) attacks wheat, barley, rye, and many grasses. The oats variety (P. graminis avenge) attacks oats and various grasses. The rye variety (P. graminis secalis) parasitises not only rye but also certain grasses. The redtop variety (P. graminis agrostidis) occurs on redtop and related grasses. The bluegrass variety (P. graminis poae) is on Kentucky bluegrass and many other bluegrasses. The timothy variety (P. graminis phlei-pratensis) attacks timothy and a few other grasses.

The wheat variety of stem rust has been found on many of the wheat-grasses (Agropyron species), orchard-grass, wild-rye grasses (Elymus species), barley grasses (Hordeum species), and a few less important grasses. The stem rust of oats occurs on tall oatgrass, sweet vernalgrass, some brome grasses, orchardgrass, a few fescues, canary-grass, timothy, a few bluegrasses, and some others of less importance. Stem rust of rye attacks many of the wheat-grasses, a few bromes, many of the wild-rye grasses, and barley grasses.

The redtop variety is practically restricted to redtop and related grasses (Agrostis species). Stem rust of blue-grasses likewise is restricted to Kentucky bluegrass and a few other of the bluegrasses. The timothy variety is virulent on tall oatgrass, orchardgrass, meadow fescue, and tall fescue. So it is apparent that three of the varieties of stem rust attack both cereals and grasses and that the other three are restricted to grasses only.

In the Southern States and other places of mild winters, stem rust usually overwinters in the red stage on grasses and on volunteer or fall-sown grain. In regions of colder winters, such overwintering is exceptional, and the rust fungus depends on the black spore stage for survival during fall and winter and for re-establishment in the following early summer. Re-establishment is impossible without intervention of a spring stage of stem rust on an entirely unrelated plant, the common barberry (species of Berberis), the alternate host.

The resulting infections on barberry bushes produce spores that set up the initial red spore stage on cereals and grasses in early summer. It is this stage that builds up in epidemic form by repeated generations during summers of frequent rains or prolonged dew.

Leaf rust (Puccinia rubigo-vera) occurs mainly on the leaves of cereals and grasses, although stem infections are encountered sometimes. Host specialisation occurs, similar to that described for stem rust, so that we have six varieties, as follows : One variety of leaf rust, Puccinia rubigo-vera agroppri, common in the Rocky Mountain region, infects many of the wheatgrasses, various bromes, and several of the wild-rye grasses. It has many alternate hosts, all in the buttercup family (clematis, buttercup, columbine, larkspur, and others).

Another rust variety, P. rubigo-vera apocrypta, also occurs on a few wheat-grasses and wild-rye grasses but has alternate hosts in the waterleaf family (e. g. Hydrophyllum) and the borage family (*e. g.* Mertensia). Of the mountainous areas. Besides wheatgrasses, it attacks bromes, meadow foxtail, alkali-grass, man-nagrass, the wild-ryes, and many others. The numerous alternate hosts are all in the buttercup family.

The wheat variety of leaf rust (P. rubigo-vera tritici) is restricted to various wheats and apparently will not infect grasses. The alternate host is meadow-rue.

The sixth variety, the leaf rust of rye (p. rubigo-vera secalis), is restricted to rye. Its alternate host is the wild forget-me-not or bugloss (Lycopsis). Leaf rust frequently is as destructive as stem rust. Occasionally infection is so heavy that the infected plants are literally dusty red with rust and may even be killed by it. As a species, leaf rust is widespread from coast to coast, although some of the varieties are limited by geography or climate.

Stripe rust (Puccinia glumarum) is quite different in appearance from stem rust and leaf rust. The individual pustules, instead of being orange red, are yel-low and arranged in long rows or stripes. A cool-weather rust, it makes its best progress in the spring and early summer before hot weather sets in. It is restricted to the Rocky Mountain States and westward.

Stripe rust is found commonly on the various wheatgrasses, wild-ryes, bromes, and barley grasses, besides wheat, barley, and rye. Some grasses, such as crested wheatgrass and blue wild-rye, frequently are infected to the extent that plants are killed outright or are rendered easy prey to drought and other diseases.

The alternate host of stripe rust is unknown. Being a cool-weather rust, it overwinters easily without an alternate host.

CROWN RUST (Puccinia coronata) is really a leaf rust and to the unaided eye looks just like true leaf rust. The difference between the two lies in the black spore stage and in its alternate hosts. Under the microscope its black spores (teliospores) are seen to bear a crown of long projections at the top (hence the name crown rust); these are lacking in the leaf rust spores. Also, it has only a few alternate hosts cascara buckthorn (chittum bark) and related species of buckthorn (Rhamnus).

Crown rust occurs on oats, wild oats, and many grasses velvetgrass, English ryegrass, Italian ryegrass, meadow fescue, tall fescue, orchardgrass, and some of the bentgrasses, among them. It is of economic significance more in the Mississippi Valley than east or west.

Bluegrass leaf rust (Puccinia poaesudeticae) is common only on grasses. It does not occur on any of the cereals. As the name indicates, the chief grasses affected are the bluegrasses, but occasionally others (such as meadow fescue and alpine timothy) are infected. It occurs chiefly in the leaves and produces numerous, minute, orange-red pustules, but frequently stem infections occur. Of the bluegrasses, Kentucky bluegrass is by far the most commonly infected. It rarely develops in epidemic proportion in lawns and pastures, however, probably because of too frequent mowing or grasing, but in nurseries or other undisturbed stands heavy developments often occur. Several other valuable bluegrasses are commonly infected, sometimes very badly. Bluegrass leaf rust does not have a known alternate host. It over-winters easily and effectively in the red spore stage.

Stem rust is controlled by eradicating the alternate host, the common barberry (Berberis vulgaris). In the interests of controlling stem rust on wheat, oats, barley, and rye, an active barberry eradication campaign has been in force in 18 States. The campaign also is in the interests of susceptible forage grasses. For example, in eastern Washington, before that State joined the eradication campaign in 1944, stem rust could be found on numerous grasses and sometimes in epidemic quantity. By 1951, with the barberry all but eliminated (except for numerous small seedlings), it was unusual to find stem rust in anything more than trace amounts.

In the States where crown rust is a problem, its alternate host, buckthorn, has been eradicated along with the common barberry. Those are the only rusts of importance on grasses that can thus be controlled to any practical degree. Elimination of the alternate host has another advantage : It is on that host that hybridisation between strains of rust takes place, often with the result of new and more virulent strains capable of attacking varieties of grasses and cereals that already were resistant.

The development or selection of grasses resistant to rust has lagged behind such projects in the cereals. Strains have been noted in timothy, orchardgrass, crested wheatgrass, blue bunchgrass, big bluegrass, and others that are resistant to their particular rusts. It is usually true, however, that those resistant strains are valid only in restricted regions and are not generally resistant to the various and numerous races or strains of rusts in all localities. The individual State colleges can best give advice on such local problems.

The fact remains that genetic resistance does exist in strains of most of our forage grasses and can be used in breeding work when necessary or desirable.

SMUTS THAT PARASITISE GRASSES

Nearly 140 different species of smuts attack approximately 300 species of grasses in the United States.

The common name smut derives from words meaning to besmirch or smudge and refers to the presence of sooty, dirty, black, or brownish masses on the affected plants. All the smuts are plant parasites. Most of them occur on grasses and cereals. As with the rusts, the average person is much more familiar with the destruction wrought by smuts on cereals than on grasses. Familiar examples of destructive cereal smuts are bunt or stinking smut of wheat, oat smuts, barley smuts, sorghum smut, and boil smut on corn. In everyday practice, the name smut is used to designate the disease as well as the fungus responsible for the disease, although the latter is more correctly spoken of as the smut fungus.

The smut fungi have much less complicated life histories than do the rust fungi and generally are easier to control. The familiar dusty black smut masses are made up mostly of millions of tiny cells or groups of cells spores, which serve the same purpose to the smut fungi as seeds do to the seed-bearing plants, namely, for reproduction and dissemination.

Some smuts destroy the flowering structure. Others are restricted only to certain parts of it. Some are confined almost exclusively to the stems of grasses. Others produce galls or tumor-like structures in various parts of their host plants.

As with the rust fungi, many smut fungi display a remarkable degree of specialisation not only to certain species of plants but also to certain varieties or strains within those host species. Likewise, there are often strains or races of the smut fungi to contend with. The common stinking smut, for example, has nearly 30 known strains or races, each capable of attacking different varieties of wheat and different strains or varieties of wheatgrasses and related grasses.

The smut fungi are not entirely obligate parasites as are the rust fungi. In fact, some smuts can easily complete their entire life cycle on an artificial medium if it contains the nutrients essential for growth. Some smuts apparently persist indefinitely in the soil or in old manure piles. They are fully capable, however, of attacking their host plants when the plants are available under the requisite conditions for infection.

The smut fungi have a more adverse effect directly (and perhaps indirectly) on their hosts than do the rust fungi. The smuts that attack all or parts of the flowering structures generally destroy the seeds entirely. The leaf smuts and the stem smuts, while only occasionally involving the flowering structures, do nevertheless generally suppress these structures and likewise result in a more or less complete loss of seed on affected plants.

The smuts that attack the vegetative structures (that is, the leaf smuts and stem smuts) have a decidedly weakening effect on their host plants and make them more susceptible to other sinister factors in their environment.

Many of the grass smuts are seed-borne. The wind carries millions of smut spores, which become lodged in or on the developing seeds of healthy Plants. When the seeds germinate, the smut spores also germinate and infect the young seedlings. Then the Plants that arise from the seedlings are smutted.

Some of the smut fungi are not limited to seedling infection. Apparently they can infect any succulent or rapidly growing part of their host plants and thereby make them smutty in time.

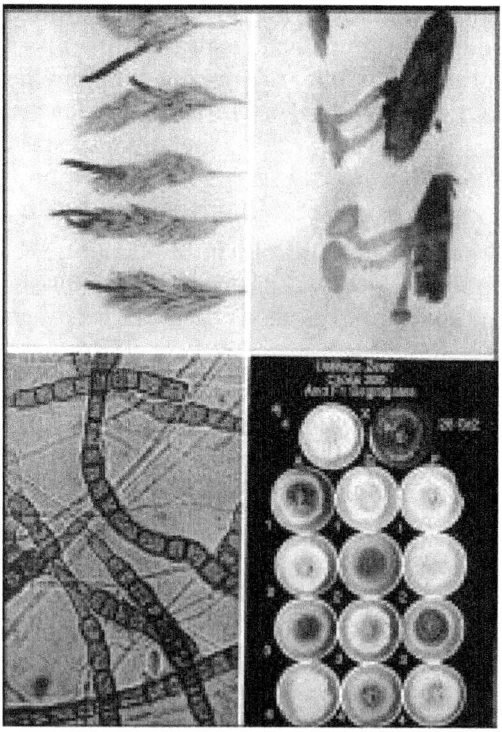

Upper left : The ergot fungus has replaced a seed in the orchardgrass flower head.

Upper right : Fruiting bodies of a fungus growing from seed of perennial ryegrass.

Lower left : Spores and mycelium of a fungus magnified by the lens of a microscope.

Lower right : Variability in progeny from a cross of two corn smut fungus cultures.

The time between infection and the appearance of smut varies rather widely. In the stripe smut of grasses, for example, seedlings often show smut within 6 weeks of the time the smut-contaminated seed is planted. Other smuts are not evident until their hosts head out. One of the stem smuts, to be described later, requires 2 to 4 years after infection before smut actually appears.

Spores of the smut fungi generally are much more durable than are those of the rust fungi. It is hard to maintain viability of rust spores for more than a year; usually it is much less than that. In the case of the smut fungi, it is unusual for viability to be maintained for less than a year. For most smuts viability is maintained for two to several years, especially if humidity is low. Some smuts have been known to retain at least some viability for 25 years. The cultivation of grasses seems to have much to do with the development of smut in them : Often cultivation practices, including harvesting and threshing, disperse the smut and contribute towards an increase in the amount of smut in succeeding crops. Any of the grass smuts therefore is a potential threat to the welfare of its grass host if the grass comes under cultivation.

Some of the grass smuts are the same as those that are destructive parasites of our cereal crops.

Seed smuts or bunt (Tilletia species) are terms used here to designate the smuts of grasses in which the normal seeds are replaced by smut"balls." The balls retain somewhat the shape of the normal seed and are even encased by the seed wall; actually, though, the inside of the seed is a solid mass of spores.

About 25 of these grass seed smuts occur in the United States. Several have actual or potential importance in cultivated grasses. One of the potentially most important is the one that on wheat is commonly called stinking smut or bunt. Three species of fungi are involved in it, including what is commonly known as dwarf smut.

These smuts are well known on wheat because of the very extensive losses they have caused for a long time the world over. Besides wheat, tall oat-grass and several of the wheatgrasses are known to be susceptible, including crested wheatgrass, slender wheatgrass, and intermediate wheatgrass. Care has to be taken to keep those grasses from falling prey to the cereal smuts.

The dwarf bunt is soil-borne. It has thus far presented a knotty problem of control except when resistant varieties are used. The other two species, being mostly seed-borne, may be controlled by seed treatment and the use of resistant varieties.

Some of the bentgrasses are susceptible to another of the seed smuts, especially on the east and west coasts where considerable smut infestation is sometimes encountered in harvested seed crops. The life antiquity of this smut is not known.

Various other groups of grasses, for example the bromes, fescues, wild-ryes, hairgrasses, velvetgrass, and others, are prey to similar seed smuts.

One of the most common of the head smuts of grasses is caused by Ustilago bullata. It complicates the cultivation and production of several of our best forage grasses. More than 60 species of grasses have been reported as hosts to it, among them wheatgrasses, bromes, fescues, wild- ryes, and barley grasses. In 1946 it was thought that head smut in mountain brome, one of our most valuable forage grasses, was under control through the release of a new variety, Bromar, which was developed jointly by the Department of Agriculture and the Washington Agricultural Experiment Station. In repeated tests the new strain had remained resistant to the various races of the head smut fungus.

In 1950, however, an entirely new and very virulent strain of head smut made its appearance and caused some very badly smutted fields of Bromar mountain brome. There are 12 known races or strains of the head smut fungus, each of which is specialised to different species of the wheatgrasses, bromes, wild-ryes, and barley grasses. None of the cereals is susceptible to it.

Several dozen other head smuts of grasses constitute a potential problem among economically important forage or range grasses. Practically nothing is known concerning the life antiquity and methods of controlling most of them.

Leaf smuts are of two types : The spot or blister smuts (Entyloma species) and the stripe smuts (Urocystis and Ustilago species).

The spot or blister smuts appear as flat or slightly raised, blackish, round or oval spots in the leaves. Although rather widely distributed, they seldom cause great losses. Once in a while spot smut develops in a bluegrass lawn to such an extent as to cause premature yellowing and death of the leaves. With the stripe smuts, however, it is quite a different matter.

The stripe smuts are evident in the leaves as black stripes, which contain the smut spores. The spores are shed and dispersed into the wind. Afterwards the affected leaves take on a shredded and curled appearance and soon wither. Affected plants quite frequently are dwarfed and contorted and produce abnormal, sterile heads, if any at all. Seedlings of grasses affected with stripe smut are predisposed to drought injury and root rot. Mature plants often are so weakened that they cannot survive severe winters.

Most of the stripe smuts of grasses cannot be told apart by the unaided eye, but they are different enough under a microscope. One of the most common and most virulent is known as flag smut. Many grasses and cultivated wheat are its hosts. Flag smut has several strains, each capable of attacking different groups of grasses.

Another of the common and destructive stripe smuts is the one that has long gone under the name of stripe smut or leaf smut. It has the same effect on its grass hosts as does flag smut and affects many of the same grasses, including some of our best forage grasses the wheatgrasses, bromes, wild-ryes, fescues, and barley grasses. None of the cereals is known to be susceptible to it. Stripe smut is carried on the seed and can be controlled by seed treatment.

Still another type of leaf smut is the"sausage" smut of the grama grasses. It produces small but conspicuous blister-like (often sausage-shaped) black Pustules on the leaves of the grama grasses, which make up a valuable component of our western ranges. It is restricted to the grama grasses, but often is widespread and undoubtedly results in reduced forage. Its life antiquity is not known. Methods of control therefore have not been developed.

Stem smuts (Ustilago species) are represented by several species that develop in significant abundance on economic grasses. All are marked by the development of conspicuous, dusty, brown or black layers of smut around the internodes of the stems. Sometimes nearly all the internodes seem to be smutty. Sometimes only the top one or two seem so. At first the smut is hidden by the leaf sheath that envelopes the stem, but as the stem elongates, the smut is exposed.

The stem smuts occur on some of our most valuable forage grasses the wheat-grasses, wild-ryes, some of the bluegrasses, the needlegrasses, Indian ricegrass, and a few of the fescues.

Quackgrass is commonly infected. It might indeed furnish a supply of smut for the infection of more desirable grasses such as crested wheat-grass, big bluegrass, and blue wild-rye. A peculiar feature about the quackgrass stem smut is that an incubation period of 2 or 3 years is required after infection before the smut appears.

The smut is perennial in the plant once it becomes infected. After the incubation period, the plant will produce a crop of smutty stems each year as long as it lives. Infection takes place in the vegetative tissues and thus spreads from plant to plant in a field.

The control of grass smuts is accomplished mainly in two ways : Chemical treatment of the seed in the case of seed-borne smuts and through the use of resistant varieties or strains.

Not all of the grass smuts can be controlled by seed treatment; but most of the head smuts and the stripe smuts can be. Among the grasses that could be expected to respond favourably to seed treatment for the control of smut are the various wheatgrasses, mountain brome, rescuegrass and related bromes, tall oatgrass, Canada wild-rye and other wild-rye grasses, big bluegrass and perhaps other of the bluegrasses, and the barley grasses. Such seed treatment for the control of smut will also help protect the seed and seedlings from seed decay and damping-off.

The procedure of treating grass seed with chemicals for the control of smut is much the same as that employed with cereal seeds and thus consists in principle of thoroughly applying a certain amount of a seed-treatment chemical to a certain weight of the seed and thoroughly mixing the two. However, comparatively little is known of the effectiveness of such control for many of the grass smuts. This is probably because intensive grass cultivation is relatively recent and the need for such knowledge has not been strongly felt.

The organic mercury dusts and some of the organic sulfur dusts generally are effective in the control of some of the grass smuts. New Improved Ceresan and Ceresan M, 1 ounce to a bushel, have given excellent control. Over-dosage, however, hurts the seed and reduces the stands.

The organic sulfur dusts, Arasan, Arasan SFX, and Tersan, at 3 to 4 ounces to a. bushel, likewise have given good control but not so consistently as organic mercury materials. They have an advantage, however, in that danger of overdosage is less. In fact, these three fungicidal dusts can be applied at maximum dosage (that is, all that can be retained by the dry seed) with little or no seed injury. This tolerance of grass seed to maximum dosage of certain of the fungicides is an advantage if small lots of seed are to be treated. Then it is not practical to attempt to weigh out minute quantities of fungicide, as needed on an ounce-per-bushel basis. Therefore, it is convenient to be able to apply an excess of the fungicide to the seed and, after mixing thoroughly, shake off the excess fungicide and not have to worry about overdosage.

Others of the newer fungicides appear promising and eventually testing may show them to be as reliable. Better results are had if the seed is stored in the treated condition for a few days. As much as 8 weeks will do no harm. It is possible therefore to treat the seed well ahead of the rush of planting time.

Comparatively little progress has been made towards the control of grasssmuts through the use of resistant varieties. In the various studies that have been made of head smut, stripe smut, stinking smut, and others, and in nursery row observations, it has been noticed repeatedly that resistant lines or strains do exist, but very few attempts have been made to combine that resistance with superior agronomic qualities by the process of hybridisation and selection. It is a neglected field of endeavor that merits attention, especially with regard to certain smut-susceptible species of forage grasses.

Certain it is that the smut diseases of grasses are going to come more and more into prominence as our grassland agriculture is extended. We should not delay, therefore, to make studies of the life histories of the grass smuts for which this information is lacking, because only on such a basis can intelligent control measures be devised.

HOW TO KEEP TURF GRASS HEALTHY

The 20 million acres of turf and 20 million lawns in the United States are prey to 100 different disease-producing organisms. Large sums are spent each year to establish good grass on golf courses, lawns, parks, and playgrounds, but the results are often disappointing. The main reason lies in the fact that the grasses that compose turf are subjected to unnatural conditions.

When man first mowed grasses closely to produce turf for specific purposes, the possibilities of disease incidence were multiplied. Turf is so commonplace that little attention has been given to its proper maintenance. However, cutting the grass blades wounds the plant and reduces its ability to manufacture food.

The lessened supply of food and repeated injury make the plant more susceptible to attack by pathogenic micro-organisms.

The grass that grows normally may be attacked by the same diseases, but it is better able to recover because it has a full leaf system. (An exception is the bentgrasses, the mowing of which makes for a healthier and better turf, while longer cutting causes matting, more disease, and actual smothering of the plants.)

Natural selection, operating over a long time among the closely mowed mixed grasses, has given us types or Species of turf grasses that are better able to survive diseases and to recover from injuries with greater vigour.

Many of the common pasture grasses, which grow more normally, never have been called on to survive repeated clipping. Sometimes a turf grass and a pasture grass may be the same, yet disease control measures applicable to lawns and golf greens are not feasible on pasture grasses that are grazed continually.

DISEASES OF TURF ARE PARASITIC OR NON-PARASITIC

Fungi cause the main parasitic diseases. Fungi usually are more destructive to grass plants that produce what is considered a good, dense turf. At their leaf tips, the rapidly growing plants secrete drops of water of guttation, which encourage the quick development of the fungi.

Moreover, the densely growing plants shade each other and tend to hold moisture longer, as there is less chance for evaporation of the surface moisture.

Because water is essential for germination and growth of the spores and other reproductive structures of fungi, conditions that favour air circulation and the removal of water films from the grass blades also help to control disease.

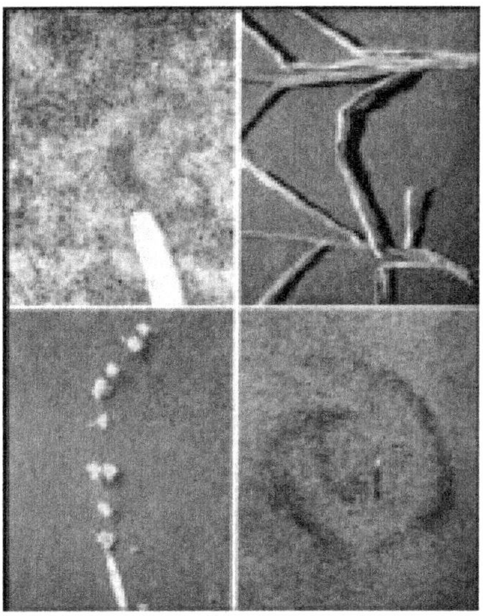

Upper left : Dollar spot injury of creeping bent, caused by Sclerotinia homeocarpa.
Upper right : Leaf spot injury of Kentucky bluegrass, caused by Helminthosporium vagans.
Lower left : Fairy ring, caused by the mushroom Marasmius oreades.
Lower right : Brown patch injury on Colonial bent, caused by Pellicularia filamentosa.

Temperature may markedly affect the development of both the grass plant and the fungus parasite. The bluegrasses and fescues grow best in the cooler temperatures of early spring and fall, for example, and grasses like Bermuda-grass and Zoysia species thrive best at summer temperatures. Similarly, the fungus that causes the destructive dollar spot disease thrives at temperatures below 80 F., but the brown patch fungus is much more injurious at temperatures above 80. The non-parasitic, or physiological, ailments injure a plant through unfavourable nutrition, moisture, light, temperature, soil reaction, or some chemical that has been added to the soil. For normal growth, plants must have the essential nutrients in proper balance.

Acidity or alkalinity of the soil must be adjusted to suit the requirements of the grasses. Frequently in alkaline (or "sweet") soil, iron is not absorbed sufficiently, so that an application of iron sulfate gives a better green colour. Heavy applications of fertilizer, especially when the grass blades are wet, often causes burning of the turf. The use of chemicals to control weeds frequently reduces the vigour of desirable turf grasses and provides a suitable medium for fungi. An oversupply of water in the soil excludes oxygen, which the roots need. Too much water also promotes the development of pathogenic fungi that attack the grass roots weakened by lack of oxygen.

Proper management, based on knowledge of the physiology of grasses and their cultural requirements, can do much to reduce the severity of diseases. Even if the recommended practices do not cure a disease, they may curb it so that chemical controls can be more effective. Because the fungi need moisture, proper drainage of turf areas to prevent excessive accumulation of moisture helps to control diseases.

Aerification of compacted soils helps water and air to reach the roots of grasses. Aerifying the soil also releases trapped air, removes excess surface water, and seems to free certain nutrients by changing the degree of acidity and alkalinity. Any practice that encourages vigour is good, for a rapidly growing, vigourous plant often is better able to withstand a disease or to recover more speedily.

Lime and fertilizers at the proper time and rate may alleviate the effects of the diseases. Technicians at the Georgia Agricultural Experiment Station learned that 2 pounds of nitrogen to 1,000 square feet, applied a month before seeding or at seeding time, reduced the occurrence of dollar spot in ryegrass. Research at Pennsylvania State College disclosed a striking inverse relationship between the application of nitrogen and the prevalence of dollar spot.

Practices that provide for good seedbed preparation, proper rate of seeding, adequate air and drainage, and the height of cut recommended for a particular species or strain of grass are other management practices that will retard disease. For example, in 1951 the workers in Georgia found that injury from diseases was much greater when ryegrass was sown at 100 pounds than at 40 pounds to 1,000 square feet.

It's a good thing to grow grasses that are resistant to disease. That is easier said than done, though, for it is hard to find and develop grasses that have high disease resistance and the required agronomic characteristics. Newly developed strains of grasses may retain resistance for only a short time, because new strains of fungi may develop.

Washington, Cohansey, and Arlington bents are much more resistant to brown patch than is Metropolitan. Seaside bent is moderately susceptible. Colonial is very susceptible. Washington bent, however, is very susceptible to dollar spot, as is the Toronto strain. Elk 16, Arlington, Congressional, Highland, Seaside, and velvet bents appear to have resistance to the dollar spot fungus. Congressional bent has shown resistance to snow mold.

The leading vegetative selections of creeping bentgrasses in use in the United States in 1953 included these strains, arranged in approximate order of their resistance to the more common diseases : Dahlgren (C-115, just as to the numbering system used by the Green Section to designate creeping bents); Arlington (C 1); Congressional (C-19); Old Orchard (C-15); Cohansey (C-7); and Toronto (C-15). Bluegrass originated from a single plant that developed from one of countless millions of bluegrass seeds sown over the years on the Merion Golf Club, Ardmore, Pa. That plant thrived and spread on the seventeenth tee until it formed a pure solid turf, which caught the eye of the golf course superintendent. A patch of the turf was sent to Arlington Farm, Va., in 1936. It was later transferred to Beltsville, Md., where it was extensively tested until 1950, when it was made available.

The consistent failure of common Kentucky bluegrass to form good turf under close mowing is an example of how mowing weakens a grass and causes a condition favourable to disease attack. The leaf spot fungus weakens and kills so many of the common bluegrass plants that open areas are produced, which afford favourable situations for crabgrass seedlings to develop rapidly. H. vagans, the fungus that is destructive to the common Kentucky bluegrass, likes cool, wet weather; it is therefore more active in spring and fall. When hot weather (which is unfavourable for bluegrass) follows, the weakened turf cannot compete with weed invasion, especially crabgrass.

Thus a fungus disease played an important part in producing conditions that made it necessary for researchers to initiate and develop a crabgrass control programme with chemicals.

Merion bluegrass turf, which is resistant to leaf spot, consistently contains less crabgrass than turf of common bluegrass grown under identical conditions. Since it is resistant to weed invasion, it produces a more attractive turf of better quality. The new bluegrass produces more vigourous rhizomes than the common bluegrass, and it can therefore spread faster. Merion bluegrass is more tolerant to heat and drought than were the other strains.

The seed of Merion bluegrass is shorter and plumper and lacks the usual brown colour found at the base of the common bluegrass seed. Seed experts are thus able to distinguish it from seed of other bluegrasses. Merion is not a"miracle grass" but it does give the home owner a better opportunity to have a lawn free of weeds, one that requires less irrigation and fewer mowings.

Since Merion bluegrass is not a heavy seed producer, it is taking somewhat longer than usual to increase the seed. It is expected that seed supplies will be ample within a few years and the prices will be much lower. Even at present prices, however, it costs very little more to establish a Merion bluegrass lawn as the recommended rate of seeding is much less than for common Kentucky bluegrass. The development of Tifton 57 Bermuda-grass by the Georgia Agricultural Experiment Station, the Department of Agriculture, and the United States Golf Association Green Section is another step towards better turf. It is a narrow-leaved strain and is resistant to a severe leaf spot and crown rot caused by a species of Helminthosporium.

U-3 Bermuda-grass is another improved selection. It is a fine-bladed, cold-hardy strain, which although discovered in Savannah, Ga., has found favour within the northern limits of the southern grass-growing areas where Bermuda is not expected to survive. It is finding increasing use in such sections as New Jersey, Maryland, Virginia, Missouri, Kansas, Oklahoma, and California. It requires close, frequent mowing and generous fertilization, but very little irrigation.

Meyer (Z-52) Zoysia is another promising disease-resistant grass that has shown up well under tests. It is an improved warm-season grass. It was developed by the United States Golf Association Green Section and co-operating agencies and named in memory of Frank N. Meyer, a Department plant explorer.

Meyer Zoysia is drought-tolerant. It needs less irrigation than do most grasses. It will grow on almost any soil. It is easy to mow when cut regularly and requires less mowing than many other grasses. It can be cut at heights of one-half inch to 4 inches without loss of vigour. It produces a firm, resilient turf which is very resistant to wear.

It also does well in association with Merion bluegrass. Meyer Zoysia is a warm-season grass so it will be more useful for planting in the southern half of the United States, although it has lived in plots as far north as Massachusetts and Michigan. It has some undesirable characteristics. Like Bermuda-grass, Meyer Zoysia loses its green colour in late fall after the first or second killing frost, but regains it in the spring about the time the weather is warm enough for people to begin using their lawns.

No seed of Meyer Zoysia is available, so vegetative planting by plugs of sod or sprigging must be done. Plugging or sprigging is done in the spring or late summer, the preferred time of planting. Brown patch, caused by the fungus Pellicularia filamentosa (Rhizoetonia solani), may attack susceptible turf wherever high temperatures and humidity prevail. It makes roundish areas that are 1 inch to 3 feet across or may extend over an entire putting green or large portions of lawns. The grass first turns dark and the leaves wilt. Gradually the area becomes light brown as the leaves die. When the fungus is growing actively, it produces grayish or black mycelium among the grass blades like a smoke ring. If the climatic conditions remain favourable and affected areas are not treated, the fungus will kill the grass to the point where reseeding may be necessary.

Brown patch has been controlled by a combination of mercury chlorides; 1 part corrosive sublimate to 2 parts calomel. Phenyl mercury acetate and lactate at dilutions of 1 1/4 ounce (10 per cent active ingredient) to 5 gallons of water per 1,000 square feet applied as sprays are partly effective. A wettable thiram known as Tersan has been widely used.

Dollar spot, once known as small brown patch, is caused by the fungus Sclerotinia homoeocarpa. The injured spots are dark and water-soaked at first. Then they turn brown. Later they become straw-coloured and are about 2 inches in diameter. The spots are usually regular in size and shape, but in cool, humid weather they may be so close together that they may merge into large, irregular areas of dead turf. When the fungus is actively growing in turf, a fine, white cobwebby mycelium may be observed in the early morning when the dew is still on the grass blades. In mild attacks of the fungus, only the uppermost grass blades may be injured. If the disease spread is checked then, the turf will recover quickly. When, however, the fungus is permitted to develop until all the leaf blades in an affected spot are destroyed, it may take many weeks for new grass to fill in the dead areas sufficiently to cover the scars of the disease.

The mercury chlorides, calomel and corrosive sublimate, are old standbys. The newer cadmium-containing fungicides, Cadminate, Puraturf 177, and Crag 531, are now recommended. Monthly applications of the cadmium compounds during the growing season are usually sufficient because of their good residual protection. The phenyl mercurials, Puraturf and PMAS, are effective but are apt to cause injury unless carefully applied.

Spot blight (grease spot or pythium blight) causes injury that appears in early morning as a circular spot surrounded by blackened grass blades, which are intertwined with mycelial threads of the causal fungus. The dark grass blades are water-soaked but soon either, becoming reddish brown, especially if the weather is sunny and windy. The spots, usually not more than 2 inches in diameter, generally occur in groups. They tend to form streaks apparently the fungus is spread by the mower or because of water in poorly drained low places. The disease can be very destructive in periods of high temperature and humidity, especially on poorly drained turf areas. Spot blight is usually more injurious to newly established turf, but if all conditions are favourable it will attack turf of any age. A sudden drop in temperature and a dry atmosphere checks its spread.

No satisfactory control measure has been developed for spot blight. Corrosive sublimate applied as a spray helps somewhat in checking it. One pound of iron sulfate in 30 gallons of water, applied to 5,000 square feet, may be worth trying. Copper sulfate at 3 ounces to 1,000 square feet gives partial control.

Copper spot makes coppery or orange areas, which usually are 1 to 3 inches in diameter. Close examination of the affected grass blades with a magnifying glass will reveal characteristic salmon-pink spore masses of the causal fungus, Gloeocercospora sorghi. The fungus seems to be active on grasses when the acidity of the soil is between pH 4.5 to 5.5.

The cadmium compounds are satisfactory in checking copper spot. The phenyl mercury formulations also are effective. Snow mold is caused by several fungi, among them Typhula itoana and Calonectria n1valls (Fusarium nivale). They grow and spread beneath the snow cover, as the snow melts, or during winter rains.

The first symptom is a dense, cottony growth of mycelium that covers patches of turf. The patches are 1 to 12 inches or more in diameter. Although it occurs as far south as Virginia, the disease is more prevalent in the northern part of the United States. Snow mold is usually more severe on greens located in low places or pockets and in places that have a deep covering of snow for long periods. The borders of greens are most often severely injured. Seaside and Highland bents are very susceptible.

Applications of mercury chlorides at the rate of 2 to 4 ounces to 1,000 square feet before the first snow is forecast, and again during a midwinter thaw, have given satisfactory control. Bluegrass leaf spot first shows small, brown, round to elliptical specks. The cause is the fungus Helminthosporizim vagans. The spots enlarge and may extend the width of the infected leaf. The tissue in the centre of the lesion dies and turns a straw colour.

A purplish-brown border may surround the straw-coloured centre. If the weather is cool and humid, spores are produced on the lesions and spread to other plants. Under favourable conditions, the lesions enlarge and may coalesce, involving most of the tissue of the leaf blade. The fungus may then spread to the leaf sheaths and to the crown of the plant, causing a crown and foot rot. The latter stage of the disease kills the grass plant, thins the turf, and leaves a bare spot for weeds to get started. Bluegrass leaf spot occurs wherever bluegrass grows. It is most severe in the spring and fall on grass that is closely clipped.

Merion bluegrass, which is resistant to leaf spot, offers a means of control. Fertilizing the bluegrass to stimulate new growth and raising the mower blades to allow longer leaves to develop will help overcome the disease.

Fertilizing the grass to stimulate rapid growth and raising the mower to permit a higher cut help the plants to withstand attacks of the disease. Metropolitan, velvet, and some of the other bentgrasses seem less injured by the disease.

Pink patch is characterised by reddish bundles of mycelial threads of the fungus, Corticium fuciforme, that mat the grass leaves together. The attacked leaves are at first water-soaked and then lose their green colour as the tissues dry. Turf is killed in distinct, isolated patches 2 to 15 inches in diameter. Pink, gelatinous branched strands (or horns) of the fungus develop during moist weather and tend to mat or bind the leaves and stems together.

The fungus bodies shrink on drying, become bright coral red, taper irregularly, and are one-eighth to 2 inches long. The brittle fragments break and spread the pathogen. The fungus attacks grasses within a wide range of temperatures, but high humidity is most conducive to its development.

The cadmium and mercury compounds seem to give satisfactory control, especially when applied as a preventive before the disease is observed. The colonial and creeping bents appear to suffer less than the velvet bents from attacks of this fungus.

Fading out : Lawns and putting greens from June to October sometimes show an off-colour that has often been attributed to drought, iron deficiency, or low fertility. Later an indefinite yellow and green dappled colour pattern develops. When the disease is severe and not controlled,the turf fades out, leaving irregular patches, seldom more than 2 to 3 inches in diameter, of dead plants.

Observations at the Rhode Island Agricultural Experiment Station have shown that species of the fungus Curvularia are always associated with grasses showing these symptoms. The disease has been found in several strains of velvet bent, piper bent, creeping bent, and annual bluegrass. Samples of infected grasses have been received from Connecticut, Massachusetts, Missouri, New Jersey, and New York.

The phenyl mercurials (PMAS, Puratised Agricultural Spray, Puraturf, Tag) when used at the equivalent of 1 1/4 ounce of 10 per cent of the active ingredient per 1,000 square feet, were effective in tests in 1949 and 1952 at the Rhode Island station.

Fairy ring : Many kinds of mushrooms or puffballs are known to make fairy rings in fairways, rough, and lawns. On greens where the turf is usually well fertilized and treated with fungicides to prevent diseases, fairy rings are seldom seen. People once believed that fairies danced in circles and caused mushrooms to grow.

Actually the ring is caused by the growth of a fungus spreading from a central point. The mycelium of the fungus grows profusely among the grass roots at the perimeter of the ring. The grass close to the fungus is often stimulated as if fertilized, producing an obvious unevenness of the turf. The soil thus invaded takes up water less rapidly than surrounding areas. In dry periods the grasses may wilt and die where the fungus has permeated the soil. Under favourable conditions the fungus produces the fruiting bodies that emerge from the soil in a ring.

Chapter 17

PLANT BREEDING

AIM OF PLANT BREEDING

The constant aim of plant breeding is to improve the quality, diversity and performance of agricultural and horticultural crops. The overriding objective is to develop plants better adapted to human needs. The evolution of plant breeding is a classic example of how improved biological understanding has been adapted to provide more effective methods of meeting the demands of a changing world. Plant breeding has existed in its most primitive form since the first farmers saved the seeds of their best plants from one season to the next more than 10,000 years ago. Over the centuries, this selection process has gradually become more scientific, bringing major improvements in the yield, quality and diversity of crops grown worldwide.

In the 19th century Gregor Mendel established the basic principles of plant genetics. Gregor Mendel in the 19th century laid the foundations for traditional plant breeding, in which selected parents are cross-pollinated to combine certain desired characteristics, such as high yield and resistance to pests and disease. He discovered that units of material, which are transferred from one generation to the next, determine inherited traits.

The plant breeder's aim is to reassemble these units of inheritance, known as genes, to produce crops with improved characteristics. In practice, this is a complex and time-consuming process. Each plant contains many thousands of genes, and the plant breeder is seeking to combine a range of desirable traits in one plant to produce a successful variety. Conventional breeding involves crossing selected parent plants, chosen because they have desirable characteristics such as high yield or disease resistance. The breeder's skill lies in selecting the best plants from the many and varied offspring. These are grown on and tested in subsequent years. Typically this involves examining thousands of individual plants for different characteristics ranging from agronomic performance to end-use quality.

Developing a new variety can take up to 15 years for wheat, 18 years for potatoes, even longer for some crops. The scope of conventional plant breeding has increased with improvements in technology. In the laboratory, chemical and mechanical techniques are used to speed up the selection process and remove natural barriers to cross-fertilization, for example between different crop species.

All forms of plant breeding involve the improvement of crops by selection. Since man began farming he has selected seeds from the best plants for the next generation. Over the centuries, this selection process has become more scientific, bringing major improvements in the yield, quality and diversity of agricultural and horticultural crops. Genes-units of hereditary material that are transferred from one generation to the next, determine these characteristics. Since the plant contains many thousands of genes, and the breeder is seeking to combine a range of traits in one plant, developing a successful variety can be an extremely lengthy process-up to 12 years in the case of cereals.

Using biological knowledge derived from cross-pollination, plant breeders have developed physical and chemically assisted ways of enhancing the speed, accuracy and scope of the selection process. Breeding systems involving grafting and hybridisation, for example, have extended breeders' capabilities at the whole crop level, while more recent cell-based techniques enable breeders to operate at the level of individual cells and their chromosomes

A major objective in modern plant breeding is the making of crop varieties with the highest possible yield potential. Yield potential is defined as "the yield of a crop when growth is not limited by water or nutrients, pests, diseases, or weeds". For farmers, whose crops are indeed limited by such constraints, this yield concept may not be seen as the most relevant objective. Relieving crops of all sorts of environmental stress, leaves light and temperature, together with varietal characteristics, as the only determinants of yield. That achieved, the same high yielding variety can be used all over a climatic zone, and target area for the variety is enormous. If stress can not be eliminated, however, adaptation to a usually site-specific environment becomes necessary. In that case, the target area for each variety will be small.

Do these different perspectives warrant different approaches to plant breeding? Conventional wisdom says no. In experiments where varieties are tested at various levels of inputs, the same high yielding variety usually comes out as top yielder at all levels. Therefore, plant breeders often claim that their varieties not only perform excellently under high-input conditions, but would also be better than traditional varieties in a low-input environment. However, when these trials are taken outside of experimental farms and the varieties are tested under local or farm conditions, researchers discover what is called 'crossover in performance'. At a certain level of stress there is a crossover point beyond which local varieties or landraces perform better than the high yielding varieties Farmers who experience such situations are not a tiny minority.

To some degree, most if not all farmers have to cope with local stress conditions. The stress-free environment is hard to achieve, and even harder to sustain. In many favourable areas, farmers abandon the high-input technology for economic reasons or because of ecological problems, thus increasing the need for locally-adapted germplasm. But modern plant breeding, whether public or private, can not supply adapted germplasm everywhere. Only a system of local seed selection can ensure that. And that means devolution of plant breeding.

Can such decentralised breeding be compatible with development needs in a changing world and meet economic aspirations in a poor society? If the answer to these questions is going to be 'yes', the decentralised breeding must be able to take advantage of the power of science as well as of the capacities of local communities. Three aspects should be considered and made mutually compatible : breeding technology, participatory research methods, and organisation at community level.

History

Plant breeding is the purposeful manipulation of plant species in order to create desired genotypes and phenotypes for specific purposes. This manipulation involves either controlled pollination, genetic engineering, or both, followed by artificial selection of progeny. *Plant breeding* often, but not always, leads to plant domestication.

Plant breeding has been practiced for thousands of years, since near the beginning of human civilization. It is now practiced worldwide by government institutions and commercial enterprises. International development agencies believe that breeding new crops is important for ensuring food security and developing practices of sustainable agriculture through the development of crops suitable for their environment.

Domestication

Domestication of plants is an artificial selection process conducted by humans to produce plants that have fewer undesireable traits of wild plants, and which renders them dependent on artificial (usually enhanced) environments for their continued existence. The practice is estimated to date back 9,000-11,000 years. Many crops in present day cultivation are the result of domestication in ancient times, about 5,000 years ago in the Old World and 3,000 years ago in the New World. In the Neolithic period, domestication took a minimum of 1,000 years and a maximum of 7,000 years. Today, all of our principal food crops come from domesticated varieties.

A cultivated crop species that has evolved from wild populations due to selective pressures from traditional farmers is called a landrace. Landraces, which can be the result of natural forces or domestication, are plants (or animals) that are ideally suited to a particular region or environment. An example are the landraces of rice, *Oryza sativa* subspecies *indica,* which was developed in South Asia, and *Oryza sativa* subspecies *japonica,* which was developed in China.

Classical plant breeding uses deliberate interbreeding (crossing) of closely or distantly related species to produce new crops with desirable properties. Plants are crossed to introduce traits/genes from one species into a new genetic background. For example, a mildew resistant pea may be crossed with a high-yielding but susceptible pea, the goal of the cross being to introduce mildew resistance without losing the high-yield characteristics. Progeny from the cross would then be crossed with the high-yielding parent to ensure that the progeny were most like the high-yielding parent, (backcrossing), the progeny from that cross would be tested for yield and mildew resistance and high-yielding resistant plants would be further developed. Plants may also be crossed with themselves to produce inbred varieties for breeding.

Classical breeding relies on homologous recombination of two genomes to generate genetic diversity. The Classical plant breeder may also makes use of a number of *in vitro* techniques such as protoplas fusion, embryo rescue or mutagenisis to generate diversity and produce plants that would not exist in nature.

Traits that breeders' have tried to incorporate into crop plants in the last 100 years include :

- Increased quality and yield of the crop
- Increased tolerance of environmental pressures (salinity, extreme temperature, drought)
- Resistance to viruses, fungi and bacteria
- Increased tolerance to insect pests
- Increased tolerance of herbicides.

Before World War II

Intraspecific hybridisation within a plant species was demonstrated by Charles Darwin and Gregor Mendel, and was further developed by geneticists and plant breeders. In the early 20th century, plant breeders realised that Mendel's findings on the non-random nature of inheritance could be applied to seedling populations produced through deliberate pollinations to predict the frequencies of different types.

In 1908, George Harrison Shull described heterosis, also known as hybrid vigour. Heterosis describes the tendency of the progeny of a specific cross to outperform both parents. The detection of the usefulness of heterosis for plant breeding has lead to the development of inbred lines that reveal a heterotic yield advantage when they are crossed. Maise was the first species where heterosis was widely used to produce hybrids.

By the 1920s, statistical methods were developed to analyse gene action and distinguish heritable variation from variation caused by environment. In 1933, another important breeding technique, cytoplasmic male sterility (CMS), developed in maise, was described by Marcus Morton Rhoades. CMS is a maternally inherited trait that makes the plant produce sterile pollen, enabling the production of hybrids and removing the need for detasseling maise plants.

These early breeding techniques resulted in large yield increase in the United States in the early 20th century. Similar yield increases were not produced elsewhere until after World War II, the Green Revolution increased crop production in the developing world in the 1960s.

After World War II

Following World War II a number of techniques were developed that allowed plant breeders to hybridise distantly related species, and artificially induce genetic diversity.

When distantly related species are crossed, plant breeders make use of a number of plant tissue culture techniques to produce progeny from otherwise fruitless mating. Interspecific and intergeneric hybrids are produced from a cross of related species or genera that do not normally sexually reproduce with each other. These crosses are referred to as *Wide crosses*. The cereal triticale is a wheat and rye hybrid. The first generation created from the cross was sterile, so the cell division inhibitor colchicine was used to double the number of chromosomes in the cell. Cells with an uneven number of chromosomes are sterile.

Failure to produce a hybrid may be due to pre- or post-fertilization incompatibility. If fertilization is possible between two species or genera, the hybrid embryo may abort before maturation. If this does occur the embryo resulting from an interspecific or intergeneric cross can sometimes be rescued and cultured to produce a whole plant. Such a method is referred to as *Embryo Rescue*. This technique has been used to produce new rice for Africa, an interspecific cross of Asian rice *(Oryza sativa)* and African rice *(Oryza glaberrima)*. Hybrids may also be produced by a technique called protoplast fusion. In this case protoplasts are fused, usually in an electric field. Viable recombinants can be regenerated in culture.

Chemical mutagens like EMS and DMSO, radiation and transposons are used to generate mutants with desirable traits to be bred with other cultivars. Classical plant breeders also generate genetic diversity within a species by exploiting a process called somaclonal variation, which occurs in plants produced from tissue culture, particularly plants derived from callus. Induced polyploidy, and the addition or removal of chromosomes using a technique called chromosome engineering may also be used.

When a desirable trait has been bred into a species, a number of crosses to the favoured parent are made to make the new plant as similar as the parent as possible. Returning to the example of the mildew resistant pea being crossed with a high-yielding but susceptible pea, to make the mildew resistant progeny of the cross most like the high-yielding parent, the progeny will be crossed back to that parent for several generations. This process removes most of the genetic contribution of the mildew resistant parent. Classical breeding is therefore a cyclical process.

It should be noted that with classical breeding techniques, the breeder does not know exactly what genes have been introduced to the new cultivars.

Some scientists therefore argue that plants produced by classical breeding methods should undergo the same safety testing regime as genetically modified plants. There have been instances where plants bred using classical techniques have been unsuitable for human consumption, for example the poison solanine was accidentally re-introduced into varieties of potato though plant breeding.

Exploiting heterogeneity and crop evolution in farmers' fields are outside the scope of most plant breeding research. One exceptional experiment, however, has shed some scientific light on the issue. It was started at the University of California (UC) in 1928. Composite cross populations of barley were produced, some of which were extremely diverse in origin of sources. These populations were exposed to continuous natural selection in current modern farming environments and became the subject of studies during the career span of several generations of UC professors.

It appears that after low yields in initial years, the composite cross populations gradually improved in performance and eventually became quite good yielders, with excellent yield stability and disease resistance. These results inspired Suneson to propose an evolutionary plant breeding method. After assessment of later generations of the same material, Soliman and Allard concluded that such evolutionary breeding "is unwarranted" if yield potential is the major goal. However, if disease resistance and yield stability are two major objectives, "the composite cross approach is an efficient method". This amounts to saying that a major part of world agriculture, many high-input systems included, could be well served by this approach.

In short, this means constructing a body of broadly diversified germplasm and exposing it to natural selection in areas of contemplated use. For those who are familiar with traditional farmers' breeding, this may sound like reinventing the wheel. In fact it is an improvement of the old wheel of plant breeding.

The first step, constructing a body of of broadly diversified germplasm, is not all that straightforward. Science has access to world collections and information sources that are unavailable to farmers. A research institute can chose relevant germplasm and make composite cross populations with an evolutionary potential, most probably far beyond that of locally available varieties.

The immediate outcome, the early generation composite cross population, will be unadapted everywhere and is likely to yield poorly. With time, however, recombinations and natural sorting will improve the adaptation, and, according to the Californian experience, narrow the gap with commercial varieties. The long-term outcome could be populations that outperform commercial varieties in disease resistance and yield stability and that may be used as a source of artificial selection for high yield.

The disease resistance appears to have evolved through the building up of polygenic complexes. Therefore, it provides a durable resistance as opposed to the monogenic and, therefore, mostly non-durable resistance usually bred into commercial pure line varieties. The stability, at least to some degree, depends on the buffering effect of crop heterogeneity. The Californian experiment shows that when the population is propagated in isolation for a very long time, diversity will start declining, resulting eventually in reduced stability.

These experimental findings into the context of current development needs, a few conclusions can be drawn. We need breeding populations with a very high evolutionary potential, and these populations must be exposed to the stress conditions of, or similar to, current farm environments. Furthermore, a certain level of diversity within populations must be maintained in order to sustain evolutionary potential and yield stability.

An Age-old Tradition

Modern plant breeding is a sophisticated, high investment business, but its origins stretch back thousands of years to primitive farmers who selected the best plants in one year to provide seed for their next crop. This selective breeding was the first human refinement of natural plant evolution. Recent scientific and technological developments have allowed a greater rate of improvement. It was Gregor Mendel who, in the mid- 19th century, first provided a scientific explanation of genetic inheritance. His conclusions on the relationship between inherited characteristics in the offspring and the genetic makeup of the parents were the theoretical basis for classical plant breeding. Mendel's work went largely unrecognised in his own lifetime, and it was not until the early 20th century that it was rediscovered to become the basis of modern scientific plant breeding.

A Modern Industry

Until the early 1960s, plant breeding in Britain was largely confined to publicly funded research. This situation changed dramatically in the mid-1960s, with the passing into UK law of the 1964 Plant Varieties and Seeds Act. This legislation introduced a system of royalty payments on individual plant varieties, known as Plant Breeders' Rights, and triggered a rapid expansion of plant breeding as a commercial enterprise in its own right. Today, much of the basic research into crop science is still conducted by public sector research organisations, but the majority of commercial plant breeding takes place within the private sector. Some

60 plant breeding companies, based in the UK, are active across the entire spectrum of plant species from the major arable crops through to ornamental garden shrubs and flowers. In total, the plant breeding sector employs around 5,000 people, and supports a further 5,000 jobs in seed production and distribution. Plant breeding remains a vital industry to keep Britain competitive on world markets. The need for new varieties, adapted to our unique growing conditions, is never ending, driven by the challenges of new disease pressures, changing market requirements and shifts in agricultural and environmental policies.

- *Maise* : Maise crops used for grain and animal fodder are derived from wild races originating in Central America

- *Sugar beet* : Modern varieties of sugar beet have been developed from wild ancestors native to Central Europe.
- *Potatoes* : Wild ancestors of the modern potato still grow in parts of South America.
- *Wheat and Barley* : Small grain cereals, the mainstay of UK crop production, are derived from wild grasses of the Middle East.
- *Oilseed Rape* : Like many crops within the brassica family, oilseed rape has its origins in wild species native to China.

Participatory Plant Breeding

Certain trends have made the world ripe for adoption of participatory plant breeding methods.

- The shielding of crops from environmental stress in high-input systems is facing increasing economic and ecological problems. Scientists are changing their attitudes, and a new paradigm is being formulated. Instead of modifying the environment to suit the requirement of high yielding varieties, the varieties need to be modified to suit the environment.
- The claim that modern varieties can be made broadly adapted and be superior across most farming environments within an ecogeographic region is being challenged.
- If relevant diversity exists in a locale, the combined action of natural and artificial selection within a local environment may be an efficient breeding method. Experiments show that this may work also in a fertilizer-intensive system.
- In recent years farmer groups working with local seeds have been organised all over the world. They are not primarily conservers of old seeds. They want their seeds to be improved, in an evolutionary, slow and steady way, and under their own control.
- Finally, participatory methods have been developed in order to facilitate the involvement of farmers together with scientists as active and equal partners in research to generate relevant farm technology. Such methods can be applied also to plant breeding.

Commercial seeds often diffuse into areas where the traditional seed supply system is still predominant. Farmers try them with an open mind and adopt or reject them according to their own criteria. If grown and multiplied in the villages, diversity will start to appear within them and local reselection will be possible. In that way commercial varieties eventually might become like landraces. It is also commonly observed that farmers change and exchange seeds. A traditional farming system rarely functions as an environment for the static preservation of old landraces.

Often practices range from neglect to very simple mass selection, but with a few scattered individuals who devote an exceptional amount of effort to the maintenance or improvement of seed quality. These exceptional persons, very often women, may be the source of good seeds for others in the community. Once such a community is organised for seed management and improvement, it becomes possible for it to establish links to scientific institutions.

Community resources for participatory approaches to seed management and breeding are not limited to indigenous culture and traditional practices. The educational status and experiences of modern farmers may also be turned into a resource for community action. In the Philippine group, a number of the members were high school graduates and a few had a university degree. And moreover, most of them had a couple of decades' experience with modern input-intensive farming. Seed activities opened their minds towards the traditional societies, towards themselves and towards the modern world. The traditional societies supplied them with their seeds, and through the seeds, they learned to appreciate the values and achievements of these societies. They discovered their own potential, and saw that the outside world could bring more than technology packets : it could bring knowledge and ideas to be exploited and further developed by themselves.

Scope of Plant Breeding

Improving Crop Performance

The dramatic gains in agricultural productivity seen in the second half of the last century are often linked uniquely to increased mechanisation and the widespread introduction of fertilizers and agrochemicals.This 'Green Revolution' would not have been possible without the huge genetic improvements made to the crops. Plant breeding has contributed around half of the threefold increase in UK wheat yields recorded from 1947 to 1992 and yields have continued to improve since then. Plant breeding can directly improve the performance of crops in different ways.

- *Productive Yield :* Developing crop varieties, which convert more of their biomass into productive yield, is the single biggest contributor to improved crop output. The introduction of shorter-strawed cereals is a striking example of how this has been achieved, by transforming more of the crop's productive energy into valuable grain.
- *Physical Characteristics :* Changing a crop's physical structure can also contribute to increased yields. For example, the development of semi-leafless varieties of field peas helped to boost intrinsic yield. It also stimulated the growth of stabilising tendrils, which significantly improved the crop's standing ability so reducing crop losses at harvest.
- *Disease Resistance :* Genetic resistance to disease enables crops to realise their yield potential–it can also mean reduced use of agrochemicals. Plant breeding has significantly improved the genetic resistance of crops against the threat of viral and fungal infection. Key examples include resistance to blight in potatoes, rhizomania in sugar beet, and barley yellow mosaic

virus in cereals. The challenge of breeding resistant varieties is constant because new strains of disease develop naturally.

- *Timing of Maturity :* Plant breeding technology has brought major improvements in the uniformity with which crops ripen ready for harvest. This not only reduces potential crop losses at harvest (as in the case of pod shatter in oilseed rape), but has also improved growers' ability to mechanise harvesting operations. In the field vegetable sector, for example, the labour intensive (and unpleasant !) task of picking crops such as cauliflower, broccoli and Brussels sprouts has been transformed by the development of improved varieties.
- *Other Agronomic Factors :* By improving crops' ability to cope with a range of other agronomic pressures, advances in plant breeding continue to underpin progress in agricultural productivity.

They include :

- Genetic resistance to pests, such as nematode resistance in potatoes
- Shorter crop life-cycle, to expand a crop's geographical growing areas
- Stress tolerance, such as frostresistance in field vegetables, to extend the seasonal availability of home-grown fresh produce.

Improving Crop Quality

Alongside progress in keeping Britain's farmers competitive through improved productivity, developments in plant breeding have also brought significant gains in crop quality. Gone are the days when Europe encouraged farmers to concentrate solely on maximising output, assured that their crops would find support under the Common Agricultural Policy.

Today, as production-based support and barriers to international trade continue to be reduced, Britain's growers must compete with the best in the world, on quality and cost of production. Consumer expectations of food are also more exacting. This is reflected in tighter quality specifications along the food chain. Plant breeders have responded with a continuous stream of new varieties, tailored to the needs of specific end-markets.

Cereals

Britain's historical dependence on North American imports of wheat for bread making has been reversed over the past 40 years thanks to improvements in baking technology and the development of high protein, hard milling varieties of wheat.

Until the mid-1960s, imported wheat typically accounted for up to 65 per cent of our bread making requirements. Today as much as 90 per cent of our bread is produced from homegrown varieties. In barley, improvements in malting quality have underpinned a fourfold increase in the volume of beer brewed per tonne, from around 2,000 liters in 1950 to nearly 8,000 liters today.

The quality of malting barley varieties grown in Britain is internationally recognised, with export markets from the near Continent to the Far East.

Oilseed Rape

Oilseed rape has become a major homegrown source of vegetable oil and animal feed within the past 30 years. The rapid expansion of this important break crop from 50,000 hectares in the mid-1970s to around 500,000 hectares today can be linked to two major breakthroughs in plant breeding. Varieties for food use were first developed in 1970. These contain reduced levels of erucic acid in the oil, making it more suitable for human consumption. This was followed in the late 1980s by 'doublelow' varieties, which also offered reduced levels of glucosinolates in the residual meal to improve quality for animal feed.

Potatoes

In the potato sector, plant breeders have responded to increasing consumer demands for quality and choice. Fresh produce counters now boast a bigger range than ever, from traditional main-crop varieties through to baby new, baking and salad potatoes. Breeding programmes have also targeted processing and catering markets, with specific varieties used to produce crisps, instant mash and chips.

Sugar Beet

Higher root yield and increased sugar content have combined to double sugar production per hectare in the past 50 years. The introduction of monogerm seed, allowing seeds to be planted individually rather than in clusters, ensures genetic improvements are fully realised in the field.

Pulses

As interest in home-grown sources of protein has increased, breeders have developed varieties of peas and beans for specialist markets–freesing and canning for human consumption, flaking for pet food, and tannin-free varieties for animal feed.

Creating a New Variety

The creation of each new variety is a complex, costly and skilled operation. It is also painstakingly slow–today's breeding programmes are already looking ten years ahead to the needs of farmers, consumers and the environment at the end of the next decade and beyond. Techniques vary between crop species, but the scientific principles of plant breeding remain unchanged from Mendel's first discovery that selected parent plants can be cross-pollinated to combine desired characteristics. Genes–units of hereditary material that are transferred from one generation to the next, determine plant characteristics. Since each plant contains many thousands of genes, and the breeder is seeking to combine a range of traits in one plant, such as high yield, quality and resistance to disease, developing a successful variety is an extremely lengthy process-up to 12 years in the case of cereals, even longer for potatoes. Plant breeding has been compared to playing a fruit machine-not with three reels, but several hundred. The skill of the plant breeder lies in improving his chances of hitting the jackpot by combining all the desired characteristics in the same variety.

METHODS IN PLANT BREEDING

Conventional Breeding

Conventional plant breeding involves crossing carefully chosen parent plants, then selecting the best plants from the resulting offspring to be grown on for further selection. For cereals, hundreds of individual crosses are carried out by hand to create seed for the first filial (or F1) generation. The resulting F1 plants are uniform, but in the following generation several hundred thousand different plants are produced. Because of the way genes work, the new combinations produced from each cross are not revealed until the second (F2) generation.

It is this enormous diversity of new gene combinations that may hold the key to a successful new variety. The plant breeder's task is to select the plants most likely to meet his breeding objectives. Seed from the best of these F2 plants is grown on in small rows or plots and the best plants again selected-this process is repeated year after year until only the very best plants remain. As promising new lines emerge, tests are conducted on each plot to assess factors such as yield, disease resistance and endues quality.

Once the best lines are purified to ensure that every plant has the same characteristics, the process of multiplying seed begins. These 'inbred' lines are then ready for entry into official trials, some six to ten years after the initial cross.

Hybrid Breeding

For some crop species, the seed supplied to growers is that produced from the first cross between selected parents. The resulting varieties, known as F1 hybrids, offer potential advantages in crop performance. Hybrid breeding is widely used to produce varieties of field vegetables, maise and oilseed rape. F1 hybrids are unique in expressing 'hybrid vigour' in the growing crops for a single year. This may result in higher yields, greater uniformity, or improvements in quality. Unlike inbred lines, F1 hybrids do not breed true year after year and their performance gains are not maintained in subsequent generations.

Improved Breeding

With increasing knowledge and improved technology, breeders have developed ways to enhance the speed, accuracy and scope of the breeding process.

There are several ways to reduce the lengthy interval between the first cross of selected parents and establishing true breeding lines of promising new varieties. For example, maintaining parallel selection programmes in northern and southern hemispheres allows two generations to be produced each year. Single seed descent produces very small plants under restricted growth conditions. Large numbers of these plants are cultivated in artificial growth rooms, with two or more generations produced in a year. In potatoes, mini-tuber breeding speeds up the slow multiplication process by producing miniature plants under greenhouse conditions.

More recent laboratory techniques enable breeders to operate at the level of individual cells and their chromosomes. In certain crop species, such as potatoes and oilseed rape, it is possible to produce new varieties in the laboratory through

Protoplast Fusion

Individual plant cells with their outer walls removed are fused, and the fused cells then induced to divide and grow in a culture medium. Whole plants are eventually regenerated containing new combinations of genes from the two parents.

Embryo Rescue and Assisted Pollination

Allow breeders to expand the range of available characters by making crosses between species, which would not produce viable offspring outside the laboratory.

Double Haploid Breeding

Enables breeders to produce genetically uniform lines within one generation. This effectively by-passes the lengthy process of self-pollination and selection normally required to produce true breeding plants. Latest developments in genetic science have greatly improved our understanding of how plants behave, offering additional ways to enhance the breeding process.

Genomics

By mapping the genetic makeup, or genome, of crop species, scientists can identify the exact position and function of individual genes. Genome mapping has revealed striking similarities in the genomes of different crop species, such as rice, wheat, barley and rye. This information is already helping to broaden the scope and precision of current breeding programmes.

Marker Assisted Breeding

Allows breeders to determine whether desired traits are present in a new variety at an early stage in the breeding programme.

Genetic Modification

Genetic modification of plants is achieved by adding a specific gene or genes to a plant, or by knocking out a gene with RNAi, to produce a desirable phenotype. The resulting plants are often referred to as transgenic plants. Genetic modification can produce a plant with the desired trait or traits faster than classical breeding because the majority of the plant's genome is not altered.

To genetically modify a plant, a genetic construct must be designed so that the gene to be added or knocked-out will be expressed by the plant. To do this, a promoter to drive transcription and a termination sequence to stop transcription of the new gene, and the gene of genes of interest must be introduced to the plant. A marker for the selection of transformed plants is also included. In the laboratory, antibiotic resistance is a commonly used marker : plants that have been successfully transformed will grow on media containing antibiotics; plants that have not been transformed will die. In some instances markers for selection are removed by backcrossing with the parent plant prior to commercial release.

The construct can be inserted in the plant genome by genetic recombination using the bacteria Agrobacterium tumefaciens or *A. rhizogenes*, or by direct methods like the gene gun or micro-injection. Using plant viruses to insert genetic constructs into plants is also a possibility, but the technique is limited by the host range of the virus. For example, Cauliflower mosaic virus (CaMV) only infects cauliflower and related species. Another limitation of viral vectors is that the virus is not usually passed on the progeny, so every plant has to be inoculated.

The majority of commercially released transgenic plants, are currently limited to plants that have introduced resistance to insect pests and herbicides. Insect resistance is achieved through incorporation of a gene from Bacillus thuringiensis (Bt) that encodes a protein that is toxic to some insects. For example, if cotton pest the cotton bollworm feeds on Bt cotton it will ingest the toxin and die. Herbicides usually work by binding to certain plant enzymes and inhibiting their action. The enzymes that the herbicide inhibits are known as the herbicides *target site*. Herbicide resistance can be engineered into crops by expressing a version of *target site* protein that is not inhibited by the herbicide. This is the method used to produce glyphosate resistant crop plants. Genetic modification of plants that can produce pharmaceuticals (and industrial chemicals), sometimes called *pharmacrops*, is a rather radical new area of plant breeding

Proteomics

Allows breeders to understand how genes behave in different parts of the plant and under different growing conditions.

Maintaining Genetic Diversity

Maintaining biodiversity is central to the process of crop improvement. It is in every breeder's interest to ensure that the gene pool from which new traits are selected remains as extensive as possible. Plant breeders created the first gene banks in the 1930s to conserve the valuable genetic diversity within past and present varieties, as well as landraces and wild relatives of cultivated crop species. Plant breeding is integral to ongoing initiatives to identify, classify and conserve existing biodiversity.

Testing Plant Varieties

Before any new crop variety can be placed on the market, it must undergo statutory testing under a process known as National Listing. Successful varieties are placed on a National List or register of varieties approved for marketing.

National Listing rules are determined on a European basis, and apply to all the major agricultural and vegetable crop species. Official trials are conducted, in most cases for a minimum of two years, to test each new variety for a range of characteristics which together determine its uniqueness, its genetic uniformity, and its value to growers and the rest of the food chain. National Listing is extremely rigourous – the majority of varieties entered do not complete the process. In winter wheat, for example, only a quarter of varieties entered for National Listing during the 1990s were finally approved.

National Listing–DUS

All varieties entered for National Listing are assessed for Distinctness, Uniformity and Stability (DUS). In the case of cereals, some 30 individual characteristics of the plant are inspected to verify that it is distinct, *i.e.* clearly distinguishable from other varieties, that its characteristics are uniform from one plant to another, and that the variety is stable in that it breeds true to type from one generation to the next.

National Listing–VCU

For agricultural crops, National Listing also involves trials to establish a variety's Value for Cultivation and Use (VCU). This provides an assurance to growers that only varieties with improved performance or end-use quality can be officially approved for sale.

Recommended and Descriptive Lists

Once a new variety has been added to the National List, it is cleared for marketing. However, its success in the market place is by no means guaranteed. Further independent trials are conducted each year to compare the performance and quality of the best varieties. These trials provide the basis for detailed information and advice to growers and their customers.

As shown in the following illustration for winter wheat, the process of statutory and commercial evaluation can take up to five years. Only a handful of varieties clear all these hurdles, which are in addition to the many years spent testing and selecting in the breeder's own trials programme.

Genetically Modified Varieties

No GM crops can be marketed until they have been assessed and approved in terms of human health, food safety and the environment. The cross-border movement of genetically modified organisms (GMOs) is regulated at a global level under the internationally agreed Biosafety Protocol.

Variety Maintenance and Identity Preservation

As well as developing new varieties, plant breeders maintain the genetic purity of existing lines and pre-commercial seed supplies year by year. This process is costly and time-consuming, but essential to maintain the quality and performance of each variety.

For cereals, variety maintenance begins after a few years of selection trials, when the promise of a variety is just emerging. At that stage, all that exists of what may become a widely grown variety is a single row containing around 100 plants. The breeder then bulks up supplies of the purified lines of breeder's seed into prebasic and then basic seed. Each year specialist seed growers are used to grow basic seed for the first generation of certified seed–or C1 seed. After one more year this becomes C2 seed, the main source of certified seed used by farmers.

The plant breeder continuously maintains breeder's seed for the process of multiplication through pre-basic, basic, C1 and C2 seed to ensure the variety's performance and quality year after year. Greater emphasis is now being placed on preserving the identity of individual varieties after harvest, both to conserve quality characteristics and to meet consumer demands for assurances about the integrity and traceability of their food.

Seed Certification

Seed of an approved variety can only be marketed if it meets strict quality criteria. Seed quality standards are laid down in UK and EU law, and policed by agencies appointed by Government. The UK's official seed certification system offers an independent assurance of quality to growers. Minimum standards apply for varietal identity, purity and germination capacity. In addition, strict limits apply to seed-borne diseases, and the presence of physical impurities such as weed seeds.

Around 9 per cent of the UK arable area is used to multiply the pure lines of seed from the plant breeder into certified seed. Several thousand individual crops are involved; each grown under specific management regimes to ensure the purity and integrity of the resulting seed is maintained. To gain certification, every seed crop must undergo crop inspection and seed testing. Seed certification underpins the health and purity status of the major arable crops in Britain. It offers an independent benchmark of quality on which buyers of seed and their customers depend.

Farm-saved Seed

For certain crop species–particularly small-grain cereals–growers can opt to save their own seed for sowing the following year provided care is taken to ensure that the crop remains healthy and free from impurities, and that the resulting seed is carefully conditioned and cleaned. Without independent testing for germination and freedom from seed borne diseases, however, farm-saved seed can harbors risks to growers that may not become apparent until well into the growing season. Many growers choose certified seed for the peace of mind that quality and performance is independently assured.

Plant Breeding in the Future

Success in any line of business depends on being able to anticipate and respond to changing market requirements. Such is the long-term nature of developing a new crop variety that plant breeders are in the unenviable position of forecasting their customers' demands at least five, and probably nearer ten years, in advance. Plant breeding objectives will continue to be shaped by changes in agricultural and environmental policy. There is no doubt that while improvements in crop yield and endues quality remain of paramount importance, the future for British agriculture lies in crop production systems which deliver both economic and environmental benefits.

Increasingly, plant breeders are targeting varieties suitable for low input and organic regimes. It is already estimated that genetic improvements in disease resistance alone save British farmers more than £100 million a year in reduced agrochemical use and improved yield. A range of other factors also determines a variety's suitability for low input, integrated farming systems. They include the crop growth cycle, straw strength and susceptibility to weed competition, as well as the variety's ability to thrive in a range of soil types, climatic conditions and rotational patterns.

Novel Crops

Breeders continue to adapt novel crops for UK growing conditions. New varieties of sunflowers and soyabeans are already creating market opportunities in the food sector. Post-BSE, attention has focused on using more home-grown fodder in animal feed. For plant breeders, this means enhancing energy and protein levels in traditional forage crops, and developing the potential of new crop species, such as lupins.

Non-food Crops

The use of crop plants for industrial, non-food applications is an area of increasing interest. Developments in plant breeding could underpin a transition from many crude, industrial processes towards renewable, field-based production of specialist chemicals, textiles and biofuels Quick-growing biomass crops such as willow coppice and miscanthus are already fuelling UK power stations, and there is huge potential to develop oilseed, starch and fibre crops for a wide range of product including lubricants, pharmaceuticals, dyes, plastics, flooring, paper and clothing.

New Traits

Pioneering research by British scientists is opening up new possibilities for crop improvement. Examples from current research include the use of genetic modification to overcome grey mould infestation in strawberries–not possible through conventional breeding because no resistance genes exist in the world strawberry germplasm. Similar techniques offer ways to control the ripening process in fruit and vegetables, to extend storage life, and improve flavour and texture for consumers. It will also increase the seasonal availability of homegrown produce, so reducing the distance food is transported. Modern biotechnology can be used to improve human nutrition. Recent research by British scientists includes the development of vegetables containing increased levels of cancer-fighting nutrients, and apples, which really do keep the dentist at, bay by protecting against tooth decay.

Patents

Recent developments in technology have also seen the emergence of patent applications in relation to plant breeding. This is particularly the case where technologies can be applied to several different varieties of the same crop, or across a range of different species. Such developments cannot be protected under the variety-specific system of Plant Breeders' Rights.

The use of patents where living organisms are concerned continues to arouse controversy, but future research will not take place unless some mechanism exists to reward and promote sharing of knowledge.

Issues and Concerns

Modern plant breeding, whether classical or through genetic engineering, comes with issues of concern, particularly with regard to food crops. The question of whether breeding can have a negative effect on nutritional value is central in this respect. Although relatively little direct research in this area has been done, there are scientific indications that, by favouring certain aspects of a plant's development, other aspects may be retarded. Reductions in calcium, phosphorus, iron and ascorbic acid were also found. The study, conducted at the Biochemical Institute, University of Texas at Austin, concluded in summary : *"We suggest that any real declines are generally most easily explained by changes in cultivated varieties between 1950 and 1999, in which there may be trade-offs between yield and nutrient content*

The debate surrounding genetic modification of plants is huge, encompassing the ecological impact of genetically modified plants and the safety of genetically modified food.

Plant breeders' rights is also a major and controversial issue. Today, production of new varieties is dominated by commercial plant breeders, who seek to protect their work and collect royalties through national and international agreements based in intellectual property rights. The range of related issues is complex. In the simplest terms, critics of the increasingly restrictive regulations argue that, through a combination of technical and economic pressures, commercial breeders are reducing biodiversity and significantly constraining individuals (such as farmers) from developing and trading seed on a regional level. Efforts to strengthen breeders' rights, for example, by lengthening periods of variety protection, are ongoing

Plant Breeding For A World Community

There is enormous potential for plant breeding to benefit less developed parts of the world, where food security is critical as populations continue to increase.

In particular, advances in genetic technology may offer more versatile solutions than conventional systems of crop improvement. For example, scientists in Britain are pioneering important research to develop salt and drought tolerance in crop plants. This may help to alleviate the devastating effects of crop failure in arid regions such as sub-Saharan Africa. Private and public sector initiatives to share knowledge and expertise are already in place. They include commitments to provide Vitamin A enhanced rice and virus resistance in sweet potatoes free of charge to developing countries.

International arrangements to recognise genetic resources, as the sovereign property of individual nation states will stimulate further programmes of technology transfer and benefit sharing between industrialised and less developed regions of the world.

Effects of Modern Plant Breeding on Environment

Some environmentalists are concerned that genes from genetically modified crops could escape and transfer to other species with unwanted consequences. For example, it is argued that herbicide-resistant crops could cross with weedy relatives to create a new strain of 'super weed'. In practice, since the reproductive systems of genetically modified and conventionally bred crops are identical, the behaviour of domesticated crop plants is unlikely to be affected by single gene changes.

In ten years of worldwide field trials there have been no adverse reports of genetically modified crops spreading in the environment. Furthermore, resistance to weed killers has been available for decades in a number of conventionally bred varieties without causing any such problem. The development conventionally bred crops-nevertheless the technology is subject to much tighter controls. In the long history of plant breeding, the strict regulations applied to the development and use of genetically modified crops is unprecedented.

In both the laboratory and the field, each genetically modified crop must go through a rigourous process of monitoring and evaluation before it can reach the customer. Extensive and ongoing evaluation of genetically modified crops has shown that the technology presents no new food safety risks. Indeed, biotechnology will enable breeders to develop food crops with improved nutritional value and better keeping qualities. Enhanced disease and pest resistance will also reduce pesticide residues.

TECHNIQUE OF CROSS-BREEDING

Here is hardly any plant-breeding project which does not involve cross-breeding. In most cases it is the object to produce variability, in order to obtain the material for selection to work upon. The quickest way to obtain variability is cross-breeding.

We should distinguish between the work of the plant breeder and that of the scientific worker. It is true that the great majority of plant-breeding problems are tackled at the larger institutes by genetically trained specialists, and yet the distinction should be made. In scientific experiments only the most painstaking precautions are good enough. We must be absolutely certain about the paternity in our hybrid plants, and for this reason both the collection of the desired pollen and the elimination of the pollen from the mother-plant must be done with the minutest precautions.

If our object is a wholly practical one, we may often relax such precautions to a certain extent. For instance, in some plants castration—the elimination of stamens—is not only extremely difficult, but the manipulation brings a risk of seeing the flower wither and die. In such plants (soybeans, oats) castration can often be dispensed with, when we choose the parents in such a way that we can recognise the hybrids from inbred plants.

In every plant-family our first step must be to study the biology of the flower, in so far as this has not been done by others. We must know at what stage of growth of the flower-bud the anthers dehisce, so as to hit upon the safe time at which we must perform the operation of removing them. We must also know when the stigma is receptive, to find the moment at which to apply the pollen. In most plants castration and the application of pollen can be done at the same time. This is the case in rice, in most of the cereals, in tobacco, Ipomoea, cocoa, peas, beans and a host of other plants. In some plants with large flowers the stamens are attached to the corolla, and it is possible to cut off the corolla with the stamens in one operation. This can some- times be done with the fingernail (apple, raspberry, straw- berry) or with a scalpel.

When a scalpel is used, it is sometimes advantageous to cover the blade almost up to the tip with adhesive tape, to avoid cutting down too deep. Where this is impossible, the stamens have to be extracted one by one. A set of fine pincers with differently curved tips should be in the possession of every worker. Very useful fine instruments can often be made to fit particular occasions. A very fine hook, made out of a bent pin, and mounted into a needle-holder can sometimes be used to great advantage, especially in castrating such minute flowers as in alfalfa or clovers.

For fine work a watch-maker's eyeglass, or even a binocular dissecting glass, may come in useful. On the whole, dispensing with castration in plants with minute flowers, and to use the method of straight cross-breeding, leaving the finding of hybrids till next season.

In some plants male and female flowers are separate. In such plants no castration is necessary; all we have to do is to protect the flowers from unwanted pollen. In cucurbits, melons, squashes, cucumbers, we found it quite easy to close unopened flower-buds, by means of small pieces of copper, or preferably lead-wire. Of course both male and female buds should be so protected. Lead-wire has the advantage of being very soft, and it can be used many times in one season.

Especially in wind-pollinated plants, care should be taken to protect the pollen used against contamination with unknown pollen. For this purpose the flowers from which pollen must be collected should be bagged. In some plants pollen is so abundant that we can use a separate anther for every flower to be pollinated, and we can hold this with fine forceps. Where the pollen is dry, it can be shaken out of the anthers into some small receptacle, for which a small gelatine capsule is just the thing. Such capsules can be closed, and the pollen contained in them can often be kept in very good condition for days, if the labelled capsules are kept cool in a well-closed bottle over calcium.

In maise both the male and female inflorescence should be bagged early. When the silks protrude they can be trimmed off short, and the pollen shaken over them from a tassel. If it is wished to obtain a cob full of kernels, it is of some advantage to insert a flowering tassel into the bag that covers the cob, and we can prolong its life and usefulness by standing this tassel in a small bottle of water hooked to the stalk.

In certain plants the operation of castration and fertilization can be performed on cut-off stalks that have been taken into the laboratory. In sugar-cane, in millet and sorghum, and even in wheat and barley, such panicles and ears, when stood in water, will ripen their seeds well, especially if only a few flowers are left on them for this purpose. Paper bags should be of ample size, and it is necessary to leave them on no longer than is absolutely essential for full protection during the few days in which the stigma is receptive. In certain cases (coffee, cocoa) a finely woven tartan may have advantages over paper; it is softer, and can be arranged around awkwardly placed flowers better than paper. Of course it is useful in certain cases to give the ripening fruit or seeds some protection against loss or damage (stone-fruit, potato, tobacco, (Enothera), and paper bags may be necessary at that stage.

For keeping small quantities of seeds, small paper or preferably cellophane envelopes can be used. Gelatin druggist capsules, however, are preferable. The labels should be en- closed with the seeds.

Labelling, Numbering and Gadgets

Every plant-breeder and every geneticist has his special methods of keeping his records. To be able to do this he must have suitable methods of numbering or labelling his plants or seeds. Any system of labelling or marking must be legible and in- destructible by weather or insects. For merely marking, as an aid to tracing a special plant or part in a field, a coloured thread may be quite sufficient.

If such a thread is used in addition to a number or a label, it helps to distinguish that label. Paper labels, even if made from very good parchment paper, tend to become discoloured.

To depend on such labels only, let us say in marking pea-pods, is courting trouble, for they closely resemble dried leaves or empty pods. The well-known Vilmorin red woollen thread is of very great value; it may even serve the double purpose of protecting the cross-fertilized ear of wheat or barley and of making it easy to find it again. To mark seed capsules, fruits and such things, we found that lead-wire was much better than string. We can string a glass or china bead on a small piece of very soft lead-wire, and if we need a few dozen distinguishing marks, combinations of coloured beads will be excellent. We used this method in marking cross-fertilized squash blossoms and melon blossoms. If we make a loop of sufficient width, we can squeeze one end of this against the fruit-stalk, and it will accompany the fruit until harvesting time, or even beyond.

A very good label that will never lose its legibility is a poultry leg-band with a perforated number. These are made out of thin aluminium, and can be opened and adjusted before closing. Such labels are just the thing to mark plants that must be taken up and wintered in a cellar, like dahlias.

Most seedsmen use stout painted stakes for carrying the name and the number of rows and individual plants, the name facing the row it marks. To obviate mistakes it is safer to have duplicate stakes, or to have the name and number of each lot written on a separate stake. To save time and worry, good use can be made of poultry leg-bands or of the newer coloured plastic numbered wing-shields. These are numbered on both faces, and can be bought in consecutively numbered hundreds or thousands.

They can be hung from wire stakes with eyelets, and they accompany such plants as cereals or beans when they are bundled for harvesting. In some countries one has to guard against theft and mischief. A European blackbird pull out dozens of numbered yellow wooden labels and scatter them all over the place. A system of book-keeping which allows one to find the missing numbers from the book is indispensable if we want to avoid mistakes.

Numerous methods have been invented for protecting flowers and branches against insects or wind-borne pollen. Paper bags come first; they must be strong, rainproof and light. Some cotton-wool may come in very useful in making the connection of bag and stalk pollen-proof. Glassine bags and lengths of seamless sausage casing are being widely used in the United States. They allow one to see the inside of the bag, and so may save much time and error.

In some sympetalic plants flower-buds may be closed hermetically by the application of a length of very soft wire (lead- wire) wrapped round the bud, and used again after fertilization. We have used this method with squashes, melons and Ipomoea, The American plant-breeding journals and publications abound with methods for castrating and cross-breeding plants. Removing the pollen from the stigmas by washing it off or killing it by heat are methods but they sound excellent. Gelatin capsules, made in two pieces, of which one fits over the other, are handy receptacles for keeping very small quantities of seeds or of pollen, for slightly larger quantities sachets for internal use may be used; they are cheap, and can be written upon, but if we close them we can never use them again. Paper bags must be used a good deal in storing seeds and similar objects (bulbils, bulbs, corms). They can be closed in different ways. The simplest system is the method of the French seed firms, in which each bag is closed by three folds, in such a way that it will never open spontaneously, even if submitted to considerable shaking.

The gadgets used in castration and cross-pollination are very numerous, and must be adapted anew to every new plant. Of course, it is necessary to kill adherent pollen with alcohol, and extra scalpels, forceps or scissors are of great help. In working with very minute flowers a good binocular microscope is invaluable.

Avoid castrating minute flowers as much as possible. Everyone doing this sort of work occasionally regrets he was born with only two hands and without a prehensile tail. A ring to which has been soldered a minute cup that holds a small gelatin capsule with pollen, and that can be worn on the index-finger of one hand, is almost as useful. This is another American invention.

Hard seeds can be treated in different ways to make them grow. Some seeds, like those of canna or musa, can be scratched with a file, others may be rubbed on stone. Enothera seeds can be soaked in water; some seeds will only take up water rapidly when we exhaust the air above the water in which we keep them and then readmit the air. Seeds of apricots, cherries and plums can be extracted after carefully cracking the stone. The extracted embryos (and the same is true for embryos of apples, pears and citrus) must be protected against bacteria and mould. Thymol solutions, even weak ones, will do this without injuring the seedlings.

Speeding-up Plant-breeding Work

Plant-breeding work involving crosses generally takes a great number of generations before the superior new plant has been obtained. In the routine work of a large plant-breeding station so many breeding projects arc under way simultaneously that it matters little how much time elapses between the initial cross and the final purification of a promising commercial variety. Every year there are many projects being started, several are under way, and a few arc in their last stages.

It often pays to cut down the time required for work of this kind. One of the most promising means of doing this is to grow more than one generation of plants in one year. This usually means that one of those generations of plants is grown in a season in which the plant is not ordinarily grown, and that for this reason a correct evaluation of the quality of the individual plants is out of the question.

In Europe or in America growing an extra generation on the spot means growing plants in the greenhouse. To grow cereals or beans in winter in greenhouse conditions not only requires heat; we must also have the means of regulating the length of day. The simplest method is to use large electric lights to furnish both the heat and the light, to lengthen the day by those means, and to let the temperature go down during the dark part of the night. Under those conditions cereals and beans will flower and set seed very well in the winter months.

The electricity required per square yard is roughly that used by lamps that require a kilowatt. It is clear that under such conditions the winter generation will not show the normal qualities of the plants in that group which are grown in summer-time in the field, but there are many cases in which this is not of much importance.

If we want to proceed by the method of making some hybrids and then growing a second and third inbred generation from this stock, we can do our crossing in the field, grow our hybrids in the winter, and then grow another crop of second-generation plants in normal field conditions. When we are using the method of mating back hybrid stock to some pure kind of plants we want to improve it matters but little if we cannot very well select the best plants during some of the generations involved. It is evident that we must grow a generation under normal field conditions whenever we want to make our selections.

In recent years a few seed-firms have hit upon the perfect scheme of growing two generations of annual plants. This does not involve any greenhouse generations. They work in two different localities with very similar climate, but so chosen that one of those localities is situated in the Southern Hemisphere.

There are regions of Chili where Californian firms can find conditions of plant-growth that are identical with conditions at home, with the exception that the growing season comes during the Californian winter. In those circumstances it is quite possible to grow two absolutely normal generations during one year. Our present greatly improved methods of communication make such schemes perfectly feasible.

With many tropical plants one generation yearly is grown of plants that need special conditions of either a wet or a dry season. Here it is easy enough to interpolate an additional generation, by looking for a suitable spot, or by providing the necessary irrigation. Sometimes it takes a very long time to bring the plants to a condition where we can evaluate them. A good example is that of the fruit trees. It takes many years for an apple seedling or a seed-grown cherry to come into bearing. Here we can save a great deal of time by budding or top-grafting parts of the seedlings upon mature trees in full bearing. By this method we may save more than half the number of years normally required for bringing the seedlings to maturity.

In this connection we might also treat of those cases where certain seeds take a very long time to germinate. Storing the seed in suitable conditions of moisture, and especially of temperature, will speed-up germination in such plants as gooseberries and celery. In stone-fruits it has been found possible to crack open the stone, to extract the kernel and grow the seedlings under aseptic or even under antiseptic conditions (thymol solution).

Shows and Showing

The exhibiting of plants and seed at the shows is a very curious thing. It pays extremely well for a seed-firm to employ some gardeners who know just exactly when and' how to get enough plants full of luscious beans or covered with flowers on the exact day for each show. Florists and lovers of gardening see the quality of such plants at the shows, and many people who admire them there try in their turn to win the coveted prises at their local shows the next year. This show game is certainly very nice and pleasant; it gives the gardeners something to strive after; it may even help to keep some of the lads in the country.

On the other hand, there is very, very little connection between the qualities judged at the shows and actual value of the material shown. The few selected mangolds, or the mammoth pumpkins, or the bundle of wheat- ears, or even of the maise-ear contests of the American agricultural shows.

All along the line, in horticulture as well as in agriculture, the selected samples shown give no indication of the value of the seed or plant stock the producers have for sale. Of course we can see that those few mangolds have the ideal shape and size, but the seed-firm must give us some guarantee about the percentage of mangolds that will grow to this size and shape from the seed they sell, just as the owner of a prize-winning bull must give some guarantee of this animal's ability to give profitable daughters, in addition to the blue ribbon awarded to the beautifully coloured and shaped prize-winner.

In agriculture this is beginning to be so well realised that the classes for beautiful corn cobs or for the largest mangolds are becoming very rare indeed, even if we still see stock-judging in dairy cows and similar farces. In the horticultural shows much interest is still shown in regard to flowers and vegetables.

It seems evident that when we strive for beauty, beauty contests for roses and begonias, for asparagus and delphiniums and dahlias are indicated. The show is a shop-window. The seller must show if his competitors show, if only to keep his name before the public. To the public the show is only just a pleasant occasion for a half-day outing. To the buyer the show is much nicer to see than a description in a catalogue.

Long lists of novelties are bought at the shows by people with an experimental turn of mind. We see the representatives at the shows kept busy noting down long lists of orders. Showing certainly pays the seller. A very much more sensible system of showing a great many varieties is that of the demonstration gardens and experimental stations, both stations in which the agricultural authorities try to compare promising new and good old varieties with the aid of objective standards, and such gardens as show-plots arranged by rose-lovers or by societies of growers of gladioli or dahlias. Here the varieties can be judged throughout the growing season in favourable conditions.

The Annual Self-fertilizing Crops

The number of plant species now being cultivated by man is really enormous, while only relatively few animal species are kept under domestication. It is possible for the author of a book on animal breeding to take those species almost one by one, and to give the specialist breeders a few hints over and above those which are contained in the general book. This would be quite out of the question in a book on plant breeding. One way out of the difficulty would be to give one example of each of the three groups – self-fertilized plants, cross-fertilizers and plants that are usually propagated by vegetative means. These plants in two sections – grains and legumes – giving some hints to the breeders of wheat, barley, oats, sorghum, rice, etc., and going into some details about the breeding of beans, soybeans, cow-peas, ground-nuts, peas, etc., separately. But, after all, the self-fertilizing annuals have so much in common from the point of view of plant-breeding methods.

Self-fertilization causes isolation. Each plant, as far as reproduction is concerned, is as self-contained and isolated in a crowded field as it would be if grown by itself in a greenhouse in a city. The inbreeding caused by self-fertilization has several results. It makes all heritable variability disappear in every line, it makes every group consist of parallel lines, in which each plant descends from only one parent plant, so that the usual complicated network of ancestry is simplified into a system of isolated parallel lines. Where inherited variability exists, such as happens after a deliberate artificial cross or when accidental cross-fertilization occurs, it will automatically disappear.

That is to say, it will disappear in so far as each separate line is concerned, for Mendelian segregation will make the lines, descending from any hybrid heterozygous for a great many genes, diverse in their hereditary make-up. The result is that a field of self-fertilizing annuals that has not been grown from just one pure line by a plant breeder always consists of a mixture of a number of genetically different lines. Even if accidental cross-fertilization or irregularities in crossing-over (mutation) are very rare, they do occur, and cause this state of things. This genetic variability, however, is wholly different from what we find in other plants and animals. We are dealing with mixtures of pure lines, and almost every plant we take from any field is pure, homozygous for all its genes, and will, if its seeds are sown, give a very homogeneous descendance. From all this it follows that enormous progress can often be made in all self-fertilized plants by simply picking out a great number of promising-looking individual plants and comparing their descendance.

Many valuable widely grown wheats, tobaccos, peanuts and beans have been just found. They "just growed ", like Topsy. The first step in plant-breeding in the self-fertilizing plants must always be to look for those ready-made pure lines. The next is to find whether a mixture of two of the very best pure lines can be discovered which will give us still higher production (10 per cent, higher production from a mixture of two strains as compared with that of the best one is quite common).

We have then to produce some more genetic variability by deliberate cross-breeding. This generally means that we try to combine the good qualities of two different strains in one new one. Here there are two wholly different schools of thought, and two wholly different methods. One of them is to analyse second-generation plants, and to find those individuals that do combine, let us say, the disease resistance of line A with the large quantity of big seeds of line B. This is a method in- vented by the experimental geneticists who are trying to combine practical plant breeding with purely scientific gene- analysis. As a theoretical geneticist, interested in genes and their action.

It is great fun playing with a hundred different hybrid lots in one species, growing a few dozen second-generation plants in each lot and tabulating the results. But this is neither sound genetics nor sound plant breeding. We are dealing with so many unanalysed and unanalysable genes which all affect production and quality at the same time, that the chance of finding something really worth while in a few hundred plants is simply not good enough.

The other method is based upon the realisation that hundreds of genes are commonly involved, and upon the fact that purification—reduction of the potential variability—is rapid and automatic in this material. It consists of growing the greatest possible number of seeds from first-generation hybrid plants. We can then proceed in two different ways. One is an analytical method, which means that we are seeking for the very best-looking plants (for instance, by weighing their seeds), and progeny testing those plants by sowing a large number of seeds from each, in order to find the most profitable families. The other method is what Baur and Nilsson-Ehle called a "Ramsch" method.

This consists simply of growing the descendance of a hybrid in a vast mixture, refraining from any analysis, and continuing this hybrid lot as a miscellaneous collection for five or six generations, sowing as much of the mixed seed of each harvest as we have room for. If we do this, natural selection will weed out most of the unproductive and undesirable types, such as those plants t1 at ripen after harvesting time, all weaklings and all plants susceptible to diseases and pests. After five or six years that lot will consist of a mixture of pure lines, every line different from every other, with the successful lines represented by more individuals than the unsuccessful ones. After this we will treat the mixture just as if we were dealing with any other mixture, growing separate rows of beds each from one likely-looking plant.

This is a very sound method, which saves a great deal of un-necessary labour and time. It is extremely difficult to find just how much better this method is as compared with that of analysis in every generation, starting from the second generation. An analytical method could be made to work very well, provided we did all we could to make the number of plants of the second and third generations exceedingly large (in the cereals we can get almost as many seeds as in tobacco by splitting up the hybrid plant repeatedly before it starts to grow any stalks).

In actual practice there are only a few crosses that really produce anything worth while, and the general practice seems to be to make a great many attempts and grow a large number of first-generation hybrids. This renders it imperative to cut down on the number of plants per experiment. It is a vicious circle. This is all very wrong and wasteful. One cross every year and work with it, both analytically and by means of a "Ramsch ", doing the final selection after six years for one lot in every season, than waste my time castrating flowers and producing numerous hybrids that would have no really good chance of showing what they could do.

In all the self-fertilizing annuals we must be continually on the look-out for qualities that may be of special merit. Mostly of adaptations to special conditions. In wheaa on irrigated ground, resistance to drowning may be as important as disease resistance or quality of the grain. Conversely, rice could be grown much more extensively if we would take the trouble to do some plant-breeding work on dry-land rices, which abound in the mountain villages of all the tropical Asian islands. In ground-nuts we should look for lines that are specially adapted to mechanical harvesting, as well as for lines adapted to the production of a first-quality product in the hands of tens of thousands of small cultivators. In all the plants of this group it is possible not only to improve the yield and the quality, but also to extend the area in which profitable production is possible. Tobacco, rice, soybeans and cow-peas are just as important in this respect as wheat, barley and oats.

Plant breeding with self-fertilizing plants is comparatively easy. It is even possible for the real amateur plant breeder to do some extremely interesting and useful work with some of the less well-known plants of this group—let us say with beans or cow-peas. Such work needs room, but very little special technical knowledge or talent for higher arithmetic. Frequently a few plants will stand out from an otherwise quite even field of barley, beans or sorghum. Such "high birds "—climbing plants in bush beans, high, scmiawned plants in wheat—are often accidental hybrids. If we sow a few of their seeds, we can see immediately whether they are hybrids or admixtures, for hybrids will have very variable offspring. If then we sow all the seeds from such hybrids, we can start a series of experiments that may easily give us something worth while, even when the father of the hybrid is unknown.

HYBRIDING PLANTS

The majority of named clones of fruit-trees, potatoes, strawberries, have been obtained by cross-breeding. This means that somebody has crossed two different kinds of plants and has found the new valuable clone among the hybrids. When we do this cross-breeding work we almost always obtain an enormously variable group, and this shows that the plants used in the cross were heterozygous for a very large number of genes. This is not surprising if we remember that those parents themselves have also been obtained by cross-breeding.

In work with plants grown from seeds, obtaining a hybrid is regarded as only the first step in our attempts to produce something new and very good. We know that we can never expect to obtain the full assortment of recombination's of the genes that went into the cross before we have seen the second generation, grown from the hybrids. Why is it that the breeders work in such a different way when they are out to improve their roses, apples or poplars?

In some cases this must be due to the fact that it frequently takes a long time for the hybrid plants to come to maturity. Another cause for this difference in method is certainly found in the great variability in the first hybrid generation, due to heterozygosity of the material.

Whatever the reason for the fact, it is easy to see that the plant breeders who are continually making new hybrids, and then look no farther than the first-generation hybrids, must very often discard hybrid plants that have no value in them-selves, but that would be potential parents of excellent new plants of the second generation. Even if many of the valuable qualities in this material are dominant, it stands to reason that some of the most valuable characters in every group are recessive; and we have not a very good chance of obtaining a full sample of the potential variability that results from a certain cross if we look no farther than the first hybrid generation. We might even say, without grossly exaggerating, that most named clones in such plants are the result of a cross that has been made with a definite object which has not been attained by the breeder. When we cross two different plants with the object of combining some good quality of one of them with those of the second plant, our work is only half done when we have obtained the hybrids.

To a certain extent we might say that the qualities of the first-generation hybrids are not very important—we must simply regard them as the parents of the variable second generation. Now, in certain groups of plants self-incompatibility is the rule, so that it is impossible to obtain a self-fertilized second generation. In that case we need more than one first-generation hybrid to produce a second generation. But this self-sterility is the exception rather than the rule.

When in plants of this group we sow a great many seeds obtained by self-fertilization, or, by mating together hybrids from the same cross, it is quite possible that we actually realise the hope which set some other breeder working to obtain those hybrids. Very often we find that some roses, ornamental shrubs or potatoes are setting seed spontaneously. It is frequently asked whether sowing this seed gives the breeder any chance of obtaining something worth while.

To answer this question we must remember that sowing seed with the object of obtaining a plant with good qualities—a valuable novelty—is always something of a lottery. It is certainly not true that our chance of drawing a winner is made very much better when the seed sown is obtained by crossing two different sorts, unless again we follow through the cross by growing an extensive second generation. As breeders generally are not in the habit of doing this, almost all the commercial clones have been obtained as hybrids, so that it will very often be worth our while to follow up their descendents. If a plant breeder, working with vegetatively reproducing plants, has the habit of sowing only cross-bred seeds, he is certain to ascribe his success to this crossing. But we must realise that in so far as valuable new recessive qualities are concerned, we only find those among the hybrids because both parents were heterozygous. Groups in which cross-breeding has always been the rule— apples, lilac, roses, potatoes—sowing inbred seeds on an extensive scale is a method very much worth while.

In some plants, as in the potato, many valuable clones have been obtained from sowing "spontaneously" formed (and presumably inbred) seeds. The method can be especially recommended in those cases where, for some technical reason, such as minute size of the flowers, difficulties in castration, etc., it is extremely difficult to obtain cross-bred seeds. Sugar-cane is a case in point.

F1- Hybrids

In the autogamous plants where genetic purity, once obtained, is easily conserved, it is worth while concentrating our efforts to produce pure-breeding superior lines. But in other cases this may be so difficult that other working methods are preferable. One of those methods is to produce first-generation hybrids for direct use. In such plants as maise or beets we often find that groups produced by crossing two distinct lines are superior in some way—for instance in yield—to the lines used in the cross. We must begin by giving an explanation of this phenomenon. If we could make a complete genetic analysis of a variable family of plants, we should see that in respect to a very large number of genes the group was pure.

All the individuals of such a group are alike in respect to a great number of genes, for which every individual is homozygous. On the other hand, there are other genes in regard to which such a line is impure, and when we would inbreed in this line, a number of offspring would be produced, that would differ in those genes from the majority. In a variable lot we should find that to a certain extent plants will differ in the set of genes for which they are either pure or impure. The same is true when we are com-paring separate lines.

When we make a cross between two different plants, and especially when we cross two plants of different lines, the result will be that the group of hybrids will be much more uniform than the descendants of either line. The explanation is very simple. The hybrids will inherit all the genes that are common to both parents. In addition to this, they will inherit all the genes for which only one of the parents was homozygous. If one parent had the composition AaBB and the other was AABb, the hybrids will all inherit both A and B.

The result will be that the hybrids are more likely to have valuable dominant qualities than inbred offspring of either line. Where dominance exists, however, and where there is no appreciable difference between homozygous and heterozygous plants (AA and Aa), individual selection will not quickly bring about purity. In other words, it is likely that good individual plants are still heterozygous for valuable genes, whose presence is desired for good quality.

We can understand why, at least in certain combinations of lines, first-generation hybrid lots of plants are superior in quality to and more uniformly good than, pure-bred lots from the same material. In other words, we could say that the homozygosity of one plant for certain genes will cover up the impurity of another plant in respect to those same genes. We can now understand why hybrid lots are frequently better than inbred lots, and why we find this superiority of the hybrids especially in certain combinations of individuals and of strains. The simplest instances are those in which we combine two pure inbred lines, or even two clones, to produce the commercial seed. Some of the azalea specialists have found combinations of species that, when crossed, give hybrid seed from which beautiful large-flowered hybrid plants can be grown.

A second example is that of the Japanese hybrid egg-plants, and that of hybrid tomato seed. In Java a few tobacco estates have frequently used hybrid seed produced from a cross between quite different lines.

This practice of using first- generation hybrids is commercially possible only if producing such seed is not excessively difficult or expensive. In the cases enumerated, each fertilization gives us a very large number of seeds. The best-known example of hybrid seed for the commercial production of first-generation hybrid plants is that of maise. Here it will suffice to say that it was found with maise that very often the first-generation hybrid plants produced by crossing different breeds gave a very high yield. The same was found true for hybrids between different inbred lines in one variety. Starting from small-scale genetical experiments, a very important hybrid-corn seed industry has grown up, first in the U.S.A. and later in other maise-growing countries.

In some cultivated plants (but not in others) inbreeding will have an adverse effect on vigour; and, conversely, crossing between different lines of plants will often give us remarkably vigourous hybrid offspring. Both from a purely scientific standpoint and from a practical one, this subject is of the greatest interest. The facts have been well known for a long time. Darwin made numerous inbreeding experiments, and found that some plants were much more subject to degeneration as a result of it than others.

Many different theories have been proposed to explain both the degeneration due to inbreeding and the great vigour due to crossing. The pre-Mendelian explanations do not concern us much. One of the chief difficulties in the way of a simple genetic explanation has certainly been the old conception of genes as determinants, each calling forth a special discrete unit character.

So long as most geneticists had this conception of genes and of their action, the mere sampling effect of Mendelian segregation of genes and the addition of genes by cross- breeding were thought to be insufficient to account either for the degeneration caused by inbreeding or for the very great special vigour of cross-bred individuals. It is necessary to go into the different theories that have been proposed and have then been abandoned. The hypothesis that, to my mind, best explains all the facts.

If we look for evidence of loss of vigour by inbreeding, we are first struck by the fact that certain classes of plants are apparently exceptions to the general rule. In the habitually self-fertilized plants, such as tobacco, wheat, barley, and garden beans, the strictest kind of inbreeding—self-fertilization— does not lower the vigour or the yield.

On the other hand, in all the allogamous plants that have been tried—in beets, in maise, in rye—inbreeding brings along a rapid decline in almost all the desirable qualities; and in the same plants, cross-breeding such deteriorated stocks immediately restores full vigour. Now, if we start cross-breeding in the habitually self-fertilizing plants, we often find that the first-generation hybrids are extremely vigourous and productive, and if this happens we notice a very large general reduction in vigour in subsequent inbred generations. This shows that we must not look for a mysterious special immunity to degeneration in some plants, but that resistance to inbreeding degeneration simply spells homozygosity.

In full accord with this is the circumstance that in animals the facts run wholly parallel to what we find in plants. Those groups in which cross-breeding has been the rule show a decline in quality after inbreeding, and those groups that have been regularly inbred—such as fancy pigeons—are inbreeding- resistant. Again, just as in plants, it has been possible both with chickens and with ducks to produce great purity by a system of inbreeding, while we retain full fertility and vigour as a result of strict selection for those very points.

Of course, the examples of the very fertile and vigourous highly inbred wheats and fowls show that the vigour we see in hybrid stock is not in any way due to the fact that the hybrids are heterozygous in respect to a great many genes, but only to the fact that they possess a favourable combination of genes. In theory it might be possible, by careful work, to make a group of plants (or animals) that would be homozygous in respect to that ideal combination of co-operating genes. As a matter of fact, that the circum-stance that we do not attain this desirable end is most often due to faulty principles of selection. Essential mistakes are frequently made that render it almost impossible to attain the much-desired results. To understand the nature of inbreeding-degeneration we must remember that dominant good qualities, as far as the effects of selection are concerned, are in a class apart from the recessive virtues. If in a variable group some individuals have some unwanted dominant quality, it is quite simple to eliminate that dominant fault from the material. If we prefer white maise to yellow, it is quite simple to produce a pure-breeding, white- seeded maise. But the reverse—to produce a pure-breeding, yellow-seeded maise from the same material—is much less simple, as some of the yellow-seeded plants chosen may again be heterozygous.

What is true for one monofactorial difference is true for all of them. If we subject a group of plants to individual selection, or even to some sort of selection by means of progeny tests (*e.g.*, ear to row in maise), and we do not have recourse to inbreeding in order to purify the genotype, the result will be that the superior strains of beets or maise we have built up in this way will be chockful of individuals heterozygous for very important genetic factors. As soon as we start inbreeding, we reduce the potential variability of such material. In every succeeding generation from self-fertilization the number of genes for which the material was still heterozygous is halved. In every generation the plants become purer and purer.

Let us take the case of just one gene, A. If we inbreed a heterozygote, Aa, 25 per cent, of its descendants will be AA9 25 per cent, will be aa and 50 per cent, will again be heterozygous, Aa. Now, if A is an important gene, the AA and Aa individuals will be somewhat better than the aa's. If we were concerned with this gene only, we could, by merely eliminating the aa plants, work towards better plants, and the proportion of pure-breeding AA plants would improve. But now we must remember that we are concerned with a multitude of genes, for all of which our original material may have been heterozygous.

If we were dealing with only three separate genes, A, B and C, we might retain only the 75 per cent. A plants (AA + Aa) and of these only the three-fourth B plants (AB -f Bb), and of these only the plants with C (again three-fourths). But we must remember that the fraction three-fourth power n becomes rapidly smaller and smaller with an increase of n (the number of desirable genes for which the stock is impure). For this reason even strict selection for vigour while inbreeding will not prevent a loss of desirable genes from the material in an inbred series. And, of course, blind inbreeding, without any selection, will very rapidly bring down the proportion of plants with dominant desirable qualities, when the material was heterozygous for many genes that helped to build up those desirable dominants.

The mutual independence of genes scattered over the chromosomes of our plants will make it a matter of chance for which desirable genes each inbred line will become homozygous, and which ones will be dropped from the line. It is certain that if we start a number of parallel inbred lines—let us say in beets or in maise—by starting every generation from just one single plant, those lines will become homozygous each for its own genotype. But the genotype of every line will become different from that of every other line. If we are lucky, and are running a good many lines, we may in the end still possess all the valuable genes that are responsible for the good qualities of superior plants. This does not mean that in one line we may have fixed the plants in respect to one desirable dominant virtue, and that in another we may find another separate good quality.

For we must realise that such good qualities do not depend each on a separate gene, but rather on the correct combination of genes. To a certain extent it is possible that we shall find good qualities in some of our lines—disease resistance in one, good quality of the product in another one. But if the foundation stock was highly cross-bred we shall seldom find highly inbred lines of sufficient all-round quality to make it worth while growing them commercially.

What happens when we cross-breed such inbred lines? As a general rule the vigour and all-round quality of hybrids between two such lines are very much better than those of the lines themselves. Sometimes, but not always, such hybrids are as vigourous and as good as the original material before we started our inbreeding.

The genetic explanation of this can only be that this is a result of the genetic difference between those lines. The loss of quality by an actual loss of genes partly responsible for desirable dominant qualities. Especially in blind inbreeding, chance determines which of such genes will be eliminated in a given line, and which will be consolidated in the line (for which the line would become pure).

If we think of the good quality and good vigour of the original strain as partly caused by the right combination of a great many genes, and of the inferior quality of inbred lines as resulting from incomplete sets of such genes, we can see that the hybrids, which inherit a combination of the genes remaining in inbred strain X with those for which strain Y happens to be homozygous, will be superior to both those strains.

In this superior quality of the hybrid generation between inbred lines we are concerned with the genotype, with the combination of useful genes present in the plants, and our present-day explanation is that it is this good combination of genes only which confers good quality upon those plants, and that the fact that they are heterozygous for many of them does not help. It would be very desirable if we could devise means for making a strain of plants that would be homozygous for all those important genes.

Hybrids of Maise and Beets

An Example of Utilising "Hybrid Vigour" from a small beginning in the experimental plots of the geneticists (G. Shull, East and Jones), the work with hybrid vigour in maise has grown to enormous proportions. The greater part of the corn harvest in the U.S. A. is now produced from hybrid seed, and the yield per acre has been much improved thereby.

In the older well-known maise varieties that formerly were always grown from open fertilized plants the potential variability was very great. Many plants are heterozygous for very many different genes, and the best vigour and the best yield are found in those plants which happen to possess the most complete combination of those genes.

If we start inbreeding in series of lines, so that in every line we take one plant and force it to pollinate itself, we quickly reduce the potential variability in each of those lines. The plants will be homozygous for many genes for which the original mother-plant was still impure, and, on the other hand, we shall lose some of the useful dominant qualities. The general quality and vigour of the inbred lines will decline, especially so if we use blind inbreeding without selection in continuing the lines. If, after some generations of inbreeding, we now cross those inbred lines, we shall find that, in some combinations more than in others, the resulting hybrids will be of very good quality. We have explained this by the hypothesis that those hybrids again have the full set of valuable genes. Now, the question may be asked why it should be advantageous to go to all this trouble of purifying those separate lines in order to reconstitute the original set of genes by cross-breeding them. Why should the hybrids between a good combination of those lines be more valuable than plants from the original non-inbred strain? There are at least two reasons for this. In the original lot—in the strain, before it was inbred—the potential variability was very high.

A large proportion of the plants—probably all of them—were heterozygous for several of the genes that help to make for quality. The germ-cells produced in the field were of the most diverse kinds, and, as a result, a certain percentage of the plants of the next generation always had undesirable recessive traits. Maise and sugar- beets are full of the most diverse undesirable recessive characters, that crop up in every generation. The inbred strains may not have a full complement of the desirable genes, but in any case they are homozygous for those they do carry. If, for this reason, we are able to find a set of two inbred lines that "click" because together they happen to have a full complement of the desirable genes, the result will be that the hybrids grown from this combination will all carry the whole set of those genes. For this reason the hybrid lot will be much less variable than the original open-fertilized variety. In other words, the method of establishing many different inbred lines is a method of cutting down the heterozygosity in the material. For practical reasons, this reduction of the variability is not attempted in one single series.

The breeders know that it would be impracticable with our present methods to produce just one inbred strain, homozygous for all the desirable genes, so they do the work in parallel series, fixing the genotype of several lines by inbreeding, and hoping that after several generations of this they will discover a few inbred lines whose combination will give them the desired hybrid seed that will produce vigourous high-yielding plants.

It is obvious that the quality of such first-generation hybrids depends on the combination of genes present, but that the plants themselves are again heterozygous for all those genes. Those hybrid plants are meant to be harvested for their pro- duce—seed from such hybrid lots would give as variable a supply of plants as the original strain, or almost so. From the point of view of the seed firms this circumstance is rather happy; the users of the seed are satisfied, but they have to come back next year for a fresh lot of seed.

The fact that most inbred strains of maise and beets are of very low vigour and production makes the production of such hybrid seed expensive. In many maise-strains the cobs are miserable nubbins, the seeds are small and irregular in shape and size, and poorly adapted to the use of sowing by machinery. There are several ways out of this difficulty. Obviously the best way is to produce at least one line that will be a fairly good producer of seed.

The methods of blind inbreeding of the first years have been improved, and it is usual nowadays to combine inbreeding with selection. Sometimes this means straight selection in the inbred line, sometimes the selection is used to fit the inbred strain for its ultimate use—cross-breeding. We can do this by testing the pollen of several plants in cross-breeding (top-crossing) tests, and so choose the sub-line which is retaining most of the valuable genes. Another method which is often used is that of the triple cross, in which the hybrid plants that furnish a lot of good-looking and very even seed are fertilized by pollen of yet a third line, that has been carefully chosen for this very purpose. If the ultimate product must be very uniform—such as in sweet-corn that is grown for canning, and that must be pre-pared by machinery – the use of one single cross between two inbred lines is indicated. The use of hybrid plants as parents of the commercial seed gives too variable a crop. In the ordinary use of maise some variability can be permitted, provided the average yield is good. In certain circumstances inbred strains can be produced that are valuable for "top-crossing".

Such strains must be pure for a great many important genes, and they must be good pollen producers. The usual practice in producing hybrid seed for sale is to grow the mother-plants in a field, and grow the pollen parents throughout that field in every third or fourth row. The mother-plants are then carefully detasselled, and the field produces hybrid seed from these detasselled plants and inbred seed from the pollen-bearers. Nowadays much of the commercial maise seed is grown by the system of "double-crosses". In this system four different inbreds are used, and to grow the commercial seed, hybrids A × B are crossed with hybrids of two other strains, C × D. This, of course, makes the production of commercial seed much cheaper. It is doubtful if from the standpoint of the seed-user this system is as good as the triple cross. We must realise that the hybrid plants of the first cross are heterozygous in respect to all the genes in which the parents differ. A uniform descendance from such plants can be expected only when the pollen used is from plants homozygous for almost all those genes; and we cannot expect that the pollen of another hybrid is in that condition.

It is certain that even commercial double-cross seed, when well-chosen inbreds are used, gives a better harvest than ordinary mass-selected stock of the same variety. But the results are never as good (and could not be expected to be as good) as those obtained from single-crossed seed. A certain extent smart advertising plays a role in this subject of the hybrid-corn seed production.

In the catalogues of the seed firms we often see pictures of "typical" ears of inbred strains, of the first hybrids, and of the "double-cross ". The enormous improvement in the first hybrids over the inbred strains, but they always show the largest and finest cob as the result of the double cross. Such pictures must be taken with a grain of salt !

If we look at the production of seed maise from the commercial angle, we must realise that the process of inbreeding those sub-strains costs money. During the first generations many plants have to be enclosed in paper bags, and to be pollinated by hand.

But, on the other hand, when the difficult and painstaking labour of purifying and testing the inbreds is done, those inbred strains can be further multiplied, without any manipulation, in fields, sufficiently far removed from other fields of corn. In fact, once we have found the two suitable inbred strains that "click", or the three necessary for a triple cross, we can reduce our selection to the very minimum, for once such strains are homozygous for the important genes, they remain in that condition, and all we have to do is to make the final cross to produce the seed-crop for sale.

The farmers, especially in the U.S.A., have become "hybrid- corn conscious". They have experienced on their own farm how much more vigourous are the crops that can be obtained from the hybrid seed. The idea of the superiority of cross-bred seed has taken firm root, the name "double-cross" has more advertising value than the facts warrant. The only yields ever seen in which "double-cross" plants gave a (slightly) better harvest than their single-cross parents were taken from a cross-breeding experiment in which four different varieties were used.

With beets, especially sugar-beet, the commercial houses use a system of seed production that has great similarities to that of the hybrid maise. They generally use a system of mild inbreeding in isolated lines. Inbreeding is much less easy here, and whole plants must be enclosed in very well-constructed and ventilated cages. Next they try combinations of their selected strains, and finally they grow their commercial seed from a field in which two strains are planted in a mixture. Until recently there has been no unfailing method of producing 100 per cent, hybrid seed, as a certain percentage of the seed must come from self-fertilization or from the inter-crossing between different plants of the same line. In practice, however, we find that the vast majority of the beets in a commercial field are first-generation hybrids.

If we deliberately cross two strains of beets, in which we can easily recognise the hybrids—by their leaf-colour or by their shape—we find that, with very few exceptions, the field contains only hybrid plants. The cause of this must be sought in the great vigour of the first-generation hybrids. In sowing beets many more seeds are sown than will ultimately grow up into mature beets.

Soon after the seedlings are up, the field is thinned, the supernumerary plantlets are hoed out. As this is done by hand, the man who does the thinning always favours the strongest, most vigourous plantlets. Even if we did not consciously select the favourable combinations between lines to be combined for seed-growing, the percentage of mature beets that have been grown from cross-fertilized seed is always very high.

Further Improvement in Maise and Beets

We saw that inbreeding will reduce the potential variability of the families subjected to it, that those lines will become homozygous for some of the valuable genes and that they will lose some of the others. One question that is of the greatest theoretical, but especially practical interest, is what we could do to influence this process of purification in such a way that each line would become pure for the maximum number of useful genes.

For discussing this, we must again look at this process of automatic reduction of the variability during inbreeding. We start with a plant of maise or of sugar-beet that is heterozygous for a great number of valuable genes—genes that we would like to have in every plant. The ordinary selection work ensures that such genes remain present in the stock, even if ordinary mass selection does not quickly ensure purity in respect of those genes. It does not matter whether we know or whether we do not know all such genes separately in their action upon the end result. But we do know that a plant heterozygous for a gene A (Aa)y on being self-fertilized, will give us 25 per cent, homozygous A A plants, 50 per cent Aa's and 25 per cent aa's (which last will tend to be discarded).

The chance that we will hit upon a plant homozygous for one of those genes is one in four; if we are dealing with five genes, we have only one chance in eight hundred of finding a descendant homozygous for all of them – in fact, the chance is one in 4" (n being the number of genes involved) that we find a plant homozygous for all the genes concerned. In a limited number of inbred plants all we can hope is that we conserve some of those useful genes in the heterozygotes.

In the work with hybrids between lines we help ourselves by continuing a number of parallel inbred lines, hoping that in the end we shall have two such lines which between them carry the whole series of important genes in homozygous condition. It is evident that (theoretically) we should be able to produce inbred lines of maise or sugar-beets that were as productive as some of the best strains we now possess, and that would be homozygous and breed true.

In fact, if the number of inbred plants from a very good parent plant were only large enough, we should be able to find such a plant homozygous for all those genes for which the mother was still impure, in the very fast generation. In practice we can never know whether we would have to sow two thousand or twenty thousand plants to find that ideal one. But one thing is certain : that if we sow a few dozen plants from self-fertilized seed we shall undoubtedly have no chance whatever ! The simplest genetic calculations show that in this selection work in such plants as maise and beets the usual number of individual plants grown per generation is far and away too small ! Not only should we combine selection with inbreeding, we should see to it that the methods used gave us an adequate chance of finding what we want to select.

Of course a great difficulty lies in the fact that it is practically impossible to appraise the genetic quality of one plant by inspecting its individual quality. This difficulty is, of course, greater in beets than in maise. Nevertheless, it would pay well to start inbreeding series on a very wide front, sowing all the seeds from an excellent plant, and testing a number of very excellent daughter-plants by sowing their inbred progeny. The circumstance that inbred series almost always deteriorate rapidly in vigour and yield does not prove a thing, so long as the numbers we work with are so inadequate. A technical difficulty, of course, is that in maise it is not possible to pick out the individually best plants before they have ripened their seeds. When we must pollinate them we do not know their quality. Beets give much less trouble in this respect.

A very great difficulty with which plant breeders have to contend in the allogamous plants, lies in the large number of genes in respect to which our material is variable. Inbreeding will decrease this potential variability, but, as we have already seen, it is extremely difficult to direct the purification in such a way that the inbred lines will be of good commercial value. There is one way out of this difficulty. We know that the inbred lines are highly homozygous.

If they are not good enough as they are, we know nevertheless that such lines are homozygous for many valuable genes. It is well worth while trying to improve such pure inbred lines, and this is being widely tried in America to-day. We can do this in two ways : by crossing to another line, followed by another series of inbreeding, or by the use of repeated back-crosses. A cross with another suitable line will give us the genes lacking in our inbred lines. But the hybrids will be highly impure, and we must try to reduce this impurity while profiting by the homozygosity of the inbred line. This we can best do by a combination of repeated back-crosses to that line, and selection for vigour and yield.

In the existence of good inbred lines in maise, that can be propagated separately, we have an excellent and sure method of producing first-class commercial seed for sale. In the beets we are far from this. But in the beet there are certain possibilities that should be worth trying out. In the first place, there is male sterility. Certain families of beets have been found that have no functioning pollen. This femaleness is inherited in a very peculiar way; it follows purely maternal lines, and it seems to be due to a peculiarity of the cytoplasm (protoplasm outside the nucleus). By cross-breeding it can be brought into any strain of beets. This gives us a possibility of producing on those plants seed that is guaranteed cross-bred, by having such plants flower in a field with other plants.

The other possibility in beets is the use of vegetative reproduction. Let us suppose that we possess two plants of sugar-beet that happen to fit together so well that the hybrid seeds produced from their combination give us beets of high purity and sugar content and with an excellent yield. Now our difficulty would be to produce two different groups of beets that would have the same constitution as those two individual beets. This would be extremely difficult and tedious work. On the other hand, if we had a very good method of rapid vegetative reproduction, all we would have to do would be to grow tens of thousands of beets of clone X and as many of clone T, and then to make them flower in a mixed field in the ordinary way. There exist a few methods of vegetative reproduction in beets. Some of these methods, involving making the beets produce shoots and reproducing new beets from cuttings from those shoots, work fairly well.

If the methods can be perfected, they could lead to an avoidance of the very costly methods of selection now in use.

The Choice of Plant-breeding Material

The advantage of having at one's disposal a large collection of plant varieties, both plants of commercial importance and imported plants of no immediate practical importance, is enormous. A difficult question for the plant breeder is always what crosses to make, what plants to combine with the greatest hope of ultimate success.

In the breeding of such plants as potatoes, raspberries and roses, certain clones have the reputation of being profitable parents of hybrid lots. The Early Rose potato, the Lloyd George raspberry, the Chilean strawberry. In some special instances the choice is obvious. When our present assortment of flaxes or potatoes or roses happens to be deficient in some such quality as disease resistance or winter-hardiness, we are on the look-out for plants that will show those desirable qualities to the greatest extent. In other words, we shall, when choosing the parents of our future hybrids, try to find material that promises the possibility of a combination of the very good points of both.

This is obvious. Often crosses that seemed very little promising have given excellent results. It would seem that here no definite rule could be indicated. It is always worth while trying crosses with new material, with newly imported species – in fact with material which has not been used previously. This is especially so when, as in horti-culture, everybody is on the look-out for something new. Another rule that could be given would be that when we make very wide crosses we cannot hope to find commercially valuable strains in the second hybrid generation, unless we grow extremely large numbers of plants in that generation. When we make a very wide cross to incorporate some special quality into the existent stocks, the best thing we can do is to use the hybrids in back-cross series.

We can then strive to retain just one special quality of the new species and incorporate it into the general set of genes with which we are familiar in our older stock. In several instances very wide crosses have been successfully used, and sometimes plants have been used in cross-breeding that in themselves showed very little promise. The inedible cattle gourd that Orton crossed with his watermelons, of some of the potato crosses with Andean wild species (Salaman) and of the back-crosses to hybrids with wild cane from which all our modern resistant varieties of sugar-cane have been derived.

It is here – in the choosing of plant material for breeding work – that cytological studies and chromosomal counts are of the greatest practical value.

Methods of Evaluation and Comparison

The aim of every plant breeder is to produce a group that will be worth a place in the assortment of cultivated plants. In some cases the new plant will have only ornamental value, in others it will, for any of a dozen reasons, be better adapted to agriculture than other similar ones. It is easy enough in all classes of plants to produce novelties, to produce variability. The experimental grounds of the plant breeder would soon be cluttered up with novelties if he did not keep on weeding out the comparatively worthless. After producing our novel plants, we must judge them. Some qualities are much easier to judge than others. It is comparatively easy to weed out a lot of seedling dahlias and to pick out a few that may be worth keeping on for further testing. In the selection of our agricultural plants, in cereals, in beets, in maise, the assortment of valuable kinds is already so great that only relatively few of the newer selections are worth keeping. In the vegetatively propagated plants this work of testing novelties is comparatively easy.

In the first place, we are not bothered by considerations of purity – once we have produced a very good specimen, we can just multiply it with all its properties intact. In the second place, we can easily, for purposes of testing and comparison, produce as many individual plants of the novelty as we require. When the differences are relatively small—as in cereals and beets, where the novel plant produced must compete with excellent old stock that has been made by a process of selection and comparison during centuries—a correct system of evaluation becomes of very great importance. What we really want to test is the inherited composition of the material.

This may be difficult, as the final quality of the plants to a large extent depends on favourable or unfavourable conditions under which they grew up. It will perhaps be as well to begin with the simplest problems. In the production of novelties in such flowers as roses or dahlias the value of a new variety depends entirely on the appeal it makes to the gardening public. The man who is proud to have found a novel dahlia seedling or a new and strikingly beautiful rose sends it to a show and hopes that it will take the judge's eye and win him the much-"prised ribbon or certificate. But of course the insider knows that many highly commended novelties will never be heard of again, or that they will drop out of cultivation in a few years. There is a difference between the fact that the grower can bring half a dozen perfect blooms to the show on time, and the question whether that same rose or chrysanthemum will ever be popular in the gardens, or even be worth while growing for the commercial producer of cut flowers. A dahlia that wins the medal may be a very weak grower, or it may be very susceptible to rain.

The strikingly beautiful rose may be very susceptible to mildew, or its flowering season may be extremely short. If we visit the growers of greenhouse roses, or the gardens of the man who produces cut flowers for the market, we find that the number of varieties that have real commercial value—varieties "with bread in them ", as our gardeners express it—is extremely small. We shall probably find only three or four of the relatively older dahlias worth growing for cut flowers, and it is seldom that a novelty is added to that list.

In testing out agricultural plants, both during the process of selection and after they have finally been introduced, many methods are in use. The direct, simple examination of a number of small plots or rows will show us the qualities of those lots, but this method will only serve to find the groups with definitely inferior quality. It is not good enough to find exceptional merit, as circumstances may have favoured those qualities we appreciate. The only way to avoid reaching the wrong conclusions is to see to it that environmental conditions do not vitiate our results. The best way to do this is to compare the quality of a number of plants with that of other plants that were grown in wholly comparable conditions.

One of the best ways to do this is to grow the plots of our plant between similar plots of an older variety, that can be used as a standard of comparison. The choice of this standard is of great importance, and we will revert to this subject when treating of the breeding of special plant groups. As an example we can here take the evaluation of a newly produced (or of an imported) potato clone, or pure line of wheat or barley. In a case like this the standard may be some line or clone that is of commercial importance, against which the tested plant will be called upon to compete. If there is sufficient material, it is wise to repeat the comparison—that is to say, we should grow the test plot between standard plots in different parts of the field, or preferably in different fields. To equalise accidental conditions in small parts of the field, oblong plots are generally better than square ones. If we are testing a number of experimental lots, we can grow them side by side, and interpolate standard plots to be used for purposes of comparison.

The experimental garden will need an efficient system of book-keeping. We must note down anything of value about the experimental and the standard plots, date of maturity, disease resistance, quality, yield. We can calculate average yield for the sum of the plots of one kind, and then compare those figures with the average computed yield of the standard plots or with the average yield of the whole field; and a comparison of the deviation of the yields in the different plots with calculated mean errors will give us valuable data about the question in how far the superiority or inferiority of each separate number is significant as an indication of merit.

The practice of plant breeding in different parts of the world. A multitude of miserably small checkerboard fields being weighed at harvest time to produce the figures for an office staff to juggle with. A successful establishments where the field-work was under the supervision of men who were thoroughly conversant with the material, and where excellent results were obtained with a minimum of calculating machine work. A method of annotating the results obtained in experimental plots which gives the plant breeder an excellent insight into the merits of his new material is that of graphical registration. This method can be adapted to the most diverse projects; it is very simple, and gives us our results at a glance. The method consists of registering the quality (yield, height, resistance) of each plot, both standard plots and testing plots, on ruled paper.

If we do this, the line that can be drawn through the points of the standard plots will probably rise in one part of the field, and descend in some other part of it, and by drawing a line that joins those points we obtain a basis upon which we can found our judgement of the results obtained in the experimental plots. If we want to test our experimental material in two or three different qualities, we must construct separate graphs for each of those qualities, or sometimes we can use the same graph, employing different colours for separate qualities (weight, sugar percentage, purity in beets, etc.).

Reading those graphs will now show us the quality of our plant plots, in comparison with the standards, and we can use this relative value as a basis for our selection work. If we compare half a dozen groups in one experiment, and we find that one or two of them will in every duplication remain well below the standard in quality, this may be a very good reason for discarding them. On the other hand, if we find one or two numbers that all through the experiment are well above the standard line in quality, we have a very good reason for preferring them. It may be that in the case of one or two numbers the results are very variable, so that in one part of the field the number is above and in other parts it is below the standard in quality. Then we may well retain this number for another trial in the next season.

The influence of one row of plants on those growing close to it may be considerable, and we must guard against this. If we can grow three rows and use only the middle one for our records, we take a step in the right direction. If we have enough seed to plant six rows, and we examine only the two middle ones, discarding two rows on each side, this is still better. The greatest difficulty arises in those cases where we are dealing with very few plants of each number; or, worse, where, as in the second generation of a cross, every plant is probably different from every other. It then becomes very important to equalise conditions for each plant as far as this is possible. A very good plan is to grow the plants in rows, and alternate those experimental rows with rows of some pure standard variety. Each plant may then have two diverse neighbours in its own row, but it will touch four other plants that will be of the same kind as four out of six neighbour plants in the case of every other plant in the experiment.

This means that to save time we must make a preliminary choice out of the many numbers in those early generations, and it is very important that we shall make the cultural conditions as nearly equal as possible for every plant. This is the place to warn against using apparent correlative qualities in judging the value of our experimental plants. In the breeding of animals even to-day we often see that almost as much attention is paid to correlative qualities as to real merit.

Some of this is certainly due to the fact that in some material real merit can only be judged in one of the two sexes. There is no good reason to give any attention to correlative qualities. In so far as a real correlation exists, selection for real quality will automatically bring along the correlative character, and in as far as we are mistaken (and this will nearly always be so) we are simply led astray by paying attention to apparent correlations. In plant-breeding selection according to qualities believed to be indirect indications of good quality used to be very frequently attempted.

This has gone out of fashion, and we only meet traces of it in very special fields (sugar-cane, apples). In our selection work we should always consider actual practical merit, and we should care-fully study in how far the qualities we can examine, measure and weigh are indications of this real merit.

The Importance of Collections

Most plant-breeding institutes own a collection of plants, sometimes consisting of imported and old varieties, sometimes completed with a collection of wild species. Such collections are of the utmost importance to plant-breeding. We must remember that by selection we cannot create new genes, we cannot even remove existing genes that are present in all the plants. For this reason the possibilities for plant-breeding work are restricted to recombination of the material present in our collection.

The importance of a collection of old varieties is sometimes seen in the historical interest, and occasionally old breeds are retained for sentimental reasons. This is wrong, for at any time somebody responsible for the collection may have other ideas; he may be less of a sentimentalist than his predecessors, or he may think the value of all those old varieties is over-rated now that we have so many superior new ones.

The truth is that the collections furnish material for serious future plant-breeding work. Every new gene added to the complement of genes in our present material doubles the possibilities of recombinations. Every novel (relative) gene-absence does the same.

It is some-times possible to use very important qualities of intrinsically worthless old kinds of plant. A garden pea, called "Acacia". In this pea each leaf consists of an uninterrupted series of pairs of leaflets like a Robinia leaf, tendrils being wholly absent. The plants therefore cannot attach themselves to supports or neighbouring plants. One of the difficulties in growing field peas in wet summers in some countries lies in the circumstance that the young plants intertwine themselves into veritable mats before the time for hoeing the weeds is past. To give the plants another cultivation in May would mean tearing apart the tendrils; in most years this would be quite out of the question.

For this reason the "Acacia" variety with a good, green, round field pea, and by selecting among the plants of the third and fourth generation succeeded in remaking a green field pea with a good yield and a high resistance to mildew, but without tendrils, a plant which could be given a last hoeing fully three weeks after the ordinary peas in neighbouring fields had grown together inextricably. When we want to extend the region in which we can grow some group of plants it is often possible to find plant varieties in an extensive collection which carry qualities that are the very thing we want.

It is uneconomical to duplicate and triplicate such collections; it is much better to have one huge central "ecole" of cultivated plants than to have a number of incomplete collections in the hands of private plant breeders and plant-breeding firms. Moreover, keeping up such a collection costs money, and for this reason the collection, or some of its less-important-looking species, always runs the risk of being suppressed some day. It is evident that the Government should take over the task of keeping up such plant collections. The American Government has acted upon this idea, and its Plant Importation Bureau has done an enormous amount of very valuable work in this connection.

The keeping of a collection and the importation of foreign plants should go hand in hand. Especially in backward countries a wealth of plants is often present, and we sometimes find a great many curious plants well adapted to special conditions of season and climate. Plants as the Hopi maise, which comes up from almost two feet underground, the pea that buries its pods like a peanut, and the cauliflower-like sugar-cane of some of the islands around Java. Hull-less oats and hull-less barleys may be out of fashion in some countries just now, nevertheless they carry extremely valuable characteristics, well worth preserving. Think of the extremely hardy peaches and apricots of the higher Himalayan valleys, of blight-free Chinese chestnuts, of the enormous seedless tangerines from Java.

Keeping a collection is no work for a geneticist; it needs its own specialists with a very special turn of mind—a zest for systematic work and for collecting. Perhaps philatelists might make excellent keepers for such collections ! The chief source of plants in such collections must be importations from newly explored countries. Sometimes, however, it might be worth while to add some of the curious plants that turn up in the descendants of hybrids—plants carrying absolutely novel qualities, new for the group. Having an important collection of cultivated plants on hand makes it unnecessary to start hunting for suitable material whenever we want to take the amelioration of some plant group in hand. Most plant-breeding firms and plant-breeding institutes keep some sort of a collection, mostly of plants that have or that have had some economic significance. It is of the utmost importance to centralise this work in every country.

POLYPLOID BREEDING

Cytological study of polyploid species shows why most of them behave in bteeding like the hybrid polyploid of the Primula kewensis type. Whether they are tetraploids, hexaploids or octoploids, their chromosomes usually unite in pairs at reduction. This rule is particularly strict in those plants which depend on seed fertility for their reproduction, *e.g.*, the cereals. In these the differentiation is sharper even than in Primula kewensis. It is more like that in the Raphanus-Brassica tetraploid, where no pairing takes place between corresponding chromosomes of the two species even in the diploid hybrid. They are, as it were, in two watertight compartments. The species behave on self-fertilization almost exactly as though they were diploids. Yet, apart from their chromosome numbers, there are three ways in which these species show us that they are polyploids.

In the first place, a particular character in a polyploid may be given by the presence of any one of several different factors. These factors are probably in corresponding chromosomes of the different diploid ancestors of the polyploid. In the hexaploid bread wheat, Tritkum vulgare, Nilsson-Ehle found that some varieties may have three independent factors affecting the red colour of the grain.

In the hexaploid oats, Akerman found similar factors affecting the presence of green leaf-colour. Other characters are affected by-two factors. We do not know that these multiple factors are due to the plant being a polyploid, but since they are not usually found in diploid plants, it is very probable. The occurrence of two independent factors for black grain-colour has been turned to advantage in breeding oats in Sweden. Old black varieties of oats had a single factor which may be called BXBX for black colour. These varieties used to sport so as to give a hundred or so white grains in every bushel, owing to the corresponding chromosomes in the second set carrying the recessive factor b9 and occasional pairing taldng place between the B and b chromosomes. The white grains, of course, reduced the value of the oats by giving it unfairly an appearance of impurity. Akerman x therefore crossed two black varieties, and from the F2 a new type doubly black BXBXB2B2 was raised, which never sports to white seeds.

Secondly, when diploid species are X-rayed their offspring change in every direction; polyploids do not vary anything like so much as their diploid relatives. While tetraploid species of oats and wheat did mutate a little under the same doses as gave big changes in the diploids, the hexaploid bread wheat and cultivated oats showed no change whatever. What is true of change induced by X-rays is probably equally true of natural change. The extra chromosome sets act as a buffer against the effects of changes in the chromosomes, so that in polyploids we can have very little idea of what is going on inside the chromosomes, as compared with diploids. The doubly black-seeded variety of oats is an example of this; the second pair of factors is there for safety and if mutation of Bx to bx occurs, we shall not notice it.

But we can occasionally see the results of other changes, which provide a third kind of evidence of polyploidy. Diploid pure lines sometimes breed true, and for generations show no variation to the closest scrutiny : such are known in the Triticum monococcum, of de Vilmorin, which has bred true for eighty years, and in Johanssen's pure lines of broad bean, Vicia Faba.

But the hexaploid cereals cannot be depended upon. Awnless cultivated oats of as pure race as can be got always throw a proportion—varying with the variety and particularly high in new varieties—of individuals having awns on their flowering glumes. And when these are selfed, their progeny (which will appear in cultivation unless oats for seed are severely rogued) are of quite a distinct and inferior type, indistinguishable, in fact, from the wild oat, Avena fatua. These are called "fatuoid" oats. They are characterised by having both the spikelets provided with a twisted, kneed awn, and each grain marked by a prominent articulation surrounded by a tuft of hairs.

The chromosome behaviour at reduction that the "mutants" are due to the same cause as the appearance of the mealy offspring of P. kswensis, which also resemble a related species, P. verticillata. In both species of Avena, A. fatua and A. sativa each chromosome type is represented by three pairs. Taking two of these pairs—we may consider that A. fatua has the constitution FF9FF9 while A. Sativa has the constitution FF9ff9 and regularly produces gametes Ff which reproduce it truly. Both, as a rule, breed true, but occasionally in A. sativa (particularly in new varieties which have been disturbed by hybridisation) F pairs with/, or the four chromosomes FFff form a ring at reduction. The result is the formation of germ-cells that are FF or ff instead of the normal Ff.

When these meet the normal germ-cell the result is the production of two new types—FFFf and Ffff. The first of these is our slightly awned type, which when selfed gives, together with the normal and its own type, the pure fatuoid, FFFF, in the proportions.

The same result is given by the first cross between the wild oat and cultivated oat, so that we know in this case roughly what makes the difference between the two.

This property of giving away its hybrid origin is probably very common in new polyploid species, for the bread wheats do just the same thing. They throw a sport called a "speltoid" in exactly the same way as the oats throw a fatuoid. Again, this form resembles a related species, Triticum Spelta, in the character of its spike and grain. Such behaviour shows that rogueing is a much more serious business with polyploids than with diploids. All the materials are there for them to produce a whole array of sports, and if they are left to themselves although they are continually self-fertilized they will none the less gradually become more and more adulterated by their own throw-backs, which will mainly be of a character unwanted in cultivation.

Crossing Polyploid Species

If this is the way polyploid species behave when they ate continually inbred, something rather complicated is to be expected when they are crossed. Diploid species-crosses very often give sterile hybrids. Polyploid species, if they have the same number of chromosomes, cross more often to give fertile hybrids. Polyploid species are therefore often more difficult to separate than diploid. Since it is not known how the polyploid species themselves are made up (except that they are hybrid), it cannot be predicted how they will behave when they are bred together; the same combination may be broken up in many different ways.

One of the exceptional results they give is what Engle-dow has called "shift." In the offspring of a cross between Triticum polonicum and T. durum he found no segregation of the long-glume character of the Polish wheat. All the progeny for several generations remained intermediate in character. In the same way, in a cross between a rough-chaffed rivet wheat, T. turgidum, and T. polonicum, Biffen never recovered the grey chaff colour on the rivet, although progeny were grown to the number of 100,000. The same suppression of a character by crossing has been found by Backhouse and Vavilov in the tetraploid cereals, and in other cases they have found the reverse effect of a character being accentuated.

Take the case of the cultivated strawberry, whose history have been well described by Bunyard and Pearl. The wood strawberry, Fragaria vesca, was cultivated for its fruits in the fifteenth century. It produced no variations of importance;, except an ever-flowering "alpine strawberry" analogous to the "All Saints" cherry and the "Lloyd George" raspberry. The fruits always remained small.

This species is diploid. Another species that came into cultivation at this time in France and England was Fragaria elatior, the Hautbois strawberry, also a native of Europe. This species differs from the former in holding its ripe fruits above the leaves, and in the musky flavour of the fruits. Like the wood strawberry, it has small fruits, and although still occasionally grown, it likewise has shown no considerable improvement in size in the course of four hundred years of cultivation. This species is usually dioecious, having male and female flowers on different plants. F. elatior is hexaploid, and therefore will not yield fertile hybrids with F. vesca.

In the seventeenth century, the scarlet strawberry, F.virginiana, from North America was introduced by John Tradescant. The plant was distinct in vegetative characters and its fruit was different in shape and flavour from that of the two European species, but it was scarcely any bigger, about half-an-inch long. It produced more varieties than the older strawberries, but no marked improvement. This species is octoploid and does not cross with either of the first two.

The last important species to be introduced to Europe was the South American F. chiloensis which came to France in 1712. But the plants introduced were all females, and consequently set no seeds until planted near other species with pollen. F. chiloensis—very distinct from the others, having stout scapes with few flowers, and fruits twice their size—is octoploid. From all accounts it is certain that the few plants of it that were introduced afforded the first possibility of raising the large-sised fruits familiar in our modern cultivated varieties.

But since the plants were female, it is evident that the first varieties to occur in cultivation were the results of natural crossing with other species. All the modern large-fruited varieties examined are octoploid, and it is quite clear that they arose from crossing with the other octoploid species, F. virginiana.

Great numbers of improvements, like this in strawberries, have been found to occur following the introduction of species into cultivation, and. they have always been attributed to cultivation. If "cultivation" means "opportunity for hybridisation," the popular idea is correct. No other effect of cultivation has ever been shown to have any importance in plant improvement. It is curious that in spite of improved varieties of strawberries being raised from crossing, they were not sufficiently improved to suggest the advantage of further crossing. The only crosses, as we have seen, were accidental. In fact, we may say they could hardly have been helped. It was only a hundred years later, with the fashion set by Knight in England of raising plants from seeds, that new strawberry seedlings appeared. Soon after, by systematic crossing of the derivatives of these early seedlings, Laxton produced the first strawberries having the qualities required in cultivation to-day. The same kind of change is undoubtedly responsible for the improvement of the Chilean wild potato. But here part of the improvement was probably carried out by the natives in South America before introduction into Europe, and only part after.

In this way it is clear that almost as rapid improvement can be made by crossing polyploid species with the same chromosome numbers as by crossing diploids and raising polyploids from them. The result, too, will not depend on chance, for polyploidy is there to start with. When two tetraploid species are crossed, four different chromosome complements originally derived from four different species are thrown into the melting pot. Not only is the result often fertile, which might well not be, were the parents diploid, but an enormously increased number of combinations can be made from the sorting out of the chromosomes derived from the four ancestral species. Among these combinations the breeder can select the new and improved forms that he wants, and propagate them vegetatively as in the case of strawberries and potatoes; or, by inbreeding, he can extract new strains that will breed true, as in the cereals.

Triploidy

So far, we have dealt chiefly with examples in which polyploidy has given advantageous results, but we must now mention some of the difficulties that polyploids offer to the breeder—difficulties which may, however, be guarded against by reference to a list of chromosome numbers. When polyploids with the same numbers of chromosomes are crossed, the result is a hybrid which, as a rule, is of moderate fertility; but when their chromosome numbers are different, difficulties are to be expected at reduction with consequent loss of fertility. Take the case of a cross between a diploid with a single set of chromosomes in its gametes (like Prunus cerasifera, the myrobalan plum) and a hexaploid with three sets (like the "Jefferson" variety of P. domestica), such as that made by Crane.1 Irregularities of reduction might be expected. But four chromosome sets are brought together in the hybrid just as in the cross between Tritkum durum (4.x) and T. polonicum (4.x).

Three of these are derived from one parent and one from the other, instead of two from each. Does this matter, since in any case the four sets are originally derived from four different species ? So far as behaviour of the chromosomes is concerned it certainly does not. The hybrid behaves just like any other tetraploid ami even gives seedlings when back-crossed with the hexaploitl plum patent. The same kind of result has been achieved in Papaper, Crepis and Digitalis hybrids.

This is well enough where there is an even number of sets of chromosomes in the hybrid, but where the number is odd, reduction can never be regular; chromosomes of the odd set have nothing corresponding to pair with, and they pass at random to one pole or the other. The result is the production of germ-cells with all kinds of chromosome numbers, and we know from other instances with odd chromosomes what this means. Here the case is worse, for a triploid ornamental cherry like Prunus avium var. nana (a cross between the diploid sweet cherry and the tetraploid sour cherry) has eight extra chromosomes. It produces germ-cells with every number between eight and sixteen, and among these the balanced ones with eight and sixteen only come once in over a hundred times. The result is that such a triploid is highly infertile. This, however, has its uses. The triploid watercress is sterile. It is eaten in winter and possibly is sturdier on account of its seed-sterility. The cherry in question is an ornamental variety, and if the cherries of the Japanese group are examined, it is found that nearly all of them are either triploid or double-flowered. Both these types are sterile. Clearly, gardeners have chosen for ornamental purposes those varieties which produce least fruit, for in trees which are grown for a display of flowers, fruit is often a disadvantage. Therefore, if one is raising ornamental trees which can be propagated readily by cuttings or grafts, tetraploids should be crossed with diploids to get, not merely something new, but also something with the advantage of sterility. Actually the triploid Japanese cherries have arisen accidentally, and not by design. Their sterility depends simply on the presence of an odd set which has been Introduced without hybridisation; the odd set is the result of a germ-cell with the unreduced diploid number on one side fusing with a haploid germ-cell on the other.

This is what has happened in the garden tulips. Several important varieties such as Keiserskroon, Massenet, Pottebakker and Pink Beauty, are triploid. They arose from the fusion of a diploid egg with a haploid pollen grain. Consequently, unlike the products of hybridisation with non-giant tetraploid species, such triploids usually show a suggestion of gigantism which has led to their selection.

This is best exemplified by the hyacinths, but it must be pointed out that the hyacinths are altogether exceptional in not being badly affected by having odd chromosome numbers. Therefore their triploids are not sterile, but give rise to seedlings having various numbers, which, although less vigourous than the triploids themselves, are occasionally worth growing. The best varieties of garden hyacinths are triploids. The first to appear was Grand Maitre in 1870.

They are all larger in bulb and flower than either the diploids which preceded them or the odd-numbered forms to which they have given rise as seedlings. They proliferate freely, their bulbils maturing in four years, that is more rapidly than those of the diploids.

Secondary Polyploidy

The apple and pear group offer another interesting peculiarity in their chromosome properties, and therefore in their inheritance. It will be seen that the basic number 17 is not paralleled in any of the other examples that have been given. The question arises, is this really a basic number, or does it conceal the evidence of a more complicated origin from a lower basic number?

Close study of germ-cell formation in the apples and pears and medlars has shown that 17 is not the primary number, but is really derived from an earlier set of seven, like that found in roses and strawberries. This earlier set at a remote period was doubled in part and triplicated in the other part. If the earlier set of 7 was ABCDEFG, then the new set of 17 is AAA9 MB, CCC, DDy EE, FF9 GG. It is unbalanced relative to the old set of seven, and it may be that the apple group owes its special characteristics, more particularly its fruits, to this change in balance. Another species which probably arose as an unbalanced polyploid and owes its special qualities to its new balance is Dahlia Merckii. This species differs from other dahlias in having thirty-six chromosomes instead of thirty-two or sixty-four. It is not in the eight-series, but has two extra pairs of chromosomes. It differs in form from all the other Dahlia species with multiples of eight by having fertile ray florets. It is another case of unbalance being a good balance.

CONCLUSION

We have attempted to find, for the guidance of the plant-breeder, the materials that transmit the hereditary qualities of the plant from one generation to the next; we have concluded that these materials lie in the chromosomes. Study of the behaviour of the chromosomes has shown how their constancy is preserved and how they vary. Particularly we see that in their reduction the chromosomes provide the mechanism of Mendelian segregation and at fertilization the mechanism of Mendelian recombination.

But in the various irregularities of their behaviour are to be seen the means by which types of inheritance can be developed much more complicated than any expressed in the simple Mendelian scheme, although they actually arise out of the same mechanism. Among these are the examples of true-breeding hybrids, such as the ring-forming evening primrose and the hexaploid bread wheat.

We also know now how sports arise from the presence of an extra chromosome and how the ever-sporting habit may be preserved by the loss of a piece of chromosome. We can see how many important improvements in plants were brought about by polyploids. We know how such plant breeders as Thomas Laxton and Thomas Andrew Knight got the results they did, although working relatively in the dark.

And we can therefore imitate them in other fields which appear, in the light of a knowledge of the chromosome numbers, as promising as those they worked in. The occasional seedlings of sterile diploids can be raised in order to obtain fertile polyploids. Where a sterile hybrid is no disadvantage diploids can be crossed with polyploids. We can see, moreover, that chromosome studies are particularly instructive in relation to fertility. Generational sterility is merely a symptom of segregation.

It is due to the pairing and separating of chromosomes that are not sufficiently like one another (as in the diploid hybrid Primula kewnsis), or to a less extent to chromosomes that are like one another pairing and failing to separate (as in tetraploid tomatoes). Further, there are many plants that are sterile, not on account of any hybridity in the ordinary sense, but simply because one of their parental germ-cells had two sets of chromosomes instead of one, so that they have an odd set which cannot pair normally at reduction.

The degree of sterility to be expected in a plant can therefore be calculated to some extent from the pairing of its chromosomes and the pairing can be predicted from a know, ledge of the parents' chromosomes. The plant-breeder therefore is in a position to decide before making a cross whether the results are likely to be fertile or sterile, and this is of great importance in the work of raising new economic plants. Finally, we can understand a great deal more of the nature of species, a knowledge of which is as important to the plant-breeder, from one point of view, as it is to the systematise from another. A study of the chromosomes shows at once that there are some species which are homosygous, uniform and true-breeding. These are the simple-pairing diploids. There are others, equally true-breeding and uniform, which are of hybrid origin and have certain properties, more or less clearly shown; of hybrids. These are the ring-forming diploids and hybrid polyploids. There are others again that contain several inter-sterile groups owing to polyploidy having arisen within them without hybridisation.

And finally there are some that are seed-sterile hybrids maintaining themselves as uniform types by vegetative means such as bulb formation, apomixis, and so on. Cytological examination at once shows which of these groups a species falls in, and is a necessary preliminary to any systematic breeding work. While chromosome studies therefore do not give the plant-breeder any greater control over his material, they enable him to direct his efforts into the right channels for obtaining and preserving the results likely to prove most profitable to him.

MICRO-ORGANISMS

The importance of cultivated micro-organisms in agriculture and medicine is rapidly increasing. Most of them are unicellulars, and only in a few cases (yeasts) is it possible to do any cross-breeding to enhance the potential variability. Winge, in Copenhagen, has done a great deal of work on yeasts, and has proved that controlled cross-breeding under microscopical control is quite possible. In most bacteria and uni-cellular algae, even in some protozoa, vegetative multiplication goes on for long series of generations, even if occasionally wholly new variability may be the result of what in higher organisms we should call cross-breeding. In the second place, true mutations may occur, and it is even possible that in this material, where the cytoplasm is of great importance compared to the nucleus (or where there may be no nucleus at all), quantitative shifts in the relative proportion between the substances present may be passed on from one generation to many subsequent ones ("Dauermodifikation "). Of course the possibility of adapting the material to our needs is due to the fact that the cultures are variable mixtures. In this respect "wild" cultures are always furnishing very valuable material for selection if we want novel qualities or if we want to improve existing qualities.

We must always bear in mind that almost always two sets of desiderata exist: the actual quality of the strain for its ultimate purpose, and the ease with which it is multiplied commercially. To give a striking example. It used to be necessary in the production of penicillin to grow the moulds on the surface of sheets of culture medium. After a while, strains were found that would grow when submerged in a liquid medium, like yeasts, and this made a continuous system of manufacturing possible, and materially reduced the cost of the methods, and so of the final product.

Once a valuable strain is obtained it is of great importance to keep it pure. The ideal method should be the isolation of one single spore, or one single cell. The usual method of the bacteriologists—that of isolating a single colony from a very dilute culture on a semi-solid plate—is also good enough in practice to bring down the potential variability. If we start new cultures continually from very small beginnings we can get round the difficulty of the unavoidable introduction of "wild" strains into our lots.

In vaccination with living microbes, such as in the treatment of young individuals with harmless strains of dreaded infections (contagious abortion in heifers, tuberculosis in human babies), it is evident that only bacteriological methods of isolation and reproduction are good enough. In the commercial production of antibiotics and yeast this is much less pressing. The same is true of nodule-bacteria in legumes.

The preparation of pure cultures for the production of Yoghourt and cheeses that must carry special qualities and aroma is certainly the province of the bacteriologist-specialist. The use of living cultures of micro-organisms in the fight against diseases is only in its infancy. In many cases where what is called "a mild infection" prepares the animals against "heavy" attacks (coccidiosis in poultry and in rabbits and fur-animals is a good example) veterinary science is bound to progress towards the use of strains of low aggressiveness. Some micro-organisms can be used to transform cheap material into foods of very high value. Yeasts are capable of making use of predigested wood pulp, and of producing feeding stuffs rich in many vitamins and proteins of very high food value. The solution of the green-food problem in animals and birds in close confinement can be solved by the plant breeder who develops suitable strains of unicellular algae (Chlorella and others).

The same methods of selection and of progeny testing can be applied to the culture and the breeding of micro-organisms that our breeders of domestic animals and of cultivated plants apply in their work. This is a very new field, and we cannot see where its development may lead us. But one thing is quite apparent, that it promises to become of enormous practical importance.

Drugs and Spices

Spices and drugs have the double advantage for the plant breeder that they have not been worked with, and that the product generally brings a very good price. Even plants that are habitually collected from the wild may very well repay cultivation; such plants as Denis, foxglove and valerian, for which there is a steady demand.

It will often pay to take such a plant into cultivation, to study its biology and try to produce it in favourable conditions, sometimes in a foreign country. Plants, compared with animals, have the advantage under conditions of cultivation, of staying where we put them; we can devote all our energy to improving their quality and productiveness.

Drugs and spices vary enormously in quality, and it is obvious that it must be the task of the plant breeder to improve both yield and quality. In this matter we are severely handicapped by the peculiarities of the trade. In every plant we must pass through a stage in which the "wild" product brings a much better price than the plantation product. This can be a matter of plain prejudice, in which the buyers know the looks and the smell of the wild stuff, and are suspicious when anything of a new kind is offered. Sometimes matters are reversed after a while (plantation quinine bark) when the trade begins to appreciate the content in an actual, measurable active substance; but, even so, appearance counts for much (lichen and moss on cinchona bark and colour of the pods in vanilla). Breeding for high quality as expressed in chemical content will always pay in the end if we can find a market which is selling the active principle pure, or that is selling a standardised article (Denis and Digitalis).

A good example is the production of tobacco for the manufacture of nicotin. Tobacco species differ enormously in yield per acre : some T. rustica lines are remarkable yielders, both in regard to quantity and nicotin content. It is understood that some drug-plants cannot be compared by actual objective standards. They may have no actual value at all, and the growers of ginseng have no hope of ever getting as good a price for their product as for wild ginseng. Their only hope lies in selecting their cultivated plants and treating them during growth in such a way that they can be palmed off as wild. Nothing is as fickle as the trade in drugs and spices. The manufacture of synthetic vanillin or alizarin may wipe out the cultivation of the two plants almost entirely, and later on the demand for real vanilla may be so much revived that the cultivation of this orchid again becomes profitable.

Certain drugs and condiments have the advantage of being eminently suited for an export trade because they are both light and expensive, like black pepper or saffron. This group of plants merits the fullest attention of the plant-breeders.

Chapter 18

TISSUE CULTURE TECHNIQUES

Fungi and bacteria grow on the plant surface and contaminate the culture media when they are not adequately eliminated. The process of explant introduction into *in vitro* conditions depends mainly on the disinfecting phase. Problems in handling produce contamination of the *in vitro* plant and so, a similar disinfection process to that of the *in vitro* introduction technique is recommended.

However, it is possible to use antibiotics included in the medium, but just temporarily, while the plantlets are growing. Among the bacterial contaminants are Bacillus sp., Erwin/a sp., Pseudomonas sp., etc. In the case of systemic infections, the use of disinfectants is not effective, because the pathogens are located in the vascular system. In this case, it is recommended to use specific antibiotics and meristems or bud cuttings.

On the other hand, the process of *in vitro* introduction involves the use of explants in different physiological stages, as dormant buds, which require growth regulators for example, Gibberellic acid) to stimulate and accelerate bud growth. During the process of *in vitro* introduction there is a close connection between the greenhouse and the laboratory; that is why the necessary precautions should be taken to avoid the entrance of contaminants into the laboratory.

- In the greenhouse
  - Take cuttings from the mother plant, or buds from greenhouse tuber sprouts.
- In the laboratory
  - Immerse the cuttings for 10 minutes in a beaker conveniently labelled with the accession name, and containing a solution of 0.5 per cent acaricide with three drops/I of Tween 20.
  - Throw away the acaricide solution and rinse the nodes with tap water.
  - Prepare the laminar flow chamber.

- Immerse the cuttings in 700/o alcohol for 30 seconds and take the beakers immediately to the laminar flow chamber.

- In the laminar flow chamber
  - Eliminate the alcohol and replace it with a 2.50/o solution of calcium hypochlorite. Keep the cuttings immersed during 15 minutes.
  - Throw away the hypochlorite, and rinse the cuttings three times with sterile water.
  - Keep the nodes immersed in sterile water until bud extraction.
  - Dissect the buds on a sterile plate and place them in the culture medium.

MICROPROPAGATION

In vitro plantlets, which are free of pathogens, are used as initial material for potato and sweetpotato seed programmes. The methods used in these micropropagation programmes mainly depend on their production volume and the available infrastructure.

In the case of potato micropropagation, the basic methods have been already described. They have been verified in many institutions and they are based on the rapid growth of individual node cuttings, or stems with multiple node cuttings. Afterwards, the basic micropropagation methods are described.

Node Micropropagation

This method is based on the principle that the node of an *in vitro* plantlet placed in an appropriate culture medium will induce the development of the axillary bud, resulting in a new *in vitro* plantlet. This type of propagation promotes the development of a pre-existent morphological structure. The nutritional and hormonal condition of the medium breaks the dormancy of the axillary bud and promotes its rapid development.

Callus formation and plant regeneration must be avoided because they tend to affect the genetic stability of the genotype. Under room-controlled conditions micropropagation is fast. Each node planted in a propagation medium will produce a plantlet which will occupy the full length of the test tube, after approximately four weeks for potato, and six weeks for sweetpotato. The resultant *in vitro* plantlets may be transplanted to *in vitro* conditions in small pots in the greenhouse.

Micropropagation by Node Cuttings in a Liquid Medium

This technique is applied both with potato and sweetpotato to produce a large number of nodes rapidly. Stem cuttings with 5 to 8 nodes are prepared by removing both the apex and the root of the *in vitro* plant to be propagated. The stems are placed in the corresponding propagation liquid medium. It is also possible to use isolated nodes : the nodes will sprout and new plantlets will develop over a period of 3 to 4 weeks.

Micropropagation Procedure

- Sterilise petri dishes (placed in paper bags or comets) and prepare the laminar flow chamber by disinfecting the internal surfaces with alcohol.

Sterilise the tools with an instrument steriliser and place them on a sterile dish.

- Open the tube, take off the plantlet and place it on a petri dish with the help of forceps.
- Remove the leaves and cut the nodes.
- Open a tube containing fresh sterile medium and place a node inside, trying to plunge it slightly into the medium with the bud up. Close the tube.
- Seal the tube with a gas-permeable plastic tape (parafilm or saram wrap) and label it correctly. It is recommended to place two explants in 16 × 125 mm tubes, three in 18 × 150 mm tubes, five in 25 × 150 mm tubes, and 20-30 in magenta vessels.

Common Problems in Micropropagation

Some problems may appear in tissue culture according to the crop or variety. To solve them, it is necessary to apply one or several preventing/solving methods such as :

Phenolisation

The explants frequently become brown or blackish shortly after isolation. When this occurs, growth is inhibited and the tissue generally dies. The young tissues are less susceptible to darkening than the more mature ones.

Prevention

The tissue darkening– mainly that of the recently isolated explants and that of the medium–may be generally prevented by :

- Removing the phenolic compounds produced by dispersion. Absorption by means of activated carbon. Absorption by polyvinilpyrolidone (PVP).
- Modifying the redox potential. Reducing agents : ascorbic acid, citric acid, L-cisteine HCL, ditriotreitol, glutation and mercaptoethanol Less availability of oxygen : stationary liquid media.
- Inactivating the phenolase enzymes. Chelating agents : NaFeEDTA, EDTA, diethyldithiocarbamate, dimethyl-dithiocarbamate
- Reducing the phenolasic activity and the availability of substrate. Low pH Darkness.

Absence of Rooting

The explants can naturally form roots during propagation, without an additional rooting stage, as with the potato. However, some wild potato species may show root production deficiency. Rooting may be induced by incorporating auxins, such as IAA, NAA, and IBA, or activated carbon to the culture medium.

POTATO *IN VITRO* TUBERISATION

Most of the potato microtubers are used in seed programmes in Europe, where large amounts (hundred of thousands) of pre-basic seed of a few varieties are produced. By means of this technique, microtubers are produced and stored, and it is possible to store thousands of them in a small area (in humid containers at 4°C) for long periods of time. The microtuber induction is produced through a stress effect by the CCC (chlorocholine chloride), BAP (B-benzylaminopurine) and Sucrose, which under darkness will produce from 3 to 4 microtubers per plant, according to the variety.

In the beginning, microtubers were used as an alternative for germplasm distribution and *in vitro* conservation. Curing the plantlet *in vitro* multiplication process, usually there is a multiplication of higher amounts to those required in the greenhouse and so, after the transference, some magentas are left over. These plantlets may be used for microtuber induction.

In accordance with the indicated procedures, the induction medium is added to the magentas and then they are placed in the dark room. After three months the microtubers will be produced. These may be harvested and transferred to sterile containers (at 4°C), where they can be maintained up to 10 months, or used in the next campaign instead of *in vitro* plantlets.

These microtubers can also be used as a reserve, in case *in vitro* material is contaminated or dies, because of temperature or handling effects. The high number and bigger size of the microtubers increases the success rates for the transfer to the greenhouse.

MEDIUM TERM GERMPLASM MAINTENANCE

The *in vitro* germplasm maintenance under normal growth conditions requires a series of transfers of the plantlets into a fresh medium. This leads to a consumption of time, increases the possibility of losses because of material contamination during successive sub-cultures, causes a loss of material by human error or failure of some equipment, and demands more labour. A way to avoid these problems is through limitation, restriction or inhibition of growth. This approach consists of growth speed reduction by modifying the physical or chemical conditions of the culture, and it is effective for a short or medium term period.

In vitro Conservation Methods

These consist in maintaining the cultures (buds, plantlets derived from nodes or directly from meristems) under physical (environmental factors] or chemical (culture medium composition) stressed conditions that make it possible to extend, as much as possible, the interval of transference into the fresh media, without affecting the viability of the cultures. The methods to reduce the *in vitro* cultivated plant growth include the reduction of temperature and light during storage, the incorporation of growth retardants in the medium, and the induction of osmotic stress in the medium, or a combination of all these.

- *Temperature* : Temperature reduction has been the most used way to curb culture growth. Most of the *in vitro* cultures are maintained at temperatures between 12 and 200C : at lower temperatures, the growth rate decreases but the reduction depends on the species.
- *Nutrients Concentration :* The reduction of the carbohydrate concentration and the nitrogenous components of the nutritive medium may affect the growth rates. In addition, the continuous absorption of these nutrients during plantet growth will bring a nutritional defficiency, which could produce a premature death of the plants.
- *Use of Growth Regulators :* The abscisic acid (ABA), phosphon-D, maleic hydrazide, and succinic acid are some of the most frequently used growth regulators.
- *Osmotic Concentration :* Growth limitation caused by osmotic concentration is due to the reduction of the water and nutrients absorbed from the medium. For example, at high concentrations, sucrose acts osmotically and it is highly metabolised. No metabolised osmotics, such as Manitol and Sorbitol, are possibly more effective than sucrose in culture growth limitation.
- *Evaluation of* in vitro *Preserved Material :* To evaluate material under these conditions, some important facts about *in vitro* maintenance such as genetic viability and stability should be considered. The viability evaluation of the *in vitro* cultures must be systematic. In slow growth conditions, when the sub-culture or transfer period extends during months or years, the frequency of the culture evaluation increases. The most important characteristics to be evaluated in the low-growth cultivars storage of apical buds are : contamination, leaf senescence, the number of green sprouts, the number of viable nodes in relation to the stem length, the presence or not of roots, and callus formation.

Maintenance of Accessions in Seed Programme

In a seed programme, a group of accessions free of virus is maintained for pre-basic seed production : however, most of them are not propagated in the greenhouse so they are maintained *in vitro* to be used in the future. The continuous propagation of the *in vitro* plantlets damages the material, mainly if we consider that environmental conditions are not adequate. The alternative is to maintain the accessions in conservation media. Each accession must be maintained in test tubes with five replications to avoid possible losses. The maintenance conditions have been tested in CIP's potato collection that consists of more than 5,000 accessions, by means of which it is possible to assure the normal recovery of the material after using stress producers.

The plantlets used in each campaign must come from the maintenance phase. Then, they will be propagated in normal media where their growth will be re-established, and the elected procedures for plantlet multiplication in the greenhouse will continue. Each season must be initiated with this material to start off with strong plants. In addition, the plantlets taken for multiplication must be replaced, trying to maintain 5 tubes per accession in the conservation media.

The sub-cultures in the conservation medium are carried out approximately every one or two years. The conservation medium renovation must pass through a previous sub-culture in the propagation media to rejuvenate the explants.

Virus Eradication through Meristems Culture and Thermotherapy

If a healthy plant is sown in the field, it is exposed to infections caused by pathogens as nematodes, fungi, bacteria, phytoplasms, virus and viroids, which have a negative effect on yield, and in some cases may kill the plants. However, not all the plant cells may become infected. A group of cells, which are in a continuous non-differentiated multiplication, are virus-free : the meristem. The *in vitro* meristems culture, together with growth at high temperatures produces potato plantlets free from viruses in more than 90 per cent of planted meristems. This routine method was established in the International Potato Centre to obtain virus-free plantlets for national and international distribution. The *in vitro* maintenance of virus-free plants provides the possibility to maintain all the time a bank of healthy and more vigourous plants, and with a more accelerated growth, than the infected ones. In a seed programme, it is essential to initiate this work with virus-free plantlets since this will affect the seed quality and the yield as well. The virus cleaning procedure is too long and expensive : that is why CIP, through its germplasm distribution programme, has provided a list of the principal virus-free potato varieties to all users.

Procedure

- Approximately 18 to 20 plantlets (virus-infected) are propagated in magentas.
- After a growth period of 20 to 25 days, or when the plantlets are 4-5 cm high, they are placed in the thermotherapy chamber. The growth conditions are : 16 hours of light 34°C; 8 hours of darkness 32°C.
- The magentas are maintained in the thermotherapy chamber for one month.
- Afterwards, the magentas are taken out of the chamber, and the outside is cleaned with 980/o alcohol. Then they are introduced to the culture room. Then they are introduced to the culture room.
- Meristems are obtained as follows :

Cut the apical portion and remove the leaves that cover the meristem (approximately 3 to 4 leaves); the meristem is observed with a prominent leaf primordium. Remove the meristem with part of the leaf primordium; cut only the translucent portion. Use a new knife.

Place the menistem in the culture medium. Be sure that the meristem is in the tube.

- Evaluate meristem growth and transfer them to fresh media if necessary.
- Each meristem that originates a plant is called <'line», which will be labelled according to the accession it belongs to. For example, Yellow Line 1, Yellow Line 2, Yellow Line 3, etc.
- Five tubes with several plants are propagated to evaluate the potato virus PSTVd, PVT, to determine the host range, the morphologic evaluation, and the *in vitro* maintenance.
- The results of each evaluation are obtained, and the infected material is replaced with clean material.

Temperature

The growth of the plantlets at temperatures higher than 20°C is accelerated (Do not use temperatures higher than 30°C).

CULTURED CELLS AND TISSUES

The large scale commercial propagation of plant material based on plant tissue culture was pioneered in the USA. During the last thirty years, tissue culture-based plant propagation has emerged as one of the leading global agro-technologies. Between 1986 and 1993, the worldwide production of tissue cultured plants increased 50 per cent. In 1993, the production was 663 million plants. By 1997, production had risen to 800 million plants. During 1990–1994, the micropropagation industry declined in Europe, mainly due to production shifting to developing countries, but since then because of the demand for high quality and number, production in European countries has increased.

Since 1995, production has increased by 14 per cent in Asian countries, mainly due to the market entry of China, while the increase in South and Central America was from production in Cuba. More recently, some companies from Israel, the USA and UK have shifted their production requirements to Costa Rica and India.

Tissue cultured plants have as yet to reach many growers and farmers in the developing countries. The primary advantage of micropropagation is the rapid production of high quality, disease-free and uniform planting material. The plants can be multiplied under a controlled environment, anywhere, irrespective of the season and weather, on a year-round basis. Production of high quality and healthy planting material of ornamentals, and forest and fruit trees, propagated from vegetative parts, has created new opportunities in global trading for producers, farmers, and nursery owners, and for rural employment. Micropropagation technology is more expensive than the conventional methods of plant propagation, and requires several types of skills. It is a capital-intensive industry, and in some cases the unit cost per plant becomes unaffordable. The major reasons are cost of production and know-how. During the early years of the technology, there were difficulties in selling tissue culture products because the conventional planting material was much cheaper.

Now this problem has been addressed by inventing reliable and cost effective tissue culture methods without compromising on quality. This requires a constant monitoring of the input costs of chemicals, media, energy, labour and capital. In the industrialised countries, labour is the main factor that contributes to the high cost of production of tissue-cultured plants.

To reduce such costs, some steps can be partially mechanised, *e.g.* use of a peristaltic pump for medium dispensing, and of dishwashers for cleaning containers. In the less developed countries of Africa, Asia, and Latin America, where labour is relatively cheaper, consumables such as media, culture containers, and electricity make a comparatively greater contribution to production costs.

For example, the cost of medium preparation (chemicals, energy and labour) can account for 30–35 per cent of the micropropagated plant production. However, automated production processes based on pre-sterilised membrane capsules, bioreactors, mechanised explant transfer, and container sealing are not commercially viable propositions in many developing countries. Therefore, low cost alternatives are needed to reduce production cost of tissue-cultured plants.

Many of the low cost technology options described in this publication can be incorporated in various steps of plant micropropagation. The occasional tissue culturegenerated variants (somaclones) and rare spontaneous bud mutants as well as those obtained from induced mutations can also be propagated by deployment of low cost techniques described. Micropropagation, in combination with radiation-induced mutations, speeds up the recovery, multiplication and release of improved varieties in vegetatively propagated plants. Hence, low cost technology will be of great value for large scale plant multiplication of mutants of many fruits, shrubs, flowers and forest trees that are conventionally vegetatively propagated.

PLANT TISSUE CULTURE

Plant tissue culture refers to growing and multiplication of cells, tissues and organs of plants on defined solid or liquid media under aseptic and controlled environment. The commercial technology is primarily based on micropropagation, in which rapid proliferation is achieved from tiny stem cuttings, axillary buds, and to a limited extent from somatic embryos, cell clumps in suspension cultures and bioreactors. The cultured cells and tissue can take several pathways. The pathways that lead to the production of true-to-type plants in large numbers are the preferred ones for commercial multiplication. The process of micropropagation is usually divided into several stages *i.e.*, prepropagation, initiation of explants, sub-culture of explants for proliferation, shooting and rooting, and hardening. These stages are universally applicable in large-scale multiplication of plants. The delivery of hardened small micropropagated plants to growers and market also requires extra care.

Plant tissue culture refers to growing and multiplication of cells, tissues and organs on defined solid or liquid media under aseptic and controlled environment. Plant tissue culture technology is being widely used for large-scale plant multiplication. The commercial technology is primarily based on micropropagation, in which rapid proliferation is achieved from tiny stem cuttings, axillary buds, and to a limited extent from somatic embryos, cell clumps in suspension cultures and bioreactors.

EXPLANT SOURCE

Plant tissue cultures are initiated from tiny pieces, called explants, taken from any part of a plant. Practically all parts of a plant have been used successfully as a source of explants. In practice, the "explant" is removed surgically, surface sterilised and placed on a nutrient medium to initiate the mother culture, that is multiplied repeatedly by sub-culture. The following plant parts are extensively used in commercial micropropagation. Shoot-tip and meristem-tip culture : Shoots develop from a small group of cells known as shoot apical meristem. The apical meristem maintains itself, gives rise to new tissues and organs, and communicates signals to the rest of the plant. Shoot-tips and meristem-tips are perhaps the most popular source of explants to initiate tissue cultures. The shoot apex explant measures between 100 to 500 μ m and includes the apical meristem with 1 to 3 leaf primordia. The apical meristem of a shoot is the portion lying distal to the youngest leaf primordium, and is ca.100 μ??m in diameter and 250μ?m in length with 800-1200 cells. In practice, shoot-tip explants between 100 to 1000 μ m are cultured to free plants from viruses. Even explants larger than 1000 μ m have been frequently used. The term "meristem-tip culture" has been suggested to distinguish the large explants from those used in conventional propagation.

Nodal or axillary bud culture : This consists of a piece of stem with axillary bud culture with or without a portion of shoot. When only the axillary bud is taken, it is designated as "axillary bud" culture. Floral meristem and bud culture : Such explants are not commonly used in commercial propagation, but floral meristems and buds can generate complete plants. Other sources of explants : In some plants, leaf discs, intercalary meristems from nodes, small pieces of stems, immature zygotic embryos and nucellus have also been used as explants to initiate cultures.

Cell suspension and callus cultures : Plant parts such as leaf discs, intercalary meristems,-stem-pieces, immature embryos, anthers, pollen, microspores and ovules have been cultured to initiate callus. A callus is a mass of unorganised cells, which in many cases, upon transfer to suitable medium, is capable of giving rise to shoot-buds and somatic embryos, which then form complete plants. Such calli on culture in liquid media on shakers are used for initiating cell suspensions. Liquid suspension cultures maintained on mechanical shakers achieve fast and excellent multiplication rates. However, in commercial micropropagation, calli are cultured mostly in bottles and flasks kept on semi-solid or liquid media. To a limited extent, bioreactors have become popular for somatic embryogenic cultures. It is considered that some day robotics could be adapted to bioreactorbased micropropagation.

PATHWAYS OF CULTURED CELLS AND TISSUES

The cultured cells and tissue can take several pathways to produce a complete plant. Among these, the pathways that lead to the production of true-to-type plants in large numbers are the popular and preferred ones for commercial multiplication. The following terms have been used to describe various pathways of cells and tissue in culture :

Regeneration and Organogenesis

In this pathway, groups of cells of the apical meristem in the shoot apex, axillary buds, root tips, and floral buds are stimulated to differentiate and grow into shoots and ultimately into complete plants. In many cases, the axillary buds formed in the culture undergo repetitive proliferation, and produce large number of tiny plants.

The plants are then separated from each other and rooted either in the next stages of micropropagation or *in vivo* (in trays, small pots or beds in glasshouse or plastic tunnel under relatively high humidity). The explants cultured on relatively high amounts of auxin (*e.g.* (2,4-D, 2,4-dichlorophenoxyacetic acid) form an unorganised mass of cells, called callus. The callus can be further sub-cultured and multiplied.

The callus shaken in a liquid medium produces cell suspension, which can be sub-cultured and multiplied into more liquid cultures. The cell suspensions form cell clumps, which eventually form calli and give rise to plants through organogenesis or somatic embryogenesis. In some cases, explants *e.g.* leaf-discs and epidermal tissue can also generate plants by direct organogenesis and somatic embryogenesis without intervening callus formation, *e.g.* in orchardgrass. Dactylis glomerata L.. In organogenesis the cultured plant cells and cell clumps (callus) and mature differentiated cells (microspores, ovules) and tissues (leaf discs, internodal segments) are induced to differentiate into complete plants to form shoot buds and eventually shoots, and rooted to form complete plants.

Somatic Embryogenesis

In this pathway, cells or callus cultures on solid media or in suspension cultures form embryo-like structures called somatic embryos, which on germination produce complete plants. The primary somatic embryos are also capable of producing more embryos through secondary somatic embryogenesis. Although, somatic embryogenesis has been demonstrated in a very large number of plants and trees, the use of somatic embryos in large-scale commercial production has been restricted to only a few plants, such as carrot, date palm, and a few forest trees.

Somatic embryos are produced as adventitious structures directly on explants of zygotic embryos, from callus and suspension cultures. Somatic embryos and synthetic seeds (embryos encapsulated in artificial endosperm) hold potential for large-scale clonal propagation of superior genotypes of heterogeneous plants.

They have also been used in commercial plant production and for the multiplication of parental genotypes in large-scale hybrid seed production. In many species, somatic embryos are morphologically similar to the zygotic embryos, although some biochemical, physiological and anatomical differences have been documented.

The synthetic auxin, 2,4-D is commonly used for embryo induction. In many angiosperms, *e.g.*, carrot and alfalfa, sub-culture of cells from 2,4-D containing medium to auxin-free medium is sufficient to induce somatic embryogenesis. The process can be enhanced with the application of osmotic stress, manipulation of medium nutrients, and reducing humidity. Selection of embryogenic cell lines has also been successfully used. For example, selection for unique morphotypes in grapevine cultures allows production of high quality embryos with predicable frequency.

A major problem in large-scale production of somatic embryos is culture synchronisation. This is achieved through selecting cells or pre-embryonic cell clusters of certain size, and manipulation of light and temperature, temporary starvation or by adding cell cycle synchronising chemicals to the medium. Cytokinins seem to play a key role in cell cycle synchronisation and embryo induction, proliferation and differentiation. Abscisic acid is crucial in all the stages of somatic development, maturation and hardening.

Synthetic Seeds

The concept of production and utilisation of synthetic seeds (somatic embryo as substitutes for true seeds) was first suggested by Murashige in 1977. Synthetic seeds can be produced either as coated or non-coated, desiccated somatic embryos or as embryos encapsulated in hydrated gel (usually calcium alginate).

Successful utilisation of synthetic seeds as propagules of choice requires an efficient and reproducible production system and a high percentage of post-planting conversion into vigourous plants. Artificial coats and gel capsules containing nutrients, pesticides and beneficial organisms have long been thought as substitutes for seed coat and endosperm. However, this technology is still in the developmental stage, and currently cannot compete with the other methods of commercial plant propagation.

Process of Micropropagation

The process of plant micro-propagation aims to produce clones (true copies of a plant in large numbers).

The process is usually divided into the following stages :

Stage 0 – Pre-propagation step or selection and pre-treatment of suitable plants.

Stage I – Initiation of explants-surface sterilisation, establishment of mother explants.

Stage II – sub-culture for multiplication/proliferation of explants.

Stage III – shooting and rooting of the explants.

Stage IV – Weaning/hardening.

These stages are universally applicable in large-scale multiplication of plants. The individual plant species, varieties and clones require specific modification of the growth media, weaning and hardening conditions.

A rule of the thumb is to propagate plants under conditions as natural or similar to those in which the plants will be ultimately grown *ex-vitro*. For example, if a chrysanthemum variety is to be grown under long day-length for flower production, it is better to multiply the material under long-day length at stages III and IV. There is a wide option to undertake production of plant material up to a limited number of stages. For example, many commercial tissue culture companies undertake production up to Stage III, and leave the remaining stages to others.

Pre-Propagation Stage

The pre-propagation stage (also called stage 0) requires proper maintenance of the mother plants in the greenhouse under disease- and insect-free conditions with minimal dust. Clean enclosed areas, glasshouses, plastic tunnels, and net-covered tunnels, provide high quality explant source plants with minimal infection. Collection of plant material for clonal propagation should be done after appropriate pre-treatment of the mother plants with fungicides and pesticides to minimise contamination in the *in vitro* cultures. This improves growth and multiplication rates of *in vitro* cultures.

The control of contamination begins with the pretreatment of the donor plants. They may be prescreened for diseases, isolated and treated to reduce contamination. The explants are then brought to the production facility, surface sterilised and introduced into culture. They may at this stage be treated with antibiotics and fungicides as well as anti-microbial formulations, such as PPM. The explants are then culture indexed for contamination by standard microbiological techniques, which are occasionally supplemented with tests based on molecular biology or other techniques.

Stage-I

This stage refers to the inoculation of the explants on sterile medium to initiate aseptic culture. Initiation of explants is the very first step in micropropagation. A good clean explant, once established in an aseptic condition, can be multiplied several times; hence, explant initiation in an aseptic condition should be regarded as a critical step in micropropagation. More than often, explants fail to establish and grow, not due to the lack of a suitable medium but because of contamination. The explants are transferred to *in vitro* environment, free from microbial contaminants. The process requires excision of tiny plant pieces and their surface sterilisation with chemicals such as sodium hypochlorite, ethyl alcohol and repeated washing with sterile distilled water before and after treatment with chemicals. After a short period of culture, usually 3 to 5 days, the contaminated explants are discarded.

The surviving explants showing growth are maintained and used for further sub-culture. In herbaceous plants *e.g.* potato, chrysanthemum, carnation, streptocarpus, strawberry, and African violet; the explant sources are meristems, apical- and axillary buds, young seedlings, developing young leaves and petioles, and unopened floral buds. The following low cost options can be adapted to initiate explants :

Sterile Instrument Technique

This method assumes that most of the deep-seated meristems and those covered by leaves or other integuments (*e.g.* floral bracts) are sterile. In this procedure, the explant is washed with sterile water, rinsed in ethanol, and instruments are sterilised every time they touch the surface of the explant, and the explant is moved to a new location on the dissection stage.

Surface Sterilisation Technique

This is by far the most commonly used method. The explants are washed in sterile water, rinsed in ethanol, and surface sterilisation is achieved by using chemicals with chlorine base. Calcium or sodium hypochlorite based solutions, 1-3 per cent (v/v) are usually used for soft herbaceous materials. A cheap and readymade sterilant is 5-7 per cent solution of 'Domestos' – a toilet disinfectant which contains 10.5 per cent v/v sodium hypochlorite, 0.3 per cent sodium carbonate, 10.0 per cent sodium chloride and 0.5 per cent (w/v) sodium hydroxide and a patented thickener). The explants are washed in sterile distilled water before and after sterilisation. Other surface sterilants used include mercuric chloride (avoid its use as far as possible, since it is highly toxic), hydrogen peroxide, and potassium permanganate.

For Soft Tissues

- Wash explants from perennial plants for 1-2 hr in tap water. Eliminate this step for material from glasshouse grown plants.
- Wash in sterile distilled water three to four times for 5 to 10 minutes each.
- Dip in 95 per cent ethanol for 3 to 5 seconds.
- Wash once again with sterile distilled water for 5 minutes.
- Surface-sterilise in 5 per cent 'Domestos' (v/v) for 20-25 minutes.

- Wash with sterile distilled water three times for 10 minutes each.
- Drain water droplets by placing on pre-sterilised blotting paper.
- Transfer explants singly to the medium.

N.B. : Sterilise forceps each time to transfer explants to avoid cross-contamination.

For woody stems (e.g. roses, hardy shrubs, and trees) :

- Collect stems, shoots, buds and store at 5 °C till needed.
- Rinse in ethanol for 3 to 5 seconds.
- Rinse in 1- per cent sodium hypochlorite (20 per cent bleach) for 10 minutes.
- Place lower parts of stems in flasks in 2 per cent sucrose and 200 PPM 8-hydroxyquinoline citrate at 23+2 oC. For items collected in September/ October, add 50-PPM GA3.

After that 10 PPM GA will help break the dormancy :

- Re-cut the bottom of stem and replace the solution after 2 days.
- Excise the softwood from the developed shoots and use material for explants or for rooting.
- Surface-sterilise as in the above protocol.

Do not forget to sterilise forceps and scalpel every time for the transfer of explants to fresh solutions. Use sterile containers in the protocol of surface sterilisation. If explants become brown or pale at the end of the protocol, reduce the strength of 'Domestos' to 2.5 per cent. Alternatively, dip explants in 10 per cent 'Domestos' for 2 minutes and then proceed to surface sterilise with 3-5 per cent 'Domestos' for 20 minutes. If basal contamination is observed after 2-3 days of culture, explants can sometimes be rescued by removing the basal end by making a single cut with a sharp scalpel and re-culturing on fresh medium.

Stage-II

Stage-II is the propagation phase in which the explants are cultured on the appropriate media for multiplication of shoots. The primary goal is to achieve propagation without losing the genetic stability. Repeated culture of axillary and adventitious shoots, cutting with nodes, somatic embryos and other organs from Stage-I leads to multiplication of propagules in large numbers. The propagules produced at this stage can be further used for multiplication by their repeated culture. Sometimes it is necessary to sub-culture the *in vitro* derived shoots onto different media for elongation.

Stage-III

The *in vitro* shoots obtained at Stage-II are rooted to produce complete plants. If the proliferated material consists of bud-like structures (*e.g.* orchids) or clumps of shoots (banana, pineapple), they should be separated after rooting and not before. Many plants (*e.g.* banana, pineapple, roses, potato, chrysanthemum, strawberry, mint, several grasses and many more) can be rooted on half-strength-MS medium without any growth-regulators. Good sturdy well-rooted plants are essential for high survival during weaning and later transfer to soil. This stage is labour intensive and expensive. The process of *in vitro* rooting has been estimated to account for approximately 35–75 per cent of the total cost of production. Efforts should be made to combine rooting and acclimatisation stages.

Stage-IV

At this stage, the *in vitro* micropropagated plants are weaned and hardened. This is the final stage of the tissue culture operation after which the micropropagated plantlets are ready for transfer to the greenhouse. Steps are taken to grow individual plantlets capable of carrying out photosynthesis. The hardening of the tissue-cultured plantlets is done gradually from high to low humidity and from low light intensity to high intensity conditions. If grown on solid medium, most of the agar can be removed gently by rinsing with water.

Plants can be left in shade for 3 to 6 days where diffused natural light conditions them to the new environment. The plants are then transferred to an appropriate substrate (sand, peat, compost, etc.), and gradually hardened. Low-cost options include the use of plastic domes or tunnels, which reduces the natural light intensity and maintains high relative humidity during the hardening process. If the plants are still joined together after rooting, these should be planted as bunches in the soil and separated after 6 to 8 weeks of growth.

DELIVERY TO THE GROWERS

The delivery of the rooted and hardened small micropropagated plants to growers and market requires extra care. In some cases, plant losses can occur during shipment and handling by growers. This is particularly true when the plants are not fully hardened and rooted or not grown for sufficient duration after transfer to soil. Growers should be given clear instructions how to handle the material provided. Apart from the economic loss, poor survival of planted material erodes the confidence of growers in the technology. The transfer of individual plants to soil in black plastic or polythene bags is widely used as a low-cost option to provide fully-grown banana plants directly to farmers in many developing countries.

Chapter 19

PLANT TRANSFORMATION

Plant cells have a variety of characteristics that distinguish them from animal cells. These characteristics include the presence of a large central vacuole and a cell wall, and the absence of entioles, which play a role in mitosis, meiosis, and cell division. Along with these physical differences, another factor distinguishes plant cells from animal cells, which is of great significance to the scientist interested in biotechnology : Many varieties of full-grown adult plants can regenerate from single, modified plant cells called protoplasts-plant cells whose cell walls have been removed by enzymatic digestion.

More specifically, when some species of plant cells are subjected to the removal of the cell wall by enzymatic treatment, they respond by synthesising a new cell wall and eventually undergoing a series of cell divisions and developmental processes that result in the formation of a new adult plant. That adult plant can be said to have been cloned from a single cell of a parent plant.

Crop genetic engineering includes :

- DNA isolation
- Gene cloning
- Gene design
- Transformation
- Plant breeding.

Plants that can be cloned with relative ease include carrots, tomatoes, potatoes, petunias, and cabbage, to name only a few. The capability to grow a whole plant from a single cell means that researchers can engage in the genetic manipulation of the cell, let the cell develop into a completely mature plant, and examine the whole spectrum of physical and growth effects of the genetic manipulation within a relatively short period of time. Such a process is far more straightforward than the parallel process in animal cells, which cannot be cloned into full-grown adults. Therefore, the results of any genetic manipulation are usually easier to examine in plants than in animals.

DNA CLONING

DNA cloning a method to manipulate DNA *in vitro* and *in vivo* or is a logical extension of the ability to manipulate DNA molecules with restriction endonucleases and ligases. Imagine that an plant gene has been obtained as a single restriction fragment after digestion of a larger molecule with the restriction enzyme *Bam*HI, which leaves 5-GATC-3 sticky ends. Imagine also that a small *E. coli* plasmid has been purified and treated with *Bam*HI, which cuts the plasmid in a single position.

The circular plasmid has therefore been converted into a linear molecule, again with 5-GATC-3 sticky ends. Mix the two DNA molecules together and add DNA ligase. Various recombinant ligation products will be obtained, one of which comprises the circularised plasmid with the animal gene inserted into the position originally taken by the *Bam*HI restriction site.

If the recombinant plasmid is now re-introduced into *E. coli*, and the inserted gene has not disrupted its replicative ability, then the plasmid plus inserted gene will be replicated and copies passed to the daughter bacteria after cell division. More rounds of plasmid replication and cell division will result in a colony of recombinant *E. coli* bacteria, each bacterium containing multiple copies of the animal gene.

Vector That Works with Plant Cells

Not all aspects of the genetic manipulation of plant cells are readily accomplished. Not only do plants usually have a great deal of chromosomal material and grow relatively slowly as compared with single cells grown in the laboratory, but few cloning vectors can successfully function in plant cells. While researchers working with animal cells can choose among a wide variety of cloning vectors to find just the right one, plant cell researchers are currently limited to just a few basic types of vectors.

Perhaps the most commonly used plant-cloning vector is the "Ti" plasmid, or tumor-inducing plasmid. This plasmid is found in cells of the bacterium known as Agrobacterium tumefaciens, which normally lives in soil. The bacterium has the ability to infect plants and cause a crown gall, or tumorous lump, to form at the site of infection. The tumor-inducing capacity of this bacterium results from the presence of the Ti plasmid.

The Ti plasmid itself, a large, circular, double-stranded DNA molecule, can replicate independently of the A. tumefaciens genome. When these bacteria infect a plant cell, a 30,000 base-pair segment of the Ti plasmid-called T DNA-separates from the plasmid and incorporates into the host cell genome. This aspect of Ti plasmid function has made it useful as a plant cloning vector. The Ti plasmid can be used to shuttle exogenous genes into host plant cells.

This type of gene transfer requires two steps :

- The endogenous, tumor-causing genes of the T DNA must be inactivated and,

- Foreign genes must be inserted into the same region of the Ti plasmid.

The resulting recombinant plasmid, carrying up to approximately 40,000 base pairs of inserted DNA and including the appropriate plant regulatory sequences, can then be placed back into the A. tumefaciens cell. That cell can be introduced into plant cell protoplasts either by the process of infection or by direct insertion.

Once in the protoplast, the foreign DNA, consisting of both T DNA and the inserted gene, incorporates into the host plant genome. The engineered protoplast-containing the recombinant T DNA-regenerates into a whole plant, each cell of which contains the inserted gene. Once a plant incorporates the T DNA with its inserted gene, it passes it on to future generations of the plant with a normal pattern of Mendelian inheritance. One of the earliest experiments that involved the transport of a foreign gene by the Ti plasmid involved the insertion of a gene isolated from a bean plant into a host tobacco plant. Although this experiment served no commercially useful purpose, it successfully established the ability of the Ti plasmid to carry genes into plant host cells, where they could be incorporated and expressed.

A. Limition of A. Tumifecian

The fact that only certain types of plants were naturally susceptible to infection with the host bacterial organism initially limited the usefulness of the Ti plasmid as a cloning vector. In nature, A. tumefaciens infects only dicotyledons or "dicots"-plants with two embryonic leaves. Dicotyledenous plants, divided into approximately 170,000 different species, include such plants as roses, apples, soybeans, potatoes, pears, and tobacco. Unfortunately, many important crop plants, including corn, rice, and wheat, are monocotyledons-plants with only one embryonic leaf-and thus could not be easily transfected using this bacterium.

Overcoming the Limition

Research efforts in the past few years have reduced the limitations of A. tumefaciens. Scientists discovered that by using the processes of micro-injection, electroporation, and particle bombardment, naked DNA molecules can be introduced into plant cell types that are not susceptible to A. tumefaciens transfection. Micro-injection involves the direct injection of material into a host cell using a finely drawn micro-pipette needle.

Electroporation uses brief pulses of high voltage electricity to induce the formation of transient pores in the membrane of the host cell. Such pores appear to act as passageways through which the naked DNA can enter the host cell. Particle bombardment actually shoots DNA-coated microscopic pellets through a plant cell wall. These developments, important in the commercial application of plant genetic engineering, render the valuable food crops of corn, rice, and wheat susceptible to a variety of manipulations by the techniques of recombinant DNA and biotechnology.

Plant Transformation

Genetic engineering of plants is much easier than that of animals.

There are several reasons for this :

- There is a natural transformation system for plants (the bacterium *Agrobacterium tumefaciens*),
- Plant tissue can redifferentiate (a transformed piece of leaf may be regenerated to a whole plant), and
- Plant transformation and regeneration are relatively easy for a variety of plants.

The soil bacterium *Agrobacterium tumefaciens* ("tumefaciens" meaning tumor-making) can infect wounded plant tissue, transferring a large plasmid, the Ti plasmid, to the plant cell. Part of the Ti (tumor-inducing) plasmid apparently randomly integrates into the chromosome of the plant.

The integrated part of the plasmid contains genes for the synthesis of :

- Food for the bacterium,
- Plant hormones.

Genes from the Ti plasmid that are integrated in the plant chromosome are expressed at high levels in the plant. Overproduction of the plant hormones leads to continuous growth of the transformed cells, causing plant tumors. Rapid, cancerous growth of the transformed plant tissue obviously is advantageous to the bacterium : more food gets produced.

The Ti plasmid has been genetically modified ("disarmed") by deleting the genes involved in the production of bacterial food and of plant hormones, and inserting a gene that can be used as a selectable marker. Selectable marker genes generally are coding for proteins involved in breakdown of antibiotics, such as kanamycin. Any gene of interest can be inserted into the Ti plasmid as well. In principle, one can thus transform any plant tissue, and select transformants by screening for antibiotic resistance.

However, unfortunately, there are some complications :

- It has proven difficult to transform some monocots (grasses, etc.) by *Agrobacterium*,
- Regeneration of plants from tissue culture or leaf discs is not always possible.

A number of genetically engineered plant varieties have been developed. Traits that have been introduced by transformation include herbicide resistance, increased virus tolerance, or decreased sensitivity to insect or pathogen attack. Traditionally, most of such genetically engineered plants were tobacco, petunia, or similar species with a relatively limited agricultural application. However, during the past decade it now has become possible to transform major staples such as corn and rice and to regenerate them to a fertile plant. Increasingly, the transformation procedures used do not depend on *Agrobacterium tumefaciens*. Instead, DNA can be delivered into the cells by small, μm-sised tungsten or gold bullets coated with the DNA.

The bullets are fired from a device that works similar to a shotgun. The modernised device uses a sudden change in pressure of He gas to propel the particles, but the principle of "shooting" the DNA into the cell remains the same. This DNA-delivery device is nicknamed "gene gun", and has been shown to work for DNA delivery into chloroplasts as well. Over the last several years, use of the "gene gun" has become a very common method to transform plants, and has been shown to be applicable to virtually all species investigated. For example, transformation of rice by this method is now routine. This is a very important development as rice is the most important crop in the world in terms of the number of people critically dependent on it for a major part of their diet.

Another method to get foreign genes into cereals is by electroporation : a jolt of electricity is used to puncture self-repairing holes in protoplasts (*i.e.*, the cell without the cell wall), and DNA can get in through these holes. However, it is often very difficult to regenerate fertile plants from protoplasts of cereals. Nonetheless, significant advances in overcoming these practical difficulties have been made over the years. Now even transgenic trees have been created : for example, the gene for a coat protein of the plum pox virus has been introduced into apricot. The plum pox virus leads to the feared Sharka disease, for which there is no cure. The resulting transgenic tree shows a markedly decreased sensitivity to this virus. The reason why continuous exposure of the tree to the viral coat protein leads to tolerance against viral infection is not yet understood, however.

Thus, now there are a number of different techniques to introduce foreign genes into plants. Essentially all major crop plants can be (and have been or are being) genetically engineered, the procedures are now routine and the frequency of success is very high. Even though genetically engineered crops are more costly than the usual ones, US farmers have rather readily accepted them provided that tangible benefits can be demonstrated. However, it is questionable whether the farmer in poorer countries can come up with the funds to "try out" and use the new crops. Another issue in this respect is how genetically engineered crops are perceived by the consumer. Even though in the US there is little resistance to such crops as long as the products can be shown to be safe and advantageous, in other countries genetically modified foods are received poorly by the consumer. It is unlikely that there is a rationally sound basis for this rather hostile reaction of the consumer, as most of the crops are the results of human manipulation (such as centuries of breeding) and may have been treated with harmful herbicides and pesticides.

Time and education will need to be invested to provide consumers and consumer advocates with a balanced opinion on the acceptability of the origin of their foods. On the other hand, as so many plants (soybean, corn, etc.) are genetically modified and the nature of the genetic modification is not necessarily easy to explain, it may be simpler to label those foods that are guaranteed free of "genetically modified organisms" or their products. However, keep in mind that traditional breeding has genetically modified essentially all agricultural products, so it may be difficult to define what is actually free of genetically modified organisms.

Transgenic Crops

H- Transgenic crops : The varieties created through the use of a biotechnological approach commonly known as recombinant DNA, genetic engineering (GE), transgenic modification, or genetic modification (GM). The products of genetic engineering are often called genetically modified organisms, or GMOs. All these terms refer to methods of recombinant DNA technology by which biologists splice genes from one or more species into the DNA of crop plants to transfer chosen genetic traits. Genes are segments of DNA that controlled the information that determines the structure and function of a living organism. Genetic engineers are molecular biologists who manipulate this information, typically by taking genes from one species—an animal, plant, bacterium, or virus—and inserting them into another species, such as an agricultural crop.

With the advent of genetic engineering of plants around 1983, it appeared that this new biotechnology would benefit and even revolutionise agriculture. The transfer of desirable genetic traits across species barriers has shown promise for solving problems in the management of agricultural crops. Potential benefits include reduced toxic pesticide use, improved weed control resulting in less tillage and soil erosion, and water conservation.

Ecological Issues

Gene flow to neighbouring crops and to related wild species. Ecological scientists have little doubt that gene flow from transgenic fields into conventional crops and related wild plants will occur. Gene flow from transgenic to conventional crops is of concern to farmers because of its potential to cause herbicide resistance in related conventional crops. For example, in western Canada, three different herbicide-resistant canola varieties have cross-pollinated to create canola plants that are resistant to all three types of herbicide. Through what is called gene-stacking, this new triple resistance has turned volunteer canola into a significant weed problem. Gene flow from transgenic crops to wild relatives creates a potential for wild plants or weeds to acquire traits that improve their fitness, turning them into "super weeds." For example, if jointed goatgrass—a weedy relative of wheat—acquires the herbicide-tolerant trait of Roundup Ready wheat, it will thrive in crop fields unless applications of other herbicides are made. There is already evidence of such outcrossing from herbicide-resistant wheat to jointed goatgrass. Frank Young and his colleagues at Washington State University found that imidazolone-resistant wheat (not a transgenic variety) outcrossed to goatgrass in one season. Other traits that wild plants could acquire from transgenic plants that would increase their weediness are insect and virus resistance.

Because of their experience with classically bred plants, few scientists doubt that genes will move from crops into the wild : seven of the world's thirteen most important crop weeds have been made weedier by genes acquired from classically bred crops. Because gene flow has the potential to affect farmers' crop and pest management, crop marketability, and liability, more research needs to be done to determine the conditions under which gene flow from transgenic plants is likely to be significant.

Pesticide Resistance in Insect Pests

Bt has been widely used as a microbial spray because it is toxic only to caterpillars. In fact, it is a pest management tool that organic farmers depend on—one of the few insecticides acceptable under organic rules. Unlike the commercial insecticide spray, the *Bt* engineered into crop plants is reproduced in all, or nearly all, the cells of every plant, not just applied on the plant surface for a temporary toxic effect. As a result, the possibility that transgenic Bt crops will accelerate insect pests' development of resistance to Bt is a serious concern. Pest resistance to Bt would remove this valuable and environmentally benign tool from farmers' and forest managers' pest control toolbox. Antibiotic resistance

The use of antibiotic-resistant marker genes for the delivery of a gene package into a recipient plant carries the danger of spreading antibiotic-resistant bacteria. The likely result will be human and animal health diseases resistant to treatment with available antibiotics. Research is needed on antibiotic resistance management in transgenic crops. Already the European Commission's new rules governing transgenic crops stipulate phasing out antibiotic-resistant marker genes by the end of 2004.

Effects on Beneficial Organisms

Evidence is increasing that transgenic crops—either directly or through practices linked to their production—are detrimental to beneficial organisms. New studies are finding that Bt crops exude Bt in concentrations high enough to be toxic to some beneficial soil organisms. In addition, a study by University of Arkansas scientists has shown that "root development, nodulation and nitrogen fixation are impaired" in some varieties of Roundup Ready soybeans. The reason is that the beneficial rhizobium responsible for nitrogen fixation in soybeans is sensitive to Roundup. It also appears that disruption of beneficial soil organisms can interfere with plant uptake of phosphorus, an essential plant nutrient.

Insecticidal crops in two ways can affect beneficial insects that prey on insect pests. First, the Bt in transgenic insecticidal crops has been shown in some laboratory studies to be toxic to ladybird beetles, lacewings, and monarch butterflies. The extent to which these beneficials are affected in the field is a matter of further study. Second, because the insecticidal properties of Bt crops function even in the absence of an economic threshold of pests, Bt crops potentially can reduce pest populations to the point that predator species are negatively affected.

Reduced Crop Genetic Diversity

As fewer and larger firms dominate the rapidly merging seed and biotechnology market, transgenic crops may continue the trend towards simplification of cropping systems by reducing the number and type of crops planted. In addition, seed saving, which promotes genetic diversity, is restricted for transgenic crops.

Food Safety

Because food safety is predominantly a concern of consumers, However, because food safety issues do impact marketing and international trade, here is a short list of consumer safety concerns about transgenic food :

- Possibility of toxins in food
- Possibility of new pathogens
- Reduced nutritional value
- Introduction of human allergens
- Transfer of antibiotic resistance to humans
- Unexpected immune-system and genetic effects from the introduction of novel compounds.

Farm Management Issues

The most widely planted transgenic crops on the market today can simplify short-term pest management for farmers and ranchers. In the case of herbicide-tolerant crops, ideally farmers can use a single broad-spectrum herbicide for all their crop weeds, though they may need more than one application in a season. By planting insecticidal crops, farmers can eliminate the need to apply pesticides for caterpillar pests like the European corn borer or the cotton bollworm, though they still have to contend with other crop pests.

While these crops offer handy pest control features, they may complicate other areas of farm management. Farmers who are growing both transgenic and conventional varieties of the same crop will need to segregate the two during all production, harvesting, storage, and transportation phases if they are selling into differentiated markets or plan to save their own seed from their conventional crops.

To minimise the risk of gene flow from transgenic to adjacent conventional crop fields, federal regulations require buffer strips of conventional varieties around transgenic fields. Different transgenic crops require different buffer widths. Because the buffer strips must be managed conventionally, producers have to be willing to maintain two different farming systems on their transgenic fields. Crops harvested from the buffer strips must be handled and marketed as though they are transgenic. Planted refuges—where pest species can live outside fields of insecticidal and herbicide-tolerant transgenic crops—are also required to slow the development of weed and insect pest resistance to Bt and broad-spectrum herbicides. These refuges allow some individuals in the pest population to survive and carry on the traits of pesticide susceptibility. Requirements governing the size of refuges differ according to the type of transgenic crop grown.

Farmers growing herbicide-tolerant crops need to be aware that volunteer crop plants the following year will be herbicide-resistant. This herbicide resistance makes no-till or direct-seed systems difficult because volunteers can't be controlled with the same herbicide used on the rest of the crop. In a no-till system that relies on the same broad-spectrum herbicide that the volunteer plants are resistant to, these plants will contaminate the harvest of a following conventional variety of the same crop, a situation farmers need to avoid for two reasons.

First, the contamination means a following conventional crop will have to be sold on the transgenic market. This leads to the second reason. If farmers grow and market a transgenic crop for which they do not have a technology agreement and did not pay royalty fees, they face possible prosecution by the company that owns the transgenic variety. Many farmers have been charged with "theft" of a company's patented seed as a result of contamination in the field Farmers growing insecticidal crops need to recognise that insect pressure is difficult to predict and may not warrant the planting of an insecticidal variety every year. In a year when pest pressure is low, the transgenic seed becomes expensive insurance against the threat of insect damage.

Farmers growing transgenic crops need to communicate with their neighbours to avoid contaminating their fields and to ensure that buffers are adequate. In Maine, farmers growing transgenic crops are now required by law to be listed with the state agriculture department, to help identify possible sources of cross-contamination when it occurs. The new law, L.D. 1266, also "requires manufacturers or seed dealers of genetically engineered plants, plant parts or seeds to provide written instructions to all growers on how to plant, grow, and harvest the crops to minimise potential cross-contamination of non-genetically engineered crops or wild plant populations". Farm management issues common to all transgenic crops include yield, cost, price, profitability, management flexibility, sustainability, market acceptance, and liability.

Crop Yield, Costs, and Profitability

Some farmers will get higher yields with a particular transgenic crop variety than with their conventional varieties, and some will get lower yields. The same is true for costs. Some producers will see their overall profitability rise, and some will see it drop. If the results of research studies so far on yield, chemical use, costs, crop prices, and profitability of these crops seem contradictory, it is because farmers and farms are not all alike. Neither are the conventional varieties used for comparison. Some yield, cost, and profitability trends do appear to be emerging from the growing body of research data for transgenic crops, however. As noted in the Wallace Center report, Roundup Ready soybeans were designed simply to resist a particular chemical herbicide, not to increase yields. In contrast, Bt corn and cotton, by resisting insect pests, may result in higher yields from reduced pest pressure.

Yield : Herbicide-tolerant Crops—soybeans, Cotton, Canola

Herbicide-tolerant soybeans appear to suffer what's referred to as a "yield drag." Again, in some areas and on some farms this tendency of Roundup Ready soybean varieties to yield less than their comparable, conventional counterparts varies, but overall, they appear to average yields that are five to ten per cent lower per acre.. Drought conditions worsen the effects. The bacterium that facilitates nodulation and nitrogen fixation in the root zone apparently is sensitive to both Roundup and drought.

Yield : Insecticidal Crops—Corn, Cotton

Bt cotton, especially, outpaces yields of conventional cotton by as much as 9 to 26 per cent in some cases, though not at all in others. Yield increases for Bt corn have not been as dramatic. Time will tell whether farmers can expect yield increases or decreases in the long run with these and other transgenic crop varieties.

Changes in Chemical Pesticide Use

One of the promises of biotechnology is that it will reduce pesticide use and thereby provide environmental benefits and reduce farmers' costs. The herbicide-tolerant and insecticidal varieties are designed specifically to meet these goals. Increased future pesticide use resulting from the buildup of resistance to heavily used herbicides is a long-term concern. Pesticide use depends on the crop and its specific traits; weather; severity of pest infestations; farm management; geographic location of the farm; and other variables. As a result, conclusions drawn by various studies analysing pesticide use on transgenic crops remain controversial. According to the Wallace Center report, in a review of the data available up through 2000, crops engineered to contain Bt appear to have decreased the overall use of insecticides slightly, while the use of herbicide-resistant crops has resulted in variable changes in overall herbicide use, with increases in use of some herbicides in some places and decreases in others

Profitability

Farmers need to consider all the factors that determine profitability. No single factor can tell the whole story. Transgenic crop seeds tend to be more costly, and farmers have the added expense of a substantial per-acre fee charged by the owners of transgenic varieties. These costs have to be considered along with input cost changes—whether herbicide or insecticide use and costs go down, go up, or stay the same. Market price is another factor : Prices for some transgenic crops in some markets are lower than prices for comparable conventional crops, though rarely are they higher. Farmers need to watch the markets. Some buyers will pay a premium for a non-transgenic product, though as transgenic seeds find their way into conventional transportation, storage, and processing streams, these premiums may disappear along with confidence that "GMO-free" products are in fact truly free of engineered genes.

Marketing and Trade

Buyer acceptance is a significant marketing issue for farmers raising transgenic crops. Farmers need to know before they plant what their particular markets will or won't accept. Since most grain handlers cannot effectively segregate transgenic from non-transgenic crops in the same facility, many companies are channeling transgenic crops into particular warehouses. Farmers need to know which ones and how far away those are.

While the European Union officially lifted its two-year moratorium on the introduction of new transgenic crops in February 2001, during the debate over labelling and traceability regulations the moratorium remains in effect. Under the proposed new EU requirements, "all foods and animal feed derived from GMOs have to be labelled and, in the case of processed goods, records have to be kept throughout the production chain allowing the GMO to be traced back to the farm".

Liability

The need to separate transgenic crops from conventional and organic ones opens farmers to liability for their product at every step from seed to table. Effective systems for segregation do not exist at present, and will be costly to develop and put into place. Farmers may well end up bearing the added costs of crop segregation, traceability, and labelling.

In the meantime, farmers who grow transgenic varieties—and, ironically, those who do not—are liable for transgenic seeds ending up where they aren't wanted : in their own non-transgenic crop fields, in neighbours' fields, in truckloads of grain arriving at the elevator, in processed food products on retail shelves, and in ships headed overseas. Farmers who choose not to grow transgenic varieties risk finding transgenic plants in their fields anyway, as a result of cross-pollination via wind, insects, and birds, which may bring pollen from transgenic crops planted miles away.

Besides pollen, sources of contamination include contaminated seed and seed brought in by passing trucks or wildlife. Those farmers whose conventional or organic crops are contaminated, regardless of the route, risk lawsuits filed against them by the companies that own the proprietary rights to seed the farmer didn't buy.

Likewise, farmers who grow transgenic crops risk being sued by neighbours and buyers whose non-transgenic crops become contaminated.

Because contamination by transgenic material has become so prevalent in such a short time, all farmers in areas of transgenic crop production are at risk. Insurance, the most common recourse for minimising potential losses because of liability, is not available to the nation's farmers for this risk because insurance companies do not have enough information to gauge the potential losses.

A full risk assessment and legal clarification of the distribution of liability among farmers, seed companies, grain handlers, processors, and retailers is needed before farmers can rest assured that transgenic crops won't result in lawsuits against them.

For farmers who have adopted transgenic technology—and for those who have not—in the words of Brian Leahy, executive director of California Certified Organic Farmers, "This technology does not respect property rights". In essence, farmers who grow transgenic crops on some of their fields, and farmers who grow none, risk bearing tremendous liability. This situation won't change until farmers either gain legal protection or until they all grow transgenic crops on all their fields.

Influences on Public Research

While transgenic crop varieties are generally the property of private corporations, those corporations often contract with public-sector agricultural research institutions for some of their development work. In fact, private investment in agricultural research, including germplasm development, has surpassed public investment in recent years. With this shift in funding sources, the following questions become important : Is the private sector unduly influencing the public research agenda? Are corporations directing public research in less socially valuable directions while research on, for instance, sustainable agriculture goes wanting? Are the outcomes of corporate-funded transgenic research and development by our public institutions equitable across the food and agricultural sectors? Is equity even a consideration of our public institutions when they accept this work?

Because intellectual property rights (patents) apply to living organisms, making them private property, the free flow of scientific information that has historically characterised public agricultural research is being inhibited. What are the implications for the future of agriculture and society of the secrecy that now surrounds so much of what was formerly shared public knowledge? For a brief history of intellectual property rights as they apply to living organisms.

These and other questions need to be addressed by citisens and their public institutions. These issues are of particular concern to farmers and consumers who would benefit from research into alternative technologies that are less costly (in every way), less risky, and more equitable. Equity requires that the economic benefits and risks of transgenic technology be fairly distributed among technology providers, farmers, merchants, and consumers.

For long-term sustainability, farmers need research that focuses on farms as systems, with internal elements whose relationships can be adjusted to achieve farm management goals. In contrast, transgenic crop research so far has focused on products that complement toxic chemical approaches to control of individual pest species. These are products that can be commercialised by large agribusiness or agri-chemical interests and that farmers must purchase every year. This research orientation only perpetuates the cost-price squeeze that continues to drive so many out of farming.

Industry Concentration and Farmers' Right to Save Seed

The broadening of intellectual property rights in 1980 to cover living organisms, including genes, has resulted in a flurry of mergers and acquisitions in the seed and biotech industries.

According to the Wallace Center report, "Relatively few firms control the vast majority of commercial transgenic crop technologies. These firms have strategically developed linkages among the biotechnology, seed, and agri-chemical sectors to capture as much market value as possible. However, these tightly controlled linkages of product sectors raise serious issues of market access, product innovation, and the flow of public benefits from transgenic crops.

Unlike Plant Variety Protection—which does not allow for the patenting of individual genes, but only of crop varieties—Intellectual Property Rights prohibit farmers from saving their seed and undertaking their own breeding programme, and prohibit plant breeders from using the material to create a new generation of varieties adapted to specific regions or growing conditions.

Farmers can't save proprietary seed for planting and so must purchase new seed every year. In addition, farmers choosing transgenic varieties must sign a contract with the owner of the variety and pay a substantial per-acre technology fee, or royalty.

The anticipated commercial introduction of transgenic wheat represents a dramatic shift in an industry in which farmers still widely save their own seed. As non-transgenic varieties become contaminated with transgenic ones, even those farmers who choose to stick with conventional varieties will lose the right to replant their own seed. This loss has already occurred in Canada's canola industry, with Monsanto winning its court case against farmer Percy Schmeiser for replanting his own canola variety that had become contaminated with Monsanto's Roundup Ready canola.

The adoption of transgenic crop varieties has brought with it an increasing prevalence of contract production. While contract production can lead to increased value and reduced risk for growers, farmers are justified in their concern about their loss of control when they sign a contract with a private company. Issues associated with contract production of transgenic crops must be considered within the broader context of a sustainable agriculture to include ownership, control, and social equity.

Chapter 20

Plant Reproduction

The plant kingdom is divided into many classes and divisions. In 1883, Eichler separated the plant kingdom (Plantae) into two sub-kingdoms, the first being Cryptogamae (bacteria, algae, fungi, molds, liverworts, moss and ferns); the second being Phanerogamae (including gymnosperms and angiosperms). There have been many changes to the classification system, as we learn more about the genetics of different species. More recent taxonomy created new kingdoms that separated fungi, algae, and molds from what we commonly consider as plants. Plantae was separated into two divisions : Bryophytes (non-vascular) and Tracheophytes (vascular). Of the vascular plants, angiosperms (the flowering plants) make up approximately 300,000 of the 312,000 known number of living species. The number of 312,000 is probably a gross under-estimation. For anyone interested there is an estimated 20,000 species of non-vascular plants.

There are many different methods of reproduction that plants have developed over the centuries. If a plant's method of reproduction makes it easier to cross species or hybridised, it is more likely that its offspring will be better adapted to a changing environment For more than 8,000 years, human beings have also been affecting plant development by purposefully breeding new characteristics into plants, creating new species. The most important fact regarding reproduction is that every living organism has, in its cell's nucleus, a set of chromosomes. Chromosomes carry genes that have all the information necessary to make new individual organisms.

DIAGRAM OF A DICOT PLANT

One of the simplest methods of reproduction is when one cell divides and becomes two equal halves that will grow large enough to split again. With plants and animals that reproduce this way, each generation is identical to the one before.

CELL DIVIDING DIAGRAM WITH ONE PARENT CELL

Plant reproduction may be sexual, in which two parents produce a genetically different individual; or asexual, involving the propagation of plants that are genetically identical to the parent.

THE SEXUAL PLANT LIFE CYCLE

Sexual reproduction is important in providing genetic variability. All plants that reproduce sexually must go through meiosis. Meiosis is a unique kind of cell division during which the paired sets of chromosomes present in sexually mature plants, called sporophytes, are halved. Because they have the two sets of chromosomes, one from the male parent and the other from the female, the cells of sporophytes are called *diploid.*

During this process, pairs of homologous, or identical, chromosomes, one from each parent, line up together. Crossing-over, or the exchange of genetic material between these homologous chromosomes, may occur at this time. Crossing-over is critical for producing some of the genetic variability in resulting offspring. Meiosis typically produces four haploid cells, each with one set of chromosomes, from a single diploid cell. These haploid cells become gametes-eggs and sperm. Without meiosis, sexually reproducing organisms would not have a mechanism for reducing the total chromosome number by half so that genetic variability can be introduced via crossing-over. Meiosis is a critical step that must occur prior to the fusion of sperm and egg, which restores the diploid chromosome number.

Alternating Generations

Plants undergo a two-phase cycle of sexual reproduction known as the *alternation of generation.* This sexual life cycle involves alternation of the diploid sporophyte generation with the haploid gametophyte generation. The sporophyte stage begins with fusion of an egg and sperm, which produces a zygote. The diploid zygote develops into a sporophytic plant. This sexually mature plant eventually produces meiocytes, cells that will undergo meiosis, typically resulting in four haploid cells. These haploid cells, in turn, develop into gametophytes, which eventually form sex organs where eggs and sperm are produced. The sporophyte and gametophyte stages usually are easy to tell apart; one is typically parasitic or dependent upon the other. Which generation is dominant differs in different plants. Generally, the gametophytes of primitive land plants such as mosses are dominant and larger than the sporophytes. In higher, or more advanced plants, the reverse is true. In other words, the alternation of generations is the alternation between meiocytes undergoing meiosis and the fusion of gametes that produces a zygote.

The Gametophyte Stage

The gametophyte generation normally begins with a spore (sexual spore or meiospore) and ends with a gamete. The cells of the gametophyte are usually haploid.

The Sporophyte Stage

The sporophyte generation usually begins with a zygote and ends with a meiocyte. The cells of the sporophyte are usually diploid.

Reproduction in Fern

Sporophyte development is very rapid in common garden and woodland ferns. The zygote develops an embryonic root, leaf, stem, and foot. The foot apparently aids in absorbing nutrients from the gametophyte. The stem develops slowly, but ultimately produces leaves. The first few leaves of juvenile ferns differ from those of the mature sporophyte. As growth continues, the leaves finally begin to resemble those of the adult species.

Life Cycle of a Fern Diagram

When the minute sporophyte has become established, the gametophyte dies. The gametophyte generation in ferns, as in other seedless vascular plants, is relatively short-lived and simple. The sporophytic phase is dominant in the life cycle. As in all vascular plants, what we recognise as "the plant" is the sporophytic generation, and at maturity it develops sporangia, the spore-producing organs.

In the common ferns, sporangia are borne on the undersides of fronds, or on modified fronds. The first fronds produced in spring are sterile—that is, they lack reproductive structures. The fertile fronds arise later in the season. In some species, such as lady fern (*Athyrium filix-femina*) the sterile and fertile fronds are similar. In other species-sensitive fern (*Onoclea sensibilis*), for example-the sterile and fertile fronds are very different. The fertile regions on fronds that bear sporangia are known as *receptacles*, and the group of sporangia on a single receptacle is called a sorus. In many species, but by no means in all, the sorus is covered during development by a flap of tissue called an *indusium*. The shape and arrangement of the sori are different in different species.

Spores arise by meiosis from sporocytes. For the most part, immature sori are a pale, whitish colour, although there are exceptions, such as autumn fern (*Dryopteris erythrosora*), whose sori, which are produced in the fall, are bright red. As the spores inside each sporangium mature, they typically get darker, until they are a deep brown or black. Not all spores mature to a deep brown or black, however. Polypodiums are ripe when they are buttercup yellow; osmundas, when they are green. The sporangia of ferns is typically a thin-walled case, usually on a stalk, and surrounded by a ring of thick-walled cells known as the *annulus*, which aids in opening the sporangium when the spores are mature.

The indusium, if present, ultimately shrivels when the spores have matured, exposing the sporangia. The annulus breaks near the base of one side, tearing the sporangium apart, and arches backward. Then the annulus snaps forward abruptly, catapulting the spores into the air. Ferns produce tremendous numbers of spores, but the special requirements of most species for moisture and shade effectively reduce the number of gametophytes that ultimately develop. Spores that are deposited by air currents on sufficiently moist soil and rocks germinate within five to six days, and the gametophyte begins to develop. Most fern spores also require light for germination; those of the bracken fern (*Pteridium aquilinum*) are an exception. Early in germination, colourless rhizoidal cells form at the base of the filament of green cells. These rhizoids, or root-like hairs, absorb and carry water and nutrients to the developing gametophyte. Eventually, the gametophyte becomes heart-shaped, and in some species may be as big as half an inch in diameter. The fern gametophyte was long ago called the *prothallus* or *prothallium*, because it was known to be the precursor of the fern plant, even before its sexual function was clearly understood.

As well-nourished, bisexual gametophytes mature they develop antheridia and archegonia, normally on their undersides. The female, egg-producing organs, or archegonia, occur near the notch of the heart-shaped prothallus. The male, sperm producing organs, or antheridia, occur on the "wings" and opposite the notch. Each archegonium has a chimney-like protuberance that is flared at the top to receive the sperm. Each antheridium is a minute, capsule-like sac where the sperm grow. The cells inside the neck of a mature archegonium dissolve, creating a moist passageway through which the sperm can swim to the mature eggs. The eggs of several archegonia may be fertilized, but usually only one of the zygotes develops into a juvenile sporophyte.

It is interesting to note that while bisexual prothalli may be capable of fertilizing themselves, experiments have shown that the resulting sporophytes often fail to thrive. This suggests that cross-fertilization, the union of sperm and egg from different prothalli, necessary for increasing genetic variability within species, is promoted in at least some ferns.

ASEXUAL REPRODUCTION

Asexual propagation can occur in various ways. Many plants produce genetically identical copies of themselves, through a mechanism referred to as asexual reproduction. However, botanists more properly refer to this mechanism as "sexual propagation" or "vegetative propagation" because many hold that only sexual reproduction should be referred to as true reproduction, since this is the only kind of propagation that results in the production of new, genetically unique individuals.

Another type of asexual propagation occurs when plants develop underground stems (or rhizomes) which grow outward, or new shoots which grow upward to form new shoots that are genetically identical to the parent. One such example is the trembling aspen (*Populus tremuloides*), which sometimes develops entire stands of trees growing out of the ground as seemingly individual stems, but are actually genetically identical and interconnected below-ground. Another example is the strawberry (*Fragaria virginiana*), although this species has its vegetative runners above-ground.

Other plants develop bulbils on their stems, which can detach, fall to the ground, and sprout to develop new plants that are genetically identical to the original one. One familiar species that does this is the tiger lily (*Lilium tigrinum*. Other plants can propagate from twigs or stem pieces that fall from the parent, then lodge into a suitable site and develop into a new plant. The crack willow (*Salix fragilis*) can spread itself along watercourses in this manner (as well as by disseminating seeds). Other non-sexual means of propagation include the production of underground bulbs, corms, and tubers that split into parts, each of which is capable of developing into a new plant. Plants that reproduce in this manner include irises and daffodils. All plant organs have been used for asexual reproduction, but stems are the most common.

Stems

In some species, stems arch over and take root at their tips, forming new plants. The horizontal above-ground stems (called stolons) of the strawberry (shown here) produce new daughter plants at alternate nodes.

Underground Stems :

- Rhizomes
- Bulbs
- Corms and
- Tubers.

Irises and day lilies, for example, spread rapidly by the growth of their rhizomes.

Leaves

This photo shows the leaves of the common ornamental plant Bryophyllum (also called Kalanchoë). Mitosis at meristems along the leaf margins produce tiny plantlets that fall off and can take up an independent existence.

Roots

Some plants use their roots for asexual reproduction. The dandelion is a common example. Trees, such as the poplar or aspen, send up new stems from their roots. In time, an entire grove of trees may form—all part of a clone of the original tree.

Plant Propagation

Commercially-important plants are often deliberately propagated by asexual means in order to keep particularly desirable traits (*e.g.*, flower colour, flavour, resistance to disease). Cuttings may be taken from the parent and rooted. Grafting is widely used to propagate a desired variety of shrub or tree.

All apple varieties, for example, are propagated this way. Apple seeds are planted only for the root and stem system that grows from them. After a year's growth, most of the stem is removed and a twig (scion) taken from a mature plant of the desired variety is inserted in a notch in the cut stump (the stock). So long the cambiums of scion and stock are united and precautions are taken to prevent infection and drying out, the scion will grow. It will get all its water and minerals from the root system of the stock. However, the fruit that it will eventually produce with be identical (assuming that it is raised under similar environmental conditions) to the fruit of the tree from which the scion was taken.

Apomixis

Citrus trees and many other species of angiosperms use their seeds as a method of asexual reproduction; a process called apomixis.

- In one form, the egg is formed with 2n chromosomes and develops without ever being fertilized.
- In another version, the cells of the ovule (2n) develop into an embryo instead of—or in addition to—the fertilized egg.

Hybridisation between different species often yields infertile offspring. But in plants, this does not necessarily doom the offspring. Many such hybrids use apomixis to propagate themselves. The many races of Kentucky bluegrass growing in lawns across North America and the many races of blackberries are two examples of sterile hybrids that propagate successfully by apomixis.

- The female cones of their own species (rare) or
- Those of a much more common species of cypress.

Breeding Apomictic Crop Plants

Many valuable crop plants (*e.g.*, corn) cannot be propagated by asexual methods like grafting. Agricultural scientists would dearly love to convert these plants to apomixis : making embryos that are genetic clones of themselves rather than the product of sexual reproduction with its inevitable gene reshuffling. After 20 years of work, an apomictic corn (maise) has been produced, but it does not yet produce enough viable kernels to be useful commercially.

FACTORS AFFECTING PLANT REPRODUCTION

Climate, soil moisture, soil nutrients, disturbances (such as defoliation, fire, cultivation, and disease) are some of the more common things that affect the reproductive ability of plants. Tolerance versus Requirements : not all plants respond the same to stress. For example, both Agrostis tenuis and Festuca ovina have the same tolerance limits for soil pH, but in a stand of Agrostis and Festuca, Agrostis had a maximum abundance at pH 4-6 and Festuca had a maximum abundance at pH 3-4 and 6.5-7.5. There is also a difference within species. For example, 2 ecotypes of Dactylis glomerata (Norwegian and Portugese) have different temperature requirements for reproduction. In Portugal, it is a winter-grown crop (lower temps) and in Norway it is a summer crop (higher temps) and they respond differently when grown in each other's typical environment.

Plant Community Development

Plant Community — "a group of plants living under relatively similar conditions in a definite area"

Development — "to expand or bring out the potentialities; to advance from a lower to higher state" If these two are combined, then plant community development could be considered "to expand or bring out the potentialities a group of plants living under relatively similar conditions in a definite area".

Stages of Community Development

There are numerous papers on the stages of community development : Succession. Nudation, migration, ecesis (adjustment of a plant to a new habitat), competition, reaction and stabilisation.

Disturbance

When discussing community development, it implies that there has been some sort of disruption in the vegetation of the community. Grime defined disturbance as "the mechanisms, which limit the plant biomass by causing its partial or total destruction". Disturbance should always be specified according to spatial extent, temporal extent and magnitude.

Type of Disturbance

Grubb summarised information for different types of plants under different disturbances. While he thought it possible to generalise about type of disturbance and plant life form it was difficult to generalise life characteristics and high disturbance levels. For example, demographic characteristics of communities exposed to grasing reveal a high number of perennial plants capable of both sexual and vegetative reproduction. Annuals and biennials are generally located in interstices and fluctuate in relative abundance.

Grasing intensity tends to increase the dominance of clonal plants. The number of species with vegetative reproduction were significantly different under grasing than those with vegetative reproduction. In the same study, seed dispersal was also analysed. Mobile and adhesion, as methods of dispersal, were both significantly different under grasing. Wind dispersal was significantly different under soil disturbance.

On a denuded glacial foreland, pioneer species were plants whose dominant reproductive form was vegetative. Unlike early colonising species of secondary successions, pioneers of primary succession are usually long-lived perennials that are relatively slow growing.

Research by Vose and White in a burned Ponderosa pine (Pinus ponderosa) community indicated that buried seed did not contribute significantly to seedling recruitment, and seed rain was more important than buried seed. Most regeneration was from seed rain, and the only vegetative reproduction was from the shrub Ceanothus fendleri. It should be noted that this shrub also reproduced by seed. Allogenic disturbances reveal a question to consider : can we take research from autogenic disturbances and extrapolate them to allogenic disturbances?

Intensity and Frequency of Disturbance

Key characteristics suited to intensity of disturbance are very different. Grubb thought special attention should be placed on plants that were able to increase their range after a disturbance and more study was required on life form, phenolgy, and requirements and tolerances.

Vose and White found that intense heat from a pole-stand burn reduced seed rain compared to sites that burned with less intensity. Burning prevented germination of any buried seed. In their study, the authors speculated that post-fire environmental conditions might have been unfavourable to germination and establishment. It is commonly thought that ruderals, with high seed production, generally dominate areas of frequent, high disturbance. Seeds are usually dormant, and these plants are largely opportunistic of the increases in light supply and nutrients in a disturbed site. A summary of data by Grubb in 1985 supported this theory. In natural areas of yearly disturbance, (for example, scour of drift lines by floods), there is a high number of annuals that use the abundant nutrient supply and set seed in the first season. The seeds of these annuals have no specialised means of dispersal and rely on dispersal though the drift material.

Low Temperatures

On the tundra, there are very low reproductive rates and predominance of vegetative proliferation. Most invading species are either self-pollinated, or invest massive resources in vegetative propagules there is a higher representation of species that possess some mode of uniparental reproduction.

Chapin calculated that the carbon cost of producing a tiller of Dupontia fisheri from seed at Pt. Barrow, Alaska was 10,000 times greater that that of producing tiller vegetatively. A single Betula pubescens in Swedish Lapland produces 40,000,000 seeds thoughout its life, only one of which is necessary to replace the original tree for a stable population. Are these seeds only important in the colonisation of new areas outside the forest or after disturbance such as fire within the forest? Recruitment from seed is important in open habitats such as fell-fields, but early mortality rates are high. Vegetative reproduction has low recruitment levels but high survival rates due to the physiological interdependence of modules. (fell fields are open communities of scattered individuals aggregated into islands). The plants are dwarfed and prostrate, and the habitat is characterised by little winter snow cover, water shortage, primitive soils and frequent soil movement due to freeze-thaw cycles).Chambers cautioned in diminishing the importance of sexual reproduction in tundra environments. And though compared to temperate climes, sexual role is reduced, for tundra it is still very important.

Callaghan and Emanuelsson indicate that sexual reproduction may be adapted to the exceptional and not too normal circumstances. Environmental conditions that select for sexual reproduction may be very different from those that select for vegetative growth.

Low Nutrients

On the tundra and in alpine disturbances, distinct species colonisation patterns can be observed. Chambers noted that species with nitrogen-fixing symbionts are frequent colonisers on mineral soils low in nitrogen, including those exposed by retreating glaciers or snowbanks. In alpine tundra areas of severe soil disturbance, where the resultant soils are poorly developed and have low nutrient status, colonisation is most frequently from dissemules, mostly seeds dispersed from adjacent communities. In these conditions, it is necessary for plants to produce many seeds that can easily be dispersed

Importance of Soil and Water

Prach and Pysek studied the role of clonal plants in plots that were 10 to 50 years old, grown under different types of habitats. They had acidic emergent bottoms, peaty soils, sandy, reclaimed, abandoned sand pit, ruderal sites ranging from poor to rich and old fields that were wet to xeric. From their observations of old fields, clonal plants dominated under xeric conditions, while non-clonal dominated under mesic conditions. These sites were 50+ years old. This indicates that more long-term studies in old fields are needed under various environmental conditions.

The study they did for plots that were 10 years old concluded the data set was limited and not able to reveal a functional relationship between the participation of clonal plants and habitats.

Stocklin and Baumler found that seeds found in areas around a glacier (primary to secondary successional areas) were extremely dependent on water availability for germination and establishment. Seedling success was dependent on "safe" sites, which were slight indentations or areas covered by dead plant material. Chambers found that in windy, soil exposed areas (tundra/alpine) relationships between soil characteristics and seed attributes significantly affected the spatial distribution of dispersed seeds. This may be one of the primary determinants of species colonisation patterns.

Chambers found that with small soil particles, small seeds and seeds with adhesive coats were trapped, but most large seeds moved across the surface; at large particle size, larger seeds were trapped, with larger particles, small seeds moved further downward in the soil as well. In the same study, surface soil temperatures were measured and on dark covered organic soils, the temperature ranged from 35-45 °C on the surface and 20-25 °C at 5 cm depth. On an adjacent site with exposed, loamy sands, midday soil temperatures were about 5 °C lower. Soil temperature can result in higher rates of mineralisation and decomposition. Most tundra species had higher germination rates under higher temperatures than what is normally found in the field.

Competition

Small clones spread laterally and are not overtopped by larger clones. Dominance and suppression by non-clonal Genetic variability in plants

Temporal Considerations

Noble and Slatyer, listed the following three plant attributes as most important in vegetation replacement :

- The method of arrival or persistence of the species at the site and after disturbance.
- The ability to establish and grow to maturity in the developing community.
- The time for the species to reach critical life stages.

Epp and Aarssen found interactions between species in recently established vegetation (1 year) are affected more by differences in attributes that are relevant to early life, *i.e* seed weight. This relates to the competitive ability of seeds in establishing themselves in an open area. They also discovered that the older a stand became, other attributes became important in a plant's ability to remain dominant. In a 1 year old plot, lateral spread only accounted for 0.3 per cent of the variation, while in the 11 year old stand, it accounted for 26 per cent of the variation in species diversity. Seed weight (number of seeds was not measured) accounted for 13 per cent of the variability in the 1-year-old plot, but only 0.03 per cent in the 11 year old plot.

Non-clonal plants had maximum coverage after 4 years, and the clonal plants had maximum coverage in years 6 and 7. Coverage for the clonal plants decreased after year 7. They also determined a difference between guerilla and phalanx plants. Phalanx had a peak of 70 per cent coverage in year 5 and that decreased to 28 per cent coverage in year 10. Guerilla plants peaked at year 6, at 72 per cent and in year 10 were 61 per cent (this was found to be significantly different).

Species capable of guerilla growth are often more successful in the later stages of succession than plants with phalanx growth; this may be explained by the guerilla's ability to produce widely spaced modules that have a greater chance of penetrating into more or less closed vegetation. Differences in clonal growth forms were also found by Oborny in simulation models of different types of vegetatively reproducing plants. Clonal plants that had a rigid structure decreased in number over time vs spp that had plastic growth form.

Vegetative reproduction becomes of more importance in later stages, when dense cover and a thick litter layer limit establishment from seed. But there are limitations to vegetative reproduction as well. Belsy found that populations of Sporobolus fimbriatus and Pennisetum mexianum began to replace Digitaria maroblephara and Sporobolu ioclados after 5 years of non-grasing. This is due to the fact that S. fimbriatus and P. mexianum are both tall rhizomatous species, whereas Digitari maroblephara and Sporobolus ioclados are both stoloniferous. The build up of the litter layer suspended the stolons above the ground. This was also true for species that reproduced by seed. The seeds could not germinate with the heavy litter layer.

Factors That Influence Colonisation : Specifically in Regards to Sexual and Vegeta Reproduction

It has been hypothesised that the role of clonal dispersal in succession is evident both in early stages, due to the rapid capture of space, and in late-successional stages, due to the rapid filling of gaps in more or less closed vegetation cover. In essentially closed cover, seeds do not have the opportunity to germinate, nor do seedlings have the chance to establish. Competition for nutrients, water and light is just too great. This would imply that eventually all plant communities should be dominated by plants that rely heavily on vegetative reproduction. Why is it then, that there are many plant communities that appear to stabilise with a combination of plants that use both seed (sexual) and vegetative (asexual) reproduction? Research on the role of clonal species is lacking, and an understanding of their behaviour is based to a large extent on speculation. The factors that inhibit or promote one reproductive strategy over another are not well understood.

Chapter 21

PROKARYOTES AND PHYTOHORMONE MUTANTS IN THE HORMONAL RESPONSES IN PLANTS

Prokaryotic systems have been important in phytochrome studies on several different levels. Bilins from cyanobacterial phycobiliproteins have allowed the production of recombinant holophytochrome and have provided insights into the attachment and functioning of the chromophore, while the recent discovery of functional phytochromes in the cyanobacterium Synechocystis and other prokaryotes has catalysed work in the field.

Synechocystis phytochrome is useful experimentally and, by making the modular structure and potential biochemical functions of phytochromes clearer, has provided an improved focus and new viewpoints for research. Some of the earliest studies of photobiology concerned phenomena in cyanobacteria : the complementary chromatic adaptation (CCA) of photosynthetic pigments to the light environment was first described in Engelmann's laboratory in Berlin a century ago. Numerous other effects such as phototaxis, photoperiodism, cell division, and differentiation are also regulated by light in cyanobacteria.

Plant plastids probably evolved from endosymbiotic cyanobacteria whose genes gradually moved to the host nucleus. There is thus every reason to expect evolutionary relationships between photoperception systems in cyanobacteria and plants. Detailed information about cyanobacterial photoreceptors was lacking, however, until genomic sequencing revealed a cyanobacterial phytochrome : ironically, the ease with which molecular methods can be used in prokaryotes has now turned the tables, with the cyanobacterial model providing a wealth of new ideas about the origins of phytochrome and its mode of action. Here we review the different ways in which cyanobacteria and other prokaryotes have contributed to research into plant photomorphogenesis and the phytochrome system.

Phytochrome is an ubiquitous plant photoreceptor that was first characterised in the late 50s in relation to its peculiar photochromic behaviour in red and far-red light. Phytochromes carry an open-chain tetrapyrrole (bilin) chromophore, which the apoprotein autocatalytically attaches to a conserved C residue our alignment2) via a Schiff base. In darkness, this autoassembly produces the redlight absorbing form Pr (lmax ' 660 nm). In red light this is photoisomerised to another form, Pfr, which absorbs maximally in far-red light (lmax ' 730 nm). In far-red light, Pfr is in turn converted back to Pr. Both forms are thermodynamically stable and can be interconverted by any number of photocycles. Because even tiny amounts of Pfr have major physiological effects, it is generally accepted that this is the active form of phytochrome, while Pr seems to be physiologically inactive.

In plants, phytochromes control a variety of developmental processes such as seed germination, stem elongation, construction of the photosynthetic apparatus, chloroplast movements, shade avoidance, and photoperiodic induction of flowering. In lower plants they are also involved in sensing light direction.

CYANOBACTERIA AND PLANTS CONTAIN FIVE DIFFERENT BILIN CHROMOPHORES

Based on the first spectral measurements, it was correctly argued that the phytochrome chromophore might be a bilin similar to those of phycobiliproteins in cyanobacteria and red algae. Phycobiliproteins bear four types of bilin, namely phycocyanobilin (PCB), the chromophore of phycocyanin (PC, which is thus generally the most abundant), phycoerythrobilin (PEB), phycoviolobilin (PVB), and phycourobilin (PUB). Apophytochromes autoassemble with PCB to form holophytochrome photoreceptors—a useful feature, as PCB can be prepared rather easily from commercially available cyanobacteria, enabling the preparation of functional phytochromes by recombinant methods.

However, in all land-plant phytochromes examined so far, another member of the family, phytochromobilin (PFB), is the natural chromophore. Phytochrome chromophores undergo a characteristic Z 3 E isomerisation around the C15AC16 double bond between rings C and D during Pr 3 Pfr conversion. PEB also assembles with apophytochrome *in vitro*, but the product between pyrroles A and D by heme oxygenase to form the linear tetrapyrrole (bilin) biliverdin IXa.

Genes for this enzyme are known from the cyanobacterium Synechocystis PCC6803, from the plastome of red algae, and from the genomes of Arabidopsis and several animals. In Arabidopsis heme oxygenase is encoded by the nuclear gene HY. The ptr116 phototropic mutant of the moss Ceratodon can be rescued by exogenous biliverdin by micro-injecting cells with mammalian heme oxygenase enzyme or by overexpressing HY1, showing that all three functionally complement the defective ptr116 gene.

The subsequent steps are not fully understood biochemically, nor have any further associated genes been described, although the Arabidopsis HY2 locus is a likely candidate.

Further sequences from the Synechocystis genome must also be involved in the production of PCB and other bilins is not photochromic because the C15-C16 bond is saturated. The five bilins (PCB, PFB, PVB, PEB, and PUB) above differ only in single double bonds and are derived from biliverdin, the first open-chain tetrapyrrole in this biosynthetic pathway.

ONLY THE FIRST GENE FOR THE BILIN SYNTHETIC PATHWAY IS KNOWN

In seed plants, the enzymes for bilin synthesis are located in the plastids but are nuclear encoded. The pathway begins with the opening of the heme tetrapyrrole ring between pyrroles A and D by heme oxygenase to form the linear tetrapyrrole (bilin) biliverdin IXa. Genes for this enzyme are known from the cyanobacterium Synechocystis PCC6803, from the plastome of red algae, and from the genomes of Arabidopsis and several animals.

In Arabidopsis heme oxygenase is encoded by the nuclear gene HY1. The ptr116 phototropic mutant of the moss Ceratodon can be rescued by exogenous biliverdin by micro-injecting cells with mammalian heme oxygenase enzyme or by overexpressing HY1, showing that all three functionally complement the defective ptr116 gene.

The subsequent steps are not fully understood biochemically, nor have any further associated genes been described, although the Arabidopsis HY2 locus is a likely candidate. Further sequences from the Synechocystis genome must also be involved in the production of PCB and other bilins.

CYANOBACTERIAL MODELS FOR PHYTOCHROME

Early physiological studies indicated that cyanobacteria might harbor useful information about the phytochrome system.

Action spectroscopy revealed photoreversible effects analogous to those of plant phytochrome but with the interesting distinction that, while in plants the responses are maximally induced by red light and reverted by far-red light, most photoreversible effects in cyanobacteria respond to red light (lmax ' 650 nm) and green light (lmax 520 nm).

The most intensely studied effect here is CCA. Unlike plants, cyanobacteria possess phycobilisome structures that funnel energy into the photosynthetic system. The principle accessory pigments involved are the blue-green (red-absorbing) PC and allophycocyanin and the red (bluegreen-absorbing) phycoerythrin (PE). Some species are able to use CCA to adjust the ratio of these pigments according to environmental conditions.

In green light PE dominates, whereas in red light PC dominates. In this way, the lmax of photosynthesis is shifted to the lmax of the light environment.

When a series of green (' 540 nm) and red (' 650 nm) pulses was given to a culture of the cyanobacterium Fremyella diplosiphon and the culture kept in darkness, the last light pulse determined the dominant accessory pigment formed.

This kind of photoreversibility points to a photoreceptor with photochromic properties, an unusual spectral feature that allowed plant phytochrome to be isolated and characterised. However, this approach was less successful in cyanobacteria, principally because, unlike angiosperms, they do not etiolate.

Although a photoreversible pigment showing difference maxima at 520 and 650 nm has been described, it was not characterised further. Later, the a-sub-unit of the minor phycobilisome component phycoerythrocyanin (PEC) was shown to be photochromic but with difference maxima at 500 and 570 nm.

The physiological role of PEC is unknown, although it might have a role as a photoreceptor as well as acting as a photosynthesis antenna. The PECa chromophore is PVB, which undergoes a Z 3 E photoisomerisation analogous to phytochrome. Moreover, the PECa and PCa sequences are approximately 65 per cent identical and the molecules have similar three-dimensional structures—yet only the former is photochromic.

While one might therefore suppose that the differences between these biliproteins could provide a master key to unlock the secrets of photochromicity at the atomic level, whether this key would fit phytochrome is questionable.

First, there is no sequence homology between phycobiliproteins and phytochrome. Second, isolated phycobiliproteins show far stronger fluorescence than phytochrome, implying that their chromophores are much more tightly held and/or have a very different photochemistry.

Third, relative to the dark state, the conformational changes in PECa are associated with a blue shift, whereas a red shift is seen with phytochrome.

Interestingly, although phycobiliprotein apoproteins are generally capable of autocatalytically attaching bilin chromophores in vitro, it has been shown that specific lyases mediate the assembly *in vivo*, accelerating the reaction and ensuring correct bilin attachment. Phytochrome autoassembles *in vitro* too, but whether a discrete phytochrome bilin lyase exists is simply not known.

TWO-COMPONENT SIGNALING ENTERS THE FRAY

A quite different line of investigation also connects plant phytochromes with prokaryotic systems. Schneider. Poetsch *et. al.* drew attention to a significant amphiphilic sequence similarity between the phytochrome C terminus and the transmitter module of bacterial sensory His protein kinases (HPKs). HPKs are a group of proteins responsible for the first step in the so-called two component signal transduction pathways that provide the prokaryotic cell with its capacity for perception and response.

The activity of each HPK is regulated by an associated sensory module whose conformation changes in response to an environmental signal such as an interaction with a specific ion or molecule. The HPK is a dimer and, upon sensor activation, each sub-unit phosphorylates the other at a conserved H target residue within the transmitter module. The phosphate is then transmitted to a conserved D residue in the receiver module on the second component of the transduction system, the response regulator.

This then does as its name suggests, either by activating transcription of specific genes itself or by interacting with other proteins to bring about specific physiological changes in the cell appropriate to the environmental signal.

These two-component signal transduction systems seem to be the primary regulatory connections between prokaryotic metabolism and the environment.

Schneider-Poetsch's suggestion that phytochrome might represent a plant sensory HPK was enhanced by the discovery of eukaryotic HPK homologs SLN1 in yeast and ETR1 in plants a year or so later, and indeed the idea that phytochrome might be a light-dependent kinase was nothing new.

A seductive aspect was that it provided phytochrome with its long-sought reaction partner : This should be a response regulator homolog. Many were not convinced by the HPK/phytochrome homology, however, and there was one very large problem : although all of the functional sub-domains characteristic of HPK transmitters are recognisable in phytochromes, the all-important target residue itself is poorly conserved.

It seemed that the affair was over when Boylan and Quail showed that even in phytochromes in which was conserved, it could be mutated without noticeable physiological effect.

A, Basic scheme. The HPK dimer is activated by conformation changes induced by stimuli perceived by the sensor module, usually N-terminal. The transmitter module of each sub-unit then transfers a phosphate (red dot) from ATP to a conserved His residue (circle) of the other sub-unit. The phosphate group is transferred to a conserved Asp residue (square) of the cognate response regulator, which is thereby activated. An autophosphatase activity returns the response regulator to its inactive state in the phosphorylation cycle (PC) – the rate of hydrolysis is sometimes regulated by interactions with the "inactive" form of the same HPK or with another molecule.

B, Cph1/Rcp1 system. Pr, The ground state of the phytochrome in darkness or far-red light is the active HPK; irradiation with red light converts the molecule to Pfr, in which the HPK activity of the transmitter module is inhibited. Following autophosphorylation at, Pr transfers the phosphate to the response regulator Rcp1. The biochemical functions of Pfr and the response regulator are unknown.

POSSIBLE PHOTORECEPTORS IN CYANOBACTERIA

Work with the cyanobacterial CCA perception system, however, continued independently. Complementation methods were used to clone genes involved in the regulation of chromatic adaptation in F. diplosiphon. RcaE encodes a 74-kD polypeptide with an approximately 150-residue portion towards the N terminus showing homologies to several regions around the chromophore-binding domain of plant phytochromes, as well as C-terminal motifs typical of transmitter modules. Intriguingly, RcaE also bears two sub-domains showing homology to the plant ethylene receptor ETR1. Although the biochemistry has yet to be demonstrated,

RcaE probably phosphorylates the response regulator RcaF, which in turn phosphorylates RcaC. The latter bears a transmitter- and two receiver-like modules, as well as a DNA-binding motif thought to mediate differential transcription of the PC and PE gene complexes. In the similar cyanobacterium Calothrix, RcaD and RcaA act as phosphorylation-dependent activators of the PC and PE gene clusters, respectively.

The 155-kD conceptual gene product of PlpA in Synechocystis PCC6803 also shows regions of similarity to phytochromes (hence the name Plp for phytochrome-like protein) and two component modules. Indeed, BLAST searches show that RcaE has approximately 25 per cent amino acid identity (40 per cent similarity) to PlpA, although the latter has a long N-terminal extension. This particular Synechocystis strain does not show CCA. It does, however, change the stochiometry between PS1 and PS2 according to the irradiance and spectral distribution of the light environment : In plpA2 knockouts the balance between the photosystems is disturbed. Furthermore, in contrast to the wild type, the plpA2 mutant cannot grow photoautotrophically in blue light.

Although both PlpA and RcaE are clearly important in cyanobacterial photoperception, it has proven difficult to demonstrate that they are photoreceptors. The sensory function could be fulfilled by a separate molecule, as in many two-component systems. At least on the basis of homology to the phytochrome N terminus, there is little reason to expect either RcaE or PlpA gene products to be *bona fide* phytochromes : If one assumes that the chromophore is thioether-linked to a C residue, as in phycobiliproteins and plant phytochromes, the alignment in this region is constrained to C198#380 and C784#380 for RcaE and PlpA, respectively.

The surrounding sub-domain is quite different from that seen in phytochromes, where it is well conserved and changes generally lead to a complete loss of function. However, although difficulties with overexpressing PlpA and RcaE in Escherichia coli have hampered in vitro studies, it has now been reported that both do seem to be capable of attaching bilins.

A PROKARYOTIC PHYTOCHROME CREATES EXCITEMENT IN THE FIELD

Quite separately, the entire 3.57-Mbp chromosome of Synechocystis PCC6803 was sequenced in a singularly efficient project at the Kazusa Institute in Japan providing the scientific community with a wealth of valuable new data. Among the 3,168 open reading frames identified, a phytochrome-like sequence (slr0473) was recognise.

The N-terminal moiety showed patchy but unmistakable similarity to phytochromes, including the all-important chromophore binding region, whereas the 30-kD C-terminal moiety was clearly homologous to typical two-component transmitter modules with characteristically conserved H-, N-, G1-, F-, and G2-boxes. The question nevertheless remained : is it a phytochrome?

The 85-kD gene product was further analysed simultaneously by Lagarias's group at University of California-Davis and by our laboratory.

The apoprotein overexpressed in E. coli autocatalytically attached PCB chromophore in vitro to form a blue-green photochromic pigment, clearly establishing that it encodes a bona fide cyanobacterial phytochrome, Cph1. The discovery of this phytochrome caused an immediate paradigm shift in the field.

Schneider-Poetsch's suggestion that the unknown mechanism of primary signal transduction could be related to the well-established twocomponent system in bacteria suddenly became a very hot topic. Moreover, the utility of a prokaryotic phytochrome system in biochemical and molecular-genetic studies opens new experimental possibilities. In particular, the efficiency with which highly soluble recombinant phytochrome can be prepared from E. coli overexpressors offers fresh hope that the three-dimensional structure of this class of photoreceptors could be resolved via NMR and x-ray diffraction analysis of phytochrome crystals.

Residue numbering is that from the alignment. The N-terminal bilin-bearing sensor module (blue) is recognisable in all phytochromes, while the C-terminal transmitter module (yellow) is common to phytochromes and most HPKs of the two-component type. The sensory modules of other HPKs (brown) are different and are sometimes carried on separate polypeptides. A PAS module (green), important for plant phytochrome signal transduction, is absent from Cph1. The chromophore attachment site, two PAS repeats, the signal transduction core, and the H-, N-, G1-, F-, and G2-boxes are highly conserved. Plant B-type phytochromes generally have an N-terminal extension, otherwise the N terminus of all phytochromes is Ser/Thr rich.

CPH1 *IN VITRO*

Cph1 has been subjected to a range of optical and biophysical studies in our laboratory and in those of our colleagues. Recombinant apoprotein (Cph1°) is expressed remarkably efficiently in E. coli, accumulating to approximately 30 per cent of cytosolic protein. Moreover, equipped with a C-terminal oligohistidine tag, it can be purified almost to homogeneity in a single Ni21-affinity chromatographic step. E. coli does not support bilin synthesis, and thus autoassembly does not occur *in vivo*. If Cph1° is added to PCB in vitro, however, a dramatic blue to blue-green colour change occurs within seconds. This results from two processes. Initially, recombinant holoprotein (Cph1) is formed as Pr, whereby the PCB red absorbance peak at 610 nm is shifted to 658 nm as the helical form of the free bilin becomes unwound in the protein environment. Thereafter, if observed in daylight, the Pr is photoconverted to Pfr, whose absorbance peak is shifted even further to 702 nm.

Both photochromic forms are quite stable in darkness. The yield of pure Cph1* is routinely about 20 mg per liter of culture. It can be concentrated to above 15 mg/

mL quite easily, satisfying an important further precondition for many biophysical and physicochemical studies, including crystallisation. Like other HPKs and plant phytochromes, Cph1* behaves as a dimer in solution. As one would expect for a photoreceptor, the extinction coefficient of Cph1* is very high (approximately 100 mm21 cm21 for Pr at lmax) and the quantum conversion efficiency is about 0.16 in both directions—values similar to those for plant phytochromes.

Cph1° can also be assembled with other chromophores : The PFB adduct shows a red shift of about 15 nm for both Pr and Pfr, as seen with plant phytochromes. PEB adducts cannot photoconvert because of their missing C15AC16 double bond; the quantum energy is released as fluorescence and their absorbance maximum is blueshifted to 579 nm.

Fourier-transform Raman-resonance (FTRR) and flash photolysis, low-temperature fluorescence, and Fourier-transform IR absorbance spectroscopic methods have also been used. Despite the considerable differences in the peptide sequences, Cph1* shows remarkably similar physicochemical properties to those of B-type phytochromes.

FTRR is a sensitive probe for the status of the chromophore in biliproteins, revealing in this case many similarities between the chromophores of native oat phytochrome A and the equivalent PFB adduct of Cph1.

For both phytochromes, spectral differences between the Pr and the Pfr form reflect the Z 3 E isomerisation of the chromophore and changes in its hydrogen bonding with the protein. Moreover, as in plant phytochromes, subtle differences between the PCB and the PFB adduct of Cph1* can be attributed to the ring D side chain (vinyl group for PFB versus ethyl group for PCB). FTRR also indicated different torsions around methane bridges within the chromophore and differences in chromophore/protein interactions between Cph1 and oat phytochrome.

As for other phytochromes, the formation of intermediates during Pr 3 Pfr photoconversion of Cph1 was readily observed by flash photolysis and fast spectroscopy. The first photoproduct detected (lumi-R) of Cph1* appeared substantially more quickly than that for plant phytochromes and was followed by a novel intermediate whose kinetics were delayed almost 2-fold by 2H exchange, implying that a protonation/deprotonation is involved at this point.

FTIR difference spectra also indicate 2H effects, and a photoreversible pH shift seems to confirm that proton extrusion accompanies Pfr formation. Fluorescence measurements at low temperature address the photoconversion from a different point of view. Whereas at ambient temperature phytochrome fluorescence yields are very low, these rise dramatically upon cooling; Pr 3 Pfr photoconversion is inhibited, although photoconversion into intermediate forms is sometimes possible.

For plant PHYA at 70 K, up to 50 per cent of the Pr can convert into lumi-R, whereas this conversion is not possible for plant PHYB. PCB and PFB Cph1* adducts are also unable to form lumi-R at this temperature, implying that

Cph1 is more related to PHYB than to PHYA photochemically. Different activation barriers for the photoreaction are thought to explain the differences between phytochrome types. The only intermediate photoproduct found after allowing the temperature to rise seemed to be rather different from the lumi-R of plant phytochromes.

Cph1 is A Light-dependent His Protein Kinase

The Lagarias group analysed the biochemistry of Cph1 regarding its apparent homology to two-component systems. Cph1 autophosphorylates at the expected H538#995, but it was a great surprise that the active kinase was not Pfr but Pr.

This flew in the face of most plant physiological data, which implied that Pfr was the active form. There was more to come, however. Unlike eukaryotes, prokaryotes often group biochemically related genes together in a single cistron, thereby keeping the job of co-ordinating expression simple while obligingly providing the scientist with clues to unknown biochemical associations. While biochemists had long sought the primary reaction partner(s) for plant phytochrome, the likely reaction partner for Synechocystis phytochrome was clear from the Kazusa chromosome map. Fifteen bases downstream of the Cph1 stop codon begins a short open reading frame, slr0474, unmistakably coding for a 17-kD response regulator of the two-component type.

Yeh *et. al.* overexpressed this gene in yeast and demonstrated that the autophosphorylated Pr form of Cph1 promptly transmitted its phosphate to the expected Asp residue D68 of the putative response regulator. slr0474 was thus the first primary reaction partner for phytochrome to be identified and was named response regulator for cyanobacterial phytochrome, Rcp1. Here again, Pr was more active than Pfr. All of the known enzymatic activities of phytochrome (bilin ligase, His autokinase, His-Asp transphosphorylase, and, as we shall see, Ser/Thr kinase) were first described by the Lagarias laboratory.

Although many response regulators are DNA-binding proteins and act as transcriptional activators, the Synechocystis model is not quite so simple. Rcp1 has no DNAbinding motifs and thus presumably acts as an intermediate phosphocarrier in a more complex relay. This is also seen in other two-component systems such as Spo and Rca. The epithet refers to the transmitter module of the kinase and the receiver module of the response regulator, whereas the transduction system as a whole can be considerably more extensive, with pathways converging and branching to form a sophisticated control network.

So with what does Rcp1 interact? Unfortunately, in Synechocystis only Cph1 and Rcp1 are co-transcribed, the flanking genes being read in the opposite direction. As the next partner cannot simply be deduced from the genome map, finding the rest of the transduction chain will prove more difficult.

CPH1 *IN VIVO* REMAINS A MYSTERY

Although Synechocystis certainly contains abundant PCB, the native chromophore of Cph1 is not known. CCA in other cyanobacteria shows well-separated maxima in the blue-green and red regions, whereas the absorbance maxima of PCB and PFB adducts of Cph1* are poorly separated and are at longer wavelengths. If a Cph1 homolog is indeed the photoreceptor for CCA, then the blue shift might result from the use of a different chromophore.

Alternatively, a quite separate photoreceptor might be involved. Measuring difference spectra in extracts of Synechocystis was unsuccessful because of masking pigments, even when using PC- deletion mutants. In an attempt to overcome this, homologous recombination was used to replace the wild-type chromosomal gene with a sequence extended to provide an oligohistidine-tagged translation product similar to that in the E. coli overexpression clones, thereby allowing the photoreceptor to be purified by affinity methods.

However, despite this technology, the extracts have yielded only tiny amounts of the modified native Cph1, indicating a very low expression level. Interestingly, the purified fraction showed not only the expected red/far-red difference spectrum with maxima around 650 and 700 nm, but also a red/green difference with maxima at 530 and 650 nm, close to the maxima for CCA. Whether the red/green reversible signal relates to a co-purified protein or directly to Cph1 remains to be determined.

Ironically, despite all the studies of Cph1 in vitro, its physiological function is unknown : What aspect of the light environment does it perceive and what response does it mediate? This gap in our knowledge is all the more surprising because the efficient homologous recombination available in Synechocystis allows knockout mutants to be created with some ease.

Indeed, both Cph12 and Cph12/ Rcp12 knockouts have been generated in several labs, but an associated phenotype has yet to be found. Perhaps the effects are masked by another photoreceptor system, or the Cph1-Rcp1 pathway might lead to a physiological dead end in this particular strain. One might expect Cph1 to regulate CCA—but, unfortunately, PCC6803 and most other strains of Synechocystis lack PE entirely and thus could not show CCA even if they wanted to. However, Synechocystis PCC6701 shows classical CCA, carrying a PE gene cluster closely homologous to that in Fremyella and replacing PC with PE in blue-green light. The Cph1 homolog in CCA-active Calothrix has also now been cloned. It will be interesting to see if a cph12 knockout in one of the CCAactive types shows the Rca2 phenotype.

OTHER PROKARYOTIC PHYTOCHROMES HAVE ALSO BEEN IDENTIFIED

Several other Synechocystis genes also show similarities to phytochromes. It seems that the true homolog of RcaE is not PlpA (sll1124), as implied above, but, rather, is represented in Cyanobase by two pseudogenes separated by a transposon (sll11473–sll1475); in other PCC6803 cultures the RcaE homolog is intact. sll0821 is also intriguing as it shows two regions with homology to the phytochrome chromophore sub-domain, both of which bind PCB *in vitro*.

Even further removed from plant phytochromes are the BphP (bacterial phytochrome photoreceptor) genes recently found on the chromosomes of Deinococcus radiodurans and Pseudomonas aeruginosa. The former autoassembles with bilin chromophores *in vitro* to yield a phytochrome-like red/far-red light photochromic product. This result is surprising because, although a region resembling the phytochrome chromophore sub-domain is apparent, residue #380 is M rather than C. While we assume that the chromophore attachment site of Cph1 is C259#380, this has yet to be demonstrated chemically.

FROM PROKARYOTES TO PLANT PHYTOCHROME

Most of the above discussion concerns phytochrome in prokaryotes, but how does that help the plant physiologist? Most importantly, it provides conceptual links. First, the alignment of Cph1 to plant phytochromes and HPKs provided a new and clearer view of phytochrome molecular architecture. Second, while the initial cloning of phytochrome was a great technical achievement in itself, the sequence did not provide us with beguiling homologies to molecules of known function. Cph1 provides a link not only to bacterial two-component systems, but also to several other eukaryotic homologs including SLN1, DHKA and DHKB, ETR1, and CKI1, all thought to take part in phosphorelay-mediated signaling. This rejuvenated the idea that phytochromes might be light-dependent protein kinases.

PLANT PHYTOCHROMES POSSESS A PAS MODULE INVOLVED IN SIGNAL TRANSDUCTION

Cph1 alignments revealed an additional approximately 300-residue module peculiar to plant phytochromes, placed between the sensory and transmitter modules. The module is also missing from the Deinococcus and Pseudomonas homologs. This region of the plant phytochrome sequence had already aroused interest since it contains a repeated motif related to the PAS3 domain family. It would seem that this PAS module was added to a Cph1-like progenitor early in plant evolution, perhaps even before eukaryotes arose, bringing with it a set of biochemical features probably including a new signaling mechanism.

PAS domains are found in diverse proteins throughout the living world; particularly interesting is the apparent PAS homology of the bacterial photoreceptor PYP (photoactive yellow protein. PAS domains often bind ligands and are involved in protein-protein interactions including signal transduction.

There is ambivalent evidence that the PAS repeats S599#675 to L683#766 and L685#768 to R815#901 are involved in the dimerisation of phytochrome A. Furthermore, random mutagenesis studies indicate that the PAS module is crucial to the plant phytochrome signaling mechanism. Yeast two-hybrid studies identified several phytochrome interacting factors that seem likely to bind to the PAS module.

One of these is involved in phytochrome signaling *in vivo*, is nuclear localised, and even possesses a DNA-binding domain, offering a remarkably – if not deceptively – simple picture of plant phytochrome action, given that newly formed Pfr migrates to the nucleus. Although the plant phytochrome PAS module is missing from Cph1, a Hidden Markov model detects PAS-domainrelated structures in Cph1, RcaE, and PlpA at different positions. A weak but significant similarity between HPK modules and the PAS domain has also been pointed out, providing the latest twist to an unfinished story.

KINASE AND KINASE-RELATED FUNCTIONS

As far as kinase function is concerned, the conceptual framework is not simple. As we have seen, athough the H538#995 target in Cph1 and its homologs in other sensory HPKs in both prokaryotes and eukaryotes is essential for autokinase and phosphorelay function, the homologous residue in plant phytochromes is neither conserved nor functional. While it is nevertheless possible that plant phytochromes could act as HPKs (for example, some PHYAs show an H-box-like [L/V][A/P] SHELQ[Q/H]AL#961-#970 motif at the PAS/transmitter module boundary) and response-regulator homologs certainly exist in plants, we emphasise that there is no evidence that any plant phytochrome functions as an HPK. Nevertheless, as we shall see, the HPK transmitter domain seems to be very much involved in signal transduction.

The two-component paradigm might help us to understand plant phytochrome function independently of the prokaryotic kinase action; the H-box is by no means the best conserved of the two-component transmitter sub-domains in plant phytochromes. The structures might have been retained for some purpose other than autophosphorylation and phosphotransfer. The unusual architecture of the chemotaxis HPK CheA suggests two possibilities.

First, the three-dimensional structure of CheA shows a relict H-box in the conventional position, which, along with downstream residues, comprises the K290#981 to R354#1055 dimerisation site. All HPKs seem to form stable dimers with sub-micromolar dissociation constants as a result of sub-unit binding in this region. Plant phytochromes are also dimers but the domains involved are uncertain. The CheA dimerisation domain is made up of two antiparallel, highly amphiphilic a-helices with hydrophobic residues exposed on the sub-unit surface; PHD predicts that this region in plant phytochromes is also largely helical with an amphiphilic pattern.

Second, recent studies in Arabidopsis indicate that the N-terminal 100-residue fragment of phytochrome B binds the response regulator homolog ARR4, whereas the equivalent phytochrome A fragment does not. B-type phytochromes generally bear a characteristic N-terminal extension relative to other family members, so it is possible that the extension mediates the interaction.

This would be analogous to the unconventional H-target sub-domain at the N terminus of CheA, although little sequence homology is apparent and there is no evidence that the phytochrome is involved in a phosphotransfer. The system seems to connect to a two-component system involved in hormone signaling. Missing a conserved H#995-target residue in plant phytochromes, it was suggested that the perfectly conserved Y#991 nearby might have taken over the acceptor function. Tyr and Ser/Thr protein kinases (YPKs and S/TPKs, respectively) work differently from HPKs; after autophosphorylation, rather than donating their own single phosphates, they phosphorylate their substrates with phosphate groups from free ATP4.

There are, however, precedents for protein kinases showing a different substrate specificity from that implied by their primary structure. Moreover, immunological methods suggest a light-regulated Y phosphorylation of oat PHYA.

Although the residue involved is not known, motifs around Y#319 and Y#1055 in various phytochromes resemble the phosphotyrosine-binding site of SH2 domains. While this appears to be the sum of current evidence for phytochrome YPK function, the possible relationship between two-component systems and YPK/ Ras GTPase signaling should encourage a careful search for related mechanisms in the case of phytochrome.

PLANT PHYTOCHROME IS A DIFFERENT KIND OF KINASE

Even before the sequence of oat PHYA was published, Quail and co-workers drew attention to its peculiarly S/Trich N terminus as a possible kinase substrate. This was perhaps born of necessity as it was the only feature of the sequence that hinted at a function. While phytochrome N-terminal sequences are not well conserved, S and T residues predominate—also in Cph1.

Indeed, S8#45 in oat PHYA is phosphorylated; however, the physiological significance of this is unclear, as the level of phosphorylation is similar for Pr and Pfr. Interestingly, mutation of the N-terminal Ser residues in PHYA increases rather than decreases its physiological potency in transgenic plants, so the phospho-Ser modification might serve to attenuate phytochrome action, analogously to arrestinmediated quenching of rhodopsin. In the case of phytochrome it is uncertain whether this is an autophosphorylation event or whether a separate kinase is involved.

The idea that phytochrome might be a S/T kinase bothered biochemists for many years, but recent evidence using recombinant systems *in vitro* and in vivo indicates that plant phytochromes can indeed autophosphorylate S/T residues and phosphorylate other proteins, including Rcp1, in a light-dependent manner. Major differences from the Cph1 system should be made clear, however. First, plant Pfr becomes more strongly labeled than Pr, implying that the assembled sensory module in its ground state represses the autokinase activity—the opposite of Cph1. Second, although plant phytochrome phosphorylated the Rcp1 response regulator *in vitro*, the target was not the D68 used by Cph1. As histones too were effective substrates, the relevance of this observation might be called into question. However, it seems that the phytochrome kinase substrate PKS1 is phosphorylated by Pfr both *in vitro* and *in vivo*, with overexpression leading to repression of phytochrome action.

Third, rather than H538#995, one or more unknown S/T residue(s) in plant phytochrome are autophosphorylated. S599#675 of oat PHYA—within the PAS module but N-terminal of the first repeat—shows Pfr-enhanced phosphorylation *in vivo* and would thus seem to be an obvious candidate. However, it is not conserved and the S599K#675 mutant still autophosphorylates and phosphorylates PKS1—although light regulation is lost. In relation to domain function it is interesting to note that, while it is probably not a functional HPK, the plant phytochrome transmitter module alone is sufficient for PSK1 binding.

Three-dimensional structural studies with kinases and their allies are already advanced, providing useful background information regarding possible functions in phytochrome. Most of the residues directly responsible for ATP binding in HPKs and gyrases are well conserved in phytochromes-interestingly deviant are N#1118 and D#1168. We look forward to the day when phytochrome will contribute to studies of kinase function in general.

PR VERSUS PFR

For the plant physiologist, perhaps the most intriguing aspect of Cph1 is that Pr is the active kinase, while in plants Pr is thought to be inactive, Pfr being the "active form of phytochrome". The red-light-induced formation of tiny amounts of Pfr from the Pr pool in the cytoplasm of imbibed seeds or dark-grown seedlings leads to the profound physiological changes associated with germination or deetiolation. Physiological responses do correlate quite well with the Pfr concentration of PHYA in etiolated tissues, although in green tissues this is less certain because spectroscopic measurements are hampered by strong chlorophyll fluorescence and the approximately 100-fold lower amounts of phytochrome. Genetic studies seem to have settled the issue, however, as phy2 mutants phenocopy Pr.

There is also an attitude problem. For the physicist, photoreceptors are in their ground state in darkness and are excited by light—thus the ground state of Cph1 is Pr and the excited state is Pfr. But for the Cph1 biochemist, the active kinase is Pr—mirroring the behaviour of the Rhizobium HPK oxygen sensor, FixL, in which ligand binding represses kinase activity. For the biologist, on the other hand, prolonged darkness is equivalent to starvation for a photosynthetic organism—"ground state" is hardly an appropriate description.

As discussed above, there is no reason to suppose that plant phytochromes act as HPKs—even if that was the original function of Pr. On the other hand, it seems now that plant phytochromes act as Pfr-active S/T protein kinases. This leaves a question open : What is the biochemical function of Pfr in cyanobacteria? Once again, the twocomponent paradigm provides possible answers.

Many HPKs are known to be bi-functional, phosphorylating or promoting de-phosphorylation of the response regulator according to their conformation as determined by the sensor module. Both activities are important as they allow the transduction system to differentiate rather than integrate the input signals from the sensor module, a principle that also applies to eukaryotic G-protein-coupled signaling, as in the rhodopsin/arrestin system.

The photochromic nature of Cph1 offers the attractive possibility that the excited Pfr form might play the opposite role to that of Pr, promoting the dephosphorylation of Rcp1. Of course, cyanobacterial Pfr might have a quite separate biochemical activity or it may simply be inactive. Whatever the role of Pfr in Cph1 is, it might be retained in plant phytochromes.

CONCLUSIONS

As we have discussed, photochromic detection systems and chromophores associated with the cyanobacterial phycobilisome have made crucial contributions to the study of phytochrome at the molecular and conceptual levels. The unexpected discovery of phytochromes in other prokaryotes both answers and poses many questions. Phytochrome apparently appeared before eukaryotes, evolving over vast tracts of time and under changing selection pressures to glean and then transmit pertinent information about the light environment to allow the organism to respond appropriately. Perhaps most surprising, then, is the clarity of the homologies between prokaryotic and plant phytochromes. The origin of the PAS module involved in plant phytochrome signaling can also be traced to prokaryotes. Thus, a variety of prokaryotic models are acting catalytically in studies of phytochrome, the active sites being the modes of signal transduction and the photochromic mechanism itself. The reaction products should prove most interesting.

Chapter 22

SIGNAL TRANSDUCTION BY PLANT HORMONES

PHOTORECEPTORS IN SIGNAL TRANSDUCTION

Sunlight is the origin of nearly all the metabolic energy that drives life processes, and plants and algae are by and large responsible for the conversion of solar energy into chemical forms that are used by all organisms.

These simple truths are perhaps obvious to plant biologists, but I think they merit enunciation every once in a while, if only to remind zoologists that, ultimately, plants rule. Even the fossil fuels that support much of modern human culture are owed to the photosynthetic activity of plants. Of course, plants have not evolved to photosynthesise on behalf of the less self-reliant inhabitants of the planet, but rather to provide for their own nourishment, growth, and development. Thus, virtually every aspect of plant life, from germination through reproduction and senescence, is affected by light.

The biology of many non-photosynthetic organisms is also linked to sunlight. The reflective and refractive properties of light are visually perceived by many animals so that they may gauge within their varied and changeable environments the availability of food, water, shelter, and even reproductive partners, as well as evade the threat of predators and other dangers. Vertebrate sleep cycles, and sleep-like states even in invertebrates, are also linked to the sun's daily periodicity. Circadian rhythms, in fact, appear to be basic to organisms as diverse as fungi, plants, and mammals.

Not all aspects of sunlight, however, are conducive to life, and it is certain that the evolution of photoperception arose in large part from the selective advantage of being able to avoid the deleterious effects of solar radiation, as opposed to exploiting solar energy and periodicity. UV light is indeed one of the most common mutagens within the biosphere. Accordingly, a range of biomolecular mechanisms that mitigate the harmful effects of UV light occur throughout the biological kingdom.

One protective strategy relies on the photolyases, chromophore-bearing enzymes that repair UV-induced pyrimidine adducts within the genomes of bacteria, plants, and animals. An intriguing aspect of photolyases is that they themselves undergo photoactivation. By virtue of a light-harvesting chromophore—either a pterin or a deazaflavin derivative—that absorbs light from the blue portion of the visible spectrum, photolyases utilise light energy to reverse DNA damage such as thymine dimers.

Specifically, the excitation energy imparted to the light-harvesting chromophore is transferred to an active site-bound molecule of (reduced) flavin adenine dinucleotide (FAD). In a free-radical mechanism, a single excited electron from FAD can be transiently transferred to a thymine dimmer lesion, thereby cleaving the adduct, before returning back to regenerate the fully reduced FAD moiety. In this way, organisms exploit for purposes of DNA repair the very same source (*i.e.*, the sun) that induces mutagenic lesions.

The existence in photosynthetic organisms of molecular events that are light-regulated by virtue of the ATP and reducing equivalents that accumulate during photosynthesis has been recognised for some time. Enzyme activation, transcription, and translation are all subject to the photosynthetic energy charge of the plant cell. But photolyase is distinctively instructive as a specific example of how light cues can directly trigger a series of biomolecular events that are not directly related to photosynthetic metabolism.

Admittedly, the series of events occur in the case of photolyase within a single enzymatic context : light energy is collected by a chromophore (*i.e.*, a photoreceptor); the light energy is transferred transiently to a distinct molecular species (*e.g.*, FAD); and a biomolecular reaction is effected (in this case, DNA repair). In other cases, however, activation of photoreceptors can result in a cascade of intermolecular events that qualify as *bona fide* mechanisms of signal transduction.

The phytochromes, photoreceptors that respond to red and far-red light, are the most familiar non-photosynthetic plant photoreceptors, but the importance of blue and UV light-absorbing photoreceptors—the cryptochromes, in particular—is currently gaining wide recognition. As in the case of the phytochrome family of proteins, cryptochromes appear throughout the plant kingdom and arise from multigene families : two cryptochrome-encoding genes have been identified in Arabidopsis, and three have heretofore been isolated in the fern, Adantium capillus-veneris. Sequence analyses, moreover, show the cryptochromes to be remarkably similar to the blue light-dependent pho tolyases discussed above. Cryptochromes also resemble photolyases in that both categories of blue light-absorbing protein extend beyond the plant kingdom; indeed, cryptochromes have been implicated in the entrainment of circadian behaviour in Drosophila, humans, and mice.

Although no cryptochrome to date has been found to possess DNA repair activity, the cryptochromes are presumed to have evolved from the photolyases in such a way that enzyme activity was discarded and photoreceptor activity was elaborated.

In Arabidopsis, both CRY1 and CRY2 have been implicated in developmental functions such as inhibition of hypocotyls elongation and cotyledon expansion, and CRY1 has been linked to additional processes of development, circadian cycle, and light-dependent gene regulation. Current research into the photobiology of plants places emphasis on further specifying the roles of individual cryptochromes, with the ultimate goal being to elucidate varying cascades of blue light signal transduction. In this regard, the fascinating relationship between the photolyases and cryptochromes is sure to be informative.

Sequence comparisons can speak not only to matters of phylogeny *per se*, but they also can potentially offer mechanistic insights into the structure–function bases of cryptochromes as signal transducers.

Capillusveneris, bringing the total number of discovered cryptochrome genes in this fern to five and thereby establishing the largest family of cryptochromes to be yet defined. Besides possessing a set of variant cryptochromes that can be compared within the biological context of a single plant species, A. capillus-veneris represents a promising experimental organism in that several aspects of its development are regulated by blue light.

Previous studies in the fern, moreover, have suggested a correlation between blue light responses and intracellular localisation of the pertinent photoreceptor(s). By analogy to the emerging picture in which distinct phytochromes are differentially translocated into the nucleus during red light photoperception and signal transduction, the intracellular localisation of cryptochromes and the factors that may control localisation are of particular interest. Imaizumi *et. al.*

Have examined the intracellular distribution of the five fern cryptochromes by creating functional cryptochrome– b-glucuronidase fusion proteins. In this way, a tendency for two of the cryptochromes, CRY3 and CRY4, to accumulate in the nucleus has been observed; CRY3 does so specifically in response to red light. The nuclear localisation signal for both proteins appears to reside within the proteins' C-terminal domains; interestingly, the C termini of cryptochromes are not related to photolyases, the presumed evolutionary forerunners that function as enzymes but not as signal transducers proper.

The translocation of cryptochrome species into the nucleus echoes data confirming that phytochromes undergo translocation into the nucleus in response to red ligh.Whether other aspects of cryptochrome signaling will prove to mirror phytochrome-mediated signal transduction remains to be seen. It is in fact only recently that mechanisms and components of phytochrome signal transduction, beyond the prosthetic tetrapyrrole chromophore, have been identified.

In addition to the involvement of classic signal transducing agents such as G proteins, calcium, and calmodulin, for example, it now appears that at least one phytochrome species has kinase activity. In Arabidopsis, moreover, a kinase that is related to a tumor suppressor in animals has been implicated in phytochrome signaling, as has been a nuclear protein that binds specifically to the C terminus of phytochromes. Whether transduction cascades that initiate with cryptochromes involve signal transducers that are similar to those recruited by phytochrome remains to be seen.

Nevertheless, Imaizumi *et. al.* have made an important step towards refining our understanding of cryptochrome- regulated developmental events. Specifically, they have examined the age-dependent expression profiles of the five A. capillus-veneris CRY genes and their responses to light. Whereas CRY, CRY2, and CRY3 are transcribed in a regulated manner throughout the A. capillus-veneris life cycle, expression of the newly described CRY4 and CRY5 genes differs strikingly over the course of development. CRY4, as opposed to CRY5, for example, is expressed preferentially in spores and is strongly downregulated in response to light.

CRY5 transcripts, on the other hand, are virtually undetectable in spores but appear during other stages of the fern life cycle and are induced by light during spore germination.

The analysis of these light effects in terms of blue and red/far-red wavelengths shows that both regions of the spectrum are linked to expression of the cryptochrome genes.

Intriguingly, crosstalk between cryptochrome and phytochrome signal transduction pathways has been suggested previously in Arabidopsis, and a case can indeed be made for the involvement of phytochromes in the regulation of both CRY4 and CRY5 gene expression in A. capillus-veneris. This fern system can thus be regarded as an attractive model for further studying the dynamics of photoreceptor signaling in multiple aspects of plant development.

SIGNALING IN PLANTS

Two major advances in plant biology have been critical for the analysis of signal transduction pathways. First, the ability to produce and analyse large numbers of mutant plants has allowed scientists to identify numerous alleles that are involved in signal transduction pathways. The Arabidopsis plant has desirable biological and genetic qualities such that tens of thousands of mutant plants can be screened to identify plants that have modified responses to a signal.

Facile screens have been developed that precisely identify mutant plants that are either response-deficient, and fail to respond to a stimulus, or response-constitutive, and respond in the absence of the signal. Genetic analyses have been performed to elucidate both the order and interactions of pathway components, and suggest that a complex network of information input and interacting components are transmitted to the nucleus to direct changes in gene expression.

Second, biotechnological developments and genome project efforts in Arabidopsis have been critical for the success of the molecular genetic approach. Mutational techniques have been developed and include chemical mutagenesis and insertional mutagenesis, in which specific DNA sequences cause mutations by insertion into the genome, thereby "tagging" the mutated gene with a novel DNA sequence. In addition, this plant has a relatively small genome ('120 million base pairs, 25 times smaller than humans) with only five chromosomes.

The Arabidopsis genome has been cloned as very large fragments in yeast artificial chromosomes and bacterial artificial chromosomes with insert sizes ranging from 100,000 to 750,000 base pairs. These large genomic fragments are being ordered into contiguous regions (contigs) so that the genetic positions of markers becomes integrated with their physical positions in the contigs.

Large numbers of cloned DNA sequences have been used to produce a physical map of the chromosomes such that the proximity of a new mutant allele to a unique DNA sequence can be established. This now allows new mutant alleles to be genetically mapped to a physically defined region of cloned DNA that greatly facilitates the isolation of the mutant gene; this process constitutes "map-based cloning."Genetic complementation analysis in which the wild-type gene is introduced into the mutant plant can then be used to demonstrate genetic proof that the gene that causes the mutation has been identified.

The complete sequence of the entire Arabidopsis genome is expected to be available by 2004. Signal transduction pathway from light absorption by photoreceptors to the light-dependent developmental changes that prepare a plant for photosynthetic activity (photomorphogenesis). Plants utilise at least three systems to perceive light : UV photoreceptors; blue light photoreceptors; and the red light photoreceptor known as phytochrome. Phytochrome senses the relative intensities of photosynthetically active red light (660 nm) and non-photosynthetic near far-red light (730 nm), thus sensing information about light quality and the suitability for photosynthesis.

Dark-grown plants develop as etiolated plants with characteristic features : absence of chlorophyll, suppression of cotyledon and leaf expansion, hypocotyls elongation, and the downward folding of cotyledons.

Exposure of plants to red or blue light initiates developmental programmes that prepare the plant to perform photosynthesis : chlorophyll and photosynthetic apparatus are produced, hypocotyls and stem elongation are suppressed, and the cotyledons and leaves unfold and expand.

Mutant screens for loss of the photomorphogenetic response have been used to identify plants that no longer respond to red, blue, or near far-red light, and these plants show characteristic features of etiolation when grown in the various wavelengths of light. These plants are effectively "blind" to the light !

This approach has led to the identification of a number of Arabidopsis light-responsive mutants, including mutants in the biosynthesis of the phytochrome chromophore, two of the five phytochrome genes, and a blue light receptor gene. Individual mutations result in different degrees of deviation from normal photomorphogenesis; however, the combination of two lightresponsive mutations generally demonstrate exaggerated effects.

These results emphasise the importance of multiple signal inputs from various types and members of receptor families, and suggest that the systems are probably partially redundant and very complicated. Screening for mutants in light-response deficiency led to important advances in the understanding of sensory input information at the level of light receptors, but unfortunately has not led to much detailed information on the signal transduction pathways beyond the photoreceptors.

The phytochrome signal transduction pathway. Signal perception by phytochrome, the light receptor, is transmitted to the nucleus by complex interactions shown inside the rectangle. Mutational analysis and molecular cloning have been used to identify and order the components of the signal transduction pathway.

Mutant screens of Arabidopsis plants for photomorphogenetic response in the absence of light have been used to identify plants that constitutively show photomorphogenesis and are considered deetiolated (det) or constitutively photomorphogenetic (cop). These plants are delusional—behaving as though they have experienced light exposure in complete darkness !

Six mutant genes have been cloned from this class of mutation by insertional mutagenesis or by map-based cloning. The COP1 protein has features of both transcription factors and G proteins, suggesting a general role for this protein in gene expression. The det2 gene has recently been characterised by the Chory lab as a steroid biosynthetic enzyme (steroid 5 alpha reductase), and has resulted in the rediscovery of a new class of phytohormones (3). Exogenous application of brassinolide to det2 mutant Arabidopsis plants produces the wild-type phenotype, suggesting that this chemical may represent a class of hormones with an important role in lightregulated development of higher plants.

Joseph Ecker, Associate Professor at the University of Pennsylvania, discussed how plants use various chemical signals to co-ordinate physiological processes. Plants use many different chemicals to send signals. Auxins, cytokinins, abscisic acid, gibberellic acid, and ethylene are "classical" phytohormones; however, a number of very recently discovered phytohormones include brassinolide (3); systemin, a peptide hormone; methyl jasmonate; and salicyclic acid.

The ethylene signal transduction pathway is ripe for analysis. The biochemical pathway and many of the genes for ethylene biosynthesis are well characterised. Exposure of dark-grown seedlings to ethylene leads to a set of clear morphological changes known as the "triple response."

These changes include (i) inhibition of hypocotyl and root elongation, (ii) radial swelling of the hypocotyl, and (iii) a diageotrophic (horizontal) growth habit. The triple response has been used to identify several types of Arabidopsis mutants (4). Ethylene-overproduction mutant plants produce high endogenous amounts of ethylene and consequently display the triple-response phenotype even in the absence of exogenous ethylene. A more interesting phenotype is shown by the constitutive tripleresponse mutant (ctr1). These plants do not overproduce ethylene, but develop as though they have been exposed to high levels of ethylene gas. The ctr1 mutant is defective in the ethylene signal transduction pathway, apparently as a result of the loss of function of a negative regulator of ethylene response.

The CTR1 gene has been cloned and encodes a protein kinase related to the Raf family.

Ethylene-insensitive mutants (*ein, etr, ain*) are ethylene response-deficient plants and represent a phenotype defined by numerous genes. These genes can be placed in a relative sequence of action by genetic analyses.

One of these very early acting genes has now been cloned and reveals a most interesting surprise. The ETR1 polypeptide shows sequence similarity with bacterial and fungal histidine kinases such as SLN1, a component of the osmotic-stress transducing pathway in yeast. Interestingly, SLN1 acts through a phospho-relay system to activate a downstream protein kinase cascade. More recent biochemical studies have now shown that ETR1 has the capacity to generate ethylene binding sites when expressed in yeast, strongly supporting the notion that ETR1 encodes an ethylene receptor.

As additional molecular components are put in place in the light and hormone signal transduction pathways, the apparent complexity of the process grows, and the increased potential for controlling plant growth and development becomes evident. Ethylene co-ordinates the agronomically important process of fruit ripening, and prospects of controlling food production by modification of ripening processes have already been explored by Calgene (Davis, CA) with the Flavr-Savr tomato and by scientists who have regulated ethylene production in tomato. Applications of signal transduction as well as basic understanding of plant developmental processes will have a very important role in future agricultural applications.

G-PROTEIN-LINKED SIGNAL TRANSDUCTION IN PLANTS

All cells, whether belonging to unicellular or multi-cellular organisms, have the capacity to communicate with their surroundings by detecting and responding to a wide range of stimuli. Extracellular signals such as hormones, neurotransmitters, growth factors and even light are detected by interaction with specific receptors present on the plasma membrane of the target cell, triggering biochemical processes that produce intracellular events.

In most transmembrane signalling systems present in animals it is possible to distinguish three functional classes of proteins : (a) the receptor – the protein that interacts physically with the extracellular signal *e.g.* ?-adrenergic receptor, rhodopsin; (b) the effector – the protein that alters the level of second messengers (cAMP, Ca2+,dyacylglycerol, etc.). This group includes enzymes such as adenylate cyclase, phospholipase C and ion channels; (c) the transducer – the protein that transduces the signal between the receptor and the effector. In this signalling system the transducer is a GTP-binding protein (G-protein). The class of signal transducing G-proteins includes heterotrimeric GTP-binding proteins involved in transmembrane signalling and monomeric low-molecular-weight GTP-binding proteins, also called small G-proteins. molecular masses of 35-36 and 7-8 kDa respectively, are tightly associated and function as a Gβγunit.

In the basal state, the Gα sub-unit is associated with the Gβγ complex and its guanine nucleotide binding site is occupied by GDP. The binding of an agonist to a G-protein coupled receptor (GPCR) causes a change of conformation in the latter.

The interaction between the agonist occupied receptor and the G-protein promotes the increase in the rate of exchange of bound GDP for GTP. The binding of GTP reduces the affinity of the Gα sub-unit for the receptor and for Gβγ, causing dissociation of the complex. The free G α sub-unit, bearing GTP, and in some cases the Gβγ heterodimer, can then modulate the activity of effectors which can alter levels of second messengers. The life time of the active Gα sub-unit is determined by its GTPase activity, which converts GTP to GDP, releasing inorganic phosphate. In its GDP bound form, Gα has high affinity for Gβγ, so the inactive heterotrimer is again formed and the system returns to the basal state.

Special analogues of GTP, such as GTPγS (guanosine-5′-O-(3-thiotriphosphate) and GppNHp (Guanylyl (βγ-imido) diphosphate) can bind to G α sub-units but can not be hydrolysed leading to an irreversible activation of G-proteins. In opposition, a GDP analogue, GDPβS (guanosine 5′-O-(thiodiphosphate), is a G-protein inhibitor that binds irreversibly to G-protein trimers preventing their dissociation. The bacterial exotoxins cholera toxin and pertussis toxin can also modulate the activity of G-proteins. Cholera toxin catalyses the ADP-ribosylation of some G α sub-units and consequent inhibition of their GTPase activity. This causes the permanent activation of the G-proteins. On the other hand, pertussis toxin ADP-ribosylates G-proteins on their trimeric state keeping them permanently inactive. Both the guanine nucleotide analogues and the bacterial toxins are frequently used in experiments on G-proteins.

Cell Signalling in Plants

Plant cells respond to a wide variety of environmental signals such as light, gravity, plant hormones, microbial elicitors, wind, minerals, water, gases and soil structure. The mechanisms by which plant cells transduce external information into intracellular signals is often unclear. However, in many cases there are similarities to animal signal transducing pathways.

For example, all the elements of a Ca2+ based signal transduction are found in plant cells. In animal cells, extracellular signals through plasma membrane receptors can activate phospholipase C which hydrolyses phosphatidylinositol 4,5-bisphosphate, producing two second messengers, inositol 1,4,5 trisphosphate (IP3) that releases Ca2+ from internal stores and diacylglycerol that activates protein kinase C. Protein kinase C, which is a Ca2+-activated phospholipid-dependent enzyme, regulates the activity of several enzymes by phosphorylation. In plants, the phosphoinositide system also appears to be present. IP3 was shown to be involved in mobilisation of vacuolar Ca2+ and to stimulate stomatal closure and the reversible inactivation of K+ channels whilst diacylglycerol was shown to be involved in stomatal opening. Ca2+-dependent protein kinases and phospholipases C have also been shown to be present in plant cells.

Cyclic AMP (cAMP) is a very important second messenger in animal tissues and in many micro-organisms but its occurrence in higher plants and its role in the regulation of their metabolism is still unclear. Adenylate cyclase, which is the enzyme that converts ATP into cAMP has been detected in plants. The presence of cAMP in cells of several plant species has also been reported. However its concentration was shown to be very low (less than 0.5 nmol/kg of tissue) compared to animal cell levels (50-250 nmol/kg of tissue).

Evidence for the Presence of Heterotrimeric G-proteins in Plants

Several different approaches have been used to demonstrate the presence of G-proteins in plants. The presence of GTPase activity, high affinity GTP-binding associated with plant membranes and the ADP-ribosylation of plant membrane proteins, catalysed by cholera and pertussis toxins have been reported. Putative plant G-proteins have also been immunologically detected. In addition, plant genes encoding sequences similar to G-protein sub-units of other organisms have been isolated. The involvement of G-proteins in plant cell signal transduction pathways has also been studied by observing the modulation of G-protein activity by external stimuli. Alternatively, the effects of G-protein modulators, *e.g.* and GTP analogues and bacterial exotoxins, on plant biochemical processes have also been tested.

GTP-binding Activity

The presence of GTP-binding proteins in plants has been demonstrated by binding assays using the radiolabelled GTP analogue [35S] GTPγS. Membrane proteins of Lemna paucicostata, Pisum sativum, Anabaema cylindrica and Avena sativa were separated by gel permeation chromatography and various fractions shown to possess GTP-binding activity. However, in these communications the authors reported that the GTP-binding proteins showed affinities to ATP comparable to GTP. Ohmori *et. al.*, (1988) proposed the existence in plants of ATP-GTP-binding proteins that might be the prototype of signal transducing GTP-binding proteins.

In comparison, in membranes of Cucurbita pepo, Nicotiana tabacu, and Oryza sativa the GTP-binding activity detected was very specific for GTP and GTP analogues and ATP was not an effective competitor. The presence of Mg2+ increased GTPγS binding in C. pepo microsomal and plasma membranes although GTPγS binding to plasma membranes did not appear to have an absolute requirement for free Mg2+ unlike the G-protein based activity in animal systems.

Immunodetection of G-proteins

Putative plant G-proteins have also been detected immunologically using antibodies directed against whole G-protein sub-units from other organisms or using antibodies directed against synthetic peptides corresponding to sequences conserved among animal G-proteins (anti G αcommon) or specific to a particular G-protein sub αtype (Gαo, Gαi, Gαs). Antibodies raised against the Arabdopsis thaliana G α sub-unit, GPα1 have also been used.

The molecular masses of the cross-reacting polypeptides are in a range of 24-68 kDa which is different from the data obtained for animal Gα. The studies reported differed in the controls used to demonstrate specificity of the immunological cross-reaction. The most appropriate control when using anti-peptide antibodies is the removal of the cross-reactivity by pre-incubation of antibodies with the appropriate synthetic peptide. Other kinds of immunoblotting controls have been used, such as probing the blots with pre immune serum, pre-incubation of antibodies with carrier protein, and probing blots without the primary antibodie.

However, in some studies controls were not used at all and some of the results obtained almost certainly represented non-specifically cross-reacting proteins.

GPA1ara2 is raised against the C-terminal sequence of A. thaliana GPα1. Gat (transducin) and Gβγwere raised against whole protein. Controls were : a, pre-incubation of antibodies with specific antigen; b, probing blots with pre-immune serum; c, pre-incubation of antibodies with carrier protein; d, probing of blots in absence of primary antibodies; e, no control. PM, plasma membranes; MM, microsomal membranes.

Weiss *et. al.* carried out extensive immunolocalisation of the A. thaliana GPα1 during development. They found that GPα1 is present through all stages of development, in all organs tested with the exception of mature seeds. It was also shown that the level of GPα1 is higher in immature than in mature organs.

Modulation of Intracellular Levels of Second Messengers and Ions

If plants have G-protein linked signal transduction, it is expected that G-protein activators and inhibitors can modulate levels of second messengers in plant cells. In fact, addition of GTP was found to induce the rapid release of Ca2+ from C. pepo microsomal membranes while GTPγS was shown to promote phosphatidyl inositide turnover in Acer pseudoplatanus membranes.

This stimulation was found to be specific for GTPγS and GTP and was inhibited by G-protein inhibitor GDPγS. In Daucus carota membranes, the G-protein activators GTPγS and NaF were found to release the phospholipase C product IP3 in a dose-dependent manner and GDP and GDP?S inhibited this release.

There is evidence that G-proteins are associated with the regulation of plant K+ channels. In Vicia faba guard cell protoplasts, GDPbS caused an increase in the inward K+ current through K+ selective ion channels while GTPγS and treatment with cholera toxin had the opposite effect.

These results indicate that the activation of a particular G-protein inhibits inward K+channels. Similar results were reported for outward K+ channels of V.

faba mesophyl cells where GTPγS and cholera toxin catalysed ADP-ribosylation negatively regulate the magnitude of outward K+ flux. It must be mentioned, however, that the concentrations of guanine nucleotides used in both these studies (500 M) were over 1000 times the apparent Kd values estimated for plant G-proteins. In addition, GTPγS is able to act as an effective substrate for many protein kinases. It is possible, therefore, that other kinds of proteins able to bind GTP may contribute to the effects observed.

G-proteins and Photosynthesis

Spinach thylakoid membranes were able to bind GTP and GTPγ in a specific and light-dependent fashion and the bound nucleotide caused inhibition of the ATP-induced decrease in chlorophyll fluorescence due to the thylakoid light harvesting complex II kinase, indicating a possible involvement of G-proteins in photosynthesis.

It was also reported that cholera toxin catalysed ADP-ribosylation of a 60 kDa polypeptide in P. sativum thylakoid membranes. The ADP-ribosylation was modulated by light and was strongly correlated with inhibition of GTPase activity.

G-proteins and Phytochrome

The best characterised plant photoreceptor is phytochrome. Phytochrome is reversibly interconvertible between its inactive Pr (red absorbing) form and active Pfr (far red absorbing) form by sequential absorption of red and far-red photons of light.

Both red and far red light were found to inhibit GTPγS binding in L. paucicostata. In contrast, in A. sativa microsomal membranes and in Medicago sativa plasma membranes, red light stimulated GTPγS binding while far red light reversed the effect.

Neuhaus *et. al.* working with phytochrome deficient tomato cells (aurea cells) also showed the involvement of G-proteins in the phytochrome response. Micro-injection of phytochrome A into aurea cells restored phytochrome mediated effects. Injection of Pertussis toxin or GDPγS (which keep the G-proteins in their trimeric inactive form) with phytochrome A blocked the response.

Injection of high GTPγS concentrations (30-100μM) or coinjection of cholera toxin and low GTPγS concentrations (1μM) produced an intracellular response indistinguishable from that mediated by phytochrome A. Although heterotrimeric G-proteins appear to be involved in the phytochrome signal transduction pathway, the mechanism of interaction between the molecules is not clear. Phytochrome is a cytosolic protein and it is not structurally related to G-protein coupled receptors which are integral membrane proteins. It is possible that an intermediate may transduce the signal from Pfr to G-proteins.

G-proteins and Blue Light Photoreceptors

Blue light regulates many aspects of growth and development in higher plants, such as increase in stomatal aperture, change in gene expression, suppression of stem elongation and curvature. A plant blue light photoreceptor has not yet been definitely characterised at the molecular level but there is evidence that a flavin is the chromophore involved in one of the blue light perception signal pathways. However, other chromophores like the carotenoid zeaxanthin may be involved in blue light responses.

In apical bud plasma membranes of P. sativum, Warpeha *et. al.* (1991) demonstrated that GTPγS binding and GTPase activity could be induced by blue light but not by red light. Both pertussis and cholera toxins were able to ADP-ribosylate a 40 kDa protein from these membranes. Blue light prevented pertussis toxin catalysed ADP-ribosylation while the cholera toxin effect was detected only when either blue light or GTP were supplied, indicating that blue light leads to activation of the G-protein. Anti-G-protein antibodies also recognised a 40 kDa protein in the plasma membrane. Finally, a 40 kDa protein covalently bound a non-hydrolysable photoactivated analogue of GTP only after irradiation with blue light. Compounds that quench flavin excited states, such as phenylacetic acid and potassium iodide inhibited the blue light enhanced GTPγS binding to the plasma membranes.Therefore, the receptor coupled to this G-protein was considered to be a flavoprotein.

G-proteins and Plant Defence Response

Plant cells release reactive oxygen species such as H_2O_2 or O_2- during stimulation by pathogenic elicitors, *e.g.* polygalacturonic acid. This oxidative burst may constitute the first line of defence against pathogen invasion. Legendre *et. al.* (1992) have studied the role of G-proteins in the oxidative burst in cultured soybean cells. They showed that internalisation of the antigen-binding fragment (Fab) of the anti-Gacommon antibody by soybean cells enhanced up to 10-fold the rapid oxidative burst mediated by polygalacturonic acid. In the absence of elicitors, the addition of mastoparan (a peptide from wasp venom that stimulates G-proteins) to soybean cells also induced the response. In Western blotting experiments, a 45 kDa soybean protein was immunologically detected by anti-G?common antibodies.

G-proteins and Plant Hormones

The action of plant hormones at the cellular level is still poorly understood. However, some hormone receptors have been characterised and there is evidence that G-proteins could be involved in plant hormone mediated signalling. Auxin increased GTPγS binding to microsomal membranes of Daucus carota and to rice aerobic coleoptiles. Auxin was also shown to increase GTPase activity in rice coleoptile membranes. Ca2+-dependent swelling of mesophyll protoplasts from dark-grown wheat was shown to be induced by auxin, gibberelic and abscisic acid. The swelling could be inhibited by GDPγS, indicating the involvement of G-proteins in the process.

Isolation of Plant G asubunit Genes

Strong evidence for the presence of G-proteins in plants is the isolation of putative plant Gα sub-unit genes. Genes encoding G-protein αsubunits from A. thaliana (GPA1), tomato (TGA1), soybean and from rice (RGA1) have been isolated. The proteins encoded by these genes, GPα1, TGα1, SGα1 and RGα1 respectively, have molecular masses around 44-45 kDa. The proteins have the consensus regions for a GTP-binding protein but show low homology to mammalian α sub-units in the regions that interact with receptors and effectors. DNA blot hybridisation experiments showed that GPA1 and TGA1 are single-copy genes. These results suggest that the number of species of Gα in plants may be small in contrast to the large numbers (more than 16Gα) found in animals.

Identification of Gb and Gg Sub-units

There are very few reports on the detection of Gβ and Gγ sub-units in plants. A 37 kDa Gβ-like polypeptide was detected immunologically in M. sativa? and several proteinsβ cross-reacted to antibodies directed against whole β sub-units in P. sativum apical bud plasma membranes. On the other hand anti-Gβ antibodies did not detect any protein in A. sativa β membranes, S. bicolor, Z. maysγ or P. sativum. A gene encoding aGβ-like polypeptide was cloned from the unicellularβalga Chlamydomonas reinhardtii.

This Gβ -like protein is about the same size as and shares a fairly low (25 per cent) amino acid identity with animal Gβ1.?However, analysis of its primary structure showed it to possess segmented internal repeats (WD-40 repeats)?which are characteristic of Gβ sub-units. Ishida *et. al.*? isolated from tobacco cells an auxin-responsive gene, arcA, which encoded a 35.8?kDa protein 25 per cent identical to the mammalian Gβ sub-units and containing the WD-40 repeats. Finally, the COP1 gene of A. thaliana was isolated by Deng *et. al.*, and was found to encode a 75 kDa protein? that contains a zinc finger domain and a Gβ -like domain containing four WD-40 repeat segments. The COP1 gene appears to govern suppression of photomorphogenesis in the absence of light. The predicted protein appeared to be hydrophylic and may represent a novel bifunctional Gβ which is able to couple signals at the cell surface to direct effects on gene expression.

Purification of Plant G-proteins

Very few reports describe the purification of G-proteins from plant tissues. Bilushi *et. al.* (1991) reported the purification of a GTP-binding protein from Z. mays plasma membranes using gel filtration and ion exchange HPLC. The purified fraction, which showed GTPγS binding activity, contained a 61 kDa protein composed of two polypeptides of 34 and 27 kDa. The authors claimed these proteins to be the α and β sub-units of a G-protein but no immunological detection, amino acid sequence data or specificity of GTPγS binding was reported. Purification of plant G-proteins by affinity method using GTP ligands has also been attempted. De Boer *et. al.* (1994) reported the affinity purification of GTP-binding proteins from oat root plasma membranes using a GTP-biotin bifunctional ligand. Several polypeptides were present in the purified fraction and proteins of 67 and 38 kDa cross-reacted with anti-Gγc antibodies. In these studies no control was used for the immunological detection. It must be borne in mind that plant proteins other than heterotrimeric G-proteins can bind GTP.

White *et. al.* (1993a), in attempts to purify G-proteins from P. sativum microsomal proteins by ion exchange HPLC, showed that most of the GTPγS binding activity measured was due to the presence of nucleoside diphosphate kinase. Nucleoside diphosphate kinases are enzymes that catalyse the transfer of terminal phosphate of various nucleoside 5′-triphosphates to acceptor nucleoside diphosphates. The apparent binding constants determined for the native and recombinant P. sativum nucleoside diphosphate kinase, 44 nM and 53 nM respectively, were of an order the magnitude appropriate for a Gγsubunit. However, the apparent binding reflected a formation of the nucleoside diphosphate kinase-thiophosphate enzyme intermediate rather than reversible binding of GTPγS. More recently, the purification of a 40 kDa polypeptide from Sorghum bicolor that cross-reacted with anti Gαo serum was accomplished using Cibacron Blue affinity resin and anion exchange HPLC. Affinity labelling studies with GTP showed that the 40 kDa protein was labelled in a GTP-specific fashion. Finally, microsequencing of the protein yielded limited amino acid sequence that was consistent with its assignment as G β sub-unit.

G-protein Coupled Receptors in Plants

All known animal G-protein coupled receptors (GPCRs) are proteins composed of a single, often glycosylated chain, which possesses seven hydrophobic regions of sufficient length to span the plasma membrane. It was shown above that plant G-proteins may interact with plant receptors such as phytochrome, and auxin binding proteins. In fact, these receptors clearly do not belong to the family of GPCRs. To date no GPCR has been isolated from plant tissues but indirect evidence for their existence will be shown bellow.

Studies with the peptide toxin mastoparan have suggested the presence of GPCRs in plants. It is well established that mastoparan is able to stimulate GTP-binding and GTP hydrolysis by G-proteins in a manner similar to G-protein coupled receptors. White *et. al.* have demonstrated that a mastoparan analogue, mas 7, stimulated GTP?S binding to microsomal and plasma membranes from P. sativum and Z. mays whereas an inactive mastoparan analogue, mas CP, was ineffective at promoting binding. In addition, Legendre *et. al.*, have shown that mastoparan was able to induce the oxidative burst response in soybean cells. More recently, a 37 kDa putative GPCR from Zea mays was isolated by affinity chromatography using both an oligopeptide corresponding to the C-terminal portion of

A. thaliana GPα1 (which is predicted to contain the receptor interaction region) and recombinant GPα1 as affinity ligands.

Additional evidence for plant GPCRs was provided by immunological studies. Microsomal membranes prepared from S. bicolor, N. tabacum, Z. mays and P. sativum contained polypeptides which crossreacted with antisera against squid rhodopsin.

Presently, the amount of plant DNA sequence data is increasing very fast because of efforts in sequencing completely the genomes of key plant species. Computational analysis of sequence database entries can be used to find plant GPCR based on the presence of transmembrane regions similar to animal GPCRs. Using this approach, candidate GPCRs from A. thaliana and Ricinus were be found.

PHYTOCHROME SIGNAL TRANSDUCTION PATHWAYS REGULATED BY RECIPROCAL CONTROL MECHANISMS

Light is an important environmental stimulus for plants. It controls many aspects of plant morphogenesis and provides the energy for photosynthesis. The initial event of light perception is carried out by at least three families of photoreceptors that detect different wavelengths within the visible spectrum : the phytochromes (which detect red and far red light), the blue light photoreceptors, and the ultraviolet (UV) light photoreceptors. Of these, the best characterised to date are the phytochromes. By genetic and other approaches, attempts are being made to elucidate the signal transduction events downstream of phytochrome that lead to various plant responses.

We have been interested in understanding specifically how phytochrome controls gene expression and have developed an experimental system that employs micro-injection to deliver signaling intermediates into the cells of a tomato mutant known as aurea, which has greatly reduced phytochrome levels (chiefly because of reduced PhyA activity).

Using this system, we have found that PhyA acts through heterotrimeric G protein activation to stimulate gene expression that results in chloroplast development and anthocyanin biosynthesis. Downstream of G-protein activation we have identified three different signal transduction pathways that are dependent on cyclic GMP (cGMP) and calcium. The activation of genes encoding anthocyanin biosynthetic enzymes is dependent only on cGMP, whereas calcium (and subsequently calcium-activated calmodulin can stimulate the development of immature chloroplasts that lack two of the photosynthetic complexes, photosystem I (PSI) and cytochrome bgf (Cyt. bgf). A third pathway, dependent on both calcium and cGMP, operates to control the synthesis of these two complexes to produce a fully mature (and presumably operational) chloroplast.

It is clearly of interest that these phytochrome signal transduction pathways are divided along functional lines, for example, photosystem II (PSII) synthesis and assembly is dependent on a signaling pathway different from that utilised for PSI. The relative inputs into these different transduction pathways may therefore be a means by which plant cells can adjust their biochemical status, particularly for controlling the levels of photoprotectants (such as anthocyanins) and the relative amounts of different photosynthetic complexes. If this were the case, one would predict the existence of cross talk mechanisms between the different phytochrome pathways, which would allow the activities of each pathway to be known by the others and would allow regulatory signals to pass between them. The study presented here was designed to detect such mechanisms by specifically manipulating signal flow through the different pathways.

Regarding end products of the signal transduction pathways, we have followed changes in the expression of genes known to be regulated by cGMP, calcium, or both in a photomixotrophic soybean cell culture, SB-P. In addition, we have performed equivalent experiments using our micro-injection approach in auiea cells.

Experiments with the SB-P cells have the advantage that they produce quantitative data about gene expression, whereas micro-injection in aurea allows analysis in a phytochrome-compromised background and allows the direct manipulation of the components of the signal transduction pathways themselves. As expected, our results suggest the existence of cross-talk mechanisms that control the activities of these signal transduction pathways and ultimately gene expression. In particular, we provide evidence for a novel mechanism of control in which a positive regulator of one signal transduction pathway can negatively regulate another pathway, and *vice versa,* a phenomenon that we propose to call reciprocal control.

Chs Gene Expression Displays Desensitisation and Adaptation

We and others have shown previously that several genes are induced in soybean SB-P cultures exposed to light and that phytochrome is a mediator of this inducibility. To examine further the signaling mechanisms that control this regulation, we chose to study representative genes induced by each of the phytochrome signal transduction pathways characterised previously. As a target of the cGMP pathway we studied chalcone synthase [chs, a gene encoding an anthocyanin biosynthetic enzyme), as a target of the calcium pathway we used cab (a gene encoding a PSII chlorophyll a,b-binding protein from the light-harvesting complex (LHCII), and as a target of the calcium/cGMP pathway we studied expression of the gene encoding the PSI component ferredoxin NADP"^ oxidoreductase [fnr). When 3-day-dark-adapted cultures were exposed to white light for extended periods of time the light induction profiles of the three genes were very different.

The cab gene was induced strongly up to 7 hr and then decreased gradually before rising again after 21 hr. Such a pattern of expression is similar to that observed when etiolated seedlings are exposed to light, and the second peak is thought to be controlled by the circadian clock. This similarity suggests that cab gene regulation in the soybean culture may also be under circadian control. chs and fnr mRNA levels clearly do not cycle, however.

In response to light, fnr mRNA abundance increases steadily and then plateaus, whereas chs mRNA increases very rapidly and reaches a maximum after ~3 hr. Subsequently, transcript levels decline sharply and by 7 hr chs mRNA is again at levels found in dark-adapted cultures. Typically, it does not rise again.

Such a transient pattern is similar to that observed previously for chs mRNA levels and for chs gene transcription in irradiated parsley cell cultures. Furthermore, the synthesis of anthocyanin biosynthetic enzymes and the expression of genes encoding them (such as chs] follow similar transient patterns in germinating seedlings of some plant species. This has been interpreted as an adaptive phenomenon for photoprotection in a newly irradiated plant.

Gene photo-induction in dark-adapted cultures. Three-day-dark-adapted SB-P cells were exposed to light (50 ixmoles/meter per sec) for the times indicated. Note that cab mRNA levels appear to display a circadian rhythm, whereas chs mRNA is induced rapidly and then returns to dark levels. Adaptation of gene expression to light of different intensities. Dark-adapted cells were exposed initially to low-intensity white light (2.5 (xmoles/meter per sec); (Low) for 10 hr and were then divided into two. For a further 10 hr, one was maintained at the low light intensity while the other was shifted to a higher light intensity (50 ixmoles/meter per sec); (High). Times are indicated in hours.

Additionally, this transient pattem of expression is highly reminiscent of the desensitisation phenomena described in animal cells, in which continued exposure to a stimulus results in alterations in the sensitivity and responsiveness to it. In many cases, these alterations are brought about by modulating the flow of information through signal transduction pathways, which can eventually switch themselves off in spite of continued stimulation. If the decrease in chs mRNA observed in SB-P cells after 3 hr light irradiation was a result of desensitisation, we would predict that when a culture that has been adapted to one light intensity was exposed to a higher intensity there would be another peak of expression followed by subsequent adaptation to the new intensity.

This was tested by exposing dark-adapted cultures to low-intensity light for 10 hr and then shifting them to a higher light intensity, chs expression in response to light of low intensity was typical in that mRNA abundance was maximal at 3 hr and then decreased rapidly, but subsequent exposure to the higher light intensity resulted in another peak of expression that decreased normally, cab and fni were induced weakly by low-intensity light and were then induced more strongly by the higher light intensity.

Modulation of cGMP Levels Controls Chs Gene Expression

The phytochrome-stimulated induction of chs and the whole anthocyanin biosynthetic pathway have been shown to be mediated by cGMP. Furthermore, we have found that if a membrane-permeable non-hydrolysable cGMP analog (8-Br-cGMP) is used as the agonist of chs expression instead of light, chs expression is maintained at high levels indefinitely. This indicates that light may normally affect chs expression by initially stimulating an increase in cGMP concentrations (leading to chs induction) and then a decrease (leading to chs desensitisation).

If 8-Br-cGMP (50 \XM] was added to dark-adapted cultures that were subsequently exposed to light, chs induction was stronger than usual and mRNA levels no longer decreased after 3 hr. Furthermore, addition of 8-Br-cGMP at the 3-hr peak of chs expression resulted in the maintenance of high mRNA levels for at least another 8 hr. We also found that if 8-Br-cGMP was added to a light-adapted culture (defined here as a 3-daydark- adapted culture that has been exposed to light for 10 hr or more), in which chs mRNA levels are minimal, it was able to stimulate chs expression again and mRNA levels did not subsequently decrease.

These data thus reinforce our hypothesis that cGMP is sufficient to stimulate chs expression and show that it is able to do so in both dark- and light-adapted cultures as well as in light-induced cultures. Because 8-Br-cGMP is non-hydrolysable, the observation that it stimulates chs gene expression that does not attenuate would indicate that reductions in cGMP concentrations mediated by light are responsible for chs desensitisation under normal conditions. Interestingly, the increases in chs expression mediated by 8-Br-cGMP were reflected by a decrease in the cab mRNA levels in these cultures. This was also found to be the case for fnr mRNA (data not shown). This suggested that in addition to being a positive regulator of chs, cGMP may also have a role in down-regulating the calcium pathway.

Note that in all three cases, the increase in chs mRNA levels is accompanied by a parallel decrease in cab mRNA. Times are indicated in hours. In each case, the experimental design is schematised above the actual results.

Although these and our previous data identify cGMP as a likely candidate for controlling the phytochrome-induced expression pattern of chs, they do not indicate whether phytochrome acts by regulating cGMP synthesis [via guanylyl cyclase (GC)], hydrolysis [via cGMP-dependent phosphodiesterase (PDE)], or both. In a first attempt to distinguish between these possibilities we examined the effects of compounds, characterised previously in animal cells, which are able to modulate the activities of these enzymes and result in increased levels of cGMP. Sodium nitroprusside (SNP) generates nitric oxide (NO), which activates GC, whereas zaprinast (M -I- B 22,948) is a highly specific inhibitor of PDE.

In dark-adapted cultures we found that SNP (100 JJLM) could stimulate chs expression in the dark in much the same way as 8-Br-cGMP; induced levels were similar, and the kinetics of induction and maintenance of expression levels were generally equivalent. Zaprinast, however, was not able to promote a response.

In light-adapted cultures the effects of 8-Br-cGMP and SNP were again highly similar, but in this case zaprinast (100 | XM) could promote an induction. The kinetics of zaprinast induction were different to the inductions mediated by 8-Br-cGMP and SNP, in that 10 hr were typically required for induction, whereas 8-Br-cGMP and SNP could stimulate expression within 3 hr in light-adapted cultures. However, levels induced by zaprinast have been formd to be maintained for up to 24 hr, like 8-Br-cGMP and SNP.

Assuming the specificity of zaprinast, the observation that it was only effective in light-adapted and not in dark-adapted cultures would suggest that PDE (and consequently GC) is only active in the former, that is, that there is no cGMP current in the dark. However, the observation that zaprinast requires 10 hr to stimulate chs in a light-adapted culture (in contrast to 3 hr for 8-BrcGMP and SNP) may suggest that the activity of the cyclase becomes greatly decreased in light-adapted cells (compared with light-induced cells) and that a long period of time is required to build up cGMP to signal-transducing levels.

Assuming that SNP acts by modulating GC activity, the observation that it can stimulate chs gene expression in both light- and dark-adapted cultures would suggest that the cyclase is present in both, that is, that it does not have to be synthesised *de novo* in the light, but light somehow stimulates and then modulates its activity. Whether NO is the physiological activator of plant GC obviously remains to be determined.

As noted previously, the increases in chs expression mediated by these compounds were accompanied by decreases in cab mRNA levels. fni mRNA followed similar patterns. This was particularly apparent in light-adapted cultures, because cab mRNA at the beginning of the experiment was much more abundant than it was in dark-adapted cultures. In this case, the effects of these compounds on cab mRNA reflect either a higher decay rate or an increased efficiency of dark repression of the cab gene or both (because the incubations with the inhibitors were done in the dark). Note that the kinetics of the reciprocal effects on chs and cab are parallel, for example, the reduction in cab mRNA was only apparent after 10 hr in the presence of zaprinast, whereas it was already clear after 3 hr when 8-Br-cGMP or SNP were used.

Induction of Chs is Sensitive to Genistein

In addition to modulating cGMP concentrations we also attempted to use compounds with known pharmacological effects in animal cells to identify inhibitors that would specifically affect the cGMP pathway and not the calcium pathway. One such inhibitor that we found to be highly specific in its effect was genistein, characterised as being an inhibitor of tyrosine and histidine kinases in animal cells. These results would thus identify the genisteinsensitive step as being far downstream in the cGMP pathway, because it does not affect fnr photoinducibility.

It was also of interest to determine whether genistein could inhibit the 8-Br-cGMP-, SNP-, and zaprinast-mediated inductions of chs.

Furthermore, v^e found that it could inhibit the inductions of chs mediated by 8-Br-cGMP and SNP in dark-adapted cultures (data not shown).

These results indicate that the genistein-sensitive component is essential for the chs induction mediated by all of these compounds, as it is for the lightmediated induction, suggesting that the same signal transduction pathway is being activated in each case. Furthermore, the fact that it inhibits chs induction by 8-Br-cGMP indicates that the genistein-sensitive component is downstream of cGMP.

In a, either 8-BrcGMP (50 fjLM) (cGMP), SNP (100 JJLM) (SNP), or zaprinast (100 | J.M) (Zap.) was added to dark-adapted cultures and samples were taken after a further 3 hr or 10 hr in darkness. In b, the experiment was identical except that light-adapted cells were used as starting material. As seen previously, in light-adaped cultures the increases in chs mRNA are reflected by decreases in cab mRNA.

Genistein (75 [LM] (Gen.) was added to dark-adapted cultures 30 min prior to their exposure to light. Times are indicated in hours. The 8-Br-cGMP-, SNP-, and zaprinast- mediated inductions of chs are also sensitive to genistein. Genistein (75 [LM] (Gen.) was added to light-adapted cultures together with DMSO (-), 50 jxM 8-Br-cGMP (cGMP), 100 JJLM SNP (SNP), or 100 p,M zaprinast (Zap.). The - and + denote the absence or presence of genistein, respectively. Following the appropriate additions, cultures (30 ml) were placed in the dark and samples were taken for RNA isolation after 10 hr. The inhibition of chs induction by genistein is accompanied by an inhibition of the increase in the dark decay rate of cab mRNA mediated by the various agonists, indicating that the negative regulation of the calcium pathway by the cGMP pathway also requires the genistein-sensitive component.

Again, as noted previously, cab mRNA levels were decreased in the presence of these compounds. Interestingly, however, this decrease was inhibited by genistein, indicating that the genistein-sensitive component of the cGMP pathway (or something downstream of it) is also necessary for the observed down-regulation of the calcium pathway. If this is the case, it would mean that cGMP per se was not able to modulate the calcium pathway directly, but that it acts through a downstream target which is sensitive to genistein. Again, as noted previously, changes in fni mRNA abundance were highly similar to those shown for cab mRNA.

To test the specificity of genistein in an independent system, we utilised the aurea mutant of tomato in micro-injection experiments similar to those reported previously. Using ^-glucuronidase (Gus)-based cab-gus, chs-gus, and fnrgus as reporters for phytochrome responses we determined whether genistein was able to inhibit the normal induction of these reporter genes in auiea hypocotyl cells when they were coinjected with PhyA. The PhyA-stimulated induction of chs-gus, but neither cab-gus nor fni-gus, was inhibited in the presence of genistein, consistent with our RNA data from SB-P cells.

Furthermore, the activation of chs-gus by the G protein activators GTP7S and cholera toxin (CTX) was also inhibited by genistein, whereas cab-gus and fni-gus were unaffected. To confirm that the genistein-sensitive component was distal of cGMP we incubated cGMP-injected material with genistein. As predicted from the SB-P data, the cGMP-mediated induction of chs-gus was inhibited by genistein. These results are thus consistent with those obtained from the soybean cells, suggesting that the importance of the genistein-sensitive component for the cGMP-mediated phytochrome signal transduction pathway has been well conserved in plants.

The importance of the genistein-sensitive component for phytochrome signaling is also underlined by the fact that it was able to block the phytochrome-mediated stimulation of the whole anthocyanin biosynthetic pathway in auiea.

However, as predicted from the soybean experiments, genistein had no effect on PhyAstimulated chloroplast development : Immunofluorescence analysis demonstrated that all of the major photosynthetic complexes [PSI, PSII, Cyt. b^f, ATP synthase, and ribulose 1,5-bisphosphate carboxylase (RUBISCO)] were present in the plastids of cells injected with PhyA and subsequently incubated in genistein. We also found that GTP^S-injected cells responded in the same way to genistein. Taken collectively, these data confirm that the genistein-sensitive component is downstream of the bifurcation between the cGMP- and the calcium/cGMP-dependent pathways.

Identification of Inhibitors that Block Calciumdependent Gene Expression

The results presented above suggest that a downstream component of the cGMP pathway is able to negatively regulate the calcium pathway. It seemed possible, therefore, that one (or several) component!s) of the calcium pathway may be able to regulate the cGMP pathway in a similar fashion. To examine this, it was of interest to identify specific inhibitors that were able to block calcium – but not cGMP-dependent gene expression. Obvious compounds to test first v^ere those that have been found to affect calcium-dependent processes in animal cells.

This compound has been shown previously to inhibit cab photo-induction in SB-P cultures. At concentrations of 25\X,M, we confirmed that trifluoperazine was able to effectively inhibit the photo-induction of cab and that it could also inhibit the photo-induction of fni, a result we would expect from a compound that inhibits Ca/CaM action. Interestingly, not only was chs photo-induction not inhibited but it was strongly superinduced, suggesting that inhibition of the calcium pathway somehow leads to stimulation of the cGMP pathway.

Similar results were observed with nifedipine, which blocks animal and plant L-type calcium channels. At a concentration of 60 | XM it could totally inhibit the light induction of cab and fnr, whereas chs induction was again superinduced. Another compound, staurosporine, which is a non-specific inhibitior of animal protein kinase, was foimd to give similar results when added to the cultures at a concentration of 50 nM. Staurosporine has been used previously with suspension-cultured tomato cells, where it was found to inhibit a variety of elicitor-stimulated responses, including phosphorylation of several proteins.

In all three cases, then, the inhibitor-sensitive step would appear to be downstream of G-protein activation but upstream of the branchpoint between the calcium and the calcium/cGMP pathways. To confirm these locations we performed equivalent experiments with these inhibitors in the micro-injection experimental system using auiea.

All three compounds were able to block the PhyA- and GTP7S-stimulated expression of cab-gus and fni-gus, but not chs-gus, corroborating the RNA data from SB-P cells. In addition, whereas trifluoperazine and staurosporine inhibited both calcium- and Ca/CaM-stimulated expression of cabgus, nifedipine could not block cab-gus expression stimulated by either of them. This would indicate that the nifedipine-sensitive component lies between the G protein and calcium and that the trifluoperazineand staurosporine-sensitive components are at or downstream of Ca/CaM but upstream of the bifurcation between the two calcium-dependent pathways.

We also observed that the development of chloroplasts induced by injection of PhyA or GTP-yS was completely inhibited in the presence of trifluoperazine, nifedipine, or staurosporine but that these compounds had no effect on the stimulation of anthocyanin pigment production by PhyA or GTP7S. With PhyA-injected cells that were subsequently incubated in trifluoperazine, nifedipine and staurosporine. Negative regulation of the cGMP pathway by the calcium pathway in addition to inhibiting calcium-dependent gene expression, trifluoperazine, nifedipine, and staurosporine are able to superinduce chs gene expression in response to light.

The results of experiments using trifluoperazine. Trifluoperazine (25 [XM) is able to augment the level of chs expression at the normal 3 hr peak, whereas cab gene expression in response to light remains fully inhibited.

In addition to blocking calcium-dependent responses, this compound therefore appears to increase chs expression (but does not appear to affect desensitisation, because chs mRNA decreases after 3 hr). Its effects on chs (but not cab or fnr) gene expression are inhibited by genistein, showing again that the genistein-sensitive component of the cGMP pathway is critical for chs gene induction. The superinductions of chs mediated by nifedipine and staurosporine were similar to that of trifluoperazine and were also sensitive to genistein.

Inhibition of cab and fnr photoinduction by 25 |xM trifluoperazine (Trif.), by 60 |xM nifedipine, and by 20 nM and 50 nM staurosporine (St.). Samples were taken for RNA analysis after 5 hr exposure to light, a time at which the mRNAs of all three of the genes studied were present. In all cases, the inhibition of cab and fnr is accompanied by an increase in chs mRNA levels. (d) The trifluoperazine-mediated superinduction of chs is sensitive to genistein. Thirty minutes prior to exposure to light, dark-adapted cells were either untreated (-), treated with 25 JJLM trifluoperazine (Trif.), or with both 25 |xM trifluoperazine and 75 p,M genistein (Trif. -f Gen.). Times are indicated in hours.

We were then interested to learn whether trifluoperazine could affect the 8-Br-cGMP-mediated induction of chs. This was examined in both light- and dark-adapted cultures. In neither culture was the addition of trifluoperazine alone able to induce chs gene expression, in clear contrast to the effects of 8-Br-cGMP.

Furthermore, again unlike 8-Br-cGMP, in light-adapted cultures placed in the dark, it did not increase the decay rate of cab mRNA either, indicating that its effects on steady-state cab mRNA levels in the light are probably not a result of mRNA destabilisation but more likely reflect an interaction that affects gene expression. Interestingly, the addition of 8-Br-cGMP, together with trifluoperazine to light-adapted cultures, led to a superinduction of chs and an increased decay rate of cab mRNA, but in dark-adapted cultures trifluoperazine affected neither the 8-Br-cGMP-mediated induction of chs nor the 8-Br-cGMP-mediated decrease in cab mRNA levels.

This suggests that the trifluoperazine-sensitive component of the phytochrome-activated calcium pathway (perhaps Ca^"^/CaM), or something down Stream of it, in addition to being a positive regulator of the calcium pathway, is also a negative regulator of chs expression. Nifedipine and staurosporine have equivalent effects to trifluoperazine in light- and dark-adapted cultures (data not shown), indicating that the effector must be either the most downstream target for these three compoimds or even farther downstream in the calcium pathway. fnT mRNA levels showed variations much like those shown for cab (data not shown).

Cultures were either untreated (-), treated with 50 JJLM 8-Br-cGMP (cGMP), 25 M-M trifluoperazine (Trif.), or with both 8-Br-cGMP and trifluoperazine (cGMP + Trif.) for the times indicated (in hours), [a] effect of these additions to a light-adapted culture; [b] effect of these compounds in a dark-adapted culture. Trifluoperazine is only able to superinduce the 8-Br-cGMP-mediated induction of chs in light-adapted cultures. In all cases, the abundance of chs mRNA is reflected as a mirror image in the abundance of cab mRNA.

Utilisation of aurea in micro-injection experiments allowed the study of these control mechanisms by directly manipulating the levels of the signaling molecules themselves, rather than working indirectly with inhibitors that block their action. For these experiments we wanted to keep the level of one signaling molecule constant but to vary the concentration of another from a different pathway. Initially, we chose to work with constant cGMP (75 |XM; estimated final intracellular concentration) and vary the levels of Ca/CaM, because Ca/ CaM could be used over a wide concentration range : Whereas 5000 molecules of Ca/CaM are sufficient to stimulate cab-gus expression, we have found that injection of as much as 100,000 molecules can be used without detriment to the cells.

As shown previously, injection of cGMP (—75 |XM) could stimulate only chs-gus, whereas injection of -10,000 molecules of Ca/CaM could stimulate only cab-gus. Also consistent with previous observations, when these amoimts were coinjected we found that all three reporter genes were activated. This was also true when 5000 molecules of Ca/CaM were coinjected with cGMP, although 2000 molecules were not sufficient to provoke any response above the normal cGMP response.

Interestingly, coinjection of higher amounts of Ca/CaM (100,000 molecules and also 50,000 molecules, to some extent) with cGMP had the effect of inhibiting chs-gus, in spite of continued expression of cab-gus. This result is therefore a direct demonstration that Ca/CaM (or something downstream of it), in addition to positively regulating the calcium pathway, is also able to negatively regulate the cGMP pathway. Calmodulin not activated with calcium (CaM) had no stimulatory or inhibitory effects in these experiments, demonstrating clearly the specificity of the responses for the active form of calmodulin (Ca/CaM). The results also reveal that there is no negative regulation by Ca/CaM (or the downstream component) acting on the third calcium/ cGMP pathway, because fni-gus was still expressed in the presence of high levels of Ca/CaM.

Furthermore, immunofluorescence analysis of cells coinjected with cGMP and high levels of Ca/ CaM revealed that all of the major photosynthetic complexes were present in the plastids, although anthocyanin fluorescence was not observed.

Negative Regulation of the Calcium Pathway by the cGMP Pathway

We then attempted to demonstrate negative regulation acting in the opposite direction, that is, to see whether high cGMP levels could inhibit signal transduction through the calcium pathway. cGMP was only able to stimulate chs-gus within a narrow concentration range 1994). We therefore coinjected different amounts of cGMP with the lowest amounts of Ca^"^/CaM, which we have found to stimulate cab-gus expression (—3000 molecules).

Injection of —3000 molecules of Ca^"^/CaM was sufficient to stimulate cab-gus expression, and when this was coinjected with 75 |XM CGMP (estimated final intracellular concentration) all three reporter genes were activated, as expected. However, when it was coinjected with 110 |XM CGMP, although the cells remained viable, cab-gus expression was now inhibited whereas chs-gus expression was activated normally. Coinjection of Ca/CaM with 130 |XM CGMP was toxic.

These results therefore provide a direct demonstration for negative regulation of the calcium pathway by the cGMP pathway, corroborating our data on gene expression from the SB-P cells. Furthermore, these results indicate that high levels of cGMP are also able to inhibit activity of the calcium/cGMP pathway, because fnr-gus expression was inhibited, like cab-gus. Cells injected with 3000 molecules of Ca/CaM and 110 JJLM cGMP were found to lack all of the major photosynthetic complexes, although anthocyanin fluorescence was still observed.

These results are fully consistent with data from the SB-P cells, which indicated that the decrease in cab mRNA levels in the presence of 8-Br-cGMP was also observed with fni mRNA these independent experiments also indicated that it is not cGMP itself that mediates negative regulation but, rather, a component at or downstream of the genistein-sensitive step.

cGMP Requirements of the Calcium/cGMP Pathway

The observation that high levels of cGMP could inhibit fnr-gus but not chs-gus expression led us to speculate whether fnr-gus expression may have different concentration requirements for cGMP than does chs-gus. To this end, we coinjected Ca /CaM with low amounts of cGMP.

Although 5 JJIM CGMP (estimated final concentration) could not stimulate chs-gus under any conditions, when it was combined with Ca^"*"/CaM it was clearly able to activate fnr-gus, demonstrating that its expression required less cGMP than did chs-gus. We reported previously that chs-gus requires at least 30

[IM cGMP for activation, so these results demonstrate that fnr-gus is able to be activated by concentrations at least sixfold lower. These low levels of cGMP combined with Ca/CaM could stimulate the production of fully mature chloroplasts (by immunological criteria) but not anthocyanin pigments.

cGMP as a Regulator of chs Gene Expression

Our photobiological observations using dark-adapted SB-P cells have shown that chs mRNA levels increase up to a 3-hr peak and then decline rapidly, most likely because of desensitisation of chs photo-induction as a result of decreasing cGMP levels.

The exogenous addition of a membrane-permeable cGMP analog, 8-BrcGMP, can both prevent the chs desensitisation normally observed after 3 hr in continuous light and can release chs from desensitisation when added to a light-adapted culture. In each case, the fact that the expression stimulated by 8-Br-cGMP was not subsequently desensitised is also consistent with our hypothesis, because, unlike the endogenous activator cGMP, 8-Br-cGMP is resistant to hydrolysis.

During the normal response to light, cGMP levels could be modulated either by controlling the activity of GC or PDF, or both. The effects of the GC activator SNP and the PDF inhibitor zaprinast would indicate that there is no cGMP synthesis in dark-adapted cells and that light increases cGMP levels primarily by stimulating GC activity. Following the stimulation of cyclase activity by light, it seems reasonable to believe that cGMP levels will rise until a point is reached when desensitisation sets in.

Desensitisation may be mediated by activation of PDF [because zaprinast was found to be effective in light-adapted cultures], but may also be a result of reduced photosensory capacity, because PhyA levels are known to be drastically reduced following exposure of dark-grown/adapted cells to light. Although we have not studied which phytochrome is responsible for mediating the observed gene inductions in response to light, PhyA is probably the most likely candidate.

Concerning cGMP, direct evidence for its role in mediating phytochrome responses can only come from the demonstration that cGMP levels rise initially and then fall in response to light and that SNP and zaprinast can augment cGMP levels. We have made exhaustive attempts to measure changes in cGMP in SB-P cells using a conventional radio-immunoassay but have so far been unable to detect any convincing changes. The problem is that in the absence of real-time imaging techniques (such as are available for calcium and cAMP), it is impossible to follow changes in cGMP concentration that are relevant to signal transduction (*i.e.*, presumably rapid and cytoplasmic and/or nuclear).

Notwithstanding, an increase in chloroplastic cGMP in response to light has been reported. The significance of such an increase for regulating nuclear gene expression is, however, not clear at this time.

Negative Regulation of Phytochrome Signal Transduction Pathways

The expression pattern of chs indicates how changes in cGMP levels may be used for both positive (for induction) and negative regulation (for desensitisation) of the same target gene. In addition to these "intra-pathway" control mechanisms, we would also expect the existence of "inter-pathway" or cross-talk regulation, to allow full coordination of phytochrome responses.

To identify such mechanisms, we have performed experiments in which signal flow through the different pathways has been manipulated. This has been done by identifying pharmacological agents that are able to block specifically either the calcium- or the cGMP-dependent pathways, and by manipulating directly the amounts of the signaling molecules themselves by micro-injection.

Genistein specifically blocks the cGMP pathway downstream of cGMP, while trifluoperazine, nifedipine and staurosporine can block all calcium- dependent responses (*i.e.*, the calcium pathway and the calcium/cGMP pathway). The location of the inhibitor-sensitive components within the phytochrome signaling pathways can be mapped and their positions, based on current information.

Although tempting to speculate on the targets of these compounds [it is interesting, for example, that the nifedipine- and trifluoperazine-sensitive steps map exactly to where we would predict for a calcium channel and for Ca/CaM, respectively], it is currently too early to know anything about them.

Nonetheless, this should in no way undermine their value nor should it complicate our interpretation, because, if nothing else, their specificity is absolutely clear at the level of their effects on the different phytochrome signal transduction pathways, and, furthermore, their effects in both SB-P cells and in aurea are entirely consistent. As such, it is now possible to examine the consequences of specific knockouts of individual pathways.

For example, a calcium pathway knockout has the effect of superstimulating chs expression, presumably by increasing signal transduction through the cGMP pathway. We believe that the inhibition of the calcium pathway is related to the superstimulation of the cGMP pathway because trifluoperazine can augment both the normal light induction of chs and the 8-Br-cGMP-mediated induction, and because it can affect an 8-Br-cGMP-induction only in a light-adapted but not in a dark-adapted culture, that is, only when there is a calcium current.

Direct evidence for this has been obtained from micro-injection experiments in aurea, in which we showed that high levels of Ca / CaM can inhibit the cGMP-mediated induction of chsgus. However, while cGMP levels appear to be the primary factor for controlling chs gene expression (induction and desensitisation), the negative regulation of the cGMP pathway by the calcium pathway would appear to affect primarily the induction process, because trifluoperazine and the other calcium pathway inhibitors do not appear to affect chs desensitisation.

Interestingly, negative regulation also operates in the opposite direction : In all cases when chs gene expression was stimulated artificially (*e.g.*, by 8-Br-cGMP,

SNP or zaprinast), we observed accompanying decreases in cab mRNA levels. cab and chs mRNA abundance are very often mirror images of one another. We have also confirmed these effects in micro-injection experiments, by showing that the induction of cab-gus observed following injection of cGMP and Ca/CaM can be inhibited if higher amounts of cGMP are used, whereas chs-gus is expressed as normal.

Which signaling intermediates could mediate these negative regulatory phenomena? It is difficult to envisage calcium and cGMP in these roles because the down-regulation of cab mRNA by cGMP is sensitive to genistein and because the superstimulation of chs can be mediated by trifluoperazine, nifedipine, and staurosporine.

This is confirmed further by the fact that injection of high amounts of Ca/CaM can inhibit the cGMP-mediated induction of chs-gus in auiea. Consequently, we would predict that either Ca / CaM, something downstream of it, or both, would mediate the cross talk in one direction and that the genistein-sensitive component, something downstream of it, or both, would act in the opposite direction.

Because negative regulation is only manifested in the presence of high amounts of signaling intermediates, to us the most reasonable mechanistic model of negative regulation is that it is mediated by components that have only low binding affinities for the effector molecules. In this way, negative regulation will only begin to operate when signal transduction through one pathway reaches an excessive level. Its mode of action may be either by inactivation of a positive regulator or by activation of a negative element.

Another interesting observation from these micro-injection experiments is that like the calcium pathway, the calcium/cGMP pathway controlling fni-gus expression does not appear to be subject to negative regulation in the presence of high amounts of Ca/CaM, although cGMP is able to act negatively on it when present in higher amounts. Again, the genistein-sensitive component (or something downstream of it, or both) is responsible for this.

Most probably, therefore, the target(s) for negative regulation lies between Ca/CaM and the bifurcation between the calciumand calcium/cGMP-dependent pathways.

Reciprocal Control and Its Physiological Significance

Concerning the calcium- and the cGMP-dependent pathways, an increase in the activity of one appears to have a negative effect on the activity of the other in both cases. Because of this, we have termed this kind of regulation reciprocal control, which we define as the negative control of one pathway by a positively acting factor from another pathway, and *vice versa.*

For phytochrome, its physiological relevance is likely to be at its strongest when the activity of one pathway is much greater than the activity of the other. In SB-P cells, this situation appears to occur after 3-hr (high cGMP/low Ca) and

7-hr (high Ca + /low cGMP) light irradiation. The following scenario of events would be consistent with all of our observations.

At the early time point, negative regulation may function to stimulate the production of photoprotectants and to concomitantly suppress the buildup of photosynthetic components, ensuring that photosynthesis does not begin in the absence of sufficient photoprotective pigments.

This would explain why negative regulation by the cGMP pathway acts on both the calcium- and the calcium/cGMP-dependent pathways, both of which are needed for the synthesis of photosynthetic components. After 3 hr, cGMP levels presumably begin to fall (switching off chs expression), by some intra-pathway mechanism, and Ca begins to rise, a phenomenon that we would expect to further attenuate chs expression.

The rise in Ca^ "^ does not, however, act negatively on the calcium/cGMP-dependent pathway, ensuring that all photosynthetic components can be synthesised unhindered. It is well known that anthocyanin accumulation precedes chloroplast development in germinating seedlings of several species, and this is also reflected in the levels of mRNAs encoding different components. This is a phytochrome-mediated response. The block points are indicated by two parallel lines drawn across each arrow in the pathway. (b] Negative regulation of the cGMP pathway by the calcium pathway. (c) Negative regulation of the calcium and calcium/ cGMP pathways by the cGMP pathway. In b and c, the broken lines indicate the negative interactions, and brackets indicate the positions of the components that are responsible, based on current information.

At first sight, the requirement of the calcium/cGMP dependent pathway for cGMP raises problems in this model, because we would expect cGMP levels to be low at the time when Ca^"^ levels are high. This problem could be overcome if its threshold for activation by cGMP is lower than it is for the cGMP pathway itself. Remarkably, this was found to be the case; coinjection of cGMP levels insufficient for chs-gus expression and anthocyanin biosynthesis, together with Ca^"^/CaM, resulted in fnr-gus expression and in the development of chloroplasts containing all of the photosynthetic complexes.

The requirement of anthocyanin biosynthesis for high levels of cGMP would ensure their synthesis more during high irradiance conditions and would allow efficient desensitisation of the genes encoding the biosynthetic enzymes as soon as they are no longer needed and when cGMP levels begin to fall. This is supported by our observations that in SB-P cells, several other genes encoding anthocyanin biosynthetic components require constant irradiation for expression, and of all the light-regulated genes studied they are the only ones that are desensitised. A lower threshold requirement of the PSI and Cyt.bgf components for cGMP would presumably permit their synthesis even when the cells are light adapted, ensuring that the genes encoding these components could be activated during conditions when the genes for anthocyanin biosynthetic enzymes are not.

In summary, the data presented here have revealed several interesting feedback mechanisms for controlling phytochrome signal transduction pathways. The relative activities of the cGMP- and the calcium-dependent pathways during different light conditions and at different times are probably key factors that determine the activities of these mechanisms. Overall, the relative push and pull between the two pathways could potentially control factors as diverse as anthocyanin-mediated photoprotection, changes in photosystem stoichiometry, and circadian rhythms, together with other light-regulated processes. Continued study of these elegant regulatory phenomena will clearly reveal novel mechanisms of cross talk between signal transduction pathways and will lead to an increased understanding of how phytochrome controls such a complex array of plant responses.

Soybean Cell Cultuie and RNA Analysis

The soybean [Glycine max) cell culture SB-P was first described by Horn *et. al.*. For all experiments it was grown photomixotrophically in constant light (50 ixmoles/meter per sec) in KNl medium containing 5 grams/liter of sucrose. Ten- to- 14-day-old cultures (500 ml) were dark-adapted for 2.5-3 days prior to experimental manipulation. Such cultures are described in the text as dark-adapted cultures. Cultures that were subsequently exposed to light (50 jjimoles/meter per sec) for at least 10 hr are described as light-adapted cultures.

All manipulations with dark-adapted cells were performed under green safe-light conditions. Samples were collected on filters by suction filtration and immediately frozen in liquid nitrogen. Although we are primarily interested in phytochrome responses, the experiments presented here have been done with white light rather than with red light. This is because we have foxmd white light to be equivalent to red light for all the responses shown.

Interestingly, signal transduction initiated by blue/UV photoreceptors is clearly different to what we have described here (manuscript in preparation). RNA was prepared from frozen cells essentially as described by Lam *et. al.*. RNA gels (10-30 (jig/lane) were blotted onto nylon filters (Stratagene) and hybridised with randomprimed probes prepared according to the manufacturers recommendations (Amersham). Soybean cDNAs encoding cab, chs, and fnr were used as probes. Filters were routinely stripped and rehybridised to different probes to control loading and blotting.

Treatment with Chemical Agonists and Antagonists

To facilitate uptake of the different compounds into the cells, we have found that acid loading greatly improves reproducibility of the reported effects. This is especially true for 8-Br-cGMP, as we reported previously. Acid loading was performed by adding sufficient 50 |JLM HCl to lower the pH of the culture to between pH 4.0 and pH 4.5. The starting pH of a dark-adapted culture was typically about pH 6.0, whereas that of a light-adapted culture was about pH 5.5.

Conveniently, we found that the cells could buffer themselves back to their starting pH after 30 min-1 hr. 8-Br-cGMP was dissolved in 10 per cent DMSO and added to the cells at 1/1000 dilutions. In the experiments reported here we used a final concentration of 50 JJLM, which we found to give a maximal response. However, all concentrations between 20 and 100 JJLM 8-Br-cGMP could stimulate chs gene expression to some extent. Concentrations above 100 ixM were never tested.These concentrations are highly similar to the concentrations found to be effective in micro-injection experiments and are very similar to the endogenous concentration of cGMP in amphibian photoreceptor cells, estimated to be 60 (iM. SNP was obtained from Sigma and was dissolved in water at a concentration of 100 IUM.

Zaprinast was a gift of Monsanto Co. and was dissolved in DMSO at 100 mM concentration. Both were used at final concentrations of 100 | xM, which we fovmd to be most effective, as has also been reported in animal cells. Once added, the cultures were maintained in darkness for the remainder of the experiment, even if light-adapted cultures were used. This was because both compounds appear to be light sensitive.

Genistein, trifluoperazine, nifedipine, and staurosporine were dissolved in DMSO and diluted to 1/1000 when added to the cultures. In appropriate experiments with darkadapted cells, the compounds were added under green safe-light conditions and a further 30 min in darkness was allowed for entry of the compound into the cells before exposing them to hght. Each of these inhibitors was tested at several concentrations; although lower levels generally had no effect, higher levels were either toxic or gave non-specific inhibition of all the phytochrome pathways. All stock solutions were prepared fresh before each experiment, and each experiment was repeated at least five times to produce essentially the same result in all cases.

All standard techniques used for the micro-injection and analysis of aurea hypocotyls, including descriptions of the reporter genes and preparation of injection solutions, have been described previously. For inhibitor experiments, injected seedlings were incubated in petri dishes with 2 ml of 50 | xM Tris-HCl (pH 7.0), containing the appropriate inhibitor (100 \x.u, genistein; 200 | xM, trifluoperazine; 125 | xM, nifedipine, 60 nM, staurosporine), and subsequently the seedlings were vacuum infiltrated for 1 min. The infiltrated seedlings were then incubated in light for 48 hr before being examined for Gus expression, chlorophyll fluorescence, anthocyanin fluorescence, or immunostaining. Genistein, nifedipine, trifluoperazine, and staurosporine were all purchased from Calbiochem and dissolved in DMSO before dilution (1/1000) to the appropriate concentrations.

DISSECTING THE PHYTOCHROME A-DEPENDENT SIGNALING NETWORK

Sessile plants have adopted a high degree of developmental plasticity to optimise their growth and reproduction in response to their ambient environments. Light is one of the major environmental signals that influence plant growth and development. Not only is light the primary energy source for plants but also it provides them with positional information to modulate their developmental processes, including seed germination, seedling de-etiolation, gravitropism, phototropism, chloroplast movement, shade avoidance, circadian rhythms and flowering time.

Plants can detect almost all facets of light, including its direction, duration, quantity and wavelength, using three major classes of photoreceptors. Phytochromes predominantly absorb the red and far-red wavelengths (600–750 nm), whereas cryptochromes and phototropins perceive blue and ultraviolet A (UV-A) wavelengths (320–500 nm), and unidentified UV-B photoreceptors absorb UV-B (282–320 nm). These photoreceptors perceive and interpret ambient light signals and transduce them to modulate plant growth and development.

The light control of seedling development has been conveniently used as a model for dissecting the signaling mechanisms of these photoreceptors. In this article, we focus on recent advances in our understanding of the signaling network of phytochrome A (phyA). We attempt to outline the recent conceptual breakthroughs, the standing issues and possible future directions in phyA-signalingmechanism research.

Molecular Properties and Functional Roles of PhyA

Historically, phyA was the first phytochrome discovered in higher plants and all higher plants have a distinct phyA among phytochrome families of variable sizes. All known phyAs from higher plants are abundant in dark-grown plants and are rapidly degraded upon light exposure. They are therefore classified as a light-labile phytochrome. The purified phyA molecule is a soluble, dimeric chromoprotein that consists of two,125-kDa polypeptides with a single covalently attached tetrapyrrole chromophore, phytochromobilin. The photosensory activity of the molecule results from its capacity to undergo a light-induced, reversible switch between two conformers : the red-light-absorbing Pr form and the far-red-light-absorbing Pfr form. Phytochrome is synthesised in the Pr form in dark-grown seedlings. Upon exposure to red light, the Pr form is converted to the Pfr form, and exposure to far-red light reverts the Pfr form to the Pr form.

There are five distinct phytochromes in Arabidopsis, designated phyA to phyE. These photoreceptors have unique, sometimes partially redundant or antagonistic, roles in different photomorphogenic responses. PhyB to phyE predominantly regulate light responses under continuous red and white light. Most of their known responses can be grouped into the classical red–far-red photoreversible phytochrome responses, the so-called low-fluence responses (LFRs, fluence requirement 1–1000 mmol m2). The classical example of a phytochrome-mediated LFR is red-light-induced germination of lettuce seeds. This induction can be inhibited by subsequent far-red light treatment. The seeds can be repeatedly treated by sequential red or far-red light, and the ultimate germination response depends only on the last light treatment. Thus, photoreversibility is one characteristic feature of LFR responses. Another distinguishing feature of LFR is that it conforms to the law of reciprocity – a response depends onthe totalnumber of photon sreceive dirrespective of the duration of exposure.

PhyA is unique among all phytochromes because it is solely responsible for the very-low-fluence response (VLFR) and for the far-red-light-dependent response (HIR). The VLFR includes light effects on the expression of some genes, seed germination and the gravitropic control of hypocotyl growth, and it can be induced with extremely low photon fluences of 0.001–1.000 mmol m2 of either red or far-red light pulses. Although the VLFR is not photoreversible, it does obey the law of reciprocity (*i.e.* the VLFR senses photon fluences).

The HIR requires relatively high photon fluence rates and a long duration of irradiation, and it is fluence rate and not total fluence that defines this type of responses. Typical HIRs include inhibition of hypocotyl elongation, opening of the apical hook, expansion of the cotyledons, accumulation of anthocyanin and a far-red light preconditioned block of greening during seedling development. The central dogma for phytochrome action (that Pfr is the biologically active form) applies to the LFRs mediated by phyB–phyE and the VLFRs mediated by phyA. However, for phyA-mediated far-red-light HIRs, a short-lived intermediate generated during Pfr-to-Pr photoconversion was suggested to be the physiologically active form.

PhyA Control of Gene Expression

In response to the far-red light signal, it is thought that phyA leads to photomorphogenic development largely by controlling far-red-light-responsive nuclear gene expression. Traditional approaches have revealed several dozen individual genes whose expression is regulated by light. Recently, DNA-microarray technology has been used to investigate genome-wide gene-expression profiles controlled by phyA.

These studies provided substantial evidence for the notion that phyA is the photoreceptor responsible for far-red-light-induced genome expression. Furthermore, many cellular metabolic and regulatory pathways are co-ordinately regulated by light. Some (including all photosynthetic genes, glycolysis and the tricarboxylic acid cycle) are activated by light, whereas others [such as cell-wall-loosening enzymes and water-channel proteins (aquaporins)] are repressed by light.

By measuring the time-dependent global gene expression profiles, a large proportion of the early phyAregulated genes have been found to be known or putative transcriptional regulators, whereas most structural and metabolic genes belong to the late-response class. The data suggest that the massive change in gene expression induced by phyA activation is probably a result of a transcriptional cascade.

Light-induced Nuclear Import of phyA

Because phytochromes are synthesised in the cytoplasm, we must address the question of how the photo-activated phytochromes control light-responsive gene expression in the nucleus. Early studies using an immunohistological approach and cell fractionation assays supported the notion that phytochromes are predominantly localised outside the nucleus. Recently, it was shown that upon photoconversion of Pr to Pfr, all phytochrome species tagged with green fluorescent protein (GFP) can translocate from the cytoplasm into the nucleus, where they form intranuclear speckles.

For phyA, brief irradiation with red or far-red light can induce rapid (a few minutes) nuclear import, and intranuclear speckle formation of phyA–GFP is preceded by even faster cytosolic-spot formation by the fusion protein when irradiated with red (but not far-red) light, a phenomenon reminiscent of SAP (sequestered area of phytochrome) formation. Additionally, nuclear translocation of phyA is also induced by continuous far-red but not by continuous red light treatment. Thus, the nuclear translocation of phyA is mediated by both the VLFR and the far-red HIR, indicating that phyA regulates its own nuclear import. Although the exact physical nature of the nuclear speckles is currently unknown, a recent study demonstrated that they are required for phyA function. Wild-type seedlings undergo photomorphogenesis and are characterised by short hypocotyls and open, expanded green cotyledons. By contrast, phyA mutant seedlings are blind to farred light and exhibit skotomorphogenic development or etiolation, which is characterised by elongated hypocotyls, closed cotyledons and apical hooks. Short arrows indicate the junctions of hypocotyl and root.

At the current stage, little is known about the molecular machinery and factors modulating the nucleocytoplasmic partitioning of phytochromes.

It has been speculated that the Pr form of phyA is anchored or retained in the cytosol, whereas the Pfr form does not interact with the anchoring proteins and is thus subject to nuclear import. Identification of the phyA nuclear localisation signal and nuclear export signal should help us to understand how lightregulated changes in phyA localisation are brought about.

PhyA as a Light-regulated Kinase

How does photo-activated phyA transfer the light signal to its downstream targets? Does phyA have enzymatic activity? Recent studies using recombinant DNA technology have added much convincing evidence to the longstanding view of phytochromes as light-regulated kinases.

Furthermore, the recent discovery of phytochrome-like photoreceptors in bacteria has also provided strong evolutionary evidence in support of such a view. Bacteriophytochromes can perceive light signals and relay information via a His-kinase signaling cascade, probably through a response regulator. Higher-plant phytochromes and bacteriophytochromes share sequence similarity in their C-termini and it has been demonstrated that, unlike their bacterial counterparts, higher plant phytochromes autophosphorylate on Ser/Thr rather than His/Asp.

The claim that higher-plant phytochromes are Ser/Thr kinases has gained support from studies that have mapped putative phosphorylation sites for oat phyA, and targeted mutagenesis studies have implicated them in regulating phyA signaling.

In addition, several phyA-interacting factors or putative in vivo substrates of phyA kinase activity have been identified in recent years. For example, phytochrome-interacting factor 3 (PIF3), phytochrome kinase substrate 1 (PKS1), nucleoside diphosphate kinase 2 (NDPK2), cryptochromes (both CRY1 and CRY2) and the AUX/IAA proteins were found to be able to interact with phyA; in some cases, their phosphorylation was enhanced by the conversion of phyA to the active Pfr form.

However, it should be realised that, even if it has been documented, no physiological relevance of the phosphorylation of these molecules has been demonstrated *in vivo*. Therefore, in spite of the compelling evidence for the kinase activity of phyA, the connection between phyA kinase activity and a phyA photosignal transfer event still remains to be established.

It is known that reversible protein phosphorylation is essential for controlling the flow of many signaling processes. Thus, it is anticipated that protein phosphatases might also participate in the regulation of phyA signaling. A recent study has identified such a phosphatase, flower-specific, phytochrome-associated protein phosphatase (FYPP), which dephosphorylates phyA in a light-dependent manner and modulates phyA-mediated light signals in the control of flowering time.

Genetically Identified phyA-specific Signaling Intermediates

Detailed genetic analyses have suggested that phyA signal transduction has at least two branches, which correspond to VLFR and HIR signaling. The Columbia (Col) ecotype does not respond to very low light fluences. Two quantitative trait loci, VLF1 and VLF2, are responsible for the expression of VLFR but their molecular nature is currently unknown. Mutants for several components affected in the HIR branch of phyA signaling pathway have been identified, including FHY1, FHY3, FIN2, SPA1, FAR1, FIN219, PAT1, EID1, HFR1/RSF1/REP1, LAF1, LAF6 and FHY4.

Several phytochrome A (phyA)-specific signaling molecules were identified by their involvement in far-red-light high-irradiance responses. These components presumably act upstream of the COP/DET/FUS genes, thus controlling the accumulation of crucial transcription factors (*e.g.* HY5 and potentially LAF1 and HFR1) and the degree of photomorphogenic development. The genes encoding all these components except FIN2 and FHY4 have been cloned. Arrows indicate a positive action, whereas the bars indicate a repressive effect. A '?' next to a component indicates its unverified placement in the signaling network model

The fhy1, fhy3, fhy4, fin2, fin219, far1, laf1, laf6 and hfr1/rep1/rsf1 mutants show less sensitivity in continuous far-red light, indicating that their genes encode positive regulators of the phyA signaling pathway. Among the identified positive regulators of phyA signaling, the loss-of-function mutants (mostly null alleles) exhibit only partial defects of various degrees in phyA signaling under far-red light.

This suggests that phyA signaling involves multiple branches or parallel pathways controlling overlapping yet distinctive sets of far-red-light-mediated HIRs. It has been shown that at least FHY1 is also involved in mediating the VLFR.

Several phyA signaling intermediates have been characterised at the molecular level. For example, LAF6 is a plastid-localised ATP-binding-cassette protein involved in co-ordinating intercompartmental communication between plastids and the nucleus. PAT1 is a new member of the GRAS (GAI, RGA, SCR) family, whereas FIN219 is a GH3-like protein whose expression is rapidly induced by auxin. Both PAT1 and FIN219 are cytoplasmic proteins, whereas FAR1, FHY3, SPA1, HFR1, LAF1 and EID1 are all nuclear. Interestingly, FAR1 and FHY3 encode two closely related proteins that constitute one branch of a small gene family. FHY1 is a novel protein, whereas HFR1 is an atypical basic helix- loop–helix (bHLH) transcription factor closely related to PIF3, and LAF1 is a Myb-type transcription activator.

It is worth mentioning that the jasmonate response locus JAR1 maps to the same locus as FIN219. Intriguingly, multiple loss-of-function alleles of jar1 mutants respond normally to far-red light (measured by hypocotyl growth) and, unlike the jar1 alleles, fin219 mutants show no obvious change in responses to jasmonate. This discrepancy might be explained by the fact that the mutation in fin219 is epigenetic, involving aberrant promoter methylation.

Biochemical analysis indicated that JAR1 (FIN219) belongs to the acyl adenylate-forming firefly luciferase superfamily, and it displays adenylation activity specific for jasmonic acid, suggesting that JAR1 could activate the carboxyl groups of certain substrates for their subsequent biochemical modification. In contrast to the many mutants with reduced phyA responses, only two mutants (spa1 and eid1) show enhanced phyA-specific light responses.

Genetic and physiological results indicate that the encoded proteins function as negatively acting components of phyA signaling and are involved in different but interacting phyA-dependent signal-transduction chains. SPA1 functions as a negative effector in both the VLFR and the HIR, whereas EID1 is only involved in signaling cascades regulating the HIR. SPA1 contains WD-40 repeats similar to COP1 and exhibits some weak homology with protein kinases, whereas EID1 is an F-box protein and a likely component of so-called SCF (Skp1/ Cdc53/F-box protein) complexes.

The biochemical function of most known phyA signaling molecules (such as PAT1, FAR1, FHY3 and SPA1) remains largely unknown and is likely to be a primary focus of future research. In addition, earlier micro-injection and pharmacological studies have implicated the cytosolic heterotrimeric G proteins, Ca2þ–calmodulin and cGMP as candidate second messengers in phyA signaling. We still do not know how these cytosolic signaling components work together with the defined nuclear factors to mediate phyA signaling.

PhyA can Directly Target Light Signals to Light-responsive Promoters

The finding that PIF3, a member of the bHLH superfamily of transcriptional regulators, can interact directly with both phyA and phyB suggests that phytochromes might regulate nuclear gene expression via direct interaction with transcription factors. In support of this notion, phyB can bind specifically and photoreversibly to PIF3 that is already bound to its cognate DNA-binding site (the light-responsive G-box DNA sequence CACGTG).

Furthermore, the expression of CCA1 and LHY1, which have G-box motifs in their promoters and encode Myb-like transcription factors known to be involved in regulating light- and clock-related responses, is reduced in transgenic Arabidopsis seedlings with reduced PIF3 levels. These and other data are consistent with a model in which light induces the phytochrome molecule to translocate into the nucleus, where it binds in its active Pfr form to promoterbound PIF3 and facilitates transcriptional activation of specific target genes.

Furthermore, HFR1, a genetically identified positive regulator specific to phyA signaling, encodes a bHLH protein closely related to PIF3 that can form a homodimer as well as a heterodimer with PIF3. These findings suggest that phyA signaling might directly regulate a diverse set of genes through an integrated transcriptional network orchestrated by an array of interacting bHLH proteins. There are,135 bHLH proteins in Arabidopsis and it is known that bHLH proteins can form homodimers and/or heterodimers. Future studies are needed to explore the possible combinations of these bHLH proteins and to sort out their downstream target genes.

Upon photo-conversion to the Pfr form, phytochrome A (phyA) translocates into the nucleus, where it interacts with the G-box-bound PIF3 and activates the expression of the primary target genes (such as the Myb-class transcription factors CCA1 and LHY).

The encoded primary target gene products (many of which are transcriptional regulators) are, in turn, responsible for orchestrating the expression of the secondary target genes (*e.g.* the induction of LHCB expression by CCA1), thus generating a transcriptional network controlling different aspects of phyA physiology.

Other phyA-interacting factors (PIFs; *e.g.* PKS1, NDPK2, cryptochromes, auxins/indole- acetic acids) and their signaling intermediates might function as modifiers to participate in such a direct light-signaling pathway.

Moreover, COP1 and the COP9 signalosome could also interact (directly or indirectly) with the transcriptional machinery, PIFs or other signaling intermediates, and regulate their abundance through a light-regulated proteolytic process. Notice that the far-red-light-absorbing Pfr could be the physiologically active form (for very-low-fluence responses) or a short-lived intermediate could be generated during the Pfr to Pr photoconversion (for far-red-light-dependent high-irradiance responses) to activate gene expression. Abbreviations : FR, far-red light; Pfr, far-red-light-absorbing conformer of phyA; PIC, pre-initiation complex; Pr, red-light-absorbing conformer of phyA; R, red light; TATA, TATA box.

It should be mentioned that some phytochromemediated responses occur quickly (within minutes; *e.g.* the unrolling of the primary leaf wrapped within oat coleoptiles), posing the question of whether nuclear gene expression regulation would be sufficiently quick. Such quick responses are possibly best explained by assuming rapid alterations in inter- or intracellular ion balances; presumably, phytochromes can mediate ionic concentrations within cell compartments that modulate extension growth.

However, electrophysiological evidence for the involvement of ion fluxes in these responses is still lacking.

Regulated Proteolysis in phyA Signaling

There is accumulating evidence that regulated proteolysis plays a crucial role in controlling the timing and amplitude of phyA signaling processes. First, phyA itself is rapidly degraded upon photo-conversion to the Pfr form and this degradation involves the ubiquitin–proteasome pathway. This downregulation of the intracellular phyA levels in response to light probably provides an adaptation mechanism to the ambient light environment.

Far-red-light activation of phyA also regulates proteolysis of downstream transcription factors such as HY5 and HYH, bZIP transcription factors that act to promote photomorphogenesis. For example, HY5 levels are,20 times higher in light-grown seedlings than in dark-grown seedlings, and HY5 levels quantitatively affect the degree of photomorphogenesis.

The degradation of HY5 in the dark involves a protein named Constitutively Photomorphogenic 1 (COP1), a negative regulator of photomorphogenesis.

COP1 is a RING-finger protein with WD-40 repeats. Studies with a functional b-glucuronidase–COP1 fusion protein indicate that COP1 acts within the nucleus to suppress photomorphogenic development in darkness, and that inactivation of COP1 by light was accompanied by reduced COP1 abundance in the nucleus. In the dark, COP1 directly interacts with and targets HY5 for degradation via a 26S proteasomemediated process. Thus, COP1 presumably functions as a ubiquitin ligase.

The finding that SPA1, a negative regulator of phyA signaling, has a WD-40 repeat domain (which is frequently used in E3 complexes for recruiting substrates) with homology to that of COP1 suggests that SPA1 might be also involved in regulated proteolysis of various phyA signaling intermediates. In support of this possibility, a direct physical interaction has been reported between COP1 and SPA1.

This observation raises the possibility that these two proteins function together in targeting substrate proteins (*e.g.* HY5, LAF1 and HFR1) for degradation by the proteasome, with SPA1 as a cofactor specific to the phyA signaling pathway to modulate COP1 function. The recent report that EID1, another negative regulator of phyA signaling, encodes an F-box protein and a likely component of a SCFEID1 ubiquitin-ligase complex underscores a possible role for targeted protein degradation in phyA signaling.

In addition to COP1, genetic screens have identified nine other pleiotropic COP/DET/FUS loci whose gene products act as negative regulators of photomorphogenesis and function downstream of multiple photoreceptors including phyA.

Among these loci, six (CSN1-CSN4, CSN7 and CSN8) encode six of the eight sub-units of the COP9 signalosome protein complex. Strikingly, the COP9 signalosome shares sub-unit-by-sub-unit similarity to the lid sub-complex of the 26S proteasome and is also involved in regulated proteolysis.

This notion is further supported by the recent finding that COP10 is an E2 ubiquitin conjugating enzyme variant. Further characterisation of the functional relationships between the various phyA signaling molecules and the COP/DET/FUS proteins is likely to enhance our understanding of the integration of phyA-specific signaling events with this group of general repressors of photomorphogenesis.

How are the COP Proteins Inactivated?

It is generally assumed that light-induced photomorphogenic development requires the inactivation of these COP/DET/FUS proteins. However, little is known about how the light-activated photoreceptors regulate the activities of those downstream COP/DET/FUS proteins to bring about the physiological responses.

Although it has been shown that nuclear abundance of COP1 is regulated by light acting through these photoreceptors, the mechanisms governing this process are largely unknown. Recent studies have shown that photoactivated cryptochromes can interact physically with COP1 and inhibit COP1 E3 ubiquitin-ligase activity, leading to HY5 accumulation and photomorphogenic development, but the same mechanism might not apply to phyA.

In addition, it has been proposed that these protein-protein interactions might be responsible for the rapid, initial inactivation of COP1 activity, but that longer-term inactivation of COP1 is achieved by subsequent depletion of the molecule from the nucleus.

Conclusions and Prospects

Dramatic progress has been made during the past decade in understanding phyA signaling mechanism, largely owing to the use of molecular genetics in the model plant Arabidopsis. However, it has also become clear that our information about the nature and extent of the phyAmediated signaling network is still limited and fragmented. Much remains to be learned to build a connected phyA signaling network with an understanding of how the various nodes of the network interact to transmit the light signals.

With the completion of the Arabidopsis genome project, new technologies have emerged that allow global and systematic approaches to dissecting signaling pathways. For example, a recent study attempted to analyse the genetic relationships among the various genetically defined phyAsignaling mutants using cDNAmicroarray technology. Clustering analysis of the genome expression profiles supports the notion that phyA signaling might entail a network with multiple paths controlling overlapping yet distinct sets of gene expression.

FHY1, FAR1 and FHY3 probably act upstream in the phyA signaling network, close to the phyA photoreceptor itself. FIN219, SPA1 and REP1 probably act somewhere more downstream in the network and control the expression of smaller sets of genes. Although such microarray-based expression profiling has been valuable in dissecting cellular responses and identifying potential downstream targets of the various signaling components, it has its limitations. First, for a particular signaling component, not all downstream target genes' expression is affected by its activity. Second, changes in some gene expression could be secondary effects of the changed developmental processes. Future application of the recently developed chromatin immunoprecipitation chip technology, together with expression profiling, will help to identify the direct target genes of various transcription regulators involved in phyA signaling.

Most gene function is largely manifested by the activity of its encoded protein. Thus, analysis of the activities and interactions of proteins will be essential for understanding the phyA signaling network. Several innovative proteomic approaches have recently emerged and are likely to make their way into phyA signaling research. For example, affinity-capture-based techniques (*e.g.* tandem affinity purification) in conjunction with enhanced mass spectrometry for characterising *in vivo* multi-protein complexes, and high-throughput protein microarray could be valuable tools in dissecting the phyA signaling network.

FHY1, FHY3 and FAR1 probably act upstream in the phyA signaling network, whereas SPA1, REP1 and FIN219 probably act further downstream in the network. All these components except SPA1 are positive regulators; SPA1 (in bold) is a negative regulator of the signaling pathway. These components could directly or indirectly regulate overlapping and distinct sets of gene expression and phyA-mediated photomorphogenic responses.

Chapter 23

PLANT MICROPROPAGATION

INTRODUCTION

Through the use of biotechnology, desirable genetic traits can be transferred from one organism to another by transfer of DNA. In plants, the DNA of interest is transferred into the new plant by using *Agrobacterium tumefaciens*, a bacterium that can infect plant tissues and incorporate part of its DNA into the DNA of the host plant. Alternatively, a particle gun can be used to directly "shoot" the DNA into plant cells. In both methods, the target for the foreign DNA is a small piece of plant tissue or a small mass of plant cells. Once the DNA has been transferred, new plants must be regenerated from the small pieces of transformed plant tissue using micropropagation (tissue culture) techniques.

Micropropagation is the aseptic culture of cells, pieces of tissue, or organs. It is possible to regenerate new plants from small pieces of plant tissue because each cell of a given plant has the same genetic makeup and is totipotent, that is, capable of developing along a "programmed" pathway leading to the formation of an entire plant that is identical to the plant from which it was derived. In addition to its biotechnological applications, micropropagation is used commercially to asexually propagate plants.

Using micropropagation, millions of new plants can be derived from a single plant. This rapid multiplication allows breeders and growers to introduce new cultivars much earlier than they could by using conventional propagation techniques, such as cuttings. Micropropagation also can be used to establish and maintain virus-free plant stock. This is done by culturing the plant's apical meristem, which typically is not virus-infected, even though the remainder of the plant may be. Once new plants are developed from the apical meristem, they can be maintained and sold as virus-free plants.

Micropropagation differs from all other conventional propagation methods in that aseptic conditions are essential to achieve success.

The process of micropropagation can be divided into four stages :

1. *Initiation stage* : A piece of plant tissue (called an explant) is
   - Cut from the plant,
   - Disinfested (removal of surface contaminants), and
   - Placed on a medium.

   A medium typically contains mineral salts, sucrose, and a solidifying agent such as agar. The objective of this stage is to achieve an aseptic culture. An aseptic culture is one without contaminating bacteria or fungi.
2. *Multiplication stage* : A growing explant can be induced to produce vegetative shoots by including a cytokinin in the medium. A cytokinin is a plant growth regulator that promotes shoot formation from growing plant cells.
3. *Rooting or preplant stage* : Growing shoots can be induced to produce adventitious roots by including an auxin in the medium. Auxins are plant growth regulators that promote root formation. For easily rooted plants, an auxin is usually not necessary and many commercial labs will skip this step.
4. *Acclimatisation* : A growing, rooted shoot can be removed from tissue culture and placed in soil. When this is done, the humidity must be gradually reduced over time because tissue-cultured plants are extremely susceptible to wilting.

BACKGROUND

Preparation for the plant micropropagation exercise should begin at least 24 hours in advance of the laboratory.

The following supplies can be provided to the class in groups of two students :

- 2 fresh African Violet leaves
- 4 culture medium plates
- Aseptic work surface/hood
- 1 aluminum foil packet containing 1 sterile forceps
- 1 aluminum foil packet containing 2 sterile razor blades (or 1 sterile #3 scalpel handle with sterile #10 blade attached)
- 2 sterile petri dishes
- 1 litre sterile water
- 1 500-ml or 1-litre jar with screw-cap lid (for disinfesting the leaf)
- Alcohol or Bunsen burner
- Spray bottle containing 70 per cent ethanol (a spray bottle of the type used to mist plants or spray window cleaning solution works very well).

The following supplies can be shared by four students :

- *500-ml bottle (for sterilising medium)
- *500-ml or 1-litre plastic or glass beaker (for preparing medium).

The teacher should have available for the entire class :

- Commercially-purchased water for medium preparation
- 70 per cent ethanol (dilute 700 ml of 95 per cent ethanol with 250 ml distilled water)
- 0.1 per cent detergent (add 1 ml detergent to 1 litre of water)
- 10 per cent bleach solution plus a few drops of dishwashing detergent
- pH metre or pH paper
- A few mls of the following solutions for adjusting pH :
  - 1.0 M NaOH
  - 1.0 M HCl
  - 0.1 M NaOH
  - 0.1 M HCl
- Parafilm for wrapping plates
- Scissors
- Sharpie marking pens
- 5-10 extra sterile scalpel blades
- 5-10 extra sterile petri plates
- Dishwashing detergent.

PREPARATION OF PLANT TISSUE CULTURE MEDIUM

The culture medium is formulated so that it provides all of the compounds needed for plant growth and development, including certain compounds that can be made by an intact plant, but not by an isolated piece of plant tissue. Typically, the medium contains mineral nutrients, organic compounds such as sucrose and vitamins, and plant growth regulators (plant hormones). Agar is added if a solid medium is desired.

Prepare 250 ml per group of four students. In this instruction guide, a prepackaged African violet medium from Carolina Biological Supply Company that includes all necessary components except agar and sucrose, will be used. The instructions that follow are for making 250 ml of medium.

Step 1 : Add about 100 ml [500 ml] of deionised or distilled water to a 500-ml or 1-litre [2-litre] beaker.

Step 2 : Add 1.16 grams [one package] of the prepackaged mix to the beaker.

Step 3 : Add 7.5 grams [30 grams] of sucrose to the beaker.

Step 4 : Bring the solution up to about 200 ml [800 ml] by adding deionised or distilled water.

Step 5 : Stir the solution until all components are dissolved.

Step 6 : Adjust the pH to about 5.6 to 5.8 with a pH metre, using NaOH and HCl as necessary. If a pH metre is not available, use pH indicating paper that can detect pH 6.0 ± 0.5 pH units. It is preferable to have the pH at slightly less than 6.0 rather than at slightly greater than 6.0.

At pH < 5.0, the ability of the explant to take up nutrients will be adversely affected. At pH > 7.0, iron will precipitate out of the medium, becoming unavailable for uptake. (When using the prepackaged mix and commercial distilled water, it will take approximately 3 drops [13 drops] of 1.0 NaOH and 6 drops [4 drops] of 0.1 NaOH to raise the pH of 250 ml [1 litre] of medium to 5.6 to 5.8.)

Step 7 : Bring the solution up to 250 ml [1 litre] by adding deionised or distilled water, and pour into a 500-ml bottle [1.5 to 2.0-litre bottle] to autoclave.

Step 8 : Add 2.0 grams [8.0 grams] agar. (Note : the agar will not dissolve until the medium is heated.)

Step 9 : Keep the cap loose on the vessel containing the medium while sterilising.

Sterilise the bottle of medium by placing it in one of the following :

- Boiling water for 30 minutes;
- Pressure cooker at 15 pounds for 15 minutes; or
- An autoclave for 15 minutes at 121ƒ C (250ƒ F) at 1.1 kg/cm^2 (15 psi).

Step 10 : After removal from the water bath, pressure cooker, or autoclave, gently swirl the medium to mix the agar. When the agar is completely dissolved and mixed, the medium should appear clear and not turbid.

Step 11 :If pouring plates, cool the medium to 45-55ƒ C (110ƒ F), then pour plates on a clean work surface (spray down with 70 per cent alcohol), working near a small Bunsen burner or alcohol lamp. Use about 25 ml per plate (about one-third to one-half full). If plates are not being used immediately, put them back into the plastic sleeve, tape the sleeve shut, and store plates at 4ƒ C (39ƒ F).

STERILISATION OF SUPPLIES

1. Sterilisation of packets of packets of forceps and razor blades can be accomplished by wrapping each item in aluminum foil, labelling the contents with a marking pen, and :
   - Baking them in an oven at 350ƒ F for 15 minutes;
   - Putting them in a pressure cooker at 15 pounds for 15 minutes; or
   - Placing them in an autoclave for 15 minutes.

The pressure cooker and autoclave should be at the desired pressure for the 15-minute period. After the packets have cooled, they should be stored unopened at room temperature. The students should be instructed when opening the packets to touch that part that will not come in contact with the plant material or petri dishes.

2. Sterilisation of the sterile water needed for the rinses can be accomplished as described for sterilisation of the medium.

Note : Teachers may decide for themselves which parts of the procedure can be successfully completed by the student. Middle school students enjoy transferring disinfested explants previously prepared by the instructor to the plates with medium. Older students may be included in parts of the medium preparation and/or the disinfestation procedure. All students will enjoy seeing the growth of shoots from the leaf surface of the African violet. Seeing this regeneration process will make it easier for the teacher to start a discussion of the totipotency of plant cells and the practical application of plant propagation.

ASEPTIC TRANSFER OF CULTURES TO FRESH MEDIUM

Micropropagation is the use of tissue culture methods to propagate plants. Micropropagation if a form of clonal propagation that differs from all other conventional propagation methods in one important aspect : aseptic conditions are necessary to achieve success with micropropagation. In commercial and research laboratories, aseptic transfer is usually done in a hood that excludes particles larger than 3 microns from the work area. Contaminants of plant tissue culture are bacteria that reside on dust particles and fungi, including spores. A hood will effectively filter out all of these contaminants.

It is possible to transfer aseptically without a hood when you take all possible precautions to keep dust and spores from falling in the sterile medium. One simple, but fairly effective method for accomplishing this, is to spray a smooth counter top with alcohol and spray all other possible sources of contamination (*e.g.*, hands, instruments, and culture vessels) before transferring. A second method used by plant pathologists is to work around a small Bunsen burner, which creates an updraft to prevent spores and dust from falling into the open dish. Extreme caution is required with this method because of the danger of burns.

A third method involves spraying hands, instruments, and vessels with alcohol and working in a large new clear plastic bag. Because newly manufactured plastic is fairly sterile, working inside this sterile bag, which contains still air, will avoid contamination. An old aquarium or cardboard box lined with aluminum foil can be sprayed with alcohol and used to cut down on drafts. With all methods, it is best to keep traffic around the transfer area to a minimum.

Step 1 : Put on a pair of sterile gloves. Clean your gloves with 70 per cent ethanol initially, and reclean them if you touch any non-sterile surface or material.

Step 2 : Spray your work surface with 70 per cent ethanol.

Step 3 : Spray vessels and all tool packets (forceps and razor blades) with 70 per cent ethanol before placing them on the work surface.

Step 4 : Arrange the objects on the work surface so that there is a cleared work area for the transferring process. Avoid placing any object "up wind" of your work area.

Step 5 : For African violets, use a fresh, healthy leaf. The disinfestation process (removal of surface contaminants) is as follows :

- Quickly rinse the leaf under cool tap water, then put the whole leaf into a closed container. A jar with a screw-cap lid is best. Depending on the size of the leaves, either a 500- or 1000-ml jar can be used. Wash the leaf in water with 0.1 per cent detergent for 3-4 minutes. Gently agitate the leaf every 20-30 seconds during this washing step.
- Rinse off the leaf and rinse out the jar with cool tap water.
- Gently agitate the leaf in 10 per cent bleach solution for 10 minutes. The jar should be filled three-fourths full with the bleach solution. *(The bleach solution disinfests the plant tissue material, killing most fungal and bacterial organisms.)*
- Pour off the bleach while keeping the lid loosely in place over the container. (Be careful not to spill bleach solution on clothing because it will leave white spots after laundering.) At this point, the leaf is considered sterile. All subsequent rinses should be done with sterile water, and all manipulations of the leaf performed with sterile instruments and supplies. Open one container at a time and never leave the lid off of any container longer than necessary.
- Take the container to the sterile counter space. Spray the container with 70 per cent ethanol before placing it in the sterile environment. Remove the lid and pour sterile water over the leaves. Fill the jar approximately half-way with the sterile water.
- Replace the lid and gently agitate for 2-3 minutes to rinse the bleach off the leaves.
- Pour off the rinse water in the sink, as in Step 5, d.
- Rinse with sterile water a total of 4 times.

Step 6. Using the sterile razor blade and forceps, cut off bleach-damaged portions in an empty, sterile petri dish. With the empty dish open, hold the leaf with forceps and cut into strips 1.5-cm x 1.5-cm wide. Place 2 or 3 explants into a petri dish that contains the fresh medium. Gently press on the explants with a forceps to ensure that they make contact with the medium. Do two dishes per student. *The tissue culture medium contains all the compounds needed for plant growth, including mineral nutrients, sucrose, vitamins, and plant growth regulators that result in shoot or root production.*

Step 7 : Replace the cover of the petri dish and wrap with parafilm. Use one 1-inch x 4-inch piece per plate. The parafilm will stretch to seal the dish. Plant tissue cultures require much longer culture times than microbial cultures, so this additional precaution is usually worth the small extra cost and time.

Step 8 : Place the vessel under lights for 16 hours per day if possible. Keep the temperature between 24 and 26ƒ C (about 75ƒ F). Do not put cultures in direct sunlight, but make sure the area where they are placed is well-lighted by fluorescent bulbs. Growth is slower at lower temperatures.

Step 9 : Look for small green shoots forming on or near cut surfaces. It may take 2-3 weeks or more before any visible evidence of new growth is noticeable. You may need a microscope to see these structures as they begin to grow, but after 5-6 weeks, they should be visible to the unaided eye.

If a sterile environment was not maintained, contamination will be obvious within 3 to 4 days. Materials contaminated by fungus will have a fuzzy growth on them, whereas material contaminated by bacteria will have slimy growth on them. Discard contaminated petri dishes promptly to avoid spreading plant diseases to other uncontaminated cultures.

The medium used in this exercise contains a cytokinin, and is specifically formulated to favour the production and multiplication of shoots. Once a sufficient number of shoots has been generated, portions of the explant that contain one or more shoots could be transferred to a medium that contains a higher concentration of an auxin, resulting in root production. Once roots have formed, the plantlets are transferred to pots containing a soil-based or soilless medium, and gradually exposed to conditions of lower humidity and greater light.

TESTING FOR VIRUS AND QUALITY

A large number of viruses infect horticultural, plantation and forest plants which are being micropropagated by the tissue culture industry. Most important of these are aphid-transmitted poty-, cucumo- and luteoviruses, whitefly transmitted gemini-viruses, thrip-transmitted tospo-viruses, mealy bug transmitted ilar-viruses.

Generation of virus-free planting material is an ideal strategy to confine these viruses and also to facilitate the movement of materials across the domestic and international boundaries. Tissue culture is an useful approach for generating virus-free planting material. In order to minimise the risk of inadvertent propagation of virus infected plants and introduction of somaclonal variability, the tissue culture raised plants need to be thoroughly indexed for freedom from viruses and checked for quality. Careful indexing based on recent biotechnological methods such as imunoprobes, nucleic acid probes and polymerase chain reaction (PCR) would ensure phytosanitary safety during the movement of planting materials. Similarly, molecular testing will ensure quality control.

The National Facility for Virus Diagnosis and Quality Control of Tissue Culture Raised Plants (NFVD and QC), is aimed at providing virus testing and quality testing support to the tissue culture industry to ensure that only :

- Virus-free mother plants are used for micropropagation, and
- Virus-free and quality tested tissue culture plants are supplied to the growers.

Hence, not only the risk of inadvertent introduction and subsequent spread of viruses would be minimised but also the quality of plants would be assured. This will be achieved by developing a biophysical, immunological and molecular technique based virus diagnosis programme and molecular based quality control programme of the plant species of interest. The testing protocols will not only detect the known viruses, but also the unknown viruses.

The test for virus and quality will have direct impact on developing a vibrant tissue culture industry as the demand for certified tissue culture raised plants will grow exponentially.

VIRUS TESTING

The country has expertise to undertake diagnosis of plant viral diseases through the application of biophysical, immunological and molecular techniques. Immunodiagnosis is the most useful technique using polyclonal and monoclonal antibodies raised against viruses. Polyclonal antibodies are now available in the country for various groups of viruses like potyviruses, tospoviruses, tobamoviruses, potex-viruses, luteo-viruses, badna-viruses, clostero-viruses, etc. These polyclonal antibodies have been very useful in the specific identification of viruses and their strains, and for plant virus diagnosis. The technology for the production of monoclonal antibodies is also available in the country, but their production has not caught up to the desired level for plant viruses due the lack of adequate facilities.

Various immunological and molecular techniques like ELISA, immuno-electron-microscopy, DIBA, Western blotting, immunofl-uorescence, nucleic acid hybridisation using radio-labelled probes and PCR have been used successfully for the diagnosis of virus infections in a variety of plant species. Of these techniques, ELISA, immuno electron microscopy, dot-blot and PCR are the most sensitive and reliable techniques which can be very effectively used for the purpose of diagnostic services to tissue culture industry in the country.

For nucleic acid hybridisation, c-DNA probes are also available for the gemini-, poty-, tobamo-, and badna-viruses. Electrophoresis and direct electron-microscopy are also being commonly used particularly for detecting viroids and unknown viruses.

The viruses for which diagnostic probes (antibodies and cDNA probes) have been developed are known to infect a variety of plant species like : banana, brinjal, cassava, citrus, Frenchbean, garlic, gladioli, groundnut, lilies, mungbean, onion, rice, soybean, sugarcane, tobacco, tomato, urdbean, ornamental plants etc.

Citrus (Citrus spp.) is the third largest fruits crop grown in India, with a great potential for export if raised from desirable virus free planting material to increase production and maintain quality. Its cultivation in India extends to over 0.37 million hectares with an annual production of 2.98 million tonnes. The productivity of citrus fruits in India is comparatively low owing to many biotic stresses of which viruses play a significant role.

In India, more than 15 virus and virus like diseases occur on citrus. Of these, three viral diseases namely tristeza, ringspot and mosaic, and citrus greening disease caused by a BLO are widely distributed in the country. While tristeza and greening present worldwide the ringspot and mosaic diseases are of specific concern in India. Tristeza disease is caused by a closterovirus which has been detected from all parts of the country.

The severe strain of CTV is most commonly distributed and result in tree decline. Mild strains of the virus are also reported, but they were not found useful for cross protection capillo-virus is associated with the ringspot disease, and a double stranded DNA virus belonging to badna-virus group is associated with the mosaic disease. Citrus greening disease, caused by a fastidious bacterium, Liberobacter asiaticum (LA) is present throughout the country and has been the major cause in nearly wiping out citrus orchards in Coorg, Karnataka. Often these pathogens are found in mixed infections. Banana is another important commercial fruit being grown in all parts of the country. The estimated area under banana is 3.92 m ha with production of 10.4 m tonnes/annum. This crop is adversely affected by the bunchy top disease which has spread throughout the country.

The causal virus has been cloned for developing c-DNA diagnostic probes. Another important viral disease of banana is infectious chlorosis which is caused by cucumber mosaic virus. This disease is also widely distributed throughout the country causing serious economic losses. Other two viral diseases namely the bract mosaic and streak mosaic are localised and are of minor importance, but these are spreading fast in the absence of appropriate diagnostic systems and healthy planting material certification programmes.

With the discovery of viroids as potential plant pathogens several diseases which were earlier attributed to viral infections are now known to be caused by viroids. Viroids are single stranded circular RNA molecules with low molecular weight, devoid of proteins. Since they are free infectious RNAs it is not possible to develop protein based diagnostic reagents. To date about 43 of them were found in vegetatively propagated crops like citrus, potato, grapes, hops, apple and avocado etc.

In India, research on viroids started only recently and the diseases caused by viroids have been recorded in tomato, potato, citrus and coleus. In India, viral diseases of ornamental plants are common. Most of these plants are infected by the viruses belonging to cucumo-virus, poty-virus and tobamo-virus. In orchids, infection by CyMV is very common.

QUALITY CONTROL

With the advent of the PCR technique and DNA based molecular markers, some laboratories in India have developed expertise in these areas and are using them for assessment of genetic diversity, phylo-genetic analysis, verietal typing and establishing true to type of tissue cultured plants. Somaclonal variants of pigeonpea were analysed using RAPD and the polymorphic fragment scored for an identifiable locus like white seed, high seed mass, low Helicoverpa incidence etc. RFLPs were reported in studies on regenerated maise, rice, potato, coffee, populus and others.

Microsatellite primed-Polymerase Chain Reactions (MP-PCR) were used to study genetic fidelity of regenerated shoots of Eucalyptus, populus and coffee species. Polymorphisms could not be detected in bread wheat genotypes using 23 simple sequence repeats indicating the inadequacy of the molecular analysis (Varshney *et. al.*, 1998). In Populus deltoides six of the off-types were reported through RAPD analysis of 32 randomly selected plants from a population of 500 individual plants. The plants were derived from nodal segments. Similarly, in sugar beet Munthali *et. al.* reported 0.05 per cent somaclonal polymorphism. Existence of distinct genetic differences among several variants of peach was reported by Hashmi *et. al.* Rani and Raina reported minor variation in RAPD banding pattern among the regenerated plants of Eucalyptus tereticornis and Eucalyptus camadulensis derived from enhanced axillary branching.

From the contrasting reports on the use of molecular marker techniques for analysis of micropropagated plants it is evident that the marker system chosen for analysis needs to be evaluated for its suitability for the application before it can be further used. TERI is a leading group to standardise the technique of AFLP in India and utilise it for studying the genetic diversity in Azadirachta indica. Attempts have also been made to compare the various techniques of genetic typing such as RAPD, RFLP and AFLP. Results have demonstrated that even though RFLP and RAPD markers are useful in studying both intra and inter-specific variation, the extent of polymorphism generated by AFLP makers is several times higher. Hence, AFLPs may be the method of choice for discriminating very closely related genotypes (such as the micropropagated plants) as compared to other molecular markers. Commercial micropropagation industry is globally operating for two major applications. One is in ornamental plants and another in crop species of agricultural importance.

Imperatives of both applications are different. For ornamental plants, major emphasis is on rapid propagation of latest cultivers of commercial importance. In crop species, significant impovement in net profits per unit area as compared to the expenses on tissue culture plants is the major consideration. In fact, although tissue culture of ornamentals got commercialised first, research in tissue culture has been a centre of attraction mainly for ability of this technology to overcome limitations of breeding programmes for improvement in crop productivity. However, what is common for both is the importance of quality control tests and systems.

Present global trade in ornamentals is hardly based on anything more than visual inspection of cultures or ex-agar plants. Situation is not any way different in crop plants. There are no specific quality tests on which the trade is based.

- A good quality culture base will provide a very sound foundation to improve process efficiency and profitability.
- Further, it will lend stable foundation for the trade. It is very easy to arrive at conflicting interpretations on quality of plants supplied in case of a dispute. In such a situation, no industry can prosper. In our opinion, this is a major global impediment in growth of this industry.
- The relationship of producer and buyer should be based on mutually accepted quality attributes, a system of process quality control based on agreed quality tests and a third party inspection based quality assurance system.
- In case of crop plants, such tests are vitally needed to base regulatory mechanisms.
- Many developing countries have become aware and their Governments are very active in promoting application of plant tissue culture to crop plants. However, doing this in absence of meaningful regulatory mechanisms is like inviting impediments in implimentation.

This article is an attempt to put together all the known principles and practices which have direct bearing on production and determination of quality of micropropagated plants and to present a coherent concept which will fill up above gap.

Here we consider only the most popular application *i.e.* multiplication of released and elite variety of a vegetatively cultivated crop of economic importance.

Strategic Imperatives for the Quality Tests

- They should be rapid.
- They should aim at identification of variety/cultivars ignoring point mutations/bud mutations which continue to occur in normal population also as a natural phenomenon. In fact, natural frequency of bud mutations is a desirable factor as it is one of the most useful survival tool for a variety.
- Definition of variety should be clear *i.e.* it is an assemblage of closely similar genotypes, not a group of "uniform" genotypes as is often misunderstood. Thus, the method should be able to make a distinction between two varieties and do not produce confusion where within variety variability is involved.

Potential Tests for Quality Control

Following are the kinds of tests, which would reveal the quality and will be consistent with the imperatives of quality tests :

- Endophytic bacterial count : Based on microbiological as well as DNA Fingerprinting based methods.
- Total virus content – assay or determination : Based on DNA Fingerprinting based methods.
- Genotypic stability/varietal identification by DNA Fingerprinting methods : DNA Fingerprinting based methods
- Metaphase chromosome count and karytotype : In polyploids, issue could be of making it sure that chromosomal aberrations have not taken place. This can be ascertained by taking metaphase chromosome counts and karyotyping of representative samples.
- Anatomy of explant, regenerating explant and multiplication cultures : Microtomy base methods to establish whether regeneration is through enhanced axillary proliferation, through adventive bud formation or through callus formation.

Requirement

- Services on standardisation of tests specific for plants of the client's interest
- Jobwork testing on routine samples in large numbers
- Culture indexing programme and Assurance system and culture management system based on above quality tests
- Culture indexing sevice based on routine analysis on jobwork basis.

IMPORTANCE OF THE WORK

The banana viruses primarily spread through the use of infected suckers but the secondary spread also occurs through aphid vectors. Field spread is much faster if virus-free suckers are not used as initial planting material. An annual loss of about Rs. 40 million has been reported due to BBTV in the Kerala state alone. In the hills of lower Palanis, BBTV destroyed 60 per cent of the banana crop. BBTV is also spreading fast in Maharashtra and Karnataka causing serious economic losses. Infectious chlorosis has been reported to be severe in Maharashtra and South India causing heavy losses in yields. The incidence of infectious chlorosis in Tamil Nadu has been reported to be 13.4 per cent. BBTV results in a yield reduction of about 40 per cent of bunch weight in cultivars Nendran, Poovan, Robusta, Red Banana, Bluggoe and Pisang Awak resulting in upto 68 per cent reduction in yield.

In India, banana streak virus severely infects 'Mysore' and Cavendish group of bananas. Many plants remain under symptomless normal condition, but, under stress like high temperature, nutrients they develop severe symptoms. The yield loss due to severe infection of BSV is reported to be more than 48 per cent. The BSV infected plants have poorly filled fingers, sometimes seediness has also been observed. Seeded fruits are unfit for marketing. The disease is assuming a major threat to banana industry owing to its fast spread, the tissue cultured plants supplied by some of the firms were found to have cent-per cent infection in Maharashtra and Tamil Nadu. Sensitive and reliable techniques for the detection of banana viruses in planting material are not available. It is necessary to develop virus specific diagnostic kits for the reliable and sensitive detection of these most common viruses so that tissue culture raised plants could be tested.

Like banana, other plant species, which are being micropropagated by the tissue culture industry in the country, are also affected by a number of viruses. For value addition, it is important that all the tissue culture raised plants are certified for freedom from viruses. Plants like citrus and oil palm are also being propagated through tissue culture intervention. Gradually, more and more plant species will be taken up for micropropagation through tissue culture.

Several plants are produced annually through tissue culture techniques. These plants are planted in large numbers in the fields either for evaluation purposes or for routine plantation. In either case it is important to maintain the genetic fidelity of the plants. Since sophisticated techniques like the AFLP have been developed and found useful, in this study we propose to use them to maintain a strict quality check of the plants produced at the MTP. This service will be made available not only to R and D centres developing tissue culture protocols but also to the established tissue culture industries for certification of planting material. Establishing clonal uniformity for horticultural species is very important as for most species it takes 3-4 years to bear fruits and certification in terms of disease free nature and expected yields are very important for the grower.

UTTINGS

Cuttings involve removing a piece from the parent plant and that piece then regrows the lost parts or tissues. Both woody and herbaceous plants are asexually propagated by cuttings of stems, leaves and roots. New plants can be grown from parts of plants because each living plant cell contains the ability to duplicate all plant parts and functions. Mature cells can change into MERISTEMATIC cells that are found at rapid growth sites like buds. There are many types of cuttings. Often, a plant can be propagated by more than one method of cutting. Some plants will reproduce readily from cuttings and others take a considerable amount of time and care.

Stock plants are the parent plants used in asexual propagation. Stock plants must be in excellent health and should possess characteristics desirable for production of new plants. Herbaceous cuttings are those taken from non-woody plants, such as perennials and houseplants. Softwood cuttings are pieces of new growth taken from woody stock plants. These cuttings must be taken before the new growth starts to harden. Hardwood cuttings are taken from tissue which has become woody. Other forms of cuttings are leaf cuttings and root cuttings.

The gardener must try to duplicate the conditions needed for a plant to root from a cutting. High humidity, indirect light and soil temperatures of 70 to 80 degrees F are best for most cuttings. These conditions may be created by keeping cuttings enclosed under glass or in plastic bags in dappled shade. Cuttings must be shielded from direct sunlight, especially if they are under glass or plastic.

STEM CUTTINGS

When a cutting is made, injured xylem and phloem cells plug the tubes so that precious fluids are not lost. Usually a CALLUS forms at the cut. Cells near the callus area reorganise to form adventitious roots.

Stem cuttings are the most commonly used method to produce houseplants. Select vigourous, new growth with no flower buds. Stem sections should be free of diseases and insects. Each cutting should be 2 to 4 inches long and have 2 or 3 leaves attached.

Make a cut 1/4 inch below a leaf node and pull off the leaves that are at the nodes that will be below the surface of the rooting medium. ROOTING HORMONE helps to stimulate rooting, but is optional. Pour a small amount of the rooting hormone into a clean container to prevent contamination of all of your rooting hormone. Dip the base of the stem, including the node area, into the rooting powder. The stem should be dry when dipped. Commercial rooting products often include a fungicide. This is a good idea given the damp conditions required for rooting success. Tap off excess powder, since too much hormone can inhibit rooting. Poke a hole in the medium before inserting the cutting to avoid loss of the rooting hormone. Insert treated cutting in a moist rooting medium. A suitable rooting medium is half perlite and half sphagnum peat moss. Any disinfested container with drainage is acceptable for use.

Cover container and cutting with a plastic bag tent to maintain high humidity. Place unit in a warm area with indirect light. Check the rooting medium every few days to make sure it remains moist. Rooting can take from a few days up to several months.

After a few weeks, test for rooting by gently tugging at the cutting. If there is resistance, rooting has started and the plastic cover may be removed.

LEAF CUTTINGS

In this method, a leaf blade or leaf with petiole is used to propagate new plants. The same steps are followed as for stem cuttings. Choose a healthy leaf from a vigourously growing plant. Cut it close to the stem with a sharp, disinfested razor or knife. Trim off 1/4 of the leaf and dip into rooting hormone, if desired. Insert the leaf into rooting medium so that 1/3 of the leaf is below the surface.

One or many new small plants form at the base of the leaf. African violet leaves will produce many new plants. Begonia leaves can be divided into segments for propagation; however each leaf piece must contain a major vein. With leaf cuttings, the original leaf is not a part of the new plant and is usually discarded. Many succulent plants, such as sedum, jade, and peperomia, can be propagated by leaf cutting. Some plants such as Kalanchoe pinnata or K. bryophyllum, piggyback or air plants, form their own new plantlets in this fashion.

ROOT CUTTINGS

Cultivation of root cuttings probably started after gardeners observed new plants growing from pieces of root accidentally left behind in the soil. Take cuttings from newer root growth. Make cuttings 1 to 4 inches long from roots that are 1/4 to 1/2 inch in diametre. Be sure that the roots collected are from the chosen plant and not neighbouring plants. Cuttings should be taken during the dormant season when roots have large carbohydrate supplies. However, they also may be taken throughout the growing season. Cut straight through the end of the root closest to the stem. Cut the other end on a slant. This allows you to remember which end is the top (the straight cut) and which is the bottom (the diagonal cut).

Store cuttings from dormant roots for 3 weeks in moist rooting medium at 40 degrees F. Remove from storage and plant upright in the growing medium. Keep moist and warm, in a bright location until growth and weather permit acclimatising to the outdoors.

If root cuttings are taken during active growth, skip the storage period and place cuttings directly in the rooting medium. For smaller plants, take 1- to 2-inch sections. Place cuttings horizontally a half inch below the surface of the rooting medium. These cuttings should be handled indoors or in a HOTBED. The fine roots of many perennials are used for propagation. Root cuttings of some variegated plants will lose their variegation.

SOFTWOOD AND HARDWOOD CUTTINGS

Softwood cuttings are taken from first-year branches that have not yet become woody. Flowering shrubs are often propagated by softwood cuttings. Late spring and early summer are the best times for success with this method. Take cuttings after it rains or water is applied in the cool morning. Make a diagonal cut. The larger diagonal cut gives more area to develop roots. Keep cuttings in water before moving them into rooting medium.

Make cuttings 2 to 10 inches long. Larger cuttings produce larger plants sooner. Make cuts slightly below a leaf node. Follow the same instructions for disinfested containers, rooting medium and light and moisture conditions used by other cutting methods. Rooting hormones may be used with this method. The base of the cutting should be dry before dipping it into rooting hormone powder.

Hardwood cuttings are taken once the tissue becomes woody and the plant is dormant. Cuttings can be taken anytime from late fall after a killing frost until late winter. Select healthy wood that was produced the previous summer. Several cuttings can be made from the same branch of some shrubs. Make cuts at a slant, 5 to 12 inches long. Basal cuts should be just below a node, while the upper cut should be slightly above a bud. Dip the basal end in rooting hormone. Mark the most terminal end of each cutting with a tag. Bury cuttings vertically in moist vermiculite or sand. Cuttings should not freeze, but must remain cool. A callus will form on the lower cut end during storage. Callus formation indicates that cuttings are ready to root. In spring, remove the cuttings from storage. Plant in a hotbed or other protected site with morning sun exposure or filtered light. Leave 1 to 2 inches of cutting above ground. Keep cuttings moist until a root system forms. Transplant the cuttings the following spring while they are still dormant.

LAYERING

Layering causes roots to develop on shoots that are still attached to the parent plant. The stem is not cut from the main plant until it has rooted. Use this method when other propagation methods are unsuccessful. Layering is a good propagation choice when only a few plants are needed. Simple layering is done by bending a branch to the ground and burying a portion of it. The tip remains uncovered. A light soil increases rooting success, as will wounding or girdling the buried section. Treatment with rooting hormone is helpful.

Hold the branch in place with a rock or peg. Layering is done in early spring while plants are still dormant or in late summer on wood that has not become woody. Plants with flexible branches are particularly suited to this method. Other types of layering include serpentine or compound layering, continuous or trench layering, and mound or stool layering.

DIVISIONS

Division is the cutting or breaking up of a crown or clump of suckers into segments. Each segment must have a bud and some roots. These segments are replanted and grow into new plants identical to the parent. Most perennials should be lifted and divided when they become overgrown and begin to lose vigour. Vigourous growth in most perennials occurs on the outer segments of the clump. Old growth in the centre of the clump is discarded. Carefully dig the plant, loosening the roots and lifting the plant from the soil. Split apart the main clump with two spades or forks or chop with a shovel or hatchet if the clump is firmly massed. In some cases outside segments of the plant can be removed and replanted without disturbing the rest of the plant. A good rule of thumb is to divide fall-flowering perennials in spring and spring- and summer-flowering perennials in fall. Some shrubs that form clumps or crowns are suitable for division. Divide in early spring while dormant. Shrubs used for division should have several clumps that are two years old or older.

BULBS AND CORMS

Bulbs can be propagated by removing small bulblets or offsets that form at the base of the parent bulb. These small bulbs take 2 or 3 years to mature into plants that flower. Place offsets in rich, light soil for their development. This same procedure should be followed for plants which form from corms, such as gladiolus.

Many lilies can be multiplied by removing scales from the mature bulb. Dust the scale with a fungicide and place, base end down, in a moist growing medium in a warm, protected area. Bulblets will form at the base of the scale. In 1 to 4 years these bulblets will grow and be ready to flower.

TUBERS AND RHIZOMES

Tuberous plants can be dug up and the TUBERS separated. In separating the tubers, each must have a segment of the crown that contains at least one eye or bud. Rhizomes grow and develop buds along their length. The rhizomes can be dug and cut into sections that each contain at least one eye or bud.

GRAFTING

Grafting involves the joining of different segments of two different plants of the same species. A branch or bud is joined to an UNDERSTOCK. In grafting, the cambium layers of the two different segments are aligned and grow together. Some plants graft naturally. Ivy commonly grafts itself. Natural grafting occurs where two branches are in close contact over several years.

Grafting allows gardeners to produce plants identical to a parent plant. It also allows growers to control size and shape of a tree or shrub. Apples are grafted onto root stock that regulates the ultimate size of the tree. Grafting can produce more vigourous, earlier fruiting, or two varieties can be grown on the same tree to provide pollination as with apples. On the negative side, some grafting attempts will be rejected. This is called graft incompatibility. This might not occur in the first year, but may in successive years. Wind or winter cold can damage a GRAFT UNION. Some grafted trees or plants produce large numbers of SUCKERS. These can crowd out the desired plant or tree and are unsightly. If the graft union is planted below the soil surface, the SCION will root and the rootstock will die.

Grafting is usually done in the spring. It involves collecting small branches called scion wood. These branches are 1/4 to 1/2 inch in diametre. Gather them several weeks before grafting is to occur. Collect wood that grew the preceding year. Select only wood with leaf buds, not flower buds. Scion wood should be gathered in winter when wood is dormant, but not frozen. Use sharp clippers so that wood is not crushed. New growth over 1 foot in length is usually best.

Discard the wood at both ends of the branch and use the middle section. End growth is too succulent for grafting and wood closer to the base will graft more slowly. Label the scion wood, wrap it in moist paper towels or sphagnum peat, enclose it in an airtight, plastic container and place it in the refrigerator.

Scion wood must be joined to the understock in spring when buds swell. It is critical that the two pieces are nearly the same size and that sap has begun to flow. The day before actually grafting, remove scions from the refrigerator and snip off the bottom ends. Place the clipped scions in a pail of water overnight. It is critical that the cambium layer on the scion precisely matches that of the understock.

For precise directions on joining scion to understock. Grafts can be placed on a root or stem and be joined by any of several methods. The union is held firmly in place using wax, tape, plastic wrap, rubber bands or plastic tubing.

The grafted area must be protected from anything that will move the scion out of alignment. Bracing some grafts is advised. Side sprouts should be removed the first summer. Sucker shoots that sprout from below the graft also should be removed. If growth of new graft is satisfactory, do not fertilize the plant during the first year. There is danger that top growth will be very brittle and it could fail to harden off before the first frost.

BUDDING OR BUD GRAFTING

Bud grafting is faster, easier and less messy than other forms of grafting. Cambium layers do not need to be aligned. Bud grafting is done from early July through early August. This method uses a newly developed LATENT BUD, taken from under a live leaf. Budwood is collected from healthy branches that grew since spring. Young trees offer good budwood because they produce a large amounts of new growth. Use buds from the middle section of the branch.

Remove leaves, but keep 1/4 inch of the leaf's petiole for grasping. The bud is cut from the branch and inserted into a T-shaped slice made in the bark of the understock. Budding should only be done when the bark slips easily away from the tree. The bud is held in place with special tape or wrap.

MICROPROPAGATION OR TISSUE CULTURE

Each plant cell has the potential to grow into a new plant exactly like the parent. This fact coupled with technical advances, specialised equipment and sterile laboratory conditions has produced modern tissue culture. In tissue culture, individual or small groups of plant cells are manipulated so they each produce a new plant. A tiny piece of bud, leaf or stem can produce incredible numbers of new plants in a small space in a short time.

The advantages of tissue culture, in addition to speed and efficiency of propagation, include production of disease-free plants. New plants can be made available to the public more quickly because of tissue culture. However, there are some problems with spontaneous mutations which naturally occur. In tissue culture, the incidence of these mutations is greatly increased.

Conditions for tissue culture are very exacting. Absolutely sterile conditions must be maintained. Temperature, light, humidity and atmosphere are strictly controlled with electronic sensors and computerised controls. Such costly equipment rules this out for most home gardeners.

Chapter 24

Pest Control in Plant

FUNGICIDE

Fungicides are biocidal chemical compounds or biological organisms used to kill or inhibit fungi or fungal spores. Fungi can cause serious damage in agriculture, resulting in critical losses of yield, quality and profit. Fungicides are used both in agriculture and to fight fungal infections in animals. Chemicals used to control oomycetes, which are not fungi, are also referred to as fungicides as oomycetes use the same mechanisms as fungi to infect plants.

Fungicides can either be contact, translaminar or systemic. Contact fungicides are not taken up into the plant tissue, and only protect the plant where the spray is deposited; translaminar fungicides redistribute the fungicide from the upper, sprayed leaf surface to the lower, unsprayed surface; systemic fungicides are taken up and redistributed through the xylem vessels. Few fungicides move to all parts of a plant. Some are locally systemic, and some move upwardly.

Most fungicides that can be bought retail are sold in a liquid form. A very common active ingredient is sulphur, present at 0.08 per cent in weaker concentrates, and as high as 0.5 per cent for more potent fungicides. Fungicides in powdered form are usually around 90 per cent sulphur and are very toxic. Other active ingredients in fungicides include neem oil, rosemary oil, jojoba oil, the bacterium *Bacillus subtilis*, and the beneficial fungus *Ulocladium oudemansii*.

Fungicide residues have been found on food for human consumption, mostly from post-harvest treatments. Some fungicides are dangerous to human health, such as vinclozolin, which has now been removed from use. A number of fungicides are also used in human health care.

Natural Fungicides

Plants and other organisms have chemical defences that give them an advantage against micro-organisms such as fungi.

Some of these compounds can be used as fungicides :

- Tea tree oil
- Cinnamaldehyde
- Cinnamon essential oil
- Jojoba oil
- Neem oil
- Oregano oil
- Rosemary oil
- Monocerin
- Milk
- Ampelomyces quisqualis AQ10, CNCM I-807.

Whole live or dead organisms that are efficient at killing or inhibiting fungi can sometimes be used as fungicides :

- The bacterium Bacillus subtilis
- The fungus Ulocladium oudemansii
- Kelp (powdered dried kelp is fed to cattle to protect them from fungi in grass).

Resistance

Pathogens respond to the use of fungicides by evolving resistance. In the field several mechanisms of resistance have been identified. The evolution of fungicide resistance can be gradual or sudden. In qualitative or discrete resistance a mutation (normally to a single gene) produces a race of a fungus with a high degree of resistance. Such resistant varieties also tend to show stability, persisting after the fungicide has been removed from the market. For example, sugar beet leaf blotch remains resistant to azoles years after they were no longer used for control of the disease. This is because such mutations often have a high selection pressure when the fungicide is used, but there is low selection pressure to remove them in the absence of the fungicide.

In instances where resistance occurs more gradually a shift in sensitivity in the pathogen to the fungicide can be seen. Such resistance is polygenic—an accumulation of many mutation in different genes each having a small additive effect. This type of resistance is known as quantitative or continuous resistance. In this kind of resistance the pathogen population will revert to a sensitive state if the fungicide is no longer applied. Little is known about how variations in fungicide treatment affect the selection pressure to evolve resistance to that fungicide. Evidence shows that the doses that provide the most control of the disease also provide the largest selection pressure to acquire resistance, and that lower doses decreased the selection pressure.

In some cases when a pathogen evolves resistance to one fungicide it automatically obtains resistance to others–a phenomenon known as cross resistance.

These additional fungicides are normally of the same chemical family or have the same mode of action, or can be detoxified by the same mechanism. Sometimes negative cross resistance occurs, where resistance to one chemical class of fungicides leads to an increase in sensitivity to a different chemical class of fungicides. This has been seen with carbendazim and diethofencarb. There are also recorded incidences of pathogens evolving multiple drug resistance-resistance to two chemically different fungicides by separate mutation events. For example *Botrytis cinerea* is resistant to both azoles and dicarboximide fungicides.

There are several routes by which pathogens can evolve fungicide resistance. The most common mechanism appears to be alteration of the target site, particular as a defence against single site of action fungicides. For example Black Sigatoka, an economically important pathogen of banana, is resistant to the QoI fungicides, due to a single nucleotide change resulting one amino acid (glycine) being replaced by another (alanine) in the target protein of the QoI fungicides, cytochrome b. This presumably disrupts the binding of the fungicide to the protein, rendering the fungicide ineffective.

Upregulation of target genes can also render the fungicide ineffective. This is seen in DMI resistant strains of *Venturia inaequalis*. Resistance to fungicides can also be developed by efficient efflux of the fungicide out of the cell. *Septoria tritici* has developed multiple drug resistance using this mechanism. The pathogen had 5 ABC type transporters with overlapping substrate specificities that together work to effectively pump toxic chemicals out of the cell. In addition to the mechanisms outlined above, fungi may also develop metabolic pathways that circumvent the target protein, or acquire enzymes that enable metabolism of the fungicide to a harmless substance.

Fungicide Resistance Management

The fungicide resistance action committee (FRAC) has several recommended practices to try to avoid the development of fungicide resistance, especially in at-risk fungicides including *Strobilurins* such as azoxystrobin. Products should not be used in isolation but rather as mixture, or alternate sprays, with another fungicide with a different mechanism of action. The likelihood of the pathogen developing resistance is greatly decreased by the fact that any resistant isolates to one fungicide will hopefully be killed by the other—in other words two mutations would be required rather than just one. The effectiveness of this technique can be demonstrated by Metalaxyl, a phenylamide fungicide. When used as the sole product in Ireland to control potato blight (*Phytophthora infestans*) resistance developed within one growing season. However in countries like the UK where it was only ever marketed as a mixture resistance problems developed more slowly.

Fungicides should only be applied when absolutely necessary, especially if they are in an at-risk group. Lowering the amount of fungicide in the environment lowers the selection pressure for resistance to develop. Manufacturers' doses should always be followed. These doses are normally designed to give the right balance between controlling the disease and limiting the risk of resistance development. Higher doses increase the selection pressure for single site mutations that confer resistance, as all strains but those that carry the mutation will be eliminated, and thus the resistant strain will propagate. Lower doses greatly increase the risk of polygenic resistance, as strains that are slightly less sensitive to the fungicide may survive.

It is also recommended that where possible fungicides are only used in a protective manner, rather than to try to cure already infected crops. Far fewer fungicides have curative/eradicative ability than protectant. Thus fungicide preparations advertised as having curative action may only have one active chemical; a single fungicide acting in isolation increases the risk of fungicide resistance.

It is better to use an integrative pest management approach to disease control, rather than relying on fungicides alone. This involves the use of resistant varieties and hygienic practises, such as the removal of potato discard piles and stubble on which the pathogen can over-winter, greatly reducing the titre of the pathogen and thus the risk of fungicide resistance development.

HERBICIDE

Herbicides, also commonly known as weedkillers, are pesticides used to kill unwanted plants. Selective herbicides kill specific targets, while leaving the desired crop relatively unharmed. Some of these act by interfering with the growth of the weed and are often synthetic mimics of natural plant hormones. Herbicides used to clear waste ground, industrial sites, railways and railway embankments are not selective and kill all plant material with which they come into contact. Smaller quantities are used in forestry, pasture systems, and management of areas set aside as wildlife habitat. Some plants produce natural herbicides, such as the genus *Juglans* (walnuts), or the tree of heaven; such action of natural herbicides, and other related chemical interactions, is called allelopathy. Herbicides are widely used in agriculture and landscape turf management. In the US, they account for about 70 per cent of all agricultural pesticide use.

History

Prior to the widespread use of chemical herbicides, cultural controls, such as altering soil pH, salinity, or fertility levels, were used to control weeds. Mechanical control (including tillage) was also (and still is) used to control weeds. Herbicides were first researched and discovered in both the UK and the US in the context of clandestine war time efforts to create chemical warfare agents in 1941. The first herbicide, 2,4-D, was co-discovered independently by two teams; one at ICI and one at Rothamsted Experimental Station. The new chemicals ability to kill weeds was entirely incidental to the aim of the research in either country. Because research into chemical warfare was illegal under the Geneva Protocols, the research operated under the plausible cover story of agricultural research.

Experimentation had become quite common (in the fashion of British team at Rothamsted Experimental Station, under the leadership of Judah Hirsch Quastel. Some discoveries were able to be eventually used to increase crop yields, although 2-4D was not used until 1946 near the end of the WWII. When it was commercially released in 1946, it became the first successful selective herbicide and allowed for greatly enhanced weed control in wheat, maise (corn), rice, and similar cereal grass crops, because it kills dicots (broadleaf plants), but not most monocots (grasses). The low cost of 2,4-D has led to continued usage today, and it remains one of the most commonly used herbicides in the world. Like other acid herbicides, current formulations use either an amine salt (often trimethylamine) or one of many esters of the parent compound. These are easier to handle than the acid.

The triazine family of herbicides, which includes atrazine, were introduced in the 1950s; they have the current distinction of being the herbicide family of greatest concern regarding groundwater contamination. Atrazine does not break down readily (within a few weeks) after being applied to soils of above neutral pH. Under alkaline soil conditions, atrazine may be carried into the soil profile as far as the water table by soil water following rainfall causing the aforementioned contamination. Atrazine is thus said to have "carryover", a generally undesirable property for herbicides.

Glyphosate, frequently sold under the brand name Roundup, was introduced in 1974 for non-selective weed control. Following the development of glyphosate-resistant crop plants, it is now used very extensively for selective weed control in growing crops. The pairing of the herbicide with the resistant seed contributed to the consolidation of the seed and chemistry industry in the late 1990s.

Many modern chemical herbicides for agriculture are specifically formulated to decompose within a short period after application. This is desirable, as it allows crops which may be affected by the herbicide to be grown on the land in future seasons. However, herbicides with low residual activity (*i.e.*, that decompose quickly) often do not provide season-long weed control.

Health and Environmental Effects

Herbicides have widely variable toxicity. In addition to acute toxicity from high exposure levels, there is concern of possible carcinogenicity, as well as other long-term problems, such as contributing to Parkinson's disease. Some herbicides cause a range of health effects ranging from skin rashes to death. The pathway of attack can arise from intentional or unintentional direct consumption, improper application resulting in the herbicide coming into direct contact with people or wildlife, inhalation of aerial sprays, or food consumption prior to the labelled pre-harvest interval. Under extreme conditions, herbicides can also be transported via surface run-off to contaminate distant water sources. Most herbicides decompose rapidly in soils via soil microbial decomposition, hydrolysis, or photolysis. Phenoxy herbicides are often contaminated with dioxins such as TCDD; research has suggested such contamination results in a small rise in cancer risk after exposure to these herbicides. Triazine exposure has been implicated in a likely relationship to increased risk of breast cancer, although a causal relationship remains unclear.

Herbicide manufacturers have at times made false or misleading claims about the safety of their products. Chemical manufacturer Monsanto Company agreed to change its advertising after pressure from New York attorney general Dennis Vacco; Vacco complained about misleading claims that its spray-on glyphosate-based herbicides, including Roundup, were safer than table salt and "practically non-toxic" to mammals, birds, and fish. Roundup is toxic and has resulted in death after being ingested in quantities ranging from 85 to 200 ml, although it has also been ingested in quantities as large as 500 ml with only mild or moderate symptoms. The manufacturer of Tordon 101 (Dow Agro-Sciences, owned by the Dow Chemical Company) has claimed Tordon 101 has no effects on animals and insects, in spite of evidence of strong carcinogenic activity of the active ingredient Picloram in studies on rats.

The risk of Parkinson's disease has been shown to increase with occupational exposure to herbicides and pesticides. The herbicide paraquat is suspected to be one such factor. All commercially sold, organic and non-organic herbicides must be extensively tested prior to approval for sale and labelling by the Environmental Protection Agency. However, because of the large number of herbicides in use, concern regarding health effects is significant. In addition to health effects caused by herbicides themselves, commercial herbicide mixtures often contain other chemicals, including inactive ingredients, which have negative impacts on human health. For example, Roundup contains adjuvants which, even in low concentrations, were found to kill human embryonic, placental, and umbilical cells *in vitro.* One study also found Roundup caused genetic damage, but the damage was not caused by the active ingredient.

Ecological Effects

Commercial herbicide use generally has negative impacts on bird populations, although the impacts are highly variable and often require field studies to predict accurately. Laboratory studies have at times overestimated negative impacts on birds due to toxicity, predicting serious problems that were not observed in the field. Most observed effects are due not to toxicity, but to habitat changes and the decreases in abundance of species on which birds rely for food or shelter. Herbicide use in silviculture, used to favour certain types of growth following clearcutting, can cause significant drops in bird populations. Even when herbicides which have low toxicity to birds are used, they decrease the abundance of many types of vegetation on which the birds rely. Herbicide use in agriculture in Britain has been linked to a decline in seed-eating bird species which rely on the weeds killed by the herbicides. Heavy use of herbicides in neo-tropical agricultural areas has been one of many factors implicated in limiting the usefulness of such agricultural land for wintering migratory birds.

Scientific Uncertainty

The health and environmental effects of many herbicides is unknown, and even the scientific community often disagrees on the risk. For example, a 1995 panel of 13 scientists reviewing studies on the carcinogenicity of 2,4-D had divided opinions on the likelihood 2,4-D causes cancer in humans. As of 1992, studies on phenoxy herbicides were too few to accurately assess the risk of many types of cancer from these herbicides, even though evidence was stronger that exposure to these herbicides is associated with increased risk of soft tissue sarcoma and non-Hodgkin lymphoma. Furthermore, there is some suggestion that herbicides can play a role in sex reversal of certain organisms that experience temperature-dependent sex determination, which could theoretically alter sex ratios.

Resistance

Scientists generally agree selection pressure applied to weed populations for a long enough period of time eventually leads to resistance. Plants have developed resistance to atrazine and to ALS-inhibitors, and more recently, to glyphosate herbicides. Marestail is one weed that has developed glyphosate resistance.

Classification

Herbicides can be grouped by activity, use, chemical family, mode of action, or type of vegetation controlled.

By Activity :

- Contact herbicides destroy only the plant tissue in contact with the chemical. Generally, these are the fastest acting herbicides. They are less effective on perennial plants, which are able to regrow from rhizomes, roots or tubers.
- Systemic herbicides are translocated through the plant, either from foliar application down to the roots, or from soil application up to the leaves. They are capable of controlling perennial plants and may be slower-acting, but ultimately more effective than contact herbicides.

By Use :

- Soil-applied herbicides are applied to the soil and are taken up by the roots and/or hypocotyl of the target plant. The three main types are :

i. Preplant incorporated herbicides are soil applied prior to planting and mechanically incorporated into the soil. The objective for incorporation is to prevent dissipation through photodecomposition and/or volatility.

ii. Pre-emergent herbicides are applied to the soil before the crop emerges and prevent germination or early growth of weed seeds.

iii. Postemergent herbicides are applied after the crop has emerged.

Their classification by mechanism of action (MOA) indicates the first enzyme, protein, or biochemical step affected in the plant following application.

The main mechanisms of action are :

- *ACCase Inhibitors Compounds Kill Grasses :* Acetyl coenzyme A carboxylase (ACCase) is part of the first step of lipid synthesis. Thus, ACCase inhibitors affect cell membrane production in the meristems of the grass plant. The ACCases of grasses are sensitive to these herbicides, whereas the ACCases of dicot plants are not.

- *ALS Inhibitors :* The acetolactate synthase (ALS) enzyme (also known as acetohydroxyacid synthase, or AHAS) is the first step in the synthesis of the branched-chain amino acids (valine, leucine, and isoleucine). These herbicides slowly starve affected plants of these amino acids, which eventually leads to inhibition of DNA synthesis. They affect grasses and dicots alike. The ALS inhibitor family includes sulfonylureas, imidazolinones, triazolopyrimidines, pyrimidinyl oxybenzoates, and sulfonylamino carbonyl triazolinones. The ALS biological pathway exists only in plants and not animals, thus making the ALS-inhibitors among the safest herbicides.
- *EPSPS Inhibitors :* The enolpyruvylshikimate 3-phosphate synthase enzyme EPSPS is used in the synthesis of the amino acids tryptophan, phenylalanine and tyrosine. They affect grasses and dicots alike. Glyphosate (Roundup) is a systemic EPSPS inhibitor inactivated by soil contact.
- Synthetic auxins inaugurated the era of organic herbicides. They were discovered in the 1940s after a long study of the plant growth regulator auxin. Synthetic auxins mimic this plant hormone. They have several points of action on the cell membrane, and are effective in the control of dicot plants. 2,4-D is a synthetic auxin herbicide.
- Photosystem II inhibitors reduce electron flow from water to NADPH2+ at the photochemical step in photosynthesis. They bind to the Qb site on the D1 protein, and prevent quinone from binding to this site. Therefore, this group of compounds causes electrons to accumulate on chlorophyll molecules. As a consequence, oxidation reactions in excess of those normally tolerated by the cell occur, and the plant dies. The triazine herbicides (including atrazine) and urea derivatives (diuron) are photosystem II inhibitors.
- Photosystem I inhibitors steal electrons from the normal pathway through FeS–Fdx–NADP leading to direct discharge of electrons on oxygen. As a result, reactive oxygen species are produced and oxidation reactions in excess of those normally tolerated by the cell occur, leading to plant death.

Bipyridinium herbicides (such as diquat and paraquat) hit the "Fe-S–Fdx step" while diphenyl ether herbicide (such as nitrofen, nitrofluorfen, and acifluorfen) hit the "Fdx–NADP step".

Organic Herbicides

Recently, the term "organic" has come to imply products used in organic farming. Under this definition, an organic herbicide is one that can be used in a farming enterprise that has been classified as organic. Commercially sold organic herbicides are expensive and may not be affordable for commercial farming.

Depending on the application, they may be less effective than synthetic herbicides and are generally used along with cultural and mechanical weed control practices.

Homemade organic herbicides include :

- Corn gluten meal (CGM) is a natural pre-emergence weed control used in turfgrass, which reduces germination of many broadleaf and grass weeds.
- Spices are now effectively used in herbicides.
- Vinegar is effective for 5–20 per cent solutions of acetic acid, with higher concentrations most effective, but it mainly destroys surface growth, so respraying to treat regrowth is needed. Resistant plants generally succumb when weakened by respraying.
- Steam has been applied commercially, but is now considered uneconomical and inadequate. It kills surface growth but not underground growth and so respraying to treat regrowth of perennials is needed.
- Flame is considered more effective than steam, but suffers from the same difficulties.
- D-limonene (citrus oil) is a natural degreasing agent that strips the waxy skin or cuticle from weeds, causing dehydration and ultimately death.
- Saltwater or salt applied in appropriate strengths to the rootzone will kill most plants.
- Monocerin produced by certain fungi will kill certain weeds such as Johnson grass.

Application

Most herbicides are applied as water-based sprays using ground equipment. Ground equipment varies in design, but large areas can be sprayed using self-propelled sprayers equipped with long booms, of 60 to 80 feet (18 to 24 m) with flat-fan nozzles spaced about every 20 inches (510 mm). Towed, handheld, and even horse-drawn sprayers are also used. Synthetic organic herbicides can generally be applied aerially using helicopters or airplanes, and can be applied through irrigation systems (chemigation).

A new method of herbicide application involves ridding the soil of its active weed seed bank rather than just killing the weed. Researchers at the Agricultural Research Service have found the application of herbicides to fields late in the weeds' growing season greatly reduces their seed production, and therefore fewer weeds will return the following season. Because most weeds are annual grasses, their seeds will only survive in soil for a year or two, so this method will be able to "weed out" the weed with only a few years of herbicide application.

Weed-wiping may also be used, where a wick wetted with herbicide is suspended from a boom and dragged or rolled across the tops of the taller weed plants. This allows treatment of taller grassland weeds by direct contact without affecting related but desirable plants in the grassland sward beneath.

Misapplication

Herbicide volatilisation or drift may result in herbicide affecting neighbouring fields or plants, particularly in windy conditions. Sometimes, the wrong field or plants may be sprayed due to error.

Terminology

- Control is the destruction of unwanted weeds, or the damage of them to the point where they are no longer competitive with the crop.
- Suppression is incomplete control still providing some economic benefit, such as reduced competition with the crop.
- Crop safety, for selective herbicides, is the relative absence of damage or stress to the crop. Most selective herbicides cause some visible stress to crop plants.

In Current Use

- 2,4-D is a broadleaf herbicide in the phenoxy group used in turf and no-till field crop production. Now, it is mainly used in a blend with other herbicides to allow lower rates of herbicides to be used; it is the most widely used herbicide in the world, and third most commonly used in the United States. It is an example of synthetic auxin (plant hormone).
- Aminopyralid is a broadleaf herbicide in the pyridine group, used to control weeds on grassland, such as docks, thistles and nettles. It is notorious for its ability to persist in compost.
- Atrazine, a triazine herbicide, is used in corn and sorghum for control of broadleaf weeds and grasses. Still used because of its low cost and because it works well on a broad spectrum of weeds common in the US corn belt, atrazine is commonly used with other herbicides to reduce the overall rate of atrazine and to lower the potential for groundwater contamination; it is a photosystem II inhibitor.
- Clopyralid is a broadleaf herbicide in the pyridine group, used mainly in turf, rangeland, and for control of noxious thistles. Notorious for its ability to persist in compost, it is another example of synthetic auxin.
- Dicamba, a postemergent broadleaf herbicide with some soil activity, is used on turf and field corn. It is another example of a synthetic auxin.
- Glufosinate ammonium, a broad-spectrum contact herbicide, is used to control weeds after the crop emerges or for total vegetation control on land not used for cultivation.
- Fluazifop (Fuselade Forte), a post emergence, foliar absorbed, translocated grass-selective herbicide with little residual action. It is used on a very wide range of broad leaved crops for control of annual and perennial grasses.

- Fluroxypyr, a systemic, selective herbicide, is used for the control of broad-leaved weeds in small grain cereals, maise, pastures, rangeland and turf. It is a synthetic auxin. In cereal growing, fluroxypyr's key importance is control of cleavers, *Galium aparine*. Other key broadleaf weeds are also controlled.
- Glyphosate, a systemic non-selective herbicide, is used in no-till burndown and for weed control in crops genetically modified to resist its effects. It is an example of an EPSPs inhibitor.
- Imazapyr a non-selective herbicide, is used for the control of a broad range of weeds, including terrestrial annual and perennial grasses and broadleaf herbs, woody species, and riparian and emergent aquatic species.
- Imazapic, a selective herbicide for both the pre- and postemergent control of some annual and perennial grasses and some broadleaf weeds, kills plants by inhibiting the production of branched chain amino acids (valine, leucine, and isoleucine), which are necessary for protein synthesis and cell growth.
- Imazamox, an imidazolinone manufactured by BASF for postemergence application that is an acetolactate synthase (ALS) inhibitor. Sold under trade names Raptor, Beyond, and Clearcast.
- Linuron is a non-selective herbicide used in the control of grasses and broadleaf weeds. It works by inhibiting photosynthesis.
- Metolachlor is a pre-emergent herbicide widely used for control of annual grasses in corn and sorghum; it has displaced some of the atrazine in these uses.
- Paraquat is a non-selective contact herbicide used for no-till burndown and in aerial destruction of marijuana and coca plantings. It is more acutely toxic to people than any other herbicide in widespread commercial use.
- Pendimethalin, a pre-emergent herbicide, is widely used to control annual grasses and some broad-leaf weeds in a wide range of crops, including corn, soybeans, wheat, cotton, many tree and vine crops, and many turfgrass species.
- Picloram, a pyridine herbicide, mainly is used to control unwanted trees in pastures and edges of fields. It is another synthetic auxin.
- Sodium chlorate, a non-selective herbicide, is considered phytotoxic to all green plant parts. It can also kill through root absorption.
- Triclopyr, a systemic, foliar herbicide in the pyridine group, is used to control broadleaf weeds while leaving grasses and conifers unaffected.

Historical Interest : 2,4,5-T

- 2,4,5-Trichlorophenoxyacetic acid (2,4,5-T) was a widely used broadleaf herbicide until being phased out starting in the late 1970s. While 2,4,5-T itself is of only moderate toxicity, the manufacturing process for 2,4,5-T contaminates this chemical with trace amounts of 2,3,7,8-tetrachlorodibenzo-p-dioxin (TCDD). TCDD is extremely toxic to humans. With proper temperature control during production of 2,4,5-T, TCDD levels can be held to about .005 ppm. Before the TCDD risk was well understood, early production facilities lacked proper temperature controls. Individual batches tested later were found to have as much as 60 ppm of TCDD.

- 2,4,5-T was withdrawn from use in the USA in 1983, at a time of heightened public sensitivity about chemical hazards in the environment. Public concern about dioxins was high, and production and use of other (non-herbicide) chemicals potentially containing TCDD contamination was also withdrawn. These included pentachlorophenol (a wood preservative) and PCBs (mainly used as stabilising agents in transformer oil). Some feel that the 2,4,5-T withdrawal was not based on sound science. 2,4,5-T has since largely been replaced by dicamba and triclopyr.
- Agent Orange was an herbicide blend used by the U.S. military in Vietnam between January 1965 and April 1970 as a defoliant. It was a 50/50 mixture of the *n*-butyl esters of 2,4,5-T and 2,4-D. Because of TCDD contamination in the 2,4,5-T component, it has been blamed for serious illnesses in many veterans and Vietnamese people who were exposed to it. However, research on populations exposed to its dioxin contaminant have been inconsistent and inconclusive. Agent Orange often had much higher levels of TCDD than 2,4,5-T used in the US. The name *Agent Orange* is derived from the orange colour-coded stripe used by the Army on barrels containing the product. It is worth noting that there were other blends of synthetic auxins at the time of the Vietnam War whose containers were recognised by their colours, such as Agent Purple and Agent Pink.

WEED CONTROL

Weed control is the botanical component of pest control, using physical and chemical methods to stop weeds from reaching a mature stage of growth when they could be harmful to domesticated plants and livestock. In order to reduce weed growth, many "weed control" strategies have been developed in order to contain the growth and spread of weeds. The most basic is ploughing which cuts the roots of annual weeds. Today, chemical weed killers known as herbicides are widely used.

A plant is often termed weed when it has one or more of the following characteristics :

- Little or no value (as in medicinal, nutritional, or energy)
- Very high growth rate and/or ease of germination
- Exhibits competition to crops, for space, light, water and nutrients

Effects on Other Plants

Weeds can compete with productive crops or pasture, or convert productive land into unusable scrub. Weeds are also often poisonous, distasteful, produce burrs, thorns or other damaging body parts or otherwise interfere with the use and management of desirable plants by contaminating harvests or excluding livestock.

Weeds tend to thrive at the expense of the more refined edible or ornamental crops. They provide competition for space, nutrients, water and light, although how seriously they will affect a crop depends on a number of factors. Some crops have greater resistance than others—smaller, slower growing seedlings are more likely to be overwhelmed than those that are larger and more vigourous. Onions are one of the crops most susceptible to competition, for they are slow to germinate and produce slender, upright stems. Quick growing, broad leafed weeds therefore have a distinct advantage, and if not removed, the crop is likely to be lost. Broad beans however produce large seedlings, and will suffer far less profound effects of weed competition other than during periods of water shortage at the crucial time when the pods are filling out. Transplanted crops raised in sterile seed or potting compost will have a head start over germinating weed seeds.

Weeds also differ in their competitive abilities, and can vary according to conditions and the time of year. Tall growing vigourous weeds such as fat hen (*Chenopodium album*) can have the most pronounced effects on adjacent crops, although seedlings of fat hen that appear in late summer will only produce small plants. Chickweed (*Stellaria media*), a low growing plant, can happily co-exist with a tall crop during the summer, but plants that have overwintered will grow rapidly in early spring and may swamp crops such as onions or spring greens.

The presence of weeds does not necessarily mean that they are competing with a crop, especially during the early stages of growth when each plant can find the resources it requires without interfering with the others. However, as the seedlings' size increases, their root systems will spread as they each begin to require greater amounts of water and nutrients. Estimates suggest that weed and crop can co-exist harmoniously for around three weeks, therefore it is important that weeds be removed early on in order to prevent competition occurring.

Weed competition can have quite dramatic effects on crop growth. Harold A Roberts cites research carried out with onions wherein *"Weeds were carefully removed from separate plots at different times during the growth of the crop and the plots were then kept clean. It was found that after competition had started, the final yield of bulbs was being reduced at a rate equivalent to almost 4 per cent per day. So that by delaying weeding for another fortnight, the yield was cut to less than half that produced on ground kept clean all the time."* (*The Complete Know And Grow Vegetables*, Bleasdale, Salter and others, OUP 1991). He goes on to record that *"by early June, the weight of weeds per unit area was twenty times that of the crop, and the weeds had already taken from the soil about half of the nitrogen and a third of the potash which had been applied"*.

Perennial weeds with bulbils, such as lesser celandine and oxalis, or with persistent underground stems such as couch grass (*Agropyron repens*) or creeping buttercup (*Ranunculus repens*) are able to store reserves of food, and are thus able to grow faster and with more vigour than their annual counterparts. There is also evidence that the roots of some perennials such as couch grass exude allelopathic chemicals which inhibit the growth of other nearby plants.

Weeds can also host pests and diseases that can spread to cultivated crops. Charlock and Shepherd's purse may carry clubroot, eelworm can be harboured by chickweed, fat hen and shepherd's purse, while the cucumber mosaic virus, which can devastate the cucurbit family, is carried by a range of different weeds including chickweed and groundsel.

However, at times the role of weeds in this respect can be overrated. As far as insect pests are concerned, often the species that live on weeds are not the same as those that attack vegetable crops; "*Tests with the common cruciferous weeds such as shepherds purse have shown that they do not act as hosts for the larvae of the cabbage root fly. One exception was found to be the wild radish, but this is not usually a weed of established vegetable gardens*" (Roberts, *The Complete Know and Grow Vegetables*). However pests such as cutworms may first attack weeds then move on to cultivated crops.

While charlock, a common weed in south-eastern USA, may be considered a weed by row crop growers, it is highly valued by beekeepers, who seek out places where it blooms all winter, thus providing pollen for honeybees and other pollinators. Its bloom is resistant to all but a very hard freeze, and even that will only kill it back briefly. By feeding an array of pollinators during a seasonal dearth, it can redound to the farmer's advantage. Many weeds are likewise highly beneficial to pollinators.

Methods

In domestic gardens, methods of weed control include covering an area of ground with several layers of wet newspaper or one black plastic sheet for several weeks. In the case of using wet newspaper, the multiple layers prevent light from reaching all plants beneath, which kills them. Saturating the newspaper with water daily speeds the decomposition of the dead plants. Any weed seeds that start to sprout because of the water will also be deprived of sunlight, be killed, and decompose. After several weeks, all germinating weed seeds present in the ground should be dead. Then the newspaper can be removed and the ground can be planted. The decomposed plants will help fertilize the plants or seeds planted later.

In the case of using the black plastic sheet, the greenhouse effect is used to kill the plants beneath the sheet. A 5–10 cm layer of wood chip mulch on the ground will also prevent most weeds from sprouting. Also, gravel can be spread over the ground as an inorganic mulch. In agriculture, irrigation is sometimes used as a weed control measure such as in the case of paddy fields. Many people find that although the black plastic sheeting is extremely effective at preventing the weeds in areas where it covers, but in actual use it is difficult to achieve full coverage.

Knowing how weeds reproduce, spread and survive adverse conditions can help in developing effective control and management strategies. Weeds have a range of techniques that enable them to thrive; Annual and biennial weeds such as chickweed, annual meadow grass, shepherd's purse, groundsel, fat hen, cleaver, speedwell and hairy bittercress propagate themselves by seeding. Many produce huge numbers of seed several times a season, some all year round. Groundsel can produce 1000 seed, and can continue right through a mild winter, whilst Scentless May weed produces over 30,000 seeds per plant. Not all of these will germinate at once, but over several seasons, lying dormant in the soil sometimes for years until exposed to light. Poppy seed can survive 80–100 years, dock 50 or more. There can be many thousands of seeds in a square foot or square metre of ground, thus and soil disturbance will produce a flush of fresh weed seedlings.

"Stale Seed Bed" Technique

One technique employed by growers is the 'stale seed bed', which involves cultivating the soil, then leaving it for a week or so. When the initial flush of weeds has germinated, the grower will lightly hoe off before the desired crop is planted. However, even a freshly cleared bed will be susceptible to airborne seed from elsewhere, as well as seed brought in by passing animals which can carry them on their fur, or from freshly imported manure. The organic solution to the problem of spreading annual weeds lies in regular, properly timed weeding, preferably just before flowering (fortuitously, this is also the time at which they will be of the most value in composting). This technique is also quite often used by farmers who let weeds germinate then return the soil before crop sowing.

Perennial weeds also propagate by seeding; the airborne seed of the dandelion and the rose-bay willow herb are parachuted far and wide. But they also have an additional range of vegetative means of spreading that gives them their pernicious reputation. Dandelion and dock put down deep tap roots, which, although they do not spread underground, are able to regrow from any remaining piece left in the ground. Removal of the complete tap root is the only sure remedy.

The most persistent of the perennials are those that spread by underground creeping rhizomes that can regrow from the tiniest fragment. These include couch grass, bindweed, ground elder, nettles, rosebay willow herb, Japanese knotweed, horsetail and bracken, as well as creeping thistle, whose tap roots can put out lateral roots. Other perennials put out runners that spread along the soil surface. As they creep along they set down roots, enabling them to colonise bare ground with great rapidity. These include creeping buttercup and ground ivy. Yet another group of perennials propagate by stolons-stems that arch back into the ground to reroot. Most familiar of these is the bramble.

All of the above weeds can be very difficult to eradicate- thick black plastic mulches can be effective to a degree, although will probably need to be left in place for at least two seasons. In addition, hoeing off weed leaves and stems as soon as they appear can eventually weaken and kill the plants, although this will require persistence in the case of plants such as bindweed. Nettle infestations can be tackled by cutting back at least three times a year, repeated over a three-year period. Bramble can be dealt with in a similar way. Some plants are said to produce root exudates that suppress herbaceous weeds. *Tagetes minuata* is claimed to be effective against couch and ground elder, whilst a border of comfrey is also said to act as a barrier against the invasion of some weeds including couch.

Use of Herbicides

The above described methods of weed control avoid using chemicals. They are often used by farmers. However, these methods may damage a fragile soil by restructuring it, hence are not always used. They are those preferred by the organic gardener or organic farmer. However weed control can also be achieved by the use of herbicides. Selective herbicides kill certain targets while leaving the desired crop relatively unharmed. Some of these act by interfering with the growth of the weed and are often based on plant hormones.

Herbicides are generally classified as follows :

- Contact herbicides destroy only that plant tissue in contact with the chemical spray. Generally, these are the fastest acting herbicides. They are ineffective on perennial plants that are able to re-grow from roots or tubers.
- Systemic herbicides are foliar-applied and are translocated through the plant and destroy a greater amount of the plant tissue. Modern herbicides such as glyphosate are designed to leave no harmful residue in the soil.
- Soil-borne herbicides are applied to the soil and are taken up by the roots of the target plant.
- Pre-emergent herbicides are applied to the soil and prevent germination or early growth of weed seeds.

In agriculture large scale and systematic weeding is usually required, often by machines, such as liquid herbicide sprayers, or even by helicopter (such as in the USA), to eliminate the massive amount of weeds present on farming lands. However there are a number of techniques that the organic farmer can employ such as mulching and carefully timed cutting of weeds before they are able to set seed.

Organic Methods

Typically a combination of methods are used in organic situations :

- *Drip Irrigation :* Rubber hoses and other methods are used to bring water directly to the roots of the desired plants. This limits weed access to water.
- *Manually Pulling Weeds :* Labourers are used to pull weeds at various points in the growing process.
- *Boiling Water :* Pour boiling water to weed, they will become more green and then die in few hours. Best for weed in cracks or other hard to reach locations.

- *Vinegar :* Vinegar kills the visible part of the weed. They will wrinkle and die next day, although the root will still be in place to continue growing.
- *Mechanically Tilling around Plants :* Tractors are used to carefully till weeds around the crop plants at various points in the growing process. Besides tilling, other mechanical weed control methods also exist
- *Ploughing :* Ploughing includes tilling of soil, intercultural ploughing and summer ploughing. Ploughing through tilling of soil uproots the weeds which causes them to die. In summer ploughing is done during deep summers. Summer ploughing also helps in killing pests.
- *Crop Rotation :* Rotating crops with ones that kill weeds by choking them out, such as hemp, *Mucuna pruriens*, and other crops, can be a very effective method of weed control. It is a way to avoid the use of herbicides, and to gain the benefits of crop rotation.
- *Weed Mat :* A weed mat is an artificial mulch, fibrous cloth material, bark or newspaper laid on top of the soil preventing weeds from growing to the surface.

Thermal Methods

There are several thermal methods known to control weed. Weed burners heat up soil quickly and destroy superficial parts of the plants. Weed seeds are often heat resistant and even react with an increase of growth on dry heat. Since the 19th century soil steam sterilisation is used as a farming technique to clean soil completely from weeds. Several research results confirm the high effectivness of humid heat against weeds and its seeds.

UK Legislation

The *Weeds Act, 1959* is described as *"Preventing the spread of harmful or injurious weeds"*, and is mainly relevant to farmers and other rural settings rather than the allotment or garden-scale grower. There are five 'injurious' (that is, likely to be harmful to agricultural production) weeds covered by the provisions of the Weeds Act.

These are :

- Spear thistle (*Cirsium vulgare*)
- Creeping or field thistle (*Cirsium arvense*)
- Curled Dock (*Rumex crispus*)
- Broad leaved dock (*Rumex obtusifolius*)
- Common ragwort (*Senecio jacobaea*) (n.b., this weed is poisonous to livestock. Livestock should not be allowed to graze where ragwort has grown until it is eradicated, and any traces have disintegrated. Ragwort should not be allowed to be harvested in hay or silage for feed).

The Department for Environment, Food and Rural Affairs (DEFRA) provides guidance for the treatment removal of these weeds from infested land. Much of this is oriented towards the use of herbicides, the majority of which may not be acceptable to the organic producer (apart from non-synthetic substances like sulphur, which in some circumstances are accepted within Soil Association standards) but in most cases there are manual techniques that can be used such as digging out the roots, mulching out or carefully timed cutting before seeds are able to spread.

Primary responsibility for weeds control rests with the occupier of the land on which the weeds are growing, therefore it is important to be alert to potential weed problems and to take prompt action. However, it should be remembered that most common farmland weeds are not "injurious" within the meaning of the Weeds Act, and many such plant species have conservation and environmental value. When dealing with complaints under the Weeds Act, DEFRA has a duty in law to try to achieve a reasonable balance among different interests.

These include agriculture, countryside conservation and the public in general. Constructive discussion about problems caused by weeds can often result in effective solutions and avoid the need for DEFRA to take official action. In addition to those weeds covered by the 1959 act, under section 14 of the *Wildlife and Countryside Act, 1981*, it can be an offence to plant or grow certain specified plants in the wild, including Giant Hogweed and Japanese Knotweed. Problems involving these plants can be referred to the local authority for the area where those weeds are growing as some local authorities have bye-laws controlling these plants. There is no statutory requirement for landowners to remove these plants from their property.

INTEGRATED PEST MANAGEMENT

Integrated pest management (IPM), also known as Integrated Pest Control (IPC) is a broad-based approach that integrates a range of practices for economic control of pests. IPM aims to suppress pest populations below the economic injury level (EIL). The Food and Agriculture Organisation of the UN defines IPM as "the careful consideration of all available pest control techniques and subsequent integration of appropriate measures that discourage the development of pest populations and keep pesticides and other interventions to levels that are economically justified and reduce or minimise risks to human health and the environment. IPM emphasises the growth of a healthy crop with the least possible disruption to agro-ecosystems and encourages natural pest control mechanisms." Entomologists and ecologists have urged the adoption of IPM pest control for many years. IPM allows for a safer means of controlling pests. This includes managing insects, plant pathogens and weeds.

Globalisation of markets and increased movements of people all over the world are allowing for increasing numbers of invasive species to be brought into countries. Appropriate responses to these pests are needed and development and implementation strategies should be arranged. It is essential that the option that poses the least risks while maximising benefits is needed and that the strategy may include all components related to integrated pest management strategies. For their leadership in developing and spreading IPM worldwide, Perry Adkisson and Ray F. Smith received the 1997 World Food Prize.

History

Shortly after World War II, when synthetic insecticides became widely available, entomologists in California developed the concept of "supervised insect control". Around the same time, some entomologists in the Cotton Belt region of the United States were advocating a similar approach. Under this scheme, insect control was "supervised" by qualified entomologists, and insecticide applications were based on conclusions reached from periodic monitoring of pest and natural-enemy populations. This was viewed as an alternative to calendar-based insecticide programmes. Supervised control was based on a sound knowledge of the ecology and analysis of projected trends in pest and natural-enemy populations.

Supervised control formed much of the conceptual basis for the "integrated control" that University of California entomologists articulated in the 1950s. Integrated control sought to identify the best mix of chemical and biological controls for a given insect pest. Chemical insecticides were to be used in manner least disruptive to biological control. The term "integrated" was thus synonymous with "compatible." Chemical controls were to be applied only after regular monitoring indicated that a pest population had reached a level (the economic threshold) that required treatment to prevent the population from reaching a level (the economic injury level) at which economic losses would exceed the cost of the artificial control measures.

IPM extended the concept of integrated control to all classes of pests and was expanded to include tactics other than just chemical and biological controls. Artificial controls such as pesticides were to be applied as in integrated control, but these now had to be compatible with control tactics for all classes of pests. Other tactics, such as host-plant resistance and cultural manipulations, became part of the IPM arsenal. IPM added the multi-disciplinary element, involving entomologists, plant pathologists, nematologists, and weed scientists.

In the United States, IPM was formulated into national policy in February 1972 when President Richard Nixon directed federal agencies to take steps to advance the concept and application of IPM in all relevant sectors. In 1979, President Jimmy Carter established an interagency IPM Co-ordinating Committee to ensure development and implementation of IPM practices.

Applications

IPM is used in agriculture, horticulture, human habitations, and preventive conservation.

Principles

An American IPM system is designed around six basic components :

1. *Acceptable Pest Levels* : The emphasis is on *control*, not *eradication*. IPM holds that wiping out an entire pest population is often impossible, and the attempt can be expensive and environmentally unsafe. IPM programmes first work to establish acceptable pest levels, called action thresholds, and apply controls if those thresholds are crossed. These thresholds are pest and site specific, meaning that it may be acceptable at one site to have a weed such as white clover, but at another site it may not be acceptable. By allowing a pest population to survive at a reasonable threshold, selection pressure is reduced. This lowers the chance of pests developing resistance to chemicals, because if many of the pests are killed then any that have resistance to the chemical will form the genetic basis of the future population. By not killing all the pests there should be unresistant pests left that will dilute the prevalence of any resistant genes that appear. Similarly, the repeated use of a single class of chemicals will create pest populations that are more resistant to that particular class, whereas alternating between different classes helps prevent this.
2. *Preventive Cultural Practices* : Selecting varieties best for local growing conditions, and maintaining healthy crops, is the first line of defence, together with plant quarantine and 'cultural techniques' such as crop sanitation (*e.g.* removal of diseased plants, and cleaning pruning shears to prevent spread of infections). Beneficial fungi and bacteria are added to the potting media of horticultural crops vulnerable to root diseases, greatly reducing the need for fungicides.
3. *Monitoring* : Regular observation is the cornerstone of IPM. Observation is broken into two steps, first; inspection and second; identification. Visual inspection, insect and spore traps, and other measurement methods and monitoring tools are used to monitor pest levels. Accurate pest identification is critical to a successful IPM programme. Record-keeping is essential, as is a thorough knowledge of the behaviour and reproductive cycles of target pests. Since insects are cold-blooded, their physical development is dependent on the temperature of their environment. Many insects have had their development cycles modeled in terms of degree days. Monitor the degree days of an environment to determine when is the optimal time for a specific insect's outbreak. Plant pathogens also follow similar patterns of response to weather and seasonal changes.
4. *Mechanical Controls* : Should a pest reach an unacceptable level, mechanical methods are the first options to consider. They include simple hand-picking, erecting insect barriers, using traps, vacuuming, and tillage to disrupt breeding.
5. *Biological Controls* : Natural biological processes and materials can provide control, with minimal environmental impact, and often at lower cost.

The main focus here is to promote beneficial insects that eat or parasitise target pests. Biological insecticides, derived from naturally occurring micro-organisms (*e.g.* : *Bt*, entomopathogenic fungi and entomopathogenic nematodes), also fit in this category.

6. *Responsible Pesticide Use* : Synthetic pesticides are used as required and often only at specific times in a pests life-cycle. Many of the newer pesticide groups are derived from plants or naturally occurring substances (*e.g.* : nicotine, pyrethrum and insect juvenile hormone analogues), but the toxophore or active component may be altered to provide increased biological activity or stability. Further 'biology-based' or 'ecological' techniques are under evaluation. Applications of pesticides must reach their intended targets. Matching the application technique to the crop, the pest, and the pesticide is critical. The use of low-volume spray equipment reduces overall pesticide use and labour cost.

An IPM regime can be quite simple or sophisticated. Historically, the main focus of IPM programmes was on agricultural insect pests. Although originally developed for agricultural pest management, IPM programmes are now developed to encompass diseases, weeds, and other pests that interfere with the management objectives of sites such as residential and commercial structures, lawn and turf areas, and home and community gardens.

Process

IPM is the intelligent selection and use of pest control actions that will ensure favourable economic, ecological and sociological consequences and is applicable to most agricultural, public health and amenity pest management situations. Reliance on knowledge, experience, observation, and integration of multiple techniques makes IPM a perfect fit for organic farming (sans artificial pesticide application). For large-scale, chemical-based farms (conventional), IPM can reduce human and environmental exposure to hazardous chemicals, and potentially lower overall costs of pesticide application material and labour.

Risk assessment is usually characterised by four issues :

1. Characterisation and identification of biological control agents,
2. Health risks,
3. Environmental risks, and
4. Efficacy.
   - *Proper Identification of Pest* : What is it? Cases of mistaken identity may result in ineffective actions. If plant damage is due to over-watering, it could be mistaken for fungal infection, since many fungal and viral infections arise under moist conditions. This could lead to spray costs, but the plant would be no better off.
   - *Learn Pest and Host Life Cycle and Biology* : What does it need? At the time you see a pest, it may be too late to do much about it except

maybe spray with a pesticide. Often, there is another stage of the life cycle that is susceptible to preventative actions. For example, weeds reproducing from last year's seed can be prevented with mulches and pre-emergent herbicide.

- *Monitor or sample environment for pest population :* How many are here? Where do we look? Preventative actions must be taken at the correct time if they are to be effective. For this reason, once the pest is correctly identified, monitoring must begin *before* it becomes a problem. For example, in school cafeterias where roaches may be expected to appear, sticky traps are set out before school starts. Traps are checked at regular intervals so populations can be monitored and controlled before they get out of hand. Some factors to consider and monitor include : Is the pest present/absent? What is the distribution-all over or only in certain spots? Is the pest population increasing, decreasing or remaining constant? This is done through crop scouting. Monitoring should include the soil/planting media fertility and water quality. Overall plant health and resistance to pests is greatly influenced by the water's pH, alkalinity,the make up of disolved minerals,and Oxygen Reduction Potential. Many diseases are water borne and spread directly by irrigation water and indirectly by splashing.
- *Establish action threshold (economic, health or aesthetic) :* How many are too many? In some cases, there is a standardised number of pests that can be tolerated. Soybeans are quite tolerant of defoliation, so if there are a few caterpillars in the field and their population is not increasing dramatically, there is not necessarily any action necessary. Conversely, there is a point at which action *must* be taken to control cost. For the farmer, that point is the one at which the cost of damage by the pest is *more* than the cost of control. This is an economic threshold. Tolerance of pests varies also by whether or not they are a health hazard (low tolerance) or merely a cosmetic damage (high tolerance in a non-commercial situation).

  Different sites may also have varying requirements based on specific areas. White clover may be perfectly acceptable on the sides of a tee box on a golf course, but unacceptable in the fairway where it could cause confusion in the field of play.
- Use resources to keep up to date on IPM developments. Researchers are always discovering new techniques, and ways to improve old techniques. Keeping up to date gives you the best options available to when using IPM.
- Choose an appropriate combination of management tactics. For any pest situation, there will be several options to consider. Options include, mechanical or physical control, cultural controls, biological controls and chemical controls. Mechanical or physical controls

include picking pests off plants, or using netting or other material to exclude pests such as birds from grapes or rodents from structures. Cultural controls include : keeping an area free of conducive conditions by removing or storing waste properly, removing diseased plants properly, late water floods, sanding, and the use of disease-resistant varieties. Biological controls are numerous. They include : conservation of natural predators or augmentation of natural predators, Sterile insect technique (SIT).

Augmentation, inoculative release and inundative release are different methods of biological control and all affect the target pest in different ways. Augmentative control includes the periodic introduction of naturally occurring predators in sufficient numbers to keep pest damage below economoic damaging levels. They can be either an inundative or inoculative release. Inundative release is where beneficials are collected, mass-reared and periodically released in large numbers into the pest area. This is used for an immediate reduction in host populations and is generally used in annual crops but is not suitable in the long run. Inoculative release is where only a limited number of the beneficial (helpful) organisms are introduced into a pest area at the start of the season. This is to give long term control as the parasitoid progeny will continue to affect the host populations throughout the season, such as in orchards. Seasonal inoculative release requires the beneficials to be collected, mass-reared and released seasonally into annual crops as to maintain the beneficial population. This is commonly used in greenhouses. In America and other western countries, inundative releases are the predominant programme while Asia and the USSR countries have put emphasis on inoculation and occasional introductions.

The SIT is an Area-Wide IPM that introduces sterile male pests into the pest population to act as birth control. The biological controls mentioned above should only be used in extreme cases, because in the introduction of new species, or supplementation of naturally occurring species can have detrimental effect to the ecosystem. Biological controls can be used to stop invasive species or pests, or they can be they route by which new pests are introduced.

Chemical controls would include horticultural oils or the application of pesticides, such as : insecticides and herbicides. A Green Pest Management IPM programme would use pesticides derived from plants, such as botanicals, or other naturally occurring materials. Pesticide classes are divided by their modes of action. Alternating between modes of action minimises pest resistance. When using any type of chemical control make sure that your pesticide applicator has the appropriate safety equipment. Their certification should be up to date, and that your equipment is well maintained and calibrated to ensure proper application.

- Evaluate results : How did it work? Evaluation is often one of the most important steps. This is the process to review an IPM programme and the results it generated. Asking the following questions is useful : Did actions have the desired effect? Was the pest prevented or managed to farmer satisfaction? Was the method itself satisfactory? Were there any unintended side effects? What can be done in the future for this pest situation? Understanding the effectiveness of the IPM programme allows the site manager to make modifications to the IPM plan prior to pests reaching the action threshold and requiring action again.

INSECTICIDE

An insecticide is a pesticide used against insects. They include ovicides and larvicides used against the eggs and larvae of insects respectively. Insecticides are used in agriculture, medicine, industry, and general home use. The use of insecticides is believed to be one of the major factors behind the increase in agricultural productivity in the 20th century. Nearly all insecticides have the potential to significantly alter ecosystems; many are toxic to humans; and others are concentrated in the food chain.

The classification of insecticides is done in several different ways :

- *Systemic* insecticides are incorporated by treated plants. Insects ingest the insecticide while feeding on the plants.
- *Contact* insecticides are toxic to insects brought into direct contact. Efficacy is often related to the quality of pesticide application, with small droplets (such as aerosols) often improving performance.
- *Natural* insecticides, such as nicotine, pyrethrum and neem extracts are made by plants as defences against insects. Nicotine-based insecticides are still being widely used in the US and Canada though they are barred in the EU.
- *Plant-incorporated protectants* (PIPs) are insecticidal substances produced by plants after genetic modification. For instance, a gene that codes for a specific *Baccilus thuringiensis* biocidal protein is introduced into a crop plant's genetic material. Then, the plant manufactures the protein. Since the biocide is incorporated into the plant, additional applications, at least of the same compound, are not required.
- *Inorganic* insecticides are manufactured with metals and include arsenates, copper compounds and fluorine compounds, which are now seldom used, and sulphur, which is commonly used.
- *Organic* insecticides are synthetic chemicals which comprise the largest numbers of pesticides available for use today.
- *Mode of Action*—how the pesticide kills or inactivates a pest—is another way of classifying insecticides. Mode of action is important in predicting whether an insecticide will be toxic to unrelated species, such as fish, birds and mammals.

Classes of Insecticides

Organochlorides

The insecticidal properties of the best known representative of this class of insecticides, DDT, was made by the Swiss Scientist Paul Müller. For this discovery, he was awarded the Nobel Prize for Physiology or Medicine in 1948. DDT was introduced on the market in 1944. The contemporary rise of the chemical industry facilitated the large-scale production of DDT and related chlorinated hydrocarbons. DDT functions by opening the sodium channels in the nerve cells of the insect.

Organophosphates and Carbamates

Another large class of synthetic insecticides are the organophosphates. Again these target the insect's nervous system. Organophosphates interfere with the enzymes acetylcholinesterase and other cholinesterases, disrupting nerve impulses, killing or disabling the insect. Organophosphate insecticides and chemical warfare nerve agents (such as sarin, tabun, soman and VX) work in the same way. Organophosphates have an accumulative toxic effect to wildlife, so multiple exposures to the chemicals amplifies the toxicity. Carbamate insecticides have similar toxic mechanisms to organophosphates, but have a much shorter duration of action and are thus somewhat less toxic.

Pyrethroids

To mimic the insecticidal activity of the natural compound pyrethrum another class of pesticides, pyrethroid pesticides, has been developed. These compounds are non-persistent sodium channel modulators, and are much less acutely toxic than organophosphates and carbamates. Compounds in this group are often applied against household pests.

Neo-nicotinoids

Neo-nicotinoids are synthetic analogues of the natural insecticide nicotine (with a much lower acute mammalian toxicity and greater field persistence). These chemicals are nicotinic acetylcholine receptor agonists. Broad-spectrum – systemic insecticides, they have a rapid action (minutes-hours). They are applied as sprays, drenches, seed and soil treatments – often as substitutes for organophosphates and carbamates. Treated insects exhibit leg tremors, rapid wing motion, stylet withdrawal (aphids), disoriented movement, paralysis and death. Imidacloprid may be the most commonly used neo-nicotinoid. It has recently come under scrutiny for its deleterious effects on honeybees, and its potential to increase the susceptibility of rice to planthopper attacks.

Ryanoids

Ryanoids are synthetic analogues with the same mode of action as ryanodine, a naturally occurring insecticide extracted from *Ryania speciosa* (Flacourtiaceae).

They bind to calcium channels in cardiac and skeletal muscle, blocking nervous transmission. Only one such insecticide is currently registered, Rynaxypyr, generic name chlorantraniliprole.

Insect Growth Regulators

Insect growth regulator (IGR) is a term coined to include insect hormone mimics and an earlier class of chemicals, the benzoylphenyl ureas, which inhibit chitin (exoskeleton) biosynthesis in insects. Diflubenzuron is a member of the latter class, used primarily to control caterpillars which are pests. The most successful insecticides in this class are the juvenoids (juvenile hormone analogues). Of these, methoprene is most widely used. It has no observable acute toxicity in rats, and is approved by WHO for use in drinking water cisterns to combat malaria.

Most of its uses are to combat insects where the adult is the pest, including mosquitoes, several fly species, and fleas. Two very similar products, hydroprene and kinoprene are used for controlling species such as cockroaches and white flies. Methoprene has been registered with the EPA since 1975, and there are virtually no reports of resistance. A more recent type of IGR is the ecdysone agonist tebufenozide (MIMIC), which is used in forestry and other applications for control of caterpillars, which are far more sensitive to its hormonal effects than other insect orders.

Biological Insecticides

Many plants exude substances to prevent insects from eating. Premier examples are substances activated by the enzyme myrosinase. This enzyme converts glucosinolates to a variety of compounds that are toxic to herbivorous insects. One product of this enzyme is allyl isothiocyanate, the pungent ingredient in horseradish sauces.

The myrosinase is only released upon crushing the flesh of horseradish by the herbivore (or preparer of horseradish sauce). Since allyl isothiocyanate is harmful to the plant as well as the insect, it is stored in the harmless form of the glucosinolate, separate from the myrosinase enzyme. In general, tree rosin is considered a natural insecticide. Specifically, the production of oleoresin by conifer species is a component of the defence response against insect attack and fungal pathogen infection.

Bacterial Insecticides

Bacillus thuringiensis is a bacterial disease that affects Lepidopterans and some other insects. Toxins produced by different strains of this bacterium are used as a larvicide against caterpillars, beetles, and mosquitoes. Toxins from *Saccharopolyspora spinosa* are isolated from fermentations and sold as Spinosad. Because these toxins have little effect on other organisms, they are considered more environmentally friendly than synthetic pesticides. The toxin from *B. thuringiensis* (Bt toxin) has been incorporated directly into plants through the use of genetic engineering.

Other biological insecticides include products based on entomopathogenic fungi (*e.g. Beauveria bassiana, Metarhizium anisopliae*), nematodes (*e.g. Steinernema feltiae*) and viruses (*e.g. Cydia pomonella* granulovirus).

Environmental Effects

Effects on Non-target Species

Some insecticides kill or harm other creatures in addition to those they are intended to kill. For example, birds may be poisoned when they eat food that was recently sprayed with insecticides or when they mistake insecticide granules on the ground for food and eat it. Sprayed insecticides may drift from the area to which it is applied and into wildlife areas, especially when it is sprayed aerially.

DDT

The development of insecticides such as DDT has been motivated by desire to replace more dangerous or less effective alternatives. DDT was introduced to replace lead and arsenic-based compounds, which were in widespread use in the early 1940s. Some insecticides have been banned due to the fact that they are persistent toxins which have adverse effects on animals and/or humans. An oft-quoted case is that of DDT, an example of a widely used (and maybe misused) pesticide, which was brought to public attention by Rachel Carson's book, Silent Spring.

One of the better known impacts of DDT is to reduce the thickness of the egg shells on predatory birds. The shells sometimes become too thin to be viable, causing reductions in bird populations. This occurs with DDT and a number of related compounds due to the process of bioaccumulation, wherein the chemical, due to its stability and fat solubility, accumulates in organisms' fatty tissues.

Also, DDT may biomagnify, which causes progressively higher concentrations in the body fat of animals farther up the food chain. The near-worldwide ban on agricultural use of DDT and related chemicals has allowed some of these birds, such as the peregrine falcon, to recover in recent years. A number of the organochlorine pesticides have been banned from most uses worldwide, and globally they are controlled via the Stockholm Convention on persistent organic pollutants. These include : aldrin, chlordane, DDT, dieldrin, endrin, heptachlor, mirex and toxaphene.

Pollinator Decline

Insecticides can kill bees and may be a cause of pollinator decline, the loss of bees that pollinate plants, and colony collapse disorder (CCD), in which worker bees from a beehive or Western honey bee colony abruptly disappear. Loss of pollinators will mean a reduction in crop yields. Sublethal doses of insecticides (*i.e.* imidacloprid and other neo-nicotinoids) affect foraging behaviour of bees. However, research into the causes of CCD was inconclusive as of June 2007.

PESTICIDE RESISTANCE

Pesticide application can artificially select for resistant pests. In this diagram, the first generation happens to have an insect with a heightened resistance to a pesticide (red). After pesticide application, its descendants represent a larger proportion of the population because sensitive pests (white) have been selectively killed. After repeated applications, resistant pests may comprise the majority of the population.

Pesticide resistance describes the decreased susceptibility of a pest population to a pesticide that was previously effective at controlling the pest. Pest species evolve pesticide resistance via natural selection : the most resistant organisms are the ones to survive and pass on their genetic traits to their offspring.

Manufacturers of pesticides tend to prefer a definition that is dependent on failure of a product in a real situation, sometimes called field resistance. For example, the Insecticide Resistance Action Committee (IRAC) definition of insecticide resistance is 'a heritable change in the sensitivity of a pest population that is reflected in the repeated failure of a product to achieve the expected level of control when used according to the label recommendation for that pest species'.

Pesticide resistance is increasing in occurrence. Farmers in the USA lost 7 per cent of their crops to pests in the 1940s; over the 1980s and 1990s, the loss was 13 per cent, even though more pesticides were being used. Over 500 species of pests have evolved a resistance to a pesticide. Other sources estimate the number to be around 1000 species since 1945.Rachel Carson predicted the phenomenon in her 1962 book *Silent Spring*.

Factors

Propensity of pest populations to evolve resistance is probably caused by a number of factors. First, pest species are usually capable of producing large number of offspring. This increases the probability of random mutations and ensures the rapid build-up in numbers of resistant mutants once such mutations have occurred. Secondly, pest species have been exposed to natural toxins for a long time before the onset of human civilization. For example, many plants produce phytotoxins to protect them from herbivores. As a result, coevolution of herbivores and their host plants required development of the physiological capability to detoxify or tolerate poisons. Thirdly, humans often rely almost exclusively on insecticides for pest control. This increases selection pressure towards resistance. Pesticides that fail to break down quickly and remain in the area contribute to selection for resistant organisms even after they are no longer being applied.

In response to pesticide resistance, pest managers may resort to increased use of pesticides, exacerbating the problem. In addition, when pesticides are toxic towards species that feed on or compete with pests, the pest population will likely expand further, requiring more pesticides. This is sometimes referred to as *pesticide trap*, or a *pesticide treadmill*, since farmers are continually paying more for less benefit.

Insect preys and parasites which live on other insects generally have smaller populations and are therefore much less likely to evolve resistance than are the primary targets of the pesticides, such as mosquitoes and those that feed on plants. This can compound the pest problem because these species normally keep pest populations in check. But resistant predators of pest species can be bred in laboratories, which can help keep pest populations down.

The fewer sources of food a pest has the more likely it is to evolve resistance, because it is exposed to higher concentrations of pesticides and has less opportunity to breed with populations that have not been exposed. Other factors in the speed with which a species evolves resistance are generation time and fecundity (shorter generations and more offspring lead to resistance more quickly).

Examples

Resistance has evolved in a variety of different pest species : Resistance to insecticides was first documented by A. L. Melander in 1914 when scale insects demonstrated resistance to an inorganic insecticide. Between 1914 and 1946, 11 additional cases of resistance to inorganic insecticides were recorded. The development of organic insecticides, such as DDT, gave hope that insecticide resistance was an issue of the past. Unfortunately, by 1947 housefly resistance to DDT was documented. With the introduction of every new insecticide class–cyclodienes, carbamates, formamidines, organophosphates, pyrethroids, even *Bacillus thuringiensis*–cases of resistance surfaced within two to 20 years.

- In the US, studies have shown that fruit flies that infest orange groves were becoming resistant to malathion, a pesticide used to kill them.
- In Hawaii and Japan and Tennessee, the diamondback moth evolved a resistance to *Bacillus thuringiensis* about three years after it began to be used heavily.
- In England, rats in certain areas have evolved such a strong resistance to rat poison that they can consume up to five times as much of it as normal rats without dying.
- DDT is no longer effective in preventing malaria in some places, a fact which contributed to a resurgence of the disease.
- In the southern United States, the weed Amaranthus palmeri, which interferes with production of cotton, has evolved widespread resistance to the herbicide glyphosate.
- Colourado potato beetle has evolved resistance to 52 different compounds belonging to all major insecticide classes. Resistance levels vary greatly among different populations and between beetle life stages, but in some cases can be very high (up to 2,000-fold).

Although the evolution of pesticide resistance is usually discussed as a result of pesticide use, it is important to keep in mind that pest populations can also adapt to non-chemical methods of control. For example, the northern corn rootworm

(*Diabrotica barberi*) became adapted to a corn-soybean crop rotation by spending the year when field is planted to soybeans in a diapause.

Multiple and Cross-resistance

Multiple resistance is the phenomenon in which a pest is resistant to more than one class of pesticides. This can happen if one pesticide is used until pests display a resistance and then another is used until they are resistant to that one, and so on. Cross resistance, a related phenomenon, occurs when the genetic mutation that made the pest resistant to one pesticide also makes it resistant to other pesticides, especially ones with similar mechanisms of action.

Physiology and Behaviour

Frequently a pest becomes resistant to a pesticide because it evolves physiological changes that protect it from the chemical. In some cases, a pest may gain an increased number of copies of a gene, allowing it to produce more of a protective enzyme that breaks down the pesticide into less toxic chemicals. Such enzymes include esterases, glutathione transferases, and mixed microsomal oxidases. Alternatively, the number of biochemical receptors for the chemicals may be reduced in the pest, or the receptor may be altered, reducing the pest's sensitivity to the compound. Behavioural resistance has also been described for some chemicals; for example, some Anopheles mosquitoes evolved a preference for resting outside that prevented them from coming in contact with pesticide sprayed on interior walls. Still other mechanisms include increased rates of excretion of toxic molecules, their sequestration and storage inside of the insect body away from vulnerable tissues and organs, and decreased toxin penetration through the insect body wall.

Often, mutation in only a single gene leads to the evolution of a resistant organism. In other cases, multiple genes are involved. Resistant genes are usually autosomal. This means that they are located on autosomes (as opposed to sex chromosomes). As a result, resistance is inherited similarly in males and females. Also, resistance is usually inherited as an incompletely dominant trait. When a resistant and a susceptible individual mate with each other, their progeny has an intermediate level of resistance (more resistant than the susceptible parent, but not as resistant as the resistant parent).

Adaptation to pesticides usually decreases relative fitness of organisms in the absence of pesticides. Resistant individuals often have reduced reproductive output, life expectancy, mobility, etc. Therefore, relatively few of them persist in a population that is not exposed to a particular insecticide to which they have evolved resistance.

Blowfly maggots produce an enzyme that confers resistance to organochloride insecticides. Scientists have researched ways to use this enzyme to break down pesticides in the environment, which would detoxify them and prevent harmful environmental effects. Later they discovered a similar enzyme produced by soil bacteria that also breaks down organochloride insecticides but which works faster and remains stable in a variety of conditions. The product, called Landguard is used in Australia to decontaminate spray equipment, soil and water after pesticide spraying and spills.

Management

Pest resistance to a pesticide can be managed by reducing selection pressure by this pesticide on the pest population. In other words, the situation when all the pests except the most resistant ones are killed by a given chemical should be avoided. This can be achieved by avoiding unnecessary pesticide applications, using non-chemical control techniques, and leaving untreated refuges where susceptible pests can survive. Adopting the integrated pest management (IPM) approach usually helps with resistance management.

When pesticides are the sole or predominant method of pest control, resistance is commonly managed through pesticide rotation. This involves alternating among pesticide classes with different modes of action to delay the onset of or mitigate existing pest resistance. Different pesticide classes may have different effects on a pest. The U.S. Environmental Protection Agency (EPA or USEPA) designates different classes of fungicides, herbicides and insecticides. Pesticide manufacturers may, on product labeling, require that no more than a specified number of consecutive applications of a pesticide class be made before alternating to a different pesticide class. This manufacturer requirement is intended to extend the useful life of a product. Tankmixing pesticides is the combination of two or more pesticides with different modes of action in order to improve individual pesticide application results and delay the onset of or mitigate existing pest resistance.

GREEN PEST MANAGEMENT

Green pest management is a pest control strategy that sets pest action thresholds, monitors pest levels, takes steps to prevent pest problems and uses control methods that are organic (plant based) materials or materials of natural origin. Green pest management is an extension of integrated pest management and is similar in all regards except for the control methods. While both integrated pest management and green pest management choose the least risky pest control material, green pest management uses organic (plant based) materials or materials of natural origin.

Overview

Green pest management (GPM) is the natural extension of integrated pest management (IPM). Pest management occurs in three primary arenas, first, the agricultural arena, second, the urban or structural arena and third the turf and ornamental arena. Green Pest Management programmes and service providers can be focused on agricultural, structural, turf and ornamental pest control. A GPM service plan follows the foundations of an IPM service plan with the exception of the type of pesticide applied when necessary according to the IPM or GPM plan.

GPM service providers (pest control applicators or pest control technicians) use naturally occurring and low risk materials as the primary pest control material instead of a synthetic pesticide when a pesticide is called for as part of an integrated pest management or IPM plan. These may or may not include materials listed on the Organic Materials Review Institute (OMRI) Pest management at an agricultural setting would include a farm or garden. Pest management at an urban or structural site includes residences, either single or multi-family, apartment houses and condominiums, hospitality such as restaurants and hotels, all types of food processing and manufacturing, warehousing, and industrial sites. Turf and ornamental pest control occurs at recreational sites, such as a lawn, park, playing field or playground, and may include plants for human or animal consumption or plants for utility, such as playing turf, or decorative use, such as a rose garden.

Government Regulation

Currently there are no regulations on the use of the term "Green Pest Control". Indeed, any pest control company could advertise a "Green Pest Control Service". Efforts to move towards a standard are being made by the pest control industry. Consumer should beware that the term green pest control is still too vague to carry much weight. The terms environmentally friendly and organic are regulated terms within the pest control industry.

The government has regulation on exempt products. Exempt products comprise only materials that are on the EPA's exempt list. These materials may include materials listed by the United States Environmental Protection Agency (EPA) as biopesticides. Biopesticides or biological pesticides are materials of natural origin or certain minerals.

Pesticides

Pesticides are used to control pests, some examples are avicide, herbicide and insecticide. Green Pest Management usually deals with insecticides and herbicides. For Green Pest Management the preferred materials are of natural origin. Materials of natural origin may be either organic, such as plant oils, or inorganic, such as boric acid. Not every material of natural origin would be preferred. It must also be low risk. Some plant oils used include Rosemary, Wintergreen, Eugenol (Clove Oil) and other oils. They act as insecticides at some concentrations and as non- selective herbicides at other, higher concentrations.

These low risk materials are often known as exempt or minimum risk pesticides or "25b products." The term 25b refers to the section 25b of the Federal Insecticide, Fungicide, and Rodenticide Act or FIFRA. Due to the nature of the active and inert ingredients in the pesticide these materials are exempt from registration. Other biological controls include use of biological pest control, the use of natural mechanisms in primarily agricultural or home lawn and garden pest control. Examples include bacteria, bacillus thuringiensis, or nematodes for example, for control of grubs, such as Oriental Beetle grubs or and in the case of bacillus thuringiensis, for control of mosquitoes.

Examples of Materials of Natural Origin :

- 1.0 Plant Oils
- 1.1 Rosemary
- 1.2 Wintergreen
- 1.3 Eugenol (Clove Oil)
- 2.0 Inorganic Materials
- 2.1 Boric Acid
- 2.2 Diatomaceous Earth (Freshwater and Salt Water).

Low Impact Pest Management

Green pest management does not include traditional chemical materials. However, recent advances in chemistry have yielded materials that have extremely low mammalian toxicity. Some key materials that are used that are chemicals not of natural origin, include insect growth regulators (IGR's). IGR's have action that is usually specific to insects, or that would affect insects primarily. Hydroprene or Methoprene (linked), are two examples of Juvenile hormone Analogs or JHA's. These are chemical materials, so are not usually appropriate for organic pest management strategies. However, given the low mammalian toxicity of the JHA's, they may be appropriate for low impact structural pest management as part of a complete IPM approach.

Certification and Accreditations

There are services available to certify or offer accreditation to businesses that provide green pest management services, to sites that use green pest management or to the manufacturers of materials used to control pests. The most widely used product list for materials acceptable to be used on operations that are certified organic under the United States Department of Agriculture(USDA) National Organic Programme(NOP) is the aforementioned Organic Materials Review Institute (OMRI). Regional organic farming associations are resources for accredited providers. An example would be the Northeast Organic Farming Association's Certified Organic Landcare or Organic Dairy Programme. Other examples include certification or registration through industry associations or private certification or accreditation service providers. For structural pest control this would include industry examples such as the National Pest Management Association's (NPMA) GreenPro Accreditation or private certification resources such as GreenShield Certification.

PUSH-PULL TECHNOLOGY

The push-pull technology is a strategy for controlling agricultural pests by using repellent "push" plants and trap "pull" plants. For example, cereal crops like maise or sorghum are often infested by stem borers. Grasses planted around the perimeter of the crop attract and trap the pests, whereas other plants, like *Desmodium*, planted between the rows of maise repel the pests and control the parasitic plant *Striga*. Push-pull technology was developed at the International Centre of Insect Physiology and Ecology (ICIPE) in Kenya in collaboration with Rothamsted Research, UK and national partners.

Pests of Cereal Crops

Stemborers, parasitic striga weeds and poor soil fertility are the three main constraints to efficient production of cereals in sub-Saharan Africa. Losses caused by stemborers can reach as high as 80 per cent in some areas and an average of about 15-40 per cent in others. Losses attributed to striga weeds, on the other hand, range between 30 per cent and 100 per cent in most areas, and are often exacerbated by the low soil fertility prevalent in the region. The soils are highly degraded due to continuous cropping with limited or no external inputs to improve soil fertility. When the two pests occur together, farmers often lose their entire crops. Crop losses caused by stemborers and striga weeds amount to about US$ 7 billion annually, affecting mostly the resource-poor subsistence farmers.

Control of stemborers using pesticides is not only expensive and harmful to the environment, but also usually ineffective, as the chemicals cannot reach deep inside the plant stems where stemborer larvae reside. Similarly, the use of herbicides against striga is neither effective nor feasible among smallholders in the region for both biological and socioeconomic reasons. Preventing crop losses from stemborers and striga weeds, and improving soil fertility in eastern Africa alone could increase cereal harvests enough to feed an additional 27 million people in the region

The Pull

The approach relies on a combination of companion crops to be planted around and among maise or sorghum. Both domestic and wild grasses can help to protect the crops by attracting and trapping the stemborers. The grasses are planted in the border around the maise and sorghum fields where invading adult moths become attracted to chemicals emitted by the grasses themselves. Instead of landing on the maise or sorghum plants, the insects head for what appears to be a tastier meal. These grasses provide the "pull" in the "push-pull" strategy. They also serve as a haven for the borers' natural enemies. Good trap crops include well-known grasses such as Napier grass (*Pennisetum purpureum*) and Sudan grass (*Sorghum vulgare sudanense*). Napier grass has a particularly effective way of defending itself against the pests : once attacked by a borer larva, it secretes a sticky substance which physically traps the pest and limits its damage.

The Push

The "push" in the intercropping scheme is provided by the plants that emit chemicals (kairomones) which repel stemborer moths and drive them away from the main crop (maise or sorghum). The best candidates discovered so far with the repellent properties are species of leguminous genus *Desmodium*. *Desmodium* is planted in between the rows of maise or sorghum. Being a low-growing plant, it does not interfere with the crops' growth and, furthermore, has the advantage of maintaining soil stability, improving soil fertility through enhanced soil organic matter content and nitrogen fixation. It also serves as a highly nutritious animal feed and effectively suppresses striga weeds. Another plant showing good repellent properties is molasses grass (*Melinis minutiflora*), a nutritious animal feed with tick-repelling and stemborer larval parasitoid attractive properties.

How Push-pull Works

The push-pull technology involves use of behaviour-modifying stimuli to manipulate the distribution and abundance of stemborers and beneficial insects for management of stemborer pests. It is based on in-depth understanding of chemical ecology, agrobiodiversity, plant-plant and insect-plant interactions, and involves intercropping a cereal crop with a repellent intercrop such as *Desmodium* uncinatum (silverleaf) (push), with an attractive trap plant such as Napier grass (pull) planted as a border crop around this intercrop. Gravid stemborer females are repelled from the main crop and are simultaneously attracted to the trap crop.

Napier grass produces significantly higher levels of attractive volatile compounds (green leaf volatiles), cues used by gravid stemborer females to locate host plants, than maise or sorghum. There is also an increase of approximately 100-fold in the total amounts of these compounds produced in the first hour of nightfall by Napier grass (scotophase), the period at which stemborer moths seek host plants for laying eggs, causing the differential oviposition preference.

However, many of the stemborer larvae, about 80 per cent, do not survive, as Napier grass tissues produce sticky sap in response to feeding by the larvae, which traps them, causing their mortality. Legumes in the *Desmodium* genus (silverleaf, *D. uncinatum* and greenleaf, *D. intortum*), on the other hand, produce repellent volatile chemicals that push away the stemborer moths. These include (E)-β-ocimene and (E)-4,8-dimethyl-1,3,7-non-atriene, semiochemicals produced during damage to plants by herbivorous insects and are responsible for the repellence of *Desmodium* to stemborers.

Desmodium also controls striga, resulting in significant yield increases of about 2 tonnes/hectare (0.9 short tons per acre) per cropping season. In the elucidation of the mechanisms of striga suppression by *D. uncinatum*, it was found that, in addition to benefits derived from increased availability of nitrogen and soil shading, an allelopathic effect of the root exudates of the legume, produced independently of the presence of striga, is responsible for this dramatic reduction in an intercrop with maise.

Presence of blends of secondary metabolites with striga seed germination stimulatory, 422,522,-dihydro-5,22,42-trihydroxy-522,-isopropenylfurano-(222,322;7,6)-isoflavanone, and postgermination inhibitory, 422,522-dihydro-

22-methoxy-5,42-dihydroxy-522-isopropenylfurano- (222,322;7,6)-isoflavanone, activities in the root exudates of D. uncinatum which directly interferes with parasitism was observed. This combination thus provides a novel means of *in situ* reduction of the striga seed bank in the soil through efficient suicidal germination even in the presence of grassy host plants in the proximity. Other *Desmodium* species have also been evaluated and have similar effects on stemborers and striga weed and are currently being used as intercrops in maise, sorghum and millets.

Chapter 25

Plant Evolutionary Developmental Biology

Evolutionary developmental biology (evo-devo) refers to the study of developmental programmes and patterns from an evolutionary perspective. It seeks to understand the various influences shaping the form and nature of life on the planet. Evo-devo arose as a separate branch of science rather recently. An early sign of this occurred in 1999. Most of the synthesis in evo-devo has been in the field of animal evolution, one reason being the presence of elegant model systems like *Drosophila melanogaster*, *C. elegans*, zebrafish and *Xenopus laevis*. However, in the past couple of decades, a wealth of information on plant morphology, coupled with modern molecular techniques has helped shed light on the conserved and unique developmental patterns in the plant kingdom also.

HISTORICAL PERSPECTIVE

Before 1900

The origin of the term "morphology" is generally attributed to Johann Wolfgang von Goethe (1749–1832). He was of the opinion that there is an underlying fundamental organisation *(Bauplan)* in the diversity of flowering plants. In his book titled The Metamorphosis of Plants, he proposed that the *Bauplan* enabled us to predict the forms of plants that had not yet been discovered. Goethe also was the first to make the perceptive suggestion that flowers consist of modified leaves.

In the middle centuries, several basic foundations of our current understanding of plant morphology were laid down. Nehemiah Grew, Marcello Malpighi, Robert Hooke, Antonie van Leeuwenhoek, Wilhelm von Nageli were just some of the people who helped build knowledge on plant morphology at various levels of organisation. It was the taxonomical classification of Carolus Linnaeus in the eighteenth century though, that generated a firm base for the knowledge to stand on and expand. The introduction of the concept of Darwinism in contemporary scientific discourse also had had an effect on the thinking on plant forms and their evolution.

Wilhelm Hofmeister, one of the most brilliant botanists of his times, was the one to diverge away from the idealist way of pursuing botany. Over the course of his life, he brought an interdisciplinary outlook into botanical thinking. He came up with biophysical explanations on phenomena like phototaxis and geotaxis, and also discovered the alternation of generations in the plant life cycle.

1900 to the Present

The past century witnessed a rapid progress in the study of plant anatomy. The focus shifted from the population level to more reductionist levels. While the first half of the century saw expansion in developmental knowledge at the tissue and the organ level, in the latter half, especially since the 1990s, there has also been a strong impetus on gaining molecular information.

Edward Charles Jeffrey was one of the early evo devo researchers of the 20th century. He performed a comparative analyses of the vasculatures of living and fossil Gymnosperms and came to the conclusion that the storage parenchyma has been derived from tracheids. His research focussed primarily on plant anatomy in the context of phylogeny. This tradition of evolutionary analyses of plant architectures was further advanced by Katherine Esau, best known for her book *The Plant Anatomy*. Her work focussed on the origin and development of various tissues in different plants. Working with Vernon Cheadle, she also explained the evolutionary specialisation of the phloem tissue with respect to its function.

In the meantime, by the beginning of the latter half of 1900s, *Arabidopsis thaliana* had begun to be used in some developmental studies. The first collection of *Arabidopsis thaliana* mutants were made around 1945. However it formally became established as a model organism only in 1998.

The recent spurt in information on various plant-related processes has largely been a result of the revolution in molecular biology. Powerful techniques like mutagenesis and complementation were made possible in *Arabidopsis thaliana* via generation of T-DNA containing mutant lines, recombinant plasmids, techniques like Transposon Tagging etc. Availability of complete physical and genetic maps, RNAi vectors, and rapid transformation protocols are some of the technologies that have significantly altered the scope of the field. Recently, there has also been a massive increase in the genome and EST sequences of various non-model species, which, coupled with the Bioinformatics tools existing today, generate opportunities in the field of plant evo devo research. Cusset provided a detailed in-depth analysis of the history of plant morphology, including plant development and evolution, from its beginnings to the end of the 20th century.

ORGANISMS, DATABASES AND TOOLS

The most important model systems in plant development have been arabidopsis and maise. Maise has traditionally been the favourite of plant geneticists, while extensive resources in almost every area of plant physiology and development are available for *Arabidopsis thaliana*. Apart from these, rice, *Antirrhinum majus, Brassica,* tomato are also being used in a variety of studies. The genomes of *Arabidopsis thaliana* and rice have been completely sequenced, while the others are in process. It must be emphasised here that the information from these "model" organisms form the basis of our developmental knowledge.

While *Brassica* has been used primarily because of its convenient location in the phylogenetic tree in the mustard family, *Antirrhinum majus* is a convenient system for studying leaf architecture. Rice has been traditionally used for studying responses to hormones like abscissic acid and gibberelin as well as responses to stress. However, recently, not just the domesticated rice strain, but also the wild strains have been studied for their underlying genetic architectures.

Some people have objected against extending the results of model organisms to the plant world. One argument is that the effect of gene knockouts in lab conditions wouldn't truly reflect even the same plant's response in the natural world. Also, these supposedly *crucial* genes might not be responsible for the evolutionary origin of that character. For these reasons, a comparative study of plant traits has been proposed as the way to go now.

Since the past few years, researchers have indeed begun looking at non-model, "non-conventional" organisms using modern genetic tools. One example of this is the Floral Genome Project, which envisages to study the evolution of the current patterns in the genetic architecture of the flower through comparative genetic analyses, with a focus on EST sequences. Like the FGP, there are several such ongoing projects that aim to find out conserved and diverse patterns in evolution of the plant shape. Expressed sequence tag (EST) sequences of quite a few non-model plants like Sugarcane, Apple, Lotus, Barley, Cycas, Coffee, to name a few, are available freely online.

The Cycad Genomics Project, for example, aims to understand the differences in structure and function of genes between gymnosperms and angiosperms through sampling in the order Cycadales. In the process, it intends to make available information for the study of evolution of structures like seeds, cones and evolution of life cycle patterns. Presently the most important sequenced genomes from an evo-devo point of view include those of *A.thaliana (a flowering plant)*, Poplar *(a woody plant)*, *Physcomitrella patens (a bryophyte)*, Maise *(extensive genetic information)*, and *Chlamydomonas reinhardtii (a green alga)*. The impact of such a vast amount of information on understanding common underlying developmental mechanisms can easily be realised.

Apart from EST and genome sequences, several other tools like PCR, Yeast two hybrid system, microarrays, RNA Interference, SAGE, QTL mapping etc. permit the rapid study of plant developmental patterns. Recently, cross-species hybridisation has begun to be employed on microarray chips, to study the conservation and divergence in mRNA expression patterns between closely related species. Techniques for analysing this kind of data have also progressed over the past decade. We now have better models for molecular evolution, more refined analysis algorithms and better computing power as a result of advances in computer sciences.

EVOLUTION OF PLANT MORPHOLOGY

Overview of Plant Evolution

Evidence suggests that an algal scum formed on the land 1,200 million years ago, but it was not until the Ordovician period, around 500 million years ago, that land plants appeared. These begun to diversify in the late Silurian period, around 420 million years ago, and the fruits of their diversification are displayed in remarkable detail in an early Devonian fossil assemblage known as the Rhynie chert. This chert preserved early plants in cellular detail, petrified in volcanic springs.

By the middle of the Devonian period most of the features recognised in plants today are present, including roots, leaves and seeds. By the late Devonian, plants had reached a degree of sophistication that allowed them to form forests of tall trees. Evolutionary innovation continued after the Devonian period. Most plant groups were relatively unscathed by the Permo-Triassic extinction event, although the structures of communities changed. This may have set the scene for the evolution of flowering plants in the Triassic (~200 million years ago), which exploded the Cretaceous and Tertiary. The latest major group of plants to evolve were the grasses, which became important in the mid Tertiary, from around 40 million years ago. The grasses, as well as many other groups, evolved new mechanisms of metabolism to survive the low CO_2 and warm, dry conditions of the tropics over the last 10 million years.

Meristems

The meristem architectures differ between angiosperms, gymnosperms and pteridophytes. The gymnosperm vegetative meristem lacks organisation into distinct tunica and corpus layers. They possess large cells called Central Mother Cells in the meristem. In angiosperms, the outermost layer of cells divides anticlinally to generate the new cells, while in gymnosperms, the plane of division in the meristem differs for different cells. However, the apical cells do contain organelles like large vacuoles and starch grains, like the angiosperm meristematic cells.

Pteridophytes, like fern, on the other hand, do not possess a multi-cellular apical meristem. They possess a tetrahedral apical cell, which goes on to form the plant body. Any somatic mutation in this cell can lead to hereditary transmission of that mutation. The earliest meristem-like organisation is seen in an algal organism from group *Charales* that has a single dividing cell at the tip, much like the pteridophytes, yet simpler. One can thus see a clear pattern in evolution of the meristematic tissue, from pteridophytes to angiosperms. Pteridophytes, with a single meristematic cell; gymnosperms with a multi-cellular, but less defined organisation and finally, angiosperms, with the highest degree of organisation.

Origins of the Leaf

Leaves are the primary photosynthetic organs of a plant. Based on their structure, they are classified into two types-microphylls, that lack complex venation patterns and megaphylls, that are large and with a complex venation. It has been proposed that these structures arose independently. Megaphylls, according to the Telome hypothesis, have evolved from plants that showed a three dimensional branching architecture, through three transformations – planation, which involved formation of a planar architecture, webbing, or formation of the outgrowths between the planar branches and fusion, where these webbed outgrowths fused to form a proper leaf lamina. Studies have revealed that these three steps happened multiple times in the evolution of today's leaves.

It has been proposed that the before the evolution of leaves, plants had the photosynthetic apparatus on the stems. Today's megaphyll leaves probably became commonplace some 360mya, about 40my after the simple leafless plants had colonised the land in the early Devonian period. This spread has been linked to the fall in the atmospheric carbon dioxide concentrations in the Late Paleozoic era associated with a rise in density of stomata on leaf surface. This must have allowed for better transpiration rates and gas exchange. Large leaves with less stomata would have gotten heated up in the sun's heat, but an increased stomatal density allowed for a better-cooled leaf, thus making its spread feasible.

Factors Influencing Leaf Architectures

Various physical and physiological forces like light intensity, humidity, temperature, wind speeds etc., are thought to have influenced evolution of leaf shape and size. It is observed that high trees rarely have large leaves, owing to the obstruction they generate for winds. This obstruction can eventually lead to the tearing of leaves, if they are large. Similarly, trees that grow in temperate or taiga regions have pointed leaves, presumably to prevent nucleation of ice onto the leaf surface and reduce water loss due to transpiration. Herbivory, not only by large mammals, but also small insects has been implicated as a driving force in leaf evolution, an example being plants of the genus *Aciphylla,* that are commonly found in New Zealand. The now extinct Moas fed upon these plants, and its seen that the leaves have spines on their bodies, which probably functioned to discourage the moas from feeding on them. Other members of *Aciphylla* that did not co-exist with the moas, do not have these spines.

Genetic Evidences for Leaf Evolution

At the genetic level, developmental studies have shown that repression of the KNOX genes is required for initiation of the leaf primordium. This is brought about by *ARP* genes, which encode transcription factors. Genes of this type have been found in many plants studied till now, and the mechanism *i.e.* repression of KNOX genes in leaf primordia, seems to be quite conserved. Expression of KNOX genes in leaves produces complex leaves. It is speculated that the *ARP* function arose quite early in vascular plant evolution, because members of the primitive group Lycophytes also have a functionally similar gene. Other players that have a conserved role in defining leaf primordia are the phytohormone auxin, gibberelin and cytokinin.

One feature of a plant is its phyllotaxy. The arrangement of leaves on the plant body is such that the plant can maximally harvest light under the given constraints, and hence, one might expect the trait to be genetically robust. However, it may not be so. In maise, a mutation in only one gene called *abphyl (ABNORMAL PHYLLOTAXY)* was enough to change the phyllotaxy of the leaves. It implies that sometimes, mutational tweaking of a single locus on the genome is enough to generate diversity. The *abphyl* gene was later on shown to encode a cytokinin response regulator protein.

Once the leaf primordial cells are established from the SAM cells, the new axes for leaf growth are defined, one important (and more studied) among them being the abaxial-adaxial (lower-upper surface) axes. The genes involved in defining this, and the other axes seem to be more or less conserved among higher plants. Proteins of the *HD-ZIPIII* family have been implicated in defining the adaxial identity. These proteins deviate some cells in the leaf primordium from the default abaxial state, and make them adaxial. It is believed that in early plants with leaves, the leaves just had one type of surface—the abaxial one. This is the underside of today's leaves. The definition of the adaxial identity occurred some 200 million years after the abaxial identity was established. One can thus imagine the early leaves as an intermediate stage in evolution of today's leaves, having just arisen from spiny stem-like outgrowths of their leafless ancestors, covered with stomata all over, and not optimised as much for light harvesting.

How the infinite variety of plant leaves is generated is a subject of intense research. Some common themes have emerged. One of the most significant is the involvement of KNOX genes in generating compound leaves, as in tomato. But this again is not universal. For example, pea uses a different mechanism for doing the same thing. Mutations in genes affecting leaf curvature can also change leaf form, by changing the leaf from flat, to a crinky shape, like the shape of cabbage leaves. There also exist different morphogen gradients in a developing leaf which define the leaf's axis. Changes in these morphogen gradients may also affect the leaf form. Another very important class of regulators of leaf development are the microRNAs, whose role in this process has just begun to be documented. The coming years should see a rapid development in comparative studies on leaf development, with many EST sequences involved in the process coming online.

EVOLUTION OF FLOWERS

Flower-like structures first appear in the fossil records some ~130 mya, in the Cretaceous era. The flowering plants have long been assumed to have evolved from within the gymnosperms; according to the traditional morphological view, they are closely allied to the gnetales. However, recent molecular evidence is at odds to this hypothesis, and further suggests that gnetales are more closely related to some gymnosperm groups than angiosperms, and that gymnosperms form a distinct clade to the angiosperms. Molecular clock analysis predicts the divergence of flowering plants (anthophytes) and gymnosperms to ~300 mya.

The main function of a flower is reproduction, which, before the evolution of the flower and angiosperms, was the job of microsporophylls and megasporophylls. A flower can be considered a powerful evolutionary innovation, because its presence allowed the plant world to access new means and mechanisms for reproduction.

Origins of the Flower

The family Amborellaceae is regarded as the sister family of all living flowering plants. That means members of this family were most likely the first flowering plants. It seems that on the level of the organ, the leaf may be the ancestor of the flower, or at least some floral organs. When we mutate some crucial genes involved in flower development, we end up with a cluster of leaf-like structures. Thus, sometime in history, the developmental programme leading to formation of a leaf must have been altered to generate a flower. There probably also exists an overall robust framework within which the floral diversity has been generated.

An example of that is a gene called *LEAFY (LFY)*, which is involved in flower development in *Arabidopsis thaliana*. The homologs of this gene are found in angiosperms as diverse as tomato, snapdragon, pea, maise and even gymnosperms. Expression of *Arabidopsis thaliana* LFY in distant plants like poplar and citrus also results in flower-production in these plants. The *LFY* gene regulates the expression of some gene belonging to the MADS-box family. These genes, in turn, act as direct controllers of flower development.

Evolution of the MADS-box Family

The members of the MADS-box family of transcription factors play a very important and evolutionarily conserved role in flower development. According to the ABC Model of flower development, three zones—A,B and C—are generated within the developing flower primordium, by the action of some transcription factors, that are members of the MADS-box family. Among these, the functions of the B and C domain genes have been evolutionarily more conserved than the A domain gene. Many of these genes have arisen through gene duplications of ancestral members of this family. Quite a few of them show redundant functions.

The evolution of the MADS-box family has been extensively studied. These genes are present even in pteridophytes, but the spread and diversity is many times higher in angiosperms. There appears to be quite a bit of pattern into how this family has evolved. Consider the evolution of the C-region gene *AGAMOUS (AG)*. It is expressed in today's flowers in the stamens, and the carpel, which are reproductive organs. It's ancestor in gymnosperms also has the same expression pattern. Here, it is expressed in the strobili, an organ that produces pollens or ovules. Similarly, the B-genes' *(AP3 and PI)* ancestors are expressed only in the male organs in gymnosperms. Their descendants in the modern angiosperms also are expressed only in the stamens, the male reproductive organ. Thus, the same, then-existing components were used by the plants in a novel manner to generate the first flower. This is a recurring pattern in evolution.

Factors Influencing Floral Diversity

How is the enormous diversity in the shape, colour and sizes of flowers established? There is enormous variation in the developmental programme in different plants. For example, monocots possess structures like lodicules and palea, that were believed to be analogous to the dicot petals and carpels respectively.It turns out that this is true, and the variation is due to slight changes in the MADS-box genes and their expression pattern in the monocots. Another example is that of a plant called *Linaria vulgaris,* which has two kinds of flower symmetries-radial and bilateral. These symmetries are due to epigenetic changes in just one gene called *CYCLOIDEA.*

Arabidopsis thaliana has a gene called *AGAMOUS* that plays an important role in defining how many petals and sepals and other organs are generated. Mutations in this gene give rise to the floral meristem obtaining an indeterminate fate, and many floral organs keep on getting produced. We have flowers like roses, carnations and morning glory, for example, that have very dense floral organs. These flowers have been selected by horticulturists since long for increased number of petals. Researchers have found that the morphology of these flowers is because of strong mutations in the *AGAMOUS* homolog in these plants, which leads to them making a large number of petals and sepals. Several studies on diverse plants like petunia, tomato, Impatiens, maise etc. have suggested that the enormous diversity of flowers is a result of small changes in genes controlling their development.

Some of these changes also cause changes in expression patterns of the developmental genes, resulting in different phenotypes. The Floral Genome Project looked at the EST data from various tissues of many flowering plants. The researchers confirmed that the ABC Model of flower development is not conserved across all angiosperms. Sometimes expression domains change, as in the case of many monocots, and also in some basal angiosperms like *Amborella.* Different models of flower development like the *The fading boundaries model,* or the *Overlapping-boundaries model* which propose non-rigid domains of expression, may explain these architectures. There is a possibility that from the basal to the modern angiosperms, the domains of floral architecture have gotten more and more fixed through evolution.

Flowering Time

Another floral feature that has been a subject of natural selection is flowering time. Some plants flower early in their life cycle, others require a period of vernalisation before flowering. This decision is based on factors like temperature, light intensity, presence of pollinators and other environmental signals. We know that genes like *CONSTANS (CO), Flowering Locus C* (*FLC*) and *FRIGIDA* regulate integration of environmental signals into the pathway for flower development. Variations in these loci have been associated with flowering time variations between plants. For example, *Arabidopsis thaliana* ecotypes that grow in the cold, temperate regions require prolonged vernalisation before they flower, while the tropical varieties, and the most common lab strains, don't. We now know that this variation is due to mutations in the *FLC* and *FRIGIDA* genes, rendering them non-functional.

Quite a few players in this process are conserved across all the plants studied. Sometimes though, despite genetic conservation, the mechanism of action turns out to be different. For example, rice is a short-day plant, while *Arabidopsis thaliana* is a long-day plant. Now, in both plants, the proteins *CO* and *FLOWERING LOCUS T (FT)* are present. But in *Arabidopsis thaliana, CO* enhances *FT* production, while in rice, the *CO* homolog represses *FT* production, resulting in completely opposite downstream effects.

Theories of Flower Evolution

There are many theories that propose how flowers evolved. Some of them are described below : The *Anthophyte Theory* was based on the observation that a gymnospermic family Gnetaceae has a flower-like ovule. It has partially developed vessels as found in the angiosperms, and the megasporangium is covered by three envelopes, like the ovary structure of angiosperm flowers. However, many other lines of evidence show that Gnetophytes are not related to angiosperms.

The *Mostly Male Theory* has a more genetic basis. Proponents of this theory point out that the gymnosperms have two very similar copies of the gene *LFY* while angiosperms only have one. Molecular clock analysis has shown that the other *LFY* paralog was lost in angiosperms around the same time as flower fossils become abundant, suggesting that this event might have led to floral evolution. According to this theory, loss of one of the *LFY* paralog led to flowers that were more male, with the ovules being expressed ectopically. These ovules initially performed the function of attracting pollinators, but sometime later, may have been integrated into the core flower.

EVOLUTION OF SECONDARY METABOLISM

Structure of Azadirachtin, a terpenoid produced by the Neem plant, which helps ward off microbes and insects. Many secondary metabolites have complex structures Plant secondary metabolites are low molecular weight compounds, sometimes with complex structures that have no essential role in primary metabolism. They function in processes such as anti-herbivory, pollinator attraction, communication between plants, allelopathy, maintenance of symbiotic associations with soil flora and enhancing the rate of fertilization. Secondary metabolites have great structural and functional diversity and many thousands of enzymes may be involved in their synthesis, coded for by as much as 15–25 per cent of the genome. Many plant secondary metabolites such as the colour and flavour components of saffron and the chemotherapeutic drug taxol are of culinary and medical significance to humans and are therefore of commercial importance.

Since bacteria possess the ability to make secondary metabolites such as the antibiotic penicillin, their production began quite early during evolution. But they assume more significant roles in eukaryotic and multi-cellular organisms. In plants they seem to have diversified using mechanisms such as gene duplications, evolution of novel genes and the development of novel biosynthetic pathways. Studies have shown that diversity in some of these compounds may be positively selected for. Cyanogenic glycosides may have been proposed to have evolved multiple times in different plant lineages, and there are several other instances of convergent evolution. For example, the enzymes for synthesis of limonene—a terpene—are more similar between angiosperms and gymnosperms than to their own terpene synthesis enzymes. This suggests independent evolution of the limonene biosynthetic pathway in these two lineages.

Mechanisms and Players in Evolution

While environmental factors are significantly responsible for evolutionary change, they act merely as agents for natural selection. Change is inherently brought about via phenomena at the genetic level-mutations, chromosomal rearrangements and epigenetic changes. While the general types of mutations hold true across the living world, in plants, some other mechanisms have been implicated as highly significant.

Polyploidy is a very common feature in plants. It is believed that at least half (*and probably all*) plants are or have been polyploids. Polyploidy leads to genome doubling, thus generating functional redundancy in most genes. The duplicated genes may attain new function, either by changes in expression pattern or changes in activity. Polyploidy and gene duplication are believed to be among the most powerful forces in evolution of plant form. It is not known though, why genome doubling is such a frequent process in plants. One probable reason is the production of large amounts of secondary metabolites in plant cells. Some of them might interfere in the normal process of chromosomal segregation, leading to polypoidy.

In recent times, plants have been shown to possess significant microRNA families, which are conserved across many plant lineages. In comparison to animals, while the number of plant miRNA families is less, the size of each family is much larger. The miRNA genes are also much more spread out in the genome than those in animals, where they are found clustered. It has been proposed that these miRNA families have expanded by duplications of chromosomal regions. Many miRNA genes involved in regulation of plant development have been found to be quite conserved between plants studied.

Domestication of plants such as maise, rice, barley, wheat etc., has also been a significant driving force in their evolution. Some studies have looked at the origins of the maise plant and found that maise is a domesticated derivative of a wild plant from Mexico called teosinte. Teosinte belongs to the genus *Zea*, just as maise, but bears very small inflorescence, 5–10 hard cobs, and a highly branched and spread-out stem.

Crosses between a particular teosinte variety and maise yield fertile offspring that are intermediate in phenotype between maise and teosinte. QTL analysis has also revealed some loci that when mutated in maise yield a teosinte-like stem or teosinte-like cobs. Molecular clock analysis of these genes estimates their origins to some 9000 years ago, well in accordance with other records of maise domestication. It is believed that a small group of farmers must have selected some maise-like natural mutant of teosinte some 9000 years ago in Mexico, and subjected it to continuous selection to yield the maise plant as known today. Another case is that of cauliflower. The edible cauliflower is a domesticated version of the wild plant *Brassica oleracea*, which does not possess the dense undifferentiated inflorescence, called the curd, that cauliflower possesses.

Cauliflower possesses a single mutation in a gene called *CAL*, controlling meristem differentiation into inflorescence. This causes the cells at the floral meristem to gain an undifferentiated identity, and instead of growing into a flower, they grow into a lump of undifferentiated cells. This mutation has been selected through domestication at least since the Greek empire.

ALTERNATION OF GENERATIONS

Alternation of generations (also known as alternation of phases or metagenesis) is a term primarily used to describe the life cycle of plants (taken here to mean the Archaeplastida). A multi-cellular gametophyte, which is haploid with n chromosomes, alternates with a multi-cellular sporophyte, which is diploid with $2n$ chromosomes, made up of n pairs. A mature sporophyte produces spores by meiosis, a process which reduces the number of chromosomes to half, from $2n$ to n. Because meiosis is a key step in the alternation of generations, it is likely that meiosis has a fundamental adaptive function. The nature of this function is still unresolved but the two main ideas are that meiosis is adaptive because it facilitates repair of DNA damages and/or that it generates genetic variation.

The haploid spores germinate and grow into a haploid gametophyte. At maturity, the gametophyte produces gametes by mitosis, which does not alter the number of chromosomes. Two gametes (originating from different organisms of the same species or from the same organism) fuse to produce a zygote, which develops into a diploid sporophyte. This cycle, from gametophyte to gametophyte (or equally from sporophyte to sporophyte), is the way in which all land plants and many algae undergo sexual reproduction.

The relationship between the sporophyte and gametophyte varies among different groups of plants. In those algae which have alternation of generations, the sporophyte and gametophyte are separate independent organisms, which may or may not have a similar appearance. In liverworts, mosses and hornworts, the sporophyte is less well developed than the gametophyte and is largely dependent on it. Although moss and hornwort sporophytes can photosynthesise, they require additional photosynthate from the gametophyte to sustain growth and spore development and depend on it for supply of water, mineral nutrients and nitrogen. By contrast, in all modern vascular plants the gametophyte is less well developed than the sporophyte, although their Devonian ancestors had gametophytes and sporophytes of approximately equivalent complexity. In ferns the gametophyte is a small flattened autotrophic prothallus on which the young sporophyte is briefly dependent for its nutrition. In flowering plants, the reduction of the gametophyte is much more extreme; it consists of just a few cells which grow entirely inside the sporophyte.

All animals develop differently. A mature animal is diploid and so is, in one sense, equivalent to a sporophyte. However, an animal *directly* produces haploid gametes by meiosis. No haploid spores capable of dividing are produced, so neither is a haploid gametophyte. There is no alternation between diploid and haploid forms.

Other organisms, such as fungi, can have life cycles in which different kinds of organism alternate. The term 'alternation of generations' has also been applied to these cases. Life cycles of plants and algae with alternating haploid and diploid phases are referred to as diplohaplontic (the equivalent terms haplodiplontic, diplobiontic or dibiontic are also in use). Life cycles, such as those of animals, in which there is only a diploid multi-cellular phase are referred to as diplontic. (Life cycles in which there is only a haploid multi-cellular phase are referred to as haplontic.)

Definition

The discussion of 'alternation of generations' above treats the alternation of a multi-cellular diploid form with a multi-cellular haploid form as the defining characteristic, regardless of whether these forms are free-living or not. In some species, such as the alga *Ulva luctuca*, the diploid and haploid forms are indeed both free-living independent organisms, essentially identical in appearance and therefore said to be isomorphic. The free-swimming, haploid gametes form a diploid zygote which germinates into a multi-cellular diploid sporophyte. The sporophyte produces free-swimming haploid spores by meiosis that germinate into haploid gametophytes.

However, in some other groups, either the sporophyte or the gametophyte is very much reduced and is incapable of free living. For example, in all bryophytes the gametophyte generation is dominant and the sporophyte is dependent on it. By contrast, in all modern vascular land plants the gametophytes are strongly reduced, although the fossil evidence indicates that they were derived from isomorphic ancestors. In seed plants, the female gametophyte develops totally within the sporophyte which protects and nurtures it and the embryo sporophyte that it produces. The pollen grains, which are the male gametophytes are reduced reduced to only three cells. Here the notion of two generations is less obvious; as

Bateman and Dimichele say "[s]porophyte and gametophyte effectively function as a single organism". The alternative term 'alternation of phases' may then be more appropriate.

Alternation of Generations in Plants

Fundamental Elements

The diagram below shows the fundamental elements of the alternation of generations in plants. The many variations found in different groups of plants are described by use of these concepts later in the article.

Starting from the right of the diagram, the processes involved are as follows :

- Two single-celled haploid gametes, each containing n unpaired chromosomes, fuse to form a single-celled diploid zygote, which now contains n pairs of chromosomes, *i.e.* $2n$ chromosomes in total.
- The single-celled diploid zygote germinates, dividing by the normal process (mitosis), which maintains the number of chromosomes at $2n$. The result is a multi-cellular diploid organism, called the *sporophyte* (because at maturity it produces spores).
- When it reaches maturity, the sporophyte produces one or more sporangia (singular : sporangium) which are the organs that produce diploid spore mother cells (sporocytes). These divide by a special process (meiosis) that reduces the number of chromosomes by a half. This results in four single-celled haploid spores, each containing n unpaired chromosomes.
- The single-celled haploid spore germinates, dividing by the normal process (mitosis), which maintains the number of chromosomes at N. The result is a multi-cellular haploid organism, called the *gametophyte* (because it produces gametes at maturity).
- When it reaches maturity, the gametophyte produces one or more gametangia (singular : gametangium) which are the organs that produce haploid gametes. At least one kind of gamete possesses some mechanism for reaching another gamete in order to fuse with it.

The 'alternation of generations' in the life cycle is thus between a diploid ($2n$) generation of sporophytes and a haploid (n) generation of gametophytes.

Gametophyte of the fern *Onoclea sensibilis* (the flat thallus at the bottom of the picture) with a descendant sporophyte beginning to grow from it (the small frond at the top of the picture). The situation is quite different from that in all animals, where the fundamental process is that a diploid ($2n$) individual *directly* produces haploid (n) gametes by meiosis. Spores (*i.e.* haploid cells which are able to undergo mitosis) are not produced, so neither is a haploid multi-cellular organism. The single-celled gametes are the only entities which are haploid.

Variations

The diagram shown above is a good representation of the life cycle of some multi-cellular algae (*e.g.* the genus *Cladophora*) which have sporophytes and gametophytes of almost identical appearance and which do not have different kinds of spores or gametes. However, there are many possible variations on the fundamental elements of a life cycle which has alternation of generations. Each variation may occur separately or in combination, resulting in a bewildering variety of life cycles. The terms used by botanists in describing these life cycles can be equally bewildering. As Bateman and Dimichele say "[...] the alternation of generations has become a terminological morass; often, one term represents several concepts or one concept is represented by several terms."

Possible Variations are :

- *Relative importance of the sporophyte and the gametophyte.*
  - *Equal* (homomorphy or isomorphy). Filamentous algae of the genus *Cladophora*, which are predominantly found in fresh water, have diploid sporophytes and haploid gametophytes which are externally indistinguishable. No living land plant has equally dominant sporophytes and gametophytes, although some theories of the evolution of alternation of generations suggest that ancestral land plants did.
  - *Unequal* (heteromorphy or anisomorphy).

Gametophyte of *Mnium hornum*, a moss.

i. *Dominant gametophyte* (gametophytic). In liverworts, mosses and hornworts, the dominant form is the haploid gametophyte. The diploid sporophyte is not capable of an independent existence, gaining most of its nutrition from the parent gametophyte, and having no chlorophyll when mature.

ii. *Dominant sporophyte* (sporophytic). In ferns, both the sporophyte and the gametophyte are capable of living independently, but the dominant form is the diploid sporophyte. The haploid gametophyte is much smaller and simpler in structure. In seed plants, the gametophyte is even more reduced (at the minimum to only three cells), gaining all its nutrition from the sporophyte. The extreme reduction in the size of the gametophyte and its retention within the sporophyte means that when applied to seed plants the term 'alternation of generations' is somewhat misleading : "[s]porophyte and gametophyte effectively function as a single organism". Some authors have preferred the term 'alternation of phases'.

   - *Differentiation of the gametes.*
     - *Both gametes the same* (isogamy). Like other species of *Cladophora*, *C. callicoma* has flagellated gametes which are identical in appearance and ability to move.
     - *Gametes of two distinct sizes* (anisogamy).

i. *Both of similar motility*. Species of *Ulva*, the sea lettuce, have gametes which all have two flagella and so are motile. However they are of two sizes : larger 'female' gametes and smaller 'male' gametes.

ii. *One large and sessile, one small and motile* (oogamy). The larger sessile megagametes are eggs (ova), and smaller motile microgametes are sperm (spermatazoa, spermatozoids). The degree of motility of the sperm may be very limited (as in the case of flowering plants) but all are able to move towards the sessile eggs. When (as is almost always the case) the sperm and eggs are produced in different kinds of gametangia, the sperm-producing ones are called antheridia (singular antheridium) and the egg-producing ones archegonia (singular archegonium).

iii. *Antheridia and archegonia occur on the same gametophyte,* which is then called monoicous. (Many sources, including those concerned with bryophytes, use the term 'monoecious' for this situation and 'dioecious' for the opposite. Here 'monoecious' and 'dioecious' are used only for sporophytes.) The liverwort *Pellia epiphylla* has the gametophyte as the dominant generation. It is monoicous : the small reddish sperm-producing antheridia are scattered along the midrib while the egg-producing archegonia grow nearer the tips of divisions of the plant.

iv. *Antheridia and archegonia occur on different gametophytes,* which are then called dioicous. The moss *Mnium hornum* has the gametophyte as the dominant generation. It is dioicous : male plants produce only antheridia in terminal rosettes, female plants produce only archegonia in the form of stalked capsules. Seed plants are also dioicous. However, the gametophytes are extremely reduced in size; the archegonium consists only of a small number of cells, and the entire male gametophyte may be represented by only two cells.

- *Differentiation of the spores.*
  - *All spores the same size* (homospory or isospory). Horsetails (species of *Equisetum*) have spores which are all of the same size.
  - *Spores of two distinct sizes* (heterospory or anisospory) : larger megaspores and smaller microspores. When the two kinds of spore are produced in different kinds of sporangia, these are called megasporangia and microsporangia. A megaspore often (but not always) develops at the expense of the other three cells resulting from meiosis, which abort.

i. *Megasporangia and microsporangia occur on the same sporophyte,* which is then called monoecious. Most flowering plants fall into this category. Thus the flower of a lily contains six stamens (the microsporangia) which produce microspores which develop into pollen grains (the microgametophytes), and three fused carpels (the megasporangia) which produce megaspores which develop into ovules (the megagametophytes). In other plants, such as hazel, some flowers have only stamens, others only carpels, but the same plant (*i.e.* sporophyte) has both kinds of flower and so is monoecious.

ii. *Megasporangia and microsporangia occur on different sporophytes,* which are then called dioecious. An individual tree of the European holly (*Ilex aquifolium*) produces either 'male' flowers which have only functional stamens (microsporangia) producing microspores which develop into pollen grains (microgametophytes) or 'female' flowers which have only functional carpels (megasporangia) producing megaspores which develop into ovules (megagametophytes).

There are some correlations between these variations, but they are just that, correlations, and not absolute. For example, in flowering plants, microspores ultimately produce microgametes (sperm) and megaspores ultimately produce megagametes (eggs). However, in ferns and their allies there are groups with undifferentiated spores but differentiated gametophytes. For example, the fern *Ceratopteris thalictrioides* has spores of only one kind, which vary continuously in size. Smaller spores tend to germinate into gametophytes which produce only sperm-producing antheridia.

A Complex Life Cycle

The diagram shows the alternation of generations in a species which is heteromorphic, sporophytic, oogametic, dioicous, heterosporic and dioecious. A seed plant example might be a willow tree (most species of the genus *Salix* are dioecious). Starting in the centre of the diagram, the processes involved are :

- An immobile egg, typically remaining in the archegonium, fuses with a mobile sperm, released from an antheridium. The resulting zygote is either 'male' or 'female'.
  - A 'male' zygote develops by mitosis into a microsporophyte, which at maturity produces one or more microsporangia. Microspores develop within the microsporangium by meiosis. In a willow (like all seed plants) the zygote first develops into a seed within the ovule (megasporangium). Later the seed is shed and grows into a mature tree. A 'male' willow tree (a microsporophyte) produces flowers with only stamens, the anthers of which are the microsporangia.
  - Microspores germinate producing microgametophytes; at maturity one or more antheridia are produced. Sperm develop within the antheridia. In a willow, microspores are not liberated from the anther (the microsporangium), but develop into pollen grains (microgametophytes) within it. The whole pollen grain is moved (*e.g.* by an insect or by the wind) to an ovule (megagametophyte), where a sperm is produced which moves down a pollen tube to reach the egg.
  - A 'female' zygote develops by mitosis into a megasporophyte, which at maturity produces one or more megasporangia. Megaspores develop within the megasporangium; typically one of the four spores produced by meiosis gains bulk at the expense of the remaining three, which disappear. 'Female' willow trees (megasporophytes) produce flowers with only carpels (the megasporangia).
  - Megaspores germinate producing megagametophytes; at maturity one or more archegonia are produced. Eggs develop within the archegonia. In a willow, megaspores develop into ovules (megagametophytes) within the

carpels (megasporangia). An archegonium develops within the ovule and produces an egg. All of this happens within the carpel (the megasporangium). The whole of the gametophytic 'generation' remains within the protection of the sporophyte except for pollen grains (which have been reduced to just three cells).

Life Cycles of Different Plant Groups

The term 'plants' is taken here to mean the Archaeplastida, *i.e.* the glaucophytes, red and green algae and land plants. Alternation of generations occurs in almost all multi-cellular red and green algae, both freshwater forms (such as *Cladophora*) and seaweeds (such as *Ulva*). In most, the generations are homomorphic (isomorphic) and free-living. Some species of red algae have a complex triphasic.

Land plants all have heteromorphic (anisomorphic) alternation of generations, in which the sporophyte and gametophyte are distinctly different. All bryophytes, *i.e.* liverworts, mosses and hornworts, have the gametophyte generation as the most conspicuous. As an illustration, consider a monoicous moss. Antheridia and archegonia develop on the mature plant (the gametophyte). In the presence of water, the biflagellate sperm from the antheridia swim to the archegonia and fertilization occurs, leading to the production of a diploid sporophyte. The sporophyte grows up from the archegonium. Its body comprises a long stalk topped by a capsule within which spore-producing cells undergo meiosis to form haploid spores. Most mosses rely on the wind to disperse these spores, although *Splachnum sphaericum* is entomophilous, recruiting insects to disperse its spores.

In ferns and their allies, including clubmosses and horsetails, the conspicuous plant observed in the field is the diploid sporophyte. The haploid spores develop in sori on the underside of the fronds and are dispersed by the wind (or in some cases, by floating on water). If conditions are right, a spore will germinate and grow into a rather inconspicuous plant body called a prothallus. The haploid prothallus does not resemble the sporophyte, and as such ferns and their allies have a heteromorphic alternation of generations. The prothallus is short-lived, but carries out sexual reproduction, producing the diploid zygote that then grows out of the prothallus as the sporophyte.

In the spermatophytes, the seed plants, the sporophyte is the dominant multicellular phase; the gametophytes are strongly reduced in size and very different in morphology. The entire gametophyte generation, with the sole exception of pollen grains (microgametophytes), is contained within the sporophyte. The life cycle of a dioecious flowering plant (angiosperm), the willow, has been outlined in some detail in an earlier section (A complex life cycle). The life cycle of a gymnosperm is similar. However, flowering plants have in addition a phenomenon called 'double fertilization'. Two sperm nuclei from a pollen grain (the microgametophyte), rather than a single sperm, enter the archegonium of the megagametophyte; one fuses with the egg nucleus to form the zygote, the other fuses with two other nuclei of the gametophyte to form 'endosperm', which nourishes the developing embryo.

Evolutionary Emergence of the Dominant Diploid Phase

It has been proposed that the basis for the emergence of the diploid phase of the life cycle (sporophyte) as the dominant phase (*e.g.* as in vascular plants) is that diploidy allows masking of the expression of deleterious mutations through genetic complementation. Thus if one of the parental genomes in the diploid cells contained mutations leading to defects in one or more gene products, these deficiencies could be compensated for by the other parental genome (which nevertheless may have its own defects in other genes). As the diploid phase was becoming predominant, the masking effect likely allowed genome size, and hence information content, to increase without the constraint of having to improve accuracy of DNA replication. The opportunity to increase information content at low cost was advantageous because it permitted new adaptations to be encoded. This view has been challenged, with evidence showing that selection is no more effective in the haploid than in the diploid phases of the lifecycle of mosses and angiosperms.

PATHOGEN RECOGNITION

Plants use a vast array of signals originating from micro-organisms and the environment to recognise pathogens and elicit plant defence responses. Non-specific elicitors of biotic and abiotic origin induce host defences in a broad range of host species. Abiotic elicitors such as heavy metal ions or UV light can induce stress responses in exposed tissues, which may provide an additional barrier to invading pathogens or alternatively, increase the plant's susceptibility to infection. Biotic elicitors include cell wall fragments released from fungi and bacteria, hydrolytic enzymes of plant or pathogen origin, some peptides, glycoproteins and polyunsaturated fatty acids. These elicitors induce defence responses in a range of host species. Often, non-specific elicitors act as a general indication that the cell has been damaged in some way (for example, the release of fragments of the host's own cell wall can elicit defence responses).

Specific elicitors enable defence against a very specific pathogen, and are conditioned by *avirulence genes in that pathogen*. Avirulence genes determine the pathogen's host range, but are only able to function in the presence of another set of genes, the 'hypersensitive response and pathogenicity' (Hrp) gene cluster. Some Hrp gene products are involved in disguising the pathogen from host recognition, thus playing a role in both virulence and avirulence. For a biotroph to form a successful infection, it must establish a basic compatibility with its host. The pathogen may also produce compatibility factors that delay, avoid or negate recognition by a normally resistant host plant. Virulent strains appear to be able to suppress the resistance mechanisms of the host, but are not able to halt resistance responses once they are activated. Incompatibility between a host and a pathogen results in the recognition of the pathogen and activation of defence mechanisms, while compatibility results in infection.

Specific elicitors are encoded by avirulence genes, and these peptides are believed to bind to receptor peptides, encoded by host resistance genes. Recognition of the avirulence gene products by the host triggers signal transduction pathways that cause a massive shift in gene transcription and plant cell metabolism, and local and systemic signals are released that prime the rest of the plant against further infection. The presence of non-specific elicitors, such as the release of host and pathogen wall fragments, during this process may amplify the defence response.

Host-parasite specific resistance is determined by the interaction of between products of pathogen avirulence genes, Specific elicitors and products of host resistance genes. The defence responses of plants can be very rapid. Host gene expression begins within minutes, or even seconds, of exposure to elicitors or pathogens. A diverse range of elicitors can induce a common set of responses in the host, suggesting that second messengers are involved in the signalling pathway between pathogen attack and host response.

RAPID ACTIVE DEFENCES

Almost every host-parasite interaction is unique in the details of the activation, localisation, timing and magnitude of the defence responses.

At the Membrane

The host membrane appears to be involved in the earliest stages of pathogen recognition and signal transduction. A change in membrane permeability after exposure to a pathogen causes fluxes in ions, such as K+, H+ and Ca2+, and results in changes to gene activation and the triggering of the defence responses. Also at the membrane, the 'oxidative burst', which involves the generation of reactive oxygen species, such as hydrogen peroxide, triggers signals that affect gene expression, cross-linking in the host cell wall and initiation of later defence responses. The reactive oxygen species at the site of infection are also produced in quantities capable of killing micro-organisms directly.

At the Cell Wall

Preparations for the reinforcement of the cell wall, which can improve host resistance, begin very quickly after a pathogen attempts to penetrate a host cell. This is characterised by an intensification of cytoplasmic streaming and the accumulation of host cytoplasm around the site of attempted penetration.

The cytoplasmic aggregates are thought to contain cellular apparatus for the synthesis of cell wall fortifications. If the host cell can repair and reinforce its cell walls quickly enough, it might reduce the penetration efficiency of the pathogen. Several types of reinforcement are produced by host cells. A papilla is a deposit of callose, silicon, lignin and proteins between the cell wall and cell membrane, directly below the point of attempted penetration, while lignitubers are lignified callose reinforcements that ensheath invading hyphal tips.

Hydroxyproline-rich glycoproteins are structural cell wall proteins involved in secondary cell wall thickening. The expression of genes governing their production is activated ahead of invading hyphae, reinforcing walls. Cross-linking of hydroxyproline-rich glycoproteins caused by the release of hydrogen peroxide in the oxidative burst also reinforces cell wall compartments.

Rapid deposition of lignin and suberin following infection also increases resistance to pathogens in many plants. Lignin can also bind to hyphal tips and bacteria, physically restraining them and restricting the diffusion of their enzymes and toxins into, and the extraction of water and nutrients out of, the host cell. Cell wall reinforcements tend to be larger and more quickly formed in resistant hosts than in susceptible hosts, and inhibition of the production of callose or lignin synthesis by the pathogen enhances its penetration efficiency.

The Hypersensitive Response

Hypersensitive cell death is another widespread mechanism used by hosts to prevent the spread of a pathogen. Infected cells and those surrounding them "suicide", preventing further spread, and in some cases, killing the pathogen. It is often associated with the initiation of other responses, such as lignification and the synthesis of anti-microbial compounds. The success of hypersensitive cell death as a resistance mechanism depends on the nutritional requirements of the specific pathogen and the timing, magnitude and location of the host response.

Antibiotic Compounds

Phytoalexins are low molecular weight antibiotics produced by many (but not all) plants in response to infection. There are many biotic elicitors of phytoalexin production, such as cell wall components, as well as abiotic elicitors, such as heavy metals and ultraviolet light. Phytoalexins inhibit the growth of bacteria and fungi *in vivo* and *in vitro*, and production of these antibiotics during an infection can induce resistance to subsequent infections by that pathogen. Over 350 phytoalexins are known in over 100 plant species. They include pterocarpans, sesquiterpenes, cryptophenols, isocoumarins, isoflavenoids, and others. Phytoalexins may be produced by any part of the plant, although different phytoalexins can accumulate in different organs.

Generally, related plant species produce structurally-related phytoalexins, and many produce more than one, enabling the plant to present a toxic cocktail to invading pathogens. Phytoalexins are produced in cells surrounding an infection site and delivered to the infected cell packaged in lipid vesicles, creating a toxic micro-environment in the infected cell and, hopefully, preventing disease establishment. Phytoalexin accumulation is often associated with hypersensitive cell death, although only living cells can synthesise phytoalexins. Some plants can also sequester phytoalexins into vacuoles as stores of inactive sugar-conjugates, which can be cleaved and released quickly if initial defence responses are unsuccessful.

DELAYED ACTIVE DEFENCES

Delayed active defences include containment of the pathogen, wound repair, expression of pathogenesis-related proteins and the acquisition of systemic resistance. These mechanisms restrict the spread of the pathogen after infection is established and contain the damage to host tissues.

Physical Responses

The ability to repair wounds can help protect the plant from further infection by other, opportunistic pathogens. A secondary meristem in fleshy tissues, fruits, roots and bark, the cork cambium, can produce cork cells, which have thick, suberised walls. These cells can create a barrier to further colonisation by the pathogen and, in some cases, develop an abscission layer around the site of infection, causing the infected tissue separate from the healthy tissue. Wounded tree trunks often secrete protective gums that seal the wound from further infection. The formation of tyloses, ingrowths if the protoplasm in xylem parenchyma can also restrict the spread of pathogenic propagules in the xylem, although they also tend to reduce the movement of water through the vessels, causing water stress in the plant.

Pathogenesis-related Proteins

There is a range of novel proteins synthesised in response to infection, many of which have b-glucanase, chitinase or lysozyme activity. Some pathogenesis-related proteins disrupt pathogen nutrition. The presence of low levels of these proteins in healthy plants suggests that they might have other roles in plant growth and development aside from disease resistance. Chitinase and glucanase accumulate in the vacuoles, and glucanase is also sometimes secreted into the intercellular space. They dissolve the fungal cell wall, fragments of which then elicit hypersensitive cell death. The breakdown of the vacuole during decompartmentalisation of the cytoplasm results in a flood of hydrolytic enzymes, which have antiviral, antifungal and antibacterial activity. The accumulation of pathogenesis-related proteins peaks around 7-10 days after initial infection. The presence of these proteins *before* infection increases the plant's resistance to pathogens, as in the case of systemic acquired resistance.

Systemic Acquired Resistance

Systemic acquired resistance, or induced resistance, is characterised by the increased resistance of a plant to a wide range of pathogens following infection by one pathogen. It is therefore fundamentally different from the specific antigen-antibody mechanism of resistance seen in the immune response of mammals. Rather than providing immunity *per se*, systemic acquired resistance reduces the severity of later diseases. The development of systemic acquired resistance usually requires the development of a slowly expanding necrotic lesion and other localised responses to infection, the release of a phloem-translocated signal originating from the infection site, and the subsequent priming of the plant against further attacks, allowing a more rapid response in the case of future infections. The nature of the signal that triggers systemic acquired resistance is as yet unknown, and is likely to be a complex signal transduction pathway mediated by a number of stress signals. Salicylic acid plays a key role, via interaction with salicylic acid-binding proteins that can cause build-up of reactive oxygen species or activate gene expression. The levels of salicylic acid increase around necrotic lesions and remain high in plants that have acquired resistance. It is not, however, salicylic acid itself that acts as the signal that is translocated systemically throughout the plant.

Co-ordination of Defence Responses

The success of defence responses is increased if activated in combination. Passive mechanisms, coupled with rapid active responses and slower follow-up defences provide a broad defence front to the plant. The specific interaction between host and pathogen is of course crucial to the success of the plant's resistance or the pathogen's invasion, and is mediated by the many pathways involved in producing or detecting elicitors, enhancers, suppressors and secondary signals.

ALLELOPATHY

Casuarina equisetifolia litter completely suppresses germination of understory plants as shown here despite the relative openness of the canopy and ample rainfall (>120 cm/yr) at the location.

Allelopathy is a biological phenomenon by which an organism produces one or more biochemicals that influence the growth, survival, and reproduction of other organisms. These biochemicals are known as allelochemicals and can have beneficial (positive allelopathy) or detrimental (negative allelopathy) effects on the target organisms. Allelochemicals are a sub-set of secondary metabolites, which are not required for metabolism (*i.e.* growth, development and reproduction) of the allelopathic organism. Allelochemicals with negative allelopathic effects are an important part of plant defence against herbivory.

Allelopathy is characteristic of certain plants, algae, bacteria, coral, and fungi. Allelopathic interactions are an important factor in determining species distribution and abundance within plant communities, and are also thought to be important in the success of many invasive plants. The process by which a plant acquires more of the available resources (such as nutrients, water or light) from the environment without any chemical action on the surrounding plants is called resource competition. This process is not negative allelopathy, although both processes can act together to enhance the survival rate of the plant species.

History

The term allelopathy from the Greek-derived compounds *allelo-* and *-pathy* (meaning "mutual harm" or "suffering"), was first used in 1937 by the Austrian professor Hans Molisch in the book *Der Einfluss einer Pflanze auf die andere-Allel-*

opathie (The Effect of Plants on Each Other) published in German. He used the term to describe biochemical interactions that inhibit the growth of neighbouring plants, by another plant. In 1971, Whittaker and Feeny published a study in the journal *Science*, which defined allelochemicals as all chemical interactions among organisms. In 1984, Elroy Leon Rice in his monograph on allelopathy enlarged the definition to include all direct positive or negative effects of a plant on another plant or on micro-organisms by the liberation of biochemicals into the natural environment.

Over the next ten years, the term was used by other researchers to describe broader chemical interactions between organisms, and by 1996 the International Homeopathy Society defined allelopathy as "Any process involving secondary metabolites produced by plants, algae, bacteria and fungi that influences the growth and development of agriculture and biological systems." In more recent times, plant researchers have begun to switch back to the original definition of substances that are produced by one plant that inhibit another plant. Confusing the issue more, zoologists have borrowed the term to describe chemical interactions between invertebrates like corals and sponges.

Long before the term allelopathy was used, people observed the negative effects that one plant could have on another. Theophrastus, who lived around 300 B.C. noticed the inhibitory effects of pigweed on alfalfa. In China around the first century A.D., the author of *Shennong Ben Cao Jing* described 267 plants that had pesticidal abilities, including those with allelopathic effects. In 1832, the Swiss botanist De Candolle suggested that crop plant exudates were responsible for an agriculture problem called soil sickness.

Allelopathy is not universally accepted among ecologists and many have argued that its effects cannot be distinguished from the competition which results when two (or more) organisms attempt to use the same limited resource, to the detriment of one or both. Allelopathy is a direct negative effect on one organism resulting from the input of substances into the environment by another. In the 1970s, great effort went into distinguishing competitive and allelopathic effects by some researchers, while in the 1990s others argued that the effects were often interdependent and could not readily be distinguished.

Examples of Allelopathy

The possible application of allelopathy in agriculture is the subject of much research. Current research is focused on the effects of weeds on crops, crops on weeds, and crops on crops. This research furthers the possibility of using allelochemicals as growth regulators and natural herbicides, to promote sustainable agriculture. A number of such allelochemicals are commercially available or in the process of large-scale manufacture. For example, Leptospermone is a purported thermochemical in lemon bottlebrush (*Callistemon citrinus*). Although it was found to be too weak as a commercial herbicide, a chemical analog of it, mesotrione (tradename Callisto), was found to be effective. It is sold to control broadleaf weeds in corn but also seems to be an effective control for crabgrass in lawns. Sheeja (1993) reported the allelopathic interaction of the weeds *Chromolaena odourata* (*Eupatorium odouratum*) and *Lantana camara* on selected major crops.

Many crop cultivars show strong allelopathic properties, of which rice (*Oryza sativa*) has been most studied. Rice allelopathy depends on variety and origin : Japonica rice is more allelopathic than Indica and Japonica-Indica hybrid. More recently, critical review on rice allelopathy and the possibility for weed management reported that allelopathic characteristics in rice are quantitatively inherited and several allelopathy-involved traits have been identified.

Many invasive plant species interfere with native plants through allelopathy. A famous case of purported allelopathy is in desert shrubs. One of the most widely known early examples was *Salvia leucophylla*, because it was on the cover of the journal *Science* in 1964. Bare zones around the shrubs were hypothesised to be caused by volatile terpenes emitted by the shrubs. However, like many allelopathy studies, it was based on artificial lab experiments and unwarranted extrapolations to natural ecosystems. In 1970, *Science* published a study where caging the shrubs to exclude rodents and birds allowed grass to grow in the bare zones.

A detailed history of this story can be found in Halsey 2004. Allelopathy has been shown to play a crucial role in forests, influencing the composition of the vegetation growth, and also provides an explanation for the patterns of forest regeneration. The black walnut *(Juglans nigra)* produces the allelochemical juglone, which affects some species greatly while others not at all. The leaf litter and root exudates of some *Eucalyptus* species are allelopathic for certain soil microbes and plant species. The tree of heaven, *Ailanthus altissima*, produces allelochemicals in its roots that inhibit the growth of many plants. The pace of evaluating allelochemicals released by higher plants in nature has greatly accelerated, with promising results in field screening.

Garlic mustard is an invasive plant species in North American temperate forests. Its success may be partly due to its excretion of an unidentified allelochemical that interferes with mutualisms between native tree roots and their mycorrhizal fungi. A study of *Kochia scoparia* in northern Montana by two high school students showed that when *Kochia* precedes spring wheat (*Triticum aestivum*), it reduces the spring wheat's growth. Effects included delayed emergence, decreased rate of growth, decreased final height and decreased average vegetative dry weight of spring wheat plants. A larger study later showed that *Kochia* seems to exhibit allelopathy on various crops in northern Montana. Corn gluten meal (CGM) is a natural pre-emergence weed control used in turfgrass, which reduces germination of many broadleaf and grass weeds.

ALLOMONE

An allomone is any chemical substance produced and released by an individual of one species that affects the behaviour of a member of another species to the benefit of the originator but not the receiver. Production of allomones is a common form of defence, particularly by plant species against insect herbivores.

Many insects have developed ways to defend against these plant defences (in an evolutionary arms race). One method of adapting to allomones is to develop a positive reaction to them; the allomone then becomes a kairomone. Others alter the allomones to form pheromones or other hormones, and yet others adopt them into their own defensive strategies, for example by regurgitating them when attacked by an insectivorous insect.

A third class of allelochemical (chemical used in interspecific communication), synomones, benefit both the sender and receiver. "Allomone was proposed by Brown and Eisner (Brown, 1968) to denote those substances which convey an advantage upon the emitter. Because Brown and Eisner did not specify whether or not the receiver would benefit, the original definition of allomone includes both substances that benefit the receiver and the emitter, and substances that only benefit the emitter. An example of the first relationship would be a mutualistic relationship, and the latter would be a repellent secretion."

KAIROMONE

A kairomone (A neo-logism using the Greek êáéñüò *opportune moment*, paralleling pheromone), is a semiochemical, emitted by an organism, which mediates interspecific interactions in a way that benefits an individual of another species which receives it, without benefiting the emitter. This "eavesdropping" is often disadvantageous to the producer (though other benefits of producing the substance may outweigh this cost, hence its persistence over evolutionary time). The kairomone improves the fitness of the recipient and in this respect differs from an allomone (which is the opposite : it benefits the producer and harms the receiver) and a synomone (which benefits both parties). The term is mostly used in the field of entomology (the study of insects). Two main ecological cues are provided by kairomones; they generally either indicate a food source for the receiver, or the presence of a predator, the latter of which is less common or at least less studied.

Predators Use Them to Find Prey

An example of this can be found in the Ponderosa Pine tree (*Pinus ponderosa*), which produces a terpene called myrcene when it is damaged by the Western pine beetle. Instead of deterring the insect, it acts synergistically with aggregation pheromones which in turn act to lure more beetles to the tree. Specialist predatory beetles find bark beetles (their prey) using the pheromones the bark beetles produce. In this case the chemical substance produced is both a pheromone (communication between bark beetles) and a kairomone (eavesdropping). This was discovered accidentally when the predatory beetles and other enemies were attracted to insect traps baited with bark beetle pheromones.

Pheromones of different kinds may be exploited as kairomones by receivers. The German wasp, *Vespula germanica* is attracted to a pheromone produced by male Mediterranean fruit flies (*Ceratitis capitata*) when lekking (a lek is a gathering of males for mating display), causing the death of some. In contrast, it is the alarm pheromone (used to communicate the presence of a threat) of an ant (*Iridomyrmex purpureus*) that a spider predator is attracted to.

Prey Use Them to Adjust to Predators

Some prey make use of chemicals originating from predators, using these cues as an indicator of the level of predation risk and changing their morphology if need be. Changes in morphology caused by predator presence is known as predator-induced polyphenism, and occurs across a variety of animals. For example *Daphnia cucullata* show formation of "helmets" when exposed to predators or the water they have lived in. Their predators include cladocerans (such as *Leptodoura kindtii*) and larvae of *Chaoborus flavicans*, a midge. They respond to these kairomones by doubling the size of their helmets, a protective structure. These changes in morphology make them safer from predators.

Mice are instinctively afraid of the smell of their natural predators, including cats and rats. This occurs even in laboratory mice that have been isolated from predators for hundreds of generations. When the chemical cues responsible for the fear response were purified from cat saliva and rat urine, two homologous protein signals were identified : Fel d 4 (*Felis domesticus* allergen 4), the product of the cat *Mup* gene, and Rat n 1 (*Rattus norvegicus* allergen 1), the product of the rat *Mup13* gene. Mice are fearful of these major urinary proteins (Mups) even when they are made in bacteria, but mutant animals that are unable to detect the Mups shown no fear of rats demonstrating their importance in initiating fearful behaviour. It is not known exactly how Mups from different species initiate disparate behaviours, but mouse Mups and predator Mups have been shown to activate unique patterns of sensory neurons in the nose of recipient mice. This implies the mouse perceives them differently, via distinct neural circuits. The pheromone receptors responsible for Mup detection are also unknown, though they are thought be members of the V2R receptor class.

Applications

Like pheromones (communication chemicals used within a species), kairomones can be utilised as an 'attracticide' to lure a pest species to a location containing pesticide. However, they might also be used to lure desired species. Kairomones produced by the hosts of parasitic wasps have been used in an attempt to attract them and keep them around in crops where they reduce herbivory, but this could instead result in fewer attacks on the herbivous pest if the applied kairomone distracts them from finding real hosts.

Chapter 26

GENETICALLY MODIFIED CROPS

Genetically modified crops (GMCs, GM crops, or biotech crops) are plants, the DNA of which has been modified using genetic engineering techniques. In most cases the aim is to introduce a new trait to the plant which does not occur naturally in the species. Examples include resistance to certain pests, diseases, or environmental conditions, or resistance to chemical treatments (*e.g.* resistance to an herbicide), or the production of a certain nutrient or pharmaceutical agent. Genetic engineering techniques are much more precise than mutagenesis (mutation breeding) where an organism is exposed to radiation or chemicals to create a non-specific but stable change. Other techniques by which humans modify plants include selective breeding; plant breeding, and somaclonal variation.

Critics have objected to GM crops *per se* on several grounds, including ecological concerns, and economic concerns raised by the fact these organisms are subject to intellectual property law. GM crops also are involved in controversies over GM food with respect to whether food produced from GM crops is safe and whether GM crops are needed to address the world's food needs.

GENE TRANSFER IN NATURE AND TRADITIONAL AGRICULTURE

Scientists first discovered that DNA naturally transfers between organisms in 1946. It is now known that there are several natural mechanisms for flow of genes, or (horizontal gene transfer), and that these occur in nature on a large scale–for example, it is a major mechanism for antibiotic resistance in pathogenic bacteria, and it occurs between plant species. This is facilitated by transposons, retrotransposons, proviruses and other mobile genetic elements that naturally translocate to new sites in a genome. They often move to new species over an evolutionary time scale and play a major role in dynamic changes to chromosomes during evolution.

The introduction of foreign germplasm into crops has been achieved by traditional crop breeders by artificially overcoming fertility barriers. A hybrid cereal was created in 1875, by crossing wheat and rye. Since then important traits have been introduced into wheat, including dwarfing genes and rust resistance. Plant tissue culture and the induction of mutations have also enabled humans to artificially alter the makeup of plant genomes.

HISTORY

The first genetically modified plant had produced in 1982, using an antibiotic-resistant tobacco plant. The first field trials of genetically engineered plants occurred in France and the USA in 1986, when tobacco plants were engineered to be resistant to herbicides. In 1987, Plant Genetic Systems (Ghent, Belgium), founded by Marc Van Montagu and Jeff Schell, was the first company to develop genetically engineered (tobacco) plants with insect tolerance by expressing genes encoding for insecticidal proteins from Bacillus thuringiensis (Bt).

The People's Republic of China was the first country to allow commercialised transgenic plants, introducing a virus-resistant tobacco in 1992. The first genetically modified crop approved for sale in the U.S., in 1994, was the *FlavrSavr* tomato, which had a longer shelf life. In 1994, the European Union approved tobacco engineered to be resistant to the herbicide bromoxynil, making it the first commercially genetically engineered crop marketed in Europe. In 1995, Bt Potato was approved safe by the Environmental Protection Agency, making it the first pesticide producing crop to be approved in the USA.

The following transgenic crops also received marketing approval in the US in 1995 : canola with modified oil composition (Calgene), *Bacillus thuringiensis* (Bt) corn/maise (Ciba-Geigy), cotton resistant to the herbicide bromoxynil (Calgene), Bt cotton (Monsanto), soybeans resistant to the herbicide glyphosate (Monsanto), virus-resistant squash (Asgrow), and additional delayed ripening tomatoes (DNAP, Zeneca/Peto, and Monsanto). As of mid-1996, a total of 35 approvals had been granted to commercially grow 8 transgenic crops and one flower crop of carnations, with 8 different traits in 6 countries plus the EU. In 2000, with the production of golden rice, scientists genetically modified food to increase its nutrient value for the first time.

METHODS

Genetically engineered plants are generated in a laboratory by altering their genetic makeup. This is usually done by adding one or more genes to a plant's genome using genetic engineering techniques. Most genetically modified plants are generated by the biolistic method (particle gun) or by *Agrobacterium tumefaciens* mediated transformation. Plant scientists, backed by results of modern comprehensive profiling of crop composition, point out that crops modified using GM techniques are less likely to have unintended changes than are conventionally bred crops.

In research tobacco and *Arabidopsis thaliana* are the most genetically modified plants, due to well developed transformation methods, easy propagation and well studied genomes. They serve as model organisms for other plant species. In the biolistic method, DNA is bound to tiny particles of gold or tungsten which are subsequently shot into plant tissue or single plant cells under high pressure. The accelerated particles penetrate both the cell wall and membranes. The DNA separates from the metal and is integrated into plant genome inside the nucleus. This method has been applied successfully for many cultivated crops, especially monocots like wheat or maise, for which transformation using *Agrobacterium tumefaciens* has been less successful. The major disadvantage of this procedure is that serious damage can be done to the cellular tissue.

Agrobacteria are natural plant parasites, and their natural ability to transfer genes provides another method for the development of genetically engineered plants. To create a suitable environment for themselves, these Agrobacteria insert their genes into plant hosts, resulting in a proliferation of plant cells near the soil level (crown gall). The genetic information for tumour growth is encoded on a mobile, circular DNA fragment (plasmid). When Agrobacterium infects a plant, it transfers this T-DNA to a random site in the plant genome. When used in genetic engineering the bacterial T-DNA is removed from the bacterial plasmid and replaced with the desired foreign gene.

The bacterium is a vector, enabling transportation of foreign genes into plants. This method works especially well for dicotyledonous plants like potatoes, tomatoes, and tobacco. Agrobacteria infection is less successful in crops like wheat and maise. Introducing new genes into plants requires a promoter specific to the area where the gene is to be expressed. For instance, if we want the gene to be expressed only in rice grains and not in leaves, then an endosperm-specific promoter would be used. The codons of the gene must also be optimised for the organism due to codon usage bias. The transgenic gene products should also be able to be denatured by heat so that they are destroyed during cooking.

FLAVR SAVR TOMATO

The first commercialised genetically modified plants (Flavr Savr tomatoes) used RNAi technology, where the inserted DNA matched an endogenous gene already in the plant. When the inserted gene is expressed it can repress the translation of the endogenous protein. Host delivered RNAi systems are being developed, where the plant will express RNA that will interfere with insects, nematodes and other parasites' protein synthesis. This may provide a novel way of protecting plants from pests.

GLYPHOSATE RESISTANCE

One of the most famous kinds of GM crops are "Roundup Ready", or glyphosate-resistant. Glyphosate, (the active ingredient in Roundup) kills plants by interfering with the shikimate pathway in plants, which is essential for the synthesis of the aromatic amino acids phenylalanine, tyrosine and tryptophan. More specifically, glyphosate inhibits the enzyme 5-enolpyruvylshikimate-3-phosphate synthase (EPSPS). The shikimate pathway is not present in animals, which instead obtain aromatic amino acids from their diet.

Some micro-organisms have a version of EPSPS that is resistant to glyphosate inhibition. One of these was isolated from an *Agrobacterium* strain CP4 (CP4 EPSPS) that was resistant to glyphosate. This CP4 EPSPS gene was cloned and transfected into soybeans. The CP4 EPSPS gene was engineered for plant expression by fusing the 5′ end of the gene to a chloroplast transit peptide derived from the petunia EPSPS. This transit peptide was used because it had shown previously an ability to deliver bacterial EPSPS to the chloroplasts of other plants.

The plasmid used to move the gene into soybeans was PV-GMGTO4. It contained three bacterial genes, two CP4 EPSPS genes, and a gene encoding beta-glucuronidase (GUS) from *Escherichia coli* as a marker. The DNA was injected into the soybeans using the particle acceleration method. Soybean cultivar A5403 was used for the transformation. The expression of the GUS gene was used as the initial evidence of transformation. GUS expression was detected by a staining method in which the GUS enzyme converts a substrate into a blue precipitate. Those plants that showed GUS expression were then taken and sprayed with glyphosate, and their tolerance was tested over many generations.

TYPES OF GENETIC ENGINEERING

Transgenic plants have genes inserted into them that are derived from another species. The inserted genes can come from species within the same kingdom (plant to plant) or between kingdoms (for example, bacteria to plant). In many cases the inserted DNA has to be modified slightly in order to correctly and efficiently express in the host organism. Transgenic plants are used to express proteins like the cry toxins from *Bacillus thuringiensis*, herbicide resistant genes and antigens for vaccinations Transgenic carrots have been used to produce the drug Taliglucerase alfa which is used to treat Gaucher's disease.

In the laboratory, transgenic plants have been modified to increase their photosynthesis (currently about 2 per cent at most plants to the theoretic potential of 9–10 per cent. This is possible by changing the rubisco enzyme (*i.e.* changing C3 plants into C4 plants), by placing the rubisco in a carboxysome, by adding CO_2 pumps in the cell wall, by changing the leaf form/size. Plants have been engineered to exhibit bioluminescence which might one day be a sustainable alternative to electric lighting. Still other transgenic plants have been modified to fixate ambient nitrogen in the plant.

Cisgenic plants are made using genes found within the same species or a closely related one, where conventional plant breeding can occur. Some breeders and scientists argue that cisgenic modification is useful for plants that are difficult to crossbreed by conventional means (such as potatoes), and that plants in the cisgenic category should not require the same level of legal regulation as other genetically modified organisms.

BUSINESS OF GM CROPS

The global value of biotech seed alone was US$13.2 billion in 2011, with the end product of commercial grain from biotech maise, soybean grain and cotton valued at approximately US$160 billion or more per year. Players in agriculture business markets include seed companies, agrochemical companies, distributors, farmers, grain elevators, and universities that develop new crops and whose agricultural extensions advise farmers on best practices. The largest share of the GMO crops planted globally are from seed created by the United States firm Monsanto. In 2007, Monsanto's trait technologies were planted on 246 million acres (1,000,000 km^2) throughout the world, a growth of 13 per cent from 2006. However, patents on the first Monsanto products to enter the marketplace will begin to expire in 2014, democratising Monsanto products. Syngenta, DuPont (especially via its Pioneer Hi-Bred subsidiary, and Bayer Crop Science are also major players in the US and Europe. In addition, a 2007 report from the European Joint Research Commission predicts that by 2015, more than 40 per cent of new GM plants entering the global marketplace will have been developed in Asia.

In the corn market, Monsanto's triple-stack corn—which combines Roundup Ready 2-weed control technology with Yield Gard (Bt) Corn Borer and Yield Gard Rootworm insect control—is the market leader in the United States. U.S. corn farmers planted more than 32 million acres (130,000 km^2) of triple-stack corn in 2008, and it is estimated the product could be planted on 56 million acres (230,000 km^2) in 2014–2015. In the cotton market, Bollgard II with Roundup Ready Flex was planted on approximately 5 million acres (20,000 km^2) of U.S. cotton in 2008.

According to the International Service for the Acquisition of Agri-Biotech Applications (ISAAA), in 2010 approximately 15 million farmers grew biotech crops in 29 countries. Over 90 per cent of the farmers were resource-poor in developing countries. 6.5 million farmers in China and 6.3 million small farmers in India grew biotech crops (mostly Bacillus thuringiensis cotton). The Philippines, South Africa (biotech cotton, maise, and soybeans often grown by subsistence women farmers) and another twelve developing countries also grew biotech crops in 2009. 10 million more small and resource-poor farmers may have been secondary beneficiaries of Bt cotton in China. According to a review published in 2012 and based on data from the late 1990s and early 2000s, much of the GM crop grown each year is used for livestock feed, and increased demand for meat will lead to increased demand for GM crops with which to feed them. Feed grain usage as a percentage of total crop production is 70 per cent for corn and more than 90 per cent of oil seed meals such as soybeans. About 65 million metric tons of GM corn grains and about 70 million metric tons of soybean meals derived from GM soybean are fed to livestock each year.

USES, ACTUAL AND PROPOSED

GM crops grown today, or under experimental development, have been modified with traits intended to provide benefit to farmers, consumers, or industry. These traits include improved shelf life, disease resistance, stress resistance, herbicide resistance, pest resistance, production of useful goods such as biofuel or drugs, and ability to absorb toxins, for use in bioremediation of pollution. Due to high regulatory and research costs, the majority of genetically modified crops in agriculture consist of commodity crops, such as soybean, maise, cotton and rapeseed. Recently, some research and development has been targeted to enhancement of crops that are locally important in developing countries, such as insect-resistant cowpea for Africa and insect-resistant brinjal (eggplant) for India.

IMPROVED SHELF LIFE

The first genetically modified crop approved for sale in the U.S. was the *Flavr Savr* tomato, which had a longer shelf life. It is no longer on the market. As of 2012, an apple that has been genetically modified to resist browning, known as the Non-browning Arctic apple produced by Okanagan Specialty Fruits, is awaiting regulatory approval in the US and Canada. A gene in the fruit has been modified such that the apple produces less polyphenol oxidase, a chemical that manifests the browning.

IMPROVED NUTRITION

The GM oilseed crops on the market today offer improved oil profiles for processing or healthier edible oils. The GM crops in development offer a wider array of environmental and consumer benefits such as nutritional enhancement and drought and stress tolerance. GM plants are being developed by both private companies and public research institutions such as CIMMYT, the International Maise and Wheat Improvement Centre. Other examples include a genetically modified cassava with lower cyanogen glucosides and enhanced with protein and other nutrients, while golden rice, developed by the International Rice Research Institute (IRRI), has been discussed as a possible cure for Vitamin A deficiency. An international group of academics has generated a vitamin-enriched corn derived from South African white corn variety M37W with 169x increase in beta carotene, 6x the vitamin C and 2x folate–it is not in production anywhere, but proves that this can be done.

STRESS RESISTANCE

Plants engineered to tolerate non-biological stresses like drought, frost, high Soil salinity, and nitrogen starvation or with increased nutritional value (*e.g.* Golden rice) were in development in 2011.

HERBICIDE RESISTANCE

Tobacco plants have been engineered to be resistant to the herbicide bromoxynil. And many crops have created that are resistant to the herbicide glyphosate. As weeds have grown resistant to glyphosate and other herbicides used in concert with resistant GM crops, companies are developing crops engineered to become resistant to multiple herbicides to allow farmers to use a mixed group of two, three, or four different chemicals.

PATHOGEN RESISTANCE–INSECTS OR VIRUSES

Tobacco, corn, rice and many other crops, have been generated that express genes encoding for insecticidal proteins from Bacillus thuringiensis (Bt). Papaya, potatoes, and squash have been engineered to resist viral pathogens, such as cucumber mosaic virus which, despite its name, infects a wide variety of plants.

PRODUCTION OF BIOFUELS

Algae, both hybrid and GM, is under development by several companies for the production of biofuels. Jatropha has also been modified to improve its qualities for fuel product. Swiss-based Syngenta has received USDA approval to market a maise seed trademarked Enogen, which has been genetically modified to convert its own starch to sugar to speed the process of making ethanol for biofuel. In 2013, the Flemish Institute for Biotechnology was investigating poplar trees genetically engineered to contain less lignin so that they would be more suitable for conversion into biofuels. Lignin is the critical limiting factor when using wood to make bio-ethanol because lignin limits the accessibility of cellulose microfibrils to depolymerisation by enzymes.

PRODUCTION OF USEFUL BY-PRODUCTS

Drugs

Bananas have been developed, but are not in production, that produce human vaccines against infectious diseases such as Hepatitis B. Tobacco plants have been developed and studied, but are not in production, that can produce therapeutic antibodies.

Materials

Several companies and labs are working on engineering plants that can be used to make bioplastics. Potatoes that produce more industrially useful starches have been developed as well.

Bioremediation

Scientists at the University of York developed a weed (*Arabidopsis thaliana*) that contains genes from bacteria that can clean up TNT and RDX-explosive contaminants from the soil : It was hoped that this weed would eliminate this pollution. 16 million hectares in the USA (1.5 per cent of the total surface) are estimated to be contaminated with TNT and RDX. However the weed *Arabidopsis thaliana* was not tough enough to withstand the environment on military test grounds and research is continuing with the University of Washington to develop a tougher native grass. Genetically modified plants have also been used for bioremediation of contaminated soils. Mercury, selenium and organic pollutants such as polychlorinated biphenyls (PCBs), TNT and RDX explosive contaminants have been removed from soils by transgenic plants containing genes for bacterial enzymes.

EXTENT OF WORLDWIDE USE OF GM CROPS

In the United States, the United States Department of Agriculture (USDA) reports on the total area of GMO varieties planted. According to National Agricultural Statistics Service, the states published in these tables represent 81–86 per cent of all corn planted area, 88–90 per cent of all soybean planted area, and 81–93 per cent of all upland cotton planted area (depending on the year). USDA does not collect data for global area. Estimates are produced by the International Service for the Acquisition of Agri-biotech Applications (ISAAA) and can be found in the report, "Global Status of Commercialised Transgenic Crops : 2007". Farmers have widely adopted GM technology.

Between 1996 and 2011, the total surface area of land cultivated with GM crops had increased by a factor of 94, from 17,000 square kilometers (4,200,000 acres) to 1,600,000 km^2 (395 million acres). 10 per cent of the world's crop lands were planted with GM crops in 2010. As of 2011, 11 different transgenic crops were grown commercially on 395 million acres (160 million hectares) in 29 countries such as the USA, Brazil, Argentina, India, Canada, China, Paraguay, Pakistan, South Africa, Uruguay, Bolivia, Australia, Philippines, Myanmar, Burkina Faso, Mexico and Spain. One of the key reasons for this widespread adoption is the perceived economic benefit the technology brings to farmers.

For example, the system of planting glyphosate-resistant seed and then applying glyphosate once plants emerged provided farmers with the opportunity to dramatically increase the yield from a given plot of land, since this allowed them to plant rows closer together. Without it, farmers had to plant rows far enough apart to control post-emergent weeds with mechanical tillage. Likewise, using Bt seeds means that farmers do not have to purchase insecticides, and then invest time, fuel, and equipment in applying them. However critics have disputed whether yields are higher and whether chemical use is less, with GM crops. Land area used for genetically modified crops by country (1996–2009), in millions of hectares. In 2011, the land area used was 160 million hectares, or 1.6 million square kilometers.

In the US, by 2009/10, 93 per cent of the planted area of soybeans, 93 per cent of cotton, 86 per cent of corn and 95 per cent of the sugar beet were genetically modified varieties. Genetically modified soybeans carried herbicide-tolerant traits only, but maise and cotton carried both herbicide tolerance and insect protection traits (the latter largely the *Bacillus thuringiensis* Bt insecticidal protein). These constitute "input-traits" which are aimed to financially benefit the producers, but may have indirect environmental benefits and marginal cost benefits to consumers. The Grocery Manufacturers of America estimated in 2003 that 70–75 per cent of all processed foods in the U.S. contained a GM ingredient.

Europe has relatively few genetically engineered crops with the exception of Spain where one fifth of maise grown is genetically engineered, and smaller amounts in five other countries. The EU had a '*de facto*' ban on the approval of new GM crops, from 1999 until 2004; in a controversial move. GM crops are now *regulated* by the EU. Developing countries grew 50 percent of genetically engineered crops in 2011.

In recent years there has been rapid growth in the area sown in developing countries. A total of 29 countries worldwide grew GM crops in 2011 by approximately 16.7 million farmers and 50 per cent of GM crops grown worldwide were grown in developing countries. For example, the largest increase in crop area planted to GM crops in 2011 was in Brazil (303,000 km_2 versus 254,000 km_2 in 2010). There has also been rapid and continuing expansion of GM cotton varieties in India since 2002 with 106,000 km^2 of GM cotton harvested in India in 2011.

However the use of GM crops in India has been controversial, as discussed in detail in the GM controversies article. According to the 2011 ISAAA brief : "While 29 countries planted commercialised biotech crops in 2010, an additional 31 countries, totaling 60 have granted regulatory approvals for biotech crops for import for food and feed use and for release into the environment since 1996... A total of 1,045 approvals have been granted for 196 events (NB : an "event" is a specific genetic modification in a specific species) for 25 crops.

Thus, biotech crops are accepted for import for food and feed use and for release into the environment in 60 countries, including major food importing countries like Japan, which do not plant biotech crops. Of the 60 countries that have granted approvals for biotech crops, USA tops the list followed by Japan, Canada, Mexico, South Korea, Australia, the Philippines, New Zealand, the European Union, and Taiwan. Maise has the most events approved (65) followed by cotton (39), canola (15), potato and soybean (14 each). The event that has received regulatory approval in most countries is herbicide tolerant soybean event GTS-40-3-2 with 25 approvals (EU=27 counted as 1 approval only), followed by insect resistant maise MON810 with 23 approvals, herbicide tolerant maise NK603 with 22 approvals each, and insect resistant cotton (MON1445) with 14 approvals worldwide."

EXAMPLES OF GENETICALLY MODIFIED CROPS

Currently, there are a number of food species for which a genetically modified version is being commercially grown.

Managing Emergence of Resistance

Constant exposure to a toxin creates evolutionary pressure for pests resistant to that toxin. One method of reducing resistance is the creation of non-Bt crop refuges to allow some non-resistant insects to survive and maintain a susceptible population. To reduce the chance an insect would become resistant to a Bt crop, the commercialisation of transgenic cotton and maise in 1996 was accompanied with a management strategy to prevent insects from becoming resistant to Bt crops, and insect resistance management plans are mandatory for Bt crops planted in the USA and other countries. The aim is to encourage a large population of pests so that any resistance genes that are recessive are greatly diluted within the population.

This means that with sufficiently high levels of transgene expression, nearly all of the heterozygotes (S/s), *i.e.*, the largest segment of the pest population carrying a resistance allele, will be killed before they reach maturity, thus preventing transmission of the resistance gene to their progeny. The planting of refuges (*i. e.*, fields of non-transgenic plants) adjacent to fields of transgenic plants increases the likelihood that homozygous resistant (s/s) individuals and any surviving heterozygotes will mate with susceptible (S/S) individuals from the refuge, instead of with other individuals carrying the resistance allele. As a result, the resistance gene frequency in the population would remain low.

Nevertheless, limitations can affect the success of the high-dose/refuge strategy. For example, expression of the Bt gene can vary. For instance, if the temperature is not ideal, this stress can lower the toxin production and make the plant more susceptible. More importantly, reduced late-season expression of toxin has been documented, possibly resulting from DNA methylation of the promoter. So, while the high-dose/refuge strategy has been successful at prolonging the durability of Bt crops, this success has also had much to do with key factors independent of management strategy, including low initial resistance allele frequencies, fitness costs associated with resistance, and the abundance of non-Bt host plants that have supplemented the refuges planted as part of the resistance management strategy. Companies that produce Bt seed are addressing this as well, by introducing plants with multiple Bt proteins. Monsanto did this with Bt cotton in India, where the product was rapidly adopted.

REGULATION

The regulation of genetic engineering concerns the approaches taken by governments to assess and manage the risks associated with the development and release of genetically modified crops. There are differences in the regulation of GM crops between countries, with some of the most marked differences occurring between the USA and Europe. Regulation varies in a given country depending on the intended use of the products of the genetic engineering. For example, a crop not intended for food use is generally not reviewed by authorities responsible for food safety.

CONTROVERSY

Critics have objected to GM crops *per se* on several grounds, including ecological concerns, and economic concerns raised by the fact these organisms are subject to intellectual property law. GM crops also are involved in controversies over GM food with respect to whether food produced from GM crops is safe and whether GM crops are needed to address the world's food needs.

Chapter 27

Plant Disease Management

Plant diseases have caused severe losses to humans in several ways. Starvation and uprooting of families resulted from the Irish famine caused by potato late blight (caused by *Phytophthora infestans*). A valued resource was lost with the virtual elimination of the American chestnut by chestnut blight (caused by *Cryphonectria parasitica*).

And direct economic loss such as the estimated one billion dollars lost in one year to American corn growers from southern corn leaf blight (caused by *Cochliobolus maydis*, anamorph *Bipolaris maydis*). Many plant diseases cause less dramatic losses annually throughout the world but collectively constitute sisable losses to farmers and can reduce the aesthetic values of landscape plants and home gardens.

The goal of plant disease management is to reduce the economic and aesthetic damage caused by plant diseases. Traditionally, this has been called plant disease control, but current social and environmental values deem "control" as being absolute and the term too rigid. More multifaceted approaches to disease management, and integrated disease management, have resulted from this shift in attitude, however. Single, often severe, measures, such as pesticide applications, soil fumigation or burning are no longer in common use.

Further, disease management procedures are frequently determined by disease forecasting or disease modelling rather than on either a calendar or prescription basis. Disease management might be viewed as proactive whereas disease control is reactive, although it is often difficult to distinquish between the two concepts, especially in the application of specific measures. This topic is a general overview of some of the many methods, measures, strategies and tactics used in the control or management of plant diseases.

Specific management programmes for specific diseases are not intended since these will often vary depending on circumstances of the crop, its location, disease severity, regulations and other factors. Most states have some agency such as the

Agricultural Extension Service or State Department of Agriculture that formulates and promulgates disease management recommendations for that state. Involvement of these agencies is especially important where the practices include some regulated component such as pesticides or quarantines.

Plant disease management practices rely on anticipating occurrence of disease and attacking vulnerable points in the disease cycle (*i.e.*, weak links in the infection chain). Therefore, correct diagnosis of a disease is necessary to identify the pathogen, which is the real target of any disease management programme. A thorough understanding of the disease cycle, including climatic and other environmental factors that influence the cycle, and cultural requirements of the host plant, are essential to effective management of any disease. The many strategies, tactics and techniques used in disease management can be grouped under one or more very broad principles of action. Differences between these principles often are not clear. The simplest system consists of two principles, prevention (prophylaxis in some early writings) and therapy (treatment or cure).

The first principle (prevention) includes disease management tactics applied before infection (*i.e.*, the plant is protected from disease), the second principle (therapy or curative action) functions with any measure applied after the plant is infected (*i.e.*, the plant is treated for the disease). An example of the first principle is enforcement of quarantines to prevent introduction of a disease agent (pathogen) into a region where it does not occur.

The second principle is illustrated by heat or chemical treatment of vegetative material such as bulbs, corms, and woody cuttings to eliminate fungi, bacteria, nematodes or viruses that are established within the plant material. Chemotherapy is the application of chemicals to an infected or diseased plant that stops (*i.e.*, eradicates) the infection. Although many attempts have been made to utilise chemotherapy, few have been successful. In a few diseases of ornamental or other high value trees, chemotherapy has served as a holding action that must be repeated at intervals of one to several years. For example, antibiotics have been infused into plants to reduce severity of phytoplasma diseases of palms (lethal yellowing) and pears (pear decline) and fungicides have been injected into elms to reduce severity of Dutch elm disease (caused by *Ophiostoma ulmi*) but in all cases the chemotherapeutant must be reapplied periodically. There also are some "systemic" fungicides such as the sterol biosynthesis inhibiting (SBI) and demethylation inhibiting (DMI) fungicides that diffuse into the plant tissues to some extent and eliminate recently established infections.

One early proposal by H. H. Whetzel included four general disease control principles, exclusion, eradication, protection and immunisation (the latter principle is more appropriately called resistance since plants do not have an immune system in the same sense as animals). These principles have been expanded or altered to some extent by others. They are still valid and are detailed here but students should investigate other systems such as those proposed by Gäumann, Sharvelle, or the National Academy of Science and use the one(s) that they believe are applicable. These and other disease control principles are discussed in Maloy, Plant Disease Control (1993) cited in the general references of this lesson.

EXCLUSION

This principle is defined as any measure that prevents the introduction of a disease-causing agent (pathogen) into a region, farm, or planting. The basic strategy assumes that most pathogens can travel only short distances without the aid of some other agent such as humans or other vector, and that natural barriers like oceans, deserts, and mountains create obstacles to their natural spread. In many cases pathogens are moved with their host plants or even on non-host material such as soil, packing material or shipping containers. Unfortunately, exclusion measures usually only delay the entry of a pathogen, although exclusion may provide time to plan how to manage the pathogen when it ultimately arrives. Karnal bunt (caused by *Tilletia indica*) of wheat is an example of a pathogen originally from India that was anticipated.

Measures were established to prevent its introduction, but it finally found its way into the United States. Soybean rust (caused by *Phakopsora pachyrhizi*) has been found recently in the southeastern U.S. and precautions have been undertaken to prevent further spread. Due to its destructiveness, South American leaf blight (SALB) (caused by *Microcyclus ulei*) is a feared disease in the major rubber producing region of Indonesia, and contingency plans have been proposed to chemically defoliate rubber trees by aerial application of herbicides if the pathogen is detected. It is hoped that this would prevent establishment of the pathogen in the region.

In the United States, the Animal and Plant Health Inspection Service (APHIS), a division of the U.S. Department of Agriculture, is responsible for promulgating and enforcing plant quarantine measures. There are also state agencies that deal with local quarantines.

Internationally, eight regional plant protection organisations (PPOs) were established in 1951 by the International Plant Protection Convention sponsored by the Food and Agricultural Organisation of the United Nations. This was revised in 1997 and now includes nine regional PPOs. The European and Mediterranean Plant Protection Organisation (EPPO) is the oldest of the regional PPOs. The regional PPOs have no regulatory authority such as APHIS or other governmental agency, but function to develop strategies against the introduction and spread of pests and to co-ordinate the use of phytosanitary regulations to ensure agreement among the different member countries.

An important and practical strategy for excluding pathogens is to produce pathogen-free seed or planting stock through certification programmes for seeds and vegetatively propagated plant materials such as potatoes, grapes, tree fruits, etc. These programmes utilise technologies that include isolation of production areas, field inspections, and removal of suspect plants to produce and maintain pathogen-free stocks. Planting stock that is freed of pathogens can be increased by tissue culture and micropropagation techniques as well as be maintained in protective enclosures such as screenhouses to exclude pathogens and their vectors. Exclusion may be accomplished by something as simple as cleaning farming equipment to remove contaminated debris and soil that can harbour pathogens such as *Verticillium*, nematodes or other soilborne organisms and prevent their introduction into non-infested fields.

ERADICATION

This principle aims at eliminating a pathogen after it is introduced into an area but before it has become well established or widely spread. It can be applied to individual plants, seed lots, fields or regions but generally is not effective over large geographic areas. Two large attempts at pathogen eradication in the United States were the golden nematode (*Globodera rostochiensis*) programme on Long Island, New York and the citrus canker (caused by *Xanthomonas axonopodis* pv. *citri* and pv. *aurantifolii*) programme in Florida. However, neither of these attempts was a lasting success.

Eradication of the golden nematode involved removing infested soil, fumigating soil in infested fields and eventually abandoning infested potato fields for housing developments and other uses. Citrus canker eradication involved widespread removal and burning of diseased trees and, in some cases, destruction of entire citrus groves and nurseries. The disease appeared to be contained and the pathogen eradicated, but the disease has reappeared and new attempts at eradication are ongoing.

Eradication can also be on a more modest scale such as the removal of apple or pear branches infected by the fire blight bacterium (*Erwinia amylovora*) or pruning to remove blister rust cankers (caused by *Cronartium ribicola*) on white pine branches. Or, it can be the sorting and removal of diseased flower bulbs, corms or rhizomes. Hot water seed-treatment of cereal seeds to kill smut mycelium in the seed and heat treatment to eliminate viruses from fruit tree budwood for grafting are other examples of pathogen eradication.

Two programmes that are actually forms of protection and not pathogen eradication are barberry eradication for reducing stem rust (caused by *Puccinia graminis*) of wheat and *Ribes* eradication for preventing white pine blister rust. The strategy is that removing these alternate hosts breaks the disease cycles and prevents infection of the economically more valuable host. These two examples are mentioned here because they are frequently cited as eradication measures.

However, stem rust can readily spread from wheat to wheat in many regions by the uredinial stage although elimination of the aecial host, barberry, may deter or diminish the development of pathogenic races of the rust. The white pine blister rust fungus is perennial in the pine host and eradication of the alternate host only protects non-infected trees but does not necessarily eliminate the pathogen from the area. Eradication may also be accomplished by destroying weeds that are reservoirs of various pathogens or their insect vectors. Elimination of potato cull piles is an effective method of eradicating overwintering inoculum of the late blight pathogen.

Soil fumigation has been a widely used eradication strategy. This technology involves introducing gas-forming chemicals such as carbon disulfide, methyl bromide, or chloropicrin into soil to kill target pathogens. However, undesirable side effects such as killing beneficial organisms, contamination of groundwater, and toxicity of these chemicals have resulted in less reliance on this approach for disease management. Volatile fumigants like methyl bromide are injected into soil and sealed with a plastic film. Some water-soluble fumigants like metam-sodium can be injected into the soil and the soil simply compacted to form a seal.

Crop rotation is a frequently used strategy to reduce the quantity of a pathogen, usually soil-borne organisms, in a cropping area. Take-all of wheat (caused by *Gaeumannomyces graminis*) and soybean cyst nematode (*Heterodera glycines*) are two examples of soilborne diseases that are easily managed by short rotations of 1 and 2 years, respectively, out of susceptible crops, which may include susceptible weed hosts such as grasses in the case of take-all.

Burning is an effective means of eradicating pathogens and is often required by law to dispose of diseased elm trees affected by Dutch elm disease (DED), citrus trees infected by citrus canker or of bean fields infected by halo blight bacteria (*Pseudomonas syringae* pv. *phaseolicola*). Propane flaming can effectively destroy *Verticillium* microsclerotia in mint stems, and flaming potato stems prior to harvest may prevent tuber infection by the late blight pathogen. However, burning agricultural fields is controversial because the smoke creates human health and safety and environmental concerns.

Students may note that the principle of eradication is a good example of the conflicting concepts of some of these general principles for at least two reasons. One is that some of the examples above could be placed as readily under protection as under eradication. The second is that complete eradication of pathogens, especially from large areas is rarely accomplished.

PROTECTION

This principle depends on establishing a barrier between the pathogen and the host plant or the susceptible part of the host plant. It is usually thought of as a chemical barrier, *e.g.*, a fungicide, bactericide or nematicide, but it can also be a physical, spatial, or temporal barrier.

The specific strategies employed assume that pathogens are present and that infection will occur without the intervention of protective measures. For example, bananas are covered with plastic sleeves as soon as the fruit are set to protect the fruit from various pests including fruit decay fungi.

Protection often involves some cultural practice that modifies the environment, such as tillage, drainage, irrigation, or altering soil pH. It may also involve changing date or depth of seeding, plant spacing, pruning and thinning, or other practices that allow plants to escape infection or reduce severity of disease. Raising planting beds to assure good soil water drainage is an example of cultural management of plant diseases such as root and stem rots.

Fungicides have been used for more than a hundred years and new fungicides continue to be developed. Bordeaux mixture, a basic copper sulfate fungicide, was the first widely used fungicide and is still used today in various forms. The earliest fungicides were simple elements like sulphur or metallic compounds of copper or mercury, and these are generally classed as inorganic fungicides. In the early to mid-1900s organic fungicides such as thiram, captan, and the bisdithiocarbamates were developed. These are broad-spectrum, contact or protectant fungicides that control a wide range of fungal diseases.

Starting in the 1960s the "systemic" fungicides were developed. Most of these are not truly systemic in plants but have some limited mobility, usually translaminar, and often give some post-infection benefits. Some of the "systemic" fungicides move upward in the plant's vascular system, but currently only one (fosetyl-Al) has ambimobile distribution (both upward and downward) that would constitute a truly systemic fungicide. In addition to the SBI and DMI fungicides mentioned earlier, a recent group of systemic fungicides are the strobilurins. Some fungicides have narrow ranges of activity and are used primarily for control of specific groups of diseases such as downy mildews, rusts, smuts or powdery mildews while others are active against a wider range of diseases.

One liability of these recent narrow-range fungicides is that they often have single-site modes of action, (that is, their site-specific activity is controlled by one or a few genes), and thus are especially prone to development of fungicide resistance in the pathogen.

Several management strategies have been developed to combat fungicide resistance. These include using mixtures of single-site and multi-site fungicides, alternating applications of fungicides with different modes of action, applying fungicides only when needed instead of on either a calendar or prescription basis, and applying the recommended dosage and not attempting to cut costs by reducing the recommended amount of fungicide applied.

Fungicides can be applied by any of several methods : ground sprayers, airplanes or through irrigation systems, but to be effective applications must be done properly. First, the fungicide must be legally registered for use on the plant involved and against the target disease.

Several different chemicals may be registered for the same crop or disease. If the different fungicides are similar in effectiveness, cost, ease of application, and safety, then timing of application becomes the most critical factor. If applied too early much of the chemical will be wasted before it can be effective; if applied too late, it will be largely ineffective. The benefits of properly applied fungicides can often be striking. Distribution of the spray droplets is important; the finer the spray the more complete the coverage on the plant surface. However, very small droplets form a mist that is easily displaced by wind.

Many cultural practices can be modified to manage the occurrence, intensity or severity of plant diseases. These include selection of suitable growing sites for the crop, adequate tillage to bury pathogen-infested plant residues, rotation to non-susceptible crops, selecting pathogen-free planting stocks, orientation of plantings to improve exposure to sun and air currents, pruning and thinning to eliminate sources of infection and improve aeration in and around susceptible plants, water management on both plants and in soil, adequate nutrition, proper cultivation to improve root growth and avoid plant injury, and sanitation procedures to eliminate sources of inoculum.

Biological control involves the use of one living organism to control another, and this management technology has received much attention in recent times. However, the number of biological agents registered for use is relatively small, success has been limited, and application has been largely restricted to intensively managed, high value crops such as greenhouse plants. Two examples of effective biological control are the use of the fungus *Peniophora gigantea* to inoculate tree stumps to prevent infection of adjacent trees by the wood decay fungus *Heterobasidion annosum*, and the application of the non-pathogenic (*i.e.*, non-tumour-producing) bacterium *Agrobacterium radiobacter* to fruit trees before planting to prevent infection by the crown gall bacterium (*Agrobacterium tumefaciens*).

RESISTANCE

Use of disease-resistant plants is the ideal method to manage plant diseases, if plants of satisfactory quality and adapted to the growing region with adequate levels of durable resistance are available. The use of disease-resistant plants eliminates the need for additional efforts to reduce disease losses unless other diseases are additionally present. Resistant plants are usually derived by standard breeding procedures of selection and/or hybridisation. A few disease-resistant lines have been obtained by inducing mutations with x-rays or chemicals. There is also interest in chemicals called "plant activators" that induce plant defence responses called systemic acquired resistance (SAR) and induced resistance. Recently, resistant plants have been developed through the use of genetic engineering (*e.g.*, resistance to the *Papaya ringspot virus*).

Selection of resistant plants involves subjecting plants to high levels of disease pressure and using the surviving plants as sources of disease resistance. Plants that survive this pressure often have genetic resistance that can be utilised directly by propagation or as sources of resistance to develop resistant plants that also have the requisite qualities for that crop. Hybridisation is a tactic where a plant having the desired agronomic or horticultural qualities, but is susceptible to a disease, is crossed with a plant that is resistant but which may or may not have the other desirable characteristics such as size, yield, flavour, aesthetics, etc.

Disease escape occurs when susceptible plants do not become diseased for some reason. This may be due to some anatomical or physical character, such as the occurrence of leaf hairs, thick cuticle, or modified stomata, or they may be environmental, in which conditions are not conducive to disease development. Although disease escape based on some anatomical feature is useful occasionally, escape more often complicates the process of developing disease resistant plants.

Development of disease-resistant plants has been relatively successful with annual and biennial plants, but less so with perennials, primarily because of the longer time required to develop and test the progeny. Woody perennials, such as ornamental, forest, and orchard trees, have been especially difficult for plant breeders to develop useful disease resistance. For example, chestnut blight and Dutch elm disease have devastated two valued native trees. In both cases there have been extensive attempts to develop resistant trees, usually by creating hybrids with exotic chestnut or elm trees, and some resistant selections have resulted.

Unfortunately, these generally lack the desirable qualities, such as nut flavour or tree forms characteristic of the native trees. Another introduced disease that has impacted native trees is white pine blister rust. There has been an intense effort for more than 50 years to select and improve rust-resistant pines from the surviving population. These trees are now being planted for reforestation, but it will be another 50 or so years, when these trees have matured to produce a timber crop, before the success of this programme is known. Development of resistance has been most successful against the more specialised pathogens such as rust fungi, smut fungi, powdery mildew fungi, and viruses, but less so against general pathogens such as many blight, canker, root rot and leaf spotting pathogens.

A major problem with genetically resistant plants is that host-differentiated pathogenic races can be selected, so that many breeding programmes become continuous processes to develop disease resistant plant lines. Disease resistance conferred by a single major gene is sometimes called specific or qualitative resistance and is race-specific. This type of resistance is often unstable, and emergence of a pathogenic race that can attack that genotype can completely overcome this type of resistance. Quantitative resistance or general resistance derives from many different genes for resistance with additive effects to provide more stable (or durable) resistance to pathogens.

There are several strategies to minimise this race development and resistance failure. These include methods of gene deployment, where different genetic plant types are interspersed on a regional basis to avoid a genetic monoculture, or planting mixtures of cultivars having different genetic compositions to ensure that some component of the crop will be resistant to the disease.

A recent and controversial technique in developing disease resistant plants is the insertion of genes from other organisms into plants to impart some characteristic. For example, genes from the bacterium *Bacillus thuringiensis* have been inserted into plants to protect against insect attacks. Plants with these inserted genes are called genetically-modified organisms (GMOs), and have caused concern that unanticipated, and perhaps detrimental, characteristics, such as unforeseen allergens, may also be transferred to the new plants. However, unforeseen and undesirable qualities also can be transmitted by conventional plant breeding techniques.

The potato cultivar Lenape was developed in part because of its resistance to *Potato virus A* and resistance to late blight tuber infection. After it was released it was discovered that the tubers contained very high levels of solanine, a toxic alkaloid. The wheat cultivar Paha had resistance to stripe rust (caused by *Puccinia striiformis*) but also was very susceptible to flag smut (caused by *Urocystis agropyri*). Both of these plant cultivars, developed by conventional breeding methods, were quickly taken out of production. There is much interest in the genetic engineering of disease-resistant plants and some success has been obtained with several virus diseases, the best known of which is papaya ringspot. This approach to plant disease management will likely expand, especially for widely grown crops such as wheat, corn, soybeans, rice, and the like, as social, legal, and economic obstacles are overcome.

INTEGRATED DISEASE MANAGEMENT

Integrated Disease Management (IDM) is a concept derived from the successful Integrated Pest Management (IPM) systems developed by entomologists for insect and mite control. In most cases IDM consists of scouting with timely application of a combination of strategies and tactics. These may include site selection and preparation, utilising resistant cultivars, altering planting practices, modifying the environment by drainage, irrigation, pruning, thinning, shading, etc., and applying pesticides, if necessary. But in addition to these traditional measures, monitoring environmental factors (temperature, moisture, soil pH, nutrients, etc.), disease forecasting, and establishing economic thresholds are important to the management scheme.

These measures should be applied in a co-ordinated integrated and harmonised manner to maximise the benefits of each component. For example, balancing fertilizer applications with irrigation practices helps promote healthy vigourous plants. However, this is not always easy to accomplish, and "disease management" may be reduced to single measures exactly the same as the ones previously called "disease control." Whatever the measures used, they must be compatible with the cultural practices essential for the crop being managed.

Chapter 28

PLANT DISEASE DIAGNOSIS

A plant pathologist or a student taking plant pathology is often asked by friends or associates the following questions. What is wrong with my plant; followed by, what can I do to get rid of the problem? It may be too late to help the specific plant when the question is asked, but proper diagnosis may be extremely important in preventing the problem on other plants or in preventing the problem in the future.

How does a plant pathologist or a plant pathology student go about diagnosing plant problems? The diagnostician must have very good observation skills, and s/he also needs to be a good detective. It is important to keep an open mind until all of the facts related to the problem can be collected. The possibility of multiple causal factors must also be considered.

Control measures depend on proper identification of diseases and of the causal agents. Therefore, diagnosis is one of the most important aspects of a plant pathologist's training. Without proper identification of the disease and the disease-causing agent, disease control measures can be a waste of time and money and can lead to further plant losses. Proper disease diagnosis is therefore vital.

Often, plant pathologists have to rely on symptoms for the identification of a disease problem. For example, Dr. Shoe is asked by Ms. Green to examine azaleas in her nursery. When Dr. Shoe gets to the nursery, he observes that the azaleas in Greenhouse 1 are wilted. When he removes plants from their pots, the roots appear to be rotten. Ms. Green wants to know right away what she should do with the azaleas in Greenhouse 2 where no wilting is currently being observed. Dr. Shoe is asked to make recommendations even before he has looked at plants in the second greenhouse. Because similar symptoms can be produced in response to different causal agents, the use of symptoms alone is often an inadequate method for disease identification. The identification of the disease-causing agent may take a week or more. What can Dr. Shoe do for Ms. Green now?

One of the most important things is for Dr. Shoe to use his powers of observation. He needs to ask many questions related to the azalea's care and culture in order to eliminate or identify possible causes of the problem. He also needs to consider various environmental and cultural factors.

As a result of his questions and observations he may :

- Be able to identify a disease and disease-causing agent,
- Be able to narrow the problem down to several possibilities which will require further study in the laboratory before he can make a final diagnosis, or
- Be completely baffled by the problem.

Regardless of the outcome, Ms. Green still expects a recommendation as to what she should do now. This article presents the various steps/activities which are associated with accurate plant disease diagnosis. The process may vary with different diseases and conditions but the overall process is relatively consistent. The steps all require careful observations and questions. The steps include :

KNOW WHAT IS NORMAL

Proper plant identification. Identification of affected plants is one of the first steps in diagnosing a plant disease. Both scientific and common names of the plant should be noted. Common names should not be relied upon since some distinctly different plant species may have the same common name, and the common name used in one area may be used for a completely different species in another area. The common name "vinca" has been used to describe plants belonging to two different genera, *Vinca,* a perennial, and *Catharanthus,* an annual. Another example is "monkey grass" which is used to describe *Liriope* and *Ophiopogon* (mondo grass). An example from forestry is "cedar" which is used to describe eastern red cedar (*Juniperus*), western red cedar (*Thuja*), Port Orford cedar (*Chamaecyparis*), incense cedar (*Libocedrus*), and Atlas cedar (*Cedrus*). Obviously the use of common names can cause confusion in identification and recognition of problems.

In addition to knowing the common and scientific names of an affected plant, it is important to know the specific variety or cultivar, whenever possible. A great variation in susceptibility to a specific disease may occur within different cultivars of a plant species. For example, when we look at the susceptibility of wheat to wheat stem rust caused by *Puccinia graminis* f. sp. *tritici,* we know that all wheat cultivars are not susceptible to all races of *P. graminis.* The major control measure for this disease is based on planting wheat cultivars each year that are resistant to the pathogen races that are predicted to be present during the growing season. Tomato cultivars having the "Better Boy" genetic background are generally resistant to root-knot nematodes while those with the genetic background of the variety "Rutgers" are susceptible, so knowing the genetic background of a cultivar can be important. Knowing the cultivar and its susceptibility to various diseases can narrow down the possible disease agents to consider.

Knowing the identity of the plant species affected allows the pathologist to utilise various resources that contain lists of plant diseases associated with specific plants. These lists are very helpful in suggesting possible pathogenic agents. Once you select the plant of interest, you will see a list of bacterial, fungal, nematode, parasitic plants and viral diseases associated with the specific plant. *Westcott's Plant Disease Handbook* is useful because specific symptoms are associated with each disease. APS Press has published a list of fungal diseases and hosts. This book followed a previous publication by USDA, but the APS Press publication only includes fungal pathogens.

The original USDA publication, while now somewhat outdated, included fungal, bacterial, viral and nematode pathogens, as well as physiological problems. Other resources are available which include the APS Press Compendium series on diseases and disorders for specific plant species, such as roses or diseases for specific regions, such as Florida, USA. In some cases these lists of plant diseases may suggest potential disease possibilities or they may lead the diagnostician to rule out other diseases. One factor to keep in mind, however, is that these lists are often incomplete or the disease may be new and unreported on the plant or in a specific region. The best possible option is to utilise several different resources since one may not have a complete index of potential diseases on a specific plant.

Recognise healthy plant appearance. It is important to know the normal appearance of the plant species you are investigating. Each plant species has special growth habits, colours and growth rates. If you do not know what to expect of the plant you cannot recognise when something is wrong. Does the plant normally have new foliage that is yellow or red and becomes darker green as the foliage ages? Many ornamental shrubs have been developed and marketed for the ornamental value of such brightly coloured new growth.

These plants are highly prised for this colouration; however, if an individual does not know that this colouration is the normal appearance of the plant, s/he may think that the plant is diseased. It is important to know what the normal appearance of a plant is before you decide there is a problem. It is also important to remember that appearance can vary with different cultivars. Some plant cultivars have naturally yellow to pale green leaves (*e. g.* new hosta cultivars, herbs like golden oregano, and coleus varieties) which at first glance appear to have symptoms of under-fertilization, root stress or soil pH problems.

Once the "normal" appearance of the specific plant is determined, several comparisons can be made between the problem plants and healthy plants. Compare characteristics such as overall size, shape, and colouration; leaf shape, size, colouration, and distribution; root distribution and colouration; and bark, stem or trunk texture and colouration. It is also important to note normal events, such as leaf drop, that may occur in a healthy plant. For example, some holly species normally drop leaves in the spring.

The affected parts of the plants should also be noted. Are there symptoms on the roots, leaves, stems, flowers, or fruit? Is the entire plant involved? Is only one limb or side of a plant involved? Answers to these questions can assist in the identification of the problem.

CHECK FOR SYMPTOMS AND SIGNS

Identify characteristic symptoms. Describing the characteristic symptoms exhibited by a specimen can be very difficult to do accurately. Because of this, it is often difficult, if not impossible, to determine what is wrong with a plant when a person is describing symptoms over the phone. As a test of this you may want to take a plant exhibiting symptoms and have three different individuals describe the symptoms that they observe on a sheet of paper. Next, compare the descriptions. Do the descriptions vary significantly? Could you visualise the symptoms by the way any one of the individuals described the diseased plant?

- Underdevelopment of tissues or organs. Examples include such symptoms as stunting of plants, shortened internodes, inadequate development of roots, malformation of leaves, inadequate production of chlorophyll and other pigments, and failure of fruits and flowers to develop.
- Overdevelopment of tissues or organs. Examples include : galls on roots, stems, or leaves, witches' brooms, and profuse flowering.
- Necrosis or death of plant parts. These may be some of the most noticeable symptoms, especially when they affect the entire plant, such as wilts or diebacks. Other examples include shoot or leaf blights, leaf spots, and fruit rots.
- Alteration of normal appearance. Examples include mosaic patterns of light and dark green on leaves, and altered colouration in leaves and flowers.

Diseases also involve a progression of symptoms that can vary significantly. The progression of symptoms is one of the most important characteristics associated with problems caused by biotic agents. Diseases can result in primary and secondary symptoms. For example, decayed roots on a tree may be a primary symptom while the toppling over of the tree or windthrow is a secondary symptom. At later stages of a disease, secondary invaders may also obscure the original disease symptoms so that symptoms observed at the later stages of the disease are not typical of the symptoms developed in response to the original pathogen.

It is important to look for a progression of disease symptoms in plants exhibiting problems. In some cases, such as improper herbicide usage, symptoms observed may be similar to spots present as a result of an infectious agent. The difference is that with herbicide injury, the symptoms usually appear suddenly and there is no observable progression of symptoms. The spots may also follow spray patterns of the herbicide. Herbicides, such as 2,4-D, can cause leaf distortion which may be confused with viral diseases. However, when new leaves form, they will generally be free of symptoms, indicating a lack of symptom progression.

Identify symptom variability. Variations in symptoms expressed by diseased plants may lead to an improper diagnosis. These variations can result from a couple of factors. It is possible that there is more than one problem present, and in some cases there may be more than one pathogen infecting a plant. Symptoms associated with these infected plants may be significantly different from the symptoms expressed in response to each of the different pathogens acting separately.

The disease symptoms exhibited by multiple pathogens infecting a plant may be either more severe or less severe than if the plant were infected with just one of the pathogens. This is commonly seen in multiple infections due to viruses. The seedling on the left is infected with both *Prune dwarf virus* and *Prunus necrotic ringspot virus*. The seedling in the middle is infected with *Prune dwarf virus* alone and the seedling on the right is infected with *Prunus necrotic ringspot virus* alone.

Look for signs of biotic causal agents. Signs of plant disease agents are the observable evidence of the actual disease-causing agent. Signs may include the mycelia of a fungal agent, fungal spores, and spore-producing bodies. Indications of insects causing problems may include the actual insect, insect frass, mite webbing, and insect eggs. Signs are much more specific to disease-causing agents than are symptoms and are extremely useful in the diagnosis of a disease and identification of the agent causing the disease.

The use of a hand lens and a knife can be valuable for a diagnostician in the field. Cutting into the bark of ornamental plants and trees at the soil surface may lead to the observation of mycelial mats of root rot fungi such as *Armillaria* spp.. Bacterial ooze can be observed by cutting stems and placing them in water. Masses of different spores such as rust spores on leaves can also be important in disease diagnosis. Powdery mildews are typically diagnosed by the observation of the gray to white mycelia and conidia observed on the surface of leaves and flowers.

Dissecting and compound microscopes are useful for the observation of specific spores and spore structures, and can lead to further identification of possible disease agents. Knowledge concerning the use of microscopes and a hand lens is vitally important to the diagnostician. Signs of plant disease agents can often be overlooked unless careful observations are conducted. Signs are not visible when taking a quick ride by plants looking through the windshield of a truck and may not even be visible to the naked eye.

Identify Plant Part Affected-Are symptoms associated with specific plant parts? It is important to note if the symptoms observed are associated with specific plant parts. For example, is a wilt observed correlated with a disruption of the vascular system which may be indicated by browning of the vascular system or are the roots of the plants abnormal including rots, decreased feeder roots, etc.; are necrotic lesions observed strictly on younger leaves? The symptoms of some diseases are most commonly seen on specific plant parts and this observation can be important in diagnosis.

OBSERVE PATTERNS

Check distribution of symptoms. One of the first things that a diagnostician should note is how the diseased plants are distributed over the affected area. Are they distributed uniformly across an area or are they localised? Is there a definite pattern to the distribution? For example, does it occur only along the edges of a greenhouse near open windows, next to roadways or driveways, in low spots of a field, along a planted row, or is it affecting plants at random in a field? This distribution can be especially important in looking at the possibility of non-infectious problems, such as improper herbicide use or various soil factors. A uniform pattern on an individual plant and uniform damage patterns over a large area are generally not associated with biotic agents, but are usually due to abiotic agents.

How prevalent is the problem? Are all plants affected? Infectious problems generally occur over time and there is a progression of symptoms. Rarely will all of the plants be affected. Generally, disease problems caused by biotic agents will be observed when they are causing problems on a low percentage of plants at least at the start of the disease, unless there were extenuating circumstances, such as the use of infected seeds. Even then, rarely will 100 per cent infection be observed. When a problem appears in 100 per cent of the plants, it more commonly results from factors such as soil conditions (deficiencies or toxicities), adverse climatic factors (cold temperatures, hail, drought, etc.), or toxic chemicals (improper pesticide use, growth regulators, air pollutants, such as ozone, etc.).

What has been the progression of symptoms on plants in the affected area? If the symptoms all appeared at the same time and there has been no further development of symptoms, this would indicate a possible episodic event such as a change in temperature or possible improper chemical usage. If however, the symptoms started in one area and slowly spread to other areas and the severity of disease symptoms changed over time, this would be more indicative of the presence of a biotic agent. Biotic agents can also include insects and mammals, such as voles, which may be feeding on plants in an area.

Check for host specificity. Is the problem occurring in only one plant species or are different plant species affected? If different plant species are affected, this suggests the possibility of a non-infectious problem which could be related to cultural or environmental problems. However, Phytophthora and Pythium root rots can cause problems on many different plant species; therefore, the fact that more than one plant species is affected does not completely eliminate infectious agents. If there is more than one species of plant involved, are these plants closely related and can they be infected by a common pathogen?

ASK QUESTIONS

Review the cultural practices and growing environment. It is vital that a diagnostician question the activities that have been conducted around the affected plants. The problem may not be due to anything that the grower has done; the problem could be related to what his/her neighbour has done. Information pertaining to the growing environment to which the affected plant has been exposed is a vital piece of the puzzle. It is especially important to document changes in the environment. Environmental factors to consider include : extreme temperatures (freesing and heat), rainfall, hail, lightning, prolonged drought, temperature inversions (important in possible air pollutant damage and pesticide drift) and prevailing winds. All of these abiotic factors can be important to the problem. Site factors such as soil type, possible drainage problems, and soil pH should also be evaluated.

Cultural and maintenance activities can be significant. What pesticides or other chemicals have been applied? At what rate and when were they applied? Who applied the chemicals? What equipment was used in its application? What other activities have occurred? Has someone been mowing in the area? Has the highway department been working along the roadway, possibly applying herbicides? Have any unusual occurrences or weather patterns been noted? Many times careful investigation by the diagnostician is required because, in some cases, someone may have done something improperly and may be unwilling to admit their error.

LABORATORY TESTS

Sometimes neither symptoms nor signs provide enough specific or characteristic information to decide the cause of an infectious plant disease. In such cases, it may be necessary to bring a sample back to the laboratory for further tests to isolate and identify the causal agent. This can be a time-consuming and labour-intensive process that takes specialised skills.

Incubation of plant material. One of the first steps when getting back to the laboratory may be to place a sample of the diseased tissue under conditions that will allow an infectious agent to grow and possibly induce sporulation. This can be accomplished by placing a leaf in a moist chamber. A moist chamber can be a sterile petri dish containing a wet filter paper in the bottom of the dish and a triangle of glass tubing on which the sample is placed so that the sample is not directly on the wet filter paper but is exposed to humid conditions. This type of moist chamber will work for small and relatively flat specimens such as leaves. Plastic bags or boxes may be necessary for larger specimens. Saprophytes that are present on the specimen can also be encouraged to grow in a moist chamber and a brief surface swab with 70 per cent isopropanol or 0.1-1 per cent sodium hypochlorite may be useful in reducing these saprophytes. Moist chambers are generally incubated at room temperature.

Isolation and identification of biotic plant disease causal agents. Isolation of fungi usually requires that pieces of infected plant tissue be placed on various nutrient media. The organism that grows out of this tissue is then isolated in pure culture. Bacteria are often isolated by chopping up infected tissue in a small amount of sterile water. This water : bacteria suspension is then streaked onto a bacteriological medium such as nutrient agar. Several problems can occur when trying to isolate the plant pathogenic agent. The infected plant tissue may contain one or more saprophytes which have moved into the infected tissue. These saprophytes may outgrow the plant pathogen on the nutrient medium, obstructing accurate identification of the pathogen. In some cases where a specific plant pathogen is suspected, a medium selective for the suspected pathogen may be utilised. It is also beneficial to attempt to isolate the plant pathogen from the margins of the diseased tissue where the pathogen is more numerous or more active than saprophytes that quickly colonise the recently killed tissue.

Once an organism is isolated, is that organism the true cause of the problem? Conducting Koch's postulates, which involves the inoculation of healthy plants, may be necessary to conclusively answer this question, especially if the organism has not been previously reported as a plant pathogen on that host. Koch's postulates are seldom conducted for routine diagnoses, but may be extremely important for new diseases and for research. Inoculation of a healthy host and obtaining the symptoms originally observed in the field may be difficult. This may be due to problems in replicating the conditions through which the host was inoculated and also in reproducing the environmental conditions present when the host became infected. It is often impossible to replicate in the laboratory the original conditions present during the disease development.

Once an infectious plant pathogen is successfully isolated, the organism must be identified. There are estimated to be some 1.6 million fungal species, most of which are not infectious pathogens. Many fungi and bacteria have never been isolated and identified. The characteristics upon which their identification is based are often complex and specialised training is necessary to be able to identify these fungi and bacteria. Diagnosticians with experience are often able to identify the most commonly isolated organisms. Identification of plant pathogenic nematodes also requires a trained individual.

Diagnostic tests for identification of biotic causal agents. A major problem in identification of biotic causal agents is the inability of some infectious pathogens to grow on artificial media. Viruses, as well as some fungi (*e.g.* powdery and downy mildew causing agents) and some prokaryotes (*e.g.* phytoplasmas), require a living host in order to grow. In cases where the plant pathogen is difficult or impossible to grow on artificial media, other methods may be used for their detection, such as the use of serological tests for viruses.

Viral identification is often accomplished utilising ELISA (enzyme-linked immunosorbent assay) which is based on the binding of an antibody produced to a specific virus with the virus in the infected plant material. More tests are currently being developed using the polymerase chain reaction (PCR) for detection of specific organisms. These types of reactions take specialised equipment and reagents, and the tests are not commonly done outside diagnostic and research laboratories. Other techniques used for the identification of viruses include negative staining and electron microscopy to view the viral particles in plant tissue or suspensions.

PCR and ELISA tests, as well as other laboratory tests, may be used for organisms that will grow on artificial media. Additional tests may include analysis of fatty acids of organisms, carbohydrate utilisation (*i.e.* BIOLOG test), and enzyme activity testing (*i.e.* pectinase, isozyme patterns). Diagnostic tests for identification of abiotic plant disease causal agents. It is extremely important to look for abiotic factors that may be important in observed symptoms. Soil and water tests may be necessary to determine pH, nutrient composition, salinity, and other factors such as pesticide residues that may induce various symptoms. It may also be important to get samples of plant tissue analysed for nutrient content to determine if there are macro- or micronutrient deficiencies or toxicities.

FINAL DIAGNOSIS

Diagnosis is a form of hypothesis testing, where the hypothesis is simply the identity of the disease, and a good diagnostician goes through multiple iterations of the scientific method (seeking evidence through testing that supports or refutes the hypothesis that s/he generates). These hypotheses are generated through observations of the plant, environment, and information from the grower. When all of the information is successfully collected, literature sources should be consulted to determine what is already known about diseases and disease-causing agents associated with the identified plant.

Information can be obtained from published resources including plant disease compendia, plant disease indexes, technical notes, commodity newsletters, online resources, and personal communication with plant disease experts. When no information is available on the specific plant, information on diseases and disease causing-agents of similar plants may be useful. There may also be rare cases where no information is available related to the disease. Then, extensive testing may be necessary to determine the identification of the plant pathogen. When this type of testing is required, it may take a long time to develop research-based control recommendations and control measures may have to be based on diseases of similar etiology. If these diseases have occurred in other areas of the world, control measures that have been previously developed in other areas may be useful.

The student should keep in mind that s/he is a detective. Causal agent identification and diagnosis of plant problems is just like a detective investigating an assault or murder case, only in this case, the victim is a plant. All clues should be investigated. Some clues may lead down blind alleys while others will lead down the correct path. It is important to note that there are exceptions that exist and these exceptions must be considered. It is the compilation of the information and clues that will ultimately lead to the most accurate diagnosis. Images such as these symptoms and signs are often used in the diagnostic process. Studying these images may assist the diagnostician in narrowing down the possible diseases to consider and others which can be eliminated.

COMMONLY OBSERVED SYMPTOMS AND SIGNS

- Fungal leaf spots : Spots usually vary in size. Generally are round and occasionally elongate on stems. Zones of different colour or texture may develop giving the spot a "bull's eye" effect. Spots are not limited by leaf veins.

- Bacterial leaf spots : Spots are often angular due to limitation by leaf veins. Colour is usually uniform and no signs of plant pathogen are evident. Tissue may appear initially as being water soaked but may become papery as it dries.
- Vein banding : Vein banding occurs when there is a band of yellow tissue along the larger veins of the leaf. This symptom is observed with viral diseases and is in contrast with nutrient deficiencies which may cause a dark green band along leaf veins.
- Mosaic and Ringspot : Mosaic and ringspot are used to describe an irregular patchwork of green and yellow areas over the surface of a leaf. In some cases leaves may also become distorted. Often these symptoms are associated with viral pathogens. There is not a sharp margin between the affected and healthy areas. Distinct margins may indicate a nutritional problem or genetic variegation.
- Leaf Distortion : Leaves of the infected plant may be distorted from their normal shape and size. Leaves may be elongated, smaller size, or thickened. This type of symptom can be associated with viral, fungal or bacterial infections as well as insect and mite infestations.
- Powdery mildew : Can affect leaves, stems, flowers and fruits with a white to gray surface coating of mycelia which can be rubbed off. Black specks may later develop in the mycelia. These specks are mature cleistothecia, the overwintering fungal structures which contain ascospores. Tissue may turn yellow, reddish or remain green under the mycelia and some leaf distortion may be observed especially on actively growing tissues.
- Presence of Spores/Spore Structures-Several fungal diseases can be easily identified based on the presence of spores or spore structures on the leaf surface. Some examples of this are rusts which are often recognised by the rusty brown to black spores and smuts which are identified by the black spores which often replace the seed structure.
- Needle drop in conifers : Conifers normally retain their needles for several years but these needles will eventually be lost. This drop is gradual and production of new needles obscures the loss of older needles. Unfavourable growing conditions, such as drought, may cause an acceleration of needle drop. If the drop occurs in only older needles especially during unfavourable growing conditions, there is no need for concern. If new needles are lost then other factors may be involved such as needle cast fungus, nutrient deficiencies, or toxic chemicals.
- Chemical spray or air pollutant injury : Spots associated with injury are relatively uniform in colour and the interface between the affected and healthy area is usually sharp. Distribution on plant may be associated with where spray or pollutant comes in contact with the plant.
- Leaf or Needle Tip Death : Death at the tips of needles and tips and margins of leaves often indicate unfavourable climatic conditions, toxic chemicals or root malfunction due to poor cultural practices. Air pollutants, soil chemicals, and excess fertilizer can cause tip burn. Drought and freesing may have

a similar effect. All needles of a specific growth period are usually affected. Needles infected by foliar fungal diseases are generally more scattered and rarely are all needles of a particular growth period killed. Needles diseased by infectious agents are generally affected over varying lengths and are often straw yellow or light tan. Fungal fruiting bodies may also be observed.

- Soil or air chemical injury : Chemicals which are absorbed from the soil by roots or absorbed from air through leaves may exhibit a burning or scorching of leaf margins. If severe, islands of tissue between the veins may also be killed. Dead tissue may drop out of the leaf leaving a ragged appearance. Other chemicals may cause a distortion of leaf shape and size.
- Cankers : Cankers are localised necrotic lesions which are often sunken in appearance. Cankers can result from mechanical injury (*e.g.* trees which have been damaged by collisions with cars or lawnmowers), and various fungi or bacteria. In the spring, ooze may be observed on the surface of bacterial cankers and fruiting bodies may be observed on the surface of fungal cankers.
- Fruit Decays and Rots : Various fungi and bacteria can cause rots of fruit. These are often distinguished by the colour, lack of firmness of tissue, and signs of spores or fruiting bodies.
- Fruit Discolouration : Discolouration of fruit is often associated with viral infections. This discolouration may be similar to mosaic and ringspot symptoms observed on leaves.
- Wilts : Wilts are characterised by a general loss of turgidity of leaves or possibly entire plants due to the loss of water. The loss is most often caused by a blocking of the water flow through the xylem. This blockage can be caused by the presence of various bacteria (*Erwinia, Ralstonia*) and fungi (*Fusarium, Verticillium*) in the xylem. Wilts may also be observed when there is a destruction of the root system due to nematodes or root-rotting fungi (*Armillaria, Phytophthora, Pythium*) or an acute water shortage in the soil.
- Shoot dieback or blights : Sudden dieback of a shoot usually indicates climatic or chemical injury rather than parasitic problems. If the line between affected and healthy bark is sharp, a soil chemical should be suspected. If dieback is somewhat more gradual and there is a cracking or splitting of the bark and wood, cold injury should be suspected along with bacterial blights caused by *Pseudomonas* or *Erwinia.* A bacterial streaming test with phase and compound microscope and isolation may be required to determine if the blights/dieback is due to a bacterial agent. Gradual decline of shoots and retention of dead leaves is more indicative of a parasitic disease. The margin between the affected and healthy tissue is often irregular and sunken. There may also be small pin-like projections or bumps over the surface of dead bark. These bumps are spore-producing structures of the fungal causal agent.
- Dying branches of trees and shrubs : If scattered branches in a tree or shrub start to decline and eventually die, a canker disease or a shoot blight should be suspected. If branches die suddenly and, especially if affected branches are concentrated on one side of tree, weather conditions should be suspected

(wind, snow, etc.) or animal or mechanical damage at the trunk base; however, this is not always the case. If symptoms develop over time on one side of a tree or plant then damage of the roots associated with one side may be suspected such as root rots due to *Phytophthora* spp.

- Death of tree and shrub top : If all or a major portion of a tree or shrub dies over a period of time, the diagnostician should suspect a problem with the roots. Examples are diseases caused by Armillaria root rot and Verticillium wilt. The decline may be gradual and may eventually affect the entire tree, but in some cases the death may occur on one side of the plant initially. If the decline is sudden, a toxic chemical in soil or weather extremes such as freesing or drought should be suspected.
- Overall Stunting or Decline : These symptoms can be caused by several very different factors. Systemic viral infections can result in stunting or decline, but such viral infections are often accompanied by other aboveground symptoms such as shortened internodes. In many cases, overall stunting of a plant may be due to problems associated with the root system.
- The roots should be examined for rotting and possible mycelial growth, reduction in roots especially feeder roots, and the presence of galls. Root galls can result from fungal and fungallike agents (*Plasmodiophora brassicae)*, nematodes (*Meloidogyne* spp.-root-knot), and bacteria (*Agrobacterium* sp.). Abiotic factors such as nutritional deficiencies, soil compaction and herbicide residues can also result in overall stunting or decline.
- Damping-off : This term describes the rapid death and collapse of young seedlings. Often the seedlings will appear to be almost broken at the soil line. It may be observed in flats of plants begun in greenhouses and can result from infection of the seedlings by the fungal organisms *Fusarium, Phytophthora, Pythium, Rhizoctonia,* or *Thielaviopsis*.

Chapter 29

Effect of the Treating with Pesticides and Plant Hormones on Catalase, Peroxidase and Polyphenol Oxidase Activities in Spinach (*Spinacia Oleraceae* L.)

F. Odabaşoğlu and Ö.İ. Küfrevioğlu*

Atatürk University, Faculty of Pharmacy, Department of Biochemistry, 25240 Erzurum -TURKEY

* Atatürk University, Science and Art Faculty, Department of Biochemistry, 25240 Erzurum-TURKEY

ABSTRACT

In this study, 6-7 week old spinach *(Spinacia oleraceae* L. cv. Gladiator) seedlings were grown for 15 days after the application of pesticide, hormone and pesticide plus hormone. The enzymatic activities of treated plants and control plants were compared after the activities of CAT, POD and PPO in extracted leaves were determined.

CAT enzyme was activated by deltamethrin, thiram and diclofop while cyfluthrin and benomyl inhibited. The activity of CAT was decreased by all of plant hormone concentrations except in all concentrations of ABA, in 10^{-7} and 10^{-9} M of IAA, and in 10^{-10} and 10^{-11} M of KIN. The CAT activity increased after the treating of ABA with deltamethrin and thiram; GA with diclofop; IAA with cyfluthrin, thiram and diclofop; KIN with benomyl, thiram and diclofop. While the activity of POD was increased by benomyl and diclofop it was decreased by all other pesticides. The POD activity was decreased by 10^{-7} M IAA and KIN but it was increased in whole concentrations of other hormones. The activity of POD was increased when ABA with whole pesticides; GA with diclofop, respectively were treated. The activity of PPO was increased by all pesticides. On the other hand, PPO activity was decreased by all hormone concentrations expect 10^{-7} M and 10^{-11} M of IAA, 10^{-7} M of GA and 10^{-7} M of KIN, respectively. The PPO activity was increased by application of ABA with all pesticides expect thiram. Also, application of GA with cyfluthrin and diclofop, IAA and KIN with diclofop increased the activity of PPO. In this study it was resulted that the adverse metabolic activities caused by pesticides may be removed by applications of different concentrations of plant hormones.

Keywords

Spinacia oleraceae L., Catalase, Peroxidase, Polyphenol oxidase, Pesticides, Plant hormones.

Abbreviations

CAT = Catalase; POD = Peroxidase; PPO = Polyphenol oxidase; ABA = Abscisic acid; IAA = Indolacetic acid; GA = Gibberellic acid; KIN = Kinetin)

INTRODUCTION

Catalase (H_2O_2 oxidoreductase, EC 1.11.1.6) is a enzyme containing heme as prostatic group (Kar and Mishra, 1976; Yamaguchi and Nishimura, 1984; Murray *et al.*, 1993), and this enzyme was located in peroxisomes and glyoxysomes of leaves (Fric,1976; Aebi,1987). Peroxidase (donor: H_2O_2 oxidoreductase, EC. 1.11.1.7) catalyses the dehydrogenation of a large number of aromatic compounds such as phenols, hydroquinones and hydroquinonoid amines, especially benzidine derivates (Putter and Becker, 1984; Van Huystee, 1987; Finger, 1994). The presence of peroxidase has been demonstrated in various subcellular components *e.g.* in nucleus, mitochondria, ribosomes, cell walls as well as in cell membranes (Fric, 1976), and have heme groups (Pütter and Becker, 1984; Chibbar *et al.*, 1984). Polyphenol oxidase (o-diphenol: O_2 oxidoreductase, EC 1.10.3.1), also known as catecol oxidase, is a cupper -containing enzyme catalysing the oxidation of o-diphenols to o-diquinones. Polymerization of quinones produced into brown pigments give rise darkening of plants. The physiological function of PPO is unknown yet, although it has been associated with disease resistance (Golbeck and Cammarata, 1981; Rathjen and Robinson, 1991; Sakiroglu *et al.*, 1991; Sakiroglu, 1994). PPO has been found in chloroplasts of higher plants. The enzyme locate on the thylakoid membrane although it reported that they present a soluble form in the stroma (Golbeck and Cammarata, 1981). Many researchers have studied on CAT, POD and PPO, which these enzymes were investigated as a indicator of oxidative damage in plants under various stressed conditions. On these enzymes activities there are a lot of studies related on researching the effects of pesticides and plant hormones. In these investigations, several authors had reported that various herbicides (alachlor, carbetamide, metribuzine, and benfluralin) decreased CAT activity while they increased POD in lupine (Fernandez Pascual *et al.*, 1992); herbicides (amitrole, holoxidine, metflurazone and difunone) inhibited CAT in winter rye (Steb *et al.*, 1993); paraquat inhibited CAT in winter rye (Steb *et al.*, 1993) while it increased in pea (Hurbe-Ormaetxe *et al.*, 1998); insecticide-carbaryl decreased both CAT and POD in mustard seedlings (Sengupta *et al.*, 1989); insecticide-profenophose activated PPO while it inhibited POD in potato (Habiba *et al.*, 1992), and it increased both PPO and POD in tomato (Ismail *et al.*, 1993); fungucides (triadimephone and propiconazole) reduced POD in wheat in fected with brown rust (Lebedev *et al.*, 1989); herbicide-amitrole enhanced PPO and POD in *Zea mays* and *Vigna anguiculata* (Begam and Vivekanandan, 1990). There are many literature related the effects of ABA on CAT and POD in maize (Prasad *et al.*, 1994), on POD in cotyledon explants of loblolly pine (Neuman *et al.*, 1992), on POD and PPO in cotton fibers (Basra *et al.*, 1992); the effects of GA on CAT in sapota (39Gautam and Chundavat, 1992), on POD in barley and rosa (Pollard, 1969; Zieslin and Ben-Zaken, 1992), on PPO in wheat (Taneja and Sachar, 1974; Kabar and Kocafaliskan, 1990); the effects of KIN on CAT in sapota (Gautam and Chundavat, 1992), on CAT-POD-PPO in rice (Kar and Mishra, 1976), on PPO and POD in *Cocculus populus* (Gaur *et al.*, 1993); the effect of IAA on POD in rosa (Zieslin and Ben-Zaken, 1992), on PPO in pea (Kadioglu and Ozbay, 1990). However, any literature related to the effects of pesticides and hormone complex on CAT, POD and PPO were not found.

The aim of this study is to determine the abnormalities effects of pesticides, which were used common in agricultural field on activities of CAT, POD and PPO enzymes which protect the tissues from toxic oxygen forms (H_2O_2, O^{-}_2 HO.) and to examine whether these abnormalities effects may be removed or not by plant hormones.

MATERIALS AND METHODS

Plant Material and Treatments

Spinach seedlings were treated with pesticides (at concentrations deltamethrin: 12.5 ppin, cyfluthrin: 25 ppm, benomyl: 300 ppm, thiram: 1600 ppm and diclofop methyl: 550 ppm), plant hormones (at concentrations 10^{-7}, 10^{-9} and 10^{-11} M) and pesticide-hormone combinations (pesticide + 10^{-11} M ABA; + 10^{-7} M GA, IAA, KIN) on 6-7 weeks old spinach leaves. Leaves were harvested on 15th day of treatments and these leaves were used in the determination of enzymatic activities.

Preparation of Leaf Extracts

0.5 g. of plant material frozen in liquid N_2 were prepared as extracts for catalase. Plant material were grinded continuously in a cold mortar at 4°C with 2.5 ml of 0.05 M K-phosphate (pH 7.5) containing 2 mg DTT. The suspension was centrifuged at *27000 g* for 15 min, and the supernatant was decanted and assayed for enzymatic activity (Havir and Mc Hale, 1987). For peroxidase extracts, one gram of treated leaves were ground in a mortar at 4°C using 5 ml of 0.1 M K-phosphate buffer (pH 7) plus 0.5 M NaCI. The homogenate was filtered through Miracloth to determine of POD activity (Angelini *et al.*, 1990). Polyphenol oxidase extracts were obtained by homogenising 1 g of treated leaves in 5 ml cold buffer (0.05 M Na-phosphate, pH 7) in a cold mortar. The homogenate was centrifuged at *20000 g* at 4°C for 10 min. The supernatant was used as the source of PPO (Omidiji and Okpuzor, 1996).

Determination of Enzyme Activities

Catalase activity was determined by measuring the rate of decrease in absorbance at 240 nm of a solution of 20 mM H_2O_2 in 50 mM K-phosphate (pH 7.5) at 30°C (Luck, 1963). The amount of enzyme per assay was adjusted so that the rate of the reaction (measured on a Ultraspec III spectrophotometer) was linear for at least 2 min. One unit is defined as the amount of enzyme catalysing the decomposition of 1 μmol H_2O_2 per minute calculated from the extinction coefficient for H_2O_2 at 240 nm of 0.036 cm^2 μmol^{-1}. Peroxidase activity was analysed in a reaction volume of 3 ml containing 0.1 M K-phosphate buffer (pH 7), 15 mM guaiacol and 7.5 mM H_2O_2. The reaction was started by the addition of an appropriate volume of the above-mentioned fractions and the formation of tetra guaiacol in 470 nm was followed continuously for 3 min (Angelini *el al.*, 1990). One unit of activity is defined as a change in absorbance of 0.001 per min. Pollyphenol oxidase activity was measured as described by Omidiji and Okpuzor (1996). The assay medium contained 2.9 ml of 0.2 M Na-phosphate buffer (pH 7) containing 25 mM catecol and 0.1 ml of enzyme extract, while the reference cuvette contained 2.9 ml of the buffer-catechol mixture and 0.1 ml of the enzyme extract which had been previously boiled and cooled. Change in absorbance of 420 nm was monitored for 120 s in a Unicam III Spectrophotometer attached to an automatic cell changer. One unit of PPO activity is defined as a chance of 0.001 absorbance unit at 420 nm at room temperature (Flurkey, 1986; Omidiji and Okpuzor, 1996). All experiments were repeated three times.

RESULTS

The effect of varying pesticides on the CAT, POD and PPO activities of spinach leaves are shown on Table 1. Different pesticide concentrations were used as described for vegetables in prospectuses. It can be seen on Table 1 clearly that CAT was activated by deltamethrin (27%), thiram (127%) and diclofop (33%) while inhibited by cyfluthrin (0.6%) and benomyl (15.8%); POD was decreased by deltamethrin (18.8%), cyfluthrin (23.7%) and thiram (19.9%) while was increased by benomyl (15%) and diclofop (30.9%); PPO was enhanced in different levels by all pesticides applicated (deltamethrin : 23.2%, cyfluthrin : 65.1%, benomyl : 134.8%, thiram : 72.8%, and diclofop : 132.5%).

In this study, we investigated the influence of ABA, GA, IAA and KIN hormones in three different concentrations on the enzymatic activities of CAT, POD and PPO. The concentrations of plant hormones were 10^{-7} M, 10^{-9} M and 10^{-11} M. The results were expressed as the EU per gram leaves in Table 2. After treating of spinach leaves with different concentration of ABA while the activities of both

CAT (approx. 17%) and POD (44-76%) were enhanced the activity of PPO (11-92%) was decreased. In spinach seedlings treated with GA hormones, 10^{-9} M and 10^{-11} M of GA have reduced both CAT (33%) and PPO (67, 65%) activity while whole concentration of GA increased that activity of POD activity (21-92%), but 10^{-7}M of GA increased both CAT (39%) and POD (38%) activity. While the CAT activity (2%) was inhibited by IAA, the CAT (34%), POD (87, 114%) and PPO (111, 132%) activities were activated by other concentrations of IAA. On the other hand, the activity of CAT (4,9%), POD (36,33%), and PPO (40) was increased by the all other concentration of KIN while the activity of CAT (15%), POD (18%) were increased by 10^{-9}M KIN, 10^{-7}M KIN respectively, whereas 10^{-9}M and 10^{-11}M concentration of KIN decreased the PPO (24, 34%) activity (Table 2).

Table 1 : Activities of CAT, POD and PPO of spinach leavestreated with pesticides (EU/g. leaves).

Pesticide	Treatments	CAT *activity* mean	%	POD *activity* mean	%	PPO *activity* Mean	%
Insecticides	Deltamethrir	45.8 ± 1.4*c*	127.0	3053 ± 55*c*	81.2	5300 ±520 *a*	1 23.2
	Cyfluthrin	35.7 ± 1.5	99.4	2871 ± 50*b*	76.3	7100±100*b*	165.1
Funguicides	Benomyl	30.2 ± 0.5*a*	84.2	4326 ± 110*a*	115.0	10100 ± 100*c*	234.8
	Thiram	82.8 ± 4.6*b*	227.0	3016±35*a*	80.1	7433 ± 208*c*	172.8
Herbicide	Diclofop	48.5 ± 2.0*a*	133.7	4923 ± 58*c*	130.9	10000 ± 200c	232.5
	Control	419 ± 1.7	100.0	3760 ± 78	100.0	4300 ±173	100.0

Values are mean ± SD of three sets of experiments,
P donates the level of significance; **a** P<0.05, **b** P<0.0l, **c** P<0.001
[a]P<0.05, [b]P<0.01,[c]P<0.001

Table 2 : CAT, POD and PPO activities of spinach leaves treated with plant hormones (EU/g leaves).

Hormone	Treatments	CAT *activity* Mean	%	POD *activity* Mean	%	PPO *activity* Mean	%
	10^{-7} M	44.0 ± 1.8	116.8	5258 ± 432.8	143.9	1177 ± 35.1	24.7
ABA	10^{-9} M	44.2 ± 0.2	117.2	6433 ± 583.6	176.0	1553 ± 11.5	32.7
	10^{-11} M	44.4 ± 2.4	117.7	5597 ± 472.2	153.0	1043 ± 5.8	21.9
	10^{-7} M	52.4 ± 4.0	139.3	4055 ± 57.7	110.9	6590 ± 177.0	138.4
GA	10^{-9} M	25.1 ± 0.8	66.7	7012 ± 154.6	191.7	1580 ± 34.6	33.1
	10^{-11} M	33.6 ± 0.0	89.2	4387 ± 460.0	119.9	1660 ± 121.7	34.8
	10^{-7} M	50.6 ± 3.7	134.3	1825 ± 42.7	50.0	6320 ± 291.0	132.7
IAA	10^{-9} M	50.7 ± 5.4	134.1	7842 ± 844.0	214.5	1397 ± 35.1	29.3
	10^{-11} M	36.9 ± 2.3	98.0	6830 ± 482.0	186.8	5270 ± 167.0	110.7
	10^{-7} M	39.1 ± 0.4	103.8	2993 ± 70.2	81.9	6667 ± 75.7	140.0
KIN	10^{-9} M	31.9 ± 1.2	84.8	4990 ± 125.0	136.5	1167 ± 28.9	24.5
	10^{-11} M	39.4 ± 0.8	108.8	4873 ± 329.0	133.3	1620 ± 34.6	34.0
	Control	37.6 ± 3.4	100.0	3657 ± 32.5	100.0	4760 ± 95.4	100.0

Table 3 : CAT, POD and PPO activities of spinach leaves applied with plant hormones and pesticides (EU/g leaves).

Pest + Hormone Treatments		CAT *activity* Mean	%	POD *activity* Mean	%	PPO *activity* Mean	%
	Deltamethrin	58.0 ± 1.3*c*	125.5	5660 ±219.0*b*	150.6	7842 ± 309*c*	204.5
	Cyfluthrin	35.2 ± 2.8*a*	76.2	3808 ± 167.5	101.3	10567 ± 104*c*	275.6
ABA+	Benomyl	46.0 ± 10.6	99.6	4995 ± 70.5*a*	132.9	11192 ± 170*c*	291.9
	Thiram	48.2 ± 1.6*a*	104.3	5935 ± 218.0*a*	157.9	1550 ± 152*b*	40.4
	Diclofop	43.2 ± 0.4	93.5	6040 ± 231.0*b*	160.7	10317 ± 227*c*	269.1
	Conirol	46.2 ± 1.6	100.0	3757 ± 231.0	100.0	3833 ± 194	100.0
	Deltamethrin	23.1 ± 1.8*a*	64.9	2657 ± 87*b*	72.9	2733 ± 270	79.4
	Cyfluthrin	19.1 ± 2.0*b*	53.6	2808 ± 275*b*	77.1	4017 ± 101*b*	116.7
GA+	Benomyl	17.9 ± 1.3*b*	50.3	2288 ± 188*b*	62.8	1767 ± 52*b*	51.8
	Thiram	34.6 ± 1.8	97.2	3163 ± 45*a*	86.9	833 ± 29*b*	24.2
	Diclofop	42.3 ± 3.7	118.8	8953 ± 966*b*	245.9	9442 ± 101*c*	274.3
	Control			3640 ±105	100.0	3442 ± 163	100.0
	Deltamethrin	29.1 ± 0.6*a*	95.1	3305 ± 241	87.0	2825 ± 250*c*	54.6
	Cyfluthrin	36.8 ± 0.7*b*	120.3	2360 ± 386*a*	62.1	1258 ± 101.0*c*	24.3
1AA+	Benomyl	28.3 ± 1.3*a*	92.5	3750 ± 678	98.7	1192 ± 80.4*c*	23.0
	Thiram	43.3 ± 1.7*b*	141.5	4850 ± 198*b*	127.6	467 ± 167.0*c*	9.0
	Diclofop	38.6 ± 1.3*a*	126.1	8642 ± 472*b*	227.4	11550 ± 996*c*	225.4
	Control	30.6 ± 0.2	100.0	3800 ± 92	100.0	5167 ± 57.7	100.0
	Deltamethrin	22.6 ± 2.4*a*	73.8	2950 ± 165*a*	77.6	4442 ± 227*a*	85.9
	Cyfluthrin	28.8 ± 1.8	94.1	3240 ± 104*a*	85.3	4858 ± 38*a*	94.0
KIN+	Benomyl	38.2 ± 5.7	124.8	4945 ± 377*a*	130.1	4567 ± 290*a*	88.3
	Thiram	48.3 ± 3.1*a*	157.8	5100 ± 513*a*	134.2	2108 ± 104*c*	40.8
	Diclofop	40.3 ± 0.9*a*	131.6	10512 ±710*b*	276.6	12208 ± 406*c*	236,2
	Control	30.6 ± 0.6	100.0	3800 ± 92	100.0	5167 ± 58	100.0

The changes of CAT, POD and PPO activities after treating with hormone + pesticide complex are shown on Table 3. The presented data in Table 3 clearly indicate that CAT activity was decreased by adding ABA. It can be seen on Table 3 that all pesticides except benomyl, but CAT activity was increased after treating of ABA with deltamethrin, thiram and diclofop with previous pesticides. On the contrary. POD activity was much increased by ABA more than all pesticides while they enhanced POD activity. Also, the activities of both CAT and POD were declined when they have treated with GA - pesticides complex except thiram. On the other hand, POD and CAT activity were reduced by application of IAA with deltamethrin and benomyl whereas POD activity was declined by adding IAA and cyfluthrin together and by this application of ABA with thiram, GA with deltamethrin-benomyl-thiram. PPO activity was reduced by the application of IAA and KIN with all of pesticides except diclofob (Table 3).

DISCUSSION

These results indicate that all pesticides have effects on the enzymatic activities of CAT, POD and PPO at different levels. Such case is a natural result of the varying chemical components of pesticides. A different behavior was observed in the presence of pesticides. Whole pesticides except diclofob, increased and decreased CAT or vice versa. Hence, it can be said that pesticides have effects on the affinity of CAT and POD to H_2O_2. In fact, it is theoretically presumed that CAT decreases the H_2O_2 level in the cells, thus limiting the peroxidative reactions catalyzed by POD (Fric, 1976). The data presented in Table 1 show that the activity of either CAT and POD was continuously increased with pesticide treatments. The increasing CAT or POD activities as a result of treatment of leaves with pesticides may be related with the senescence processes. CAT and POD enzymes use H_2O_2 as substrate. H_2O_2 can be also reason of occurrence of OH. (from H_2O_2 and O^-_2 in the Huber -Weiss reaction) which it is a highly reactive and can damage significantly the cell (Anderson *et al.*, 1994). Senescence processes have been traditionally associated with variations of the enzyme levels, which protect the tissues from toxic oxygen derivatives (Lorenzo *et al.*, 1990). Also, CAT and POD are two enzymes of these enzymes group (Lorenzo *et al.*, 1990; Fernandez Pascual *et al.*, 1992). The activity level of PPO has been related to leaf senescence as CAT and POD (Kar and Mishra, 1976). In this study, we noticed that PPO activity was increased as a result of pesticides treatment.

We think this result is indication of beginning senescence processes of youngest leaves by increasing PPO activity after treatments of pesticides and pesticide treatments generate negative environment for spinach leaves. This suggestion can be supported by the Fric (1976) who reported that the PPO activity of plant tissues can be caused by adverse environmental factors and by the other workers (Kar and Mishra, 1976; Mayer, 1987; Pilar Cano *et al.*, 1995;Thipyapong, 1997).

We found that the both CAT and POD activities were elevated by ABA. This case is a natural result because of ABA hinder the growth of plants. The both CAT and POD is known antioxidant enzymes (Lorenzo *et al.*, 1990). The increased activities of CAT and POD had been related to the result of seedlings treated with ABA responding to oxidative stress generated by antioxidant enzymes which scavenge or utilize H_2O_2 (Scandilos, 1993; Prasad *et al.*, 1994). On the other hand, Neuman *et al.* (1992) reported that the number of peroxidase bands were increased with the length of exposure to ABA. Our data clearly demonstrated that the PPO activity was inhibited by phytohormone ABA. The activity of PPO inhibited by

ABA, had been associated with the with the inhibition of PPO activity by ABA which is inhibiting calmodulin synthesis or its mobilization to sites of action (Basra *et al.*, 1992). Some researchers made on maize (Anderson *et al.*, 1994), cotyledon explant of loblolly pine, wheat and cotton are supporting our results which are presented on Table 2 for ABA.

We have found that POD activity was increased by all concentration levels of GA and activity of CAT was decreased by the 10^{-9} and 10^{-7} M concentrations of GA but reduced by 10^{-11} M concentration of GA. As relating to the affects of GA hormone on the POD activity there are some reports in which it report the increasing of POD both in vitro (Zieslin and Ben Zaken, 1992) and in vivo (Pollard, 1969; Zieslin and Ben Zaken, 1992). Our data are in good agreement with these reports. However, hand our findings relating to effect of GA on activities of both CAT and PPO can be confirmed by results of CAT activity which was decreased after treating with GA in sapota (Gautam and Chundawat, 1992), and PPO activity of wheat endosperm which was declined by GA (Taneja and Sachar, 1974; Kabar and Kocacalypkan, 1990). In light of the presented data on Table 2 it can be said that the activities of CAT, POD and PPO were affected at vary levels after treating with both IAA and KIN. Stafford and Galston (1970) have been reported that nucleic acid synthesis can be increased by IAA. According to these conclusion we can suggest that the increasing of nucleic acid synthesis in spinach leaves is parallel with increasing PPO activity after spraying IAA. This result has also been supported by finding of exogenic spraying of IAA on to plant tissues. Increasing of PPO activity in nicotiana (Taneja and Sachar, 1974) and pea (Kadyodlu and Ozbay, 1990) by IAA have been reported in some papers. In literature, it has been recorded that increasing of the activity of PPO by IAA was decreased by KIN (Staffort and Galston, 1970). This conclusion can be supported by our data that the PPO activity was increased after treating with both 10^{-7} M and 10^{-11} M of IAA or was decreased after treating with both 10^{-9} M and 10^{-11} M of KIN (Table 2). Kar and Mishra (1976) have reported that 0.5×10^{-7} M of KIN enhance the level of PPO activity. The result of Kar and Mishra (1976) supports our findings about the level of PPO activity which has been enhanced significantly by 0.5×10^{-7} KIN. On the other hand it has been recorded that CAT activity was decreased by KIN in both oryza (Kar and Mishra, 1976) and sapota (Gautam and Chundawat, 1992), and POD activity was increased by KIN in oryza (Kar and Mishra, 1976) and by IAA in rosa (Zieslin and Ben Zaken, 1992). These all findings are supporting our data partially.

The increasing POD and PPO and PPO activities demonstrate that ABA has not a regulating effect on enzyme but pesticides has negative effect. Thus we adopted a natural indicator of ABA which it is a hindering hormone for growing and maturing of plants. On the other hand except diclofop, mixture of GA and pesticides have a regulating effects on the activities of CAT, POD and PPO enzymes. Moreover, IAA and KIN the mixture of containing different pesticides can be regulate the adverse effects which they are occurred by the all pesticides.

In addition, our results indicates that CAT, POD and PPO activities have partially regulated in favour of spinach leaves by treating IAA-KIN with deltamethrin, IAA with benomyl and GA with thiram, respectively. Unfortunately we could not compare finding with literature since we have not come across any related articles.

CONCLUSION

The conclusion drawn from the present study is that pesticides and plant hormones might has effect at different levels on activities of catalase, peroxidase and polyphenol oxidase, and the adverse metabolic activities, occurred by pesticides, might prevent by the some hormone concentrations added to pesticides.

According to the findings of this study, it was proposed that the using of IAA and KIN with deltamethrin, IAA with benomyl, and GA with thiram would be useful, for plant growing and hormones would be used to correct the negative effects of the pesticides.

ACKNOWLEDGEMENT

This work was financially supported by the research council of Ataturk Univers ity (AUAF 1996/27). The authors wish to thank Dr. H. Ozdemir, Dr. B. Nalbantoglu and Dr. I. Celik who are researcher in Department of Biochemistry.

REFERENCES

1. Aebi, H.E. (1987). Catalase. Methods of Enziymatic Analysis, Bergmeyer, Third Edition, VCH, Volume III, *Enzymes*, l;p. 273-282.
2. Anderson, M.D.; Prasad, T.K.; Martin, B.A. and Stewart, C.R.(1994). Differential gene expression in chilling-acclimatod maize seedlings and evidence for the involvement of abscisic acid in chilling tolerance., *Plant Physiol*, 105;p. 331-339
3. Angelini, R.; Manes, F. and Federico, R.(1990). Spatial and functional correlation between diamine-oxidase and peroxidase activities and their dependence upon deetiolation and wounding in chick-pea stems. *Planta*, 182;p. 89-96
4. Basra, A.S., Sarlach, R.S., Dhillon-Grewal, R. and Malik, C.P.(1992). Calcium-mediated changes in peroxidase and o-diphenol oxidase activities of cotton fibers *(Gossypium* spp) and its possible relationsnip to ABA. *Plant Growth Regulation*, 11;p. 159-164
5. Begam, M.N. and Vivekanandan, M.(1990). Light activation of enzymes in relation to leafage in *Vigna unguiculata* (L.) Walp and *Zea mays* L., *Proceedings of the Indian Academy of Sciences, Plant Sciences*, 100/4;p. 225-231
6. Chibbar, R.N.; Van Huystee, R.B. and Cella, R.(1984). The heme moiety in peanut peroxidase. *Can. J. Biochem*, 62;p. 1046-1050
7. Fernandez-Pascual, M.; De Lorenzo, C; Pozuelo, J.M. and De Felipe, M.R.(1992). Alterations induced by four herbicides on lupine ncdule cortex structure, protein metabolism and some senescense-related enzymes. *J. Plant Physiol*, 140;p. 385-390
8. Finger, A.(1994). In - vitro studies on the effect of polyphenol oxidase and peroxidase on the formation of polyphenols black tea constituents. *J. Sci. Food Agric*, 64;p. 293-305
9. Flurkey, W.H.(1986).Pollyphenoloxidase in higher plants., *Plant Physiol.*, 81;p.614-619

10. Fric, F. (1976). Oxidative enzymes. In Pirson, A. and Zimmerman, M.H. (Ed.): Encyclophedia of Plant Physiology, new series. 4, Springer Verlag, New York;p. 617-631.

11. Gaur, A.; Bhardwaj, L.; Merilon, J.M. and Ramavat, K.G.(1993). changes in phospholipids, fatty acids, oxidativ enzymes, phenolics and protein levels during growth of normal and habituated tissues of *Cocculus pendulus.* Indian *Journal of Experimental Biology,* 31; 987-990

12. Gautam, S.K. and Chundawat, B.S.(1992). Effect of post-harvest treatments on biochemical changes in sapota cv. 'cricket ball'. *J, Fd. Sci. Technol.,* 29 (l);p. 53-54

13. Golbeck, J.H. and Cammarata, K.V.(1981). Spinach thylakoid polyphenol oxidase. *Plant Physiol,* 67;p. 977-984

14. Habiba, R.A.; Ali, H.M. and Ismail, S.M.M.(1992). Biochemical Effects of Profenofos Residues in Potatoes../ *Agric and Food Chem.,* 40/10;p. 1852-55

15. Havir. E.A. and McHale, N.A.(1987). Biochemical and developmental characterization of multiple forms of catalase in tobacco leaves. *Plant Physiol.,* 84;p. 450-455

16. Hurbe-Ormaetxe, I.; Escuredo, P.R.; Arrese-lgor, C. and Becana,M.(1998). Oxidative Damage in Pea Plants Exposed to Water Deficit or Paraquat. *Plant Physiol.,* 116;p. 173-181

17. Ismail, S.M.M.; Ali, H.M. and Habiba, R.A.(1993). Analyses of Profenofos Residues and its Biochemical Effects in Tomatoes and Tomato Products. *J. Agric and Food Chem.,* 41/4;p. 610-615

18. Kabar, K. and Kocacaliskan, 1.(1990). Interactions among salinity (NaCl), polyphenol oxidase, and growth regulators in the germination of wheat seeds. Doga-Tr. *J. of Botany,* 14;p. 235-245

19. Kadioglu, A. and Ozbay, 0.(1990). Polyphenol oxidase and related isoenzymes in crown- gall tumors of pea. *Phyton,* 30 (2);p. 283-290

20. Kar, M. and Mishra, D.(1976). Catalase, peroxidase, and polyphenoloxidase activities d during rice leaf senescence. *Plant Physiol.,* 57;p.315-319

21. Kar, M. and Mishra, D.(1989). Catalase, peroxidase, and polyphenoloxidase activities during rice leaf senescence. *Plant Physiol.,* 57;p. 315-319

22. Lebedev, V.B.; Suslova, T.A.; Khorosheva, T.M. and Gryazeva, O.A.(1989). Leaf Peroxidase Activity and Isozyme Composition When Wheat is Infected with Brown Rust and Fungucides are Applied. *Soviet Agric. Sci.* 5,p.14-17,

23. Lorenzo, C; Lucas, M.M. and Felipe, M.R.(1990). Effect of nitrate on peroksizome ultrastructure and catalase activity in nodules *aSLupinus albus* L. cv. Multolupa. *Journal-of Experimental Botany,* 41 (233);p. 1573-1578

24. Luck, H. (1963). Catalase., *In* HI) Bergmeyer, ed., Methods of Enzymatic Analysis. VerlagChemie, Weinheim/Bergstr.;p. 885-894

25. Mayer. A.M.,(1987). Polyphenol oxidases in plant-recent progress. *Phytochemistry,* 26/l;p.ll-20.

26. Murray, R.K.; Mayes, P.A.; Granner, D.K. and Rodwell, V.W.(1993). Harper'in Biyokimyasi,Mente$, G. ve Ersoz, B. cevirisi, muhtelif kisimlan, Ban? Kitabevi, Istanbul.

27. Neuman, PR.; Fong, F.; Newton, RJ. and Sen, S.(1992). Induction of peroxidase isozymes during shoot enhancement by ABA in cotyledon explants of loblolly pine (*Pinus taeda* L.). *J. Plant Physiol,* 139;p. 343-349

28. Oinidiji, O. and Okpuzor, 3.(1996). Time course of PPO-related browning of yams. *J. Sci. Food Agric,* 70;p. 190-196

29. Pilar Cano, M.; De Anco, B. and Lobo, G.(1995). Peroxidase and polyphenol oxidase activities in papaya during post harvest ripening and after freezing/thawing., *J.Food Sci.,* 60/4;p.815-817

30. Pollard, C.J.(1969). A survey of the sequance of some effects of gibberellic acid in the metabolism of cereal grains. *Plant Physiol.,* 44: 1227-1232

31. Prasad, T.K.; Anderson, M.D. and Stewart, C.R.(1994). Acclimation, hydrogen peroxide, and abscisic acid protect mitochondria against irreversible chilling injury in maize seedlings. *Plant Physiol,* 105p. 619-627

32. Putter, J. and Becker, R. (1984). Perokxidases. Methods of Enzymatic Analysiss. Bergmeyer, third edition, VHC, V III;p.86-291.

33. Rathjen, A.H. and Robinson, S.P.(1991). Aberrant processing of polyphenol oxidase in a variegated grapevine mutant. *Plant Physiol,* 95;p. 319-325

34. Scandalios, J.G.(1993).Oxygen stress and superoxide dismutases./Vaw/ *Physiol.* 101;p.7-12

35. Sengupta, P.K.; Chakrabarti, A. and Banerjee, S.K.(1989). Alterations in the growth rate, activities of catalase, peroxidase, IAA oxidase and the level of hydrogen peroxide in mustard seedlings *(Brassica nigra)* during carbaryl toxicity. *Biologia Plantarum,* 31 (3);p. 196-201

36. Staffort, H.A. and Galston, A.W.(!970). Ontogeny and hormonal control of polyphenol oxidase isoenzyms in tobacco pith. *Plant Physio/.,* 46;p.763-767

37. Streb. P.; Michael-Knauf. A. and Feierabend, J.(1993).Preferential photoinactivation of catalase and photoinhibition of photosystem II are common early symptoms under various osmotic and chemical stress conditions. *Physiologia Plantarum,* 88;p. 590-598

38. Sakiroglu, H.; Kiifrevioglu, 0.1. ve Oktay, M.(1991). Kusburnu meyvasindan izole edilen polifenol oksidaz enziminin substrat spesifitesi. VIII. *Kimya ve Kimya Miihendisligi Sempozyumu*;p. 233-235

39. Taneja, S.R. and Sachar, R.C.(1974). Stimulation of polyphenol oxidase (monophenolase) activity in wheat endosperm by gibberellic acid, cycloheximide and actinomycin D. *Plan/a.* 116;p. 133-142.

40. Thipyopong, P. And Steffans, J.C.(1997). Tomato polyphenol oxidase. *Plant Physiol.,*115;p.409-418

41. Van Huystee, R.B.(1987).Some molecular aspects of plant peroxidase biosynthetic studies. *Annu. Rev. Plant. Physiol. "Plant Molecular Biology",* 38;p. 205-219

42. Yamaguchi, *i.* and Nishimura, M.(1984). Purification of glyoxysomal catalase and immunochemical comparison of glyoxysomal and leaf peroxisomal catalase in germinating pumpkin cotyledons. *Plant Physiol.,* 74;p. 261-267

43. Zieslin, N. and Ben-Zaken, R.(1992). Effects of applied auxin, gibberellin and cytokinin on the activity of peroxidases in the peduncles of rose flowers. *Plant Growth Regulation,* 11; p. 53-57

Chapter 30

HORMONAL REGULATION OF PLANT GROWTH AND DEVELOPMENT

William M. Gray

William M. Gray is in Plant Biology at the University of Minnesota, St. Paul, Minnesota, United States of America. E-mail: grayx051@tc.umn.edu

Phytohormones: What Are they?

Plant growth and development involves the integration of many environmental and endogenous signals that, together with the intrinsic genetic program, determine plant form. Fundamental to this process are several growth regulators collectively called the plant hormones or phytohormones. This group includes auxin, cytokinin, the gibberellins (GAs), abscisic acid (ABA), ethylene, the brassinosteroids (BRs), and jasmonic acid (JA), each of which acts at low concentrations to regulate many aspects of plant growth and development.

With the notable exception of the steroidal hormones of the BR group, plant hormones bear little resemblance to their animal counterparts (Figure 1). Rather, they are relatively simple, small molecules such as ethylene gas and indole-3-acetic acid (IAA), the primary auxin in the majority of plant species.

The concept of plant hormones originates from a classical experiment on phototropism, the bending of plants toward light, carried out by Charles Darwin and his son Francis in 1880. The Darwins were able to demonstrate that when oat seedlings were exposed to a lateral light source, a transported signal originating from the plant apex promoted differential cell elongation in the lower parts of the seedling that resulted in it bending toward the light source. This signal was subsequently shown to be IAA, the first known plant hormone.

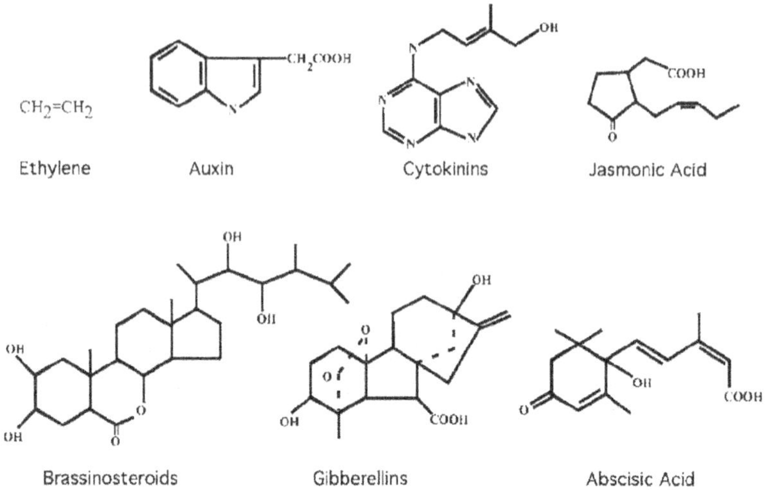

Figure 1. Chemical Structures of the Plant Hormones A partial list of the responses elicited by each hormone is provided below. Ethylene gas promotes fruit ripening, senescence, and responses to pathogens and abiotic stresses. IAA (an auxin) regulates cell division and expansion, vascular differentiation, lateral root development, and apical dominance. Cytokinins are adenine derivatives first identified by their ability to promote cytokinesis. JA is a volatile signal that modulates pollen development and responses to pathogen infection. The BRs regulate cell expansion and photomorphogenesis (light-regulated development). GAs are diterpenoid compounds that promote germination, stem elongation, and the induction of flowering. ABA promotes seed dormancy and is involved in several stress signaling pathways.

What Do They Do?

Virtually every aspect of plant growth and development is under hormonal control to some degree. A single hormone can regulate an amazingly diverse array of cellular and developmental processes, while at the same time multiple hormones often influence a single process. Well-studied examples include the promotion of fruit ripening by ethylene, regulation of the cell cycle by auxin and cytokinin, induction of seed germination and stem elongation by GA, and the maintenance of seed dormancy by ABA. Historically, the effects of each hormone have been defined largely by the application of exogenous hormone. More recently, the isolation of hormone biosynthetic and response mutants has provided powerful new tools for painting a clearer picture of the roles of the various phytohormones in plant growth and development.

How Do They Work?

Plant biologists have been fascinated by the regulatory capacity of phytohormones since the time of their discovery, and the notion that hormone levels or responses could be manipulated to improve desired plant traits has long been an area of intense interest. Perhaps the best-known example of this is the isolation of dwarf varieties of wheat and rice that led to the "green revolution" in the second half of the 20th century, which is credited with saving millions of people around the globe from starvation. These dwarf varieties have shorter stems than wild-type, making these plants less susceptible to damage by wind and rain. The molecular isolation of these "dwarfing genes" has revealed that they encode components of the GA biosynthesis and response pathways (Peng *et al.* 1999; Sasaki *et al.* 2002).

To elucidate the molecular mechanisms underlying phytohormone action, several researchers have utilized the genetically facile model plant *Arabidopsis thaliana* to isolate mutations that confer altered response to applied hormone. Molecular and biochemical analysis of the gene products defined by these mutations, coupled with expression studies aimed at identifying the downstream target genes that mediate hormonal changes in growth and development, has begun to unlock some of the mysteries behind phytohormone action. While no hormone transduction pathway is completely understood, we now have a rudimentary understanding of many of the molecular events underlying hormone action. Several reviews covering the individual hormone pathways in greater detail have recently been published (Turner *et al.* 2002; Gomi and Matsuoka 2003; Himmelbach *et al.* 2003; Kakimoto 2003; Dharmasiri and Estelle 2004; Guo and Ecker 2004; Wang and He 2004).

Common Themes

Regulation by proteolysis has emerged as a resounding theme in plant hormone signaling. The ubiquitin-mediated degradation of key regulatory proteins has been demonstrated, or is at least likely, for all of the phytohormone response pathways (Smalle and Vierstra 2004). In the case of auxin, the response pathway is normally subject to repression by a large family of transcriptional regulators called the Aux/IAA proteins (Figure 2). These proteins dimerize with members of the auxin response factor (ARF) family of transcription factors, thus preventing ARFs from activating auxin-responsive genes (Tiwari *et al.* 2004). Upon an auxin stimulus, an SCF (SKP1/Cullin/F-box protein) ubiquitin ligase (Deshaies 1999) containing the TIR1 F-box protein ubiquitinates the Aux/IAA proteins, marking them for degradation by the 26S proteasome thereby de-repressing the response pathway (Gray *et al.* 2001). The hormone promotes the Aux/IAA–TIR1 interaction; however, the molecular mechanisms behind this regulation are unclear. Most yeast and animal SCF substrates must be post-translationally modified, usually by phosphorylation, before they are recognized by their cognate F-box protein. Despite numerous efforts to identify auxin-induced modification of Aux/IAA proteins, no such signal has been discovered, raising the distinct possibility that auxin uses a novel mechanism to regulate SCF–substrate interactions.

Ethylene and cytokinin are both perceived by receptors sharing similarity to bacterial two-component regulators. Common in prokaryotes, but apparently restricted to plants and fungi in eukaryotes, these modular signaling systems involve a membrane-bound receptor containing an intracellular histidine kinase (HK) domain (Wolanin *et al.* 2002). Ligand binding activates the kinase, resulting in autophosphorylation and initiation of a series of phosphotransfer reactions that culminates with the activation of a response regulator protein that functions as the effector component of the pathway. Cytokinin signaling appears to largely follow this paradigm (Kakimoto 2003). Ethylene response, however, appears more complex (Guo and Ecker 2004).

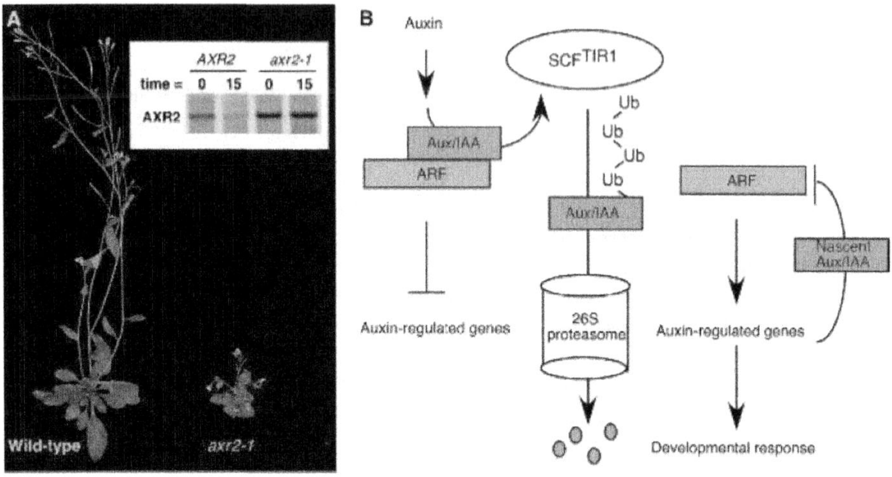

Figure 2. The Ubiquitin-Mediated Proteolysis of Aux/IAA Proteins Regulates Auxin Response (A) Wild-type Arabidopsis thaliana and the axr2-1 mutant. axr2-1 is a dominant gain-of-function mutation in an Aux/IAA gene that confers reduced auxin response. The mutant axr2-1 protein constitutively represses auxin response because it cannot be targeted for proteolysis by the SCFTIR1 ubiquitin ligase. The effect of the mutation on AXR2 stability is shown in a pulse-chase experiment (inset). Wild-type and axr2-1 seedlings were labeled with 35S-methionine and AXR2/axr2-1 protein was immunoprecipitated either immediately after the labeling period (t = 0) or following a 15-minute chase with unlabeled methionine (t = 15). (B) A simplified model for auxin response. In the absence of an auxin stimulus, Aux/IAA proteins inhibit ARF transcriptional activity by forming heterodimers. Auxin perception (by an unknown receptor) targets the Aux/IAA proteins to the SCFTIR1 complex, resulting in their ubiquitination and degradation, thereby de-repressing the ARF transcription factors. Among the ARF targets are the Aux/IAA genes themselves, which produce nascent Aux/IAA proteins that restore repression upon the pathway in a negative feedback loop.

Ethylene is perceived by a family of five receptors. ETR1 and ERS1 contain a consensus HK domain, however, the HK domains of ETR2, ERS2, and EIN4 are degenerate and lack elements necessary for catalytic activity. This fact, together with studies of "kinase-dead" mutants of *ETR1*, suggests that HK activity is not required for ethylene response. Mutations that abolish ethylene binding in any of the five receptor genes are dominant and confer ethylene insensitivity, indicating that the receptors function as negative regulators of the ethylene pathway.

Genetic and molecular studies have positioned these receptors upstream of the Raf-like MAP kinase kinase kinase, CTR1, which interacts with the receptors and also acts as a negative regulator (Figure 3). The integral membrane protein, EIN2, and the transcription factors EIN3 and EIL1 are positive regulators of ethylene

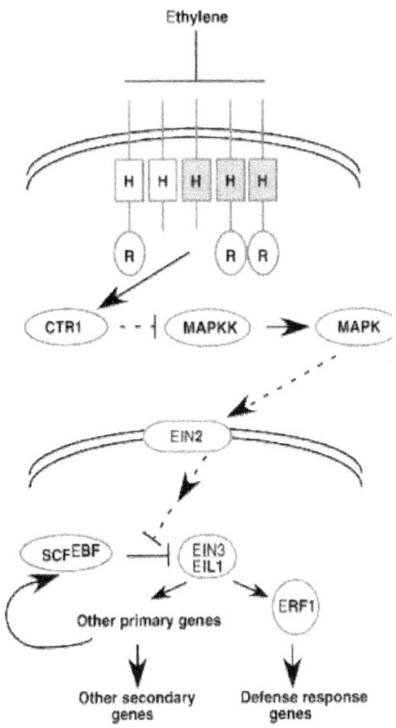

Figure 3. A Model for the *Arabidopsis* Ethylene Response Pathway Ethylene is perceived by a family of two-component receptors containing a consensus (unshaded) or degenerate (shaded) HK domain (H). Three of the receptors also contain a C-terminal receiver domain (R). The receptors negatively regulate ethylene response together with CTR1 in a complex on the endoplasmic reticulum membrane. Perception results in reduced receptor and CTR1 activities and activation of a MAP kinase kinase, which transmits the signal through the EIN2 membrane protein, ultimately resulting in the activation of a transcriptional cascade in the nucleus. The EIN3 and EIL1 transcription factors regulate primary response genes including *ERF1*, which activates a subset of secondary ethylene-induced genes involved in defense responses. EIN3/EIL1 abundance is regulated in an ethylene-dependent manner by SCF complexes containing F-box proteins encoded by the ethylene-induced genes *EBF1* and *EBF2*. Positive-and negative-acting components of the pathway are indicated in green and red, respectively. Solid lines indicate regulation that is likely to be through direct interactions. Dotted lines indicate speculative interactions based on genetic studies.

signaling downstream of CTR1. Current models propose that hormone binding inactivates the receptors, thus resulting in down-regulation of CTR1 activity. Since the identifi cation of CTR1, biologists have speculated that a MAP kinase cascade may be involved. Only recently, however, have putative MAP kinase kinase and MAP kinase components of the ethylene pathway been identified (Chang 2003). Interestingly, these kinases appear to positively regulate ethylene response, suggesting that CTR1 must inhibit their function. If so, this would represent a novel twist on the traditional MAP kinase signaling paradigm. Precisely how the ethylene signal is transduced to the EIN3 and EIL1 transcription factors remains unclear. However, the recent finding that ethylene stabilizes these transcription factors, which are targeted for degradation by an SCF complex in the absence of ethylene, clearly indicates a role for the ubiquitin pathway (Guo and Ecker 2003;

Potuschak *et al.* 2003). One of the known targets for EIN3 is the ERF1 transcription factor, which activates several genes involved in a subset of ethylene responses.

Signal Integration and Combinatorial Control

Long ago, plant physiologists noted the apparent antagonistic interactions between some of the phytohormones, such as between auxin and cytokinin in the regulation of root-shoot differentiation and between GA and ABA in germination. Other processes are synergistically regulated by multiple hormones. While it has long been obvious that hormones do not function in discrete pathways, but rather exhibit extensive cross-talk and signal integration with each other and with environmental and developmental signaling pathways, the molecular basis for such coordinated regulation has been unclear. Several recent findings have begun to elucidate the molecular details of some of these events.

One example of such signal integration was recently described for the ethylene and JA pathways (Lorenzo *et al.* 2003). Genetic studies had previously implicated both hormones as important regulators of pathogen defense responses, as well as of the wounding response and other stress-related pathways. Additionally, microarray analysis has identified a large number of genes that are responsive to both hormones. The ERF1 transcription factor was recently found to be an intersection point for these two signaling pathways (Lorenzo *et al.* 2003). Like ethylene, JA rapidly induces *ERF1* expression, and treatment with both hormones synergistically activates *ERF1*. Induction of *ERF1* by both hormones alone or in combination is dependent upon both signaling pathways, and constitutive overexpression of *ERF1* rescues the defense-response defects of both ethylene- and JA-insensitive mutants. These findings suggest that *ERF1* represents one of the first signaling nodes identified in the complex web of hormonal cross-talk.

The auxin and BR pathways also appear to converge and mutually regulate some developmental processes. Both hormones promote cell expansion, and microarray studies have revealed that as many as 40% of all BR-induced genes are also up-regulated by auxin (Goda *et al.* 2004; Nemhauser *et al.* 2004). BR is perceived by the cell surface receptor kinase BRI1 (Wang and He 2004). The SHAGGY/GSK3-type kinase BIN2 acts as a negative regulator of the pathway downstream of the receptor. In the absence of a BR signal, BIN2 phosphorylates the transcription factors BES1 and BZR1, targeting them for proteolysis by the 26S proteasome. Upon a BR stimulus, BIN2 is inactivated, allowing BES1 and BZR1to accumulate in the nucleus, where they are presumably involved in regulating BR-responsive genes.

Using combined genetic, physiological, and genomic approaches, Nemhauser and colleagues (2004) were able to demonstrate that auxin and BR regulate *Arabidopsis* hypocotyl (embryonic stem) elongation in a synergistic and interdependent fashion. Elevating endogenous auxin levels rendered plants more sensitive to BR application in hypocotyl elongation assays, and this response was dependent upon both the auxin and BR signaling pathways. Genetic studies suggest that the convergence of these two pathways occurs at a late point in hormone signaling, perhaps at the promoters of the many genes responsive to both hormones. In support of this notion, bioinformatic analysis identified distinct sequence elements that were enriched specifically in the promoters of auxin-induced, BR-induced, and auxin/BR-induced genes.

Many Unanswered Questions

While great strides have been made in recent years in understanding the molecular basis of phytohormone action, many fundamental questions remain. Receptors and other upstream signaling components remain to be identified for the majority of the phytohormones. Equally important are the elucidation of hormonal networks and the integration of these networks with the morphogenetic program, such that our understanding of hormone action can be placed in a developmental context.

ABBREVIATIONS

ABA, abscisic acid; ARF, auxin response factor; BR, brassinosteroid; GA, gibberellin; HK, histidine kinase; IAA, indole-3-acetic acid; JA, jasmonic acid; SCF, SKP1/Cullin/F-box protein

ACKNOWLEDGMENTS

The author wishes to thank members of his lab for helpful comments on this manuscript. Work in the author's laboratory on auxin response is supported by National Institutes of Health grant GM067203 and the Mcknight Foundation.

REFERENCES

Chang C (2003) Ethylene signaling: The MAPK module has finally landed. Trends Plant Sci 8: 365-368.

Deshaies RJ (1999) SCF and Cullin/Ring H2-based ubiquitin ligases. Annu Rev Cell Dev Biol 15: 435-467.

Dharmasiri N, Estelle M (2004) Auxin signaling and regulated protein degradation. Trends Plant Sci 9: 302-308.

Goda H, Sawa S, Asami T, Fujioka S, Shimada Y, *et al.* (2004) Comprehensive comparison of auxin-regulated and brassinosteroid-regulated genes in *Arabidopsis*. Plant Physiol 134: 1555-1573.

Gomi K, Matsuoka M (2003) Gibberellin signalling pathway. Curr Opin Plant Biol 6: 489-493.

Gray WM, Kepinski S, Rouse D, Leyser O, Estelle M (2001) Auxin regulates SCFTIR1-dependent degradation of AUX/IAA proteins. Nature 414: 271–276.

Guo H, Ecker JR (2003) Plant responses to ethylene gas are mediated by SCF(EBF1/EBF2)-dependent proteolysis of EIN3 transcription factor. Cell 115: 667–677.

Guo H, Ecker JR (2004) The ethylene signaling pathway: New insights. Curr Opin Plant Biol 7: 40–49.

Himmelbach A, Yang Y, Grill E (2003) Relay and control of abscisic acid signaling. Curr Opin Plant Biol 6: 470–479.

Kakimoto T (2003) Perception and signal transduction of cytokinins. Annu Rev Plant Biol 54: 605–627.

Lorenzo O, Piqueras R, Sanchez-Serrano JJ, Solano R (2003) ETHYLENE RESPONSE FACTOR1 integrates signals from ethylene and jasmonate pathways in plant defense. Plant Cell 15: 165–178.

Nemhauser JL, Mockler TC, Chory J (2004) Interdependency of brassinosteroid and auxin signaling in *Arabidopsis*. PLoS Biol 2: e258.

Peng J, Richards DE, Hartley NM, Murphy GP, Devos KM, *et al.* (1999) 'Green revolution' Open access, freely available online genes encode mutant gibberellin response modulators. Nature 400: 256–261.

Potuschak T, Lechner E, Parmentier Y, Yanagisawa S, Grava S, *et al.* (2003) EIN3-dependent regulation of plant ethylene hormone signaling by two arabidopsis F box proteins: EBF1 and EBF2. Cell 115: 679–689.

Sasaki A, Ashikari M, Ueguchi-Tanaka M, Itoh H, Nishimura A, *et al.* (2002) Green revolution: A mutant gibberellin-synthesis gene in rice. Nature 416: 701–702.

Smalle J, Vierstra RD (2004) The ubiquitin 26s proteasome proteolytic pathway. Annu Rev Plant Physiol Plant Mol Biol 55: 555–590.

Tiwari SB, Hagen G, Guilfoyle TJ (2004) Aux/IAA proteins contain a potent transcriptional repression domain. Plant Cell 16: 533–543.

Turner JG, Ellis C, Devoto A (2002) The jasmonate signal pathway. Plant Cell 14: S153–S164.

Wang ZY, He JX (2004) Brassinosteroid signal transduction—Choices of signals and receptors. Trends Plant Sci 9: 91–96.

Wolanin PM, Thomason PA, Stock JB (2002) Histidine protein kinases: Key signal transducers outside the animal kingdom. Genome Biol 3: REVIEWS3013.

INDEX

F

G

H

I

J

K

L

M

N

O

P

This page left intentionally blank.